CONTEMPORARY
CRIMINAL LAW
Third Edition

Titles of Related Interest

Criminal Procedure by Matthew Lippman

Criminal Courts by Craig Hemmens, David Brody, and Cassia Spohn

How Do Judges Decide? (2nd Edition) by Cassia Spohn

Courts: A Text/Reader (2nd Edition) by Cassia Spohn and Craig Hemmens

Corrections: A Text/Reader by Mary Stohr, Anthony Walsh, and Craig Hemmens

Community Corrections by Robert Hanser

Administration and Management in Criminal Justice by Jennifer Allen and Rajeev Sawhney

Juvenile Justice: A Text/Reader by Richard Lawrence and Craig Hemmens

Juvenile Justice (7th Edition) by Steven Cox, Jennifer Allen, Robert Hanser, and John Conrad

Preventing and Reducing Juvenile Delinquency by Buddy Howell

Introduction to Criminology (7th Edition) by Frank Hagan

Introduction to Criminology: A Text/Reader by Anthony Walsh and Craig Hemmens

Criminological Theory: A Text/Reader by Stephen Tibbetts and Craig Hemmens

Criminals in the Making by John Wright, Stephen Tibbetts, and Leah Daigle

Criminological Theory (4th Edition) by Robert Lilly, Frank Cullen, and Richard Ball

Crime Types and Criminals by Frank Hagan

Profiling Violent Crimes (4th Edition) by Ronald Holmes and Stephen Holmes

Criminal Justice Ethics (2nd Edition) by Cynthia Banks

Forensic Psychology by Curt Bartol and Anne Bartol

The Practice of Research in Criminology and Criminal Justice (3rd Edition) by Ronet Bachman and Russell Schutt

Fundamentals of Research in Criminology and Criminal Justice by Ronet Bachman and Russell Schutt

Adventures in Criminal Justice Research (4th Edition) by Kim Loggio and George Dowdall

Understanding Terrorism (3rd Edition) by Gus Martin

Essentials of Terrorism by Gus Martin

Terrorism in Perspective (2nd Edition) by Susan Mahan and Pamela Griset

Drugs and Drug Policy by Clay Mosher and Scott Akins

Race and Crime (2nd Edition) by Shaun Gabbidon and Helen Greene

Crime Analysis With Crime Mapping (2nd Edition) by Rachel Boba

Critical Issues in Criminal Justice by Mary Maguire and Daniel Okada

CONTEMPORARY CRIMINAL LAW

CONCEPTS, CASES, AND CONTROVERSIES

Third Edition

MATTHEW LIPPMAN

University of Illinois at Chicago

Los Angeles | London | New Delhi
Singapore | Washington DC

Los Angeles | London | New Delhi
Singapore | Washington DC

FOR INFORMATION:

SAGE Publications, Inc.
2455 Teller Road
Thousand Oaks, California 91320
E-mail: order@sagepub.com

SAGE Publications Ltd.
1 Oliver's Yard
55 City Road
London EC1Y 1SP
United Kingdom

SAGE Publications India Pvt. Ltd.
B 1/I 1 Mohan Cooperative Industrial Area
Mathura Road, New Delhi 110 044
India

SAGE Publications Asia-Pacific Pte. Ltd.
3 Church Street
#10-04 Samsung Hub
Singapore 049483

Acquisitions Editor: Jerry Westby
Associate Editor: Megan Krattli
Assistant Editor: Rachael Leblond
Editorial Assistant: MaryAnn Vail
Production Editor: Laureen Gleason
Copy Editor: Ellen Howard
Typesetter: C&M Digitals (P) Ltd.
Proofreader: Gretchen Treadwell
Indexer: Michael Ferreira
Cover Designer: Michael Dubowe
Marketing Manager: Terra Schultz
Permissions Editor: Karen Ehrmann

Printed in the United States of America

Library of Congress Cataloging-in-Publication Data

Lippman, Matthew Ross, 1948-

Contemporary criminal law : concepts, cases, and controversies / Matthew R. Lippman. — 3rd ed.

p. cm.
Includes bibliographical references and index.

ISBN 978-1-4522-3002-3 (pbk.)

1. Criminal law—United States. I. Title.

KF9219.L54 2013
345.73—dc23   2012012968

This book is printed on acid-free paper.

12 13 14 15 16 10 9 8 7 6 5 4 3 2 1

BRIEF CONTENTS

DETAILED CONTENTS

PREFACE

This book reflects the insights and ideas developed over the course of more than twenty years of teaching criminal law and criminal procedure to undergraduate criminal justice students. The volume combines the concepts and learning tools found in undergraduate texts with the types of challenging cases and issues that are characteristic of law school casebooks. Each chapter incorporates several features:

- **Essays.** Essays introduce and summarize the chapters and topics.
- **Cases.** Edited cases are accompanied by "Questions for Discussion."
- **Case Notes.** Following the edited case decisions, "Cases and Commentaries" and "You Decide" review exercises are provided. In the "You Decide" sections, actual cases are discussed, and readers are asked to act as judges.
- **The Model Penal Code and Discussion Boxes.** In these sections, selected statutes and the provisions of the Model Penal Code are reprinted and analyzed. Discussion boxes and graphs supplement the coverage in most chapters.
- **Learning Tools.** Learning tools summarize and reinforce the material. These include introductory vignettes, chapter outlines, questions for discussion following each case, legal equations, chapter review questions, legal terminology lists, bibliographies, and Web-based study aides.

The book provides a contemporary perspective on criminal law that encourages students to actively read and analyze the text. I hope that at the conclusion of the course, students will have mastered the substance of criminal law and have developed the ability to understand and to creatively apply legal rules. My aspiration is that students come to appreciate that criminal law is dynamic and evolutionary and is not merely a static and mechanical set of rules.

THE CASE METHOD

One of my aims is to provide a book that students find interesting and instructors consider educationally valuable. I have found that undergraduates enjoy and easily absorb material taught through the case method. In my experience, learning is encouraged when students are presented with concrete factual situations that illustrate legal rules. The case method also lends itself to an interactive educational environment in which students engage in role playing or apply legal precedents to novel factual scenarios. The case method has the additional benefit of assisting students to refine their skills in critical reading and analysis and in logical thinking.

The cases in the text are organized to enhance learning and comprehension. The decisions have been edited to emphasize the core components of the judgments, and technicalities have been kept to a minimum. Each case is divided into **Facts, Issue**, **Reasoning,** and **Holding**. I strongly believe in the educational value of factual analysis and have included a fairly full description of the facts. The textbook highlights the following:

- **Classic Cases.** The book includes various classic cases that are fundamental to the study of criminal law as well as cases that provide a clear statement of the law.

- **Contemporary Cases.** I have incorporated contemporary cases that reflect our increasingly diverse and urbanized society. This includes cases that address the issues of carjacking, computer crime, drugs, gangs, stalking, terrorism, white-collar crime, cultural diversity, and animal rights. Attention is also devoted to gender, race, domestic violence, and hate crimes.
- **Legal Issues.** The vast majority of the decisions have been selected to raise important and provocative legal issues. For instance, students are asked to consider whether the law should be expanded to provide that a vicious verbal attack constitutes adequate provocation for voluntary manslaughter.
- **Facts.** In other instances, the cases illustrate the challenge of applying legal rules. For example, decisions present the difficulty of distinguishing between various grades of homicide and the complexity of determining whether an act constitutes a criminal attempt.
- **Public Policy.** I have found that among the most engaging aspects of teaching criminal law are the questions of public policy, law, and morality that arise in various cases. The book constantly encourages students to reflect on the impact and social context of legal rules and raises issues throughout, such as whether we are justified in taking a life to preserve several other lives under the law of necessity.

CHAPTER ORGANIZATION

Each chapter is introduced by a **vignette**. This is followed by **Learning Objectives**, which helps students focus on the important points. The **Introduction** to the chapter then provides an overview of the discussion.

The cases are introduced by **essays**. These discussions clearly present the development and elements of the relevant defense, concept, or crime and also include material on public policy considerations. Each case is introduced by a **question** that directs students to the relevant issue.

At the conclusion of the case, **Questions for Discussion** ask students to summarize and analyze the facts and legal rule. These questions, in many instances, are followed by **Cases and Comments** that expand on the issues raised by the edited case in the textbook. There is also a feature titled **You Decide** that provides students with the opportunity to respond to the facts of an actual case. The "answers" are available on the book's **Web site** at **www.sagepub.com/ lippmanccl3e**.

The essays are often accompanied by an analysis of the **Model Penal Code**. This provides students with an appreciation of the diverse approaches to criminal statutes. The discussion of each defense or crime concludes with a **legal equation** that clearly presents the elements of the defense or crime.

The chapters close with a **Chapter Summary** that outlines the important points. This is followed by **Chapter Review Questions**, **Legal Terminology**, a **Web exercise**, and a **Bibliography**. A **Glossary** appears at the end of the book. Additional learning tools are included on the Web site.

Most of the chapters also include **Crime in the News**. This is a brief discussion of legal developments and cases that students have likely encountered in the media. The purpose is to highlight contemporary issues and debates and to encourage students to consider the impact of the media in shaping our perceptions. Several chapters also include **Crime on the Streets**, which employs graphs to illustrate the frequency of various criminal offenses or other pertinent information. This is intended to give students a sense of the extent of crime in the United States and to connect the study of criminal law to the field of criminal justice. The Web site provides resources that enable instructors to augment the material in the book and to assist in student learning.

ORGANIZATION OF THE TEXT

The textbook provides broad coverage. This enables instructors to select from a range of alternative topics. You will also find that subjects are included that are not typically addressed. The discussion of rape, for instance, includes "withdrawal of consent" and "rape shield statutes." Expanded coverage is provided on topics such as sentencing, homicide, white-collar crime, and terrorism.

The textbook begins with the nature, purpose, and constitutional context of criminal law as well as sentencing and then covers the basic elements of criminal responsibility and offenses. The next parts of the textbook discuss crimes against the person and crimes against property and business. The book concludes with discussions of crimes against public morality and crimes against the state.

- **Nature, Purpose, and Constitutional Context of Criminal Law.** Chapter 1 discusses the nature, purpose, and function of criminal law. This introduction to criminal law is followed by an appendix on reading legal cases. Chapter 2 covers the constitutional limits on criminal law, including due process, equal protection, freedom of speech, and the right to privacy. Chapter 3 provides an overview of punishment and sentencing and discusses the Eighth Amendment prohibition on cruel and unusual punishment.
- **Principles of Criminal Responsibility.** This part covers the foundation elements of a crime. Chapter 4 discusses criminal acts and Chapter 5 is concerned with criminal intent, concurrence, and causation.
- **Parties, Vicarious Liability, and Inchoate Crimes.** The third part of the textbook discusses the scope of criminal responsibility. Chapter 6 discusses parties to crime and vicarious liability. Chapter 7 covers the inchoate crimes of attempt, conspiracy, and solicitation.
- **Criminal Defenses.** The fourth part of the text discusses defenses to criminal liability. Chapter 8 outlines justifications and Chapter 9 encompasses excuses.
- **Crimes Against the Person.** The fifth part focuses on crimes against the person. Chapter 10 provides a lengthy treatment of homicide. Chapter 11 is concerned with criminal sexual conduct, assault and battery, kidnapping, and false imprisonment.
- **Crimes Against Habitation and Property, and White-Collar Crime.** Chapter 12 covers burglary, trespass, arson, and mischief. These crimes against property were originally conceived as protecting the safety and security of the home. Chapter 13 centers on other crimes against property, including larceny, embezzlement, identity theft, and carjacking. Chapter 14 provides an overview of white-collar crime, commercial offenses that are designed to illegally enhance an individual's income or corporate profits. This chapter covers a range of topics, including environmental crimes, securities fraud, mail and wire fraud, and public corruption.
- **Crimes Against Public Order, Morality, and the State.** Chapter 15 focuses on crimes against public order and morality that threaten the order and stability of the community. The chapter covers a number of topics including disorderly conduct, riot, vagrancy, and efforts to combat homelessness, gangs, and prostitution. Chapter 16 discusses crimes against the state, stressing counterterrorism.

THIRD EDITION

In writing the third edition, I have benefited from the insightful comments of reviewers. I also have drawn on my experience in teaching the text. The changes to the book were adopted following a thorough review of contemporary court decisions and developments. I focused my efforts on sharpening topics that caused students particular problems in previous editions. The standard was whether a modification assisted in teaching and learning. The primary changes to the text include the following:

- **Cases.** New cases have been added that illuminate important concepts. This includes decisions on criminal intent, causality, conspiracy, necessity, the insanity defense, and money laundering. A number of cases have been placed on the study site. Several cases from the first edition have been edited to highlight important aspects of the decision.
- **Reorganization.** The chapter on homicide appears earlier in the text and is more logically integrated into the book.
- **Statutory Standard.** State statutes that illustrate contemporary developments in areas such as computer crime are available on the study site.
- **New Material.** Chapters have been updated to maintain the contemporary content and theme of the book and to clarify concepts discussed in the book. The text references a number of recent U.S. Supreme Court decisions on areas ranging from juveniles and life imprisonment to terrorism.

- **You Decide.** Most chapters include several new "You Decide" sections. These problems clarify concepts, illustrate the complexity of legal analysis, and enhance the interactive character of the text. Instructors also will find additional hypothetical problems on the password-protected instructor teaching site, which can be accessed at www.sagepub.com/lippmanccl3e.
- **Student Study Site.** New material has been added to the study site to assist in student learning. You will find a number of interesting cases, author podcasts, SAGE journal articles, Answers to the "You Decide" boxes, video links, and more on the student study site, which is available at www.sagepub.com/lippmanccl3e.

ACKNOWLEDGMENTS

I am hopeful that the textbook conveys my passion and enthusiasm for the teaching of criminal law and contributes to the teaching and learning of this most fascinating and vital topic. The book has been the product of the efforts and commitment of countless individuals who deserve much of the credit.

Let me first thank those responsible for compiling the state-specific material on the Web site.

I also greatly benefited from reviewers who as noted made valuable contributions to the manuscript: Robbin Day Brooks, Arizona State University; Mark S. Brown, The University of South Carolina; M. Lisa Clayton, College of Southern Nevada; Glenn Coffey, formerly of Virginia Commonwealth University; Julie Currie, Temple University; Roger Enriquez, University of Texas at San Antonio; Jona Goldschmidt, Loyola University Chicago; Willis Geer, California State University, Chico; O. Hayden Griffin, III, University of Florida; Dan Haley, Tidewater Community College; Ray Kessler, Sul Ross State University; Thomas A. Lateano, Kean University; Donald Liddick, Penn State University, The Eberly Campus; Russell Loving, California State University, Sacramento; Yue Ma, John Jay College; Evan J. Mandery, John Jay College of Criminal Justice; Kerry Muehlenbeck, Mesa Community College; Richard Warren Perry, University of California, Berkeley; Joseph G. Sandoval, Metropolitan State College of Denver; Philip M. Stinson, Indiana University of Pennsylvania; Jennifer K. Wesely, University of North Florida; and Jane Younglove, California State University, Stanislaus.

The people at SAGE Publications are among the most skilled professionals that an author is likely to encounter. An author is fortunate to publish with SAGE, a company that is committed to quality books. Publisher Jerry Westby provided intelligent suggestions and expert direction. Associate Editor Megan Krattli made an immense contribution to conceptualizing and improving the third edition, and Assistant Editor Rachael Leblond supervised the preparation of the study site. Senior Project Editor Laureen Gleason supervised the preparation of the lengthy manuscript and was responsible for monitoring a myriad of details associated with publication of the text. A special thanks as well to marketing manager Terra Schultz. I would also like to thank all the expert professionals at SAGE in production and design, and in marketing and in sales, who contributed their talent. The text was immensely improved by the meticulous and intelligent copyediting and expertise of Ellen Howard.

I must mention colleagues at the University of Illinois at Chicago: Greg Matoesian, Dennis Judd, John Hagedorn, Lisa Frohmann, Evan McKenzie, the late Gordon Misner, Beth Richie, Laurie Schaffner, Gene Scaramella, Ana Petrovic, Eric Leafblad, Dave Williams, Dean Bette Bottoms, Dagmar Lorenz, Amie Schuck, and Dennis Rosenbaum who first proposed that I write this textbook. Paola Baldo did an admirable job in working on the peripherals, and Robbin Day Brooks of Arizona State University substantially improved the PowerPoints. A great debt of gratitude, of course, is owed to my students, who constantly provide new and creative insights.

During the writing of this text, I was the victim of a vicious criminal attack. I owe a special debt of gratitude for the support provided by the University of Illinois at Chicago administration and community.

I am fortunate to have loyal friends who have provided inspiration and encouragement. These include my dear friends Wayne Kerstetter, Deborah Allen-Baber, and Agata Fijalkowski, as well as Nan Kamen-Judd, Sharon Savinski, Mindie Lazarus-Black, the late Leanne Lobravico, Jess Maghan, Sean McConville, Oneida Mascarenas, Sheldon Rosing, Bryan Burke, Bill Lane, Annamarie Pastore, Kerry Petersen, Robin Wagner, Donna Dorney, Ken Janda, Maeve Barrett Burke, Kris Clark, Jennifer

Woodard, Dr. Peter Ivanovic, Tom Morante, and Marianne Splitter. I also must thank the late Ralph Semsker and Isadora Semsker. Dr. Mary Hallberg has been an important person in my life throughout the writing of the text, and the late Lidia Janus remains my true north and source of inspiration and the love of my life.

I have two members of my family living in Chicago. My sister, Dr. Jessica Lippman, and niece, Professor Amelia Barrett, remain a source of encouragement and generous assistance. Finally, the book is dedicated to my parents, Mr. and Mrs. S.G. Lippman, who provided me with a love of learning. My late father, S.G. Lippman, practiced law for seventy years in the service of the most vulnerable members of society. He believed that law was the highest calling and never turned away a person in need. Law, for him, was a passionate calling to pursue justice and an endless source of discussion, debate, and fascination.

My source for the Model Penal Code excerpts throughout the text is *Model Penal Code* © 1985 by the American Law Institute. Reprinted with permission. All rights reserved.

For my parents and Lidia Janus

1 THE NATURE, PURPOSE, AND FUNCTION OF CRIMINAL LAW

May the police officers be subjected to prosecution in both state and federal court?

As the videotape begins, it shows that King rose from the ground and charged toward Officer Powell. Powell took a step and used his baton to strike King on the side of his head. King fell to the ground. From the eighteenth to the thirtieth second on the videotape, King attempted to rise, but Powell and Wind each struck him with their batons to prevent him from doing so. From the thirty-fifth to the fifty-first second, Powell administered repeated blows to King's lower extremities; one of the blows fractured King's leg. At the fifty-fifth second, Powell struck King on the chest, and King rolled over and lay prone. At that point, the officers stepped back and observed King for about 10 seconds. . . . At one-minute-five-seconds (1:05) on the videotape, Briseno, in the District Court's words, "stomped" on King's upper back or neck. King's body writhed in response. At 1:07, Powell and Wind again began to strike King with a series of baton blows, and

Wind kicked him in the upper thoracic or cervical area six times until 1:26. At about 1:29, King put his hands behind his back and was handcuffed.

For a deeper look at this topic, visit the study site at www.sagepub.com/lippmanccl3e.

Learning Objectives

1. Define a crime, provide examples of criminal behavior, and distinguish between civil and criminal law.

2. Explain the difference between criminal law and criminal procedure.

3. Discuss the difference between felonies and misdemeanors and the difference between *mala in se* and *mala prohibitum.*

4. List and describe the various sources of criminal law.

INTRODUCTION

The criminal law is the foundation of the criminal justice system. The law defines the conduct that may lead to an arrest by the police, trial before the courts, and incarceration in prison. When we think about criminal law, we typically focus on offenses such as rape, robbery, and murder. States, however, condemn a range of acts in their criminal codes, some of which may surprise you. In Alabama, it is a criminal offense to promote or engage in a wrestling match with a bear or to train a bear to fight in such a match.[1] A Florida law states that it is unlawful to possess "any ignited tobacco product" in an elevator.[2] Rhode Island declares that an individual shall be imprisoned for seven years who voluntarily engages in a duel with a dangerous weapon or who challenges an individual to a duel.[3] In Wyoming you can be arrested for skiing while being impaired by alcohol[4] or for opening and failing to close a gate in a fence that "crosses a private road or river."[5] You can find criminal laws on the books in various states punishing activities such as playing dominos on Sunday, feeding an alcoholic beverage to a moose, cursing on a miniature golf course, making love in a car, or performing a wedding ceremony when either the bride or groom is drunk.[6] In Louisiana, you risk being sentenced to ten years in prison for stealing an alligator, whether dead or alive, valued at $1,000.[7]

THE NATURE OF CRIMINAL LAW

Are there common characteristics of acts that are labeled as crimes? How do we define a crime? The easy answer is that a **crime** is whatever the law declares to be a criminal offense and punishes with a penalty. The difficulty with this approach is that not all criminal convictions result in a fine or imprisonment. Rather than punishing a **defendant**, the judge may merely warn him or her not to repeat the criminal act. Most commentators stress that the important feature of a crime is that it is an act that is officially condemned by the community and carries a sense of shame and humiliation. Professor Henry M. Hart, Jr., defines crime as "conduct which, if . . . shown to have taken place" will result in the "formal and solemn pronouncement of the moral condemnation of the community."[8]

The central point of Professor Hart's definition is that a crime is subject to formal condemnation by a judge and jury representing the people in a court of law. This distinguishes a crime from acts most people would find objectionable that typically are not subject to state prosecution and official punishment. We might, for instance, criticize someone who cheats on his or her spouse, but we generally leave the solution to the *individuals involved*. Other matters are left to *institutions* to settle; schools generally discipline students who cheat or disrupt classes, but this rarely results in a criminal charge. Professional baseball, basketball, and football leagues have their own private procedures for disciplining players. Most states leave the decision whether to recycle trash to the *individual* and look to *peer pressure* to enforce this obligation.

CRIMINAL AND CIVIL LAW

How does the criminal law differ from the **civil law**? The civil law is that branch of the law that protects the individual rather than the public interest. A legal action for a civil wrong is brought by an individual rather than by a state prosecutor. You may sue a mechanic who breaches a contract to repair your car or bring an action against a landlord who fails to adequately heat your apartment. The injury is primarily to you as an individual, and there is relatively little harm to society. A mechanic who intentionally misleads and harms a number of innocent consumers, however, may find himself or herself charged with criminal fraud.

Civil and criminal actions are characterized by different legal procedures. For instance, conviction of a crime requires the high standard of proof beyond a reasonable doubt, although responsibility for a civil wrong is established by the much lower standard of proof by a preponderance of the evidence or roughly fifty-one percent certainty. The high standard of proof in criminal cases reflects the fact that a criminal conviction may result in a loss of liberty and significant damage to an individual's reputation and standing in the community.[9]

The famous eighteenth-century English jurist William Blackstone summarizes the distinction between civil and criminal law by observing that civil injuries are "an infringement . . . of the civil rights which belong to individuals . . . public wrongs, or crimes . . . are a breach and violation of the public rights and duties, due to the whole community . . . in its social aggregate capacity." Blackstone illustrates this difference by pointing out that society has little interest in whether someone sues a neighbor or emerges victorious in a land dispute. On the other hand, society has a substantial investment in the arrest, prosecution, and conviction of individuals responsible for espionage, murder, and robbery.[10]

The difference between a civil and criminal action is not always clear, particularly with regard to an action for a **tort**, which is an injury to a person or to his or her property. Consider the drunken driver who runs a red light and hits your car. The driver may be sued in tort for negligently damaging you and your property as well as criminally prosecuted for reckless driving. The purpose of the civil action is to compensate you with money for the damage to your car and for the physical and emotional injuries you have suffered. In contrast, the criminal action punishes the driver for endangering society. Civil liability is based on a preponderance of the evidence standard, while a criminal conviction carries a possible loss of liberty and is based on the higher standard of guilt beyond a reasonable doubt. You may recall that former football star O.J. Simpson was acquitted of murdering Nicole Brown Simpson and Ron Goldman but was later found guilty of wrongful death in a civil court and ordered to compensate the victims' families in the amount of $33.5 million.

The distinction between criminal and civil law proved immensely significant for Kansas inmate Leroy Hendricks. Hendricks was about to be released after serving ten years in prison for molesting two thirteen-year-old boys. This was only the latest episode in Hendricks's almost thirty-year history of indecent exposure and molestation of young children. Hendricks freely conceded that when not confined, the only way to control his sexual urge was to "die."

Upon learning that Hendricks was about to be released, Kansas authorities invoked the Sexually Violent Predator Act of 1994, which authorized the institutional confinement of individuals who, due to a "mental abnormality" or a "personality disorder," are likely to engage in "predatory acts of sexual violence." Following a hearing, a jury found Hendricks to be a "sexual predator." The U.S. Supreme Court ruled that Hendricks's continued commitment was a civil rather than criminal penalty, and that Hendricks was not being unconstitutionally punished twice for the same criminal act of molestation. The Court explained that the purpose of the commitment procedure was to detain and to treat Hendricks in order to prevent him from harming others in the future rather than to punish him.[11] Do you think that the decision of the U.S. Supreme Court makes sense?

THE PURPOSE OF CRIMINAL LAW

We have seen that the criminal law primarily protects the interests of society, and the civil law protects the interests of the individual. The primary purpose or function of the criminal law is to help maintain social order and stability. The Texas criminal code proclaims that the purpose of criminal law is to "establish a system of prohibitions, penalties, and correctional measures to deal with conduct that unjustifiably and inexcusably causes or threatens harm to those individual or public interests for which state protection is appropriate."[12] The New York criminal code sets out the basic purposes of criminal law as follows:[13]

- *Harm.* To prohibit conduct that unjustifiably or inexcusably causes or threatens substantial harm to individuals as well as to society
- *Warning.* To warn people both of conduct that is subject to criminal punishment and of the severity of the punishment
- *Definition.* To define the act and intent that is required for each offense
- *Seriousness.* To distinguish between serious and minor offenses and to assign the appropriate punishments
- *Punishment.* To impose punishments that satisfy the demands for revenge, rehabilitation, and deterrence of future crimes
- *Victims.* To insure that the victim, the victim's family, and the community interests are represented at trial and in imposing punishments

The next step is to understand the characteristics of a criminal act.

THE PRINCIPLES OF CRIMINAL LAW

The study of **substantive criminal law** involves an analysis of the definition of specific crimes (specific part) and of the general principles that apply to all crimes (general part), such as the defense of insanity. In our study, we will first review the general part of criminal law and then look at specific offenses. Substantive criminal law is distinguished from **criminal procedure**. Criminal procedure involves a study of the legal standards governing the detection, investigation, and prosecution of crime and includes areas such as interrogations, search and seizure, wiretapping, and the trial process. Criminal procedure is concerned with "how the law is enforced"; criminal law involves "what law is enforced."

Professors Jerome Hall[14] and Wayne R. LaFave[15] identify the basic principles that compose the general part of the criminal law. Think of the general part of the criminal law as the building blocks that are used to construct specific offenses such as rape, murder, and robbery.

- *Criminal Act.* A crime involves an act or failure to act. You cannot be punished for bad thoughts. A criminal act is called *actus reus.*
- *Criminal Intent.* A crime requires a criminal intent or *mens rea.* Criminal punishment is ordinarily directed at individuals who intentionally, knowingly, recklessly, or negligently harm other individuals or property.

- *Concurrence.* The criminal act and criminal intent must coexist or accompany one another.
- *Causation.* The defendant's act must cause the harm required for criminal guilt, death in the case of homicide, and the burning of a home or other structure in the case of arson.
- *Responsibility.* Individuals must receive reasonable notice of the acts that are criminal so as to make a decision to obey or to violate the law. In other words, the required criminal act and criminal intent must be clearly stated in a statute. This concept is captured by the Latin phrase *nullum crimen sine lege, nulla poena sin lege* (no crime without law, no punishment without law).
- *Defenses.* Criminal guilt is not imposed on an individual who is able to demonstrate that his or her criminal act is justified (benefits society) or excused (the individual suffered from a disability that prevented him or her from forming a criminal intent).

We now turn to a specific part of the criminal law to understand the various types of acts that are punished as crimes.

CATEGORIES OF CRIME

Felonies and Misdemeanors

There are a number of approaches to categorizing crimes. The most significant distinction is between a **felony** and a **misdemeanor**. A crime punishable by death or by imprisonment for more than one year is a felony. Misdemeanors are crimes punishable by less than a year in prison. Note that whether a conviction is for a felony or misdemeanor is determined by the punishment provided in the statute under which an individual is convicted rather than by the actual punishment imposed. Many states subdivide felonies and misdemeanors into several classes or degrees to distinguish between the seriousness of criminal acts. **Capital felonies** are crimes subject either to the death penalty or to life in prison in states that do not have the death penalty. The term **gross misdemeanor** is used in some states to refer to crimes subject to between six and twelve months in prison, whereas other misdemeanors are termed **petty misdemeanors**. Several states designate a third category of crimes that are termed **violations** or **infractions**. These tend to be acts that cause only modest social harm and carry fines. These offenses are considered so minor that imprisonment is prohibited. This includes the violation of traffic regulations.

Florida classifies offenses as felonies, misdemeanors, or noncriminal violations. Noncriminal violations are primarily punishable by a fine or forfeiture of property. The following list shows the categories of felonies and misdemeanors and the maximum punishment generally allowable under Florida law:

- *Capital Felony.* Death or life imprisonment without parole
- *Life Felony.* Life in prison and a $15,000 fine
- *Felony in the First Degree.* Thirty years in prison and a $10,000 fine
- *Felony in the Second Degree.* Fifteen years in prison and a $10,000 fine
- *Felony in the Third Degree.* Five years in prison and a $5,000 fine
- *Misdemeanor in the First Degree.* One year in prison and a $1,000 fine
- *Misdemeanor in the Second Degree.* Sixty days in prison and a $500 fine

The severity of the punishment imposed is based on the seriousness of the particular offense. Florida, for example, punishes as a second-degree felony the recruitment of an individual for prostitution knowing that force, fraud, or coercion will be used to cause the person to engage in prostitution. This same act is punished as a first-degree felony in the event that the person recruited is under fourteen years old or if death results.[16]

Mala in Se and *Mala Prohibita*

Another approach is to classify crime by "moral turpitude" (evil). ***Mala in se*** crimes are considered "inherently evil" and would be evil even if not prohibited by law. This includes murder, rape, robbery, burglary, larceny, and arson. ***Mala prohibita*** offenses are not "inherently evil" and are only considered wrong because they are prohibited by a statute. This includes offenses ranging from tax evasion to carrying a concealed weapon, leaving the scene of an accident, and being drunk and disorderly in public.

Why should we be concerned with classification schemes? A felony conviction can prevent you from being licensed to practice various professions, bar you from being admitted to the armed forces or joining the police, and prevent you from adopting a child or receiving various forms of federal assistance. In some states, a convicted felon is still prohibited from voting, even following release. The distinction between *mala in se* and *mala prohibita* is also important. For instance, the law provides that individuals convicted of a "crime of moral turpitude" may be deported from the United States.

There are a number of other classification schemes. The law originally categorized as **infamous** those crimes that were considered to be deserving of shame or disgrace. Individuals convicted of infamous offenses such as treason (betrayal of the nation) or offenses involving dishonesty were historically prohibited from appearing as witnesses at a trial.

Subject Matter

This textbook is organized in accordance with the subject matter of crimes, the scheme that is followed in most state criminal codes. There is disagreement, however, concerning the classification of some crimes. Robbery, for instance, involves the theft of property as well as the threat or infliction of harm to the victim, and there is a debate about whether it should be considered a crime against property or against the person. Similar issues arise in regard to burglary. Subject matter offenses in descending order of seriousness are as follows:

- *Crimes Against the State.* Treason, sedition, espionage, terrorism (Chapter 16)
- *Crimes Against the Person, Homicide.* Homicide, murder, manslaughter (Chapter 10)
- *Crimes Against the Person, Sexual Offenses, and Other Crimes.* Rape, assault and battery, false imprisonment, kidnapping (Chapter 11)
- *Crimes Against Habitation.* Burglary, arson, trespassing (Chapter 12)
- *Crimes Against Property.* Larceny, embezzlement, false pretenses, receiving stolen property, robbery, fraud (Chapters 13 and 14)
- *Crimes Against Public Order.* Disorderly conduct, riot (Chapter 15)
- *Crimes Against the Administration of Justice.* Obstruction of justice, perjury, bribery
- *Crimes Against Public Morals.* Prostitution, obscenity (Chapter 15)

The book also covers the general part of criminal law, including the constitutional limits on criminal law (Chapter 2), sentencing (Chapter 3), criminal acts (Chapter 4), criminal intent (Chapter 5), the scope of criminal liability (Chapters 6 and 7), and defenses to criminal liability (Chapters 8 and 9).

SOURCES OF CRIMINAL LAW

We now have covered the various categories of criminal law. The next questions to consider are these: What are the sources of the criminal law? How do we find the requirements of the criminal law? There are a number of sources of the criminal law in the United States:

- *English and American Common Law.* These are English and American judge-made laws and English acts of Parliament.
- *State Criminal Codes.* Every state has a comprehensive written set of laws on crime and punishment.
- *Municipal Ordinances.* Cities, towns, and counties are typically authorized to enact local criminal laws, generally of a minor nature. These laws regulate the city streets, sidewalks, and buildings and concern areas such as traffic, littering, disorderly conduct, and domestic animals.

- *Federal Criminal Code.* The U.S. government has jurisdiction to enact criminal laws that are based on the federal government's constitutional powers, such as the regulation of interstate commerce.
- *State and Federal Constitutions.* The U.S. Constitution defines treason and together with state constitutions establishes limits on the power of government to enact criminal laws. A criminal statute, for instance, may not interfere with freedom of expression or religion.
- *International Treaties.* International treaties signed by the United States establish crimes such as genocide, torture, and war crimes. These treaties, in turn, form the basis of federal criminal laws punishing acts such as genocide and war crimes when Americans are involved. These cases are prosecuted in U.S. courts.
- *Judicial Decisions.* Judges write decisions explaining the meaning of criminal laws and determining whether criminal laws meet the requirements of state and federal constitutions.

At this point, we turn our attention to the common law origins of American criminal law and to state criminal codes.

The Common Law

The English *common law* is the foundation of American criminal law. The origins of the common law can be traced to the Norman conquest of England in 1066. The Norman king, William the Conqueror, was determined to provide a uniform law for England and sent royal judges throughout the country to settle disputes in accordance with the common customs and practices of the country. The principles that composed this common law began to be written down in 1300 in an effort to record the judge-made rules that should be used to decide future cases.

By 1600, a number of **common law crimes** had been developed, including arson, burglary, larceny, manslaughter, mayhem, rape, robbery, sodomy, and suicide. These were followed by criminal attempt, conspiracy, blasphemy, forgery, sedition, and solicitation. On occasion, the king and Parliament issued decrees that filled the gaps in the common law, resulting in the development of the crimes of false pretenses and embezzlement. The distinctive characteristic of the common law is that it is for the most part the product of the decisions of judges in actual cases.

The English civil and criminal common law was transported to the new American colonies and formed the foundation of the colonial legal system that in turn was adopted by the thirteen original states following the American Revolution. The English common law was also recognized by each state subsequently admitted to the Union; the only exception was Louisiana, which followed the French Napoleonic Code until 1805 when it embraced the common law.[17]

State Criminal Codes

States in the nineteenth century began to adopt comprehensive written criminal codes. This movement was based on the belief that in a democracy the people should have the opportunity to know the law. Judges in the common law occasionally punished an individual for an act that had never before been subjected to prosecution. A defendant in a Pennsylvania case was convicted of making obscene phone calls despite the absence of a previous prosecution for this offense. The court explained that the "common law is sufficiently broad to punish . . . although there may be no exact precedent, any act which directly injures or tends to injure the public."[18] There was the additional argument that the power to make laws should reside in the elected legislative representatives of the people rather than in unelected judges. As Americans began to express a sense of independence, there was also a strong reaction against being so clearly connected to the English common law tradition, which was thought to have limited relevance to the challenges facing America. As early as 1812, the U.S. Supreme Court proclaimed that federal courts were required to follow the law established by Congress and were not authorized to apply the common law.

States were somewhat slower than the federal government to abandon the common law. In a Maine case in 1821, the accused was found guilty of dropping the dead body of a child into a river. The defendant was convicted even though there was no statute making this a crime. The court explained that "good morals" and "decency" all forbid this act. State legislatures reacted against these types of decisions and began to abandon the common law in the mid-nineteenth century. The Indiana Revised Statutes of 1852, for example, proclaims that "[c]rimes and misde-meanors shall be defined, and punishment fixed by statutes of this State, and not otherwise."[19]

Some states remain **common law states**, meaning that the common law may be applied where the state legislature has not adopted a law in a particular area. The Florida criminal code states that the "common law of England in relation to crimes, except so far as the same relates to the mode and degrees of punishment, shall be of full force in this state where there is no existing provision by statute on the subject." Florida law further provides that where there is no statute, an offense shall be punished by fine or imprisonment but that the "fine shall not exceed $500, nor the term of imprisonment 12 months."[20] Missouri and Arizona are also examples of common law states. These states' criminal codes, like that of Florida, contain a **reception statute** that provides that the states "receive" the common law as an unwritten part of their criminal law. California, on the other hand, is an example of a **code jurisdiction**. The California criminal code provides that "no act or omission . . . is criminal or punishable, except as prescribed or authorized by this code."[21] Ohio and Utah are also code jurisdiction states. The Utah criminal code states that common law crimes "are abolished and no conduct is a crime unless made so by this code . . . or ordinance."[22]

Professor LaFave observes that courts in common law states have recognized a number of crimes that are not part of their criminal codes, including conspiracy, attempt, solicitation, utter-ing gross obscenities in public, keeping a house of prostitution, cruelly killing a horse, public inebriation, and false imprisonment.[23]

You also should keep in mind that the common law continues to play a role in the law of code jurisdiction states. Most state statutes are based on the common law, and courts frequently consult the common law to determine the meaning of terms in statutes. In the well-known California case of *Keeler v. Superior Court*, the California Supreme Court looked to the common law and determined that an 1850 state law prohibiting the killing of a "human being" did not cover the "murder of a fetus." The California state legislature then amended the murder statute to punish "the unlaw-ful killing of a human being, or a fetus."[24] Most important, our entire approach to criminal trials reflects the common law's commitment to protecting the rights of the individual in the criminal justice process.

State Police Power

Are there limits on a state's authority to pass criminal laws? Could a state declare that it is a crime to possess fireworks on July Fourth? State governments possess the broad power to promote the public health, safety, and welfare of the residents of the state. This wide-ranging **police power** includes the "duty . . . to protect the well-being and tranquility of a community" and to "prohibit acts or things reasonably thought to bring evil or harm to its people."[25] An example of the far-reaching nature of the state police power is the U.S. Supreme Court's upholding of the right of a village to prohibit more than two unrelated people from occupying a single home. The Supreme Court proclaimed that the police power includes the right to "lay out zones where family values, youth values, the blessings of quiet seclusion, and clean air make the area a sanctuary for people."[26]

State legislatures in formulating the content of criminal codes have been profoundly influ-enced by the Model Penal Code.

The Model Penal Code

People from other countries often ask how students can study the criminal law of the United States, a country with fifty states and a federal government. The fact that there is a significant degree of agreement in the definition of crimes in state codes is due to a large extent to the **Model Penal Code**.

In 1962, the American Law Institute (ALI), a private group of lawyers, judges, and scholars, concluded after several years of study that despite our common law heritage, state criminal statutes radically varied in their definition of crimes and were difficult to understand and poorly organized. The ALI argued that the quality of justice should not depend on the state in which an individual was facing trial and issued a multivolume set of model criminal laws, *The Proposed Official Draft of the Model Penal Code*. The Model Penal Code is purely advisory and is intended to encourage all fifty states to adopt a single uniform approach to the criminal law. The statutes are accompanied by a commentary that explains how the Model Penal Code differs from existing state statutes. Roughly thirty-seven states have adopted some of the provisions of the Model Penal Code, although no state has adopted every single model law. The states that most closely follow the code are New Jersey, New York, Pennsylvania, and Oregon. As you read this book, you may find it interesting to compare the Model Penal Code to the common law and to state statutes.[27]

This book primarily discusses state criminal law. It is important to remember that we also have a federal system of criminal law in the United States.

Federal Statutes

The United States has a federal system of government. The states granted various powers to the federal government that are set forth in the U.S. Constitution. This includes the power to regulate interstate commerce, to declare war, to provide for the national defense, to coin money, to collect taxes, to operate the post office, and to regulate immigration. The Congress is entitled to make "all Laws which shall be necessary and proper" for fulfilling these responsibilities. The states retain those powers that are not specifically granted to the federal government. The Tenth Amendment to the Constitution states that the powers "not delegated to the United States by the Constitution, nor prohibited by it to the states, are reserved to the states respectively, or to the people."

The Constitution specifically authorizes Congress to punish the counterfeiting of U.S. currency, piracy and felonies committed on the high seas, and crimes against the "Law of Nations" as well as to make rules concerning the conduct of warfare. These criminal provisions are to be enforced by a single Supreme Court and by additional courts established by Congress.

The **federal criminal code** compiles the criminal laws adopted by the U.S. Congress. This includes laws punishing acts such as tax evasion, mail and immigration fraud, bribery in obtaining a government contract, and the knowing manufacture of defective military equipment. The **Supremacy Clause** of the U.S. Constitution provides that federal law is superior to a state law within those areas that are the preserve of the national government. This is termed the **preemption doctrine**. In 2012, the Supreme Court held that federal immigration law preempted several sections of an Arizona statute directed at undocumented individuals.

Several recent court decisions have held that federal criminal laws have unconstitutionally encroached on areas reserved for state governments. This reflects a trend toward limiting the federal power to enact criminal laws. For instance, the U.S. government, with the **Interstate Commerce Clause**, has interpreted its power to regulate interstate commerce as providing the authority to criminally punish harmful acts that involve the movement of goods or individuals across state lines. An obvious example is the interstate transportation of stolen automobiles.

In the past few years, the U.S. Supreme Court has ruled several of these federal laws unconstitutional based on the fact that the activities did not clearly affect interstate commerce or involve the use of interstate commerce. In 1995, the Supreme Court ruled in *United States v. Lopez* that Congress violated the Constitution by adopting the Gun Free School Zones Act of 1990, which made it a crime to have a gun in a local school zone. The fact that the gun may have been transported across state lines was too indirect a connection with interstate commerce on which to base federal jurisdiction.[28]

In 2000, the Supreme Court also ruled unconstitutional the U.S. government's prosecution of an individual in Indiana who was alleged to have set fire to a private residence. The federal law made it a crime to maliciously damage or destroy, by means of fire or an explosive, any building used in interstate or foreign commerce or in any activity affecting interstate or foreign commerce. The Supreme Court ruled that there must be a direct connection between a building and interstate commerce and rejected the government's contention that it is sufficient that a building is constructed of supplies or serviced by electricity that moved across state lines or that the owner's insurance payments are mailed to a company located in another state. Justice Ruth Bader Ginsburg explained that this would mean that "every building in the land" would fall within the reach of federal laws on arson, trespass, and burglary.[29]

In 2006, in *Oregon v. Gonzalez,* the Supreme Court held that U.S. Attorney General John Ashcroft lacked the authority to prevent Oregon physicians acting under the state's Death With Dignity Law from prescribing lethal drugs to terminally ill patients who are within six months of dying.[30]

The sharing of power between the federal and state governments is termed **dual sovereignty**. An interesting aspect of dual sovereignty is that it is constitutionally permissible to prosecute a defendant for the same act at both the state and federal levels so long as the criminal charges slightly differ. You might recall in 1991 that Rodney King, an African American, was stopped by the Los Angeles police. King resisted and eventually was subdued, wrestled to the ground, beaten, and handcuffed by four officers. The officers were acquitted by an all-Caucasian jury in a state court in Simi Valley, California, leading to widespread protest and disorder in Los Angeles. The federal government responded by bringing the four officers to trial for violating King's civil right to be arrested in a reasonable fashion. Two officers were convicted and sentenced to thirty months in federal prison and two were acquitted. Later in this chapter, you will be asked to decide whether this "double prosecution" is fair.

We have seen that the state and federal governments possess the power to enact criminal laws. The federal power is restricted by the provisions of the U.S. Constitution that define the limits on governmental power.

Constitutional Limitations

The U.S. Constitution and individual state constitutions establish limits and standards for the criminal law. The U.S. Constitution, as we shall see in Chapter 2, requires that

- a state or local law may not regulate an area that is reserved to the federal government. A federal law may not encroach upon state power.
- a law may infringe upon the fundamental civil and political rights of individuals only in compelling circumstances.
- a law must be clearly written and provide notice to citizens and to the police of the conduct that is prohibited.
- a law must be nondiscriminatory and may not impose cruel and unusual punishment. A law also may not be retroactive and punish acts that were not crimes at the time that they were committed.

The ability of legislators to enact criminal laws is also limited by public opinion. The American constitutional system is a democracy. Politicians are fully aware that they must face elections and that they may be removed from office in the event that they support an unpopular law. As we learned during the unsuccessful effort to ban the sale of alcohol during the prohibition era in the early twentieth century, the government will experience difficulties in imposing an unpopular law on the public.

Of course, the democratic will of the majority is subject to constitutional limitations. A classic example is the Supreme Court's rulings that popular federal statutes prohibiting and punishing flag burning and desecration compose an unconstitutional violation of freedom of speech.[31]

CRIME IN THE NEWS

In 1996, California became one of seventeen states to authorize the use of marijuana for medical purposes. (The other states are Arizona, Alaska, Colorado, Connecticut, Delaware, Hawaii, Maine, Michigan, Montana, New Jersey, Nevada, New Mexico, Oregon, Rhode Island, Vermont, and Washington, and the same applies in the District of Columbia. Maryland exempts medical marijuana users from jail sentences.)

California voters passed Proposition 215, the Compassionate Use Act of 1996, which is intended to ensure that "seriously ill" residents of California are able to obtain marijuana. The act provides an exemption from criminal prosecution for doctors who, in turn, may authorize patients and primary caregivers to possess or cultivate marijuana for medical purposes. The California legislation is directly at odds with the federal Controlled Substances Act, which declares it a crime to manufacture, distribute, or possess marijuana. There are more than 100,000 medical marijuana users in California, and roughly one-tenth of one percent of the population uses medical marijuana in the states that collect information on medical marijuana users.

Angel Raich and Diane Monson are two California residents who suffer from severe medical disabilities. Their doctors have found that marijuana is the only drug that is able to alleviate their pain and suffering. Raich's doctor goes so far as to claim that Angel's pain is so intense that she might die if deprived of marijuana. Monson cultivates her own marijuana, and Raich relies on two caregivers who provide her with California-grown marijuana at no cost.

On August 15, 2000, agents from the federal Drug Enforcement Administration (DEA) raided Monson's home and destroyed all six of her marijuana plants. The DEA agents disregarded objections from the Butte County Sheriff's Department and the local California District Attorney's Office that Monson's possession of marijuana was perfectly legal.

Monson and Raich, along with several doctors and patients, refused to accept the destruction of the marijuana plants and asked the U.S. Supreme Court to rule on the constitutionality of the federal government's refusal to exempt medical marijuana users from criminal prosecution and punishment. The case was supported by the California Medical Association and the Leukemia and Lymphoma Society. Raich suffers from severe chronic pain stemming from fibromyalgia, endometriosis, scoliosis, uterine fibroid tumors, rotator cuff syndrome, an inoperable brain tumor, seizures, life-threatening wasting syndrome, and constant nausea. She also experiences extreme chemical sensitivities that result in violent allergic reactions to virtually every pharmaceutical drug. Raich was confined to a wheelchair before reluctantly deciding to smoke marijuana, a decision that led to her enjoying a fairly normal life.

A doctor recommended that Monson use marijuana to treat severe chronic back pain and spasms. She alleges that marijuana alleviates the pain that she describes as comparable to an uncontrollable cramp. Monson claims that other drugs have proven ineffective or resulted in nausea and create the risk of severe injuries to her kidneys and liver. The marijuana reportedly reduces the frequency of Monson's spasms and enables her to continue to work.

The U.S. Supreme Court, in Gonzalez v. Raich in 2005,[32] held that the federal prohibition on the possession of marijuana would be undermined by exempting marijuana possession in California and other states from federal criminal enforcement. The Supreme Court explained that the cultivation of marijuana under California's medical marijuana law, although clearly a local activity, frustrated the federal government's effort to control the shipment of marijuana across state lines, because medical marijuana inevitably would find its way into interstate commerce, increase the nationwide supply, and drive down the price of the illegal drug. There was also a risk that completely healthy individuals in California would manage to be fraudulently certified by a doctor to be in need of medical marijuana. Three of the nine Supreme Court judges dissented from the majority opinion. Justice Sandra Day O'Connor observed that the majority judgment "stifles an express choice by some States, concerned for the lives and liberties of their people, to regulate medical marijuana differently."

Following the decision, Angel Raich urged the federal government to have some "compassion and have some heart" and not to "use taxpayer dollars to come in and lock us up . . . we are using this medicine because it is saving our lives." She asked why the federal government was trying to kill her. Opponents of medical marijuana defend the Supreme Court's decision and explain that individuals should look to traditional medical treatment rather than being misled into thinking that marijuana is an effective therapy. They also argue that marijuana is a highly addictive drug that could lead individuals to experiment with even more harmful narcotics.

There were over 850,000 arrests for possession or sale of marijuana in 2009, most of which were carried out by state authorities. The question is whether the federal authorities will use the Supreme Court decision as a justification for arresting individuals growing or possessing medical marijuana. The federal government under President George W. Bush adopted a policy of targeting individuals in California accused of growing 100 plants or more and raided marijuana dispensaries in the state.

The Obama administration stated in 2009 that it would take a "hands off" attitude towards medical marijuana. In 2011, the Department of Justice (DOJ) reversed course and wrote strongly worded letters to the governors of states with laws permitting medical marijuana. The letters indicated that although individuals could grow and use small amounts of medical marijuana, the DOJ would criminally prosecute growers of more than 100 plants and individuals who distributed and marketed marijuana. Medical cannabis would be tolerated but not medical "cannibusiness."

Where do you stand on the medical marijuana controversy?

Consider the following factual scenario that is taken from the U.S. Supreme Court's description of the events surrounding the beating of Rodney King.[33]

You Decide

1.1 On the evening of March 2, 1991, Rodney King and two of his friends sat in King's wife's car in Altadena, California, a city in Los Angeles County, and drank malt liquor for a number of hours. Then, with King driving, they left Altadena via a major freeway. King was intoxicated. California Highway Patrol (CHP) officers observed King's car traveling at a speed they estimated to be in excess of 100 mph. The officers followed King with red lights and sirens activated and ordered him by loudspeaker to pull over, but he continued to drive. The Highway Patrol officers called on the radio for help. Units of the Los Angeles Police Department joined in the pursuit, one of them manned by petitioner Laurence Powell and his trainee, Timothy Wind. (The officers are all Caucasian; King is African American. King later explained that he fled because he feared that he would be returned to prison after having been released four months earlier following a year spent behind bars for robbery.)

King left the freeway, and after a chase of about eight miles, stopped at an entrance to a recreation area. The officers ordered King and his two passengers to exit the car and to assume a felony prone position—that is, to lie on their stomachs with legs spread and arms behind their backs. King's two friends complied. King, too, got out of the car but did not lie down. Petitioner Stacey Koon arrived, at once followed by Ted Briseno and Roland Solano. All were officers of the Los Angeles Police Department; and, as sergeant, Koon took charge. The officers again ordered King to assume the felony prone position. King got on his hands and knees but did not lie down. Officers Powell, Wind, Briseno, and Solano tried to force King down, but King resisted and became combative, so the officers retreated. Koon then fired Taser darts (designed to stun a combative suspect) into King.

The events that occurred next were captured on videotape by a bystander. As the videotape begins, it shows that King rose from the ground and charged toward Officer Powell. Powell took a step and used his baton to strike King on the side of his head. King fell to the ground. From the eighteenth to the thirtieth second on the videotape, King attempted to rise, but Powell and Wind each struck him with their batons to prevent him from doing so. From the thirty-fifth to the fifty-first second, Powell administered repeated blows to King's lower extremities; one of the blows fractured King's leg. At the fifty-fifth second, Powell struck King on the chest, and King rolled over and lay prone. At that point, the officers stepped back and observed King for about ten seconds. Powell began to reach for his handcuffs. (At the sentencing phase, the district court found that Powell no longer perceived King to be a threat at this point.) At one-minute-five-seconds (1:05) on the videotape, Briseno, in the District Court's words, "stomped" on King's upper back or neck. King's body writhed in response. At 1:07, Powell and Wind again began to strike King with a series of baton blows, and Wind kicked him in the upper thoracic or cervical area six times until 1:26. At about 1:29, King put his hands behind his back and was handcuffed.

Powell radioed for an ambulance. He sent two messages over a communications network to the other officers that said "oops" and "I haven't [sic] beaten anyone this bad in a long time." Koon sent a message to the police station that said: "Unit just had a big time use of force. . . . Tased and beat the suspect of CHP pursuit big time." King was taken to a hospital where he was treated for a fractured leg, multiple facial fractures, and numerous bruises and contusions. Learning that King worked at Dodger Stadium, Powell said to King: "We played a little ball tonight, didn't we Rodney? . . . You know, we played a little ball, we played a little hardball tonight, we hit quite a few home runs. . . . Yes, we played a little ball and you lost and we won."

Koon, Powell, Briseno, and Wind were tried in California state court on charges of assault with a deadly weapon and excessive use of force by a police officer. The officers were acquitted of all charges, with the exception of one assault charge against Powell that resulted in a hung jury. (The jury was composed of ten Caucasians, one Hispanic, and one Asian American.) The verdicts touched off widespread rioting in Los Angeles. More than 40 people were killed in the riots, more than 2,000 were injured, and nearly $1 billion in property was destroyed.(Los Angeles Mayor Tom Bradley declared that there "appears to be a dangerous trend of racially motivated incidents running through at least some segments of the police department," and President George H.W. Bush announced in May that the verdict had left him with a deep sense of personal frustration and anger and that he was ordering the Justice Department to initiate a prosecution against the officers.)

On August 4, 1992, a federal grand jury indicted the four officers, charging them with violating King's constitutional rights under color of law. Powell, Briseno, and Wind were charged with willful use of unreasonable force in arresting King. Koon was charged with willfully permitting the other officers to use unreasonable force during the arrest. After a trial in U.S. District Court for the Central District of California, the jury convicted Koon and Powell but acquitted Wind and Briseno. Koon and Powell were sentenced to thirty months in prison. This jury was comprised of nine Caucasians, two African Americans, and one Hispanic. King later won a $3.8 million verdict from the City of Los Angeles. He used some of the money to establish a rap record business.

The issue to consider is whether Officers King and Powell may be prosecuted and acquitted in California state court and then prosecuted in federal court. This seems to violate the prohibition on **double jeopardy** in the Fifth Amendment to the U.S. Constitution, which states that individuals shall not be "twice put in jeopardy of life or limb." Double jeopardy means that an individual should not be prosecuted more than once for the same offense. Without this protection, the government could subject people to a series of trials in an effort to obtain a conviction.

It may surprise you to learn that judges have held that the dual sovereignty doctrine permits the U.S. government to prosecute an individual under federal law who has been acquitted on the state level. The theory is that the state and federal governments are completely different entities and that state government is primarily concerned with punishing police officers and with protecting residents against physical attack, while the federal government is concerned with safeguarding the civil liberties of all Americans. Each of these entities provides a check on the other to ensure fairness for citizens. The evidence introduced in the two prosecutions to establish the police officers' guilt in the King case was virtually identical, and the federal prosecution likely was brought in response to political pressure. On the other hand, the federal government historically has acted to prevent unfair verdicts, such as the acquittal of members of the Ku Klux Klan charged with killing civil rights workers during the 1960s.

Do you believe that it was fair to subject the Los Angeles police officers to the expense and emotional stress of two trials? As the attorney general to the United States, would you have advised President George H.W. Bush to bring federal charges against the officers following their acquittal by a California jury?

You can find the answer at www.sagepub.com/lippmanccl3e.

CHAPTER SUMMARY

Criminal law is the foundation of the criminal justice system. The law defines the acts that may lead to arrest, trial, and incarceration. We typically think about crime as involving violent conduct, but in fact a broad variety of acts are defined as crimes.

Criminal law is best defined as conduct that if shown to have taken place, will result in the "formal and solemn pronouncement of the moral condemnation of the community." Civil law is distinguished from criminal law by the fact that it primarily protects the interests of the individual rather than the interests of society.

The purpose of criminal law is to prohibit conduct that causes harm or threatens harm to the individual and to the public interests, to warn people of the acts that are subject to criminal punishment, to define criminal acts and intent, to distinguish between serious and minor offenses, to punish offenders, and to ensure that the interests of victims and the public are represented at trial and in the punishment of offenders.

In analyzing individual crimes, we will be concerned with several basic concerns that compose the general part of the criminal law. A crime is composed of a concurrence between a criminal act (*actus reus*) and criminal intent (*mens rea*) and the causation of a social harm. Individuals must be provided with notice of the acts that are criminally condemned in order to have the opportunity to obey or to violate the law. Individuals must also be given the opportunity at trial to present defenses (justifications and excuses) to a criminal charge.

The criminal law distinguishes between felonies and misdemeanors. A crime punishable by death or by imprisonment for more than one year is a felony. Other offenses are misdemeanors. Offenses are further divided into capital and other grades of felonies and into gross and petty misdemeanors. A third level of offenses are violations or infractions, acts that are punishable by fines.

Another approach is to classify crime in terms of "moral turpitude." *Mala in se* crimes are considered "inherently evil," and *mala prohibita* crimes are not inherently evil and are only considered wrong because they are prohibited by statute.

Our textbook categorizes crimes in accordance with the subject matter of the offense, the scheme that is followed in most state criminal codes. This includes crimes against the state, crimes against the person, crimes against habitation, crimes against property, crimes against public order, and crimes against the administration of justice.

There are a number of sources of American criminal law. These include the common law, state and federal criminal codes, the U.S. and state constitutions, international treaties, and judicial decisions. The English common law was transported to the United States and formed the foundation for the American criminal statutes adopted in the nineteenth and twentieth centuries. Some states continue to apply the common law in those instances in which the state legislature has not adopted a criminal statute. In code jurisdiction states, however, crimes only are punishable if incorporated into law.

States possess broad police powers to legislate for the public health, safety, and welfare of the residents of the state. The drafting of state criminal statutes has been heavily influenced by the American Law Institute's Model Penal Code, which has helped ensure a significant uniformity in the content of criminal codes.

The United States has a system of dual sovereignty in which the state governments have provided the federal government with the authority to legislate various areas of criminal law. The Supremacy Clause provides that federal law takes precedence over state law in the areas that the U.S. Constitution explicitly reserves to the national government. There is a trend toward strictly limiting the criminal law power of the federal government. The U.S. Supreme Court, for example, has ruled that the federal government has unconstitutionally employed the Interstate Commerce Clause to extend the reach of federal criminal legislation to the possession of a firearm adjacent to schools.

The authority of the state and federal governments to adopt criminal statutes is limited by the provisions of federal and state constitutions. For instance, laws must be drafted in a clear and nondiscriminatory fashion and must not impose retroactive or cruel or unusual punishment. The federal and state governments possess the authority to enact criminal legislation only within their separate spheres of constitutional power.

CHAPTER REVIEW QUESTIONS

1. Define a crime.

2. Distinguish between criminal and civil law. Distinguish between a criminal act and a tort.

3. What is the purpose of criminal law?

4. Is there a difference between criminal law and criminal procedure? Distinguish between the specific and general part of the criminal law.

5. List the basic principles that compose the general part of criminal law.

6. Distinguish between felonies, misdemeanors, capital felonies, gross and petty misdemeanors, and violations.

7. What is the difference between *mala in se* and *mala prohibita* crimes?

8. Discuss the development of the common law. What do we mean by common law states and code jurisdiction states?

9. Discuss the nature and importance of the state police power.

10. Why is the Model Penal Code significant?

11. What is the legal basis for federal criminal law? Define the preemption doctrine and dual sovereignty. What is the significance of the Interstate Commerce Clause?

12. What are the primary sources of criminal law? How does the U.S. Constitution limit the criminal law?

13. Why is understanding the criminal law important in the study of the criminal justice system?

LEGAL TERMINOLOGY

capital felony	federal criminal code	Model Penal Code
civil law	felony	petty misdemeanor
code jurisdiction	gross misdemeanor	police power
common law crimes	infamous crimes	preemption doctrine
common law states	infractions	reception statutes
crime	Interstate Commerce Clause	substantive criminal law
criminal procedure	*mala in se*	Supremacy Clause
defendant	*mala prohibita*	tort
double jeopardy	misdemeanor	violation
dual sovereignty		

CRIMINAL LAW ON THE WEB

Log on to the Web-based student study site at www.sagepub.com/lippmanccl3e to assist you in completing the Criminal Law on the Web exercises, as well as for additional features such as podcasts, Web quizzes, and video links.

1. A number of sites contain collections of state and federal laws and links to state criminal cases. As a first step, go to www.findlaw.com, click on Criminal Law, and read about the steps in a criminal case. This is also a good site at which to find the definitions of various crimes. Then explore the site maintained by the Cornell University Law School, and find the criminal law statutes of the state in which you live. You also might want to go to www.lawsource.com.

2. Learn more about the Rodney King case. Would you have convicted the police officers?

3. You may also want to ask yourself whether it is possible for an innocent individual to be convicted. The Innocence Project works to exonerate the wrongfully convicted.

4. Read about medical marijuana laws.

BIBLIOGRAPHY

Joshua Dressler, *Understanding Criminal Law,* 3rd ed. (New York: Lexis, 2001), pp. 1–32. A good introduction to the nature of criminal law and to the common law and statutes.

George Fletcher, *Rethinking Criminal Law* (New York: Oxford University Press, 2000). A challenging discussion of the history and philosophy of criminal law.

Lawrence M. Friedman, *Crime and Punishment in American History* (New York: Basic Books, 1993). A history of criminal justice in the United States.

Hyman Gross, *A Theory of Criminal Justice* (New York: Oxford University Press, 1979). A highly sophisticated discussion of the philosophical basis of criminal law.

Jerome Hall, *General Principles of Criminal Law,* 2nd ed. (Indianapolis, IN: Bobbs-Merrill, 1960), pp. 1–359. A detailed discussion of the theory of criminal law and the basic elements of a crime.

Wayne R. LaFave, *Criminal Law,* 3rd ed. (St. Paul, MN: West Publishing, 2000), pp. 1–198. A comprehensive discussion of the characteristics, purpose, classification, and common law foundation of criminal law.

Rollin M. Perkins and Ronald N. Boyce, *Criminal Law,* 3rd ed. (Mineola, NY: Foundation Press, 1982), pp. 1–46. A sophisticated analysis of the purpose, definition, and classification of criminal law and development of the common law.

Appendix

Reading and Briefing Cases

INTRODUCTION

A unique aspect of studying criminal law is that you have the opportunity to read actual court decisions. Reading cases will likely be a new experience, and although you may encounter some initial frustrations, in my experience students fairly quickly master the techniques of legal analysis.

The case method was introduced in 1870 by Harvard law professor Christopher Columbus Langdell and is the primary method of instruction in nearly all American law schools. This approach is based on the insight that students learn the law most effectively when they study actual cases. Langdell encouraged instructors to employ a question and answer classroom technique termed the **Socratic method**. The most challenging aspect of this approach involves posing *hypothetical* or fictitious examples that require students to apply the case material to new factual situations.

The study of cases assists you to

- understand the principles of criminal law,
- improve your skills in critical reading and thinking,
- acquaint yourself with legal vocabulary and procedures,
- appreciate how judges make decisions, and
- learn to apply the law to the facts.

The cases in this textbook have been edited to highlight the most important points. Some nonessential material has been omitted to assist you in reading and understanding the material. You may want to read the entire, unedited case in the library or online.

The cases you read are the products of an *adversary system* in which the prosecutors and defense attorneys present evidence and examine witnesses at trial. The evaluation of the facts is the responsibility of the jury or, in the absence of a jury, the judge. A case heard by a judge without a jury is termed a **bench trial**. The adversary system is premised on the belief that truth will emerge from the clash between two dedicated attorneys "zealously presenting their cause."

The lowest courts in the judicial hierarchy are *trial courts*. The proceedings are recorded in **trial transcripts** that recite the selection of jurors, testimony of witnesses, arguments of lawyers, and rulings by the judge. Individuals convicted before a trial court may appeal the guilty verdict to **appellate** (or appeal) **courts**. The cases you read in this book in most instances are the decisions issued by appellate court judges reviewing a guilty verdict entered against a **defendant** at trial. These reviews are based on transcripts and briefs. **Briefs** are lengthy written arguments submitted to the court by the prosecution and defense. The two sides may also have the opportunity to engage in an **oral argument** before the appellate court. In issuing a decision, the appellate court will accept as established those facts that are most favorable to the party that prevailed at the trial court level.

Defendants appealing a verdict by a trial court ordinarily file an appeal with the intermediate court of appeals, which in many states can provide the defendant with a new trial or **trial de novo**. The losing party may then file an additional appeal to the state supreme court. The party who is appealing is termed the **appellant**, and the second name is typically the party against whom the appeal is filed or the **appellee**. You also will notice the insertion of *"v."* between the names of the parties, which is an abbreviation for the Latin *versus*.

Individuals who have been convicted and have exhausted their state appeals may file a constitutional challenge or **collateral attack** against their conviction in federal court provided that they have raised a federal constitutional claim in their state appeals. The first name in the title is the name of the prisoner bringing the case, or the **petitioner**, and the second name, or **respondent**, is typically the warden or individual in charge of the prison in which the petitioner is incarcerated.

In a collateral attack, an inmate bringing the action files a petition for **habeas corpus** review requesting a federal court to issue an order requiring the state to demonstrate that the petitioner is lawfully incarcerated. The ability of a petitioner to compel the state to demonstrate that he or she has been lawfully detained is one of the most important safeguards for individual liberty and is guaranteed in Article I, Section 9, Clause 2 of the U.S. Constitution.

Federal courts may also preside over criminal cases charging a defendant with a violation of a federal statute. There are three levels of federal courts. First, there are ninety-four district courts, which are the trial courts. Appeals may be taken to the thirteen courts of appeals and ultimately to the U.S. Supreme Court. The U.S. Supreme Court generally may choose whether to review a case. Four of the nine justices must vote to grant a **writ of certiorari** or an order to review the decision of a lower court.

THE STRUCTURE OF CASES

A case is divided into an *introduction* and *judicial opinion.* These two sections have several components that you should keep in mind.

Introduction

The initial portion of a case is divided into title, citation, and identification of the judge.

Title

Cases are identified by the names of the parties involved in the litigation. At the trial level, this typically involves the prosecuting authority (a city, county, state, or the federal government) and the name of the defendant. On direct appeals, the first name refers to the appellant who is bringing the appeal and the second to the appellee who is defending against the appeal. On collateral attack, remember that the parties are termed petitioners and respondents. You will notice that judicial decisions often use a shorthand version of a case and refer only to one of the parties, much like calling someone by his or her first or last name.

Citation

Immediately following the names you will find the citation that directs you to the book or **legal reporter** where you can find the case in a law library. Increasingly, cases are also becoming available online. The standard form for citations of cases, statutes, and law journals is contained in *The Bluebook* published by the Harvard Law Review Association.

Judge

The name of the judge who wrote the opinion typically appears at the beginning of the case. An opinion written by a respected judge may prove particularly influential with other courts. The respect accorded to a judge may also be diminished if his or her decisions have frequently been reversed by appellate courts.

Outline

The full, unedited cases in legal reporters typically begin with a list of numbered paragraphs or **head notes** that outline the main legal points in the case. There is also a summary of the case and of the decisions of other courts that have heard the case. These outlines have been omitted from the edited cases reprinted in this book.

Judicial Opinion

The judge's legal discussion is referred to as the opinion, judgment, or decision. The opinion is usually divided into history, facts, and law. These component parts are not always neatly distinguished, and you may have to organize the material in your mind as you read the case.

History

The initial portion of a case typically provides a summary of the decisions of the lower courts that previously considered the case and the statutes involved.

Facts

Each case is based on a set of facts that present a question to be answered by the judge. This question, for instance, may involve whether a defendant acted in self-defense or whether an individual cleaning his or her rifle intentionally or accidentally killed a friend. This question is termed the *issue*. The challenge is to separate the relevant from the irrelevant facts. A **relevant** fact is a fact that assists in establishing the existence or nonexistence of a *material fact* or element of the crime that the government is required to prove beyond a reasonable doubt at trial. For instance, in the gun example, whether the defendant possessed a motive to kill the victim would be relevant in establishing the material element of whether the defendant possessed a specific intent to kill.

Law

The judge then applies the legal rule to the facts and reaches a **holding** or decision. The **reasoning** is the explanation offered by the judge for the holding. Judges also often include comments and observations (in Latin **obiter dicta**, or comments from the bench) on a wide range of legal and factual concerns that provide important background but may not be central to the holding. These comments may range from legal history to a discussion of a judge's philosophy of punishment.

Judges typically rely on **precedents** or the holdings of other courts. Precedent or "**stare decisis** *et no quieta movere*" literally translates as "to stand by precedent and to stand by settled points." The court may follow a precedent or point out that the case at hand should be distinguished from the precedent and calls for a different rule, which is called a **distinguishing precedent**.

Appellate courts are typically composed of a **multiple judge panel** consisting of three or more judges, depending on the level of the court. The judges typically meet and vote on a case and issue a **majority opinion**, which is recognized as the holding in the case. Judges in the majority may choose to write a **concurring opinion** supporting the majority, which is typically based on slightly different grounds. On occasion, a majority of judges agree on the outcome of a case but are unable to reach a consensus on the reasoning. In these instances there is typically a **plurality opinion** as well as one or more concurring opinions. In cases in which a court issues a plurality opinion, the decisions of the various judges in the majority must be closely examined to determine the precise holding of the case. You may encounter a **per curiam** opinion. This is an opinion that is not attributed to a particular judge; it is attributed as by the court.

A judge in the minority has the discretion to write a **dissenting opinion**. Other judges in the minority may also issue separate opinions or join the dissenting opinion of another judge. In those instances in which a court is closely divided, the dissenting opinion with the passage of time may come to reflect the view of a majority of the members of the court. The dissent may also influence the majority opinion. The judges in the majority may feel compelled to answer the claims of the dissent or to compromise in order to attract judges who may be sympathetic to the dissent.

You should keep in mind that cases carry different degrees of authority. The decisions of the Ohio Supreme Court possess **binding authority** on lower courts within Ohio. The decision of a lower-level Ohio court that fails to follow precedent will likely be appealed by the losing party and reversed by the appellate court. The decisions of the Ohio Supreme Court, however, are not binding on lower courts outside of Ohio, but may be considered by these other tribunals to possess **persuasive authority**. Of course, precedents are not written in stone, and courts will typically adjust the law to meet new challenges.

As you read the edited cases reprinted in this textbook, you will notice that the cases are divided into various sections. The "facts" of the case and the "issue" to be decided by the court are typically followed by the court's "reasoning" or justification and "holding" or decision. A number of questions appear at the end of the case to help you understand the opinion.

BRIEFING A CASE

Your instructor may ask you to **brief** or summarize the main points of the cases reprinted in this textbook. A student brief is a concise, shorthand written description of the case and is intended to assist you in *understanding* and *organizing* the material and *in preparing for class and examinations*. A brief generally includes several standard features. These, of course, are only broad guidelines, and there are differing opinions on the proper form of a brief. Bear in mind that a particular case that you are reading may not be easily reduced to a standard format.

1. *The Name of the Case and the Year the Case Was Decided.* The name of the case will help you in organizing your class notes. Including the year of decision places the case in historical context and alerts you to the possibility that an older decision may have been revised in light of modern circumstances.

2. *The State or Federal Court Deciding the Case and the Judge Writing the Decision.* This will assist you in determining the place of the court in the judicial hierarchy and whether the decision constitutes a precedent to be followed by lower-level courts.

3. *Facts.* Write down the relevant facts. You should think of this as a story that has a factual beginning and conclusion. The best approach is to put the facts into your own words. *Pay particular attention to*

   a. the background facts leading to the defendant's criminal conduct;

   b. the defendant's criminal act, intent, and motives; and

   c. the relevant facts as distinguished from the irrelevant facts.

4. *Criminal Charge.* Identify the crime with which the defendant is charged and the text of the relevant criminal statute.

5. *The Issue That the Court Is Addressing in the Case.* This is customarily in the form of a question in the brief and typically is introduced by the word "whether." For instance, the issue might be "whether section 187 of the California criminal code punishing the unlawful killing of a human being includes the death of a fetus."

6. *Holding.* Write down the legal principle formulated by the court to answer the question posed by the issue. This requires only a statement that the "California Supreme Court ruled that section 187 does not include a fetus."

7. *Reasoning.* State the reasons that the court provides for the holding. Note the key precedents the court cites and relies on in reaching its decision. Ask yourself whether the court's reasoning is logical and persuasive.

8. *Disposition.* An appellate court may *affirm* and uphold the decision of a lower court or *reverse* the lower court judgment. In addition, a lower court's decision may be *reversed in part and affirmed in part.* Lastly the appellate court may *reverse* the lower court and *remand* or return the case for additional judicial action. Take the time to understand the precise impact of the court decision.

9. *Concurring and Dissenting Opinions.* Note the arguments offered by judges in concurring and dissenting opinions.

10. *Public Policy and Psychology.* Consider the impact of the decision on society and the criminal justice system. In considering a court decision, do not overlook the psychological, social, and political factors that may have affected the judge's decision.

11. *Personal Opinion.* Sketch your own judicial opinion and note whether you agree with the holding of the case and the reasoning of the court.

APPROACHING THE CASE

You will most likely develop a personal approach to reading and briefing cases. You might want to keep the following points in mind:

- *Skim the case.* This will enable you to develop a sense of the issue, facts, and holding of the case.
- *Read the case slowly a second time.* You may find it helpful in the beginning to read the case out loud and write notes in the margin.
- *Write down the relevant facts in your own words.*
- *Identify the relevant facts, issues, reasoning, and holding. You should not merely mechanically copy the language of the case.* Most instructors suggest that you express the material in your own words in order to improve your understanding. You should pay careful attention to the legal language. For instance, there is a significant difference between a statute that provides that an individual who "reasonably believes" that he

or she is being attacked is entitled to self-defense, and a statute that provides that an individual who "personally believes" that he or she is being attacked is entitled to self-defense. The first is an objective test measured by a "reasonable person," and the second is a "subjective test" measured by the victim's personal perception. Can you explain the difference? You should incorporate legal terminology into your brief. The law, like tennis or music, possesses a distinctive vocabulary that is used to express and communicate ideas.

- *Consult the glossary or a law dictionary for the definition of unfamiliar legal terms, and write down questions that you may have concerning the case.*
- *The brief should be precise and limited to essential points. You should bring the brief to class and compare your analysis to the instructor's.* Modify the brief to reflect the class discussion, and provide space for insights developed in class.
- *Consider that each case is commonly thought of as "standing for a legal proposition."* Some instructors suggest that you write the legal rule contained in the case as a "banner" across the first page of the brief.
- *Consider why the case is included in the textbook and how the case fits into the general topic covered in the chapter. Remain an active and critical learner, and think about the material you are reading.* You should also consider how the case relates to what you learned earlier in the course. Bring a critical perspective to reading the case, and resist mechanically accepting the court's judgment. Keep in mind that there are at least two parties involved in a case, each of whom may have a persuasive argument. Most important, remember that briefing is a learning tool; it should not be so time-consuming that you fail to spend time understanding and reflecting on the material.
- *Consider how the case may relate to other areas you have studied.* A case on murder may also raise interesting issues concerning criminal intent, causality, and conspiracy. Thinking broadly about a case will help you integrate and understand criminal law.
- *Outline the material.* Some instructors may suggest that you develop an outline of the material covered in class. This can be used to assist you in preparing for examinations.

LOCATING CASES

The names of the cases are followed by a set of numbers and alphabetical abbreviations. These abbreviations refer to various legal reporters in which the cases are published. This is useful in the event that you want to read an unedited version in the library. An increasing number of cases are also available online. The rules of citation are fairly technical and are of immediate concern only to practicing attorneys. The following discussion presents the standard approach to citation used by lawyers. Those of you interested in additional detail should consult *The Bluebook: A Uniform System of Citation,* 18th edition (Cambridge, MA: Harvard Law Review Association, 2005).

The first number you encounter is the volume in which the case appears. This is followed by the abbreviation of the reporter and by the page number and year of the decision. State cases are available in "regional reporters" that contain appellate decisions of courts in various geographic areas of the United States. These volumes are cited in accordance with standard abbreviations: Atlantic (A.), Northeast (N.E.), Pacific (P.), Southeast (S.E.), South (S.), and Southwest (S.W.). The large number of cases decided has necessitated the organization of these reporters into various "series" (e.g., P.2d and P.3d).

Individual states also have their own reporter systems containing the decisions of intermediate appellate courts and state supreme courts. Decisions of the Nebraska Supreme Court appear in the Northwest Reporter (N.W. or N.W.2d) as well as in the Nebraska Reports (Neb.). The decisions of the Nebraska Court of Appeals are reprinted in Nebraska Court of Appeals (Neb. Ct. App.). These decisions are usually cited to the Northwest Reporter, for example, *Nebraska v. Metzger,* 319 N.W.2d 459 (Neb. 1982). New York and California cases appear in state and regional reporters as well as in their own national reporter.

The federal court reporters reprint the published opinions of federal trials as well as appellate courts. District court (trial) opinions appear in the Federal Supplement Reporter (F.Supp) and appellate court opinions are reprinted in the Federal Reporter (F.), both of which are printed in several series (F.Supp.2d; F.2d and F.3d). These citations also provide the name of the federal court that decided the case. The Second Circuit Court of Appeals in New York, for instance, is cited as *United States v. MacDonald,* 531 F.2d 196 (2nd Cir. 1976). The standard citation for U.S. Supreme Court decisions is the United States Reports (U.S.), for example, *Papachristou v. Jacksonville,* 405 U.S. 156 (1971). This is the official version issued by the Supreme Court; the decisions are also available in two privately published reporters, the Supreme Court Reporter (S. Ct.) and Lawyers edition (L. Ed.).

There is a growing trend for cases to appear online in commercial electronic databases. States are also beginning to adopt "public domain citation formats" for newly decided cases that appear on state court Web pages.

These are cited in accordance with the rules established by the state judiciary. The standard format includes the case name, the year of decision, the state's two-digit postal abbreviation, the abbreviation of the court in the event that this is not a state supreme court decision, the number assigned to the case, and the paragraph number. A parallel citation to the relevant regional reporter is also provided. *The Bluebook* provides examples of this format. The following example is for a state supreme court case: *Gregory v. Class,* 1998 SD 106, ¶ 3, 54 N.W.2d 873, 875.

LEGAL TERMINOLOGY

appellant	habeas corpus	plurality opinion
appellate courts	head notes	precedent
appellee	holding	reasoning
bench trial	legal reporters	relevant
binding authority	majority opinion	respondent
brief	multiple judge panel	Socratic method
collateral attack	*obiter dicta*	*stare decisis*
concurring opinion	oral argument	trial de novo
defendant	per curiam	trial transcript
dissenting opinion	persuasive authority	writ of certiorari
distinguishing precedents	petitioner	

2 CONSTITUTIONAL LIMITATIONS

Was the defendant discriminated against based on gender?

Wright, six feet tall and weighing 216 pounds, beat and kicked his wife Wendy on the evening of February 16, 1999. Her injuries were so severe that two of her ribs were fractured and her spleen had to be removed. Wright was indicted for criminal domestic violence of a high and aggravated nature. The aggravating factors alleged in the indictment were "a difference in the sexes of the victim and the defendant" and/or that "the defendant did inflict serious bodily harm upon the victim by kicking her in the mid-section requiring her to seek medical attention." Wright contends the judge's charge on the aggravating circumstance of a "difference in the sexes" violated his right to "equal protection."

Learning Objectives

1. Explain the rule of legality.

2. Distinguish between bills of attainder and ex *post facto laws.*

3. Understand the importance of statutory clarity and know the legal test for identifying laws that are void for vagueness.

4. Describe the three levels of scrutiny under the Equal Protection Clause.

5. Discuss the importance of freedom of expression and the categories of expression that are not protected by the First Amendment.

6. Understand the constitutional basis for the right to privacy and the type of acts that are protected within the "zone of privacy."

7. Know the meaning of the Second Amendment right to bear arms.

INTRODUCTION

In the American democratic system, various constitutional provisions limit the power of the federal and state governments to enact criminal statutes. For instance, a statute prohibiting students from criticizing the government during a classroom discussion would likely violate the First Amendment to the U.S. Constitution. A law punishing individuals engaging in "unprotected" sexual activity, however socially desirable, may unconstitutionally violate the right to privacy.

Why did the framers create a **constitutional democracy**, a system of government based on a constitution that limits the powers of the government? The founding fathers were profoundly influenced by the harshness of British colonial rule and drafted a constitution designed to protect the rights of the individual against the tyrannical tendencies of government. They wanted to ensure that the police could not freely break down doors and search homes. The framers were also sufficiently wise to realize that individuals required constitutional safeguards against the political passions and intolerance of democratic majorities.

The limitations on government power reflect the framers' belief that individuals possess natural and inalienable rights, and that these rights may be restricted only when absolutely necessary to ensure social order and stability. The stress on individual freedom was also practical. The framers believed that the fledgling new American democracy would prosper and develop by freeing individuals to passionately pursue their hopes and dreams.

At the same time, the framers were not wide-eyed idealists. They fully appreciated that individual rights and liberties must be balanced against the need for social order and stability. The striking of this delicate balance is not a scientific process. A review of the historical record indicates that at times, the emphasis has been placed on the control of crime and, at other times, stress has been placed on individual rights.

Chapter 2 describes the core constitutional limits on the criminal law and examines the balance between order and individual rights. Consider the costs and benefits of constitutionally limiting the government's authority to enact criminal statutes. Do you believe that greater importance should be placed on guaranteeing order or on protecting rights? You should keep the constitutional limitations discussed in this chapter in mind as you read the cases in subsequent chapters. The topics covered in the chapter are as follows:

- The first principle of American jurisprudence is the rule of legality.
- Constitutional constraints include the following:
  - Bills of attainder and *ex post facto* laws
  - Statutory clarity
  - Equal protection
  - Freedom of speech
  - Privacy
  - The right to bear arms

We will discuss an additional constitutional constraint, the Eighth Amendment prohibition on cruel and unusual punishment, in Chapter 3.

THE RULE OF LEGALITY

The **rule of legality** has been characterized as "the first principle of American criminal law and jurisprudence."[1] This principle was developed by common law judges and is interpreted today to mean that an individual may not be criminally punished for an act that was not clearly condemned in a statute prior to the time that the individual committed the act.[2] The doctrine of legality is nicely summarized in the Latin expression ***nullum crimen sine lege, nulla poena sine lege,*** meaning "no crime without law, no punishment without law." The doctrine of legality is reflected in two constitutional principles governing criminal statutes:

- the constitutional prohibition on bills of attainder and *ex post facto* laws, and
- the constitutional requirement of statutory clarity.

BILLS OF ATTAINDER AND *EX POST FACTO* LAWS

Article I, Sections 9 and 10 of the U.S. Constitution prohibit state and federal legislatures from passing **bills of attainder** and ***ex post facto* laws**. James Madison characterized these provisions as a "bulwark in favor of personal security and personal rights."[3]

Bills of Attainder

A bill of attainder is a legislative act that punishes an individual or a group of persons without the benefit of a trial. The constitutional prohibition of bills of attainder was intended to safeguard Americans from the type of arbitrary punishments that the English Parliament directed against opponents of the Crown. The Parliament disregarded the legal process and directly ordered that dissidents should be imprisoned, executed, or banished and forfeit their property.[4] The prohibition of a bill of attainder was successfully invoked in 1946 by members of the American Communist Party, who were excluded by Congress from working for the federal government.[5]

Ex Post Facto Laws

Alexander Hamilton explained that the constitutional prohibition on *ex post facto* laws was vital because "subjecting of men to punishment for things which, when they were done were breaches of no law, and the practice of arbitrary imprisonments, have been, in all ages, the favorite and most formidable instrument of tyranny."[6] In 1798, Supreme Court Justice Samuel Chase in *Calder v. Bull* listed four categories of *ex post facto* laws:[7]

- Every law that makes an action, done before the passing of the law, and was *innocent* when done, criminal; and punishes such action.
- Every law that *aggravates* a crime, or makes it *greater* than it was, when committed.
- Every law that *changes the punishment,* and inflicts a *greater punishment,* than the law annexed to the crime, when committed.
- Every law that alters the *legal* rules of *evidence,* and receives less, or different, testimony, than the law required at the time of the commission of the offense, *in order to convict the offender.*

The constitutional rule against *ex post facto* laws is based on the familiar interests in providing individuals notice of criminal conduct and protecting individuals against retroactive "after the fact" statutes. Supreme Court Justice John Paul Stevens noted that all four of Justice Chase's categories are "mirror images of one another. In each instance, the government refuses, after the fact, to play by its own rules, altering them in a way that is advantageous only to the State, to facilitate an easier conviction."[8]

In summary, the prohibition on *ex post facto* laws prevents legislation being applied to *acts committed before the statute went into effect.* The legislature is free to declare that in the *future* a previously innocent act will be a crime. Keep in mind that the prohibition on *ex post facto* laws is directed against enactments that disadvantage defendants; legislatures are free to retroactively assist defendants by reducing the punishment for a criminal act.

The distinction between bills of attainder and *ex post facto* laws is summarized as follows:

- A bill of attainder punishes a specific individual or specific individuals. An *ex post facto* law criminalizes an act that was legal at the time the act was committed.
- A bill of attainder is not limited to criminal punishment and may involve any disadvantage imposed on an individual; *ex post facto* laws are limited to criminal punishment.
- A bill of attainder imposes punishment on an individual without trial. An *ex post facto* law is enforced in a criminal trial.

The Supreme Court and *Ex Post Facto* Laws

Determining whether a retroactive application of the law violates the prohibition on *ex post facto* laws has proven more difficult than might be imagined given the seemingly straightforward nature of this constitutional ban.

In *Stogner v. California,* the Supreme Court ruled that a California law authorizing the prosecution of allegations of child abuse that previously were barred by a three-year statute of limitations constituted a prohibited *ex post facto* law.[9] This law was challenged by Marion Stogner, who found himself indicted for child abuse after having lived the past nineteen years without fear of criminal prosecution for an act committed twenty-two years prior. Justice Stephen Breyer ruled that California acted in an "unfair" and "dishonest" fashion in subjecting Stogner to prosecution many years after the State had assured him that he would not stand trial. Judge Anthony Kennedy argued in dissent that California merely reinstated a prosecution that was previously barred by the three-year statute of limitations. The penalty attached to the crime of child abuse remained unchanged. What is your view?

We now turn our attention to the requirement of statutory clarity.

STATUTORY CLARITY

The Fifth and Fourteenth Amendments of the U.S. Constitution prohibit depriving individuals of "life, liberty or property without due process of law." Due process requires that criminal statutes should be drafted in a clear and understandable fashion. A statute that fails to meet this standard is unconstitutional on the grounds that it is **void for vagueness**.

Due process requires that individuals receive notice of criminal conduct. Statutes are required to define criminal offenses with sufficient *clarity* so that ordinary individuals are able to understand what conduct is prohibited.

Due process requires that the police, prosecutors, judges, and jurors are provided with a reasonably clear statement of prohibited behavior. The requirement of definite standards ensures the uniform and nondiscriminatory enforcement of the law.

In summary, due process ensures clarity in criminal statutes. It guards against individuals being deprived of life (the death penalty), liberty (imprisonment), or property (fines) without due process of law.

Clarity

Would a statute that punishes individuals for being a member of a gang satisfy the test of statutory clarity? The U.S. Supreme Court, in *Grayned v. Rockford,* ruled that a law was void for vagueness that punished an individual "known to be a member of any gang consisting of two or more persons." The Court observed that "no one may be required at peril of life, liberty or property to speculate as to the meaning of [the term gang in] penal statutes."[10]

In another example, the Supreme Court ruled in *Coates v. Cincinnati* that an ordinance was unconstitutionally void for vagueness that declared that it was a criminal offense for "three or more persons to assemble . . . on any of the sidewalks . . . and there conduct themselves in a manner annoying to persons passing by." The Court held that the statute failed to provide individuals with reasonably clear guidance because "conduct that annoys some people does not annoy others," and that an individual's arrest may depend on whether he or she happens to "annoy" a "police officer or other person who should happen to pass by." This did not mean that Cincinnati was helpless to maintain the city sidewalks; the city was free to prohibit people from "blocking sidewalks, obstructing traffic, littering streets, committing assaults, or engaging in countless other forms of antisocial conduct."[11]

Definite Standards for Law Enforcement

Edward Lawson was detained or arrested on roughly fifteen occasions between March and July 1977. Lawson certainly stood out; he was distinguished by his long dreadlocks and habit of wandering the streets of San Diego at all hours. Lawson did not carry any identification, and each of his arrests was undertaken pursuant to a statute that required that an individual detained for investigation by a police officer present "credible and reliable" identification that carries a "reasonable assurance" of its authenticity and that provides "means for later getting in touch with the person who has identified himself."[12]

The U.S. Supreme Court explained in *Kolender v. Lawson* that the void-for-vagueness doctrine was aimed at ensuring that statutes clearly inform citizens of prohibited acts and simultaneously provide definite standards for the enforcement of the law. The California statute was clearly void for vagueness, because no standards were provided for determining what constituted "credible and reliable" identification, and "complete discretion" was vested in the police to determine whether a suspect violated the statute. Was a library or credit card or student identification "credible and reliable" identification? A police officer explained at trial that a jogger who was not carrying identification might satisfy the statute by providing his or her running route or name and address. Did this constitute "credible and reliable" identification? The Court was clearly concerned that a lack of definite standards opened the door to the police using the California statute to arrest individuals based on their race, gender, or appearance.

Due process does not require "impossible standards" of clarity, and the Supreme Court stressed that this was not a case in which "further precision" was "either impossible or impractical." There seemed to be little reason why the legislature could not specify the documents that would satisfy the statutory standard and avoid vesting complete discretion in the "moment-to-moment judgment" of a police officer on the street. Laws were to be made by the legislature and enforced by the police: "To let a policeman's command become equivalent to a criminal statute comes dangerously near to making our government one of men rather than laws."[13]

The Supreme Court has stressed that the lack of standards presents the danger that a law will be applied in a discriminatory fashion against minorities and the poor. In *Papachristou v. Jacksonville,* the U.S. Supreme Court expressed the concern that a broadly worded vagrancy statute punishing "rogues and vagabonds"; "lewd, wanton and lascivious persons"; "common railers and brawlers"; and "habitual loafers" failed to provide standards for law enforcement and risked that the poor, minorities, and nonconformists would be targeted for arrest based on the belief that they posed a threat to public safety.[14] The court humorously noted that middle-class individuals who frequented the local country club were unlikely to be arrested, although they might be guilty under the ordinance of "neglecting all lawful business and habitually spending their time by frequenting . . . places where alcoholic beverages are sold or served."[15]

Broadly worded statutes are a particular threat in a democracy in which we are committed to protecting even the most extreme nonconformist from governmental harassment. The U.S. Supreme Court, in *Coates v. Cincinnati,* expressed concern that the lack of clear standards in the local ordinance might lead to the arrest of individuals who were exercising their constitutionally protected rights. Under the Cincinnati statute, association and assembly on the public streets would be "continually subject" to whether the demonstrators' "ideas, their lifestyle, or their physical appearance is resented by the majority of their fellow citizens."[16]

Void for Vagueness

Judges are aware that language cannot achieve the precision of a mathematical formula. Legislatures are also unable to anticipate every possible act that may threaten society, and understandably they resort to broad language. Consider the obvious lack of clarity of a statute punishing a "crime against nature." In *Horn v. State,* the defendant claimed that a law punishing a "crime against nature" was vague and indefinite and failed to inform him that he was violating the law in raping a ten-year-old boy. An Alabama court ruled that the definition of a "crime against nature" was widely discussed in legal history and was "too disgusting and well known" to require further details or description.[17] Do you agree?

Judges appreciate the difficulty of clearly drafting statutes and typically limit the application of the void-for-vagueness doctrine to cases in which the constitutionally protected rights and liberties of people to meet, greet, congregate in groups, move about, and express themselves are threatened.

A devil's advocate may persuasively contend that the void-for-vagueness doctrine provides undeserved protection to "wrongdoers." In *Nebraska v. Metzger,* a neighbor spotted Metzger standing naked with his arms at his sides in the large window of his garden apartment for roughly five seconds.[18] The neighbor testified that he saw Metzger's body from "his thighs on up." The police were called and observed Metzger standing within a foot of the window eating a bowl of cereal and noted that "his nude body, from the mid-thigh on up, was visible." The ordinance under which Metzger was charged and convicted made it unlawful to commit an "indecent, immodest or filthy act within the presence of any person, or in such a situation that persons passing might ordinarily see the same." The Nebraska Supreme Court ruled that this language provided little advance notice as to what is lawful and what is unlawful and could be employed by the police to arrest individuals for entirely lawful acts that some might consider immodest, including holding hands, kissing in public, or wearing a revealing swim suit. Could Metzger possibly believe that there was no legal prohibition on his standing nude in his window? Keep these points in mind as you read the first case in the textbook, *State v. Stanko.*

Did the defendant know that he was driving at an excessive rate of speed?

State v. Stanko, 974 P.2d 1132 (Mont. 1998). Opinion by: Trieweiler, J.

Facts

Kenneth Breidenbach is a member of the Montana Highway Patrol who, at the time of trial and the time of the incident that formed the basis for Stanko's arrest, was stationed in Jordan, Montana. On March 10, 1996, he was on duty patrolling Montana State Highway 24 and proceeding south from Fort Peck toward Flowing Wells in "extremely light" traffic at about 8 a.m. on a Sunday morning when he observed another vehicle approaching him from behind.

He stopped or slowed, made a right-hand turn, and proceeded west on Highway 200. About one-half mile from that intersection, in the first passing zone, the vehicle that had been approaching him from behind passed him. He caught up to the vehicle and trailed the vehicle at a constant speed for a distance of approximately eight miles while observing what he referred to as the two- or three-second rule. . . . He testified that he clocked the vehicle ahead of him at a steady 85 miles per hour during the time that he followed it. At that speed, the distance between the two vehicles was from 249 to 374 feet. . . . Officer Breidenbach signaled him to pull over and issued him a ticket for violating Section 61-8-303(1), Montana Code Annotated (MCA). The basis for the ticket was the fact that Stanko had been operating his vehicle at a speed of 85 miles per hour at a location where Officer Breidenbach concluded it was unsafe to do so.

The officer testified that the road at that location was narrow, had no shoulders, and was broken up by an occasional frost heave. He also testified that the portion of the road over which he clocked Stanko included curves and hills that obscured vision of the roadway ahead. However, he acknowledged that at a distance of from 249 to 374 feet behind Stanko, he had never lost sight of Stanko's vehicle. The roadway itself was bare and dry, there were no adverse weather conditions, and the incident occurred during daylight hours. Officer Breidenbach apparently did not inspect the brakes on Stanko's vehicle or make any observation regarding its weight. The only inspection he conducted was of the tires, which appeared to be brand new. He also observed that it was a 1996 Camaro, which was a sports car, and that it had a suspension system designed so that the vehicle could be operated at high speeds. He also testified that while he and Stanko were on Highway 24 there were no other vehicles that he observed, that during the time that he clocked Stanko . . . they approached no other vehicles going in their direction, and that he observed a couple of vehicles approach them in the opposite direction during that eight-mile stretch of highway.

Although Officer Breidenbach expressed the opinion that 85 miles per hour was unreasonable at that location, he gave no opinion about what would have been a reasonable speed, nor did he identify anything about Stanko's operation of his vehicle, other than the speed at which he was traveling, which he considered to be unsafe. Stanko testified that on the date he was arrested he was driving a 1996 Chevrolet Camaro that he had just purchased one to two months earlier and that had been driven fewer than 10,000 miles. He stated that the brakes, tires, and steering were all in perfect operating condition, the highway conditions were perfect, and he felt that he was operating his vehicle in a safe manner. He conceded that after passing Officer Breidenbach's vehicle, he drove at a speed of 85 miles per hour but testified that because he was aware of the officer's presence he was extra careful about the manner in which he operated his vehicle. He felt that he would have had no problem avoiding any collision at the speed that he was traveling. Stanko testified that he was fifty years old at the time of trial, drives an average of 50,000 miles a year, and has never had an accident.

Issue

Is Section 61-8-303(1), MCA, so vague that it violates the Due Process Clause found at article 2 section 17 of the Montana Constitution? . . . Stanko contends that Section 61-8-303(1), MCA, is unconstitutionally vague because it fails to give a motorist of ordinary intelligence fair notice of the speed at which he or she violates the law, and because it delegates an important public policy matter, such as the appropriate speed on Montana's highways, to policemen, judges, and juries for resolution on a case-by-case basis. Section 61-8-303(1), MCA, provides as follows:

A person operating or driving a vehicle of any character on a public highway of this state shall drive the vehicle in a careful and prudent manner and at a rate of speed no greater than is reasonable and proper under the conditions existing at the point of operation, taking into account the amount and character of traffic, condition of brakes, weight of vehicle, grade and width of highway, condition of surface, and freedom of obstruction to the view ahead. The person operating or driving the vehicle shall drive the vehicle so as not to unduly or unreasonably endanger the life, limb, property, or other rights of a person entitled to the use of the street or highway.

The question is whether a statute that regulates speed in the terms set forth above gave Stanko reasonable notice of the speed at which his conduct would violate the law.

Reasoning

In Montana, we have established the following test for whether a statute is void on its face for vagueness: "A statute is void on its face if it fails to give a person of ordinary intelligence fair notice that his contemplated conduct is forbidden." . . . No person should be required to speculate as to whether his contemplated course of action may be subject to criminal penalties. We conclude that, as a speed limit, Section 61-8-303(1), MCA, does not meet these requirements of the Due Process Clause of article 2 section 17 of the Montana Constitution, nor does it further the values that the void-for-vagueness doctrine is intended to protect.

For example, while it was the opinion of Officer Breidenbach that 85 miles per hour was an unreasonable speed at the time and place where Stanko was arrested, he offered no opinion regarding what a reasonable speed at that time and place would have been. Neither was the attorney general, the chief law enforcement officer for the state, able to specify a speed that would have been reasonable for Stanko at the time and place where he was arrested. . . .

The difficulty that Section 61-8-303(1), MCA, presents as a statute to regulate speed on Montana's highways, especially as it concerns those interests that the void-for-vagueness doctrine is intended to protect, was further evident from the following discussion with the attorney general during the argument of this case:

Q. Well how many highway patrol men and women are there in the State of Montana?

A. There are 212 authorized members of the patrol. Of that number, about 190 are officers and on the road.

Q. And I understand there are no specific guidelines provided to them to enable them to know at what point, exact point, a person's speed is a violation of the basic rule?

A. That's correct, your honor, because that's not what the statute requires. We do not have a numerical limit. We have a basic rule statute that requires the officer to take into account whether or not the driver is driving in a careful and prudent manner, using the speed.

Q. And it's up to each of their individual judgments to enforce the law?

A. It is, your honor, using their judgment applying the standard set forth in the statute. . . .

It is evident from the testimony in this case and the arguments to the court that the average motorist in Montana would have no idea of the speed at which he or she could operate his or her motor vehicle on this state's highways without violating Montana's "basic rule" based simply on the speed at which he or she is traveling. Furthermore, the basic rule not only permits, but requires the kind of arbitrary and discriminatory enforcement that the Due Process Clause in general, and the void-for-vagueness doctrine in particular, are designed to prevent. It impermissibly delegates the basic public policy of how fast is too fast on Montana's highways to "policemen, judges, and juries for resolution on an ad hoc and subjective basis."

. . . For example, the statute requires that a motor vehicle operator and Montana's law enforcement personnel take into consideration the amount of traffic at the location in question, the condition of the vehicle's brakes, the vehicle's weight, the grade and width of the highway, the condition of its surface, and its freedom from obstruction to the view ahead. However, there is no specification of how these various factors are to be weighted, or whether priority should be given to some factors as opposed to others. This case is a good example of the problems inherent in trying to consistently apply all of these variables in a way that gives motorists notice of the speed at which the operation of their vehicles becomes a violation of the law. . . .

Holding

We do not, however, mean to imply that motorists who lose control of their vehicles or endanger the life, limb, or property of others by the operation of their vehicles on a street or highway cannot be punished for that conduct pursuant to other statutes. . . . We simply hold that Montanans cannot be charged, prosecuted, and punished for speed alone without notifying them of the speed at which their conduct violates the law. . . . The judgment of the district court is reversed. . . .

Dissenting, *Turnage, C.J.*

This important traffic regulation has remained unchanged as the law of Montana . . . since 1955. . . . Apparently for the past forty-three years, other citizens driving upon our highways had no problem in understanding this statutory provision. Section 61-8-303(1), MCA, is not vague and most particularly is not unconstitutional as a denial of due process. . . .

Dissenting, *Regnier, J.*

The arresting officer described in detail the roadway where Stanko was operating his vehicle at 85 miles per hour. The roadway was very narrow with no shoulders. There were frost heaves on the road that caused the officer's vehicle to bounce. The highway had steep hills, sharp curves, and multiple no-passing zones. There were numerous ranch and field access roads in the area, which ranchers use for bringing hay to their cattle. The officer testified that at 85 miles per hour, there was no way for Stanko to stop in the event there had been an obstruction on the road beyond the crest of a hill. In the officer's judgment, driving a vehicle at the speed of 85 miles per hour on the stretch of road in question posed a danger to the rest of the driving public. In my view, Stanko's speed on the roadway where he was arrested clearly falls within the behavior proscribed by the statute. . . .

Questions for Discussion

1. What were the facts the police officer relied on in arresting Stanko for speeding? Contrast these with the facts recited by Stanko in insisting that he was driving at a reasonable speed.

2. The statute employs a "reasonable person" standard and lists a number of factors to be taken into consideration in determining whether a motorist is driving at a proper rate of speed. Was the decision of the Montana Supreme Court based on the lack of notice provided to motorists concerning a reasonable speed or based on the failure to provide law enforcement officers with clear standards for enforcement?

3. Why does Chief Justice Turnage refer to Section 61-8-303(1), MCA as an "important traffic regulation" and stress that this has been the law for forty-three years? Can you speculate as to why Montana failed to post speed limits on highways?

4. Do you agree with the majority opinion or with the dissenting judges?

5. The Montana state legislature reacted by establishing speed limits of "75 mph at all times on Federal . . . interstate highways outside an urban area" . . . and "70 mph during the daytime and 65 mph during the nighttime on any other public highway." Why did the legislature believe that this statute solved the void-for-vagueness issue?

Cases and Comments

Stanko's Subsequent Arrests. Stanko was arrested for reckless driving on August 13, 1996, and again on October 1, 1996. He was charged on both occasions with operating a vehicle with "willful or wanton disregard for the safety of persons or property." Two officers cited the fact that Stanko was driving between 117 and 120 miles per hour on narrow, hilly highways with the risk of encountering farm, ranch, tourist, and recreational vehicles and wildlife and placing emergency personnel at risk. Stanko possessed extraordinary confidence in his driving ability and dismissed the suggestion that he was driving in a wanton and reckless fashion.

He pointed out that he drove roughly 6,000 miles a month without an accident and that he had won several stock-car races in Oregon almost twenty years previously. The Montana Supreme Court unanimously ruled that Stanko should have reasonably understood that the manner in which he was driving posed a risk to other motorists who "do not assume the risk of driving in racetrack conditions." The Montana Supreme Court stressed that Stanko's conviction was not "based on speed alone" and dismissed his claim that the reckless driving law was unconstitutionally vague. See *State v. Stanko*, 974 P.2d 1139 (Mont. 1998).

See more cases on the study site: **State v. Metzger, www.sagepub.com/lippmanccl3e.**

You Decide

2.1 David C. Bryan was involved in a relationship with a young woman during the fall semester of 1994 at the University of Kansas. The relationship ended and Bryan allegedly repeatedly contacted the young woman, including personally approaching her in a university building. Bryan subsequently was charged under the Kansas stalking statute. The Kansas statute at the time prohibited an "intentional and malicious following or course of conduct when such following or course of conduct seriously alarms, annoys or harasses the person." The statute failed to specify whether a "following" that "alarms, annoys or harasses" was to be measured by the standard of a "reasonable person." Bryan contends that the statute is unconstitutionally vague. How should the judge rule? Could you suggest how the state legislature could clarify the law? Consider the perspectives of a female victim and male defendant. See *State v. Bryan*, 910 P.2d 212 (Kan. 1996). Another Kansas case on stalking is *State v. Rucker*, 987 P.2d 1080 (Kan. 1999).

You can find the answer at www.sagepub.com/lippmanccl3e.

EQUAL PROTECTION

The U.S. Constitution originally did not provide for the **equal protection** of the laws. Professor Erwin Chemerinsky observes that this is not surprising, given that African Americans were enslaved and women were subject to discrimination. Slavery, in fact, was formally embedded in the legal system. Article I, Section 2 of the U.S. Constitution provides for the apportionment of the House of Representatives based on the "whole number of free persons" as well as three-fifths of the slaves. This was reinforced by Article IV, Section 2, the Fugitive Slave Clause, which requires the return of a slave escaping into a state that does not recognize slavery.[19]

Immediately following the Civil War in 1865, Congress enacted and the states ratified the Thirteenth Amendment, which prohibits slavery and involuntary servitude. Discrimination against African Americans nevertheless continued, and Congress responded by approving the Fourteenth Amendment in 1868. Section 1 provides that "no state shall deprive any person of life, liberty or property without due process of law, or deny any person equal protection of the law." The Supreme Court declared in 1954 that the Fifth Amendment Due Process Clause imposes an identical obligation to ensure the equal protection of the law on the federal government.[20]

The Equal Protection Clause was rarely invoked for almost one hundred years. Justice Oliver Wendell Holmes, Jr., writing in 1927, typified the lack of regard for the Equal Protection Clause when he referred to the amendment as "the last resort of constitutional argument."[21] The famous 1954 Supreme Court decision in *Brown v. Board of Education* ordering the desegregation of public schools with "all deliberate speed" ushered in a period of intense litigation over the requirements of the clause.[22]

For an international perspective on this topic, visit the study site.

Three Levels of Scrutiny

Criminal statutes typically make distinctions based on various factors, including the age of victims and the seriousness of the offense. For instance, a crime committed with a dangerous weapon may be punished more harshly than a crime committed without a weapon. Courts generally accept the judgment of state legislatures in making differentiations so long as a law is rationally related to a legitimate government purpose. Legitimate government purposes generally include public safety, health, morality, peace and quiet, and law and order. There is a strong presumption that a law is constitutional under this **rational basis test** or **minimum level of scrutiny test**.[23]

In *Westbrook v. Alaska,* nineteen-year-old Nicole M. Westbrook contested her conviction for consuming alcoholic beverages when under the age of twenty-one. Westbrook argued that there was no basis for distinguishing between a twenty-one-year-old and an individual who was slightly younger. The Alaska Supreme Court recognized that there may be some individuals younger than twenty-one who possess the judgment and maturity to handle alcoholic beverages and that some individuals over twenty-one may fail to meet this standard. The court observed that states have established the drinking age at various points and that setting the age between nineteen and twenty-one years of age seemed to be rationally related to the objective of ensuring responsible drinking. As a result, the court concluded that "even if we assume that Westbrook is an exceptionally mature 19-year-old, it is still constitutional for the legislature to require her to wait until she turns 21 before she drinks alcoholic beverages."[24]

In contrast, the courts apply a strict scrutiny test in examining distinctions based on race and national origin. Racial discrimination is the very evil that the Fourteenth Amendment was intended to prevent, and the history of racism in the United States raises the strong probability that such classifications reflect a discriminatory purpose. In *Strauder v. West Virginia,* the U.S. Supreme Court struck down a West Virginia statute as unconstitutional that limited juries to "white male persons who are twenty-one years of age."[25]

Courts are particularly sensitive to racial classifications in criminal statutes and have ruled that such laws are unconstitutional in almost every instance. The Supreme Court observed that "in this context . . . the power of the State weighs most heavily upon the individual or the group."[26] In *Loving v. Virginia,* in 1967, Mildred Jeter, an African American, and Richard Loving, a Caucasian, pled guilty to violating Virginia's ban on interracial marriages and were sentenced to twenty-five years in prison, a sentence that was suspended on the condition that the Lovings leave Virginia. The Supreme Court stressed that laws containing racial classifications must be subjected to the "most rigid scrutiny" and determined that the statute violated the Equal Protection Clause.

The Court failed to find any "legitimate overriding purpose independent of invidious racial discrimination" behind the law. The fact that Virginia "prohibits only interracial marriages involving white persons demonstrates that the racial classifications must stand on their justification, as measures designed to maintain White Supremacy. . . . There can be no doubt that restricting the freedom to marry solely because of racial classifications violates the central meaning of the Equal Protection Clause."[27] The strict scrutiny test also is used when a law limits the exercise of "fundamental rights" (such as freedom of speech).

The Supreme Court has adopted a third, intermediate level of scrutiny for classifications based on gender. The decision to apply this standard rather than strict scrutiny is based on the consideration that although women historically have confronted discrimination, the biological differences between men and women make it more likely that gender classifications are justified. Women, according to the Court, also possess a degree of political power and resources that are generally not found in "isolated and insular minority groups." Intermediate scrutiny demands that the State provide some meaningful justification for the different treatment of men and women and not rely on stereotypes or classifications that have no basis in fact. Justice Ruth Ginsburg applied intermediate scrutiny in ordering that the Virginia Military Institute admit women and ruled that gender-based government action must be based on "an exceedingly persuasive justification. . . . The burden of justification is demanding and it rests entirely on the State."[28]

In *Michael M. v. Superior Court,* the U.S. Supreme Court upheld the constitutionality of California's "statutory rape law" that punished "an act of sexual intercourse accomplished with a female not the wife of the perpetrator, where the female is under the age of 18 years."[29] Is it constitutional to limit criminal liability to males?

The Supreme Court noted that California possessed a "strong interest" in preventing illegitimate teenage pregnancies. The Court explained that imposing criminal sanctions solely on males roughly "equalized the deterrents on the sexes," because young men did not face the prospects of pregnancy and child rearing. The Court also deferred to the judgment of the California legislature that extending liability to females would likely make young women reluctant to report violations of the law.[30]

In summary, there are three different levels of analysis under the Equal Protection Clause:

- *Rational Basis Test.* A classification is *presumed valid* so long as it is rationally related to a constitutionally permissible state interest. An individual challenging the statute must demonstrate that there is no rational basis for the classification. This test is used in regard to the "nonsuspect" categories of the poor, the elderly, and the mentally challenged and to distinctions based on age.
- *Strict Scrutiny.* A law singling out a racial or ethnic minority must be strictly necessary, and there must be no alternative approach to advancing a compelling state interest. This test is also used when a law limits fundamental rights.
- *Intermediate Scrutiny.* Distinctions on the grounds of gender must be substantially related to an important government objective. A law singling out women must be based on factual differences and must not rest on overbroad generalizations.

The next case in the textbook, *Wright v. South Carolina,* asks you to consider whether the defendant was sentenced under a statutory provision that reflects an outdated view of women. Is this a case of gender discrimination against the male defendant under the intermediate scrutiny test? You will want to refer back to this case when we discuss equal protection and sentencing in Chapter 3.

Was the defendant's prison sentence based on a statutory provision that discriminated against men?

Wright v. South Carolina, 563 S.E.2d 311 (S.C. 2000). Opinion by: Waller, J.

Todd William Wright was convicted of criminal domestic violence of a high and aggravated nature (CDVHAN) and sentenced to ten years imprisonment, suspended upon service of eight years, and five years probation. We affirm.

Facts

Wright, six feet tall and weighing 216 pounds, beat and kicked his wife Wendy on the evening of February 16, 1999. Her injuries were so severe that two of her ribs were fractured and her spleen had to be removed. Wright was indicted for criminal domestic violence of a high and aggravated nature. The aggravating factors alleged in the indictment were "a difference in the sexes of the victim and the defendant" and/or that "the defendant did inflict serious bodily harm upon the victim by kicking her in the mid-section requiring her to seek medical attention."

The offense of CDVHAN incorporates the aggravating factor of an assault and battery of a high and aggravated nature (ABHAN). The elements of ABHAN that result in a defendant receiving a harsher sentence are (1) the unlawful act of violent injury to another, accompanied by circumstances of aggravation. Circumstances of aggravation include the use of a deadly weapon, the intent to commit a felony, infliction of serious bodily injury, great disparity in the ages or physical conditions of the parties, a difference in gender, the purposeful infliction of shame and disgrace, taking indecent liberties or familiarities with a female, and resistance to lawful authority. . . .

Wright objected to the judge's charge on the aggravating circumstance of "a difference of the sexes," contending it violated equal protection. The objection was overruled; Wright was found guilty as charged.

Issue

Does the aggravating circumstance of a "difference in the sexes" violate equal protection in violation of the Fourteenth Amendment Section 1 ("No State shall . . . deny to any person within its jurisdiction the equal protection of the laws")?

Reasoning

Wright contends the judge's charge on the aggravating circumstance of a "difference in the sexes" violated his right to equal protection. We disagree. The Equal Protection Clause prevents only irrational and unjustified classifications, not all classifications. For a gender-based classification to pass constitutional muster, it must serve an important governmental objective and be substantially related to the achievement of that objective. A law will be upheld where the gender classification realistically reflects the fact that the sexes are not similarly situated in certain circumstances. See *Michael M. v. Superior Court of Sonoma County,* 450 U.S. 464, 469 (1981) (holding that as long as the rule of nature that the sexes are not similarly situated in certain circumstances is realistically reflected in a gender classification, the statute will be upheld as constitutional). In *Michael M.,* Justice Stewart wrote that "when men and women are not in fact similarly situated . . . the Equal Protection Clause does not mean that the physiological differences between men and women must be disregarded. While those differences must never be permitted to become a pretext for invidious discrimination . . . the Constitution . . . does not require a State to pretend

that demonstrable differences between men and women do not really exist."

In *State v. Gurganus,* 250 S.E.2d 668 (N.C. 1979), the North Carolina Supreme Court upheld a statute enhancing the punishment for males convicted of assault on a female stating, "We base our decision . . . upon the demonstrable and observable fact that the average adult male is taller, heavier and possesses greater body strength than the average female." We . . . think that the South Carolina General Assembly was also entitled to take note of the differing physical sizes and strengths of the sexes. Having noted such facts, the General Assembly could reasonably conclude that assaults and batteries without deadly weapons by physically larger and stronger males are likely to cause greater physical injury and risk of death than similar assaults by females. Having so concluded, the General Assembly could choose to provide greater punishment for these offenses, which it found created greater danger to life and limb, without violating the Fourteenth Amendment. . . .

Certainly some individual females are larger, stronger, and more violent than many males. The General Assembly is not, however, required by the Fourteenth Amendment to modify criminal statutes that have met the test of time in order to make specific provisions for any such individuals. The Constitution of the United States has not altered certain virtually immutable facts of nature, and the General Assembly of South Carolina is not required to undertake to alter those facts. The South Carolina statute establishes classifications by gender that serve important governmental objectives and are substantially related to achievement of those objectives. Therefore, we hold that the statute does not deny males equal protection of law in violation of the Fourteenth Amendment to the Constitution of the United States. . . .

Holding

We find that the "difference in gender" aggravator is legitimately based upon realistic physiological size and strength differences of men and women such that it does not violate equal protection. . . . We therefore affirm Wright's convictions.

Concurring, *Toal, J.*

While I concur with the majority's decision to affirm Wright's CDVHAN conviction, I disagree with the majority's conclusion that the "difference in the sexes" aggravating circumstance does not violate equal protection. I believe the "difference in the sexes" aggravating circumstance, as a gender-based classification, violates equal protection. . . .

The CDVHAN statute was designed to address violence in the home; it applies when any person harms any member of his or her household. The statute then is designed to prevent domestic violence against men, women, and children by perpetrators of both sexes (household members include spouses, former spouses, parents and children, relatives to the second degree, persons with a child in common, and males and females who are cohabiting or have previously cohabited). Having an aggravating circumstance based solely on gender does not substantially further this objective or the narrower objective of protecting women from domestic abuse. In my opinion, this gender-based classification is no different than the classification . . . in *In the Interest of Joseph T.* In that case, this court held that a statute criminalizing communication of indecent messages to females violated the Equal Protection Clause. Although the court recognized that some gender-based classifications that realistically reflect that men and women are not similarly situated can withstand equal protection scrutiny on occasion, it clarified that distinctions in the law that were based on "old notions" that women should be afforded "special protection" could no longer withstand equal protection scrutiny.

In my opinion, this "difference in gender" aggravating circumstance is a distinction that perpetuates

these "old notions." There is no logical purpose for it except to protect physically inferior women from stronger men. . . . Deterring domestic violence is more efficiently and appropriately accomplished through other aggravators, such as the "great disparity in ages or physical conditions of the parties" and "infliction of serious bodily injury" aggravators. In many cases, there may be a great disparity in strength between a male and a female, but if there is not, there is no reason why a difference in gender should serve as an aggravating circumstance to "protect" women to the detriment of men. Therefore, I would find that the "difference in the sexes" aggravating circumstance violates equal protection, because it fails to substantially relate to the government objective of preventing domestic violence. However, I would affirm Wright's conviction, because the jury also found a permissible, gender-neutral aggravating circumstance: infliction of serious bodily injury. Accordingly, I respectfully concur in result only.

Questions for Discussion

1. Explain why Wright claims that his enhanced sentence is based on gender discrimination. Would this aggravator apply to a homosexual couple or in a case in which a daughter abused her mother?

2. Why does Judge Waller reject the defendant's equal protection claim?

3. Do you agree with Judge Toal that the "difference in gender" aggravating factor reflects outdated stereotypes concerning women? What is his solution?

4. How would you rule as a judge in this case?

Cases and Comments

Detention of Japanese Americans During World War II. In *Korematsu v. United States,* the U.S. Supreme Court upheld the conviction of Fred Korematsu, an American citizen of Japanese descent, for remaining in San Leandro, California, in defiance of Civilian Exclusion Order No. 34 issued by the commanding general of the Western Command, U.S. Army. This prosecution was undertaken pursuant to an act of Congress of March 21, 1942, that declared it was a criminal offense punishable by a fine not to exceed $5,000 or by imprisonment for not more than a year for a person of Japanese ancestry to remain in "any military area or military zone" established by the president, secretary of defense, or a military commander. Japanese Americans who were ordered to leave their homes were detained in remote relocation camps. Exclusion Order No. 34 was one of a number of orders and proclamations issued under the authority of President Franklin Delano Roosevelt; it stated that "successful prosecution of the war [World War II] requires every possible protection against espionage and against sabotage to

national defense material, national defense premises and national-defense utilities." Justice Hugo Black recognized that legal restrictions that "curtail the civil rights of a single racial group are immediately suspect" and that individuals excluded from the military zone would be subject to relocation and detention without trial in a camp far removed from the West Coast. The Supreme Court nevertheless affirmed the constitutionality of the order by a vote of six to three. The majority concluded the following:

> Korematsu was not excluded from the Military Area because of hostility to him or to his race. He was excluded because we are at war with the Japanese Empire, because the properly constituted military authorities feared an invasion of our West Coast and felt constrained to take proper security measures, because they decided that the military urgency of the situation demanded that all citizens of Japanese ancestry be segregated from the West Coast

temporarily, and finally, because Congress, reposing its confidence in this time of war in our military leaders . . . determined that they should have the power to do just this. There was evidence of disloyalty on the part of some, the military authorities considered that the need for action was great, and time was short. We cannot—by availing ourselves of the calm perspective of hindsight—say that at that time these actions were unjustified.

Justice Frank Murphy questioned the constitutionality of this order, which he contended unconstitutionally excluded both citizens and noncitizens of Japanese ancestry from the Pacific Coast. He concluded that the "exclusion goes over 'the very brink of constitutional power' and falls into the ugly abyss of racism." Was this a case of racial discrimination or an effort to safeguard the United States from an attack by Japan? What is the standard of review? Would such a law be ruled constitutional today? See *Korematsu v. United States*, 323 U.S. 214 (1944).

You Decide

2.2 Jeanine Biocic was walking on the beach on the Chincoteague National Wildlife Refugee in Virginia with a male friend. Biocic wanted to get some extra sun and removed the top of her two-piece bathing suit, exposing her breasts. She was observed by a U.S. Fish and Wildlife Service officer who issued a summons charging Biocic with an "act of indecency or disorderly conduct. . . . prohibited on any national wildlife refuge." Biocic was convicted and fined $25 and appealed on the grounds that her conviction violated equal protection under law. Her claim was based on the fact that the ordinance prohibited the exposure of female breasts and did not prohibit the exposure of male breasts. How would you rule? See *United States v. Biocic*, 928 F.2d 112 (4th Cir. 1991).

You can find the answer at www.sagepub.com/lippmanccl3e.

We next look at constitutional protections for freedom of speech and privacy.

FREEDOM OF SPEECH

The **First Amendment** to the U.S. Constitution provides that "Congress shall make no law . . . abridging the freedom of the speech or of the press; or the right of the people peaceably to assemble, and to petition the Government for a redress of grievances." The U.S. Supreme Court extended this prohibition to the states in a 1925 Supreme Court decision in which the Court proclaimed that "freedom of speech and of the press . . . are among the fundamental personal rights and 'liberties' protected under the Due Process Clause of the Fourteenth Amendment from impairment by the States."[31]

The Fourteenth Amendment to the Constitution applies to the states and was adopted following the Civil War in order to protect African Americans against the deprivation of "life, liberty and property without due process" as well as to guarantee former slaves "equal protection of the law." The Supreme Court has held that the Due Process Clause incorporates various fundamental freedoms that generally correspond to the provisions of the **Bill of Rights** (the first ten amendments to the U.S. Constitution that create rights against the federal government). This **incorporation theory** has resulted in a fairly uniform national system of individual rights that includes freedom of expression.

The famous, and now deceased, First Amendment scholar Thomas I. Emerson identified four functions central to democracy performed by freedom of expression under the First Amendment:[32]

- Freedom of expression contributes to *individual self-fulfillment* by encouraging individuals to express their ideas and creativity.
- Freedom of expression insures *a vigorous "marketplace of ideas"* in which a diversity of views are expressed and considered in reaching a decision.
- Freedom of expression *promotes social stability* by providing individuals the opportunity to be heard and to influence the political and policy-making process. This promotes the acceptance of decisions and discourages the resort to violence.
- Freedom of expression ensures that there is a steady stream of innovative ideas and enables the *government to identify and address newly arising issues.*

The First Amendment is vital to the United States' free, open, and democratic society. Justice William Douglas wrote in *Terminello v. Chicago*[33] that speech

> may indeed best serve its high purpose when it induces a condition of unrest, creates dissatisfaction with the conditions as they are, or even stirs people to anger. Speech is often provocative and challenging. It may strike at prejudices and preconceptions and have profound unsettling effects as it presses for acceptance of an idea.

Justice Robert H. Jackson, reflecting on his experience as a prosecutor during the Nuremberg trials of Nazi war criminals, cautioned Justice Douglas that the

> choice is not between order and liberty. It is between liberty with order and anarchy without either. There is danger that, if the Court does not temper its doctrinaire logic with a little practical wisdom, it will convert the constitutional Bill of Rights into a suicide pact.

Justice Jackson is clearly correct that there must be some limit to freedom of speech. But where should the line be drawn? The Supreme Court articulated these limits in *Chaplinsky v. New Hampshire* and observed that there are "certain well-recognized categories of speech which may be permissibly limited under the First Amendment." The Supreme Court explained that these "utterances are no essential part of any exposition of ideas, and are of such slight social value as a step to truth that any benefit that may be derived from them is clearly outweighed by the social interest in order and morality."[34] The main categories of speech for which *content is not protected by the First Amendment* and that may result in the imposition of criminal punishment are as follows:

- *Fighting Words.* Words directed to another individual or individuals that an ordinary and reasonable person should be aware are likely to cause a fight or breach of the peace are prohibited under the **fighting words** doctrine. In *Chaplinsky v. New Hampshire,* the Supreme Court upheld the conviction of a member of the Jehovah's Witnesses who, when distributing religious pamphlets, attacked a local marshal with the accusation that "you are a God damned racketeer" and "a damned Fascist and the whole government of Rochester are Fascists or agents of Fascists."
- *Incitement to Violent Action.* A speaker, when addressing an audience, is prohibited from **incitement to violent action**. In *Feiner v. New York,* Feiner addressed a racially mixed crowd of seventy-five or eighty people. He was described as "endeavoring to arouse" the African Americans in the crowd "against the whites, urging that they rise up in arms and fight for equal rights." The Supreme Court ruled that "when clear and present danger of riot, disorder, interference with traffic upon the public streets, or other immediate threat to public safety, peace, or order, appears, the power of the State to prevent or punish is obvious."[35] On the other hand, in *Terminello v. Chicago,* the Supreme Court stressed that a speaker could not be punished for speech that merely "stirs to anger, invites dispute, brings about a condition of unrest, or creates a disturbance."[36]
- *Threat.* A developing body of law prohibits threats of bodily harm directed at individuals. Judges must weigh and balance a range of factors in determining whether a statement constitutes a political exaggeration or a **true threat**. In *Watts v. United States,* the defendant proclaimed to a small gathering following a public rally on the grounds of the Washington Monument that if inducted into the army and forced to carry a rifle that "the first man I want to get in my sights is L.B.J. [President Lyndon Johnson] . . . They are not going to make me kill my black brothers." The onlookers greeted this statement with laughter. Watts's conviction was overturned by the U.S. Supreme Court, which ruled that the government had failed to demonstrate that Watts had articulated a true threat and that these types of bold statements were to be expected in a dynamic and democratic society divided over the Vietnam War.[37]
- *Obscenity.* Obscene materials are considered to lack "redeeming social importance" and are not accorded constitutional protection. Drawing the line between obscenity and protected speech has proven problematic. The Supreme Court conceded that obscenity cannot be

defined with "God-like precision," and Justice Potter Stewart went so far as to pronounce in frustration that the only viable test seemed to be that he "knew obscenity when he saw it."[38] The U.S. Supreme Court was finally able to agree on a test for obscenity in *Miller v. California*. The Supreme Court declared that **obscenity** was limited to works that when taken as a whole, in light of contemporary community standards, appeal to the prurient interest in sex; are patently offensive; and lack serious literary, artistic, political, or scientific value. This qualification for scientific works means that a medical textbook portraying individuals engaged in "ultimate sexual acts" likely would not constitute obscenity.[39] Child pornography may be limited despite the fact that it does not satisfy the *Miller* standard.[40] (Obscenity and pornography are discussed in Chapter 15.)

- *Libel.* You should remain aware that the other major limitation on speech, **libel**, is a civil law rather than a criminal action. This enables individuals to recover damages for injury to their reputations. In *New York Times v. Sullivan,* the U.S. Supreme Court severely limited the circumstances in which public officials could recover damages and held that a public official may not recover damages for a defamatory falsehood relating to his or her official conduct "unless . . . the statement was made with 'actual malice'—that is, with knowledge that it was false or with reckless disregard of whether it was false or not."[41] The Court later clarified that this "reckless disregard" or actual knowledge standard applied only to "public figures" and that states were free to apply a more relaxed, simple negligence (lack of reasonable care in verifying the facts) standard in suits for libel brought by private individuals.[42]

Speech lacking First Amendment protection shares several common characteristics:

- The expression lacks social value.
- The expression directly causes social harm or injury.
- The expression is narrowly defined in order to avoid discouraging and deterring individuals from engaging in free and open debate.

Keep in mind that these are narrowly drawn exceptions to the First Amendment's commitment to a lively and vigorous societal debate. The general rule is that the government may neither require nor substantially interfere with individual expression. The Supreme Court held in *West Virginia v. Barnette* that a student may not be compelled to pledge allegiance to the American flag. The Supreme Court observed that "if there is any fixed star in our constitutional constellation, it is that no official, high or petty, can prescribe what shall be orthodox in politics, nationalism, religion or other matters of opinion or force citizens to confess by word or action their faith therein." This commitment to a free "marketplace of ideas" is based on the belief that delegating the decision as to what "views shall be voiced largely into the hands of each of us" will "ultimately produce a more capable citizenry and more perfect polity and . . . that no other approach would comport with the premise of individual dignity and choice upon which our political system rests."[43]

Overbreadth

The doctrine of **overbreadth** is an important aspect of First Amendment protection. This provides that a statute is unconstitutional that is so broadly and imprecisely drafted that it encompasses and prohibits a substantial amount of protected speech relative to the coverage of the statute. In *New York v. Ferber,* the U.S. Supreme Court upheld a New York child pornography statute that criminally punished an individual for promoting a "performance which includes sexual conduct by a child less than sixteen years of age." Sexual conduct was defined to include "lewd exhibition of the genitals." Justice Byron White was impatient with the concern that although the law was directed at hardcore child pornography, "[s]ome protected expression ranging from medical textbooks to pictorials in the National Geographic would fall prey to the statute." White doubted whether these applications of the statute to protected speech constituted more than a "tiny fraction of the materials" that would be affected by the law, and he expressed confidence that prosecutors would not bring actions against these types of publications. This, in short, is the "paradigmatic case of state statute whose legitimate reach dwarfs its arguably impermissible applications."[44]

Hate Speech

Hate speech is one of the central challenges confronting the First Amendment. This is defined as speech that denigrates, humiliates, and attacks individuals on account of race, religion, ethnicity, nationality, gender, sexual preference, or other personal characteristics and preferences. Hate speech should be distinguished from hate crimes or penal offenses that are directed against an individual who is a member of one of these "protected groups."

The United States is an increasingly diverse society in which people inevitably collide, clash, and compete over jobs, housing, and education. Racial, religious, and other insults and denunciations are hurtful, increase social tensions and divisions, and possess limited social value. This type of expression also has little place in a diverse society based on respect and regard for individuals of every race, religion, ethnicity, and nationality.

Regulating this expression, on the other hand, runs the risk that artistic and literary depictions of racial, religious, and ethnic themes may be deterred and denigrated. In addition, there is the consideration that debate on issues of diversity, affirmative action, and public policy may be discouraged. Society benefits when views are forced out of the shadows and compete in the sunlight of public debate.

The most important U.S. Supreme Court ruling on hate speech is *R.A.V. v. St. Paul.* In *R.A.V.*, several Caucasian juveniles burned a cross inside the fenced-in yard of an African American family. The young people were charged under two statutes, including the St. Paul Bias Motivated Crime Ordinance (St. Paul Minn. Legis. Code § 292.02), which provided that "whoever places on public or private property a symbol, object, . . . including and not limited to, a burning cross or Nazi swastika, which one knows or has reasonable grounds to know arouses anger, alarm or resentment . . . on the basis of race, color, creed, religion or gender commits disorderly conduct . . . shall be guilty of a misdemeanor."[45]

The Supreme Court noted that St. Paul punishes certain fighting words, yet permits other equally harmful expressions. This discriminates against speech based on the content of ideas. For instance, what about symbolic attacks against a greedy real estate developer?

A year later, in *Wisconsin v. Mitchell,* in 1993, the Supreme Court ruled that a Wisconsin statute that enhanced the punishment of individuals convicted of hate crimes did not violate the defendant's First Amendment rights. Todd Mitchell challenged a group of other young African American males by asking whether they were "hyped up to move on white people." As a young Caucasian male approached the group, Mitchell exclaimed "there goes a white boy; go get him" and led a collective assault on the victim. The Wisconsin court increased Mitchell's prison sentence for aggravated assault from a maximum of two years to a term of four years based on his intentional selection of the person against "whom the crime . . . is committed . . . because of the race, religion, color, disability, sexual orientation, national origin or ancestry of that person."[46]

Mitchell creatively claimed that he was being punished more severely for harboring and acting on racially discriminatory views in violation of the First Amendment. The Supreme Court, however, ruled that Mitchell was being punished for his harmful act rather than for the fact that his act was motivated by racist views. The enhancement of Mitchell's sentence was recognition that acts based on discriminatory motives are likely "to provoke retaliatory crimes, inflict distinct emotional harms on their victims, and incite community unrest." Mitchell also pointed out that the prosecution was free to introduce a defendant's prior racist comments at trial to prove a discriminatory motive or intent and that this would "chill" racist speech. The Supreme Court held that it was unlikely that a citizen would limit the expression of his or her racist views based on the fear that these statements would be introduced one day against him or her at a prosecution for a hate crime.

In 2003, in *Virginia v. Black,* the U.S. Supreme Court held unconstitutional a Virginia law prohibiting cross burning with "an intent to intimidate a person or group of persons."[47] This law, unlike the St. Paul statute, did not discriminate on the basis of the content of the speech. The Court, however, determined that the statute's provision that the jury is authorized to infer an intent to intimidate from the act of burning of a cross without any additional evidence "permits a jury to convict in every cross burning case in which defendants exercise their constitutional right not to put on a defense." This provision also makes "it more likely that the jury will find an intent to intimidate regardless of the particular facts of the case." The Virginia law failed to distinguish between cross burning intended to intimidate individuals and cross burning intended to make a political statement by groups such as the Ku Klux Klan that view the flaming cross as a symbolic representation of their political point of view.

In the next case in the text, *George T. v. California,* the California Supreme Court was asked to determine whether a student who wrote a poem that stated that he may be the "next kid to bring guns to kill students at school" constituted a "true threat."

Was George's poem a criminal threat?

George T. v. California, *93 P.3d 1007 (Cal. 2004). Opinion by: Moreno, J.*

Issue

We consider in this case whether a high school student made a criminal threat by giving two classmates a poem labeled "Dark Poetry," which read in part,

> I am Dark, Destructive, & Dangerous. I slap on my face of happiness but inside I am evil!! For I can be the next kid to bring guns to kill students at school. So parents watch your children cuz I'm BACK!!

Facts

Fifteen-year-old George T. (minor) had been a student at Santa Teresa High School in Santa Clara County for approximately two weeks when on Friday, March 16, 2001, toward the end of his honors English class, he approached fellow student Mary S. and asked her, "Is there a poetry class here?" Minor then handed Mary three sheets of paper and told her, "Read these." Mary did so. The first sheet of paper contained a note stating, "These poems describe me and my feelings. Tell me if they describe you and your feelings." The two other sheets of paper contained poems. Mary read only one of the poems, which was labeled "Dark Poetry" and entitled "Faces":

> Who are these faces around me? Where did they come from? They would probably become the next doctors or loirs [sic] or something. All really intelligent and ahead in their game. I wish I had a choice on what I want to be like they do. All so happy and vagrant. Each original in their own way. They make me want to puke. For I am Dark, Destructive, & Dangerous. I slap on my face of happiness but inside I am evil!! For I can be the next kid to bring guns to kill students at school. So parents watch your children cuz I'm BACK!!
> by: Julius AKA Angel

Minor had a "straight face," not "show[ing] any emotion, neither happy or sad or angry or upset," when he handed the poems to Mary. Upon reading the "Faces" poem, Mary became frightened, handed the poems back to minor, and immediately left the campus in fear. After she informed her parents about the poem, her father called the school, but it was closed. Mary testified she did not know minor well, but they were on "friendly terms." When asked why she felt minor gave her the poem to read, she responded, "I thought maybe because the first day he came into our class, I approached him because that's the right thing to do" and because she continued to be nice to him.

After Mary handed the poems back to minor, minor approached Erin S. and Natalie P., students minor had met during his two weeks at Santa Teresa High School. Erin had been introduced to minor a week prior and had subsequently spoken with him on only three or four occasions, whereas Natalie considered herself minor's friend and had come to know him well during their long afterschool conversations, which generally lasted from an hour to an hour and a half and included discussions of poetry. Minor handed Erin a "folded up" piece of paper and asked her to read it. He also handed a similarly folded piece of paper to Natalie, who was standing with Erin. Because Erin was late for class, she only pretended to read the poem to be polite but did not actually read it. She placed the unread poem in the pocket of her jacket.

The next day, Saturday, Mary e-mailed her English teacher William Rasmussen to report her encounter with minor. A substitute teacher had been teaching the class on the day that Mary received the note. She wrote,

> I'm sorry to bother you over the weekend, but I don't think this should wait until Monday. During 6th period on Friday, 3/16, the guy in our class called Julius (actually his name is Theodore?) gave me two poems to read. He explained to me that these poems "described him and his feelings," and asked if I "felt the same way." I was surprised to find that the poems were about how he is "nice on the outside," and how he's "going to be the next person to bring a gun to school and

kill random people." I told him to bring the poems to Room 315 to Ms. Gonzalez because [she] is in charge of poetry club. He said he would but I don't know for sure if he did.

Mary remained in fear throughout the weekend, because she understood the poem to be personally threatening to her, as a student. Asked why she felt the poem was a threat, Mary responded,

It's obvious he thought of himself as a dark, destructive, and dangerous person. And if he was willing to admit that about himself and then also state that he could be the next person to bring guns and kill students, then I'd say that he was threatening.

She understood the term "dark poetry" to mean "angry threats; any thoughts that aren't positive." Rasmussen called Mary on Sunday regarding her e-mail. Mary sounded very shaken during the conversation, and based on this and on what she stated about the contents of the poem, Rasmussen contacted the school principal and the police. He read "Faces" for the first time during the jurisdictional hearing and, upon reading it, felt personally threatened by it, because, according to Rasmussen, "He's saying he's going to come randomly shoot." His understanding of "dark poetry" was that it entailed "the concept of death and causing and inflicting a major bodily pain and suffering. . . . There is something foreboding about it."

On Sunday, March 18, 2001, officers from the San Jose Police Department went to minor's uncle's house, where minor and his father were residing. An officer asked minor, who opened the door when the officers arrived, whether there were any guns in the house. Minor "nodded." Minor's uncle was surprised that minor was aware of his guns, and handed the officers a .38-caliber handgun and a rifle. When asked about the poems disseminated at school, minor handed an officer a piece of paper he took from his pocket. The paper contained a poem entitled, "Faces in My Head," which read as follows:

Look at all these faces around me. They look so vacant. They have their whole lives ahead of them. They have their own individuality. Those kind of people make me wanna puke. For I am a slave to very evil masters. I have no future that I choose for myself. I feel as if I am going to go crazy. Probably I would be the next high school killer. A little song keeps playing in my head. My daddy is worth a dollar not even 100 cents. As I look at these faces around me I wonder why r [sic] they so happy. What do they have that I don't. Am I the only one with the messed up mind. Then I realize, I'm cursed!!

As with the poem entitled "Faces," this poem was labeled "dark poetry," but it was not shown or given to anyone at school. Minor had drafted "Faces in My Head" that morning in an attempt to capture what he had written in "Faces," because he wanted a copy for his poetry collection. Minor was taken into custody.

Police officers went to the school the following Monday to investigate the dissemination of the poem. Erin was summoned to the vice-principal's office and asked whether minor had given her any notes. She responded in the affirmative, realized that the poem was still in the pocket of her jacket, and retrieved it. The paper contained a poem entitled "Faces," which was the same poem given to Mary. Upon reading the poem for the first time in the vice-principal's office, Erin became terrified and broke down in tears, finding the poem to be a personal threat to her life. She testified that she was not in the poetry club and had no interest in the subject.

Natalie, who testified on behalf of minor, recalled that minor said, "Read this" as he handed her and Erin the pieces of paper. The folded-up sheet of paper Natalie received contained a poem entitled, "Who Am I." When a police officer went to Natalie's home to inquire about the poem minor had given her on Friday, Natalie was not completely cooperative and truthful, telling the officer that the poem was about water and dolphins and that she believed it was a love poem. The police retrieved the poem from Natalie's trash can and although it was torn, some of it could still be deciphered:

. . . I created? . . . cause it really . . . feel as if . . . stolen from . . . of peace . . . Taken to a place that you hate. Your locked up and when your let out of your cage it is to perform. Not able to be yourself and always hiding & thinking would people like me if I behaved differently? by Julius AKA Angel.

Natalie did not feel threatened by the poem; rather it made her "feel sad," because "it was kind of lonely." She testified that "dark poetry is . . . relevant to like pure emotions, like sadness, loneliness, hate or just like pure emotions. Sometimes it tells a story, like a dark story." Based on her extended conversations with minor, Natalie found him to be "mild and calm and very serene" and did not consider him to be violent.

Minor testified the poem "Faces" was not intended to be a threat, and, because Erin and Natalie were his friends, he did not think they would have taken his poems as such. He thought of poetry as art and stated that he was very much interested in the subject, particularly as a medium to describe "emotions instead of acting them out." He wrote "Faces" during his honors English class on the day he showed it to Mary and Erin. Minor was having a bad day as a consequence of having forgotten to ask his parents for lunch money and having to forgo lunch that day, and because he was unable to locate something in his backpack. He had many thoughts going through his head, so he decided to write them down as a way of getting them out. The poem "Who Am I," which was given to Natalie, was written the same day as "Faces," but was written during the lunch period. Neither poem was intended to be a threat. Instead they were "just creativity."

Minor and his friends frequently joked about the school shootings at Columbine High School in Colorado (where, in 1999, two students killed twelve fellow students and one faculty member). They would jokingly say, "I'm going to be the next Columbine kid." Minor testified that Natalie and Erin had been present when he and some of his friends had joked about Columbine, with someone stating that "I'll probably be the next Columbine killer," and indicating who would be killed and who would be spared. Given this history, minor believed Natalie and Erin would understand the poems as jokes.

The poems were labeled "dark poetry" to inform readers that they were exactly that, and minor testified,

> If anybody was supposed to read this poem, or let's say if my mom ever found my poem or something of that nature, I would like them to know that it was dark poetry. Dark poetry is usually just an expression. It's creativity. It is not like you're actually going to do something like that, basically.

Asked why he wrote, "for I can be the next kid to bring guns to school and kill students," minor responded,

> The San Diego killing [on March 5, 2001, a student at Santana High School shot and killed two students and wounded thirteen others] was about right around this time. So since I put the three Ds—dark, destructive, and dangerous—and since I said—"I am evil," and since I was talking about people around me—faces—how I said, like, how they would make me want to—did I say that?—well, even if I didn't—yeah, I did say that. Okay. So, um, I said from all these things, it sounds like, for I can be the next Columbine kid, basically. So why not add that in? And so, "Parents, watch your children, because I'm back," um, I just wanted to—kind of like a dangerous ending, like a—um, just like ending a poem that would kind of get you, like,—like, whoa, that's really something.

Minor stated that he did not know Mary and did not give her any poems. However, he was unable to explain how Mary was able to recount the contents of the "Faces" poem.

On cross-examination, minor conceded that he had had difficulties in his two previous schools, including being disciplined for urinating on a wall at his first school and had been asked to leave his second school for plagiarizing from the Internet. He explained that the urination incident was caused by a doctor-verified bladder problem. He denied having any ill will toward the school district, but he conceded when pressed by the prosecutor that he felt the schools "had it in for me."

An amended petition under Welfare and Institutions Code section 602 was filed against minor, alleging minor made three criminal threats in violation of Penal Code section 422. The victims of the alleged threats were Mary (count 1), Erin (count 3), and Rasmussen (count 2).

Following a contested jurisdictional hearing, the juvenile court found true the allegations with respect to Mary and Erin but dismissed the allegation with respect to Rasmussen. At the hearing, the court adjudicated minor a ward of the court and ordered a 100-day commitment in juvenile hall. Minor appealed, challenging the sufficiency of the evidence to support the juvenile court's finding that he made criminal threats. Over a dissent, the court of appeal affirmed the juvenile court in all respects with the exception of remanding the matter for the sole purpose of having that court declare the offenses to be either felonies or misdemeanors. We granted review and now reverse.

Reasoning

We made clear that not all threats are criminal and enumerated the elements necessary to prove the offense of making criminal threats under section 422. The prosecution must prove

> (1) that the defendant "willfully threaten[ed] to commit a crime which will result in death or great bodily injury to another person," (2) that the defendant made the threat "with the specific intent that the statement . . . is to be taken as a threat, even if there is no intent of actually carrying it out," (3) that the threat—which may be "made verbally, in writing, or by means of an electronic communication device"—was "on its face and under the circumstances in which it [was] made, . . . so unequivocal, unconditional, immediate, and specific as to convey to the person threatened, a gravity of purpose and an immediate prospect of execution of the threat," (4) that the threat actually caused the person threatened "to be in sustained fear for his or her own safety or for his or her immediate family's safety," and (5) that the threatened person's fear was "reasonabl[e]" under the circumstances.

Minor challenges the juvenile court's findings that he made criminal threats in violation of section 422 and contends that his First Amendment rights were infringed by the court's conclusion that his poem was a criminal threat.

In cases raising First Amendment issues, [it has] repeatedly held that an appellate court has an obligation to "make an independent examination of the whole record in order to make sure that the judgment does not constitute a forbidden intrusion on the field of free expression." The current version of section 422 was drafted with the mandates of the First Amendment in mind, incorporating language from a federal appellate court true-threat decision:

> . . . to describe and limit the type of threat covered by the statute. Independent review is particularly important in the threat's context, because it is a type of speech that is subject to categorical exclusion from First Amendment protection, similar to obscenity, fighting words, and incitement of imminent lawless action. "What is a threat must be distinguished from what is constitutionally protected speech."

As discussed above, this court . . . enumerated five elements the prosecution must prove in order to meet its burden of proving that a criminal threat was uttered. Minor challenges the findings with respect to two of the five elements, contending that the poem "was [not] 'on its face and under the circumstances in which it [was disseminated] so unequivocal, unconditional, immediate, and specific as to convey to [Mary and Erin] a gravity of purpose and an immediate prospect of execution of the threat'" and that the facts fail to establish he harbored the specific intent to threaten Mary and Erin.

With respect to the requirement that a threat be "so unequivocal, unconditional, immediate, and specific as to convey to the person threatened a gravity of purpose and an immediate prospect of execution of the threat," we explained that the word "so" in Section 422 meant that "'unequivocality,' 'unconditionality,' 'immediacy' and 'specificity' are not absolutely mandated, but must be sufficiently present in the threat and surrounding circumstances. . . . The four qualities are simply the factors to be considered in determining whether a threat, considered together with its surrounding circumstances, conveys those impressions to the victim." A communication that is ambiguous on its face may nonetheless be found to be a criminal threat if the surrounding circumstances clarify the communication's meaning.

With the above considerations in mind, we examine the poem at issue—"Faces." What is readily apparent is that much of the poem plainly does not constitute a threat. "Faces" begins by describing the protagonist's feelings about the "faces" that surround him:

> Where did they come from? They would probably become the next doctors or loirs [sic] or something. All really intelligent and ahead in their game. I wish I had a choice on what I want to be like they do. All so happy and vagrant. Each original in their own way. They make me want to puke.

These lines convey the protagonist's feelings about the students around him and describe his envy over how happy and intelligent they appear to be, with opportunities he does not have. There is no doubt this portion of the poem fails to convey a criminal threat, as no violent conduct whatsoever is expressed or intimated. Neither do the next two lines of the poem convey a threat: "For I am Dark, Destructive, & Dangerous. I slap on my face of happiness but inside I am evil!!" These lines amount to an introspective description of the protagonist, disclosing that he is "destructive," "dangerous," and "evil." But again, such divulgence threatens no action.

Only the final two lines of the poem could arguably be construed to be a criminal threat: "For I can be the next kid to bring guns to kill students at school. So parents watch your children cuz I'm BACK!!" Mary believed this was a threat, but her testimony reveals that her conclusion rested upon a considerable amount of interpretation:

> I feel that when he said, "I can be the next person," that he meant that he will be, because also he says that he's dark, destructive, and dangerous person. And I'd describe a dangerous person as someone who has something in mind of killing someone or multiple people.

The juvenile court's finding that minor threatened to kill Mary and Erin likewise turned primarily on its interpretation of the words, "For I can be the next kid to bring guns to kill students at school" to mean not only that minor could do so, but that he would do so. In other words, the court construed the word "can" to mean "will." But that is not what the poem says. However the poem was interpreted by Mary and Erin and the court, the fact remains that "can" does not mean "will." While the protagonist in "Faces" declares that he has the potential or capacity to kill students given his dark and hidden feelings, he does not actually threaten to do so. While perhaps discomforting and unsettling, in this unique context this disclosure simply does not constitute an actual threat to kill or inflict harm.

As is evident, the poem "Faces" is ambiguous and plainly equivocal. It does not describe or threaten future conduct, because it does not state that the protagonist plans to kill students, or even that any potential victims would include Mary or Erin. Such ambiguity aside, it appears that Mary actually misread the text of the poem. In her e-mail to Rasmussen, she stated that the poem read, "He's 'going to be the next person to bring a gun to school and kill random people.'" She did not tell Rasmussen that this was her interpretation of the poem but asserted that those were the words used by minor. Given the student killings in Columbine and Santee, this may have been an understandable mistake, but it does not alter the requirement that the words actually used must constitute a threat in light of the surrounding circumstances.

The court of appeal rejected minor's contention that the protagonist in the poem was a fictional character rather than minor, because he gave the poem to Mary with a note stating that the poem described "me and my feelings." There is no inconsistency, however, in viewing the protagonist as a fictional character while also concluding that the poem reflects minor's personal feelings. And when read by another person, the poem may similarly describe that reader's feelings, as minor implied when he asked Mary if the poem also "described [her] and [her] feelings." More important, the note is consistent with the contention that the poem did nothing more than describe certain dark feelings. The note asked whether Mary had the same feelings; it did not state or imply something to the effect of, "This is what I plan to do; are you with me?" (Of course, exactly what the poem means is open to varying interpretations, because a poem may mean different things to different readers.)

As a medium of expression, a poem is inherently ambiguous. In general, "reasonable persons understand musical lyrics and poetic conventions as the figurative expressions which they are," which means they "are not intended to be and should not be read literally on their face, nor judged by a standard of prose oratory." Ambiguity in poetry is sometimes intended: "'Ambiguity' itself can mean an indecision as to what you mean, an intention to mean several things, a probability that one or the other or both of two things has been meant, and the fact that a statement has several meanings." As the court of appeal observed in a case involving a painting graphically depicting a student shooting a police officer in the back of the head, "a painting—even a graphically violent painting—is necessarily ambiguous because it may use symbolism, exaggeration, and make-believe." This observation is equally applicable to poetry, since it is said that "painting is silent poetry, and poetry painting that speaks."

In short, viewed in isolation the poem is not "so unequivocal" as to have conveyed to Mary and Erin a gravity of purpose and an immediate prospect that minor would bring guns to school and kill them. Ambiguity, however, is not necessarily sufficient to immunize the poem from being deemed a criminal threat, because the surrounding circumstances may clarify facial ambiguity. As Section 422 makes clear, a threat must "on its face and under the circumstances in which it is made, [be] so unequivocal, unconditional, immediate, and specific as to convey . . . a gravity of purpose and an immediate prospect of execution of the threat." When the words are vague, context takes on added significance, but care must be taken not to diminish the requirements that the communicator have the specific intent to convey a threat and that the threat be of such a nature as to convey a gravity of purpose and immediate prospect of the threat's execution.

Unlike some cases that have turned on an examination of the surrounding circumstances given a communication's vagueness, incriminating circumstances in this case are noticeably lacking: there was no history of animosity or conflict between the students . . . no threatening gestures or mannerisms accompanied the poem . . . and no conduct suggested to Mary and Erin that there was an immediate prospect of execution of a threat to kill. Thus the circumstances surrounding the poem's dissemination fail to show that as a threat, it was sufficiently unequivocal to convey to Mary and Erin an immediate prospect that minor would bring guns to school and shoot students.

The themes and feelings expressed in "Faces" are not unusual in literature:

> Literature illuminates who 'we' are: the repertory of selves we harbor within, the countless feelings we experience but never express or perhaps even acknowledge, the innumerable other lives we could but do not live, all those 'inside' lives that are not shown, not included in our resumes.[48]

"Faces" was in the style of a relatively new genre of literature called "dark poetry" that . . . is an extension of the poetry of Sylvia Plath, John Berryman, Robert Lowell, and other confessional poets who depict "extraordinarily mean, ugly, violent, or harrowing experiences." Consistent with that genre, "Faces" invokes images of darkness, violence, discontentment, envy, and alienation. The protagonist describes his duplicitous nature—malevolent on the inside, felicitous on the outside.

Holding

For the foregoing reasons, we hold the poem entitled "Faces" and the circumstances surrounding its dissemination fail to establish that it was a criminal threat, because the text of the poem, understood in light of the surrounding circumstances, was not "so unequivocal, unconditional, immediate, and specific as to convey to [the two students] a gravity of purpose and an immediate prospect of execution of the threat." Our conclusion that the poem was not an unequivocal threat disposes of the matter and we need not, and do not, discuss minor's contention that he did not harbor the specific intent to threaten the students, as required by Section 422.

This case implicates two apparently competing interests: a school administration's interest in ensuring the safety of its students and faculty versus students' right to engage in creative expression. Following Columbine, Santee, and other notorious school shootings, there is a heightened sensitivity on school campuses to latent signs that a student may undertake to bring guns to school and embark on a shooting rampage. Such signs may include violence-laden student writings. For example, the two student killers at Columbine had written poems for their English classes containing "extremely violent imagery." Ensuring a safe school environment and protecting freedom of expression, however, are not necessarily antagonistic goals.

Minor's reference to school shootings and his dissemination of his poem in close proximity to the Santee school shooting no doubt reasonably heightened the school's concern that minor might emulate the actions of previous school shooters. Certainly, school personnel were amply justified in taking action following Mary's e-mail and telephone conversation with her English teacher, but that is not the issue before us. We decide here only that minor's poem did not constitute a criminal threat.

For the foregoing reasons, we reverse the judgment of the court of appeal.

Concurring, *Baxter, J.*

Applying the independent review standard proper for cases implicating First Amendment interests, I agree the evidence does not establish this specific element.

The writing, in the form of a poem, that defendant handed to Mary S. and Erin S. said that the protagonist, "Julius AKA Angel," "can be the next kid to bring guns to kill students at school." It did not say, in so many words, that defendant presently intended to do so. And the surrounding circumstances did not lend unconditional meaning to this conditional language. That said, there is no question that defendant's ill-chosen words were menacing by any common understanding, both on their face and in context. The terror they elicited in Mary S., and the concern they evoked in the school authorities, were real and entirely reasonable. It is safe to say that fears arising from a raft of high school shooting rampages, including those in Colorado and Santee, California, are prevalent among American high school students, teachers, and administrators. Certainly this was so on March 16, 2001, only eleven days after the Santee incident had occurred. That is the day defendant selected to press his violent writing on two vulnerable and impressionable young schoolmates who hardly knew him. Defendant admitted at trial that he intentionally combined the subject matter and the timing for maximum shock value. Indeed, he acknowledged, his words would be interpreted as threats by "kids who didn't know [he was] just kidding."

Under these circumstances, as the majority observe, school and law enforcement officials had every reason to worry that defendant, deeply troubled, was contemplating his own campus killing spree. The important interest that underlies the criminal-threat law—protection against the trauma of verbal terrorism—was also at stake. Accordingly, the authorities were fully justified, and should be commended, insofar as they made a prompt, full, and vigorous response to the incident. They would have been remiss had they not done so. Nothing in our very narrow holding today should be construed as suggesting otherwise.

Questions for Discussion

1. Summarize the facts in *George T.*

2. Describe the responses of Mary, Erin, and Natalie to George T.'s poem. What occurred when the police confronted George T. and conducted an investigation at George T.'s school?

3. What are the elements of the crime of a "true threat" under Section 422 of the California Penal Code?

4. Why did the California Supreme Court conclude that George T.'s poem did not constitute a clear threat? Did the court fully consider the circumstances surrounding the threat? Should the court have analyzed whether George T. intended to harm other students?

5. Do you think that the supreme court's decision was influenced by the fact that George T. was a juvenile and that the alleged threat was contained in a "poem"? Note that a number of prominent writers viewed George T.'s prosecution as a violation of artistic freedom and urged the court to dismiss the charges against George T. Would the court have ruled differently if the poem had stated clearly that George T. planned to return to school with a gun? What if George T. had expressed the sentiments in the letter directly to various students and teachers?

6. Does the California Supreme Court's focus on specific words lead the court to overlook that the poem was interpreted as a threat by Mary and Erin?

7. Do you agree with the California Supreme Court's ruling that George T.'s poem is protected speech under the First Amendment?

Cases and Comments

1. ***Flag Burning.*** In *Texas v. Johnson,* the U.S. Supreme Court addressed the constitutionality of Texas Penal Code Annotated section 42.09 (1989), which punished the intentional or knowing desecration of a "state or national flag." Desecration under the statute was interpreted as to "efface, damage, or otherwise physically mistreat in a way that the actor knows will seriously offend one or more persons likely to observe or discover his action."

Johnson participated in a political demonstration during the Republican National Convention in Dallas in 1984. The purpose was to protest the policies of the Reagan administration and certain Dallas-based corporations and to dramatize the consequences of nuclear war. The demonstrators gathered in front of the Dallas City Hall, where Johnson unfurled an American flag, doused the flag with kerosene, and set it on fire. The demonstrators chanted "America, the red, white, and blue, we spit on you" as the flag burned. None of the participants was injured or threatened retribution.

Justice Brennan observed that the Supreme Court had recognized that conduct may be protected under the First Amendment where there is an intent to convey a particularized message and there is a strong likelihood that this message will be understood by observers. Justice Brennan observed that the circumstances surrounding Johnson's burning of the flag resulted in his message being "both intentional and overwhelmingly apparent." In those instances in which an act contains both communicative and noncommunicative elements, the standard in judging the constitutionality of governmental regulation of *symbolic speech* is whether the government has a substantial interest in limiting the nonspeech element (the burning).

The Supreme Court rejected Texas's argument that the statute was a justified effort to preserve the flag as a symbol of nationhood and national unity. This would permit Texas to "prescribe what is orthodox by saying that one may burn the flag . . . only if one does not endanger the flag's representation of nationhood and national unity." In the view of the majority, Johnson was being unconstitutionally punished based on the ideas he communicated when he burned the flag. See *Texas v. Johnson,* 491 U.S. 397 (1989).

In 1989, the U.S. Congress adopted the Flag Protection Act, 19 U.S.C. § 700. The act provided that anyone who "knowingly mutilates, defaces, physically defiles, burns, maintains on the floor or ground, or tramples upon" a U.S. flag shall be subject to both a fine and imprisonment for not more than one year. This law exempted the disposal of a worn or soiled flag. The U.S. government asserted an interest in preserving the flag as "emblematic of the Nation as a sovereign entity." In *United States v. Eichman,* Justice Brennan failed to find that this law was significantly different from the Texas statute in *Johnson* and ruled that the law "suppresses expression out of concern for its likely communicative impact." Justice Stevens, in a dissent joined by Justices Rehnquist, White, and O'Connor, argued that the government may protect the symbolic value of the flag and that this does not interfere with the speaker's freedom to express his or her ideas by other means. He noted that various types of expression are subject to regulation. For example, an individual would not be free to draw attention to a cause through a "gigantic fireworks display or a parade of nude models in a public park." See *United States v. Eichman,* 496 U.S. 310 (1990).

2. ***Picketing Military Funerals.*** The American embrace of freedom of speech was tested in the 2011 case of *Snyder v. Phelps,* where the U.S. Supreme Court overturned a judgment against the Westboro United Church for the civil tort of the intentional infliction of emotional distress. The case was brought by Al Snyder, the father of Lance Corporal Matthew Snyder who had been killed in the line of duty in Iraq.

Members of the Westboro Church picketed Corporal Snyder's funeral on public land adjacent to the burial site. The picketing was designed to call attention to the belief of church members that the United States had angered God by tolerating homosexuality and that God had retaliated by allowing the killing of American soldiers. The church had picketed more than 600 military funerals over the last six years. Chief Justice John Roberts, writing for the eight-judge majority, overturned the verdict against Westboro United Church, reasoning that the members of the congregation

> [had] addressed matters of public import on public property, in a peaceful manner, in full compliance with the guidance of local officials. The speech . . . did not itself disrupt that funeral, and Westboro's choice to conduct its picketing at that time and place did not alter the nature of its speech.

> Speech is powerful. It can stir people to action, move them to tears of both joy and sorrow, and—as it did here—inflict great pain. On the facts before us, we cannot react to that pain by punishing the speaker. As a Nation we have chosen a different course—to protect even hurtful speech on public issues to ensure that we do not stifle public debate. That choice requires that we shield Westboro from tort liability for its picketing in this case.

Snyder v. Phelps, ___ U.S. ___, 131 S.Ct. 1207, 179 L.Ed.2d 172 (2011).

In reaction to the picketing of military funerals, the U.S. Congress passed the Respect for America's Fallen Heroes Act (RAFHA). Roughly 29 states have adopted antipicketing statutes or have broadened their laws to impose restrictions on the picketing of funerals. These laws regulate the time, place, and manner of demonstrations at funerals and do not restrict the content of the demonstration.

See more cases on the study site: **Snyder v. Phelps, Virginia v. Black, People v. Rokicki, Stevens v. United States, www.sagepub.com/Lippmanccl3e.**

You Decide

2.3 Lori MacPhail, a peace officer in Chico, California, assigned to a high school, observed Ryan D. with some other students off campus during school hours. She conducted a pat down, discovered that Ryan possessed marijuana, and issued him a citation.

Roughly a month later, Ryan turned in an art project for a painting class at the high school. The projects generally are displayed in the classroom for as long as two weeks. Ryan's painting pictured an individual who appeared to be a juvenile wearing a green hooded sweatshirt discharging a handgun at the back of the head of a female peace officer with badge No. 67 (Officer MacPhail's number) and the initials CPD (Chico Police Department). The officer had blood on her hair and pieces of her flesh and face were blown away. An art teacher saw the painting and found it to be "disturbing" and "scary," and an administrator at the school informed Officer MacPhail.

An assistant principal confronted Ryan, who stated the picture depicted his "anger at police officers" and that he was angry with MacPhail and agreed that it was "reasonable to expect that Officer MacPhail would eventually see the picture." Ryan was charged with a violation of section 422 and brought before juvenile court.

How would you rule? See *In re Ryan D.*, 123 Cal. Rptr. 2d 193 (Cal. App. 2002).

You can find the answer at www.sagepub.com/lippmanccl3e.

PRIVACY

The idea that there should be a legal right to **privacy** was first expressed in an 1890 article in the *Harvard Law Review* written by Samuel D. Warren and Louis D. Brandeis, who was later appointed to the U.S. Supreme Court. The two authors argued that the threats to privacy associated with the dawning of the twentieth century could be combated through recognition of a civil action (legal suit for damages) against individuals who intrude into individuals' personal affairs.[49]

In 1905, the Supreme Court of Georgia became the first court to recognize an individual's right to privacy when it ruled that the New England Life Insurance Company illegally used the image of artist Paolo Pavesich in an advertisement that falsely claimed that Pavesich endorsed the company.[50] This decision served as a precedent for the recognition of privacy by courts in other states.

For a deeper look at this topic, visit the study site.

The Constitutional Right to Privacy

A constitutional right to privacy was first recognized in *Griswold v. Connecticut* in 1965. The U.S. Supreme Court proclaimed that although privacy was not explicitly mentioned in the U.S. Constitution, it was implicitly incorporated into the text. The case arose when Griswold, along with Professor Buxton of Yale Medical School, provided advice to married couples on the prevention of procreation through contraceptives. Griswold was convicted of being an accessory to the violation of a Connecticut law that provided that any person who uses a contraceptive shall be fined not less than $50 or imprisoned not less than sixty days nor more than one year or be both fined and imprisoned.[51]

Justice William O. Douglas noted that although the right to privacy was not explicitly set forth in the Constitution, this right was "created by several fundamental constitutional guarantees." According to Justice Douglas, these fundamental rights create a "zone of privacy" for individuals. In a famous phrase, Justice Douglas noted that the various provisions of the Bill of Rights possess "penumbras, formed by emanations from those guarantees . . . [that] create zones of privacy." Justice Douglas cited a number of constitutional provisions that together create the right to privacy.

The right of association contained in the penumbra of the First Amendment is one; the Third Amendment in its prohibition against the quartering of soldiers "in any house" in time of peace without the consent of the owner is another facet of that privacy. The Fourth Amendment explicitly affirms the "right of the people to be secure in their persons, houses, papers, and effects, against unreasonable searches and seizures." The Fifth Amendment's Self-Incrimination Clause "enables the citizen to create a zone of privacy that Government may not force him to surrender to his detriment." The Ninth Amendment provides that "[t]he enumeration in the Constitution of certain rights shall not be construed to deny or disparage others retained by the people."

In contrast, Justice Arthur Goldberg argued that privacy was found within the Ninth Amendment, and Justice Harlan contended that privacy is a fundamental aspect of individual "liberty" within the Fourteenth Amendment.

We nevertheless should take note of Justice Hugo Black's dissent in *Griswold* questioning whether the Constitution provides a right to privacy, a view that continues to attract significant support. Justice Black observed that "I like my privacy as well as the next one, but I am nevertheless compelled to admit that government has a right to invade [my privacy] unless prohibited by some specific constitutional provision."

The right to privacy recognized in *Griswold* guarantees that we are free to make the day-to-day decisions that define our unique personality: what we eat, read, and watch; where we live and how we spend our time, dress, and act; and with whom we associate and work. In a totalitarian society, these choices are made by the government, but in the U.S. democracy, these choices are made by the individual. The courts have held that the right to privacy protects several core concerns:

- *Sanctity of the Home*. Freedom of the home and other personal spaces from arbitrary governmental intrusion
- *Intimate Activities*. Freedom to make choices concerning personal lifestyle and an individual's body and reproduction
- *Information*. The right to prevent the collection and disclosure of intimate or incriminating information to private industry, the public, and governmental authorities
- *Public Portrayal*. The right to prevent your picture or endorsement from being used in an advertisement without permission or to prevent the details of your life from being falsely portrayed in the media[52]

In short, as noted by Supreme Court Justice Louis Brandeis, "The makers of our Constitution undertook to secure conditions favorable to the pursuit of happiness. . . . They conferred as against the Government, the right to be let alone—the most comprehensive of rights and the right most valued by civilized men."[53]

There are several key Supreme Court decisions on privacy.

In *Eisenstadt v. Baird*, in 1972, the Supreme Court extended *Griswold* and ruled that a Massachusetts statute that punished individuals who provided contraceptives to unmarried individuals violated the right to privacy. Justice William Brennan wrote that "if the right to privacy means anything, it is the right of the individual, married or single, to be free from unwarranted governmental intrusion into matters so fundamentally affecting a person as the decision whether to bear or beget a child."[54]

The Supreme Court, in *Carey v. Population Services International*, next declared a New York law unconstitutional that made it a crime to provide contraceptives to minors and for anyone other than a licensed pharmacist to distribute contraceptives to persons over fifteen. Justice Brennan noted that this imposed a significant burden on access to contraceptives and impeded the "decision whether or not to beget or bear a child" that was at the "very heart" of the "right to privacy."[55]

In 1973, in *Roe v. Wade*, the U.S. Supreme Court ruled unconstitutional a Texas statute that made it a crime to "procure an abortion." Justice Blackmun wrote that the "right to privacy . . . is broad enough to encompass a woman's decision whether or not to terminate her pregnancy."[56] The Supreme Court later ruled that Pennsylvania's requirement that a woman obtain her husband's consent unduly interfered with her access to an abortion.[57]

The zone of privacy also was extended to an individual's intellectual life in the home in 1969 in *Stanley v. Georgia*. A search of Stanley's home for bookmaking paraphernalia led to the seizure of three reels of film portraying obscene scenes. Justice Marshall concluded that "whatever the power of the state to control public dissemination of ideas inimical to the public morality, it cannot constitutionally premise legislation on the desirability of controlling a person's private thoughts."[58]

The Constitutional Right to Privacy and Same-Sex Relations Between Consenting Adults in the Home

Privacy, however appealing, lacks a clear meaning. Precisely what activities are within the right of privacy in the home? In answering this question, we must balance the freedom to be let alone against the need for law and order. The issue of sodomy confronted judges with the question of whether laws upholding sexual morality must yield to the demands of sexual freedom within the home.

In 1986, in *Bowers v. Hardwick,* the Supreme Court affirmed Hardwick's sodomy conviction under a Georgia statute. Justice White failed to find a fundamental right deeply rooted in the nation's history and tradition to engage in acts of consensual sodomy, even when committed in the privacy of the home. He pointed out that sodomy was prohibited by all thirteen colonies at the time the constitution was ratified, and twenty-five states and the District of Columbia continued to criminally condemn this conduct.[59]

Bowers v. Hardwick was reconsidered in 2003, in *Lawrence v. Texas.* In *Lawrence,* the Supreme Court called in doubt the historical analysis in *Bowers* and noted that only thirteen states currently prohibited sodomy and that in these states, there is a "pattern of nonenforcement with respect to consenting adults in private." The Court held that the right to privacy includes the fundamental right of two consenting males to engage in sodomy within the privacy of the home.[60]

You can find Lawrence v. Texas *at the study site www.sagepub.com/lippmanccl3e.*

Cases and Comments

Voyeurism. On April 26, 1999, Sean Glas used a camera to take pictures underneath the skirts of two women working at the Valley Mall in Union Gap, Washington. In one instance, Inez Mosier was working the women's department at Sears and saw a light flash out of the corner of her eye. She turned around to discover Glas squatting on the floor a few feet behind her. She noticed a small, silver camera in his hand. The police later confiscated the film and discovered photos of the undergarments of Mosier and another woman. Richard Sorrells, in a separate case, was apprehended after using a video camera to film the undergarments of women and young girls at the "Bite of Seattle" at the Seattle Center. Both Glas and Sorrells were convicted of voyeurism for taking photos underneath women's skirts ("upskirt" voyeurism). The Washington voyeurism statute (Wash. Rev. Code § 9A.44.115(2)(a)) reads,

A person commits the crime of voyeurism if, for the purpose of arousing or gratifying the sexual desire of any person, he or she knowingly views, photographs, or films: another person without that person's knowledge and consent while the person being viewed, photographed, or filmed is in a place where he or she would have a reasonable expectation of privacy.

The statute defines a place in which a person would have a reasonable expectation of privacy as a place where a "reasonable person would believe that he or she could disrobe in privacy, without being concerned that his or her undressing was being filmed by another," or as a "place where one may reasonably expect to be safe from casual or hostile intrusion or surveillance." The Washington Supreme Court interpreted a location where an individual may "disrobe in privacy" to include the bedroom, bathroom, dressing room, or tanning salon. A location in which an individual may reasonably expect to be safe from intrusion or surveillance includes the other rooms in an individual's home as well as locations where someone would not normally disrobe, but would not expect others to intrude, such as a private suite or office.

The court acquitted the two defendants, ruling that although Glas and Sorrells engaged in "disgusting and reprehensible behavior," Washington's voyeurism statute "does not apply to actions taken in purely public places and hence does not prohibit the 'upskirt' photographs" taken by Glas and Sorrells. Do you agree that the women had no expectation of privacy? See *Washington v. Glas,* 54 P.3d 147 (Wash. 2002).

You can find more cases on the study site: Utah v. Holm, ***www.sagepub.com/lippmanccl3e.***

You Decide

2.4 The plaintiffs allege that the Florida law requiring motorcyclists to wear helmets violates their right to privacy under the U.S. Constitution. Are they correct? See *Picou v. Gillum*, 874 F.2d 1519 (11th Cir. 1989).

You can find the answer at www.sagepub.com/lippmanccl3e.

THE RIGHT TO BEAR ARMS

The American people historically have considered the handgun to be the quintessential self-defense weapon. Handguns are easily accessible in an emergency and require only a modest degree of physical strength to use and cannot easily be wrestled away by an attacker. In the past several decades, various cities and suburbs have placed restrictions on the right of Americans to possess handguns, even for self-defense. The constitutionality of these limitations on the possession of handguns was addressed by two recent U.S. Supreme Court decisions.

The Second Amendment to the U.S. Constitution provides that "A well regulated Militia being necessary to the security of a free State, the right of the people to keep and bear Arms shall not be infringed."

The meaning of the Second Amendment has been the topic of considerable debate. Courts historically focused on the first clause of the amendment that recognizes the importance of a "well regulated Militia" and held that the Amendment protects the right of individuals to possess arms in conjunction with service in an organized government militia. In 1939 in *United States v. Miller*, the U.S. Supreme Court upheld the constitutionality of a federal law prohibiting the interstate shipment of sawed-off shotguns, reasoning that the Second Amendment protections are limited to gun ownership that has "some reasonable relationship to the preservation or efficiency of a well regulated militia."[61]

Gun rights activists contended that the Second Amendment protection of the "right of the people to keep and bear Arms" is not limited to members of the militia. They argued that the Second Amendment also protects individuals' right to possess firearms "unconnected" with service in a militia. The founding fathers, according to gun activists, viewed gun ownership as essential to the preservation of individual liberty. A state or federal government could abolish the state national guard and leave citizens unarmed and vulnerable. The framers concluded that the best way to safeguard and to protect the people was to guarantee individuals' right to bear arms.

In *District of Columbia v. Heller*, the U.S. Supreme Court adopted the view of gun rights activists. The Court majority held that the Second Amendment protects the right of individuals to possess firearms.[62]

Dick Heller, a special police officer, was authorized to carry a handgun while on duty at the Federal courthouse in the District of Columbia (D.C.) and applied for a registration certificate from the D.C. government for a handgun that he planned to keep at home for self-defense. A D.C. ordinance prohibited the possession of handguns and declared that it was a crime to carry an unregistered firearm. A separate portion of the D.C. ordinance authorized the Chief of Police to issue licenses for 1-year periods. Lawfully registered handguns were required to be kept "unloaded and dissembled or bound by a trigger lock or similar device" when not "located" in a place of business place or used for lawful recreational activities.

Justice Anton Scalia writing for a five-judge majority held that the D.C. ordinance was unconstitutional because the regulations interfered with the ability of law-abiding citizens to use a firearm for self-defense in the home, the "core lawful purpose" of the right to bear arms. "Undoubtedly some think that the Second Amendment is outmoded in a society where our standing army is the pride of our Nation, where well-trained police forces provide personal security, and where gun violence is a serious problem. That is perhaps debatable, but what is not debatable is that it is not the role of this Court to pronounce the Second Amendment extinct."

The Court decision noted that while D.C. could not constitutionally ban the possession of firearms in the home, the right to bear arms is subject to limitations. The Court did not limit the ability of states to prohibit possession of firearms by felons and the mentally challenged, to prohibit the carrying of firearms in "sensitive places" such as schools and government buildings, to regulate the commercial sale of arms, to ban the possession of dangerous and unusual weapons, or to require the safe storage of weapons.

Heller, although important for defining the meaning of the Second Amendment, applied only to D.C. and to other federal jurisdictions. In 2010, in *Chicago v. McDonald,* residents of Chicago, Illinois, and the Chicago suburb of Oak Park, Illinois, challenged local ordinances that were almost identical to the law that the Court struck down as unconstitutional in the federal enclave of Washington, D.C. The Supreme Court addressed whether the Second Amendment right of individuals to bear arms extended to state as well as to the federal government.[63]

The Fourteenth Amendment had been adopted following the Civil War to insure former African American slaves equal rights, and the Supreme Court in a series of cases had ruled that most of the Bill of Rights was applicable to the States and protected individuals against the State as well as the Federal governments. The Second Amendment was one of the few amendments in the Bill of Rights that had not been incorporated into the Fourteenth Amendment and made applicable to the states. The result was that even after *Heller,* the right to possess firearms was not considered a fundamental right protected by the Fourteenth Amendment, and state governments were free to restrict or even to prohibit the possession of firearms.

The Fourteenth Amendment prohibits a state from denying an individual life, liberty, or property without due process of law. The question in *McDonald v. Chicago* was whether the right to keep and to bear arms was a liberty interest protected under the Due Process Clause of the Fourteenth Amendment. Justice Samuel Alito wrote that self-defense is a "basic right, recognized by many legal systems from ancient times to the present day." He concluded that Second Amendment right to possess firearms in the home for the purpose of self-defense is incorporated into the Fourteenth Amendment and is applicable to the states. The right to keep and bear arms for purposes of self-defense is "among the fundamental rights necessary to our system of ordered liberty," which is "deeply rooted in this Nation's history and tradition." A number of state constitutions already protected the right to own and to carry arms. The incorporation of the Second Amendment into the Fourteenth Amendment clearly established that the right to bear arms for the purpose of self-defense is a fundamental right that may not be infringed by state governments.

The precise meaning of the decisions in *Heller* and *McDonald* will not be clear until various state gun control laws are reviewed by the courts. Chicago has modified its gun control law to allow guns to be stored inside the home although residents may not possess "more than one handgun in operating order at any given time." Residents in homes with children are required to keep handguns in lock boxes or to equip the firearms with trigger locks. Gun owners must take a safety class and certain categories of offenders are prohibited from owning a firearm. The question remains whether these limitations on the right to own arms will be upheld as constitutional under the Second Amendment.

You might be interested in the fact that Utah and Colorado permit students on college campuses to carry concealed weapons. Will state laws that prohibit guns on campus be held to violate students' Second Amendment rights?

You can find McDonald v. Chicago *at the study site www.sagepub.com/lippmanccl3e.*

CHAPTER SUMMARY

The United States is a constitutional democracy. The government's power to enact laws is constrained by the constitution. These limits are intended to safeguard the individual against the passions of the majority and the tyrannical tendencies of government. The restrictions on government also are designed to maximize individual freedom, which is the foundation of an energetic and creative society and dynamic economy. Individual freedom, of course, must be balanced against the need for social order and stability. We all have been reminded that "you cannot yell 'fire' in a crowded theater." This chapter challenges you to locate the proper balances among freedom, order, and stability.

The rule of legality requires that individuals receive notice of prohibited acts. The ability to live your life without fear of unpredictable criminal punishment is fundamental to a free society. The rule of legality provides the philosophical basis for the constitutional prohibition on bills of attainder and *ex post facto* laws. Bills of attainder prohibit the legislative punishment of individuals without trial. *Ex post facto* laws prevent the government from criminally punishing acts that were innocent when committed. The constitutional provision for due process insures that individuals are informed of acts that are criminally condemned and that definite standards are established that limit the discretion of the police. An additional restriction on criminal statutes is the Equal Protection Clause. This prevents the government from creating classifications that unjustifiably disadvantage or discriminate against individuals; a particularly heavy burden is imposed on the government to justify distinctions based on race or ethnicity. Classifications on gender are subject to intermediate scrutiny. Other differentiations are required only to meet a rational basis test.

Freedom of expression is of vital importance in American democracy, and the Constitution protects speech that some may view as offensive and disruptive. Courts may limit speech only in isolated situations that threaten social harm and instability. The right to privacy protects individuals from governmental intrusion into the intimate aspects of life and creates "space" for individuality and social diversity to flourish. The U.S. Supreme Court has held that the Second Amendment protects the right of individuals to possess handguns for the purpose of self-defense in the home. The full extent of the Second Amendment "right to bear arms" has yet to be determined.

This chapter provided you with the constitutional foundation of American criminal law. Keep this material in mind as you read about criminal offenses and defenses in the remainder of the textbook. We will look at the Eighth Amendment prohibition on cruel and unusual punishment in Chapter 3.

CHAPTER REVIEW QUESTIONS

1. Explain the philosophy underlying the United States' constitutional democracy. What are the reasons for limiting the powers of state and federal government to enact criminal legislation? Are there costs as well as benefits in restricting governmental powers?

2. Define the rule of legality. What is the reason for this rule?

3. Define and compare bills of attainder and *ex post facto* laws. List the various types of *ex post facto* laws. What is the reason that the U.S. Constitution prohibits retroactive legislation?

4. Explain the standards for laws under the Due Process Clause.

5. Why does the U.S. Constitution protect freedom of expression? Is this freedom subject to any limitations?

6. What is the difference among the "rational basis," "intermediate scrutiny," and "strict scrutiny" tests under the Equal Protection Clause?

7. Where is the right to privacy found in the U.S. Constitution? What activities are protected within this right?

8. Write a short essay on the constitutional restrictions on the drafting and enforcement of criminal statutes.

9. As a final exercise, consider life in a country that does not provide safeguards for civil liberties. How would your life be changed?

LEGAL TERMINOLOGY

bill of attainder

Bill of Rights

constitutional democracy

equal protection

ex post facto law

fighting words

First Amendment

hate speech

incitement to violent action

incorporation theory

intermediate level of scrutiny

libel

minimum level of scrutiny test

nullum crimen sine lege, nulla poena sine lege

obscenity

overbreadth

privacy

rational basis test

rule of legality

strict scrutiny

true threats

void for vagueness

CRIMINAL LAW ON THE WEB

Log on to the Web-based student study site at www.sagepub.com/lippmanccl3e to assist you in completing the Criminal Law on the Web exercises, as well as for additional features such as podcasts, Web quizzes, and video links.

1. Read about flag burning and freedom of speech.
2. Consider the debate over the meaning of the Second Amendment and the right to bear arms.

BIBLIOGRAPHY

Erwin Chemerinsky, *Constitutional Law Principles and Policies,* 2nd ed. (New York: Aspen, 2002), pp. 641–1137. A clear and well-researched discussion of various provisions of the Bill of Rights to the U.S. Constitution.

Joshua Dressler, *Understanding Criminal Law,* 3rd ed. (New York: Lexis, 2001), pp. 33–62. An overview of various constitutional constraints on criminal legislation by a leading legal scholar.

Thomas I. Emerson, *The System of Freedom of Expression* (New York: Vintage Books, 1970). A classic volume on the theory and law of the First Amendment.

Alexander Hamilton, "Federalist No. 51," in A. Hamilton, J. Madison, & J. Jay, *The Federalist Papers* (New York: New American Library, 1961), pp. 320–325. James Madison confronts critics who advocate the inclusion of a Bill of Rights in the U.S. Constitution.

Wayne R. LaFave, *Criminal Law,* 3rd ed. (St. Paul, MN: West Publishing, 2000), pp. 97–113, 175–195. A sophisticated discussion of the constitutional constraints on criminal statutes.

Arnold Loewy, *Criminal Law* (St. Paul, MN: West Publishing, 2003), pp. 291–306. A concise discussion of vagueness, *ex post facto* laws, and other limitations.

Herbert Packer, *The Limits of the Criminal Sanctions* (Palo Alto, CA: Stanford University Press, 1968), pp. 149–248. A classic discussion on the challenge of striking the balance between individual rights and societal safety and security.

3 PUNISHMENT AND SENTENCING

May Missouri legally execute seventeen-year-old murderer Christopher Simmons?

Simmons and Benjamin entered the home of the victim, Shirley Crook, after reaching through an open window and unlocking the back door. Simmons turned on a hallway light. Awakened, Mrs. Crook called out, "Who's there?" In response Simmons entered Mrs. Crook's bedroom, where he recognized her from a previous car accident involving them both. Simmons later admitted this confirmed his resolve to murder her.

Using duct tape to cover her eyes and mouth and bind her hands, the two perpetrators put Mrs. Crook in her minivan and drove to a state park. They reinforced the bindings, covered her head with a towel, and walked her to a railroad trestle spanning the Meramec River. There they tied her hands and feet together with electrical wire, wrapped her whole face in duct tape and threw her from the bridge, drowning her in the waters below. . . . Simmons, meanwhile, was bragging about the killing, telling friends he had killed a woman "because the bitch seen my face. . . ."

Learning Objectives

1. Know the factors that are considered in determining whether a statute imposes criminal punishment or a civil penalty. Understand why it is significant whether a law imposes criminal punishment or a civil penalty.

2. Discuss the purposes of punishment, various types of punishments, and the four different approaches to sentencing offenders.

3. Understand the development, purpose, and significance of sentencing guidelines.

4. Know about truth in sentencing, and victims' rights.

5. Understand the relationship between the Eighth Amendment and criminal punishment.

6. Summarize the role of the Equal Protection Clause in criminal sentencing.

INTRODUCTION

One of the primary challenges confronting any society is to ensure that people follow the legal rules that protect public safety and security. This is partially achieved through the influence of families, friends, teachers, the media, and religion. Perhaps the most powerful method to persuade people to obey legal rules is through the threat of criminal punishment. Following a defendant's conviction, the judge must determine the appropriate type and length of the sentence.

The sentence typically reflects the purpose of the punishment. A penalty intended to exact revenge will result in a harsher punishment than a penalty designed to assist an offender to "turn his or her life around."

An American judge in colonial times and during the early American republic was able to select from a wide array of punishments, most of which were intended to inflict intense pain and public shame. A Virginia statute of 1748 punished the stealing of a hog with twenty-five lashes and a fine. The second offense resulted in two hours of pillory (public ridicule) or public branding. A third theft resulted in a penalty of death. False testimony during a trial might result in mutilation of the ears or banishment from the colony. These penalties were often combined with imprisonment in a jail or workhouse and hard labor. You should keep in mind that minor acts of insubordination by African American slaves resulted in swift and harsh punishment without trial. Between 1706 and 1784, 550 African slaves were sentenced to death in Virginia alone.[1]

We have slowly moved away from most of these physically painful sanctions. The majority of states followed the example of the U.S. Congress, which in 1788 prohibited federal courts from imposing whipping and standing in the pillory. Maryland retained corporal punishment until 1953, and Delaware repealed this punishment only in 1972. Delaware, in fact, subjected more than 1,600 individuals to whippings in the twentieth century.[2] This practice was effectively ended in 1968, when the Eighth Circuit Court of Appeals ruled that the use of the strap "offends contemporary concepts of decency and human dignity and precepts of civilization which we profess to possess."[3] In 1994, President Bill Clinton and twenty-four U.S. senators wrote the president of Singapore in an unsuccessful effort to persuade him to make an "enlightened decision" and to halt plans to subject an American teenager charged with vandalism to four lashings with a rattan rod.[4]

In the United States, courts have attempted to balance the need for swift and forceful punishment with the recognition that individuals are constitutionally entitled to fair procedures and are to be free from cruel and unusual punishments. You should be familiar with several central concerns when you complete the study of this chapter:

- the definition of punishment,
- justifications for punishment,
- the types of sentences that may be imposed by judges,
- the considerations employed to evaluate the merits of sentencing schemes,
- the approaches to sentencing in federal and state courts, and
- the constitutional standards that must be met by criminal sentences.

You should also come away from this chapter with an understanding that sentencing policies have evolved over time. Disillusionment with flexible sentences and rehabilitation led to the development of sentencing guidelines and determinate sentences that are intended to ensure uniform and fixed sentences that fit the crime. The federal and state governments also adopted "truth in sentencing" laws that assure the public that defendants are serving a significant portion of their prison terms. These developments have been accompanied by a growing concern for victims.

The central point that you should appreciate is that the United States is witnessing a revolution in sentencing. Keep the following points in mind:

- *Purpose of Punishment.* The emphasis is on deterrence, retribution, incapacitation, education, and treatment of offenders rather than on rehabilitation.
- *Judicial Discretion.* Judicial discretion in sentencing is greatly reduced. The federal government and states have introduced sentencing guidelines and mandatory minimum sentences, illustrated by Three Strikes and You're Out legislation and drug laws.
- *Truth in Sentencing.* The authority of parole boards to release prisoners prior to the completion of their sentences and the ability of incarcerated individuals to accumulate "good time" is vastly reduced as a result of truth in sentencing legislation. As a consequence, offenders are serving a greater percentage of their sentences.
- *Victims.* Victims are being provided a greater role and more protections in the criminal justice process.
- *Death Penalty.* The death penalty does not violate the Eighth Amendment. Capital punishment, however, is subject to a number of constitutional limitations under the Eighth Amendment intended to ensure that death is a penalty proportionate to the offender's crime.
- *Terms of Years.* Courts have deferred to the decisions of state legislatures and the Congress in regard to sentencing decisions and generally have held that prison sentences are proportionate to the offender's crime.
- *Equal Protection.* Courts have ruled that sentencing decisions and statutes based on race or gender violate the Equal Protection Clause.

The larger point to consider as you read this chapter is whether we have struck an appropriate balance among the interests of society, defendants, and victims in the sentencing process. You should make an effort to develop your own theory of punishment.

PUNISHMENT

Professor George P. Fletcher writes that the central characteristic of a criminal law is that a violation of the rule results in punishment before a court. Whether an act is categorized as a criminal as opposed to a civil violation is important, because a criminal charge triggers various constitutional rights, such as a right against double jeopardy, the right to a lawyer, the right not to testify at trial, and the right to a trial by a jury.[5] Would the quarantine of individuals during a flu pandemic be considered a civil disability or a criminal penalty? The U.S. Supreme Court has listed various considerations that determine whether a law is criminal.[6]

- Does the legislature characterize the penalty as civil or criminal?
- Has the type of penalty imposed historically been viewed as criminal?
- Does the penalty involve a significant disability or restraint on personal freedom?
- Does the penalty promote a purpose traditionally associated with criminal punishment?
- Is the imposition of the penalty based on an individual's intentional wrongdoing, a requirement that is central to criminal liability?
- Has the prohibited conduct traditionally been viewed as criminal?

Whether a law is considered to impose criminal punishment can have important consequences for a defendant. For instance, in *Smith v. Doe,* the U.S. Supreme Court was asked to decide whether Alaska's sex registration law constituted *ex post facto* criminal punishment.

In 1994, the U.S. Congress passed the Jacob Wetterling Crimes Against Children and Sexually Violent Offender Registration Act that makes certain federal criminal justice funding dependent on a state's adoption of a sex offender registration law. By 1996, every state, the District of Columbia, and the federal government had enacted some type of **Megan's Law**. These statutes were named in memory and honor of Megan Kanka, a seven-year-old New Jersey child who had been sexually assaulted and murdered in 1994 by a neighbor who, unknown to Megan's family, had prior convictions for sexual offenses against children.[7]

Alaska adopted a retroactive law that required both convicted sex offenders and child kidnappers to register and keep in contact with local law enforcement authorities. Alaska provided nonconfidential information to the public on the Internet, including an offender's crime, address, place of employment, and photograph.

Supreme Court Justice Anthony Kennedy, in his majority opinion in *Smith v. Doe,* agreed with Alaska that this statute was intended to protect the public from the danger posed by sexual offenders through the dissemination of information, and that the law was not intended to constitute and did not constitute unconstitutional *ex post facto* (retroactive) criminal punishment.[8]

Justice Ruth Ginsburg dissented from the majority judgment affirming the constitutionality of the Alaska statute. She observed that placing a registrant's face on a Web site under the label "Registered Sex Offender" was reminiscent of the shaming punishment that was employed during the colonial era when individuals were branded or placed in stocks and subjected to public ridicule.

Justice Ginsburg pointed out that John Doe I, one of the individuals bringing this case, had been sentenced to prison for sexual abuse nine years before the passage of the Alaska statute. He successfully completed a rehabilitation program and gained early release on supervised probation. John Doe subsequently remarried, established a business, and gained custody of one of his daughters based on a judicial determination that he no longer posed a threat. The Alaska version of Megan's Law now required John Doe I "to report personal information to the State four times per year," and permitted the state publicly to label him a registered sex offender for the rest of his life.

Justice Ginsburg's argument that Megan's Law constitutes *ex post facto* punishment would render Alaska helpless to alert citizens to the continuing danger posed by sex offenders convicted prior to the passage of Megan's Law. Does Justice Ginsberg overlook the fact that, although registrants must inform authorities of changes in appearance, employment, and address, these individuals remain free to live their lives without restraint or restriction and can hardly claim to have been punished? On the other hand, there was evidence that registrants were scorned by the community, experienced difficulties in employment and housing, and encountered hostility. However, this resulted from the acts of members of the public rather than the government.

The next section briefly outlines the purposes or goals that are the basis of sentencing in the criminal justice system. These purposes include retribution, deterrence, rehabilitation, incapacitation, and restoration.

PURPOSES OF PUNISHMENT

In the United States, we have experienced various phases in our approach to criminal punishment. We continue to debate whether the primary goal of punishment should be to assist offenders to turn their lives around or whether the goal of punishment should be to safeguard society by locking up offenders. Some rightly point out that we should not lose sight of the need to require offenders to compensate crime victims. In considering theories of punishment, ask yourself what goals should guide our criminal justice system.

Retribution

Retribution imposes punishment based on **just deserts**. Offenders should receive the punishment that they deserve based on the seriousness of their criminal acts. The retributive philosophy is based on the familiar biblical injunction of "an eye for an eye, a tooth for a tooth." Retribution assumes that we all know right from wrong and are morally responsible for our conduct and should be held accountable. The question is what punishment is "deserved": a prison term, a fine, or confinement? How do we determine the appropriate length of a prison sentence and in what type of institution the sentence should be served? This is not always clear, because what an individual "deserves" may depend on the circumstances of the crime, the background of the victim, and the offender's personal history.

Deterrence

The theory of **specific deterrence** imposes punishment to deter or discourage a defendant from committing a crime in the future. Critics note that the recidivism rate indicates that punishment rarely deters crime. Also, we once again confront the challenge of determining the precise punishment required to achieve the desired result, in this case to deter an individual from returning to a life of crime. **General deterrence** punishes an offender as an example to deter others from violating the law. Critics contend that offenders have little concern or awareness of the punishment imposed on other individuals and that even harsh punishments have little general deterrent effect. Others reply that swift and certain punishment sends a powerful message, and that a credible threat of punishment constitutes a deterrent. There are also objections to punishing an individual as an example to others, because this may result in a harsher punishment than is required to deter the defendant from committing another crime.

Rehabilitation

The original goal of punishment in the United States was to reform the offender and to transform him or her into a law-abiding and productive member of society. **Rehabilitation** appeals to the idealistic notion that people are essentially good and can transform their lives when encouraged and given support. However, studies cast doubts on whether prison educational and vocational programs are able to rehabilitate inmates. Reformers, on the other hand, point out that rehabilitation has never been seriously pursued and requires a radically new approach to imprisonment.

Incapacitation

The aim of **incapacitation** is to remove offenders from society to prevent them from continuing to menace others. This approach accepts that there are criminally inclined individuals who cannot be deterred or rehabilitated. The difficulty with this approach is that we lack the ability to accurately predict whether an individual poses a continuing danger to society. As a result, we may incapacitate an individual based on a faulty prediction of what he or she may do in the future rather than for what he or she did in the past. **Selective incapacitation** singles out offenders who have committed designated offenses for lengthy incarceration. In many states, a conviction for a drug offense or a second or third felony under a Three Strikes and You're Out law results in a lengthy prison sentence or life imprisonment. There is continuing debate over the types of offenses that merit selective incapacitation.

Restoration

Restoration stresses the harm caused to victims of crime and requires offenders to engage in financial restitution and community service to compensate the victim and the community and to "make them whole once again." The restorative justice approach recognizes that the needs of victims are often overlooked in the criminal justice system. This approach is also designed to encourage offenders to develop a sense of individual responsibility and to become responsible members of society.

This discussion of the purposes of punishment is not mere academic theorizing. Judges, when provided with the opportunity to exercise discretion, are guided by these purposes in determining the appropriate punishment. For example, in a New York case, the court described Dr. Bernard Bergman as a man of "unimpeachably high character, attainments and distinction" who is respected by people around the world for his work in religion, charity, and education. Bergman's desire for money apparently drove him to fraudulently request payment from the U.S. government for medical treatment that he had not provided to nursing home patients. He entered guilty pleas to fraud charges in both New York and federal courts and argued that he should not be imprisoned, because he did not require "specific deterrence."

Judge Marvin Frankel recognized in his judgment that there was little need for incapacitation and doubted whether imprisonment could provide useful rehabilitation. Nevertheless, he imposed a four-month prison sentence, explaining that this is "a stern sentence. For people like Dr. Bergman who might be disposed to engage in similar wrongdoing, it should be sufficiently frightening to serve the . . . [purpose] of general deterrence." Judge Frankel also explained that the four-month sentence served the interest in retribution and that "for all but the profoundly vengeful, [the sentence] should not depreciate the seriousness of his offenses." [9] Do you agree with Judge Frankel's reasoning and sentence?

SENTENCING

Various types of punishments are available to judges. These punishments often are used in combination with one another:

- *Imprisonment.* Individuals sentenced to a year or less are generally sentenced to local jails. Sentences for longer periods are typically served in state or federal prisons.
- *Fines.* State statutes usually provide for fines as an alternative to incarceration or in addition to incarceration.
- *Probation.* Probation involves the suspension of a prison sentence so long as an individual continues to report to a probation officer and to adhere to certain required standards of personal conduct. For instance, this may entail psychiatric treatment or a program of counseling for alcohol or drug abuse. The conditions of probation are required to be reasonably related to the rehabilitation of the offender and the protection of the public.
- *Intermediate Sanctions.* This includes house arrest with electronic monitoring, short-term "shock" incarceration, community service, and restitution. Intermediate sanctions may be imposed as a criminal sentence, as a condition of probation, following imprisonment, or in combination with a fine.
- *Death.* Thirty-three states and the federal government provide the death penalty for homicide. The seventeen other states and Washington, D.C., provide life without parole.

We should also note that the federal government and most states provide for **assets forfeiture** or seizure pursuant to a court order of the fruits of illegal narcotics transactions (along with certain other crimes) or of the instrumentalities that were used in such activity. The burden rests on the government to prove by a preponderance of the evidence that instrumentalities (vehicles), profits (money), or property are linked to an illegal transaction. In *United States v. Ursery,* the U.S. Supreme Court held that the seizure of money and property did not constitute double jeopardy, because forfeitures do not constitute punishment.[10]

The U.S. Department of Justice reported in 2004 that state and federal courts convicted almost 1,145,000 adults of felonies. State courts accounted for 1,079,000 of these convictions. Seventy percent of individuals convicted in state courts were sentenced to prison, and thirty percent were sentenced to probation with no jail time. The average sentence for felons sentenced to state prisons was fifty-seven months; the average probation sentence was thirty-eight months. Individuals sentenced to local jails, on average, received a six-month sentence. Thirty-three percent of convicted felons were ordered to pay fines, and roughly eighteen percent were required to pay restitution. Thirty percent were required to undergo some form of treatment, perform community service, or satisfy some other requirement.

EVALUATING SENTENCING SCHEMES

As you read about various approaches to sentencing in the remainder of the chapter, keep several considerations in mind that might prove useful in evaluating the merits of a particular approach.

- *Proportionality.* A sentence should fit the crime.
- *Individualism.* A sentence should reflect the offender's criminal history and the threat posed to society.
- *Disparity.* The sentences for a particular offense should be uniform; like cases should be treated alike.
- *Predictability and Simplicity.* The sentence to be imposed for a particular offense should be clear and definite and should not be dependent on the personality or biases of the judge. It should be relatively easy for a judge to determine the appropriate sentence.
- *Excessiveness.* A sentence should not inflict unnecessary and needless pain and suffering.
- *Truthfulness.* An offender's sentence should reflect the actual time served in prison.
- *Purpose.* A sentence should be intended to achieve one or more of the purposes of punishment.

Clearly no single approach can achieve each of these goals.

APPROACHES TO SENTENCING

The approach to sentencing in states historically has shifted in response to the prevailing criminal justice thought and philosophy. The federal and state governments generally follow four different approaches to sentencing offenders. Criminal codes may incorporate more than a single approach.

- **Determinate Sentences.** The state legislature provides judges with little discretion in sentencing and specifies that the offender is to receive a specific sentence. A shorter or longer sentence may be given to an offender, but this must be justified by the judge.
- **Mandatory Minimum Sentences.** The legislature requires judges to sentence an offender to a minimum sentence, regardless of mitigating factors. Prison sentences in some jurisdictions may be reduced by good-time credits earned by the individual while incarcerated.
- **Indeterminate Sentences.** The state legislature provides judges with the ability to set a minimum and maximum sentence within defined limits. In some jurisdictions, the judge possesses discretion only to establish a maximum sentence. The decision to release an inmate prior to fully serving his or her sentence is vested in a parole board.
- **Presumptive Sentencing Guidelines.** A legislatively established commission provides a sentencing formula based on various factors, stressing the nature of the crime and the offender's criminal history. Judges may be strictly limited in terms of discretion or may be provided

with some flexibility within established limits. The judge must justify departures from the presumptive sentence on the basis of various aggravating and mitigating factors that are listed in the guidelines. Appeals are provided in order to maintain reasonable sentencing practices in those instances in which a judge departs from the presumptive sentence in the guidelines.

An individual convicted of multiple crimes may be given **consecutive sentences**, meaning that the sentences for each criminal act are served one after another. In the alternative, **concurrent sentences** are served at the same time.

Governors and, in the case of federal offenses, the President of the United States may grant an offender **clemency**, resulting in a reduction of an individual's sentence or in a commutation of a death sentence to life in prison. A **pardon** exempts an individual from additional punishment. The U.S. Constitution, in Article II, Section 2, authorizes the president to pardon "offenses against the United States." In 2004, former Illinois Governor George Ryan concluded that the problems in the administration of the death penalty risked the execution of an innocent person and responded by pardoning four individuals on death row and commuting the sentences of over one hundred individuals to life in prison. In another example, in 2010 Florida Governor Charlie Crist pardoned the deceased lead singer of the Doors, Jim Morrison, who had been convicted in 1970 for lewd behavior during a Miami concert.

See a discussion of Morrison v. Dade County *on the study site, www.sagepub.com/lippmanccl3e.*

SENTENCING GUIDELINES

At the turn of the twentieth century, most states and the federal government employed indeterminate sentencing. The legislature established the outer limits of the penalty, and parole boards were provided with the authority to release an individual prior to the completion of his or her sentence in the event the offender demonstrated that he or she had been rehabilitated. This approach is based on the belief that an individual who is incarcerated will be inspired to demonstrate that he or she no longer poses a threat to society and deserves an early release. The disillusionment with the notion of rehabilitation and the uncertain length and extreme variation in the time served by offenders led to the introduction of determinate sentences.

In 1980, Minnesota adopted sentencing guidelines in an effort to provide for uniform proportionate and predictable sentences. Currently over a dozen states employ guidelines. In 1984, the U.S. Congress responded by passing the Sentencing Reform Act. The law went into effect in 1987 and established the U.S. Sentencing Commission, which drafted binding guidelines to be followed by federal judges in sentencing offenders. The Sentencing Commission is composed of seven members appointed by the president with the approval of the U.S. Senate. At least three of the members must be federal judges. The Sentencing Commission has the responsibility to monitor the impact of the guidelines on sentencing and to propose needed modifications.[11]

The Sentencing Reform Act abandoned rehabilitation as a purpose of imprisonment. The goals are retribution, deterrence, incapacitation, and the education and treatment of offenders. All sentences are determinate, and an offender's term of imprisonment is reduced only by any good-behavior credit earned while in custody.

Sentences under the federal guidelines are based on a complicated formula that reflects the seriousness and characteristics of the offense and the criminal history of the offender. The judge employs a sentencing grid and is required to provide a sentence within the narrow range where the offender's criminal offense and criminal history intersect on the grid.

Judges are required to document the reasons for criminal sentences and are obligated to provide a specific reason for an upward or downward departure. The prosecution may appeal a sentence below the presumed range and the defense any sentence above the presumed range. This process can be incredibly complicated and requires the judge to undertake as many as seven separate steps. The federal guidelines also specify that any **plea bargain** (a negotiated agreement between defense and prosecuting attorneys) must be approved by a judge to ensure that any sentence agreed upon is within the range established by the guidelines. The impact of the guidelines is difficult to measure, but studies suggest that the guidelines have increased the percentage of defendants who receive prison terms.

The federal guidelines are much more complicated than most state guidelines and provide judges with much less discretion in sentencing. Experts conclude that as a result of several recent Supreme Court cases, federal as well as state sentencing guidelines should now be considered merely advisory rather than binding on judges. These complicated and confusing legal decisions, outlined as follows, hold that it is unconstitutional to enhance a sentence based on facts found to exist by the judge by a **preponderance of the evidence** (a probability) rather than **beyond a reasonable doubt** by a jury. According to the Supreme Court, excluding the jury from the fact-finding process constitutes a violation of a defendant's Sixth Amendment right to trial by a jury of his or her peers. In *Apprendi v. New Jersey,* the U.S. Supreme Court explained that to "guard against . . . oppression and tyranny on the part of rulers, and as the great bulwark of [our] . . . liberties, trial by jury has been understood to require that 'the truth of every accusation . . . should . . . be confirmed by the unanimous suffrage of twelve of [the defendant's] equals and neighbors.'"[12]

In *Blakely v. Washington,* decided in 2004, Blakely pled guilty to kidnapping his wife. The judge followed Washington's sentencing guidelines and found that Blakely had acted with "deliberate cruelty" and imposed an "exceptional" sentence of ninety months rather than the standard sentence of fifty-three months. The U.S. Supreme Court ruled that a judge's sentence is required to be based on "the facts reflected in the jury verdict or admitted by the defendant" and that a judge may not enhance a sentence based on facts that were not determined by the jury to exist.[13]

The decisions in *Apprendi* and in *Blakely* were relied on by the Supreme Court in *Cunningham v. California* to hold unconstitutional California's determinate sentencing law (DSL). Cunningham was tried and convicted of the continuous sexual abuse of a child under the age of fourteen. Under the DSL, the offense is punishable by imprisonment for a lower term of six years, a middle term of twelve years, or an upper term sentence of sixteen years. The judge was obligated to sentence Cunningham to the twelve-year middle term unless the judge found one or more additional facts in aggravation. The trial judge found six aggravating circumstances by a preponderance of the evidence that outweighed the single mitigating factor, and Cunningham was sentenced to sixteen years. The Supreme Court held that "fact finding to elevate a sentence . . . falls within the province of the jury employing a beyond-a-reasonable-doubt standard . . . [B]ecause the DSL allocates to judges sole authority to find facts permitting the imposition of an upper term sentence, the system violates the Sixth Amendment."[14]

In 2005, in *United States v. Booker,* the U.S. Supreme Court held that the enhancement of sentences by a judge under the federal sentencing guidelines unconstitutionally deprives defendants of their right to have facts determined by a jury of their peers. Booker was convicted of possession with intent to distribute at least fifty grams of crack cocaine. His criminal history and the quantity of drugs in his possession required a sentence of between 210 and 262 months in prison. The judge, however, concluded by a preponderance of the evidence that Booker had possessed an additional 556 grams of cocaine and that he also was guilty of obstructing justice. These findings required the judge to select a sentence of between 360 months and life. The judge sentenced Booker to thirty years in prison. The Supreme Court ruled that the trial judge had acted unconstitutionally and explained that Booker had, in effect, been convicted of possessing a greater quantity of drugs than was charged in the indictment and that the determination of facts was a matter for the jury rather than for the judge. Justice Breyer concluded that the best course under the circumstances was for judges to view the guidelines as advisory rather than as requiring the selection of a particular sentence. Why? An advisory system enables judges to formulate a sentence without consulting with a jury. On the other hand, mandatory guidelines under the Supreme Court's decisions require the jury to find each fact on which a sentence is based beyond a reasonable doubt. A number of federal judges had publicly criticized the guidelines as unduly complicated and as limiting their discretion to impose more lenient sentences on deserving defendants and likely silently rejoiced over the Supreme Court's pronouncement that the guidelines should be considered as advisory rather than as binding.[15]

More recently, in *Rita v. United States, Gall v. United States,* and *Kimbrough v. United States,* the U.S. Supreme Court once again addressed the federal guidelines and explicitly held that the guidelines are advisory.[16] In these judgments, the Court held that an appellate court should examine whether a judge's sentencing decision, whether inside or outside the sentencing range in the guidelines, is reasonable. In other words, a trial court judge does not have to satisfy an extraordinarily high standard on appeal to justify a sentence that departs from the guidelines.

TRUTH IN SENTENCING

Whatever the fate of federal sentencing guidelines is, keep in mind that in 1984 the U.S. government moved from indeterminate to determinate sentencing. This was part of a general trend away from rehabilitation. A federal prisoner currently serves his or her complete sentence, reduced only by good-time credits earned while incarcerated. This replaces a system in which good-time credits and parole reduced a defendant's incarceration to roughly one-third of the sentence. Crime victims complained in frustration that the criminal justice system favored offenders over victims.

As part of this more open and honest approach to sentencing, the U.S. Congress championed **truth in sentencing laws**. What does this mean? The indeterminate sentencing model resulted in the release of prisoners prior to the completion of their sentences who succeeded in persuading parole boards that they had been rehabilitated. Truth in sentencing ensures that offenders serve a significant portion of the sentence. In the Violent Crime Control and Law Enforcement Act of 1994, Congress authorized the federal government to provide additional funds for prison construction and renovation to states that guarantee violent offenders serve eighty-five percent of their prison sentences. Roughly forty states have some form of truth in sentencing legislation and have qualified for funding. The result is that over seventy percent of violent offenders are serving longer sentences than they did prior to truth in sentencing.

VICTIMS' RIGHTS

Early tribal codes viewed criminal attacks as offenses against the victim's family or tribe. The family had the right to revenge or compensation. By the late Middle Ages, crime came to be viewed as an offense against the "King's Peace," which is the right of the monarch to insist on social order and stability within his realm. Government officials now assumed the responsibility to apprehend, prosecute, and punish offenders. The victim's interest was no longer of major consequence. In 1964, California passed legislation to assist victims, and today every state as well as the District of Columbia provides monetary payments to various categories of crime victims. The plans typically cover compensation for physical and emotional injuries and also provide restitution for medical care, lost wages, and living and burial expenses. Most states have statutes that authorize courts to require offenders to provide this restitution as part of their criminal sentence. Forty-three states have adopted so-called **Son of Sam** laws, named after a New York law directed at serial killer David Berkowitz. These laws prohibit convicted felons from profiting from books, films, or television programs that recount their crimes; instead, these laws make such funds available to victims.[17]

In 1986, the U.S. Congress passed the Victims of Crime Act (VOCA). This provides for a compensation fund and establishes the Office for Victims of Crime (OVC), which is responsible for coordinating all victim-related federal programs. President George W. Bush also signed the Crime Victims Rights Act of 2004, which proclaims various rights for crime victims, including the right to be informed of all relevant information involving the prosecution, imprisonment, and release of an offender as well as the right to compensation and return of property. California, along with nineteen other states, has adopted constitutional amendments protecting victims.

Another important development is the U.S. Supreme Court's approval of **victim impact statements** in death penalty cases. In *Payne v. Tennessee,* the defendant stabbed to death Charisse Christopher and her two-year-old daughter in front of Charisse's three-year-old son Nicholas. This was a particularly brutal crime; Charisse suffered eighty-four knife wounds and was left helplessly bleeding on the floor. The Supreme Court ruled that the trial court had acted properly in permitting Charisse's mother to testify during the sentencing phase of the trial that Nicholas continued to cry for his mother. The Court explained that the jury should be reminded that "just as the murderer should be considered as an individual, so too the victim is an individual whose death represents a unique loss to society and in particular to his family." The federal government and an estimated twenty states have laws that authorize direct victim involvement at sentencing for criminal offenses, and all fifty states and the District of Columbia provide for some form of written submissions.[18]

In the next portion of the chapter, we will see that criminal sentences must satisfy the constitutional requirements of the Cruel and Unusual Punishment Clause of the Eighth Amendment and meet the requirements of equal protection that we discussed in Chapter 2.

CRUEL AND UNUSUAL PUNISHMENT

The **Eighth Amendment** to the U.S. Constitution is the primary constitutional check on sentencing. The Eighth Amendment states that "[e]xcessive bail shall not be required, nor excessive fines imposed, nor cruel and unusual punishments inflicted." The prohibition on cruel and unusual punishment received widespread acceptance in the new American nation. In fact, the language in the U.S. Bill of Rights is taken directly from the Virginia Declaration of Rights of 1776, which in turn was inspired by the English Bill of Rights of 1689. The English document significantly limited the powers and prerogatives of the British monarchy and recognized certain basic rights of the English people.[19]

The U.S. Supreme Court has ruled that the prohibition against cruel and unusual punishment applies to the states as well as to the federal government, and virtually every state constitution contains similar language. Professor Wayne LaFave lists three approaches to interpreting the clause: (1) it limits the *methods* employed to inflict punishment, (2) it restricts the *amount of punishment* that may be imposed, and (3) it *prohibits* the criminal punishment of certain acts.[20]

Methods of Punishment

Patrick Henry expressed concern during Virginia's consideration of the proposed federal Constitution that the absence of a prohibition on cruel and unusual punishment would open the door to the use of torture to extract confessions. In fact, during the debate in the First Congress on the adoption of a Bill of Rights, one representative objected to the Eighth Amendment on the grounds that "villains often deserve whipping, and perhaps having their ears cut off."[21]

There is agreement that the Eighth Amendment prohibits punishment that was considered cruel at the time of the amendment's ratification, including burning at the stake, crucifixion, breaking on the wheel, drawing and quartering, the rack, and the thumbscrew.[22] The Supreme Court observed as early as 1890 that "if the punishment prescribed for an offense against the laws of the state were manifestly cruel and unusual as burning at the stake, crucifixion, breaking on the wheel, or the like, it would be the duty of the courts to adjudge such penalties to be within the constitutional prohibition."[23] In 1963, the Supreme Court of Delaware held that whipping was constitutionally permissible on the grounds that the practice was recognized in the state in 1776.[24]

The vast majority of courts have not limited cruel and unusual punishment to acts condemned at the time of passage of the Eighth Amendment and have viewed this as an evolving concept. The U.S. Supreme Court in *Trop v. Dulles* stressed that the Eighth Amendment "must draw its meaning from the evolving standards of decency that mark the progress of a maturing society."[25] *Trop* is an example of the application of the prohibition on cruel and unusual punishment to new situations. In *Trop,* the U.S. Supreme Court held that it was unconstitutional to deprive Trop and roughly 7,000 others convicted of military desertion of their American citizenship. Chief Justice Earl Warren wrote that depriving deserters of citizenship, although involving "no physical mistreatment," was more "primitive than torture" in that individuals are transformed into "stateless persons without the right to live, work or enjoy the freedoms accorded to citizens in the United States or in any other nation."

The death penalty historically has been viewed as a constitutionally acceptable form of punishment.[26] The Supreme Court noted that punishments are "cruel when they involve torture or a lingering death; but the punishment of death is not cruel within the meaning of that word as used in the constitution. [Cruelty] implies there is something inhuman and barbarous—something more than the mere extinguishment of life."[27]

The Supreme Court has rejected the contention that death by shooting[28] and electrocution are cruel and barbarous, noting in 1890 that the newly developed technique of electricity was a "more humane method of reaching the result."[29] In *Louisiana ex rel. Francis v. Resweber,* Francis was strapped in the electric chair and received a bolt of electricity before the machine malfunctioned. The U.S. Supreme Court rejected the claim that subjecting the petitioner to the electric chair a second time constituted cruel and unusual punishment. The Court observed that there was no intent to inflict unnecessary pain, and the fact that "an unforeseeable accident prevented the prompt consummation of the sentence cannot . . . add an element of cruelty to a subsequent execution."[30]

Judges have actively intervened to prevent barbarous methods of discipline in prison. In *Hope v. Pelzer*, in 2002, the U.S. Supreme Court ruled that Alabama's use of a "hitching post" to discipline inmates constituted "wanton and unnecessary pain." During Hope's seven-hour ordeal on the hitching post in the hot sun, he was painfully handcuffed at shoulder level to a horizontal bar without a shirt, taunted, and provided with water only once or twice and denied bathroom breaks. There was no effort to monitor the petitioner's condition despite the risks of dehydration and sun damage. The ordeal continued despite the fact that Hope expressed a willingness to return to work. The Supreme Court determined that the use of the hitching post was painful and punitive retribution that served no legitimate and necessary penal purpose.[31]

In 2011, in *Brown v. Plata*, Justice Anthony Kennedy affirmed a lower court judgment requiring California prisons to release roughly 46,000 inmates to relieve prison overcrowding. The California system housed twice as many prisoners as the institutions were designed to hold. Justice Anthony Kennedy concluded that the overcrowding of California prisons constituted unconstitutional cruel treatment because the prison system lacked the resources to provide adequate health and mental health care to the large prison population. Overcrowding also had led to rising tension and to violence. The Supreme Court held that "[a] prison that deprives prisoners of basic sustenance, including adequate medical care, is incompatible with the concept of human dignity and has no place in civilized society."[32]

In judging whether a method of criminal punishment or prison discipline is cruel and unusual, courts consider the following:

- *Prevailing Social Values*. The punishment must be acceptable to society.
- *Penological Purpose*. The punishment must be strictly necessary to the achievement of a valid correctional goal, such as deterrence, rehabilitation, or incapacitation.
- *Human Dignity*. Individuals subject to the punishment must be treated with human respect and dignity.

There is an argument that individuals convicted of crimes have forfeited claims to humane treatment and that courts have gone too far in coddling criminals and in handcuffing state and local criminal justice professionals. According to individuals who adhere to this position, judges are too far removed from the realities of crime to appreciate that harsh penalties are required to deter crime and to control inmates. The debate over appropriate forms of punishment will likely continue as society moves toward utilizing alternative forms of social control. A number of states already authorize the chemical castration of individuals convicted of sexual battery, and some statutes also provide individuals with the option of surgically removing their testes.[33]

You can find Brown v. Plata *on the study site, www.sagepub.com/lippmanccl3e.*

The next section explores whether capital punishment constitutes cruel and unusual punishment.

The Amount of Punishment: Capital Punishment

The prohibition on cruel and unusual punishment has also been interpreted to require that punishment is proportionate to the crime. In other words, the "punishment must not be excessive"; it must "fit the crime." Judges have been particularly concerned with the **proportionality** of the death penalty. This reflects an understandable concern that a penalty that is so "unusual in its pain, in its finality and in its enormity" is imposed in an "evenhanded, nonselective, and nonarbitrary" manner against individuals who have committed crimes deserving of death.[34]

In *Furman v. Georgia*, five Supreme Court judges wrote separate opinions condemning the cruel and unusual application of the death penalty against some defendants while others convicted of equally serious homicides were sentenced to life imprisonment. Justice Byron White reviewed the cases before the Supreme Court and concluded that there was "no meaningful basis for distinguishing the few cases in which it [the death penalty] is imposed from the many cases in which it is not." Justice Potter Stewart observed in a concurring opinion that "these death sentences are cruel and unusual in the same way that being struck by lightning is cruel and unusual. . . . The Eighth

For a deeper look at this topic, visit the study site.

and Fourteenth Amendments cannot tolerate the infliction of a sentence of death under legal systems that permit this unique penalty to be so wantonly and so freakishly imposed."[35]

Justice Douglas controversially concluded in *Furman* that the death penalty was being selectively applied against the poor, minorities, and uneducated at the same time privileged individuals convicted of comparable crimes were sentenced to life in prison. Justice Douglas argued that the United States' system of capital punishment operated in practice to exempt anyone making over $50,000 from execution, although "blacks, those who never went beyond the fifth grade in school, those who make less than $3,000 a year or those who were unpopular or unstable [were] the only people executed."

States reacted to this criticism by adopting mandatory death penalty laws that required that defendants convicted of intentional homicide receive the death penalty. The U.S. Supreme Court ruled in *Woodson v. North Carolina* that treating all homicides alike resulted in death being cruelly inflicted on undeserving defendants. The Court held that a jury "fitting the punishment to the crime" must consider the "character and record of the individual offender" as well as the "circumstances of the particular offense." The uniform system adopted in North Carolina treated "all persons convicted of a designated offense not as uniquely individual human beings, but as members of a faceless, undifferentiated mass to be subject to the blind infliction of the penalty of death."[36]

In *Gregg v. Georgia,* in 1976, the U.S. Supreme Court approved a Georgia statute designed to ensure the proportionate application of capital punishment. The Georgia law limited the discretion of jurors to impose the death penalty by requiring jurors to find that a murder had been accompanied by one of several aggravating circumstances. This evidence was to be presented at a separate sentencing hearing and was to be weighed against any and all mitigating considerations. Death sentences were to be automatically reviewed by the state supreme court, which was charged with ensuring that the verdict was supported by the facts and that capital punishment was imposed in a consistent fashion. This system was intended to ensure that the death penalty was reserved for the most severe homicides and was not "cruelly imposed on undeserving defendants."[37]

Were there offenses other than aggravated and intentional murder that merited the death penalty? What of aggravated rape? In *Coker v. Georgia,* in 1977, the U.S. Supreme Court ruled that death was a grossly disproportionate and excessive punishment for the aggravated rape of an adult and constituted cruel and unusual punishment.[38] Thirty-one years later, in *Kennedy v. Louisiana* (discussed in Crime in the News), the Supreme Court held that imposition of capital punishment for the rape of a child constituted cruel and unusual punishment.[39]

In 2008, the U.S. Supreme Court addressed the constitutionality of the execution of individuals through the use of lethal injection. In *Baze v. Rees,* the Court upheld the constitutionality of Kentucky's lethal injection protocol.[40] In 1977, Oklahoma passed the first lethal injection law. The law was motivated by the desire to find a less expensive and more humane method of execution. All of the thirty-three death penalty states along with the federal government presently provide for lethal injection. Seventeen states provide that lethal injection is the only method of execution. In sixteen states, lethal injection is the primary method of execution, but other methods are available (in nine, electrocution; in four, gas chamber; in two, hanging; and in one, firing squad). Between 1976 and 2006, 838 of the 1,016 executions in the United States were carried out by lethal injection. Three were carried out by the federal government and the remainder by the states. Thirty state correctional agencies employ the identical three-drug sequence of sodium thiopental, pancuronium bromide, and potassium chloride used by Oklahoma.

Opponents of lethal injection claim that the individuals who administer the protocol lack the training to safely administer the drugs. The anesthesia level from sodium thiopental at times fails to sufficiently insulate the inmate from pain, inmates may experience suffocation from the pancuronium bromide (which causes death by asphyxiation) and excruciating pain from the potassium chloride (which results in cardiac arrest), and the equipment on some occasions has malfunctioned. Pancuronium bromide also can prevent an inmate from communicating that he or she is suffering pain. There are stories of veins collapsing and needles popping out of an inmate's arm and blocked tubes preventing the administration of the anesthesia. In December 2006, Florida executed Nieves Dias for murder. Diaz remained alive in obvious pain for twenty minutes following the administration of the first dose, and after thirty-five minutes a second lethal dose was administered. The medical examiner determined that the chemicals accidentally had been injected into soft tissue rather than into the vein. Governor Jeb Bush temporarily suspended executions in the state and appointed a commission to evaluate the humanity and legality of lethal injections. Florida reintroduced lethal injection eighteen months later. In June 2008, an Ohio lower court judge held that the state's three-drug protocol ran the risk of causing unnecessary pain and was unconstitutional in light of the Ohio statute that "death by lethal injection must be caused quickly and painlessly."

The Juvenile Death Penalty

The next case in the book involves the issue of whether the capital punishment of juvenile offenders constitutes cruel and unusual punishment.

In 1966, in *Kent v. United States,* the U.S. Supreme Court limited the broad authority exercised by state and local judges in waiving juveniles over for criminal prosecution as adults.[41] The U.S. Supreme Court was next asked and refused on several occasions during the 1980s to rule on the constitutionality of the juvenile death penalty. In *Eddings v. Oklahoma,* in 1982, the Supreme Court declined to rule on the constitutionality of the death penalty against juveniles, but held that a defendant's youth and psychological and social background must be considered in mitigation of punishment.[42]

You can find Baze v. Reese on the study site, www .sagepub.com/ lippmanccl3e.

In *Thompson v. Oklahoma,* in 1988, the Supreme Court ruled that the execution of a young person who was under the age of sixteen at the time of his or her offense constituted cruel and unusual punishment. Justice John Paul Stevens wrote that "inexperience, less education, and less intelligence make the teenager less able to evaluate the consequences of his or her conduct while at the same time he or she is much more apt to be motivated by mere emotion or pressure than is an adult."[43]

In *Stanford v. Kentucky,* in 1989, the U.S. Supreme Court finally addressed the issue of the application of the death penalty against individuals under the age of eighteen and ruled that there was no national consensus against the execution of individuals sixteen or seventeen years of age and that the imposition of capital punishment could not be considered either cruel or unusual. Justice Scalia relied on the objective fact that of the thirty-seven states that provided for capital punishment, only fifteen declined to impose it on sixteen-year-olds and twelve did not extend the death penalty to seventeen-year-old defendants.[44]

The petitioners in *Stanford* pointed to the fact that of the 2,106 sentences of death handed out between 1982 and 1988, only fifteen were imposed against individuals who were under sixteen at the time of their crimes and only thirty against individuals who were seventeen at the time of the crime. Actual executions for crimes committed by individuals under age eighteen constituted only about two percent of the total number of executions between 1642 and 1986. Justice Scalia explained that the statistics merely indicated that prosecutors and juries shared the view that there was a select but dangerous group of juveniles deserving of death.

In *Roper v. Simmons,* the U.S. Supreme Court once again considered whether the execution of individuals who are sixteen or seventeen years of age constitutes cruel and unusual punishment.

Did sentencing seventeen-year-old Christopher Simmons to death for murder constitute cruel and unusual punishment?

Roper v. Simmons, 543 U.S. 551 (2005). Opinion by: Kennedy, J.

This case requires us to address . . . whether it is permissible under the Eighth and Fourteenth Amendments to the Constitution of the United States to execute a juvenile offender who was older than fifteen but younger than eighteen when he committed a capital crime.

Facts

At the age of seventeen, when he was still a junior in high school, Christopher Simmons, the respondent here, committed murder. About nine months later, after he had turned eighteen, he was tried and sentenced to death. There is little doubt that Simmons

was the instigator of the crime. Before its commission Simmons said he wanted to murder someone. In chilling, callous terms he talked about his plan, discussing it for the most part with two friends, Charles Benjamin and John Tessmer, then aged fifteen and sixteen respectively. Simmons proposed to commit burglary and murder by breaking and entering, tying up a victim, and throwing the victim off a bridge. Simmons assured his friends they could "get away with it" because they were minors.

The three met at about 2 a.m. on the night of the murder, but Tessmer left before the other two set out. (The state later charged Tessmer with conspiracy, but dropped the charge in exchange for his testimony against Simmons.) Simmons and Benjamin entered the home of the victim, Shirley Crook, after reaching through an open window and unlocking the back door. Simmons turned on a hallway light. Awakened, Mrs. Crook called out, "Who's there?" In response Simmons entered Mrs. Crook's bedroom, where he recognized her from a previous car accident involving them both. Simmons later admitted this confirmed his resolve to murder her.

Using duct tape to cover her eyes and mouth and bind her hands, the two perpetrators put Mrs. Crook in her minivan and drove to a state park. They reinforced the bindings, covered her head with a towel, and walked her to a railroad trestle spanning the Meramec River. There they tied her hands and feet together with electrical wire, wrapped her whole face in duct tape and threw her from the bridge, drowning her in the waters below.

By the afternoon of September ninth, Steven Crook had returned home from an overnight trip, found his bedroom in disarray, and reported his wife missing. On the same afternoon fishermen recovered the victim's body from the river. Simmons, meanwhile, was bragging about the killing, telling friends he had killed a woman "because the bitch seen my face." The next day, after receiving information of Simmons' involvement, police arrested him at his high school and took him to the police station in Fenton, Missouri. They read him his *Miranda* rights. Simmons waived his right to an attorney and agreed to answer questions. After less than two hours of interrogation, Simmons confessed to the murder and agreed to perform a videotaped reenactment at the crime scene.

The state charged Simmons with burglary, kidnapping, stealing, and murder in the first degree. As Simmons was seventeen at the time of the crime, he was outside the criminal jurisdiction of Missouri's juvenile court system. He was tried as an adult. At trial the state introduced Simmons' confession and the videotaped reenactment of the crime, along with testimony that Simmons discussed the crime in advance and bragged about it later. The defense called no witnesses in the guilt phase. The jury having returned a verdict of murder, the trial proceeded to the penalty phase.

The state sought the death penalty. As aggravating factors, the state submitted that the murder was committed for the purpose of receiving money; was committed for the purpose of avoiding, interfering with, or preventing lawful arrest of the defendant; and involved depravity of mind and was outrageously and wantonly vile, horrible, and inhuman. The state called Shirley Crook's husband, daughter, and two sisters, who presented moving evidence of the devastation her death had brought to their lives.

In mitigation Simmons' attorneys first called an officer of the Missouri juvenile justice system, who testified that Simmons had no prior convictions and that no previous charges had been filed against him. Simmons' mother, father, two younger half brothers, a neighbor,

and a friend took the stand to tell the jurors of the close relationships they had formed with Simmons and to plead for mercy on his behalf. Simmons' mother, in particular, testified to the responsibility Simmons demonstrated in taking care of his two younger half brothers and of his grandmother and to his capacity to show love for them.

During closing arguments, both the prosecutor and defense counsel addressed Simmons' age, which the trial judge had instructed the jurors they could consider as a mitigating factor. Defense counsel reminded the jurors that juveniles of Simmons' age cannot drink, serve on juries, or even see certain movies, because "the legislatures have wisely decided that individuals of a certain age aren't responsible enough." Defense counsel argued that Simmons' age should make "a huge difference to [the jurors] in deciding just exactly what sort of punishment to make." In rebuttal, the prosecutor gave the following response: "Age, he says. Think about age. Seventeen years old. Isn't that scary? Doesn't that scare you? Mitigating? Quite the contrary I submit. Quite the contrary."

The jury recommended the death penalty after finding the state had proved each of the three aggravating factors submitted to it. Accepting the jury's recommendation, the trial judge imposed the death penalty. . . . After these proceedings in Simmons' case had run their course, the Supreme Court held that the Eighth and Fourteenth Amendments prohibit the execution of a mentally retarded person (*Atkins* v. *Virginia*, 536 U.S. 304 (2002)). Simmons filed a new petition for state postconviction relief, arguing that the reasoning of *Atkins* established that the Constitution prohibits the execution of a juvenile who was under 18 when the crime was committed. The Missouri Supreme Court agreed that "a national consensus has developed against the execution of juvenile offenders." . . . On this reasoning it set aside Simmons's death sentence and resentenced him to "life imprisonment without eligibility for probation, parole, or release except by act of the Governor."

Issue

The Eighth Amendment provides, "Excessive bail shall not be required, nor excessive fines imposed, nor cruel and unusual punishments inflicted." The provision is applicable to the States through the Fourteenth Amendment. As the court has explained, the Eighth Amendment guarantees individuals the right not to be subjected to excessive sanctions. The right flows from the basic "precept of justice that punishment for crime should be graduated and proportioned to [the] offense." By protecting even those convicted of heinous crimes, the Eighth Amendment reaffirms the duty of the government to respect the dignity of all persons.

The prohibition against "cruel and unusual punishments," like other expansive language in the Constitution, must be interpreted according to its text, by considering history, tradition, and precedent, and with due regard for its purpose and function in the constitutional design. To implement this framework we have established the propriety and affirmed the necessity of referring to "the evolving standards of decency that mark the progress of a maturing society" to determine which punishments are so disproportionate as to be cruel and unusual. . . . We now reconsider the issue . . . whether the death penalty is a disproportionate punishment for juveniles.

Reasoning

The evidence of national consensus against the death penalty for juveniles is similar, and in some respects parallel, to the evidence *Atkins* held sufficient to demonstrate a national consensus against the death penalty for the mentally retarded. When *Atkins* was decided, thirty states prohibited the death penalty for the mentally retarded. This number comprised twelve that had abandoned the death penalty altogether, and eighteen that maintained it but excluded the mentally retarded from its reach. By a similar calculation in this case, thirty states prohibit the juvenile death penalty, comprising twelve that have rejected the death penalty altogether and eighteen that maintain it but, by express provision or judicial interpretation, exclude juveniles from its reach.

Atkins emphasized that even in the twenty states without formal prohibition, the practice of executing the mentally retarded was infrequent. In the present case, too, even in the twenty states without a formal prohibition on executing juveniles, the practice is infrequent. Since *Stanford,* six states have executed prisoners for crimes committed as juveniles. In the past ten years, only three have done so: Oklahoma, Texas, and Virginia. In December 2003, the Governor of Kentucky decided to spare the life of Kevin Stanford, and commuted his sentence to one of life imprisonment without parole, with the declaration that "we ought not to be executing people who, legally, were children." . . . By this act the Governor ensured Kentucky would not add itself to the list of States that have executed juveniles within the last ten years even by the execution of the very defendant whose death sentence the Court had upheld in *Stanford* v. *Kentucky.*

There is, to be sure, at least one difference between the evidence of consensus in *Atkins* and in this case. Impressive in *Atkins* was the rate of abolition of the death penalty for the mentally retarded. Sixteen states that permitted the execution of the mentally retarded at the time of *Penry v. Lynaugh,* 492 U.S. 302 (1989) [finding no national consensus against execution of mentally challenged individuals] had prohibited the practice by the time we heard *Atkins.* By contrast, the rate of change in reducing the incidence of the juvenile death penalty, or in taking specific steps to abolish it, has been slower. Five states that allowed the juvenile death penalty at the time of *Stanford* have abandoned

it in the intervening fifteen years—four through legislative enactments and one through judicial decision.

Though less dramatic than the change from *Penry* to *Atkins* . . . we still consider the change from *Stanford* to this case to be significant. As noted in *Atkins,* with respect to the States that had abandoned the death penalty for the mentally retarded . . . "it is not so much the number of these States that is significant, but the consistency of the direction of change." In particular we found it significant that, in the wake of *Penry,* no state that had already prohibited the execution of the mentally retarded had passed legislation to reinstate the penalty. The number of States that have abandoned capital punishment for juvenile offenders since *Stanford* is smaller than the number of States that abandoned capital punishment for the mentally retarded after *Penry;* yet we think the same consistency of direction of change has been demonstrated. Since *Stanford,* no state that previously prohibited capital punishment for juveniles has reinstated it. This fact, coupled with the trend toward abolition of the juvenile death penalty, carries special force in light of the general popularity of anticrime legislation, and in light of the particular trend in recent years toward cracking down on juvenile crime in other respects. Any difference between this case and *Atkins* with respect to the pace of abolition is thus counterbalanced by the consistent direction of the change.

The slower pace of abolition of the juvenile death penalty over the past fifteen years, moreover, may have a simple explanation. When we heard *Penry,* only two death penalty states had already prohibited the execution of the mentally retarded. When we heard *Stanford,* by contrast, twelve death penalty states had already prohibited the execution of any juvenile under eighteen, and fifteen had prohibited the execution of any juvenile under seventeen. If anything, this shows that the impropriety of executing juveniles between sixteen and eighteen years of age gained wide recognition earlier than the impropriety of executing the mentally retarded. In the words of the Missouri Supreme Court, "It would be the ultimate in irony if the very fact that the inappropriateness of the death penalty for juveniles was broadly recognized sooner than it was recognized for the mentally retarded were to become a reason to continue the execution of juveniles now that the execution of the mentally retarded has been barred." Congress considered the issue when enacting the Federal Death Penalty Act in 1994, and determined that the death penalty should not extend to juveniles.

As in *Atkins,* the objective indicia of consensus in this case—the rejection of the juvenile death penalty in the majority of States; the infrequency of its use even where it remains on the books; and the consistency in the trend toward abolition of the practice—provide sufficient evidence that today our society views juveniles, in the words *Atkins* used respecting the mentally retarded, as "categorically less culpable than the average criminal."

A majority of States have rejected the imposition of the death penalty on juvenile offenders under eighteen, and we now hold this is required by the Eighth Amendment.

. . . Capital punishment must be limited to those offenders who commit "a narrow category of the most serious crimes" and whose extreme culpability makes them "the most deserving of execution." This principle is implemented throughout the capital sentencing process. States must give narrow and precise definition to the aggravating factors that can result in a capital sentence. In any capital case a defendant has wide latitude to raise as a mitigating factor "any aspect of [his or her] character or record and any of the circumstances of the offense that the defendant proffers as a basis for a sentence less than death." There are a number of crimes that beyond question are severe in absolute terms, yet the death penalty may not be imposed for their commission. These rules vindicate the underlying principle that the death penalty is reserved for a narrow category of crimes and offenders.

The general differences between juveniles under eighteen and adults demonstrate that juvenile offenders cannot with reliability be classified among the worst offenders. First, as any parent knows and as the scientific and sociological studies respondents cite tend to confirm, "A lack of maturity and an underdeveloped sense of responsibility are found in youth more often than in adults and are more understandable among the young. . . ." In recognition of the comparative immaturity and irresponsibility of juveniles, almost every state prohibits those under eighteen years of age from voting, serving on juries, or marrying without parental consent.

The second area of difference is that juveniles are more vulnerable or susceptible to negative influences and outside pressures, including peer pressure. ("Youth is more than a chronological fact. It is a time and condition of life when a person may be most susceptible to influence and to psychological damage.") This is explained in part by the prevailing circumstance that juveniles have less control, or less experience with control, over their own environment. The third broad difference is that the character of a juvenile is not as well formed as that of an adult. The personality traits of juveniles are more transitory, less fixed.

These differences render suspect any conclusion that a juvenile falls among the worst offenders. The susceptibility of juveniles to immature and irresponsible behavior means "their irresponsible conduct is not as morally reprehensible as that of an adult." Their own vulnerability and comparative lack of control over their immediate surroundings mean juveniles have a greater claim than adults to be forgiven for failing to escape negative influences in their whole environment. The reality that juveniles still struggle to define their identity means it is less supportable to conclude that even a heinous crime committed by a juvenile is evidence of irretrievably depraved character.

From a moral standpoint it would be misguided to equate the failings of a minor with those of an adult, for a greater possibility exists that a minor's character deficiencies will be reformed. Indeed, "the relevance of youth as a mitigating factor derives from the fact that the signature qualities of youth are transient; as individuals mature, the impetuousness and recklessness that may dominate in younger years can subside."

. . . Once the diminished culpability of juveniles is recognized, it is evident that the penological justifications for the death penalty apply to them with lesser force than to adults. We have held there are two distinct social purposes served by the death penalty: "retribution and deterrence of capital crimes by prospective offenders." As for retribution, . . . whether viewed as an attempt to express the community's moral outrage or as an attempt to right the balance for the wrong to the victim, the case for retribution is not as strong with a minor as with an adult. Retribution is not proportional if the law's most severe penalty is imposed on one whose culpability or blameworthiness is diminished, to a substantial degree, by reason of youth and immaturity.

As for deterrence, it is unclear whether the death penalty has a significant or even measurable deterrent effect on juveniles. . . . Here . . . the absence of evidence of deterrent effect is of special concern because the same characteristics that render juveniles less culpable than adults suggest as well that juveniles will be less susceptible to deterrence. . . . To the extent the juvenile death penalty might have residual deterrent effect, it is worth noting that the punishment of life imprisonment without the possibility of parole is itself a severe sanction, in particular for a young person.

Certainly it can be argued, although we by no means concede the point, that a rare case might arise in which a juvenile offender has sufficient psychological maturity, and at the same time demonstrates sufficient depravity, to merit a sentence of death. . . . The differences between juvenile and adult offenders are too marked and well understood to risk allowing a youthful person to receive the death penalty despite insufficient culpability. An unacceptable likelihood exists that the brutality or coldblooded nature of any particular crime would overpower mitigating arguments based on youth as a matter of course, even where the juvenile offender's objective immaturity, vulnerability, and lack of true depravity should require a sentence less severe than death. . . . Drawing the line at eighteen years of age is subject, of course, to the objections always raised against categorical rules. The qualities that distinguish juveniles from adults do not disappear when an individual turns eighteen. By the same token, some under eighteen have already attained a level of maturity some adults will never reach.

Our determination that the death penalty is disproportionate punishment for offenders under eighteen finds confirmation in the stark reality that the United

States is the only country in the world that continues to give official sanction to the juvenile death penalty. This reality does not become controlling, for the task of interpreting the Eighth Amendment remains our responsibility. Yet . . . the laws of other countries and . . . international authorities are instructive in interpreting the Eighth Amendment's prohibition of "cruel and unusual punishments." Respondent does not contest, that only seven countries other than the United States have executed juvenile offenders since 1990: Iran, Pakistan, Saudi Arabia, Yemen, Nigeria, the Democratic Republic of Congo, and China. Since then each of these countries has either abolished capital punishment for juveniles or made public disavowal of the practice. In sum, it is fair to say that the United States now stands alone in a world that has turned its face against the juvenile death penalty. . . .

Holding

The Eighth and Fourteenth Amendments forbid imposition of the death penalty on offenders who were under the age of eighteen when their crimes were committed. The judgment of the Missouri Supreme Court setting aside the sentence of death imposed upon Christopher Simmons is affirmed.

Dissenting, *O'Connor, J.*

The Court's decision today establishes a categorical rule forbidding the execution of any offender for any crime committed before his eighteenth birthday, no matter how deliberate, wanton, or cruel the offense. . . . The rule decreed by the Court rests, ultimately, on its independent moral judgment that death is a disproportionately severe punishment for any seventeen-year-old offender. I do not subscribe to this judgment. Adolescents as a class are undoubtedly less mature, and therefore less culpable for their misconduct, than adults. But the Court has adduced no evidence impeaching the seemingly reasonable conclusion reached by many state legislatures: that at least some seventeen-year-old murderers are sufficiently mature to deserve the death penalty in an appropriate case. Nor has it been shown that capital sentencing juries are incapable of accurately assessing a youthful defendant's maturity or of giving due weight to the mitigating characteristics associated with youth.

Questions for Discussion

1. Summarize the data Justice Kennedy reviews in concluding that capital punishment for juveniles is disproportionate punishment.

2. What are the similarities and differences in statistics relating to the execution of the mentally retarded compared to the data concerning juveniles? Is there a clear consensus against capital punishment for individuals under eighteen?

3. Why does Justice Kennedy conclude that juveniles are not among the "worst offenders who merit capital punishment"? What does Justice Kennedy write about the interests in retribution and deterrence in regard to juveniles?

4. Explain why Justice Kennedy refers to other countries. Is this relevant to a decision of the U.S. Supreme Court?

5. The jurors at trial concluded that Simmons deserved the death penalty. Would it be a better approach to permit each state to remain free to determine whether to impose the death penalty for juveniles under eighteen? Is life imprisonment without parole a proportionate penalty for a juvenile convicted of the intentional killing of another person?

6. How would you rule in *Simmons*?

Cases and Comments

Juveniles and Life Without Parole. In 2010, in *Graham v. Florida*, the U.S. Supreme Court held that sentencing a juvenile to life imprisonment without parole for a nonhomicide offense violated the Eighth Amendment prohibition on cruel and unusual punishment. Terrance Jamar Graham's parents were addicted to crack cocaine, and in elementary school he was diagnosed with attention deficit hyperactivity disorder. He began drinking alcohol and using tobacco at age 9 and smoked marijuana at age 13. At age 16, Graham was arrested and charged as an adult with armed burglary and with attempted armed robbery. He pled guilty and received a 3-year term of probation, the first six months of which he served in the county jail. Roughly six months following Graham's release, he was arrested along with two accomplices for home invasion robbery following a high speed chase. Three firearms were found in his automobile.

The trial court judge found that Graham had violated his probation by committing a home invasion robbery, possessing a firearm, and by associating with individuals involved in criminal activity. The court sentenced Graham to life imprisonment without parole for the earlier armed burglary and 15 years for the armed robbery. The judge explained that "[g]iven your escalating pattern of criminal conduct, it is apparent to the court that you have decided that is the way you are going to live your life and that the only thing I can do now is to try and protect the community from your actions."

The U.S. Supreme Court considered Graham's claim that his sentence of life without parole constituted cruel and unusual punishment. The Court determined that there was a national consensus against life imprisonment for juveniles convicted of nonhomicide offenses.

There were 129 juvenile nonhomicide offenders serving life without parole sentences. Seventy-seven of these offenders were incarcerated in Texas and the other fifty-two inmates were imprisoned in 10 states and in the federal system. Twenty-six states and D. C. had not imposed life imprisonment without parole despite statutory authorization. The Court concluded that considering the large number of juveniles who may be eligible for life imprisonment based on having committed aggravated assault, forcible rape, robbery, burglary, and arson, the sentence of life imprisonment without parole for nonhomicide offenses is infrequently imposed.

The Court stressed that "community consensus," although entitled to great weight, is not determinative whether life imprisonment for juveniles constitutes cruel and unusual. The important step is to evaluate the degree of responsibility of juvenile offenders for their crimes. An additional consideration is whether the sentence serves legitimate penological goals.

Offenders. Roper v. Simmons established that juveniles lack maturity, have an underdeveloped sense of responsibility, and are susceptible to outside pressures. As a consequence, juvenile offenders cannot be considered to be the "worst of the worst" and cannot be considered as morally responsible as an adult.

Nature of crime. The taking of the life of another person results in the loss of human life and is more serious than a nonhomicide felony.

Nature of punishment. Life without parole is the second most severe punishment authorized under law. The offender is deprived of basic liberties and is incarcerated for the remainder of his or her life. A juvenile offender will serve more years in prison than an adult offender.

Penological justification. The interest in retribution does not justify the imposition of life imprisonment on juveniles because they are not as responsible for their actions as are adults. Juveniles are impulsive and emotional and are unlikely to be deterred by the threat of punishment. The protection of society in most instances does not require the incapacitation of juvenile offenders for the remainder of their lives, and there is no justification for dismissing the possibility of rehabilitation.

The Supreme Court held that a state is required to provide defendants like Graham "some meaningful opportunity to obtain release based on demonstrated maturity and rehabilitation." The Eighth Amendment, however, does not "foreclose the possibility that juveniles convicted of nonhomicide crimes "will remain behind bars for life." *See Graham v. Florida* ___ U.S. ___, 130 S.Ct. 2011, 176 L.Ed.2d 825 (2010).

Do you agree with the Supreme Court's ruling? In light of *Graham,* will life imprisonment for juveniles for homicide offenses be upheld as constitutional?

Life imprisonment for juveniles for homicide offenses. In June 2012, in *Miller v. Alabama,* the Supreme Court in a 5-to-4 decision held that the Eighth Amendment prohibits mandatory sentencing schemes that require life in prison without the possibility for parole for juvenile offenders under the age of 18 convicted of homicide. Justice Elena Kagan noted that "youth matters" and mandatory sentencing schemes by making "age irrelevant" pose "too great a risk of disproportionate punishment."

States are not precluded from sentencing juveniles to life imprisonment without parole in homicide cases, although given juveniles' "diminished culpability" and "capacity for change," this "harshest possible penalty [should be] uncommon." A "sentencer" before imposing life imprisonment on a juvenile is required to consider mitigating factors, and the sentence is to be based on "individualized consideration[s]," such as the juvenile's age, the juvenile's background, the juvenile's development, the nature of the juvenile's involvement in the crime, the juvenile's capacity to assist his or her attorney, and the potential for rehabilitation. Would you sentence the two 14-year-old offenders in *Miller* to life imprisonment? *See Miller v. Alabama,* ___ U.S. ___ (2012).

You can find Graham v. Florida *and* Miller v. Alabama *on the study site, www.sagepub.com/lippmanccl3e.*

CRIME IN THE NEWS

In 2008 in *Kennedy v. Louisiana,* the U.S. Supreme Court ruled by a vote of five to four that the imposition of the death penalty on a defendant convicted of the rape of a child constituted cruel and unusual punishment. The case almost immediately became embroiled in controversy and is one of the Court's most controversial decisions in recent memory.

On March 2, 1998, Patrick Kennedy called 911 and reported that his stepdaughter, L.H., had been dragged from the garage and raped by two neighborhood boys who fled on their bikes. The police found L.H. in bed wrapped in a bloody blanket bleeding profusely. At the hospital, an expert in pediatric forensic medicine reported that L.H.'s injuries were the most severe that he had witnessed from a sexual assault, and L.H. was rushed into surgery. A laceration to the left wall of the vagina had separated her cervix from the back of her vagina, causing her rectum to protrude into the vaginal structure. L.H.'s entire perineum was torn from the posterior fourchette to the anus. Both L.H. and Kennedy told investigators that L.H. had been raped by two neighborhood boys, and L.H. repeated this account during a lengthy examination by a psychologist.

Eight days following the rape, Kennedy was arrested and charged with the aggravated rape of a child under twelve. A number of factors led the police to question whether L.H. had been raped by two neighborhood boys. For example, Kennedy had called his employer three hours prior to the time that he allegedly discovered that L.H. had been raped and reported that he would not be at work. He then called a fellow employee to ask how to get blood out of a carpet because his daughter had "just become a lady." An hour later, Kennedy phoned a carpet cleaning service and asked for emergency assistance in removing bloodstains from the carpet, and thirty minutes later he called 911 to report the rape.

In December 1999, twenty-one months following the assault, L.H. revealed that Kennedy had raped her. The jury unanimously convicted Kennedy. At the sentencing stage, a cousin of Kennedy's former wife reported that Kennedy had abused her when she was eight, and the jury sentenced Kennedy to death. The verdict was affirmed by the Louisiana Supreme Court, which noted that other than first-degree murder, there is no other nonhomicide crime more deserving of the death penalty.

Kennedy, an African American, was the first person to receive the death penalty for rape under Louisiana's 1995 death penalty law. He was an eighth-grade dropout with an IQ of 70 whose record was marked by a prior conviction for attempting to cash five worthless checks. In 2007, Richard Davis joined Kennedy on death row when he was sentenced to death for the aggravated rape of a five-year-old. Louisiana stands alone among the states in having sentenced a defendant to death for the rape of a child.

U.S. Supreme Court Justice Anthony Kennedy authored the majority opinion of the Supreme Court in *Kennedy* and asserted that the death penalty for the rape of child is a disproportionate punishment as measured by the evolving standards of contemporary society. Justice Kennedy rested this conclusion on the "objective indicator" of state legislation and practice.

In 1925, eighteen states, the District of Columbia, and the federal government authorized the death penalty for the rape of a child or of an adult. Between 1930 and 1964, 455 people were executed for this offense. The last person executed for the rape of a child was Ronald Wolfe in 1964 in Missouri.

The landscape of the death penalty changed in 1972 with *Furman v. Georgia*, in which the Supreme Court invalidated most state death penalty statutes. Louisiana reintroduced the death penalty for the rape of a child in 1995. Louisiana law at the time of Kennedy's prosecution provided that anal or vaginal intercourse with a child

under twelve constitutes aggravated rape and is punishable by death. Five states followed Louisiana's example: Georgia, Montana, Oklahoma, South Carolina, and Texas. Four of these states' statutes are narrower than Louisiana's and limit the death penalty to offenders with a previous rape conviction. Georgia requires a finding of aggravating circumstances such as a prior conviction for a designated offense.

Justice Kennedy pointed out that forty-four states did not authorize the death penalty for child rape and that in 1994 Congress expanded the number of federal crimes punishable by death and yet failed to include child rape or abuse. The forty-four states that did not provide for capital punishment for the rape of a child is greater than the number of jurisdictions that prohibited capital punishment for the mentally challenged (thirty) and juveniles (thirty) at the time that the Supreme Court held that the execution of these types of individuals constituted cruel and unusual punishment. It is roughly the same as the number of states that in 2002 prohibited the execution of an individual who participated in a robbery and who was not responsible for the killing (forty-two) at the time that the Supreme Court held the execution of these individuals to constitute cruel and unusual punishment.

Justice Kennedy also noted that no defendants had been executed for childhood rape in recent years. The death penalty also had been infrequently imposed on juveniles (five executions) and mentally challenged defendants (five executions) despite the fact that twenty states had provided for the imposition of this penalty on these individuals. Six individuals between 1954 and 1983 had been sentenced to death for participation in a robbery.

Justice Kennedy dismissed the argument that in the last thirteen years there had been a movement toward making child rape a capital offense. It was true that six states had adopted statutes providing for the death penalty for the rape of a child since 1995 and that three of these laws had been passed in the last two years. The trend of legislation, however, had been far stronger at the time of the Supreme Court decisions declaring that it was cruel and unusual punishment to execute mentally challenged individuals and juveniles. The six states that declared child rape a capital offense was comparable to the number of states (eight) that imposed the death penalty for involvement in robbery.

Louisiana contended that any analysis of the number of states that provided for the death penalty for the rape of the child should consider the confusion that resulted from the Supreme Court's 1976 decision in *Coker v. Georgia*, in which the U.S. Supreme Court held that the death penalty for rape was disproportionate and excessive under the Eighth Amendment. Justice Kennedy responded that the Court had stressed in *Coker* that the decision was limited to an act of rape against an "adult woman" and found no indication that state legislatures and state courts had misinterpreted

Coker to stand for the proposition that the death penalty for child rape is unconstitutional. The fact that only five states had adopted statutes providing for capital punishment for child rape, in Justice Kennedy's view, could not be explained by a misunderstanding of *Coker*.

Justice Kennedy recognized that rape results in a permanent psychological, emotional, and often a physical impact on a child. This, however, does not mean that the imposition of death is proportionate to the offense. The Supreme Court has limited the use of capital punishment to crimes in which there has been an intentional taking of a victim's life. Other offenses may be "devastating in their harm," but they cannot be compared to first degree murder in their "severity and irrevocability." Justice Kennedy argued that the imposition of capital punishment for the rape of a child would result in a significant extension of the use of the death penalty that is contrary to "evolving standards of decency."

Number of executions. In 2005, there were 5,702 incidents of vaginal, anal, or oral rape of a child under twelve. This is almost twice the number of intentional murders committed during the same period. Only roughly 2.2 percent of these murderers are sentenced to death. The authorization of the death penalty for the rape of a child would significantly expand the application of the death penalty.

Standards. The Supreme Court had developed a set of aggravating and mitigating factors that a jury may weigh and balance in deciding whether a defendant deserves the death penalty. This imprecise process is acceptable where the victim dies but should not be expanded to other types of offenses.

Justice Kennedy questioned whether a death penalty prosecution will comfort the child victim who is required to testify at trial and to relive his or her brutalization. The unreliability of childhood testimony also may result in false convictions and call into question whether capital punishment is advancing the goals of retribution and incapacitation of dangerous child molesters.

The extension of the death penalty for the rape of a child may undermine the goal of bringing offenders to justice. Children in most cases know their abusers, and their families tend to circle the wagons and in many instances do not report the abuse to the police. Relatives may prove even more reluctant to report molestation when the penalty is death. Perpetrators confronting death for the rape of a child also will have an incentive to murder their victims and to eliminate the person who is often the sole witness to the crime.

Justice Samuel Alito wrote the dissenting opinion, in which he was joined by Chief Justice Roberts and by Justices Scalia and Thomas. Justice Alito criticized the broad and sweeping nature of the majority opinion, which prohibited the death sentence no matter the age of the child, the frequency and viciousness of the molestation, the number of children raped, or the length of the perpetrator's criminal record. Justice Alito asked,

"Is it really true that every person who is convicted of capital murder and sentenced to death is more morally depraved than every child rapist?" He answered his own question by asserting that "in the eyes of ordinarily Americans, the very worst child rapists . . . are the epitome of moral depravity."

Justice Alito pointed out that despite the fact that many state legislators and judges mistakenly viewed *Coker* as holding that the death penalty may not be imposed for the rape of a child, six states had adopted laws providing for capital punishment for the death of a child. A number of state legislatures currently were considering imposing the death penalty for the rape of a child, and the majority decision in *Kennedy* put the brakes on what may have proven to be a movement toward the expansion of capital punishment. The Court majority failed to provide a convincing reason why the judgment of these democratically elected legislatures should be short-circuited.

Shortly after the Supreme Court decision in *Kennedy,* it was discovered that in 2006, the lawyers in their briefs submitted to the Court had failed to inform the Supreme Court that the U.S. Congress in the National Defense Authorization Act had provided that the rape of a child when committed by a member of the military was punishable by death. Critics contended that this called into question the notion that there was no national consensus for imposing the death penalty for the rape of the child. The Court responded by taking the unusual step of announcing that it would evaluate whether to reconsider the judgment in *Kennedy.* In October 2008, the majority decided that the authorization of the death penalty in the "military sphere does not indicate that the penalty is constitutional in the civilian context" and does not "affect our reasoning or conclusions." Justice Antonin Scalia thundered in response that "the indifferent response of the majority of the Court reveals that they are imposing their own political preference rather than following a national consensus" (*Kennedy v. Louisiana,* 554 U.S. 407 (2008)).

Louisiana Governor Bobby Jindal nonetheless proclaimed that he was "outraged" by the Supreme Court decision in *Kennedy* and condemned the judgment as "incredibly absurd" and as a "clear abuse of judicial authority." Governor Jindal characterized the rape of a child as a "repugnant crime" that deserves the death penalty and observed that the majority of the Supreme Court clearly did not share the "same standards of decency as the people of Louisiana." He vowed that Louisiana officials would find ways to maintain the death penalty for the rape of a child. Alabama Attorney General Troy King condemned the decision as creating a "situation where the country is a less safe place to grow up," and Texas Republican Lieutenant Governor David Dewhurst announced that Texas would not follow a decision that placed at risk "our most precious resource—our children."

The prominent liberal legal scholar Laurence Tribe, of Harvard Law School, joined the chorus of conservative criticism and argued in the *Wall Street Journal* that people concerned about the rule of law must "cry foul" when the Supreme Court interferes with the judgment of an elected state legislature and holds that "torturers or violent rapists of young children . . . [are] constitutionally exempt from the death penalty." He noted that by restricting the death penalty to murder, the Court, in essence, had denied equal protection of the law to children victimized by rape.

The Court majority clearly was concerned that upholding the death penalty for individuals who raped a child would open the floodgates to the expansion of capital punishment to encompass a variety of reprehensible offenses. What would be next, the death penalty for individuals who severely abused or maimed a child? There was no indication that individuals tempted to rape and molest children were not already deterred by the prospect of a lengthy prison term and by the requirement that they register as sex offender when released from prison.

The U.S. already was subject to international criticism for reliance on capital punishment, and by approving of the death penalty for the rape of a child, the United States would be joining the ranks of fundamentalist Islamic regimes like Saudi Arabia. Louisiana's system of capital punishment already had come under criticism as unreliable and discriminatory. A Columbia University study found that roughly half of death penalty convictions in Louisiana had been overturned on appeal due to errors and misconduct at trial. Roughly one-third of Louisiana's residents are African American, and yet, seventy percent of the individuals on its death row are African American, and roughly one-half of the individuals who have been executed in Louisiana in the past twenty-six years are African Americans. Confidence in Louisiana's system of capital punishment was further shaken by the fact that nine individuals sentenced to death in recent years had been determined to have been falsely convicted and were released from prison.

Was the U.S. Supreme Court in *Kennedy* justified in holding that the rape of a child under all circumstances is cruel and unusual punishment under the Eighth Amendment? Should the Court have allowed the states to determine for themselves whether to impose the death penalty for the rape of a child or, in the alternative, should the Court have articulated various aggravating circumstances that would justify the death penalty for the rape of a child?

 You can find Kennedy v. Louisiana *on the study site, www.sagepub.com/lippmanccl3e.*

The Amount of Punishment: Sentences for a Term of Years

The U.S. Supreme Court has remained sharply divided over whether the federal judicial branch is constitutionally entitled to extend its proportionality analysis beyond the death penalty to imprisonment for a "term of years." The Court appears to have accepted that the length of a criminal sentence is the province of elected state legislators and that judicial intervention should be "extremely rare" and limited to sentences that are "grossly disproportionate" to the seriousness of the offense. Excessively severe sentences are not considered to advance any of the accepted goals of criminal punishment and constitute the purposeless and needless imposition of pain and suffering.

The implications of this approach were illustrated by Justice Sandra Day O'Connor's opinion in *Lockyer v. Andrade,* in 2003, in which the Supreme Court affirmed two consecutive twenty-five-year-to-life sentences for a defendant who, on two occasions in 1995, stole videotapes with an aggregate value of roughly $150 from two stores.[45] These two convictions, when combined with Andrade's arrest thirteen years earlier for three counts of residential burglary, triggered two separate mandatory sentences under California's **Three Strikes and You're Out law**. This statute provides a mandatory sentence for individuals who commit a third felony after being previously convicted for two serious or violent felonies. Stringent penalties also are provided for a second felony. Justice O'Connor held that the "gross disproportionality principle reserves a constitutional violation for only the extraordinary case" and that the sentence in *Andrade* was not "an unreasonable application of our clearly established law."[46]

In *Ewing v. California,* decided on the same day as *Lockyer,* Justice O'Connor affirmed a twenty-five-year sentence for Daniel Ewing under California's Three Strikes and You're Out law. Ewing while on parole was adjudged guilty of the grand theft of three golf clubs worth $399 apiece and had previously been convicted of several serious or violent felonies. As required by the three strikes law, the prosecutor formally alleged, and the trial court found, that Ewing had been convicted previously of four serious or violent felonies. Justice Sandra Day O'Connor ruled that the Supreme Court was required to respect California's determination that it possessed a public-safety interest in incapacitating and deterring recidivist felons like Ewing, whose previous offenses included robbery and three residential burglaries.[47]

Weems v. United States is an example of the rare case in which the Supreme Court has ruled that a punishment is grossly disproportionate to the crime and is unconstitutional. Weems was convicted under the local criminal law in the Philippines of forging a public document. He was sentenced to twelve years at hard labor as well as to manacling at the wrist and ankle. During Weems's twelve-year imprisonment, he was deprived of all legal rights and, upon his release, lost all political rights (such as the right to vote) and was monitored by the court. The U.S. Supreme Court ruled that Weems's sentence was "cruel in its excess of imprisonment and that which accompanies and follows imprisonment. It is unusual in its character. Its punishments come under the condemnation of the Bill of Rights, both on account of their degree and kind."[48]

The next case, Humphrey v. Wilson, asks you to consider whether the defendant's mandatory minimum sentence of ten years in prison under Georgia's aggravated child molestation law constitutes cruel and unusual punishment. In Rummel v. Estelle, Justice Louis Powell observed that a "mandatory life sentence for overtime parking might well deter vehicular lawlessness, but it would offend our felt sense of justice."[49] Consider Justice Powell's statement in evaluating whether Wilson's sentence is grossly disproportionate.

You can find Ewing v. California on the study site, www.sagepub.com/lippmanccl3e.

Did Wilson's sentence of ten years in prison for consensual oral sex constitute cruel and unusual punishment?

Humphrey v. Wilson, 652 S.E.2d 501 (Ga. 2007). Opinion by: Sears, J.

Facts

In February 2005, Wilson was found guilty in Douglas County for the aggravated child molestation of T. C. Wilson was seventeen years old at the time of the crime, and the victim was fifteen years old. The sexual act involved the victim willingly performing oral sex on Wilson. At the time of Wilson's trial, the minimum sentence for a conviction of aggravated child molestation was ten years in prison with no possibility of probation or parole; the maximum sentence was thirty years in prison. The trial court sentenced Wilson to eleven years, ten to serve and one year on probation. In addition to the foregoing punishment, Wilson was also subject to registration as a sex offender. In this regard . . . Wilson would be required, before his release from prison, to provide prison officials with, among other things, his new address, his fingerprints, his social security number, his date of birth, and his photograph. Prison officials would have to forward this information to the sheriff of Wilson's intended county of residence, and Wilson, within 72 hours of his release, would have to register with that sheriff, and he would be required to update the information each year for the rest of his life. Moreover, upon Wilson's release from prison, information regarding Wilson's residence, his photograph, and his offense would be posted in numerous public places in the county in which he lived and on the internet. Significantly, Wilson could not live or work within 1,000 feet of any child care facility, church, or area where minors congregate.

After the trial court denied Wilson's motion for new trial, Wilson filed a notice of appeal to this court. This court transferred the appeal to the court of appeals, and that court affirmed Wilson's conviction on April 28, 2006. On appeal . . . Wilson contended that his sentence constituted cruel and unusual punishment. The court of appeals did not address Wilson's contention that his sentence constituted cruel and unusual punishment and rejected the appeal on other grounds. In a motion for reconsideration filed on May 8, 2006, Wilson stated that, two days before the court of appeals issued its opinion, Georgia Governor Sonny Perdue signed House Bill 1059, which amended Official Code of Georgia (OCGA) section 16-6-4 effective July 1, 2006, by adding a new subsection (d)(2) to make conduct such as Wilson's a misdemeanor and which amended OCGA 42-1-12 to relieve him from having to register as a sex offender. . . . The court of appeals denied Wilson's motion for reconsideration. Wilson thereafter petitioned this court for consideration. This court subsequently denied Wilson's petition for certiorari.

On April 16, 2007, Wilson filed the present application for writ of habeas corpus, contending that his sentence constituted cruel and unusual punishment due in large part to the fact that the 2006 amendment to OCGA section 16-6-4 makes conduct such as his a misdemeanor, while the 2006 amendment to OCGA section 42-1-12 relieved him from the requirements of the sex offender registry. In this regard, the 2006 amendment to OCGA section 16-6-4 provides that, if a person engages in sodomy with a victim who "is at least 13 but less than 16 years of age" and, if the person who engages in the conduct is "18 years of age or younger and is no more than four years older than the victim," the person is guilty of the new crime of misdemeanor aggravated child molestation. Moreover, the 2006 amendment to OCGA section 42-1-12 provided that teenagers whose conduct is a misdemeanor under the 2006 amendment to OCGA section 16-6-4 do not have to register as sex offenders.

Concluding that the extraordinary changes in the law reflected in the 2006 amendments to OCGA sections 16-6-4 and 42-1-12 reflected this state's contemporary view of how Wilson's conduct should be punished, the habeas court ruled that Wilson's punishment was cruel and unusual. Finally, the habeas court, as a remedy, ruled that Wilson was guilty of misdemeanor aggravated child molestation under the 2006 amendment to OCGA section 16-6-4, and it sentenced Wilson to twelve months to serve with credit for time served. On June 11, 2007, the warden filed a notice of appeal from the habeas court's grant of relief to Wilson.

Issue

The prison warden . . . contends that the habeas court erred in ruling that Wilson's sentence constituted cruel and unusual punishment.

Reasoning

Under the Eighth Amendment to the United States Constitution and under article 1, section 1, paragraph 17 of the Georgia Constitution, a sentence is cruel and unusual if it "is grossly out of proportion to the severity of the crime." Moreover, whether "a particular punishment is cruel and unusual is not a static concept, but instead changes in recognition of the 'evolving standards of decency that mark the progress of a maturing society.'" Legislative enactments are the clearest and best evidence of a society's evolving standard of decency and of how contemporary society views a particular punishment.

In determining whether a sentence set by the legislature is cruel and unusual, this court has cited with approval Justice Kennedy's concurrence in *Harmelin v. Michigan.* Under Justice Kennedy's concurrence in *Harmelin,* as further developed in *Ewing v. California,* in order to determine if a sentence is grossly disproportionate, a court must first examine the "gravity of the offense compared to the harshness of the penalty" and determine whether a threshold inference of gross disproportionality is raised. In making this determination, courts must bear in mind the primacy of the legislature in setting punishment and seek to determine whether

the sentence furthers a "legitimate penological goal" considering the offense and the offender in question. If a sentence does not further a legitimate penological goal, it is not a rational legislative judgment that is entitled to deference and a threshold showing of disproportionality has been made.

If this threshold analysis reveals an inference of gross disproportionality, a court must proceed to the second step and determine whether the initial judgment of disproportionality is confirmed by a comparison of the defendant's sentence to sentences imposed for other crimes within the jurisdiction and for the same crime in other jurisdictions.

Before undertaking the foregoing analysis, we address the warden's contention that this court's recent decision in *Widner v. State* controls the cruel and unusual punishment issue adversely to Wilson. We conclude that *Widner* is not controlling. Widner was eighteen years old when he had oral sex with a willing fourteen-year-old girl, and he received a ten-year sentence under OCGA section 16-6-4. On appeal, we resolved Widner's claim of cruel and unusual punishment against him. However, the basis of Wilson's claim in the present case—the 2006 amendment to OCGA section 16-6-4—did not become effective until after Widner's appeal, and Widner thus did not predicate his cruel and unusual punishment contention on the 2006 amendment for that reason.

There is, however, a more significant reason *Widner* is not controlling in the present case. The 2006 amendment to OCGA section 16-6-4 did not alter the punishment for Widner's conduct. In *Widner,* the minor child turned fourteen five days before the incident in question. Widner was eighteen and a half years old at that time. Widner was thus more than four years older than the victim. The 2006 amendment to OCGA section 16-6-4 changed the punishment for oral sex with a thirteen-, fourteen-, or fifteen-year-old child when the defendant is "no more than four years older than the victim." Accordingly, the amendment does not apply to Widner's conduct and does not raise an inference of gross disproportionality with respect to his sentence.

We turn now to the threshold inquiry of disproportionality as developed in *Harmelin* and *Ewing.* In this regard, we conclude that the rationale of our decisions in *Fleming* . . . leads to the conclusion that, considering the nature of Wilson's offense, his ten-year sentence does not further a legitimate penological goal and thus the threshold inquiry of gross disproportionality falls in Wilson's favor.

Here, the legislature has recently amended OCGA section 16-6-4 to substitute misdemeanor punishment for Wilson's conduct in place of the felony punishment of a minimum of ten years in prison (with the maximum being 30 years in prison) with no possibility of probation or parole. Moreover, the legislature has relieved such teenage offenders from registering as a sex offender. It is beyond dispute that these changes represent a seismic shift in the legislature's view of the gravity of oral sex between two willing teenage participants. Acknowledging, as we must . . . that no one has a better sense of the evolving standards of decency in this state than our elected representatives, we conclude that the amendments to OCGA sections 16-6-4 and 42-1-12 reflect a decision by the people of this state that the severe felony punishment and sex offender registration imposed on Wilson make no measurable contribution to acceptable goals of punishment.

Stated in the language of *Ewing* and *Harmelin,* our legislature compared the gravity of the offense of teenagers who engage in oral sex but are within four years of age of each other and determined that a minimum ten-year sentence is grossly disproportionate for that crime. This conclusion appears to be a recognition by our General Assembly that teenagers are engaging in oral sex in large numbers; that teenagers should not be classified among the worst offenders because they do not have the maturity to appreciate the consequences of irresponsible sexual conduct and are readily subject to peer pressure; and that teenage sexual conduct does not usually involve violence and represents a significantly more benign situation than that of adults preying on children for sex. Similarly, the Model Penal Code adopted a provision in 1980 decriminalizing oral or vaginal sex with a person under sixteen years old where that person willingly engaged in the acts with another person who is not more than four years older. The commentary to the Model Penal Code explains that the criminal law should not target "sexual experimentation among social contemporaries"; that "it will be rare that the comparably aged actor who obtains the consent of an underage person to sexual conduct . . . will be an experienced exploiter of immaturity"; and that the "more likely case is that both parties will be willing participants and that the assignment of culpability only to one will be perceived as unfair."

In addition to the extraordinary reduction in punishment for teenage oral sex reflected in the 2006 amendment to OCGA section 16-6-4, the 2006 amendment to that statute also provided for a large increase in the punishment for adults who engage in child molestation and aggravated child molestation. The new punishment for adults who engage in child molestation is ten years to life in prison, whereas the punishment under the prior law was imprisonment "for not less than five nor more than twenty years." For aggravated child molestation, the punishment for adults is now twenty-five years to life, followed by life on probation, with no possibility of probation or parole for the minimum prison time of twenty-five years. The significant increase in punishment for adult offenders highlights the legislature's view that a teenager engaging in oral sex with a willing teenage partner is far from the worst offender and is, in fact, not deserving of similar punishment to an adult offender.

Although society has a significant interest in protecting children from premature sexual activity, we must acknowledge that Wilson's crime does not rise to the level of culpability of adults who prey on children and

that, for the law to punish Wilson as it would an adult, with the extraordinarily harsh punishment of ten years in prison without the possibility of probation or parole, appears to be grossly disproportionate to his crime.

Based on the foregoing factors and, in particular, based on the significance of the sea change in the General Assembly's view of the appropriate punishment for teenage oral sex, we could comfortably conclude that Wilson's punishment, as a matter of law, is grossly disproportionate to his crime without undertaking the further comparisons outlined in *Harmelin* and *Ewing.* However, we nevertheless will undertake those comparisons to complete our analysis.

A comparison of Wilson's sentence with sentences for other crimes in this state buttresses the threshold inference of gross disproportionality. For example, a defendant who gets in a heated argument and shoving match with someone, walks away to retrieve a weapon, returns minutes later with a gun, and intentionally shoots and kills the person may be convicted of voluntary manslaughter and sentenced to as little as one year in prison. A person who plays Russian Roulette with a loaded handgun and causes the death of another person by shooting him or her with the loaded weapon may be convicted of involuntary manslaughter and receive a sentence of as little as one year in prison and no more than ten years. A person who intentionally shoots someone with the intent to kill, but fails in his aim such that the victim survives, may be convicted of aggravated assault and receive as little as one year in prison. A person who maliciously burns a neighbor's child in hot water, causing the child to lose use of a member of his or her body, may be convicted of aggravated battery and receive a sentence of as little as one year in prison. Finally, at the time Wilson committed his offense, a fifty-year-old man who fondled a five-year-old girl for his sexual gratification could receive as little as five years in prison, and a person who beat, choked, and forcibly raped a woman against her will could be sentenced to ten years in prison. There can be no legitimate dispute that the foregoing crimes are far more serious and disruptive of the social order than a teenager receiving oral sex from another willing teenager. The fact that these more culpable offenders may receive a significantly smaller or similar sentence buttresses our initial judgment that Wilson's sentence is grossly disproportionate to his crime.

Finally, we compare Wilson's sentence to sentences imposed in other states for the same conduct. A review of other jurisdictions reveals that most states either would not punish Wilson's conduct or would, like Georgia now, punish it as a misdemeanor. Although some states retain a felony designation for Wilson's conduct, we have found no state that imposes a minimum punishment of ten years in prison with no possibility of probation or parole, such as that provided for by former section 16-6-4. This review thus also reinforces our initial judgment of gross disproportionality between Wilson's crime and his sentence.

At this point, the Supreme Court's decision in *Weems v. United States* merits discussion. In that case,

Weems forged signatures on several public documents. The Supreme Court found that a minimum sentence of twelve years in chains at hard labor for falsifying public documents, combined with lifetime surveillance by appropriate authorities after Weems's release from prison, constituted cruel and unusual punishment. The Court stated that, because the minimum punishment imposed on Weems was more severe than or similar to punishments for some "degrees of homicide" and other more serious crimes, Weems's punishment was cruel and unusual. According to the Court, this contrast shows more than different exercises of legislative judgment. It is greater than that. It condemns the sentence in this case as cruel and unusual. It exhibits a difference between unrestrained power and that which is exercised under the spirit of constitutional limitations formed to establish justice.

Holding

All of the foregoing considerations compel the conclusion that Wilson's sentence is grossly disproportionate to his crime and constitutes cruel and unusual punishment under both the Georgia and United States Constitutions. We emphasize that it is the "rare case in which the inference of gross disproportionality will be met" and a rarer case still in which that threshold inference stands after further scrutiny. The present case, however, is one of those rare cases. We also emphasize that nothing in this opinion should be read as endorsing attempts by the judiciary to apply statutes retroactively. We are not applying the 2006 amendment retroactively in this case. Instead . . . we merely factor the 2006 amendment into the evaluation of whether Wilson's punishment is cruel and unusual.

Today's opinion will affect only a small number of individuals whose crimes and circumstances are similar to Wilson's, i.e., those teenagers convicted only of aggravated child molestation, based solely on an act of sodomy, with no injury to the victim, involving a willing teenage partner no more than four years younger than the defendant. Wilson stands convicted of aggravated child molestation, and . . . we have determined that, under the statute then in effect, the minimum punishment authorized by the legislature for that crime is unconstitutional. . . .

Dissenting, *Carley, J.*

Because I believe that the majority's conclusion that Wilson's felony sentence constitutes cruel and unusual punishment does violence to the fundamental constitutional principle of separation of powers and is contrary to the doctrine of *stare decisis,* I respectfully dissent. . . .

It is important to note at the outset that the factual basis for Wilson's prosecution is not an act which is in any sense protected by the constitutional right of privacy. The evidence shows that a group of teenagers rented adjacent rooms at a motel and held a raucous, unsupervised New Year's Eve party. Among the participants were seventeen-year-old Genarlow Wilson, seventeen-year-old L. M., and fifteen-year-old T. C. The next morning, L. M. reported to her mother that she had been raped. Police were notified, and the motel rooms were searched. During the search, a video camera and videocassette tape were found. The tape showed Wilson having sexual intercourse with an apparently semiconscious L. M., and T. C. performing oral sex on Wilson. As a result, Wilson was charged with the rape of L. M. and with the aggravated child molestation of T. C. Acquitted of the former offense and convicted of the latter, he was given a mandatory sentence of ten years imprisonment without possibility of parole. When Wilson engaged in the very public act of oral sodomy with a fifteen-year-old child, he committed the crime of aggravated child molestation and, as a result, he received the felony sentence mandated for that offense (former OCGA section 16-6-4 (d)(1)).

Subsequently, the General Assembly did amend the statute so as to provide that the crime of aggravated child molestation committed under the factual circumstances which underlay Wilson's prosecution would only be punishable as a misdemeanor (OCGA section 16-6-4(d)(2)). However, the effective date of that change in the law was July 1, 2006, which is more than a year-and-a-half after Wilson committed the offense for which he was convicted. In amending the law to provide for misdemeanor punishment, the General Assembly not only provided generally that the change would become effective on July 1, 2006. It also specifically addressed the issue of retroactive application. The effect of this clear and unambiguous provision is to preclude giving retroactive effect to the 2006 amendment so as to "affect or abate" the status of Wilson's crime as felony aggravated child molestation punishable in accordance with the sentence authorized at the time he committed that offense. The majority fails to acknowledge this provision of the statute, presumably because to do so would completely destroy the foundation upon which it bases its ultimate conclusion that Wilson's felony sentence constitutes cruel and unusual punishment.

In connection with a claim of cruel and unusual punishment, the enactments of the General Assembly are the clearest and best evidence of a society's evolving standards of decency and of how contemporary society views a particular punishment. The majority acknowledges this tenet and purports to invoke it. However, a faithful adherence to that principle would seem to require a consideration of the totality of the law in question, which in this case certainly includes every provision of the 2006 statute. Accordingly, while I am very sympathetic to Wilson's argument regarding the injustice of sentencing this promising young man, with good grades and no criminal history, to ten years in prison without parole and a lifetime registration as a sexual offender because he engaged in consensual oral sex with a fifteen-year-old victim only two years his junior, this court is bound by the Legislature's determination that young persons in Wilson's situation are not entitled to the misdemeanor treatment now accorded to identical behavior under OCGA section 16-6-4(d)(2).

The majority does not demonstrate that an unqualified felony sentence for aggravated child molestation constituted cruel and unusual punishment at the time that Wilson committed that crime. Indeed, it cannot so demonstrate, since the law which was then in effect "provide[d] no such exception [to mandatory felony sentencing based upon the age of the defendant and victim], and, because the required punishment does not unconstitutionally shock the conscience, [such a] sentence must stand." Wilson's sentence does not become cruel and unusual simply because the General Assembly made the express decision that he cannot benefit from the subsequent legislative determination to reduce the sentence for commission of that crime from felony to misdemeanor status. To the contrary, it is because the General Assembly made that express determination that his felony sentence cannot be deemed cruel and unusual. It is for the legislature to "determine to what extent certain criminal conduct has demonstrated more serious criminal interest and damaged society and to what extent it should be punished."

In actuality . . . today's decision is rare because of its unprecedented disregard for the General Assembly's constitutional authority to make express provision against the giving of any retroactive effect to its legislative lessening of the punishment for criminal offenses. If . . . the judiciary is permitted to determine that a formerly authorized harsher sentence nevertheless constitutes cruel and unusual punishment, then it necessarily follows that there are no circumstances in which the General Assembly can insulate its subsequent reduction of a criminal sentence from possible retroactive application by courts. Wilson is certainly not the only defendant convicted of aggravated child molestation who benefits at the expense of today's judicial reduction of the General Assembly's power to legislate. At present, any and all defendants who were ever convicted of aggravated child molestation and sentenced for a felony under circumstances similar to Wilson are, as a matter of law, entitled to be completely discharged from lawful custody even though the General Assembly expressly provided that their status as convicted felons would not be affected by the very statute upon which the majority relies to free them. . . . Moreover, nothing in today's decision limits its application to cases involving minors who engage in voluntary sexual acts. Any defendant who was ever convicted in this state for the commission of any crime for which the sentence was subsequently reduced is now entitled to claim that his harsher sentence, though authorized under the statute in effect at the time it was imposed, has since become cruel and unusual and that, as a consequence, he is not only entitled to the benefit of the more lenient sentence, but should be released entirely from incarceration. . . . Accordingly, as a result of this "rare case," the superior courts should be prepared for a flood of habeas corpus petitions filed by prisoners who seek to be freed from imprisonment because of a subsequent reduction in the applicable sentences for the crimes for which they were convicted.

The courts of this state must give due regard to the authority of the legislative branch of government. The constitutional principle of separation of powers is intended to protect the citizens of this state from the tyranny of the judiciary, insuring that the authority to enact the laws will be exercised only by those representatives duly elected to serve as legislators. The General Assembly "being the sovereign power in the State, while acting within the pale of its constitutional competency, it is the province of the courts to interpret its mandates, and their duty to obey them, however absurd and unreasonable they may appear."

Questions for Discussion

1. What was Wilson's original prison sentence? How was the law changed following his conviction?

2. Why did the Georgia Supreme Court rule that Wilson had been subjected to cruel and unusual punishment?

3. Explain how, following Wilson's conviction, Georgia courts could hold that Wilson had not been subjected to cruel and unusual punishment, and then two years later reverse themselves and hold that Wilson's punishment constituted cruel and unusual punishment.

4. Had the Georgia legislature not modified the punishment for aggravated child molestation, would the Georgia Supreme Court have ruled that Wilson's sentence constituted cruel and unusual punishment?

5. Does it make sense that because Widner was more than four years older than his sexual partner, his punishment was not considered to be cruel and unusual?

6. Summarize the main points made in the dissenting opinion.

7. How would you rule in *Wilson*?

You Decide

3.1 In *United States v. Gementera*, Shawn Gementera stole letters from several mailboxes in San Francisco. He entered a plea agreement and pled guilty to mail theft. The twenty-four-year-old Gementera already had an extensive arrest record including criminal mischief, driving with a suspended license, misdemeanor battery, possession of drug paraphernalia, and taking a vehicle without the owner's consent. United States District Court Judge Vaughn Walker sentenced Gementera to two months imprisonment and three years supervised release.

Several conditions were placed on the supervised release including requiring Gementera to perform one day of community service consisting of either wearing a two-sided sandwich board-style sign or carrying a large two-sided sign stating, "I stole mail; this is my punishment." Gementera was required to display the sign for 8 hours while standing in front of a San Francisco postal facility. The prosecution and the defense attorneys jointly agreed that Gementera also would lecture at a high school and write apologies to any identifiable victims.

Do "shaming punishments" promote rehabilitation? Deter criminal conduct? Unnecessarily shame and humiliate defendants? See *United States v. Gementera*, 379 F.3d 596 (9th Cir. 2004).

You can find the answer at www.sagepub.com/lippmanccl3e.

You can find more cases on the study site: **People v. Meyer, Ewing v. California,** *www.sagepub.com/lippmanccl3e.*

The Amount of Punishment: Drug Offenses

Three Strikes and You're Out legislation is an example of determinate sentencing. Determinate sentences possess the advantage of ensuring predictable, definite, and uniform sentences. On the other hand, this "one size fits all" approach may prevent judges from handing out sentences that reflect the circumstances of each individual case.

A particularly controversial area of determinate sentencing is mandatory minimum sentences for drug offenses. In 1975, New York Governor Nelson Rockefeller initiated the controversial "Rockefeller drug laws" that required that an individual convicted of selling two ounces or possessing eight ounces of a narcotic substance receive a sentence of between eight and twenty years, regardless of the individual's criminal history. This approach in which a judge must sentence a defendant to a minimum sentence was followed by other states. The federal government joined this trend and introduced mandatory minimums in the Anti-Drug Abuse Act of 1986 and the 1988 amendments. The most debated aspect of federal law is the punishment of an individual based on the type and amount of drugs in his or her possession, regardless of the individual's criminal history.

The 2010 Fair Sentencing Act of 2010 constituted a major reform of U.S. narcotics laws. Under the previous federal law, a conviction for possession with intent to distribute 5 grams of crack cocaine and 500 grams of powder cocaine resulted in the same five-year sentence. Fifty grams of crack cocaine and 5 grams of powder cocaine triggered the same ten-year sentence. The thinking behind the law was that crack is sold in small, relatively inexpensive amounts on the street, ravages communities, and leads to street violence between street gangs competing for control of the drug trade. The law was criticized for resulting in the disproportionate arrest and imprisonment of African Americans for lengthy prison terms while Caucasian sellers and users of powder cocaine received much less severe prison terms. The sentencing reform law reduced the 100–1 ratio between crack and powder cocaine to an 18–1 ratio.

The following quantities are punishable by five years in prison under federal law:

- 100 grams of heroin
- 500 grams of powder cocaine
- 28 grams of crack cocaine
- 100 kilograms of marijuana

The following quantities are punishable by ten years in prison under federal law:

- 1 kilogram of heroin
- 5 kilograms of powder cocaine
- 280 grams of crack cocaine
- 1,000 kilograms of marijuana

Congress softened the impact of the mandatory minimum drug sentences by providing that a judge may issue a lesser sentence in those instances in which prosecutors certify that a defendant has provided "substantial assistance" in convicting other drug offenders. There also is a "safety valve" that permits a reduced sentence for defendants determined by the judge to be low-level, nonviolent, first-time offenders.

Prosecutors argue that the mandatory minimum sentences are required to deter individuals from entering into the lucrative drug trade. The threat of a lengthy sentence is also necessary in order to gain the cooperation of defendants. Prosecutors also point out that individuals who are convicted and sentenced are fully aware of the consequences of their criminal actions.

These laws, nevertheless, have come under attack by both conservative and liberal politicians and by the American Bar Association, a justice of the U.S. Supreme Court, and by the Judicial Conference, which is the organization of federal judges. An estimated twenty-two states have recently modified or are considering amending their mandatory minimum narcotics laws, including Connecticut, Louisiana, Michigan, North Dakota, and Pennsylvania. New York also has modified its Rockefeller drug laws. This trend is encouraged by studies that indicate that these laws have several flaws:

- *Inflexibility.* They fail to take into account the differences between defendants.
- *Disparities in Enforcement.* Drug kingpins are able to trade information for reduced sentences, and some prosecutors who object to harsh drug laws charge defendants with the possession of a lesser quantity of drugs to avoid the mandatory sentencing provisions.
- *Increasing Prison Population.* These laws are thought to be responsible for the growth of the state and federal prison population.
- *Disproportionate Affect on Minorities and Women.* A significant percentage of individuals sentenced under these laws are African Americans or Hispanics involved in street-level drug activity. The increase in the number of women who are incarcerated is attributed to the fact that women find themselves arrested for assisting their husbands or lovers who are involved in the drug trade.

Mandatory minimum state drug laws have been held to be constitutional by the U.S. Supreme Court. In *Hutto v. Davis,* the Supreme Court ruled that Hutto's forty-year prison sentence and $20,000 fine was not disproportionate to his conviction on two counts of possession with intent to distribute and on distribution of a total of nine ounces of marijuana with a street value of roughly $200. The Court held that the determination of the proper sentence for this offense was a matter that was appropriately determined by the Virginia legislature.[50]

The Tenth Circuit Court of Appeals upheld the fifty-five year sentence given to rap music producer Weldon Angelos. The twenty-six-year-old Angelos, who did not possess an adult criminal history, was convicted of three counts of dealing twenty-four ounces of marijuana while in possession of a firearm. The federal statute provides that a first offense carries a mandatory minimum five-year sentence, and each subsequent conviction carries a mandatory minimum of twenty-five years. Twenty-nine former federal judges and U.S. attorneys protested that Angelos's punishment violated the Eighth Amendment, pointing out that his sentence was longer than he likely would have received for various forms of murder or rape. The three-judge panel disregarded this argument and held that Angelos's punishment for possession of firearms was justified on the grounds that the weapon facilitated his drug activity by providing protection from purchasers and that Angelos's possession of a firearm endangered his neighbors. The Tenth Circuit reasoned that the federal statute furthered these interests and that Angelos's sentence was not grossly disproportionate.[51]

What do you think about the argument that such sentences are so disproportionate and impose such hardship that jurors should refuse to convict defendants charged with quantities of narcotics carrying mandatory minimum sentences?[52]

Criminal Punishment and Status Offenses

In *Robinson v. California,* the U.S. Supreme Court overturned Robinson's conviction under a California law that declared it was a criminal offense "to be addicted to the use of narcotics." The Supreme Court ruled that it was cruel and unusual punishment to impose criminal penalties on Robinson based on his conviction of the **status offense** of narcotics addiction, which a majority of the judges considered an addictive illness. Justice Potter Stewart noted that "even one day in prison would be cruel and unusual punishment for the 'crime' of having a common cold. . . . It is unlikely that any state would . . . make it a criminal offense for a person to be mentally ill,

Figure 3.1 Crime on the Streets: Incarceration Rates. The number of adults in the correctional population has been increasing.

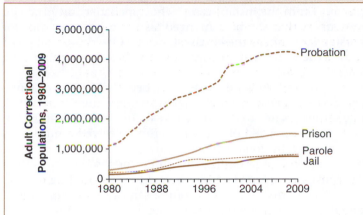

In 2009, over 7.2 million people were under some form of correctional supervision including the following:

Probation – count-order period of correctional supervision in the community generally as an alternative to incarceration. In some cases probation can be a combined sentence of incarceration followed by a period of community supervision. These data include adults under the jurisdiction of probation agency, regardless of supervision status (i.e., active supervision, inactive supervision, financial conditions only, warrant status, absconder status, in a residential/other treatment program, or supervised out of jurisdiction).

Prison – confinement in a state of federal correctional facility to serve a sentence of more than 1 year, although in some jurisdictions the length of sentence which results in prison confinement is longer.

Jail – confinement in a local jail while pending trail, awaiting sentencing, serving a sentence that is usually less than 1 year, or awaiting transfer to other facilities after conviction.

Parole – period of conditional supervised release in the community following a prison term, including prisoners released to parole either by a parole board decision (discretionary parole) or according to provisions of a statute (mandatory parole). These data include adults under the jurisdiction of a parole agency, regardless of supervision status (i.e., active supervision, inactive supervision, financial conditions only, absconder status, or supervised out of state).

Source: Bureau of Justice Statistics Correctional Surveys, U.S. Department of Justice.

or a leper, or to be afflicted with venereal disease. . . . [Such a] law . . . would . . . be universally thought to be an infliction of cruel and unusual punishment in violation of the Eighth and Fourteenth amendments."[53]

Cruel and Unusual Punishment: A Summary

The prohibition on cruel and unusual punishment in the Eighth Amendment is applicable to the states through the Due Process Clause of the Fourteenth Amendment. This prohibition has three components:

- *How.* Punishment may not be inflicted in a cruel or unusual fashion.
- *What.* Capital punishment must be imposed in a proportionate fashion and is reserved for acts of aggravated murder. Courts defer to the legislative branch and accept that sentences for a "term of years" are proportionate. A finding that a sentence for a term of years is disproportionate should be "extremely rare."
- *Who.* Punishment may not be extended to individuals based on a conviction for a status offense; a socially harmful act is required.

EQUAL PROTECTION

Judicial decisions have consistently held that it is unconstitutional for a judge to base a sentence on a defendant's race, gender, ethnicity, or nationality. In other words, a sentence should be based on a defendant's act rather than on a defendant's identity. A federal district court judge's sentence of thirty years in prison and lifetime supervision for two first-time offenders convicted of a weapons offense and two narcotics offenses was rejected by the federal Second Circuit Court of Appeals based on the trial court judge's observation that the South American defendants "should have stayed where they were. . . . Nobody tells them to come and get involved in cocaine. . . . My father came over with $3 in his pocket." The appellate court noted that it appeared that "ethnic prejudice somehow had infected the judicial process in the instant case." The appellate court observed that one of the defendant's "plaintive request that she be sentenced 'as for my person, not for my nationality,'" was completely understandable under the circumstances.[54] Could the trial court judge have constitutionally handed down the same sentence in order to deter Colombian drug gangs from operating in the United States?

Statutes that provide different sentences based on gender also have been held to be in violation of the Equal Protection Clause. In *State v. Chambers,* the New Jersey Supreme Court struck down a statute providing indeterminate sentences not to exceed five years (or the maximum provided in a statute) for women, although men convicted of the same crime received a minimum and maximum sentence, which could be reduced by good behavior and work credits.[55]

This complicated scheme resulted in men receiving significantly shorter prison sentences than women convicted of the same crime. For instance, a female might be held on a gambling conviction "for as long as five years . . . [a] first offender male, convicted of the same crime, would likely receive a state prison sentence of not less than one or more than two years." The female offender was required to serve the complete sentence, although the male would "quite likely" receive parole in four months and twenty-eight days. The New Jersey Supreme Court dismissed the argument that the "potentially longer period of detention" for females was justified on the grounds that women were good candidates for rehabilitation who could turn their lives around in prison. The court pointed out that there "are no innate differences [between men and women] in capacity for intellectual achievement, self-perception or self-control or the ability to change attitude and behavior, adjust to social norms and accept responsibility." What was the New Jersey legislature thinking when it adopted this sentencing scheme? Can you think of a crime for which the legislature might constitutionally impose differential sentences based on gender? At least one appellate court in Illinois has upheld a statute that punished a man convicted of incest with his daughter more severely than a woman convicted of incest with her son, reasoning that this furthered the interest in preventing pregnancy.

What about seemingly neutral laws that possess a discriminatory impact? *The general rule is that a defendant must demonstrate both a discriminatory impact and a discriminatory intent.* The difficulty of this task is illustrated by the Supreme Court's consideration of the discriminatory application of the death penalty in *McCleskey v. Kemp.*[56]

Warren McCleskey, an African American, was convicted of two counts of armed robbery and one count of the murder of a Caucasian police officer. He was sentenced to death on the homicide count and to consecutive life sentences on the robbery. McCleskey claimed that the Georgia capital punishment statute violated the Equal Protection Clause in that African American defendants facing trial for the murder of Caucasians were more likely to be sentenced to death. McCleskey relied on a sophisticated statistical study of 2,000 Georgia murder cases involving 230 variables that had been conducted by Professors David C. Baldus, Charles Pulaski, and George Woodworth. This led to a number of important findings, including that defendants charged with killing Caucasian victims were over four times as likely to receive the death penalty as defendants charged with killing African Americans, and that African American defendants were one and one-tenth times as likely to receive the death sentence as other defendants.

The Supreme Court ruled that McCleskey had failed to meet the burden of clearly establishing that the decision makers in his specific case acted with a discriminatory intent to disadvantage McCleskey on account of his race. What of McCleskey's statistical evidence? The Supreme Court majority observed that the statistical pattern in Georgia reflected the decisions of a number of prosecutors in cases with different fact patterns, various defense counsel, and different jurors and did not establish that the prosecutor or jury in McCleskey's specific case or in other cases were biased. McCleskey killed a police officer, a charge that clearly permitted the imposition of capital punishment under Georgia law. Where was the discrimination?

Justice William Brennan, in dissent, criticized the five-judge majority for approving a system in which "lawyers must tell their clients that race casts a large shadow on the sentencing process." Justice Brennan noted that a lawyer, when asked by McCleskey whether McCleskey was likely to receive the death sentence, would be forced to reply that six of every eleven defendants convicted of killing a Caucasian would not have received the death penalty if their victim had been African American. At the same time, among defendants with aggravating and mitigating factors comparable to McCleskey's, twenty of every thirty-four would not have been sentenced to die if their victims had been African American. This pattern of racial bias was particularly significant given the history of injustice against African Americans in the Georgia criminal justice system. The Bureau of National Justice Statistics reports that in 2004, 1,851 Caucasians and 1,390 African Americans were on death row. (Seventy-four additional individuals classified as "other" were also on death row.)

The following facts were presented to a California judge at sentencing. What punishment would you impose?

You Decide

3.2 Defendant Soon Ja Du was convicted of voluntary manslaughter in the killing of Latasha Harlins, a customer in defendant's store. Defendant was sentenced to ten years in state prison. The sentence was suspended (not enforced), and defendant was placed on probation under certain terms and conditions. The district attorney contends the court abused its discretion in granting probation and seeks a . . . legal sentence of an appropriate term in state prison.

The crime giving rise to defendant's conviction occurred on the morning of March 16, 1991, at the Empire Liquor Market, one of two liquor stores owned and operated by defendant and her family. Although Empire Liquor was normally staffed by defendant's husband and son while defendant, Ja Du, worked at the family's other store in Saugus, defendant worked at Empire on the morning of March 16, 1991, so that her son, who had been threatened by local gang members, could work at the Saugus store instead. Defendant's husband, Billy Du, was present at the Empire Liquor Market that morning; but at defendant's urging he went outside to sleep in the family van, because he had worked late the night before.

Defendant was waiting on two customers at the counter when the victim, fifteen-year-old Latasha Harlins, entered the store. Latasha proceeded to the section where the juice was kept, selected a bottle of orange juice, put it in her backpack, and proceeded toward the counter.

Defendant had observed many shoplifters in the store, and it was defendant's experience that people who were shoplifting would take the merchandise, "place it inside the bra or anyplace where the owner would not notice," and then approach the counter, buy some small items and leave. Defendant saw Latasha enter the store, take a bottle of orange juice from the refrigerator, place it in her backpack, and proceed to the counter. Although the orange juice was in the backpack, it was partially visible. Defendant testified that she was suspicious, because she expected if the victim were going to pay for the orange juice, she would have had it in her hand. Defendant's son, Joseph Du, testified that there were at least forty shoplifting incidents a week at the store.

Thirteen-year-old Lakesha Combs and her brother, nine-year-old Ismail Ali, testified that Latasha approached the counter with money ("about two or three dollars") in her hand. According to these witnesses, defendant confronted Latasha, called her a "bitch," and accused her of trying to steal the orange juice; Latasha stated she intended to pay for it. According to defendant, she asked Latasha to pay for the orange juice, and Latasha replied, "What orange juice?" Defendant concluded that Latasha was trying to steal the juice.

Defendant testified that it was Latasha's statement, "What orange juice?" that changed defendant's attitude toward the situation, since prior to that time defendant was not afraid of Latasha. Defendant also thought Latasha might be a gang member. Defendant had asked her son, Joseph Du, what gang members in America look like, and he replied "either they wear some pants and some jackets, and they wear light sneakers, and they either wear a cap or a hairband, headband. And they either have some kind of satchel, and there were some thick jackets. And he told me to be careful with those jackets sticking out." Latasha was wearing a sweater and a "Bruins" baseball cap.

Defendant began pulling on Latasha's sweater in an attempt to retrieve the orange juice from the backpack. Latasha resisted and the two struggled. Latasha hit defendant in the eye with her fist twice. With the second blow, defendant fell to the floor behind the counter, taking the backpack with her. During the scuffle, the orange juice fell out of the backpack and onto the floor in front of the counter. Defendant testified that she thought if she were hit one more time, she would die. Defendant also testified that Latasha threatened to kill her. Defendant picked up a stool from behind the counter and threw it at Latasha, but it did not hit her.

After throwing the stool, defendant reached under the counter, pulled out a holstered .38-caliber revolver, and, with some difficulty, removed the gun from the holster. As defendant was removing the gun from the holster, Latasha picked up the orange juice and put it back on the counter, but defendant knocked it away. As Latasha turned to leave defendant shot her in the back of the head from a distance of approximately three feet, killing her instantly. Latasha had $2 in her hand when she died.

Defendant's husband entered the store [upon hearing defendant's calls for help] and saw Latasha lying on the floor. Defendant leaned over the counter and asked, "Where is that girl who hit me?"

Defendant then passed out behind the counter. Defendant's husband attempted to revive her and also dialed 911 and reported a holdup. Defendant, still unconscious, was transported to the hospital by ambulance, where she was treated for facial bruises and evaluated for possible neurological damage.

At defendant's trial, she testified that she had never held a gun before, did not know how it worked, did not remember firing the gun, and did not intend to kill Latasha.

Defendant's husband testified that he had purchased the .38-caliber handgun from a friend in 1981 for self-protection. He had never fired the gun, however, and had never taught defendant how to use it. In 1988, the gun was stolen during a robbery of the family's store in Saugus. Defendant's husband took the gun to the Empire store after he got it back from the police in 1990.

David Butler, a Los Angeles Police Department ballistics expert, testified extensively about the gun, a Smith & Wesson .38-caliber revolver with a two-inch barrel. In summary, he testified that the gun had been altered crudely and that the trigger pull necessary to fire the gun had been drastically reduced. Also, both the locking mechanism of the hammer and the main spring tension screw of the gun had been altered so that the hammer could be released without putting much pressure on the trigger. In addition, the safety mechanism did not function properly. . . . The jury found defendant guilty of voluntary manslaughter (murder in the heat of passion). By convicting defendant of voluntary manslaughter, the jury impliedly found that defendant had the intent to kill and . . . rejected the defenses that the killing was unintentional and that defendant killed in self-defense.

After defendant's conviction, the case was evaluated by a Los Angeles County probation officer, who prepared a presentence probation report. That report reveals the following about defendant.

At the time the report was prepared, defendant was a fifty-one-year-old Korean-born naturalized American citizen, having arrived in the United States in 1976. For the first ten years of their residence in the United States, defendant worked in a garment factory and her husband worked as a repairman. Eventually, the couple saved enough to purchase their first liquor store in San Fernando. They sold this store and purchased the one in Saugus. In 1989, they purchased the Empire Liquor Market, despite being warned by friends that it was in a "bad area."

These warnings proved prophetic, as the store was plagued with problems from the beginning. The area surrounding the store was frequented by narcotics dealers and gang members, specifically the Main Street Crips. Defendant's son, Joseph Du, described the situation as "having to conduct business in a war zone." In December 1990, defendant's son was robbed while working at the store, and he incurred the wrath of local gang members when he agreed to testify against one of their number who he believed had committed the robbery. Soon thereafter, the family closed the store for two weeks while defendant's husband formulated a plan (which he later realized was "naive") to meet with gang members and achieve a form of truce. The store had only recently been reopened when the incident giving rise to this case occurred.

Joseph Du testified at trial that on December 19, 1990, approximately ten to fourteen African American persons entered the store, threatened him, and robbed him again. The store was burglarized over thirty times, and shoplifting incidents occurred approximately forty times per week. If Joseph tried to stop the shoplifters, "they show me their guns." Joseph further testified that his life had been threatened over thirty times, and more than twenty times people had come into the store and threatened to burn it down. Joseph told his mother about these threats every day, because he wanted to emphasize how dangerous the area was and that he could not do business there much longer.

The probation officer concluded "it is true that this defendant would be most unlikely to repeat this or any other crime if she were allowed to remain free. She is not a person who would actively seek to harm another. . . ." However, she went on to state that although defendant expressed concern for the victim and her family, this remorse was centered largely on the effect of the incident on defendant and her own family. The respondent court found, however, that defendant's "failure to verbalize her remorse to the Probation Department [was] much more likely a result of cultural and language barriers rather than an indication of a lack of true remorse."

The probation report also reveals that Latasha had suffered many painful experiences during her life, including the violent death of her mother. Latasha lived with her extended family (her grandmother, younger brother and sister, aunt, uncle, and niece) in what the probation officer described as "a clean, attractively furnished three-bedroom apartment" in south central Los Angeles. Latasha had been an honor student at Bret Hart Junior High School, from which she had graduated the previous spring. Although she was making only average grades in high school, she had promised that she would bring her grades up to her former standard. Latasha was involved in activities at a youth center as an assistant cheerleader, member of the drill team, and summer junior camp counselor. She was a good athlete and an active church member.

The probation officer's ultimate conclusion and recommendation was that probation be denied and defendant sentenced to state prison. The court sentenced defendant to ten years in state prison (six years for the base term and four for the gun use). The sentence was suspended (not enforced), and defendant was placed on probation for a period of five years with the usual terms and conditions and on the condition that she pay $500 to the restitution fund and reimburse Latasha's family for any out-of-pocket medical expenses and expenses related to Latasha's funeral. Defendant was also ordered to perform 400 hours of community service. The court did not impose any jail time as a condition of probation. The trial judge possessed the option of sentencing Du to prison for three, six, or eleven years and an additional four years for the use of a gun.

What are the objectives of sentencing? Were these goals achieved by a sentence of probation? Under California law, probation is not to be granted for an offense involving the use of a firearm other than under certain conditions. These include whether the crime was committed under unusual circumstances, such as great provocation. Other conditions are whether the carrying out of the crime indicated criminal sophistication and whether the defendant will be a danger to others in the event he or she is not incarcerated. See *People v. Superior Court*, 7 Cal. Rptr. 2d 177 (1992).

You can find the answers at www.sagepub.com/lippmanccl3e.

CHAPTER SUMMARY

The distinguishing characteristic of a criminal offense is that it is subject to punishment. Categorizing a law as criminal or civil has consequences for the protections afforded to a defendant, such as the prohibition against double jeopardy. Punishment is intended to accomplish various goals, including retribution, deterrence, rehabilitation, incapacitation, and restoration. Judges seek to accomplish the purposes of punishment through penalties ranging from imprisonment, fines, probation, and intermediate sanctions to capital punishment. Assets forfeiture may be pursued in a separate proceeding.

The federal government and the states have initiated a major shift in their approach to sentencing. The historical commitment to indeterminate sentencing and to the rehabilitation of offenders has been replaced by an emphasis on deterrence, retribution, and incapacitation. This primarily involves presumptive sentencing guidelines and mandatory minimum sentences. Several recent U.S. Supreme Court cases appear to have resulted in sentencing guidelines that are advisory rather than binding on judges. A sentence, whether inside or outside the guidelines, is required only to be "reasonable."

Truth in sentencing laws are an effort to ensure that offenders serve a significant portion of their sentences and are intended to prevent offenders from being released by parole boards who determine that offenders have exhibited progress toward rehabilitation. We also have seen the development of a greater sensitivity to victims' rights.

Constitutional attacks on sentences are typically based on the Eighth Amendment prohibition on the imposition of cruel and unusual punishment. The U.S. Supreme Court has ruled that the prohibition against cruel and unusual punishment applies to the federal government as well as to the states, and virtually every state constitution contains similar language. Professor Wayne LaFave lists three approaches to interpreting the clause: (1) it limits the *methods* employed to inflict punishment, (2) it restricts the *amount of punishment* that may be imposed, and (3) it *prohibits* the criminal punishment of certain acts. The Equal Protection Clause provides an avenue to challenge statutes and sentencing practices that result in different penalties for individuals based on their race, religion, gender, and ethnicity.

The effort to ensure uniform approaches to sentencing is exemplified by the procedural protections that surround the death penalty. Legal rulings under the Eighth Amendment have limited the application of capital punishment to a narrow range of aggravated homicides committed by adult offenders. The imposition of capital punishment on juveniles was held disproportionate in *Roper v. Simmons*. Constitutional challenges under the Eighth Amendment have proven unsuccessful against mandatory minimum sentences, as illustrated by the federal court's upholding of Three Strikes and You're Out laws and determinate penalties for drug possession. Judges have generally deferred to the decision of legislators and have ruled that penalties for terms of years are proportionate to the offenders' criminal acts. The U.S. Supreme Court has stressed that such challenges should be upheld on "extremely rare" occasions where the sentence is "grossly disproportionate" to the seriousness of the offense. The reconsideration of the disparity in treatment between crack and powder cocaine is a significant step in the lessening the harshness of drug laws.

Criminal sentences may not be based on the "suspect categories" of race, gender, religion, ethnicity, and nationality. Despite the condemnation of racial practices in the criminal justice system, the due process procedures surrounding the death penalty do not appear to have eliminated racial disparities in capital punishment. An equal protection challenge to the application of capital punishment, however, proved unsuccessful in *McCleskey v. Kemp*.

As you read the cases in the next chapters of the textbook, pay attention to the sentences handed down to defendants by the trial court. Consider what sentence you believe the defendant deserved.

CHAPTER REVIEW QUESTIONS

1. Distinguish between civil disabilities and criminal punishments. Why did the Supreme Court rule that the sanction provided for in Megan's Law was not a criminal punishment?

2. Discuss the purposes of punishment. Which do you believe should be the primary reason for criminal punishment?

3. What are some types of sentences that a court may impose?

4. List several of the criteria that are used to evaluate approaches to sentencing. Which do you believe is most important?

5. Describe the various approaches to sentencing. Contrast indeterminate and determinate sentencing. What approach do you favor?

6. Why were sentencing guidelines introduced? How were the guidelines affected by recent Supreme Court decisions?

7. Describe truth in sentencing laws.

8. What types of protections are included within victims' rights?

9. List the three protections provided under the Eighth Amendment.

10. How do courts determine whether a method of punishment is prohibited under the Eighth Amendment? Does lethal injection constitute cruel and unusual punishment?

11. Discuss the efforts of the Supreme Court to ensure that the death penalty is applied in a proportionate fashion.

12. Why did the Supreme Court rule that it is cruel and unusual punishment to execute juveniles?

13. What is the approach of courts that are asked to decide whether a sentence for a "term of years" is proportionate to the crime? Why do judges take such a hands-off approach in this area?

14. Why is it a violation of equal protection for race, gender, religion, ethnicity, or nationality to play a role in sentencing?

15. What is the legal test for determining whether a law that is neutral on its face is in violation of the Equal Protection Clause?

16. Outline the debate over whether the possession of crack cocaine should be punished more severely than the possession of powder cocaine. What is the approach under the federal sentencing guidelines and in most states?

LEGAL TERMINOLOGY

assets forfeiture	indeterminate sentencing	rehabilitation
beyond a reasonable doubt	just deserts	restoration
clemency	mandatory minimum sentences	retribution
concurrent sentences	Megan's Law	selective incapacitation
consecutive sentences	pardon	Son of Sam laws
determinate sentencing	plea bargain	specific deterrence
disparity	preponderance of the evidence	status offenses
Eighth Amendment	presumptive sentencing guidelines	Three Strikes and You're Out law
general deterrence		truth in sentencing laws
incapacitation	proportionality	victim impact statements

CRIMINAL LAW ON THE WEB

Log on to the Web-based student study site at www.sagepub.com/lippmanccl3e to assist you in completing the Criminal Law on the Web exercises, as well as for additional features such as podcasts, Web quizzes, and video links.

1. The Death Penalty Information Center provides information on capital punishment. You will also want to read about the death penalty around the world.

2. Explore various public policy issues in sentencing on the Web site of The Sentencing Project.

3. The Justice Policy Institute has reports on the Three Strikes laws that you can find on the Web site of this nonprofit research organization.

BIBLIOGRAPHY

American Law Institute, *Model Penal Code: Sentencing* (Philadelphia: American Law Institute, 2003). A discussion of the proposal to modify the indeterminate sentencing structure of the Model Penal Code to provide for sentencing guidelines. The report provides a good outline of state sentencing practices.

Lynn S. Branham, *The Law and Policy of Sentencing and Corrections in a Nutshell,* 8th ed. (St. Paul MN: West Publishing, 2010). A summary of the law of sentencing and corrections.

Nicholas N. Kittrie, Elyce H. Zenoff, and Vincent A. Eng, *Sentencing, Sanctions, and Corrections: Federal and State Law, Policy, and Practice,* 2nd ed. (New York: Foundation Press, 2002). A collection of essays and materials that covers the entire range of issues in sentencing. A comprehensive bibliography is included.

Arnold Loewy, *Criminal Law in a Nutshell,* 4th ed. (St. Paul, MN: West Publishing, 2003), pp. 1–26. A short summary of punishment and sentencing.

Jeffrie G. Murphy, *Punishment and Rehabilitation,* 3rd ed. (Belmont, CA: Wadsworth, 1995). A collection of essays on the philosophy of punishment.

4 ACTUS REUS

Did the defendant act voluntarily in killing his brother-in-law?

Family members testified to a substantial history going back to defendant's childhood of defendant's acting as if he were "in his own world." . . . Dr. Harrell clearly testified that in his opinion defendant was unable to exercise conscious control of his physical actions at the moment of the fatal shooting. He stated further, "I think he was acting sort of like a robot. He was acting like an automaton. . . . When he goes into the altered state of consciousness, . . . then he engages in a motor action." This testimony . . . if accepted by the jury, "exclude[d] the possibility of a voluntary act without which there can be no criminal liability. . . ."

Learning Objectives

1. Understand why the criminal law punishes voluntary criminal acts and does not penalize thoughts or involuntary acts.

2. List some examples of involuntary acts.

3. Know why it is unconstitutional to punish a defendant for a status or condition.

4. Describe the circumstances in which an individual may be held liable for an omission to act.

5. Know the definition of possession and the difference between actual and constructive possession.

INTRODUCTION

A crime comprises an *actus reus*, or a criminal act or omission, and a *mens rea*, or a criminal intent. Conviction of a criminal charge requires evidence establishing beyond a reasonable doubt that the accused possessed the required mental state and performed a voluntary act that caused the social harm condemned in the statute.[1]

There must a concurrence between the *actus reus* and *mens rea*. For instance, common law burglary is the breaking and entering of the dwelling house of another at night with the intent to commit a felony. A backpacker may force his or her way into a cabin to escape the sweltering summer heat and, once having entered, find it impossible to resist the temptation to steal hiking equipment. The requisite intent to steal developed following the breaking and entering, and our backpacker is not guilty of common law burglary.[2] The requirement of concurrence is illustrated by the California Penal Code, which provides that "in every crime . . . there must exist a union or joint operation of act and intent. . . ."[3]

Actus reus generally involves three elements or components: (1) a voluntary act or failure to perform an act (2) that causes (3) a social harm condemned under a criminal statute. Homicide, for instance, involves the voluntary shooting or stabbing (act) of another human being that results in (causation) death (social harm).[4] The Indiana Criminal Code provides that a "person commits an offense only if he voluntarily engages in conduct in violation of the statute defining the offense . . . [A] person who omits to perform an act commits an offense only if he has a statutory, common law, or contractual duty to perform the act."[5]

First, keep in mind that an act may be innocent or criminal depending on the context or **attendant circumstances**. Entering an automobile, turning the key, and driving down the highway may be innocent or criminal depending on whether the driver is the owner or a thief. Second, crimes require differing attendant circumstances. An assault on a police officer requires an attack on a law enforcement official; an assault with a dangerous weapon involves the employment of an instrument capable of inflicting serious injury, such as a knife or firearm. A third point is that some offenses require that an act cause a specific harm. Homicide, for instance, involves an act that directly causes the death of the victim, while false pretenses require that an individual obtain title to property through the false representation of a fact or facts. In the case of these so-called **result crimes**, the defendant's act must be the "actual cause" of the resulting harm. An individual who dangerously assaults a victim who subsequently dies may not be guilty of homicide in the event that the victim would have lived and her death was caused by the gross negligence of an ambulance driver.

This chapter covers *actus reus*. We first discuss the requirement of an act and then turn our attention to status offenses, omissions, and possession. Chapter 5 addresses criminal intent, concurrence, and causality, and we will apply these concepts to specific crimes in Chapters 10 through 15. At this point, merely appreciate that a crime consists of various "elements" or components that the prosecution must prove beyond a reasonable doubt.

CRIMINAL ACTS

Scholars have engaged in a lengthy and largely philosophical debate over the definition of an "act." It is sufficient to note that the modern view is that an act involves a bodily movement, whether voluntary or involuntary.[6] The significant point is that the *criminal law punishes voluntary acts and does not penalize thoughts*. Why?

- This would involve an unacceptable degree of governmental intrusion into individual privacy.
- It would be difficult to distinguish between criminal thoughts that reflect momentary anger, frustration, or fantasy, and thoughts involving the serious consideration of criminal conduct.
- Individuals should be punished only for conduct that creates a social harm or imminent threat of social harm and should not be penalized for thoughts that are not translated into action.
- The social harm created by an act can be measured and a proportionate punishment imposed. The harm resulting from thoughts is much more difficult to determine.

An exception to this rule was the historical English crime of "imagining the King's death." How should we balance the interest in freedom of thought and imagination against the social interest in the early detection and prevention of social harm in the case of an individual who records dreams of child molestation in his or her private diary?

A Voluntary Criminal Act

A more problematic issue is the requirement that a crime consist of a **voluntary act**. The Indiana Criminal Law Study Commission, which assisted in writing the Indiana statute on criminal conduct, explains that voluntary simply means a conscious choice by an individual to commit or not to commit an act.[7] In an often-cited statement, Supreme Court Justice Oliver Wendell Holmes, Jr., observed that a "spasm is not an act. The contraction of the muscles must be willed."[8] Professor Joshua Dressler compares an involuntary movement to the branch of a tree that is blown by the wind into a passerby.[9]

The requirement of a voluntary act is based on the belief that it would be fundamentally unfair to punish individuals who do not consciously choose to engage in criminal activity and who therefore cannot be considered morally blameworthy. There also is the practical consideration that there is no need to deter, incapacitate, or rehabilitate individuals who involuntarily engage in criminal conduct.[10]

A North Carolina court of appeals ruled that a jury hearing the case of a defendant charged with taking indecent liberties with his girlfriend's eight-year-old daughter should have been instructed that "if they found that defendant was unconscious or, more specifically, asleep, they must find the defendant not guilty."[11] In another case, the Kentucky Supreme Court ruled that a defendant, who claimed that he was a "sleepwalker," should not be convicted in the event that he was "unconscious when he killed the deceased."[12]

The criminal defense of involuntariness has been unsuccessfully invoked by individuals charged with criminal negligence while operating a motor vehicle. In the frequently cited case of *People v. Decina,* the defendant's automobile jumped a curb and killed four children. The appellate court affirmed Decina's conviction despite the fact that the accident resulted from an epileptic seizure. The judges reasoned that the statute "does not necessarily contemplate that the driver be conscious at the time of the accident" and that it is sufficient that the defendant knew of his medical disability and knew that it would interfere with the operation of a motor vehicle.[13]

Some defendants have actually managed to be acquitted by persuading judges or juries that their crimes were **involuntary acts**. A California court of appeals concluded that the evidence supported the "inference" that a defendant who had been wounded in the abdomen had shot and killed a police officer as a reflex action and was in a "state of unconsciousness."[14]

The next case in the textbook, *State v. Fields,* challenges you to determine whether the defendant's act should be considered involuntary or voluntary.

Model Penal Code

The Model Penal Code section 2.01 defines the Requirement of Voluntary Act as follows:

1. A person is not guilty of an offense unless his liability is based on conduct that includes a voluntary act or the omission to perform an act of which he is physically capable.

2. The following are not voluntary acts within the meaning of this Section:
   a. a reflex or convulsion;
   b. a bodily movement during unconsciousness or sleep;
   c. conduct during hypnosis or resulting from hypnotic suggestion;
   d. a bodily movement that otherwise is not a product of the effort or determination of the actor, either conscious or habitual.

Analysis

The Model Penal Code requires that the guilt of a defendant should be based on conduct that includes a voluntary act or omission. An individual who is aware of a serious heart defect who voluntarily drives an automobile may be liable for harm resulting from an accident he or she causes while having a heart attack, based on the fact that he or she voluntarily drove the automobile or failed to stop as he or she began to feel ill. The Model Penal Code avoids the difficulties involved in defining voluntary conduct and, instead, lists the type of conditions that are not voluntary. Section 2(d) encompasses a range of unspecified conditions.

The Legal Equation

Actus reus = A voluntary act or failure to perform an act.

Voluntary act = A bodily movement that is the product of a conscious choice.

Was the defendant entitled to a jury instruction on unconsciousness?

State v. Fields, 376 S.E.2d 740 (N.C. 1989). Opinion by: Whichard, J.

Defendant was convicted of first degree murder. . . . The trial court sentenced him to life imprisonment. We award a new trial for error in refusing a requested jury instruction.

Facts

The State's evidence, in pertinent summary, showed the following: Connie Williams, defendant's half-sister, testified that she had been dating Isaiah Barnes, the victim, for two years at the time of his death. On September 18, 1986, the couple was drinking liquor at Robert Cobb's house. . . .

Cobb testified that defendant and his girlfriend were at his house when Williams and Barnes arrived. Defendant offered Williams a drink and left soon thereafter. Before leaving, defendant "played some numbers" with Cobb. Williams and Barnes also left Cobb's house, but returned later that evening. Williams and Barnes were sitting on a trunk in Cobb's bedroom, drinking and talking. Cobb and his friend, Joyce Ann Pettaway, also were talking in the bedroom. Defendant entered the bedroom about midnight. He called Cobb by a nickname, "Snow." Defendant asked Cobb to keep the ticket for the numbers he had played, saying, "If I hit, I want you to get the money and keep it until you see me." Cobb asked why defendant could not keep it himself, and defendant answered, "You'll see."

Defendant and Barnes had not spoken to one another . . . defendant then walked around the foot of the bed, pulled a gun out of his belt, and shot Barnes. Barnes fell on the floor. Pettaway cried, "Oh, Lord have mercy. Please don't shoot that man anymore." Defendant turned toward her and said, "Shut up," then shot Barnes again as he lay on the floor gasping for breath. Cobb told defendant to get out of his house because he was calling "the law." Defendant said, "Okay, Snow," and walked out.

Wallace Fields, defendant's brother, testified on defendant's behalf. He recounted the difficult circumstances of their childhood. Their stepfather, called "Dump," drank regularly and beat the children and their mother. They had little money and were often hungry. When Dump was on a rampage, the mother and children would often sleep outside to avoid him.

One night when defendant was fourteen, Dump held a knife to defendant's mother's throat and threatened to kill her. Defendant grabbed a gun and shot Dump, killing him. Wallace Fields testified that up to the time of this incident, defendant was a normal boy who liked to play and go to school. After the shooting, defendant had nightmares and became "a different person," acting as if he were "in his own world." Defendant was extremely devoted to his mother, helping her cook and clean and giving her money.

Wallace Fields further testified that his sister, Connie Williams, had become a different person since beginning her relationship with Barnes. She often appeared bruised and beaten and cared little for her appearance.

Willard Mills, defendant's stepbrother, also testified regarding defendant's devotion to his mother. Mills stated that Connie Williams became dependent on alcohol or drugs and lost all interest in her family and appearance after she became involved with Barnes. Defendant and his brothers were worried about Connie and frequently discussed how to help her.

Mills reported that defendant had become very morose after the childhood shooting incident. As defendant grew older, Mills advised him to put it all behind him and join the service. While in service, another soldier performed a trick in which the soldier put lighter fluid in his mouth, lit it, and blew out the flames. Defendant saw his stepfather's face in the flames and ran away. He was hospitalized for several months following this episode. Mills testified that defendant was concerned about Barnes's drinking and tried to persuade him to stop, but that defendant bore Barnes no malice. Ten days after shooting Barnes, defendant called Mills. Mills picked up defendant at the bus station and took him to the Tarboro police station to turn himself in.

Agnes Williams, defendant's mother, testified that defendant's nerves had been bad ever since the incident with Dump. Defendant had a nervous breakdown in the service and was never the same afterward. Defendant's nerves were "just racked all to pieces" over Connie Williams's problems.

Dr. Evans Harrell, a psychologist, testified on defendant's behalf. . . . He stated that after defendant killed Dump, he felt very protective toward his mother and sisters. Defendant felt guilty about the family being left without a father figure, and he tried to assume that role. Defendant suffered from frequent nightmares featuring Dump and often felt Dump's presence even when awake. In Dr. Harrell's opinion, defendant suffered from post-traumatic stress disorder, and certain aspects of his behavior were characteristic of a disassociative state. Dr. Harrell described a disassociative state as a sudden temporary alteration in the state of consciousness, during which defendant would not remember what happened and did not intend to do anything, "like his mind and his body weren't connected."

Dr. Harrell recounted what defendant related to him about the killing of Isaiah Barnes. Defendant told Dr. Harrell he had tried to get his sister Connie to leave Cobb's house that night, because he was worried about her drinking. Connie's arm was bandaged from a burn, which she attributed to an accident but which defendant and the family suspected Barnes inflicted. Defendant saw Barnes reach out and grab Connie, and Connie grimaced in pain. At this point defendant pulled out the gun and shot Barnes. Defendant told Dr. Harrell he had not planned to kill Barnes, had not thought of killing Barnes, and even as he shot him, was not thinking of killing Barnes. Defendant denied any memory of firing a second shot. Instead, defendant was seeing Dump and his mother "and all of these things flashing before [him] in a blur."

Dr. Harrell testified that defendant perceived Barnes to be treating Connie the same way Dump had treated defendant's mother. . . . In Dr. Harrell's opinion, defendant did not plan or intend to shoot Barnes and was unable to exercise conscious control of his physical actions at that moment. Dr. Harrell concluded, "I think he was acting sort of like a robot. He was acting like an automaton." Dr. Harrell testified that defendant told him he had been drinking on the night of the shooting but did not tell him how much he had had to drink.

Issue

Defendant assigns error to the trial court's refusal to instruct the jury on the defense of unconsciousness. This defense, also called automatism, has been defined as connoting the state of a person who, though capable of action, is not conscious of what he is doing. It is to be equated with unconsciousness, involuntary action [and] implies that there must be some attendant disturbance of conscious awareness. Undoubtedly automatic states exist, and medically they may be defined as conditions in which the patient may perform simple or complex actions in a more or less skilled or uncoordinated fashion without having full awareness of what he is doing. . . .

Reasoning

Defendant's evidence tended to show that immediately preceding and during the killing of his victim, he was unconscious. Family members testified to a substantial history going back to defendant's childhood of defendant's acting as if he were "in his own world." In the context of this testimony, and on the basis of a personal and family history obtained from defendant and members of his family, Dr. Harrell testified that in his opinion defendant suffered from post-traumatic stress disorder and was prone to experiencing disassociative states. In Dr. Harrell's opinion, defendant was in a disassociative state when he shot the victim. Dr. Harrell testified that the defendant "was acting sort of like a robot" and "like an automaton." . . . "When Isaiah reached out and grabs Connie's arm and Connie grimaces, and his whole past life and material that is so similar in his mind to what he's seen, flashes before him, then he engages in a motor action. . . ."

This testimony, if believed, permits a jury finding that defendant was unable to exercise conscious control of his physical actions when he shot the victim. The defendant thus was entitled to the unconsciousness or automatism jury instruction stating that the defense of unconsciousness does not apply to a case in which the mental state of the person in question is

due to insanity, mental defect or voluntary intoxication resulting from the use of drugs or intoxicating liquor, but applies only to cases of the unconsciousness of persons of sound mind as, for example, sleepwalkers or persons suffering from the delirium of fever, epilepsy, a blow on the head or the involuntary taking of drugs or intoxicating liquor, and other cases in which there is no functioning of the conscious mind and the person's acts are controlled solely by the subconscious mind. . . .

Holding

The defendant's evidence . . . merited the requested instruction on unconsciousness or automatism. . . . As noted above, family members testified to a substantial history going back to defendant's childhood of defendant's acting as if he were "in his own world." . . . Dr. Harrell clearly testified that in his opinion defendant was unable to exercise conscious control of his physical actions at the moment of the fatal shooting. He stated further, "I think he was acting sort of like a robot. He was acting like an automaton. . . . When he goes into the altered state of consciousness, . . . then he engages in a motor action." This testimony, combined with the family members' testimony, if accepted by the jury, "exclude[d] the possibility of a voluntary act without which there can be no criminal liability. . . ." Therefore, an instruction on the legal principles applicable to the unconsciousness or automatism defense was required. . . . .

Questions for Discussion

1. Why is unconsciousness or automatism a criminal defense? What facts support the defendant's contention that the killing of Barnes was an unconscious automatic act?

2. What facts undermine the contention that Fields shot Barnes as an act of unconsciousness or automatism? Why, for instance, was Fields carrying a pistol? Did he have a motive for killing Barnes? Was his conduct consistent with an individual displaying "disassociation"? Are the witnesses cited by the court neutral and objective or biased? Were you persuaded by Dr. Harrell's testimony?

3. Should unconsciousness or automatism reduce the seriousness of an offense rather than constitute an absolute defense? What about holding Fields guilty for reckless homicide based on the fact that he was irresponsible for not seeking medical treatment for his severe emotional problems?

4. The North Carolina Supreme Court distinguishes unconsciousness or automatism from insanity. What is the difference between these two concepts? Under what circumstances is a defendant entitled to the jury receiving an unconsciousness or automatism jury instruction?

Cases and Comments

Sleepwalking. The first American case to recognize the defense of sleepwalking or somnambulism involved the acquittal in 1846 of Albert Tirrell for the murder of his mistress and the arson of her Boston brothel. This was followed in 1879 by the Kentucky case, *Fain v. Commonwealth*, 78 Ky. 183, in which sleepwalking was recognized as a defense for homicide. Over 120 years later, Adam Kieczykowski, age nineteen, was acquitted of the burglary of a number of dorm rooms at the University of Massachusetts Amherst and of sexual assault. In 2003, twenty-four-year-old Marc Reider was acquitted of aggravated manslaughter based on a jury's determination that he was "sleep-driving." An even more startling case was an English jury's 2005 acquittal of British bartender James Bilton for raping a woman three times.

In 2002, Timothy Stowell was convicted by a California appellate court of digital penetration and of lewd actions "upon a four-year-old female." Stowell was found in bed with Tracie and her daughter Taylor, who were spending the night with Stowell and his girlfriend LeeAnne. Stowell and LeeAnne were sleeping in the living room while Tracie and Taylor slept in the bedroom. At trial, Stowell testified that he did not know "how he came to be in bed" with Taylor and that the last thing that he recalled from that evening was watching a film on television and then "waking up to Tracie yelling, 'What the hell is going on?'"

The appellate court affirmed the trial court judge's refusal to instruct the jury that they should acquit Stowell if after reviewing the evidence "you have a reasonable doubt that the defendant was conscious at the time the alleged crime was committed." The appellate court noted that while Stowell and LeeAnne had testified that Stowell had walked in his sleep on various occasions, he did not claim that he had been sleepwalking on the night of the molestation. There also was no testimony that he had engaged in similar types of sexual behavior when he previously had been sleepwalking. At the time of his arrest, Stowell did not explain to the police that he had been sleepwalking, and he raised the sleepwalking defense for the first time at trial. The last point made by the appellate court was that while studies indicate that sleepwalkers are able to unlock doors and operate machinery, there was no expert testimony presented at trial documenting that a sleepwalker could have undressed and sexually molested a four-year-old. The appellate court concluded that Stowell's claim that he had been sleepwalking and had unconsciously molested Taylor simply was not supported by substantial evidence.

Should the criminal law recognize the "involuntariness" defense of sleepwalking? See *People v. Stowell*, 202 WL 1068259 (Cal. Ct. App. 2002).

You Decide

4.1 Brown, while drinking beer and talking with friends in the parking lot of an apartment complex, became involved in an argument with James McLean. One week earlier, Brown had been badly beaten in a fight with McLean. Brown purchased a .25 caliber handgun to protect himself and his friends from McLean, who was known to "possess and discharge firearms in the vicinity of the apartment complex" where Brown lived.

Brown, on the day in question, got into a heated exchange with McLean. Brown, who is right-handed, testified that he held the firearm in his left hand because of an injury that he had experienced to his right hand. He testified that while raising the handgun, the gun accidentally fired when he was bumped from behind by Coleman. The shot fatally wounded one of Brown's friends, Joseph Caraballo. Did Brown involuntarily fire the shot that killed Caraballo? What is your opinion? See *Brown v. State*, 955 S.W.2d 276 (Tex.1997).

You can find the answer at www.sagepub.com/lippmanccl3e.

STATUS OFFENSES

Can you be criminally convicted of being a drunk or drug addict or a common thief? Or for a violent personality? The commentary to Model Penal Code section 2.01 stresses that a crime requires an act and that individuals may not be punished based on a mere status or condition. The code cites as an example the 1962 U.S. Supreme Court decision in *Robinson v. California*, which, as you might recall from Chapter 3, held that it was cruel and unusual punishment under the Eighth and Fourteenth Amendments to convict Robinson of the **status offense** of being "addicted to the use of narcotics."[15]

In *Robinson*, Los Angeles police officers observed scar tissue and discoloration on Robinson's arms that was consistent with the injection of drugs. Robinson was convicted by a jury under a statute that declared it a misdemeanor for a person "either to use narcotics or to be addicted to the use of narcotics." The verdict was based on Robinson's status as a narcotics addict rather than for the act of using narcotics or for other illegal acts such as the manufacture, selling, transport, or purchase of narcotics. The Supreme Court reversed Robinson's criminal conviction and condemned the fact that Robinson could be held "continuously guilty . . . whether or not he ever used or possessed any narcotics within the State, and whether or not he has been guilty of any antisocial behavior [in California]."

Holding Robinson liable for his status as an addict seemed particularly unfair given that four judges viewed narcotics addiction as a disease. Justice Potter Stewart explained that it was "cruel and unusual punishment to condemn Robinson for an illness, which like being mentally or physically challenged or leprosy, may be contracted innocently or involuntarily."

Six years later, the U.S. Supreme Court issued a decision in *Powell v. Texas*. Powell was arrested for "being found in a state of intoxication in a public place."[16] The Texas law was aimed at preventing the disruptive behavior accompanying public drunkenness. Powell claimed that his conviction was based on his status as a chronic alcoholic and that the law constituted cruel and unusual punishment. The Supreme Court rejected this argument and ruled in a five-to-four decision that Powell was not convicted for being a chronic alcoholic, but for his public behavior that posed "substantial health and safety hazards, both to himself and for members of the general public." Justice White was the fifth justice in the *Powell* majority. White agreed with the dissent that alcoholism was a disease but concurred in the majority decision based on his determination that Powell was capable of "making plans to avoid his being found drunk in public." Justice White cautioned that he believed that it would constitute cruel and unusual punishment to convict and punish chronic alcoholics who were homeless, because such individuals "have no place else to go and no place else to be when they are drinking."

The Supreme Court majority in *Powell* dismissed the argument that alcoholism was a disease similar to drug addiction and that *Robinson*'s condemnation of status offenses should be expanded to encompass Powell's irresistible compulsion to appear drunk in public. A shift of Justice White's vote would have resulted in a different outcome. Would this prove a dangerous doctrine? The next step might involve sex offenders claiming that their criminal acts are an expression of a sexually dysfunctional personality or violent offenders contending that their behavior resulted from their antisocial personalities.

In *People v. Kellogg*, the Superior Court of San Diego, California, confronted an appellant who contended that his criminal convictions for public intoxication unconstitutionally punished him for his status as a homeless, chronic alcoholic. Is Kellogg's claim persuasive?

Was Kellogg convicted for the status of being a homeless, chronic alcoholic?

People v. Kellogg, 14 Cal. Rptr. 3d 507 (Cal. Ct. App. 2004). Opinion by: Haller, J.

Facts

On January 10, 2002, Officer Heidi Hawley, a member of the Homeless Outreach Team, responded to a citizen's complaint of homeless persons camping under bridges and along State Route 163. She found Kellogg sitting on the ground in some bushes on the embankment off the freeway. Kellogg appeared inebriated and was largely incoherent. He was rocking back and forth, talking to himself, and gesturing. Officer Hawley arrested Kellogg for public intoxication. He had $445 in his pocket from disability income.

In February 2001, Kellogg had accepted an offer from the Homeless Outreach Team to take him to Mercy Hospital. However, on three other occasions when Officer Hawley had offered Kellogg assistance from the Homeless Outreach Team, he had refused.

After his arrest on January 10, 2002, Kellogg posted $104 cash bail and was released. Because he was homeless, he was not notified of his court date, and he did not appear for his January thirty-first arraignment. A warrant for his arrest was issued on February 11, 2002; he was arrested again for public intoxication on February nineteenth and twenty-seventh and subsequently charged with three violations of section 647(f), prohibiting public intoxication.

After a pretrial discussion in chambers about Kellogg's physical and psychological problems, the trial court conditionally released Kellogg on his own recognizance and ordered that he be escorted to the Department of Veterans Affairs hospital (VA) by Officer Hawley. He was not accepted for admission at the hospital and accordingly was returned to county jail. Kellogg pleaded not guilty and filed a motion to dismiss the charges based on his constitutional right to be free of cruel and unusual punishment.

Psychologist Gregg Michel and psychiatrist Terry Schwartz testified on behalf of Kellogg. These experts explained that Kellogg had a dual diagnosis. In addition to his severe alcohol dependence, which caused him to suffer withdrawal symptoms if he stopped drinking, he suffered from dementia, long-term cognitive impairment, schizoid personality disorder, and symptoms of post-traumatic stress disorder. He had a history of seizure disorder and a closed head injury, and he reported anxiety, depressive symptoms, and chronic pain. He was estranged from his family. Physically, he had peripheral edema, gastritis, acute liver damage, and ulcerative colitis requiring him to wear a colostomy bag. To treat his various conditions and symptoms, he had been prescribed Klonopin and Vicodin, and it was possible that he suffered from addiction to medication.

Dr. Michel opined that Kellogg was gravely disabled and incapable of providing for his basic needs and that his degree of dysfunction was life-threatening. His mental deficits impeded his executive functioning (planning, making judgments) and memory. . . . Drs. Michel and Schwartz opined that Kellogg's homelessness was not a matter of choice but a result of his gravely disabled mental condition. His chronic alcoholism and cognitive impairment made it nearly impossible for him to obtain and maintain an apartment without significant help and support. . . . Dr. Schwartz explained that for a person with Kellogg's conditions, crowded homeless shelters can be psychologically disturbing and trigger post-traumatic stress or anxiety symptoms, causing the person to prefer to hide in a bush where minimal interactions with people would occur. Additionally, a homeless person such as Kellogg, particularly when intoxicated, might refuse offers of assistance from authorities, because he has difficulty trusting people and fears his situation, although bad at present, will worsen.

In Dr. Michel's view, Kellogg's incarceration provided some limited benefit in that he obtained medication for seizures, did not have access to alcohol, received some treatment, and was more stable during incarceration than he was when homeless on the streets. However, such treatment was insufficient to be therapeutic, and medications prescribed for inmate management purposes can be highly addictive and might not be medically appropriate.

Testifying for the prosecution, physician James Dunford stated that at the jail facility, medical staff assess the arrestee's condition and provide treatment as needed, including vitamins for nutritional needs and medication to control alcohol withdrawal symptoms or other diseases such as hypertension, seizure disorders, and diabetes. . . . Dr. Dunford opined that between March second and seventh, Kellogg's condition had improved, because his seizure medicine was restarted, his alcohol withdrawal was treated, his vital signs were stable, his colostomy bag was clean and intact, his overall cleanliness was restored, and he was interacting with people in a normal way.

After the presentation of evidence, the trial court found that Kellogg suffered from both chronic alcohol dependence and a mental disorder and was homeless at the time of his arrests. Further, his alcohol dependence was both physical and psychological and caused him to be unable to stop drinking or to engage in rational choice-making. Finding that before his arrest Kellogg was offered assistance on at least three occasions and that his medical condition improved while in custody, the court denied the motion to dismiss the charges. On April 2, 2002, the court found Kellogg guilty of one charge of violating section 647(f) arising from his conduct on January 10, 2002. At sentencing on April thirtieth, the probation officer requested that the hearing be continued for another month, so Kellogg could be evaluated for a possible conservatorship.

After expressing the difficult "Hobson's choice" whereby there were no clear prospects presented to effectively assist Kellogg, the court sentenced him to 180 days in jail, with execution of sentence suspended for three years on the condition that he complete an alcohol treatment program and return to court on June 4, 2002, for a progress review. . . .

After Kellogg's release from jail, defense counsel made extensive, but unsuccessful, efforts to place Kellogg in an appropriate program and to find a permanent residence for him. On May 25 and 28, 2002, he was again arrested for public intoxication. After he failed to appear at his June fourth review hearing, his probation was summarily revoked. Kellogg was rearrested on June twelfth. After a probation revocation hearing, Kellogg's probation was formally revoked, and he was ordered to serve the 180-day jail sentence. The court authorized that his sentence be served in a residential rehabilitation program. However, no such program was found. According to defense counsel, the VA concluded Kellogg could not benefit from its residential treatment program due to his cognitive defects. Further, his use of prescribed, addictive narcotics precluded placement in other residential treatment programs, and his iliostomy precluded placement in board and care facilities. On July 11, 2003, the appellate division of the superior court affirmed the trial court's denial of Kellogg's motion to dismiss on Eighth Amendment grounds. We granted Kellogg's request to have the matter transferred to this court for review.

Issue

Section 647(f) defines the misdemeanor offense of disorderly conduct by public intoxication as occurring when a person

> is found in any public place under the influence of intoxicating liquor . . . in such a condition that he or she is unable to exercise care for his or her own safety or the safety of others, or by reason of his or her being under the influence of intoxicating liquor . . . interferes with

or obstructs or prevents the free use of any street, sidewalk, or other public way.

Kellogg argues that this statute, as applied to him, constitutes cruel and/or unusual punishment prohibited by the Eighth Amendment to the U.S. Constitution and article 1 section 17 of the California Constitution. He asserts that his chronic alcoholism and mental condition have rendered him involuntarily homeless and that it is impossible for him to avoid being in public while intoxicated. He argues because his public intoxication is a result of his illness and beyond his control, it is inhumane for the State to respond to his condition by subjecting him to penal sanctions.

Reasoning

It is well settled that it is cruel and unusual punishment to impose criminal liability on a person merely for having the disease of addiction. In *Robinson v. California* 370 U.S. 660, 666–667 (1962), the U.S. Supreme Court invalidated a California statute that made it a misdemeanor to "be addicted to the use of narcotics." The *Robinson* Court recognized that a state's broad power to provide for the public health and welfare made it constitutionally permissible for it to regulate the use and sale of narcotics, including, for example, such measures as penal sanctions for addicts who refuse to cooperate with compulsory treatment programs. But the Court found the California penal statute unconstitutional, because it did not require possession or use of narcotics, or disorderly behavior resulting from narcotics, but rather imposed criminal liability for the mere status of being addicted. *Robinson* concluded that just as it would be cruel and unusual punishment to make it a criminal offense to be mentally ill or a leper, it was likewise cruel and unusual to allow a criminal conviction for the disease of addiction without requiring proof of narcotics possession or use or antisocial behavior.

In *Powell v. Texas*, 392 U.S. 514 (1968), the U.S. Supreme Court, in a five-to-four decision, declined to extend *Robinson's* holding to circumstances where a chronic alcoholic was convicted of public intoxication, reasoning that the defendant was not convicted merely for being a chronic alcoholic but rather for being in public while drunk. That is, the State was not punishing the defendant for his mere status, but rather was imposing "a criminal sanction for public behavior which may create substantial health and safety hazards, both for [the defendant] and for members of the general public." . . . In the plurality decision, four justices rejected the proposition set forth by four dissenting justices that it was unconstitutional to punish conduct that was " involuntary" or "occasioned by a compulsion."

The fifth justice in the *Powell* plurality, Justice White, concurred in the result only, concluding that the issue of involuntary or compulsive behavior could be pivotal to the determination of cruel and unusual punishment, but the record did not show the defendant (who had a home) suffered from any inability to refrain from drinking in public. Justice White opined that punishing a homeless alcoholic for public drunkenness could constitute unconstitutional punishment if it was impossible for the person to resist drunkenness in a public place. Relying on Justice White's concurring opinion, Kellogg argues that Justice White, who was the deciding vote in *Powell*, would have sided with the dissenting justices had the circumstances of his case (i.e., an involuntarily homeless chronic alcoholic) been presented, thus resulting in a finding of cruel and unusual punishment by a plurality of the Supreme Court.

We are not persuaded. Although in *Robinson* the U.S. Supreme Court held it was constitutionally impermissible to punish for the mere condition of addiction, the Court was careful to limit the scope of its decision by pointing out that a state may permissibly punish disorderly conduct resulting from the use of narcotics. This limitation was recognized and refined by the plurality opinion in *Powell*, where the Court held it was permissible for a state to impose criminal punishment when the addict engages in conduct that spills into public areas. . . .

Here, the reason Kellogg was subjected to misdemeanor culpability for being intoxicated in public was not because of his condition of being a homeless alcoholic but rather because of his conduct that posed a safety hazard. If Kellogg had merely been drunk in public in a manner that did not pose a safety hazard (i.e., if he were able to exercise care for his own and the public's safety and was not blocking a public way), he could not have been adjudicated guilty under section 647(f). The state has a legitimate need to control public drunkenness when such drunkenness creates a safety hazard. It would be neither safe nor humane to allow intoxicated persons to stumble into busy streets or to lie unchecked on sidewalks, driveways, parking lots, streets, and other such public areas where they could be trampled upon, tripped over, or run over by cars. The facts of Kellogg's public intoxication in the instant case show a clear potential for such harm. He was found sitting in bushes on a freeway embankment in an inebriated state. It is not difficult to imagine the serious possibility of danger to himself or others had he wandered off the embankment onto the freeway. . . .

Holding

We conclude that the California legislature's decision to allow misdemeanor culpability for public intoxication, even as applied to a homeless chronic alcoholic such as Kellogg, is neither disproportionate to the offense nor inhumane. In deciding whether punishment is unconstitutionally excessive, we consider the degree of the individual's personal culpability as compared to the amount of punishment imposed. To the extent Kellogg has no choice but to be drunk in public given the nature of his impairments, his culpability is low; however, the penal sanctions imposed on him under section 647(f) are correspondingly low. Given the state's interest in providing for the safety of its citizens, including Kellogg, imposition of low-level criminal sanctions for Kellogg's conduct does not tread on the federal or state constitutional proscriptions against cruel and/or unusual punishment.

In presenting his argument, Kellogg points to the various impediments to his ability to obtain shelter and effective treatment, apparently caused by a myriad of factors including the nature of his condition and governmental policies and resources, and asserts that these impediments do not justify criminally prosecuting him. He posits that the Eighth Amendment "mandates that society do more for [him] than prosecute him criminally and repeatedly incarcerate him for circumstances which are beyond his control." We are sympathetic to Kellogg's plight; however, we are not in a position to serve as policy maker to evaluate societal deficiencies and amelioration strategies. . . . The judgment is affirmed.

Dissenting, *McDonald, J.*

Because Kellogg is involuntarily homeless and a chronic alcoholic with a past head injury who suffers from dementia, severe cognitive impairment, and a schizoid personality disorder, and there is no evidence he was unable by reason of his intoxication to care for himself or others, other than inability inherent in intoxication, or interfered in any manner with a public way, his section 647(f) conviction solely for being intoxicated in public constitutes cruel and unusual punishment in violation of the Eighth and Fourteenth Amendments. Accordingly, the trial court erred by denying Kellogg's motion to dismiss the section 647(f) charge against him. . . .

Although the People assert that incarceration of Kellogg provides him with treatment similar to or better than he would receive were he civilly committed, the quality of his treatment in jail does not prevent his criminal conviction from constituting cruel and unusual punishment in violation of the Eighth and Fourteenth Amendments. As Justice Fortas stated in his dissenting opinion in *Powell,*

> It is entirely clear that the jailing of chronic alcoholics is punishment. It is not defended as therapeutic, nor is there any basis for claiming that it is therapeutic (or indeed a deterrent). The alcoholic offender is caught in a

"revolving door"—leading from arrest on the street through a brief, unprofitable sojourn in jail, back to the street and, eventually, another arrest. The jails, overcrowded and put to a use for which they are not suitable, have a destructive effect upon alcoholic inmates. . . .

The scope of the California Constitution's prohibition of cruel or unusual punishment is not well-defined. . . . Nevertheless, the California Supreme Court has consistently followed the principle that "a sentence is cruel or unusual as applied to a particular defendant . . . [when] the punishment shocks the conscience and offends fundamental notions of human dignity." . . . In this case, the focus is on the nature of the offense and the offender in discussing whether Kellogg's public intoxication conviction is cruel or unusual punishment under article I section 17 of the California Constitution. . . .

A section 647(f) public intoxication offense, both in the abstract and as committed by Kellogg, is a nonviolent, fairly innocuous offense. . . . It is a nonviolent offense, does not require a victim, and poses little, if any, danger to society in general. As committed by Kellogg, the offense was nonviolent, victimless, and posed no danger to society. Kellogg was found intoxicated sitting under a bush in a public area. He was rocking back and forth, talking to himself, and gesturing. The record does not show that Kellogg's public intoxication posed a danger to other persons or society in general. His motive in drinking presumably was merely to fulfill his physical and psychological compulsion as an alcoholic to become intoxicated. Because Kellogg is involuntarily homeless and did not have the alternative of being intoxicated in private, he did not have any specific purpose or motive to be intoxicated in a public place. Rather, it was his only option. . . . As an involuntarily homeless person, Kellogg cannot avoid appearing in public. As a chronic alcoholic, he cannot stop drinking and being intoxicated. Therefore, Kellogg cannot avoid being intoxicated in a public place.

Based on the nature of the offense and the offender, Kellogg's section 647(f) public intoxication conviction "shocks the conscience and offends fundamental notions of human dignity," and therefore constitutes cruel or unusual punishment in violation of . . . the . . . U.S. Constitution and . . . California Constitution. I would reverse the judgment.

Questions for Discussion

1. Summarize the opinions of the U.S. Supreme Court in *Robinson v. California* and in *Powell v. Texas.*

2. Why does Kellogg argue that his arrest and incarceration constitute unconstitutional cruel and unusual punishment? How does the California court respond to this argument? What is the basis for Judge McDonald's dissent? Is the majority or the dissent more consistent with Supreme Court precedents?

3. In terms of public policy, discuss the impact of the majority decision on the criminal justice system. How would the minority decision affect public policy?

See more cases on the study site: Powell v. Texas, **www.sagepub.com/lippmanccl3e.**

OMISSIONS

Can you be held criminally liable for a failure to act? For casually stepping over the body of a dying person who is blocking the entrance to your favorite coffee shop? The Model Penal Code, as we have seen, requires that criminal conduct be based on a "voluntary act or omission to perform an act of which [an individual] is physically capable." An **omission** is a failure to act or a "negative act."

The criminal law is generally concerned with punishing individuals who engage in voluntary acts that violate the law. The law, on occasion, imposes a duty or obligation on individuals to act and punishes a failure to act. For example, we are obliged to pay taxes, register for the draft, serve on juries, and report an accident. These duties are required in the interests of society and are limited exceptions to the requirement that a crime requires a voluntary act.

The American and European Bystander Rules

The basic rule in the United States is that an individual is not legally required to assist a person who is in peril. This principle was clearly established in 1907 in *People v. Beardsley*. The Michigan Supreme Court ruled that the married Beardsley was not liable for failing to take steps to ensure the safety of Blanche Burns, a woman with whom he was spending the weekend. The court explained that the fact that Burns was in Beardsley's house at the time she overdosed on drugs and alcohol did not create a legal duty to assist her.[17] The Michigan judges cited in support of this verdict the statement of U.S. Supreme Court Justice Joseph Stephen Field that it is "undoubtedly the moral duty of every person to extend to others assistance when in danger . . . and, if such efforts should be omitted . . . he would by his conduct draw upon himself the just censure and reproach of good men; but this is the only punishment to which he would be subjected by society."[18] Chief Justice Carpenter of the New Hampshire Supreme Court earlier had recognized that an individual did not possess a duty to rescue a child standing in the path of an oncoming train. Justice Carpenter noted that "if he does not, he may . . . justly be styled a ruthless savage and a moral monster; but he is not liable in damages for the child's injury, or indictable under the statute for its death."[19]

This so-called **American bystander rule** contrasts with the **European bystander rule** common in Europe that obligates individuals to intervene. Most Americans would likely agree that an Olympic swimmer is morally obligated to rescue a young child drowning in a swimming pool. Why then is this not recognized as a legal duty in the United States? There are several reasons for the American bystander rule:[20]

- Individuals intervening may be placed in jeopardy.
- Bystanders may misperceive a situation, unnecessarily interfere, and create needless complications.
- Individuals may lack the physical capacity and expertise to subdue an assailant or to rescue a hostage and place themselves in danger. This is the role of criminal justice professionals.

- The circumstances under which individuals should intervene and the acts required to satisfy the obligation to assist another would be difficult to clearly define.
- Criminal prosecutions for a failure to intervene would burden the criminal justice system.
- Individuals in a capitalist society are responsible for their own welfare and should not expect assistance from others.
- Most people will assist others out of a sense of moral responsibility, and there is no need for the law to require intervention.

Critics of the American bystander rule contend that there is little difference between pushing a child onto the railroad tracks and failing to intervene to ensure the child's safety and that criminal liability should extend to both acts and omissions. This also would deter crime, because offenders may be reluctant to commit crimes in situations in which they anticipate that citizens will intervene. We can see how the readiness of passengers to confront terrorists on airplanes has prevented several attacks, most notably in the case of the "shoe bomber," Richard Reed. The Good Samaritan rule also assists in promoting a sense of community and regard for others.[21]

The conflict between law and morality was starkly presented in 1964 when thirty-eight residents of New York City were awakened by the desperate screams of Kitty Genovese, a twenty-eight-year-old woman returning home from work. Kitty parked her car in a lot roughly one hundred feet from her apartment and was confronted by Winston Moseley, a married father of two young children, who later would testify that he received emotional gratification from stalking women. The thirty-eight residents of the building turned on their lights and opened their windows and watched as Moseley returned on three separate occasions over a period of thirty-five minutes to stab Kitty seventeen times. The third time Moseley returned, he found that Kitty had crawled to safety inside a nearby apartment house, and he stabbed her in the throat to prevent her from screaming, attempted to rape her, and took $49 from her wallet. One person found the courage to persuade a neighbor to call the police, who arrived in two minutes to find Kitty's dead body. This event profoundly impacted America. Commentators asked whether we had become a society of passive bystanders who were concerned only with our own welfare.[22]

The Duty to Intervene

American criminal law does not impose a general duty on the individuals witnessing the murder of Kitty Genovese to intervene. There is a duty, however, to assist another under certain limited conditions.[23] The primary requirement is that a duty must be recognized under either the common law or a statute.

- *Status.* The common law recognized that individuals possess an obligation to assist their child, spouse, or employee. In *State v. Mally,* the defendant was convicted of "hastening" the death of his wife who had fallen and broken both of her arms, precipitating severe shock and the degeneration of her kidneys. Michael Mally left his wife, Kay, alone in bed for two days, bothering only to provide her with a single glass of water. A Montana district court held that "the failure to obtain medical aid for one who is owed a duty is a sufficient degree of negligence to constitute involuntary manslaughter provided death results from the failure to act."[24]
- *Statute.* A **duty to intervene** may be created by a statute that imposes a duty of care. This may be a criminal statute requiring that a doctor report child abuse or a statute that sets forth the obligations of parents. In *Craig v. State,* the defendants followed the dictates of their religion and treated their child's fatal illness with prayer rather than medicine and were subsequently convicted of failing to obtain medical care for their now-deceased six-year-old daughter. The court ruled that the parents had breached their duty under a statute that provided that a father and mother are jointly and individually responsible for the "support, care, nurture, welfare and education of their minor children." The statute failed to mention medical care, but the court had "no hesitancy in holding that it is embraced within the scope of the broad language used."[25]
- *Contract.* An obligation may be created by an agreement. An obvious example is a babysitter who agrees to care for children or a lifeguard employed to safeguard swimmers.

In *Commonwealth v. Pestinikas,* Walter and Helen Pestinikas verbally agreed to provide shelter, food, and medicine to ninety-two-year-old Joseph Kly, who had been hospitalized with a severe weakness of the esophagus. Kly agreed to pay the Pestinikas $300 a month in return for food, shelter, care, and medicine. Kly was found dead of dehydration and starvation roughly nineteen months later. A Pennsylvania superior court ruled that although failure to provide food and medicine could not have been the basis for prosecuting a stranger who learned of Kly's condition, a "duty to act imposed by contract is legally enforceable and, therefore, creates a legal duty."[26]

- *Assumption of a Duty.* An individual who voluntarily intervenes to assist another is charged with a duty of care. In *People v. Oliver,* Oliver, knowing that Cornejo was extremely drunk, drove him from a bar to Oliver's home, where she assisted him to inject drugs. Cornejo collapsed on the floor, and Oliver instructed her daughter to drag Cornejo's body outside and hide him behind a shed. The next morning Cornejo was discovered dead. A California superior court ruled that by taking Cornejo into her home, Oliver "took charge of a person unable to prevent harm to himself," and she "owed Cornejo a duty" that she breached by failing to summon medical assistance.[27]
- *Creation of Peril.* An individual who intentionally or negligently places another in danger has a duty of rescue. In *Jones v. State,* the defendant raped a twelve-year-old girl who almost immediately jumped or fell off a bridge into a stream. The defendant waded into the water, but neglected to rescue the young woman. The court asked, "Can it be doubted that one who by his own overpowering criminal act has put another in danger of drowning has the duty to preserve her life?"[28]
- *Control.* An individual has a duty to direct and to care for those under his or her supervision and command, including employees or members of the military. A California criminal statute provides that parents or legal guardians of any person under eighteen "shall have the duty to exercise reasonable care, supervision, protection, and control over their minor child."[29] The California Supreme Court noted that this act was part of an effort to combat gangs and that the law applies to parents who intentionally or with criminal negligence fail to fulfill their duty to control their child and, as a result, contribute to child delinquency.[30]
- *Property Owner.* Property owners owe a duty of care to those invited onto their land. The defendants, in *Commonwealth v. Karetny,* operated a nightclub on a pier in Philadelphia, knowing that the pier was in imminent danger of collapse. The pier subsequently collapsed, killing three persons, and the Pennsylvania Supreme Court held that there was sufficient evidence to warrant a jury in finding the appellees' "reckless creation of a risk of catastrophe."[31]

In addition to establishing a duty, the prosecutor must demonstrate a number of other facts beyond a reasonable doubt:

- *Possession of Knowledge of the Peril.* The prosecution must establish that an individual was actually aware or should have been aware that another person was in danger. A mother cannot be held liable for her boyfriend's molestation of her child unless she knew or ought to have known that her child was being sexually mistreated.[32]
- *Acted With the Required Intent.* Most omission cases involve death and are prosecuted as either murder or manslaughter (reckless disregard).[33] As we will see in Chapter 10, homicide requires a specific intent to kill, while manslaughter requires knowledge that death is substantially certain to result. Poor judgment, a reasonable mistake, or a debatable decision is generally not sufficient to establish criminal guilt.[34]
- *Caused the Harm to the Victim.* The defendant's failure to assist the victim must have caused the harm.[35]

We can see the interrelationship between these three factors in *Craig v. State*. In *Craig,* two *parents* were held not to be *grossly negligent* in causing their daughter's death because they were found not to have possessed *knowledge* of her serious illness, which was not apparent until two or three days prior to her death. The evidence indicated at this point that medical assistance would not have saved her life. As a result, the parents *were held not to have caused* the child's death.[36]

Last, individuals are not expected to "accomplish the impossible." *The law excuses persons from fulfilling their duty in those instances in which they would be placed in peril.* Individuals, however, must take whatever action is feasible under the circumstances. In *State v. Walden,* the defendant observed the beating of her infant son by his biological father. The North Carolina Supreme Court recognized that parents cannot be expected or required to exhibit unreasonable courage and heroism in protecting their children. However, the defendant was convicted based on the fact that she neglected to take every reasonable step under the circumstances to avert the harm, such as protesting, alerting authorities, or seeking assistance.[37]

In serious cases of family abuse, duty, knowledge, intent, causality, and a failure to intervene are easily established. In *People v. Burton,* the defendants Sharon Burton and Leroy Locke were convicted of first-degree murder. On January 22, 1996, Sharon Burton passively watched Leroy Locke chase her daughter Dominique with a belt after learning that she had a "toilet training accident" on the carpet while shouting "the little bitch pissed again." Locke then filled the bathtub with water and forced Dominique's head under the water three times for fifteen seconds at a time. Dominique's body reportedly went limp in the water, and Locke and Burton left the three-year-old unattended in the bathtub for thirty minutes while they played cards. Burton, after discovering Dominique's lifeless body, called her mother rather than authorities and later falsely reported to investigators that the child had fallen off the toilet. An Illinois appellate court found that Burton possessed knowledge that Dominique was being subjected to an ongoing pattern of abuse and that there was a substantial likelihood that Dominique would suffer death or great bodily harm.[38]

In several states, a **Good Samaritan statute** requires individuals to aid individuals in peril. A Vermont law, Title 12, Chapter 23, § 519, requires individuals to provide reasonable medical assistance and, in return, relieves individuals of liability for civil damages unless their actions constitute gross negligence. Willful violation of the statute is punishable by a fine of not more than $100. This has the advantage of focusing on medical assistance and of not requiring individuals to intervene to prevent harm or to rescue individuals. Most state Good Samaritan laws do not require individuals to intervene, although people who do intervene are provided some degree of protection from civil liability.

The next case in the textbook, *Jones v. United States,* challenges you to determine whether the defendant possessed a duty of care to the children in her home.

Model Penal Code

Section 2.01. Requirement of a Voluntary Act, Omission as Basis of Liability

(1) A person is not guilty of an offense unless his liability is based on conduct that includes a voluntary act or the omission to perform an act of which he is physically capable.

(2) . . .

(3) Liability for the commission of an offense may not be based on an omission unaccompanied by action unless:

(a) the omission is expressly made sufficient by the law defining the offense; or

(b) a duty to perform the omitted act is otherwise imposed by law.

Analysis

The Model Penal Code adopts the conventional position and does not generally impose criminal liability for omissions.

Section (1) excuses an individual from liability when the intervention is beyond his or her physical capacities or would place him or her in peril.

Section (3)(a) recognizes that the definition of some crimes requires an omission. This would encompass a doctor who fails to fulfill the duty to report child abuse. Section (3)(b) provides that an omission may be committed by a failure to fulfill a legal duty. This may arise from statute or the common law. A duty, however, may not arise under a moral or religious code.

The Legal Equation

Omission of a duty = (1) A failure to act

+ (2) status, statute, contract, assume a duty, peril, control, landowner

+ (3) knowledge that the victim is in peril

+ (4) criminal intent

+ (5) possession of the capacity to perform the act

+ (6) would not be placed in danger.

Did the appellant breach a legal duty to Robert Lee Green and Anthony Lee Green?

Jones v. United States, 308 F.2d 307 (D.C. Cir. 1962). Opinion by: Wright, J.

Appellant, together with one Shirley Green, was tried on a three-count indictment charging them jointly with (1) abusing and maltreating Robert Lee Green (2), abusing and maltreating Anthony Lee Green, and (3) involuntary manslaughter through failure to perform their legal duty of care for Anthony Lee Green, which failure resulted in his death. At the close of evidence, after a trial before a jury, the first two counts were dismissed as to both defendants. On the third count, appellant was convicted of involuntary manslaughter. Shirley Green was found not guilty.

Appellant argues that there was insufficient evidence as a matter of law to warrant a jury finding of breach of duty in the care she rendered Anthony Lee. Alternatively, appellant argues that the trial court committed plain error in failing to instruct the jury that it must first find that appellant was under a legal obligation to provide food and necessities to Anthony Lee before finding her guilty of manslaughter in failing to provide them. The first argument is without merit. Upon the latter we reverse.

Facts

A summary of the evidence, which is in conflict upon almost every significant issue, is necessary for the disposition of both arguments. In late 1957, Shirley Green became pregnant, out of wedlock, with a child, Robert Lee, subsequently born August 17, 1958. Apparently to avoid the embarrassment of the presence of the child in the Green home, it was arranged that appellant, a family friend, would take the child to her home after birth. Appellant did so, and the child remained there continuously until removed by the police on August 5, 1960. Initially appellant made some motions toward the adoption of Robert Lee, but these came to naught, and shortly thereafter it was agreed that Shirley Green was to pay appellant $72 a month for his care. According to appellant, these payments were made for only five months. According to Shirley Green, they were made up to July 1960.

Early in 1959 Shirley Green again became pregnant, this time with the child Anthony Lee, whose death is the basis of appellant's conviction. This child was born October 21, 1959. Soon after birth, Anthony Lee developed a mild jaundice condition. . . . The jaundice resulted in his retention in the hospital for three days beyond the usual time, or until October 26, 1959, when, on authorization signed by Shirley Green, Anthony Lee was released by the hospital to appellant's custody. Shirley Green, after a two or three day stay in the hospital, also lived with appellant for three weeks, after which she returned to her parents' home, leaving the children with appellant. She testified she did not see them again, except for one visit in March, until August 5, 1960. Consequently, though there does not seem to have been any specific monetary agreement with Shirley Green covering Anthony Lee's support, appellant had complete custody of both children until they were rescued by the police.

With regard to medical care, the evidence is undisputed. In March 1960, appellant called a Dr. Turner to her home to treat Anthony Lee for a bronchial condition. Appellant also telephoned the doctor at various times to consult with him concerning Anthony Lee's diet and health. In early July 1960, appellant took Anthony Lee to Dr. Turner's office where he was treated for "simple diarrhea." At this time the doctor noted the "wizened" appearance of the child and told appellant to tell the mother of the child that he should be taken to a hospital. This was not done.

On August 2, 1960, two collectors for the local gas company had occasion to go to the basement of appellant's home, and there saw the two children. Robert Lee and Anthony Lee at this time were age two years and ten months respectively. Robert Lee was in a "crib" consisting of a framework of wood, covered with a fine wire screening, including the top which was hinged. The "crib" was lined with newspaper, which was stained, apparently with feces, and crawling with roaches. Anthony Lee was lying in a bassinet and was described as having the appearance of a "small baby monkey." One collector testified to seeing roaches on Anthony Lee.

On August 5, 1960, the collectors returned to appellant's home in the company of several police officers and personnel of the Women's Bureau. At this time, Anthony Lee was upstairs in the dining room in the bassinet, but Robert Lee was still downstairs in his "crib." The officers removed the children to the D.C. General Hospital, where Anthony Lee was diagnosed as suffering from severe malnutrition and lesions over large portions of his body, apparently caused by severe diaper rash. Following admission, he was fed repeatedly, apparently with no difficulty, and was described as being very hungry. His death, thirty-four hours after admission, was attributed without dispute to malnutrition. At birth, Anthony Lee weighed six pounds, fifteen ounces—at death at age ten months, he weighed seven pounds, thirteen ounces. Normal weight at this age would have been approximately fourteen pounds.

Appellant argues that nothing in the evidence establishes that she failed to provide food to Anthony Lee. She cites her own testimony and the testimony of a lodger, Mr. Wills, that she did in fact feed the baby regularly. At trial, the defense made repeated attempts to extract from the medical witnesses opinions that the jaundice, or the condition that caused it, might have prevented the baby from assimilating food. The doctors conceded this was possible but not probable, since the autopsy revealed no condition that would support the defense theory. It was also shown by the disinterested medical witnesses that the child had no difficulty in ingesting food immediately after birth, and that Anthony Lee, in the last hours before his death, was able to take several bottles, apparently without difficulty, and seemed very hungry. This evidence, combined with the absence of any physical cause for nonassimilation, taken in the context of the condition in which these children were kept, presents a jury question on the feeding issue.

Moreover, there is substantial evidence from which the jury could have found that appellant failed to obtain proper medical care for the child. Appellant relies upon the evidence showing that on one occasion she summoned a doctor for the child, on another took the child to the doctor's office, and that she telephoned the doctor on several occasions about the baby's formula. However, the last time a doctor saw the child was a month before his death, and appellant admitted that on that occasion the doctor recommended hospitalization. Appellant did not hospitalize the child, nor did she take any other steps to obtain medical care in the last crucial month. Thus there was sufficient evidence to go to the jury on the issue of medical care, as well as failure to feed.

Issue

Appellant takes exception to the failure of the trial court to charge that the jury must find beyond a reasonable doubt, as an element of the crime, that appellant was under a legal duty to supply food and necessities to Anthony Lee. . . .

Reasoning

The problem of establishing the duty to take action that would preserve the life of another has not often arisen in the case law of this country. The most commonly cited statement of the rule is found in *People v. Beardlsey*, 150 Mich. 206, 113 N.W. 1128, 1129 (1962), which provides that the

> law recognizes that . . . the omission of a duty owed by one individual to another, where such omission results in the death of the one to whom the duty is owing, will make the other chargeable with manslaughter. . . . It must be a duty imposed by law or by contract, and the omission to perform the duty must be the immediate and direct cause of death.

There are at least four situations in which the failure to act may constitute breach of a legal duty. One can be held criminally liable first, where a statute imposes a duty to care for another; second, where one stands in a certain status relationship to another; third, where one has assumed a contractual duty to care for another; and fourth, where one has voluntarily assumed the care of another and so secluded the helpless person as to prevent others from rendering aid.

It is the contention of the government that either the third or the fourth ground is applicable here. However, it is obvious that in any of the four situations, there are critical issues of fact that must be passed on by the jury—specifically in this case, whether appellant had entered into a contract with the mother for the care of Anthony Lee or, alternatively, whether she assumed the care of the child and secluded him from the care of his mother, his natural protector. On both of these issues, the evidence is in direct conflict, appellant insisting that the mother was actually living with appellant and Anthony Lee, and hence should have been taking care of the child herself, while Shirley Green testified she was living with her parents and was paying appellant to care for both children.

Holding

In spite of this conflict, the instructions given in the case failed even to suggest the necessity for finding a legal duty of care. The only reference to duty in the instructions was the reading of the indictment, which charged, inter alia, that the defendants "failed to perform their legal duty." A finding of legal duty is the critical element of the crime charged, and failure to instruct the jury concerning it was plain error.

Reversed and remanded.

Questions for Discussion

1. Why was this case remanded to the trial court?

2. Did both Jones and Shirley Green breach a duty of care?

3. Would you acquit Jones in the event that she informed Shirley Green that she no longer desired to take care of Robert and Anthony, and Shirley Green made no effort to remove the children from Jones's home?

4. Is it significant that Jones did not call Shirley Green as suggested by the doctor? Did the doctor breach a duty in this case?

5. What if Shirley Green left Anthony on Ms. Jones's porch with a note asking Ms. Jones to care for him, and Ms. Jones ignored Anthony? In the event that Anthony froze to death, would both Green and Jones be criminally liable?

Cases and Comments

1. **Broadening the Parental Duty of Care.** Defendants Darryl Stephens and Yvette Green lived together for eleven years. Eight of the ten children in their household were the product of their relationship. In November 1996, Yvette was named legal guardian of eight-year-old Sabrina Green, the daughter of Yvette's deceased, crack-addicted sister. Yvette did not continue the medication Sabrina required to control her oppositional defiant disorder. Sabrina was punished almost daily as a result of her tantrums, fights with other children, refusal to follow household rules, and bed wetting. The defendants resorted to tying Sabrina to her bed and making her sit in the hallway where she could be monitored. They disregarded the pleas of the other children to seek medical care for Sabrina. A week prior to Sabrina's death, Stephens was seen hitting her with a belt ten or twelve times. At the time of Sabrina's death in November 1997, the autopsy indicated that she suffered from a hemorrhage caused by numerous blunt impacts to the head, a third-degree burn to the hand that was left untreated until infection and gangrene set in, pneumonia, bruises that were consistent with being hit with a belt, scars from her hands being tied with a rope, and bedsores from being immobilized for many days. The injuries to her right hand were found to be consistent with being slammed by the refrigerator door as punishment for Sabrina's taking food without permission. The injuries to her head resulted from a heavy instrument, such as a baseball bat.

The prosecution's case was based on two alternative theories. The first theory was that Stephens acted together with Yvette to engage in reckless conduct that created a grave risk of serious physical injury or death to Sabrina that resulted in her death. The New York court had little difficulty in finding that this was supported by the evidence. The second approach, that Stephens failed to fulfill his duty to secure medical care for Sabrina, was more difficult to establish. Stephens argued that he did not possess either a family or guardian relationship with Sabrina and was merely a live-in boyfriend who did not possess a legal duty to ensure that she received the necessary medical care. In fact, in *People v. Myers*, 608 N.Y.S.2d 544 (N.Y. App. Div. 1994), a New York court refused to utilize the *in loco parentis* doctrine (which refers to an individual who assumes a parental role in the absence of a parent) to impose liability on a defendant charged with the negligent murder of a two-month-old child who died of severe dehydration and malnutrition. The defendant was the live-in boyfriend of the child's mother and had limited his involvement to contributing to the household finances, babysitting, and the occasional purchase of formula. The judges in *Myers* concluded that the defendant had not intended to assume responsibility for the child's welfare.

The appellate court in *Stephens*, however, distinguished this case from *Myers* and ruled that the *in loco parentis* doctrine imposed a duty on Stephens based on his having undertaken the "fundamental responsibilities that are normally those of a parent." He was determined to have taken joint responsibility for Sabrina's housing, clothing, food, and supervision and treated her with the same degree of concern as he treated the other children in the household. The court thus concluded that the evidence indicated that at the time of Sabrina's death, Stephens was responsible for Sabrina's care and that, together with Yvette, he recklessly engaged in conduct that created a grave risk of serious physical injury or death. See *People v. Stephens* (3 A.D.3d 57 [N.Y. App. Div. 2003]).

In *State v. Miranda,* the Connecticut Supreme Court recognized the doctrine of *in loco parentis* and observed that the United States was experiencing a significant increase in nontraditional alternative family arrangements. The court observed that the obligation of caretakers to safeguard children should not depend upon whether the adults had entered into a formal marital relationship. See *State v. Miranda* 715 A.2d 680 (Conn. 1998).

What factors are most important in establishing that an individual is acting *in loco parentis*? Should a live-in boyfriend or girlfriend who occasionally cares for an infant be exempted from a duty of care? Is there a danger that individuals will unfairly be held criminally liable as the duty to intervene is broadened?

2. *The Free Exercise of Religion and the Duty of Care to a Child.* On the morning of July 19, 1989, Michael and Zelia McCauley brought their eight-year-old daughter, Elisha, to the hospital. Tests revealed leukemia. The results indicated that Elisha had a hematocrit reading (percentage of red to whole blood) of fourteen and a half percent. A normal hematocrit for a young child is roughly forty percent. Further laboratory tests disclosed the presence of other symptoms consistent with the diagnosis of leukemia. The doctors determined that a bone marrow procedure was required in order to confirm the diagnosis. They, however, were unwilling to perform the operation until they raised Elisha's hematocrit to a safe clinical range in order to eliminate the risk of congestive heart failure. The only treatment available to raise the red blood cell level was through a blood transfusion. The doctors explained that in the event that the diagnosis was confirmed, the leukemia would be treated with chemotherapy and blood transfusions.

Michael and Zelia McCauley are Jehovah's Witnesses. They were baptized over fifteen years prior and attend services three days a week. A principal tenet of their religion is the belief based on an interpretation of the Bible that the reception of blood or blood products precludes resurrection and everlasting life after death. They accordingly refused to consent to the administration of blood or blood products to Elisha. The McCauleys do not object to other medical procedures.

The Massachusetts Supreme Judicial Court recognized that the free exercise of religion is a fundamental right protected under the First Amendment to the U.S. Constitution and that it includes the right of parents to raise their children in accordance with the tenets of their religion. The court, however, ruled that this religious interest is not absolute and must give way to the State's concern with protecting the welfare and life of children and in maintaining the medical profession's ethical commitment to the care and preservation of life. The tests that the doctors desire to undertake will identify the type of leukemia involved and enable doctors to determine the required chemotherapy. Absent these procedures, Elisha likely will die. The supreme court recognized the sincerity of the McCauleys' belief, but concluded that Elisha's "best interests and welfare, coupled with the strong interest of the State, must outweigh her parents' objections to the blood transfusions."

Do you agree with this decision? See *Matter of McCauley,* 565 N.E.2d 411 (Mass. 1991).

Now consider the 1972 U.S. Supreme Court decision in *Wisconsin v. Yoder.* Members of the Old Order Amish religion and the Conservative Amish Mennonite Church were convicted of violating Wisconsin's compulsory school-attendance law by declining to send their children to public or private schools beyond the eighth grade. Wisconsin law requires school attendance until the age of sixteen; the Amish contend that the law violates their free exercise of religion.

The Amish recognize that children must master basic math, reading, and writing and develop a familiarity with the "outside world." They contend that education beyond the eighth grade exposes their children to "worldly values" and, more importantly, that the children must be integrated into a communal and agrarian lifestyle and inculcated with attitudes favoring manual work, self-reliance, community welfare, and a life of faith and goodness. This reflects the biblical injunction from the Epistle of Paul to the Romans, "Be not conformed to this world. . . ." The Amish have protected and preserved their "simple and uncomplicated" communities in the United States for roughly 200 years.

The Supreme Court pronounced that Wisconsin must articulate an interest of "the highest order" to overcome a claim of the free exercise of religion. Wisconsin claims that the system of compulsory education is essential to train citizens to participate in politics and to be productive members of the economy. The Supreme Judicial Court ruled in favor of the Amish, noting that the qualities of reliability, self-reliance, and dedication to work are highly valued in the general economy and that the Amish are fully capable of fulfilling their democratic responsibilities.

Why did the U.S. Supreme Court uphold the parents' free exercise of religion in *Yoder* and the Massachusetts Supreme Court rule to limit the same constitutional right in *McCauley?* Thirty states at present recognize that a parent's religious belief constitutes a criminal defense to various types of offenses. See *Wisconsin v. Yoder,* 406 U.S. 205 (1972).

You can find the answer at www.sagepub.com/lippmanccl3e.

You Decide

4.3 In May 1997, nineteen-year-old Jeremy Strohmeyer together with his friend David Cash played video games at a Las Vegas casino while Strohmeyer's father gambled. Seven-year-old Sherrice Iverson threw a wet paper towel at Strohmeyer and a paper towel fight ensued. He followed her into the restroom to continue the game. The forty-six-pound Iverson threw a yellow floor sign at Strohmeyer and then began screaming. Strohmeyer covered her mouth and forced her into a bathroom stall. David Cash wandered into the restroom to look for Strohmeyer. He peered over the stall and viewed Strohmeyer gripping and threatening to kill Sherrice. Cash allegedly made an unsuccessful effort to get Strohmeyer's attention and left the bathroom. Strohmeyer then molested Sherrice and strangled her to suffocate the screams. As he was about to leave, Strohmeyer decided to relieve Sherrice's suffering and twisted her head and broke her neck. He placed the limp body in a sitting position on the toilet with Sherrice's feet in the bowl.

Strohmeyer confessed to Cash and, after being apprehended by the police three days later, explained that he wanted to experience death. His lawyer argued that Strohmeyer was in a "dream-like state" as a result of a combination of alcohol, drugs, and stress. In order to avoid the death penalty, Strohmeyer pled guilty to first-degree murder, first-degree kidnapping, and to the sexual assault of a minor, all of which carry a life sentence in Nevada.

Iverson's mother called for Cash to be criminally charged, but Nevada law neither required him to intervene nor to report the crime to the police. The administration at the University of California at Berkeley responded to a student demonstration calling for Cash's dismissal by explaining that there were no grounds to expel him from the institution because he had not committed a crime. Cash, who was studying nuclear engineering, refused to express remorse, explaining that he was concerned about himself and was not going to become upset over other people's problems, particularly a little girl whom he did not know.

Should David Cash be held criminally liable for a failure to rescue Sherrice Iverson? See Joshua Dressler, *Cases and Materials on Criminal Law,* 3rd ed. (St. Paul, MN: West Publishing Co., 2003), pp. 133–134.

POSSESSION

Possession is a **preparatory offense**. The thinking is that punishing possession deters and prevents the next step—a burglary, sale of narcotics, or the use of a weapon in a robbery. The possession of contraband such as drugs and guns may also provoke conflict and violence.[39]

How does the possession of contraband meet the requirement that a crime involve a voluntary act or omission? This difficulty is overcome by requiring proof that the accused knowingly obtained or received the contraband (a voluntary act) or failed to immediately dispose of the property (failure to fulfill a duty).[40]

The challenge in the crime of possession is to balance the competing values of punishing the guilty while at the same time protecting the innocent. There is little difficulty in convicting an individual who is found to have drugs in his or her pocket. Complications are created when drugs are discovered in the glove compartment of a car or in the living room of a house with four occupants. There is a temptation to charge all four with drug possession. On the other hand, there is the risk that individuals who had no knowledge of the contraband will be convicted.

Possession is typically defined as the ability to exercise "dominion and control over an object." This means that a drug dealer has the ability to move, sell, or transfer the contraband. There are several other central concepts to keep in mind.

For an international perspective on this topic, visit the study site.

- *Actual possession* refers to drugs and other contraband within an individual's physical possession or immediate reach.
- *Constructive possession* refers to contraband that is outside of an individual's actual physical control but over which he or she exercises control through access to the location where the contraband is stored or through ability to control an individual who has physical control over the contraband. A drug dealer has constructive possession over narcotics stored in his or her home or under the physical control of a member of his or her gang.

- **Joint possession** refers to a situation in which a number of individuals exercise control over contraband. Several members of a gang may all live in the home where drugs are stored. There must be specific proof connecting each individual to the drugs. The fact that a gang member lives in the house is not sufficient.
- **Knowing possession** refers to an individual's awareness that he or she is in possession of contraband. A drug dealer, for instance, is aware that marijuana is in his or her pocket.
- **Mere possession** refers to physical control without awareness of contraband. An individual may be paid by a drug dealer to carry a suitcase across international borders and lack awareness that the luggage contains drugs.

Criminal statutes punishing possession are typically interpreted to require that an individual (1) know of the presence of the item, (2) exercise actual or constructive possession, and (3) know the general character of the material. There may be individual or joint possession. An individual is required to know that the material is contraband but is not required to know the precise type of contraband involved.[41]

Hawkins v. State is an example of conviction for actual possession of a firearm by a felon. The case illustrates how courts use circumstantial, or indirect, evidence to find possession. The defendant was apprehended following a high-speed chase, and the police seized a loaded shotgun in the back seat within reach of the driver. The Texas District Court stated that possession is a voluntary act if the possessor "knowingly obtains or receives the thing possessed or is aware of his control of the thing for a sufficient time to permit him to terminate his control." The court concluded that the prosecution affirmatively established Hawkins's knowledge and control over the firearm. The gun was in plain view in the back seat and was within easy reach, and Hawkins was the sole occupant of the vehicle. His guilty state of mind was indicated by his effort to escape.[42]

The **fleeting possession** rule is a limited exception to criminal possession. This permits an innocent individual to momentarily possess and dispose of an illegal object. In *People v. Mijares,* the defendant removed and disposed of narcotics that he took from an unconscious friend whom he was driving to the hospital. The California District Court, citing a fleeing possession exception, ruled that to hold Mijares liable would "result in manifest injustice to admittedly innocent individuals."[43]

The concept of constructive possession is illustrated by the federal appellate court decision in *United States v. Byfield.* Byfield traveled from New York to Washington, D.C., with a young female who carried a tote bag. The two separated after arriving at the train station, and Byfield was alleged to have directed her movements through hand signals. The police detained and searched the young woman's bag and found that it contained men's clothing in Byfield's size and a shoe box for the brand of athletic shoes worn by Byfield, along with sixty grams of crack cocaine.

Byfield carried no luggage and yet, when questioned by the police, explained that he planned to stay in Washington, D.C., for several days. The appellate court affirmed that there was sufficient evidence establishing that Byfield had previous contact with the young woman in New York and that Byfield possessed "some stake," "power," and "dominion and control" over the crack cocaine either personally or through his female companion. The court noted that it was not unusual for juveniles to be employed as drug couriers.[44]

For a deeper look at this topic, visit the study site.

The most difficult issue for courts undoubtedly is joint possession. For example, the police searched an apartment shared by Jason Stansbury, Crisee Moore, and Anthony Webb and discovered marijuana. Only Moore and his son were present at the time. Webb arrived during the course of the search. The police seized $336 from Webb; he explained that he had been paid for babysitting Moore's son. The officers were justifiably suspicious of Webb's explanation, because he had earlier pled guilty to possession of drugs found in another apartment that he had shared with Moore. The Iowa Supreme Court ruled that where an accused such as Webb is not the only person occupying the apartment, but one of several individuals in joint possession, the knowledge and ability to maintain control over narcotics must be demonstrated by direct proof. The fact that Webb occupied the apartment was insufficient to establish possession absent additional evidence. There were no fingerprints linking Webb to the narcotics, drug paraphernalia, or firearms or bullets found on the premises, and none of these items were near or among Webb's personal belongings or in a location subject to Webb's exclusive control. A search of Webb failed to find drugs on his person, and there was no evidence that he was under the influence of narcotics.[45]

Webb starkly presents the conflict between broadly interpreting possession in order to combat narcotics traffic and the due process requirement that possession should be established beyond a reasonable doubt. Courts are clearly concerned that the "war on drugs" and "war on terrorism" will result in the conviction of individuals who have not been clearly demonstrated to have exercised control over contraband. On the other hand, requiring an unrealistic standard of proof can result in the guilty escaping criminal liability.

We should note that although Washington and North Dakota do not require knowing possession, in practice these courts have imposed a knowledge requirement to ensure fair results.[46] The importance of the knowledge requirement is illustrated by the Maryland case of *Dawkins v. State*. Dawkins was arrested in a hotel room in which the police found a tote bag containing narcotics paraphernalia and a bottle cap containing heroin residue. He claimed that the bag belonged to his girlfriend, who had asked him to carry the bag to her hotel room. Dawkins claimed to have had no idea what was in the bag and explained that he only arrived a few minutes before the police. The Maryland Supreme Court reversed the defendant's conviction and explained that in order to be guilty of possession of a controlled substance, the accused "must know of both the presence and . . . general character or illicit nature of the substance. Of course, such knowledge may be proven by circumstantial evidence, and by inferences. . . ."[47]

Model Penal Code

Section 2.01 Possession as an Act

(1) . . .

(2) Possession is an act, within the meaning of this Section, if the possessor knowingly procured or received the thing possessed or was aware of his control thereof for a sufficient period to have been able to terminate his possession.

Analysis

The Model Penal Code establishes that the voluntary procurement of contraband or knowing possession of contraband for a "sufficient period" satisfies the standard for possession. The code also clarifies that an individual is required only to be aware of the nature (e.g., drugs) of an item in his or her possession and need not be informed of the item's illegal character.

The Legal Equation

Possession = Knowledge of presence of object
+ exercise of dominion and control
+ knowledge of the character of object.

Did Toups have constructive possession of the cocaine? Was Toups guilty of cocaine possession?

State v. Toups, 859 So.2d 768 (La. Ct. App. 2003). Opinion By: Bagneris, J.

Issue

The central issue before the appellate court is to determine whether there is sufficient evidence to hold Mary Toups legally liable for drug possession.

Facts

Defendant Mary L. Toups, aka Mary Billiot ("the defendant"), was charged with possession of cocaine, a violation of La. R.S. 40:967(C). The defendant pled not guilty at her arraignment on December 10, 1999. . . . At trial on February 9, 2000, a six-member jury, found the defendant guilty as charged. . . . The defendant waived all legal delays and was sentenced to four years at hard labor, with credit for time served. On May 8, 2000, after being advised of her rights and waiving them, the defendant admitted a prior conviction and was adjudicated a second-felony habitual offender. After the defendant waived all legal delays, the trial court vacated her original sentence and resentenced her to four years at hard labor, with credit for time served. The defendant's motion for appeal was granted on that same date.

New Orleans Police Officer Dennis Bush testified that on the evening of October 18, 1999, he and five other officers executed a search warrant at a residence located at 633 N. Scott Street. They were in search of a male known as "Stan." After receiving no response at the front door, Officer Bush entered the "shotgun" residence. Officer Bush testified that the defendant and Stanley Williams were found seated on a sofa in the front living room, facing each other. In another room, police found an elderly male connected to some type of respirator. He was the only other person found in the residence. Two pieces of crack cocaine, three clear glass crack pipes and a razor blade were observed on a coffee table that was positioned directly in front of the sofa on which the defendant and Williams were sitting. Officer Bush identified the crack cocaine and the crack pipes. Officer Bush admitted on cross-examination that he had no knowledge of the defendant, and that Williams had been residing there for a period of months. He further acknowledged that even though he arrested the defendant for possession of the contraband, he had not seen her smoking out of one of the pipes. He admitted stating at an earlier hearing that he did not have any evidence to "connect" the defendant to this contraband. Officer Bush then conceded that he did not arrest the defendant for possession of eighteen other pieces of crack cocaine because he did not believe she was "connected" to them. Officer Bush testified that he observed the residence for approximately one-half hour before executing the search warrant, and he had not seen the defendant enter during that time.

New Orleans Police Detective Jeff Keating observed the defendant and Stanley Williams seated on a sofa when he entered the residence. He confirmed that three crack pipes and two rocks of crack cocaine were seized from the coffee table. Detective Keating said he also seized a plastic container containing sixteen pieces of crack cocaine that was next to Williams, although it was not in "plain view." Three hundred and four dollars was also seized from the same area.

Corey Hall, employed by the New Orleans Police Department Crime Lab, was qualified by stipulation as an expert in the analysis of controlled dangerous substances, specifically cocaine. He tested two pieces of a rock-like substance and three glass tubes related to the defendant's case, and he said that all were positive for cocaine. He also stated that seventeen pieces of rock-like substance in a plastic container tested positive for cocaine, as did a metal tube. Mr. Hall acknowledged on cross examination that it did not appear from a document presented to him by defense counsel that any of the items had been submitted for fingerprint identification.

Reasoning

The defendant was convicted of possession of cocaine, a violation of La. R.S. 40:967, which makes it unlawful for any person to knowingly or intentionally possess a controlled dangerous substance. To convict for possession of a controlled dangerous substance, the State must prove that the defendant knowingly possessed it. Guilty knowledge is an essential element of the offense of possession of a controlled dangerous substance. The State need not prove that the defendant was in actual possession of the narcotics found; constructive possession is sufficient to support conviction. "A person may be deemed to be in joint possession of a drug which is in the physical possession of a companion if he willfully and knowingly shares with the other the right to control it." However, mere presence in an area where drugs are found is insufficient to establish constructive possession. Factors to be considered in determining whether a defendant exercised dominion and control over drugs are: the defendant's knowledge that illegal drugs were present in the area; the defendant's relationship with the person in actual possession; the defendant's access to the area where the drugs were found; evidence of recent drug use; the defendant's proximity to the drugs; and evidence that the area was being frequented by drug users.

The defendant cites *State v. Bell*, where police approached a vehicle to request that the occupants, the driver and his passenger, turn down music blaring from the radio. One officer noticed a distinctively wrapped package, containing what was later determined to be cocaine, among some cassette tapes in a plastic console between the front seats. Both men were convicted of attempted possession of cocaine, which convictions were affirmed by this court. There had been no evidence presented that the passenger had any other drugs in his possession or on his person, that he appeared under the influence of narcotics, or that he had possession of any drug paraphernalia. The court also noted that the vehicle was parked only two doors from the driver's residence. The Louisiana Supreme Court found that, based on the passenger's mere presence in the car close to the sealed package, "even assuming he was aware of the contents," a rational trier of fact could not have concluded that the passenger exercised dominion and control over the package, or that he willfully and knowingly shared the right to control it with the driver.

In *State v. Jackson,* an individual named Steele threw down drugs at the sight of police, and ran into his apartment. Police followed, and saw the defendant Jackson standing at homemade bar on which were displayed a glass pipe with cocaine residue, another pipe, a mirror with trace amounts of cocaine on it, a razor blade with traces of cocaine on it, test tubes, glasses, and a bag containing a material used to cut or dilute cocaine. This court held that because there was no indication that the residue-containing pipe was warm, that the defendant's fingerprints were on any of the items, that the defendant [her blood] tested positive for the presence of cocaine, or that the defendant was anything more than a guest in the apartment, the evidence was insufficient to establish that the defendant exercised dominion and control over the objects which had cocaine residue in or on them.

In *State v. Harris*, police executing a search warrant at an apartment rented by the defendant's brother found the defendant and another individual seated at a kitchen table on top of which were a plastic bag of cocaine, a cigarette pack containing fifty-two marijuana cigarettes, two plates with cocaine residue on them, a scale, some cash, and several other items which an expert testified were used to free-base cocaine. The defendant's brother was standing at the kitchen sink free-basing cocaine. This court found that the evidence was sufficient to support a finding that the defendant possessed the cocaine and marijuana.

In *State v. Kingsmill*, police stopped a vehicle containing four men, and the front seat passenger stepped out and dropped a marijuana cigarette. A large plastic bag containing marijuana was found on the front floorboard near the passenger seat. All four men were arrested for possession of marijuana. During a subsequent search of the vehicle, an open cardboard box was found in the middle of the rear seat, containing a scale and some other drug paraphernalia, as well as a plastic bag with two smaller bags of cocaine inside. The four men were then arrested for possession of cocaine. Police testified that one defendant informed them that all four had pooled their money to purchase the marijuana and cocaine. This court found that the evidence was sufficient to support a finding that the two defendants appealing their convictions for possession of cocaine, the front seat passenger and one rear seat passenger, had exerted dominion and control over the cocaine.

In the instant case, it is undisputed that the target of the narcotics investigation was Stanley Williams, the known resident of the premises. Williams was the suspected drug dealer, and seventeen pieces of crack cocaine were found in a plastic container next to him. All of the evidence points to Williams having dominion and control over not only those seventeen pieces of cocaine in the plastic container, but also the two pieces and the three crack pipes on the coffee table. There was no evidence to suggest that the defendant lived in the residence. There was no evidence as to her relationship with Williams. While the two pieces of crack cocaine and three crack pipes were on the coffee table directly in front of the sofa on which the defendant and Williams were sitting, and the defendant had been inside of the residence for at least thirty minutes before police entered, there was no direct evidence that anyone in the residence had been smoking crack cocaine. There was no evidence of smoke, an unusual odor, or warm crack pipes. Nor was there evidence that any means of combusting crack cocaine was on the coffee table, such as matches or a cigarette lighter. There was no evidence that the defendant purchased one or both of the two rocks found on the coffee table, thus gaining dominion and control over one or both, and then she decided to linger and socialize with Williams. Considering the evidence adduced at trial, one can only speculate as to what the defendant was doing in the residence. She could have been a non-drug using member of a neighborhood church proselytizing defendant or an unrepentant crack addict preparing to prostitute herself in exchange for a rock of crack cocaine. While any rational trier of fact could have found that the defendant had knowledge of the cocaine and crack pipes on the coffee table, there was no evidence from which any rational trier of fact could have concluded beyond a reasonable doubt that the contraband was subject to the defendant's dominion and control at the point when police entered. Nor is there any evidence from which it could be concluded beyond a reasonable doubt that the defendant specifically intended to gain dominion and control over any cocaine or cocaine-tainted contraband and did or omitted to do an act toward the accomplishing of that object so as to constitute the crime of attempted possession of cocaine.

Holding

There is merit to this assignment of error. Consequently, the defendant's conviction and sentence are hereby vacated.

Questions for Discussion

1. Outline the facts in the case.

2. What facts must the prosecution establish beyond a reasonable doubt to convict Toups of narcotics possession? Outline the factors that a judge should consider in determining whether a defendant exercises "dominion and control" over narcotics.

3. Why does the appellate court vacate Toups's conviction? Is the court's ruling consistent with the precedents cited by the court?

4. As a judge how would you rule in *Toups*?

Cases and Comments

1. Constructive Possession. Ross Cashen was convicted of marijuana possession. He was a passenger in the back seat of an automobile that was stopped for a traffic violation. There were six people in the car, four of whom were in the back seat. Cashen was sitting next to a window with his girlfriend sitting on his lap. A lighter and cigarette rolling papers were found on Cashen and cigarette rolling papers and a small baggie of marijuana seeds were discovered in the pants pocket of his girlfriend. The officers also found a baggie of marijuana wedged in the rear seat on the side where Cashen and his girlfriend had been seated. The baggie was stuck in the crack between the back and bottom of the rear seat. At the jail, Cashen denied knowledge of the marijuana and later told the police that the drugs belonged to his girlfriend. She subsequently confessed to owning the drugs. Cashen was prosecuted and convicted for marijuana possession.

The Iowa Supreme Court noted that the issue was whether Cashen exercised constructive possession over the marijuana. His presence alone was ruled to be insufficient to establish possession, because he was not in exclusive possession of the automobile and did not have exclusive access to the back seat. The rolling papers, at most, demonstrated that Cashen possessed marijuana "in the past and intended to do so again in the future. However, we cannot infer from this fact alone that Cashen had authority or the ability to exercise unfettered influence of these drugs." Cashen's question to the police whether anyone had "fessed up to ownership may indicate that he had knowledge of the presence of the drugs, but does not constitute 'dominion and control over the marijuana.'"

The Iowa Supreme Court stressed that Cashen was not the owner of the automobile, the drugs were not in plain view, and the marijuana was not found among Cashen's personal effects. Cashen was completely cooperative, and the police also did not offer evidence indicating that Cashen's fingerprints were on the baggie. The other three passengers were as close to the narcotics as Cashen, and the prosecution was "required to prove facts other than mere proximity to show [Cashen's] dominion and control of the drugs."

Are appellate courts imposing too stringent a standard of proof and impeding the effort to combat drug possession and drug dealing? On the other hand, are prosecutors unjustifiably filing charges of drug possession? See *State v. Cashen,* 666 N.W.2d 566 (Iowa, 2003).

Compare *Cashen* with the U.S. Supreme Court case of *Maryland v. Pringle.* The Supreme Court affirmed Pringle's conviction for possession with intent to distribute cocaine and possession of cocaine, and he was sentenced to ten years in prison. Pringle was one of three passengers in an automobile that was stopped for speeding in the early morning hours in Baltimore. He was sitting in the front seat directly in front of the glove compartment, which contained $763. Five plastic glassine baggies of cocaine were behind him in the back seat armrest and were accessible to all three passengers. The Supreme Court concluded that it is entirely reasonable to conclude "that any or all . . . of the occupants [of this confined space] had knowledge of, and exercised dominion and control over the cocaine. . . . There was probable cause to believe Pringle committed the crime of possession of cocaine, either solely or jointly." The Court stressed that the quantity of drugs and cash indicated the "likelihood of drug dealing" and that a "dealer would be unlikely to admit an innocent person with the potential to furnish evidence against him." See *Maryland v. Pringle,* 540 U.S. 366 (2004).

2. Possession and Computer Files. Christopher Worden was convicted of four counts of possession of child pornography, and of one count each of indecent exposure and of unlawful exploitation of a minor. Worden admitted to inappropriate contact with two young juveniles. The police seized and searched two computers from Worden's home and found images of child pornography in the computer cache files.

Virgil Gattenby, a police technician who examined the computer, testified that Worden had visited certain websites containing child pornography "more than once" and that "it would have taken Worden's computer several minutes to load the images and the images recovered had loaded completely." Gattenby testified that although the images of child pornography were found among the cache files on the hard drive of Worden's computer, there was no indication that Worden "had any intent to store the images—his intent was simply to view the images on his computer screen during the time that he visited a website." Gattenby explained that when a person uses a computer to access a site, the "computer automatically stores the images from the web page in the browser cache." According to Gattenby this "enables to computer to load the web page more quickly when you revisit it, because data is accessed directly from the computer's hard drive rather than loading that data over the internet." There was no evidence that Worden possessed the type of "specialized knowledge" required to know that images were "being stored in his computer cache or that he intended to save them on his computer." Did Worden's viewing of sexual images constitute "knowing possession" of "material that visually or aurally depicts conduct [constituting] child pornography?" The Alaska Supreme Court held that the state statute prohibiting the possession of child pornography does not prohibit viewing material on a computer screen and observed that "[i]f Worden had gone to a movie depicting child pornography, it could not be said that he possessed the child pornography depicted in the movie, even though it might be clear that he had intentionally set out to view those images." See *Worden v. Alaska,* 213 P.3d 144 (Alaska Ct. App. 2009).

You can find Worden v. Alaska *on the study site, www.sagepub.com/lippmanccl3e.*

You Decide

4.4 Six Spirit Lake, Iowa, police officers executed a search warrant at an apartment shared by Bash, her husband Kevin, and their three sons. The warrant authorized the police to seize controlled substances and a safety deposit box. The officers arrested Kevin and removed him from the home. The defendant stated that she could "show [the police] where the stuff is." One of the officers followed Bash into the master bedroom where she told the officer that "it's on his nightstand in a cardboard box, that it's Kevin's stuff, that is his bong . . . sitting on the floor next to the bed."

Bash's version was somewhat different. She testified that when questioned that she stated that "[i]f there is anything here, it would be on Kevin's side of the bed." She pointed towards his nightstand, which was on the left side of the bed.

On Kevin's nightstand, the officers found a cardboard box bearing the word "Friscos." Inside the box, they found 1.37 grams of marijuana. Bash claimed that she "did not know what was in the box until after the officers opened it. However, she admitted that she knew there had been marijuana in the house, in the box, in the past."

Bash was charged with possession of a controlled substance. She was convicted and sentenced to a thirty-day suspended sentence with credit for time served and a $250 fine.

Was Bash guilty of possession of a controlled substance? See *State v. Bash,* 670 N.W.2d 135 (Iowa 2003).

You can find the answer at www.sagepub.com/lippmanccl3e.

CRIME IN THE NEWS

In 1989, Denver, Colorado, enacted an ordinance banning pit bulls from the city. The law was precipitated by dog attacks that resulted in the death of a five-year-old boy and the savage maiming of a pastor. Denver had experienced twenty such attacks over a five-year period. The Colorado legislature subsequently passed a law prohibiting counties and municipalities from enacting breed-specific bans on dogs. In December 2004, a Denver court ruled that Colorado lacked the authority to prevent the city from prohibiting any person from "owning, possessing, keeping, exercising control over, maintaining, harboring, or selling a Pit Bull in the City and County of Denver." A pit bull is defined in the ordinance as any dog that is an American Pit Bull Terrier, an American Staffordshire Terrier, a Staffordshire Bull Terrier, or any dog displaying the majority of the physical traits of any one or more of these breeds.

Animal control officers under the ordinance are authorized to confiscate pit bulls, and a determination then is made by a veterinarian as to whether the dog is one of the three "banned breeds." In the event that the animal is found to be a member of a banned breed, the owner is provided the opportunity to remove the dog from the city. A failure to remove the animal results in the dog being put to sleep. A second offense of possession results in automatic euthanization. An owner who removes his or her dog must provide a statement listing the dog's new home. The penalty for harboring an illegal pit bull is a fine of up to $1,000 and a year in jail. The ordinance permits the transportation of a pit bull through Denver so long as the dog remains in a vehicle. Since 1989, opponents of the Denver ordinance estimate that roughly 1,100 pit bulls have been seized and put down. There reportedly have been no deaths in Denver from pit bulls since the prohibition went into effect. As for national statistics, roughly thirty-eight percent of the 238 deaths of humans from dogs between 1979 and 1998 resulted from pit bulls and Rottweilers.

Denver and Miami are the largest cities to ban pit bulls, and similar laws are being considered by several states and municipalities. Thirteen states have legislation that prohibits breed-specific bans. The legislation in most states focuses on a dog's behavior rather than on a dog's breed. The typical approach is represented by Michigan, which prohibits "dangerous dogs"; such an animal is defined as a dog that "bites or attacks a person, or a dog that bites or attacks and causes serious injury or death to another dog while the other dog is on the property or under the control of its owner." An exception is made for an attack against trespassers and persons who provoke or torment the animal, or in those instances in which the animal acts to protect an individual.

The Denver ordinance is based on the belief that pit bulls tend to be inherently aggressive toward other animals and children and inflict more severe injuries than other dogs. In addition, the breed is favored by gang members and drug dealers. Defenders of the breed claim that pit bulls are no more dangerous than other dogs and that most of the pit bulls that are impounded are completely harmless. Historically, various breeds have been victims of the same form of social hysteria that is being directed at pit bulls. Linda Blair has spoken out against the pit bull ban and has argued that it is irresponsible owners who present the problem rather than the breed.

Some courts have struck down pit bull ordinances on the ground that the term *pit bull* "is vague and risks depriving owners of their pets without due process of law." The majority view, however, is that the regulation of pit bulls is a valid exercise of the state and local government's power to protect the public health and safety. A Kansas court found that pit bulls "possess a strongly developed 'kill instinct' not shared by other breeds of dogs," are "unique in their 'savageness and unpredictability,'" and are "twice as likely to cause multiple injuries as other breeds of dogs." See *Hearn v. City of Overland Park*, 772 P.2d 758 (Kan. 1989).

Would you support a pit bull ban in your local community?

CHAPTER SUMMARY

A crime involves a *concurrence* between an *actus reus* (act) and *mens rea* (intent). The act generally must have *caused* the social harm punishable under the relevant statute.

A crime is limited to acts and omissions; an individual may not be punished for "mere thoughts." This would involve an unacceptable degree of governmental intrusion into individual privacy and would result in the disproportionate punishment of individuals for ideas that ultimately may not be translated into criminal conduct. An act must be voluntary. It is fundamentally unfair to punish individuals for involuntary acts that are the product of a disease or the unconscious and are not the product of a conscious and deliberate choice. The punishment of an individual based on status is also considered "particularly obnoxious" and "cruel and unusual," because it involves punishment for a personal condition or characteristic that may not be translated into socially harmful acts.

The criminal law, with some limited exceptions, typically does not punish individuals for a failure to act. There are limited circumstances in which individuals are required to assist those in peril. These involve a status, statute, contract, assumption of duty, creation of peril, and control and ownership of property.

The possession of contraband is also subject to punishment based on a knowing and voluntary acquisition or failure to dispose of the material. Possession requires "dominion and control." This may be actual or constructive as well as either individual or joint.

CHAPTER REVIEW QUESTIONS

1. Why are individuals not punished for their thoughts?

2. What is the reason for requiring a voluntary act? Provide some examples of acts that are considered involuntary. May a defendant be criminally condemned for reckless driving despite the fact that an accident results from a stroke?

3. Why do status offenses constitute cruel and unusual punishment?

4. Is there a difference between the American and European rules on omissions? What are the reasons behind the American rule? When does a duty arise to intervene to assist an individual in peril?

5. Discuss the difference between actual and constructive possession and between sole and joint possession. What facts are important in establishing possession?

LEGAL TERMINOLOGY

actual possession	fleeting possession	omission
actus reus	Good Samaritan statute	possession
American bystander rule	involuntary act	preparatory offense
attendant circumstances	joint possession	result crime
constructive possession	knowing possession	status offense
duty to intervene	*mens rea*	voluntary act
European bystander rule	mere possession	

CRIMINAL LAW ON THE WEB

Log on to the Web-based student study site at www.sagepub.com/lippmanccl3e to assist you in completing the Criminal Law on the Web exercises, as well as for additional features such as podcasts, Web quizzes, and video links.

1. Read about post-traumatic stress and Iraq war veterans.

2. Learn more about the Kitty Genovese case. Could a similar incident happen today?

3. Learn about the history of the Denver pit bull ordinance.

BIBLIOGRAPHY

American Law Institute, *Model Penal Code and Commentaries* (Philadelphia: American Law Institute, 1985), § 2.01. A comprehensive discussion of the requirements of a criminal act.

Joshua Dressler, *Understanding Criminal Law,* 3rd ed. (New York: Lexis, 2001), Chap. 9. An understandable and thorough discussion of *actus reus* with helpful citations to relevant journal articles.

Markus D. Dubber, *Criminal Law: Model Penal Code* (New York: Foundation Press, 2002), pp. 33–42. A volume that outlines the requirements of the Model Penal Code and also includes a critical commentary that provides insight into the policy choices that were made in drafting the Model Code provision on criminal acts.

Hyman Gross, *A Theory of Criminal Justice* (New York: Oxford University Press, 1979), Chap. 4. A sophisticated discussion of the philosophical issues involved in the concept of *actus reus.*

Leo Katz, *Bad Acts and Guilty Minds: Conundrums of the Criminal Law* (Chicago: University of Chicago Press, 1987), pp. 129–153. Challenging legal cases and hypothetical problems on omissions.

Wayne R. LaFave, *Criminal Law,* 3rd ed. (St. Paul, MN: West Publishing, 2000), pp. 206–224. A comprehensive discussion of criminal acts with citations to important cases.

Arnold H. Lowey, *Criminal Law in a Nutshell,* 4th ed. (St. Paul, MN: West Publishing, 2003), Chap. 9. A concise discussion of *actus reus.*

Paul H. Robinson, *Would You Convict? Seventeen Cases That Challenged the Law* (New York: New York University Press, 1999), pp. 78–84. A case raising issues involved in the imposition of a legal duty to assist another.

Richard G. Singer and John Q. La Fond, *Criminal Law Examples and Explanations,* 2nd ed. (New York: Aspen, 2001), Chap. 3. A straightforward discussion of criminal acts under the common law and Model Penal Code with review questions.

5

MENS REA, CONCURRENCE, CAUSATION

Did the defendant know that his pet tiger cats endangered his daughter?

By June 6, 1999, the tigers were two years old. Lauren was ten. She stood fifty-seven inches tall and weighed eighty pounds. At dusk that evening, Lauren joined Hranicky in the tiger cage. Suddenly, the male tiger attacked her. It mauled the child's throat, breaking her neck and severing her spinal cord. She died instantly. . . . Hranicky testified . . . [that] he did not view the risk to be substantial because he thought the tigers were domesticated and had bonded with the family. . . . Thus, he argues, he had no knowledge of any risk.

Learning Objectives

1. Know the role of criminal intent in criminal law and punishment.

2. Understand the categories of criminal intent in the Model Penal Code.

3. Explain the difference between the criminal intents of purposely and knowingly and between the criminal intents of recklessly and negligently.

4. Know the characteristics of strict liability offenses.

5. Understand the concept of concurrence in defining crimes.

6. Appreciate the role of causality in the determination of criminal guilt.

7. In causality analysis, understand the concepts of cause-in-fact, proximate cause, coincidental intervening cause, and responsive intervening cause.

INTRODUCTION

In the last chapter, we noted that a criminal act or *actus reus* is required to exist in unison with a criminal intent or *mens rea*; and as you soon will see, these two components must combine to *cause* a prohibited injury or harm. This chapter completes our introduction to the basic elements of a crime by introducing you to criminal intent, concurrence, and causation.

One of the common law's great contributions to contemporary justice is to limit criminal punishment to "morally blameworthy" individuals who consciously choose to cause or to create a risk of harm or injury. Individuals are punished based on the harm caused by their decision to commit a criminal act rather than because they are "bad" or "evil" people. Former Supreme Court Justice Robert Jackson observed that a

system of punishment based on intent is a celebration of the "freedom of the human will" and the "ability and duty of the normal individual to choose between good and evil." Jackson noted that this emphasis on individual choice and free will assumes that criminal law and punishment can deter people from choosing to commit crimes, and those who do engage in crime can be encouraged to develop a greater sense of moral responsibility and avoid crime in the future.[1]

MENS REA

You read in the newspaper that your favorite rock star shot and killed one of her friends. There is no more serious crime than murder; yet before condemning the killer, you want to know, "What was on her mind?" The rock star may have intentionally aimed and fired the rifle. On the other hand, she may have aimed and fired the gun believing that it was unloaded. We have the same act, but a different reaction based on whether the rock star intended to kill her friend or acted in a reckless manner. As Oliver Wendell Holmes, Jr., famously remarked, "Even a dog distinguishes between being stumbled over and being kicked."[2]

As we have seen, it is the bedrock principle of criminal law that a crime requires an act or omission and a criminal intent. The appropriate punishment of an act depends to a large extent on whether the act was intentional or accidental. Law texts traditionally have repeated that *actus non facit rum nisi mens sit rea:* "There can be no crime, large or small, without an evil mind." The "mental part" of crimes is commonly termed **mens rea** ("guilty mind") or **scienter** ("guilty knowledge") or criminal intent. The U.S. Supreme Court noted that the requirement of a "relation between some mental element and punishment for a harmful act is almost as instinctive as the child's familiar exculpatory (not responsible) plea, 'But I didn't mean to.'"[3]

The common law originally punished criminal acts and paid no attention to the mental element of an individual's conduct. The killing of an individual was murder, whether committed intentionally or recklessly. Canon, or religious law, with its stress on sinfulness and moral guilt, helped to introduce the idea that punishment should depend on an individual's "moral blameworthiness." This came to be fully accepted in the American colonies; and, as observed by the U.S. Supreme Court, *mens rea* is now the "rule of, rather than the exception to, the principles . . . of American criminal jurisprudence." There are some good reasons for requiring moral blameworthiness.

- **Responsibility.** It is just and fair to hold a person accountable who intentionally chooses to commit a crime.
- **Deterrence.** Individuals who act with a criminal intent pose a threat to society and should be punished in order to discourage them from violating the law in the future and in order to deter others from choosing to violate the law.
- **Punishment.** The punishment should fit the crime. The severity of criminal punishment should depend on whether an individual's act was intentional, reckless, or accidental.

The concept of *mens rea* has traditionally been a source of confusion, and the first reaction of students and teachers has been to flee from the topic. This is understandable when it is realized that in 1972, U.S. statutes employed seventy-six different terms to describe the required mental element of federal crimes. This laundry list included terms such as *intentionally, knowingly, fraudulently, designedly, recklessly, wantonly, unlawfully, feloniously, willfully, purposely, negligently, wickedly,* and *wrongfully.* These are what Justice Jackson termed "the variety, disparity and confusion" of the judicial definition of the "elusive mental element" of crime.[4]

The Evidentiary Burden

The prosecution must establish the required *mens rea* beyond a reasonable doubt. Professor Hall observed that we cannot observe or record what goes on inside an individual's mind. The most reliable indication of intent is a defendant's confession or statement to other individuals. Witnesses may also testify that they saw an individual take careful aim when shooting or that a killing did not appear to be accidental.[5]

In most cases, we must look at the surrounding circumstances and apply our understanding of human behavior. In *People v. Conley,* a high school student at a party hit another student with a wine bottle, breaking the victim's upper and lower jaws, nose, and cheek and permanently numbing his mouth. The victim and his friend were alleged to have made insulting remarks at the party and were leaving when one of them was assaulted with a wine bottle. The attacker was convicted of committing an aggravated battery that "intentionally" or "knowingly" caused "great bodily harm or permanent disability or disfigurement." The defendant denied possessing this intent. An Illinois appellate court held that the "words, the weapon used, and the force of the blow . . . the use of a bottle, the absence of warning and the force of the blow are facts from which the jury could reasonably infer the intent to cause permanent disability." In other words, the Illinois court held that the defendant's actions spoke louder than his words in revealing his thoughts. Evidence that helps us indirectly establish a criminal intent or criminal act is termed **circumstantial evidence**.[6]

The Model Penal Code Standard

The common law provided for two confusing categories of *mens rea,* a general intent and a specific intent. These continue to appear in various state statutes and decisions.

A **general intent** is simply an intent to commit the *actus reus* or criminal act. There is no requirement that prosecutors demonstrate that an offender possessed an intent to violate the law, an awareness that the act is a crime, or that the act will result in a particular type of harm. Proof of the defendant's general intent is typically inferred from the nature of the act and the surrounding circumstances. The crime of battery or a nonconsensual, harmful touching provides a good illustration of a general intent crime. The prosecutor is only required to demonstrate that the accused intended to commit an act that was likely to substantially harm another. In the case of a battery, this may be inferred from factors such as the dangerous nature of the weapon, the number of blows, and the statements uttered by the accused. A statute that provides for a general intent typically employs terms such as *intentionally* or *willfully* to indicate that the crime requires a general intent.

A **specific intent** is a mental determination to accomplish a specific result. The prosecutor is required to demonstrate that the offender possessed the intent to commit the *actus reus* and then is required to present additional evidence that the defendant possessed the specific intent to accomplish a particular result. For example, a battery with an intent to kill requires proof of a battery along with additional evidence of a specific intent to murder the victim. The classic example is common law burglary. This requires the *actus reus* of breaking and entering and evidence of a specific intent to commit a felony inside the dwelling. Some commentators refer to these offenses as **crimes of cause and result** because the offender possesses the intent to "cause a particular result."

Courts often struggle with whether statutes require a general or specific intent. The consequences can be seen from the Texas case of *Alvarado v. State.* The defendant was convicted of "intentionally and knowingly" causing serious bodily injury to her child by placing him in a tub of hot water. The trial judge instructed the jury that they were merely required to find that the accused deliberately placed the child in the water. The appellate court overturned the conviction and ruled that the statute required the jury to find that the defendant possessed the intent to place the child in hot water, as well as the specific intent to inflict serious bodily harm.[7]

You may encounter two additional types of common law intent. A **transferred intent** applies when an individual intends to attack one person but inadvertently injures another. In *People v. Conley,* Conley intended to hit Marty but instead struck and inflicted severe injuries on Sean. Nevertheless, he was convicted of aggravated battery. The classic formulation of the common law doctrine of transferred intent states that the defendant's guilt is "exactly what it would have been had the blow fallen upon the intended victim instead of the bystander." Transferred intent also applies to property crimes in cases where, for example, an individual intends to burn down one home, and the wind blows the fire onto another structure, burning the latter dwelling to the ground.

Constructive intent is a fourth type of common law intent. This was applied in the early twentieth century to protect the public against reckless drivers and provides that individuals who are grossly and wantonly reckless are considered to intend the natural consequences of their actions. A reckless driver who caused an accident that resulted in death is, under the doctrine of constructive intent, guilty of a willful and intentional battery or homicide.

In 1980, the U.S. Supreme Court complained that the common law distinction between general and specific intent had caused a "good deal of confusion."[8] The Model Penal Code attempted to clearly define the mental intent required for crimes by providing four easily understood levels of responsibility. All crimes requiring a mental element (some do not, as we shall see) must include one of the four mental states provided in the Model Penal Code. These four types of intent, in descending order of culpability, are

For a deeper look at this topic, visit the study site.

- purposely,
- knowingly,
- recklessly, and
- negligently.

Model Penal Code

Section 2.02. General Requirements of Culpability

(1) Minimum Requirements of Culpability. . . . [A] person is not guilty of an offense unless he acted purposely, knowingly, recklessly or negligently . . . with respect to each material element of the offense.

(2) Kinds of Culpability Defined.

(a) **Purposely**.

A person acts purposely with respect to material elements of an offense when:

(i) . . . it is his conscious object to engage in conduct of that nature or to cause such a result. . . .

(b) **Knowingly**.

A person acts knowingly . . . when:

(i) If the element involves the nature of his conduct, . . . he is aware of the existence of such circumstances or he believes or hopes that they exist; and

(ii) If the element involves a result of his conduct, he is aware that it is practically certain that his conduct will cause such a result.

(c) **Recklessly**.

A person acts recklessly with respect to a material element of an offense when he consciously disregards a substantial and unjustifiable risk that the material element exists or will result from his conduct. The risk must be of such a nature and degree that, considering the nature and purpose of the actor's conduct and the circumstances known to him, its disregard involves a gross deviation from the standard of conduct that a law-abiding person would observe in the actor's situation.

(d) **Negligently**.

A person acts negligently with respect to a material element of an offense when he should be aware of a substantial and unjustifiable risk that the material element exists or will result from his conduct. The risk must be of such a nature and degree that the actor's failure to perceive it, considering the nature and purpose of his conduct and the circumstances known to him, involves a gross deviation from the standard of care that a reasonable person would observe in the actor's situation.

Analysis

- *Purposely.* "You borrowed my car and wrecked it on purpose."
- *Knowingly.* "You may not have purposely wrecked my car, but you knew that you were almost certain to get in an accident because you had never driven such a powerful and fast automobile."
- *Recklessly.* "You may not have purposely wrecked my car, but you were driving over the speed limit on a rain-soaked and slick road in heavy traffic and certainly realized that you were extremely likely to get into an accident."
- *Negligently.* "You may not have purposely wrecked my car and apparently did not understand the power of the auto's engine, but I cannot overlook your lack of awareness of the risk of an accident. After all, any reasonable person would have been aware that such an expensive sports car would pack a punch and would be difficult for a new driver to control."

We now turn our attention to a discussion of each type of criminal intent.

PURPOSELY

The Model Penal Code established *purposely* as the most serious category of criminal intent. This merely means that a defendant acted "on purpose" or "deliberately." In legal terms, the defendant must possess a specific intent or "conscious object" to commit a crime or cause a result. A murderer pulls the trigger with the purpose of killing the victim, the burglar breaks and enters with the purpose of committing a felony inside the dwelling, and a thief possesses the purpose of permanently depriving an individual of the possession of his or her property.

Did Ferino commit a hate crime?

Commonwealth v. Ferino, 640 A.2d 934
(Pa. Super. Ct. 1994). Opinion by: Popovich, J.

Issue

We are asked to review the judgment of sentence (1–2 years imprisonment, to be followed by a consecutive term of 4 years probation) for ethnic intimidation and terroristic threats by the appellant, Theresa Ferino, a/k/a Delores Cullin.

Facts

Viewing the evidence in a light most favorable to the verdict-winner, as well as all reasonable inferences to be drawn therefrom, it appears that at approximately 3:00 a.m. on the 31st day of July, 1992, Emmitt Harris (a 20-year-old black male) and his friend of some 3 years Matthew Chapman (a 17-year-old white male) arrived at the Arlington Deli on the South Side of Pittsburgh.

The Deli had closed at 9:00 p.m. and Harris, as the nighttime stock boy, had asked Chapman to assist him in cutting boxes and depositing the refuse in the dumpster in the rear of the establishment. After the two had disposed of a number of the boxes, the two were en route to the dumpster when the appellant was observed "walking up towards" them. Both had seen her patronizing the Deli and Harris knew her as a resident of the area.

As the appellant walked toward Harris and Chapman, and at a distance of about 50 yards, she extended her arms and said: "I'm going to kill you, you f—king n__, and fired two shots." Nothing was said by Harris or Chapman to the appellant prior to or after the shooting. The two were so frightened that they ran into the Deli and phoned "911." More specifically, the events during the vocalization of the appellant's statement were described by Harris as follows:

She [the appellant] was holding a gun and like taking steps towards us, coming—like she was coming up the street towards the deli, and she fired two shots. Me and Matt ran back into the store.

The police arrived within 30–45 minutes of the call and found the appellant in the street. However, no arrest was made of the appellant at that time. Rather, it was not until Harris phoned Pittsburgh police officer Gerald Watkins during the 1st week of August, 1992, out of which a report was filed and forwarded to Watkins, that he and Chapman were interviewed,

a complaint was filed and a search warrant was issued on August 12th leading to the seizure of a .38 caliber Taurus handgun from the appellant's residence (located 5–6 houses away from the Deli). Her arrest occurred on the same day as the search of her premises.

At the non-jury trial before the Honorable James F. Clarke, Harris' account of what transpired was corroborated by Chapman. For example, Chapman testified that the appellant "was walking towards us . . . and holding the gun" when she made the threatening remark. Chapman also indicated that the appellant "pointed the gun" before firing it, but he did not state at whom the weapon was aimed. Yet, Chapman, as well as Harris, became "scared" and "darted" back into the store." Further, like Harris, Chapman disclosed that he had seen the appellant in the Deli, but he never had any problems with her.

After the Commonwealth rested, the appellant took the stand and denied ever brandishing or firing a weapon at Harris or calling him a "n__." . . . At the close of testimony, the appellant was found guilty of ethnic intimidation and terroristic threats . . . This appeal ensued raising questions concerning the sufficiency and weight of the evidence, as well as a claim that trial counsel was ineffective.

On the ethnic intimidation charge, the statute defines the offense thusly:

(a) **Offense defined**.—A person commits the offense of ethnic intimidation if, with malicious intention toward the race, color, religion or national origin of another individual or group of individuals, he commits an offense under any other provision of this article or under Chapter 33 (relating to arson, criminal mischief and other property destruction) exclusive of section 3307 (relating to institutional vandalism) or under section 3503 (relating to criminal trespass) or under section 5504 (relating to harassment by communication or address) with respect to one or more members of such group or their property. . . .

(c) **Definition**.—As used in this section "malicious intention" means the intention to commit any act, the commission of which is a necessary element of any offense referred to in subsection (a) motivated by hatred toward the race, color, religion or national origin of another individual or group of individuals.

Reasoning

It is the appellant's contention that she was charged with ethnic intimidation merely because she used the word "n__" during the shooting. According to the appellant, to uphold her conviction on this charge would be to "criminalize the use of a particular word, even [though] used while in the commission of a crime." She would have us focus on the "crime itself" in determining whether it was racially motivated. By doing so, she argues, the evidence "clearly showed that the crime could only have been committed as the result of ill will between the neighbors and the black victim and his white employer. Therefore, the evidence was insufficient to show that the purported crime was directed at the victim because he was black. Without more than the use of the word 'n___' . . . the evidence was insufficient in law to prove beyond a reasonable doubt that the appellant was guilty of the crime of which she stands convicted."

In *Commonwealth v. Rink,* 574 A.2d 1078 (Pa. Super. 1990), in the only other case to date to reach the appellate courts on the charge of ethnic intimidation, this Court affirmed the judgment of sentence. In *Rink,* the victim, his wife and four children were the only black individuals living in the lower Frankford section of the City of Philadelphia for 2 years. During this time there had been no incidents of racism against the victim or his family.

On November 6, 1987, at about 10:00 P.M., Mr. Snow was driven home by his friend Al Bendzynski, a co-worker, with whom he intended to share the six pack of beer he had brought home with him. There was a crowd of teenagers across the street in front of the Snow residence playing a radio loudly. Mr. Snow asked them to hold the noise down. Moments after entering his home, Mr. Snow heard a knock at the front door. The group of teenagers was now at his front steps. Believing his wife and children to be asleep upstairs, he shut the door behind him to confront the knocker. A two-by-four hit him at the thigh, knocking him off the step. Armed with sticks, the crowd of about sixteen or seventeen white youths then started to pummel him on the arms, head, and body. They threw objects at his home, breaking windows. Appellant was urging the group to "kill the n__, get him."

When Mrs. Snow came to the door, her husband was on the ground surrounded by a group of young white males, wielding two-by-fours, some holding on to her husband, others tossing beer bottles at them and the house, cussing, yelling that they hated n__s, and "kill a couple of . . . n__s." Prominent in the group, holding a board and urging the group to "kill

the n___s" was the appellant. The appellant punched Mrs. Snow and called her a "bitch" and "n___." The crowd of youths dispersed as the police arrived on the scene. At the end of it all, Mr. Snow was bleeding, suffering from contusions of arms, legs, and body. The housefront was in shambles with both first and second floor windows shattered, including the windows of the living room located six feet behind the porch windows.

It was the appellant's position that the victim's co-worker (Mr. Bendzynski) had engaged in a friendly discussion with the appellant's group, but upon leaving Bendzynski "patted" one of the teenagers on the cheek. This offended the youth and prompted the person to go to the victim's home and seek an apology from Mr. Bendzynski. . . . [W]e found "overwhelming" evidence that the victim, and not Mr. Bendzynski, was the object of the teenagers' wrath. . . . [I]f one were to believe the appellant's view of the disgruntled teenager seeking retribution from Mr. Bendzynski, amazingly Mr. Bendzynski and his vehicle (situated outside the victim's home) remained unscathed in the assault. Even Mr. Bendzynski conceded at trial that "the teenagers did not seem to want him, but were after [the victim]." Further, no other slurs were directed at Mr. Bendzynski. . . .

Instantly, unlike in *Rink,* we had the use of a single racial word preceding the discharge of a weapon in the direction of not only Harris but his white companion (Chapman) as well. This behavior by the appellant was sufficiently intimidating that it "frightened" Chapman as well as Harris to flee the scene in tandem to seek the safety and security of the Deli.

It is to be recalled that both Harris and Chapman were consistent in their accounts painting a picture of the appellant as someone walking toward the two, with arms outstretched and weapon in hand, but neither testified at whom the weapon was directed either prior to, during or after the shooting. . . .

Initially, we read Section 2710 in a common sense fashion, so as to give effect to all of its provisions as intended by the Legislature. In the course of effectuating the commission of such a crime, the actor must manifest a malicious intent toward the intended victim and have as its motivation the hatred of the victim's "race, color, religion or national origin."

It cannot be disputed that, viewed in a vacuum (i.e., without the utterance directed at Harris' race), the appellant's aiming and firing a weapon at and in the direction of Harris would not have taken on the prohibited conduct of ethnic intimidation. Neither do we believe that the line of criminality proscribed by Section 2701 was crossed when the appellant aimed and fired a weapon in the direction of Harris and Chapman preceded by the threat ("I'm going to kill you, you f—king n__"). This was insufficient to evidence an intention malicious in nature and having as its origin racial prejudice which evoked or was the underlying cause for the prohibited behavior. In other words, the singularity of the act committed by the appellant, directed as it was against both Harris and his companion (a Caucasian), the antecedent of which was neither a harsh word, gesture nor conduct exhibited between the victim and the appellant, we do not believe rises to the proof-level sufficient to constitute a contravention of the ethnic intimidation statute. Stated otherwise, the appellant's conduct was isolated in nature, brief in its execution and unattended by any trappings consistent with a finding that the terroristic threat had an origin of malicious intent "motivated by a hatred toward race, color . . . or national origin" of the victim. . . .

Holding

Therefore, based on the unique facts at bar and for the reasons stated herein, we fail to discern such a malevolence on the part of the appellant directed specifically at the victim because of his race to justify an affirmance of the judgment of sentence for ethnic intimidation. . . .

Consistent with our determination that the evidence was insufficient to sustain the conviction for ethnic intimidation, we reverse the judgment of sentence for such offense and let stand the conviction for terroristic threats. . . .

Questions for Discussion

1. What is the prosecution required to establish to convict Ferino of ethnic intimidation?

2. Compare and contrast the facts in *Ferino* with the facts in *Rink*.

3. Should the court have given greater importance to Ferino's statement when she shot at Emmitt and Matthew?

4. Do you agree with the court's decision?

You Decide

5.1 Five African-American juveniles between the ages of eleven and fourteen were walking in the street when they heard a vehicle approaching and moved onto the sidewalk. Hennings, a Caucasian, drove past the young men and shouted at them to "get the f__ off the road." One of the young men, K.W. yelled back at Hennings that "[W]e don't have to get the f__ off the street."

Hennings exited the truck and threatened the young men with a pocket knife with a serrated blade between three and four inches long. Four of the juveniles fled, although K.W. stood his ground. K.W. challenged Hennings that "if you drop the knife," "we'll beat [your] ass." The other four boys started back toward K.W., and Hennings walked back to his truck. Hennings called the boys "f___ng __s" as he got back into his truck.

Hennings sped off and circled back around the block. As the boys were crossing a street, they saw Hennings heading toward them in his truck. He aimed the truck at A.M., and the truck's tires drove over him. Hennings left the scene.

A.M. was able to recover from potentially severe injuries, although he suffered permanent scarring and discoloration across his body, including on his face.

Hennings, when questioned by the police, referred to the young men as "monkeys" and stated if they "don't have enough sense to stay out the f__ing road, . . . they deserve to get hit." Hennings' mother, who was present during the police interrogation, asked, "Why didn't you wait for 'em to move?" Hennings responded, "When they're standing in the f__ing road like stupid monkeys?" Hennings' parents suggested the complaint was brought against Hennings because the family was not well-liked because of their opinions on race relations. Was Hennings guilty of ethnic intimidation? See *State v. Hennings*, 791 N.W.2d 828 (Iowa 2010).

You can find the answer at www.sagepub.com/lippmanccl3e.

You can find more cases on the study site:
Commonwealth v. Barnette, *www.sagepub.com/lippmanccl3e*.

KNOWINGLY

An individual satisfies the knowledge standard when he or she is "aware" that circumstances exist or that a result is practically certain to follow from his or her conduct. Examples of knowledge of circumstances are to knowingly "possess narcotics" or to knowingly "receive stolen property." It is sufficient that a person is aware that there is a high probability that property is stolen; he or she need not be certain. An illustration of a result that is practically certain to occur is a terrorist who bombs a public building knowing the people inside are likely to be maimed or injured or to die.

The commentary to the Model Penal Code uses the example of treason to illustrate the difference between purpose and knowledge. In *United States v. Haupt,* Chicago resident Hans Haupt was accused of treason during World War II based on the assistance he provided to his son, whom he knew was a German spy. The U.S. Supreme Court ruled that treason requires a specific intent (purpose) to wage war on the United States. Haupt claimed that as a loving father, he knowingly assisted his son, who unfortunately happened to be sympathetic to the German cause, and he did not possess the purpose to injure the U.S. government. The Supreme Court, however, pointed to Haupt's statements that "he hoped that Germany would win the war" and that "he would never permit his son to fight for the United States" as indicating that Haupt's "son had the misfortune of being a chip off the old block."[9]

For an international perspective on this topic, visit the study site.

In the next case in the chapter, *State v. Nations,* the defendant remained "willfully blind" or deliberately unaware of the criminal circumstances and claims that she did not knowingly violate the law. This type of situation typically arises in narcotics prosecutions in which drug couriers claim to have been unaware that they were transporting drugs.[10]

Did the defendant know the dancer's age?

State v. Nations, 676 S.W.2d 282 (Mo. Ct. App. 1984). Opinion by: Satz, J.

Issue

Defendant, Sandra Nations, owns and operates the Main Street Disco, in which police officers found a scantily clad sixteen-year-old girl dancing for tips. Consequently, defendant was charged with endangering the welfare of a child "less than seventeen years old." Defendant was convicted and fined $1,000. Defendant appeals. We reverse.

Specifically, defendant argues the State failed to show she knew the child was under seventeen and, therefore, failed to show she had the requisite intent to endanger the welfare of a child "less than seventeen years old." We agree.

Reasoning

The pertinent part of section 568.050 provides as follows:

(1) A person commits the crime of endangering the welfare of a child if:

. . . .

(2) He knowingly encourages, aids or causes a child less than seventeen years old to engage in any conduct which causes or tends to cause the child to come within the provisions of subdivision (1)(c) . . . of section 211.031, RSMo. . . .

Thus, section 568.050 requires the State to prove the defendant "knowingly" encouraged a child "less than seventeen years old" to engage in conduct tending to injure the child's welfare; and "knowing" the child to be less than seventeen is a material element of the crime.

"Knowingly" is a term of art, whose meaning is limited to the definition given to it by our present criminal code. Literally read, the code defines "knowingly" as actual knowledge—"A person 'acts knowingly,' or with knowledge, (1) with respect . . . to attendant circumstances when he is aware . . . that those circumstances exist. . . ." So read, this definition of "knowingly" or "knowledge" excludes those cases in which "the fact [in issue] would have been known had not the person willfully 'shut his eyes' in order to avoid knowing." The Model Penal Code, the source of our criminal code, does not exclude these cases from its definition of "knowingly." Instead, the Model Penal Code proposes that "[when] knowledge of the existence of a particular fact is an element of an offense, such knowledge is established if a person is aware of a high probability of its existence." . . .

The additional or expanded definition of "knowingly" proposed in section 2.02(7) of the Model Penal Code "deals with the situation British commentators have denominated **willful blindness** or connivance," the case of the actor who is aware of the probable existence of a material fact but does not satisfy himself that it does not in fact exist. . . . The inference of "knowledge" of an existing fact is usually drawn from proof of notice of substantial probability of its existence, unless the defendant establishes an honest, contrary belief. . . .

Our legislature, however, did not enact this proposed definition of "knowingly." . . . The sensible, if not compelling, inference is that our legislature rejected the expansion of the definition of "knowingly" to include willful blindness of a fact, and chose to limit the definition of "knowingly" to actual knowledge of the fact. Thus, in the instant case, the State's burden was to show defendant actually was aware the child was under seventeen, a heavier burden than showing there was a "high probability" that the defendant was aware that the child was under seventeen. . . .

Facts

The record shows that, at the time of the incident, the child was sixteen years old. When the police arrived, the child was dancing on stage for tips with another female. The police watched her dance for some five to seven minutes before approaching defendant in the service area of the bar. Believing that one of the girls appeared to be "young," the police questioned defendant about the child's age. Defendant told them that both girls were of legal age and that she had checked the girls' identification when she hired them. When the police questioned the child, she initially stated that she was eighteen but later admitted that she was only sixteen. She had no identification.

The State also called the child as a witness. Her testimony was no help to the State. She testified the defendant asked her for identification just prior to the police arriving, and she was merely crossing the stage to get her identification when the police took her into custody. Nor can the State secure help from the defendant's testimony.

She simply corroborated the child's testimony; i.e., she asked the child for her identification; the child replied she would "show it to [her] in a minute"; the police then took the child into custody.

Holding

These facts simply show defendant was untruthful. Defendant could not have checked the child's identification, because the child had no identification with her that day, the first day defendant hired the child. This does not prove that defendant knew the child was less than seventeen years old. At best, it proves defendant did not know or refused to learn the child's age. . . . Having failed to prove defendant knew the child's age was less than seventeen, the State failed to make a . . . case.

Admittedly, a person in defendant's shoes can easily avoid conviction of a crime under section 568.050 by simply refusing to check the age of dancers. This result is to be rectified, however, by the legislature, not by judicial redefinition of already precisely defined statutory language or by improper inferences from operative facts. The Model Penal Code's expanded definition of "knowingly" attracts us by its logic. Apparently, it was not as attractive to our legislature for use throughout our criminal code. . . .

Questions for Discussion

1. Why does the court conclude that the defendant is not guilty under the statute of endangering the welfare of the young dancer?

2. In your view, was the defendant aware that there was a "high probability" that the dancer was under seventeen and for that reason intentionally avoided checking her age?

3. How does the Missouri statute differ from the Model Penal Code in regard to willful blindness? What is the impact of the court decision for offenses involving the possession of narcotics?

4. How would you amend the Missouri statute to eliminate the willful blindness defense?

5. If you were a judge, how would you rule in *Nations*?

You Decide

5.2 The government indicted fifteen men for offenses arising from their participation in an illegal gambling enterprise. Nicholas Janis was convicted of conducting an illegal gambling business or aiding and abetting its conduct and was sentenced to 60 days.

The head of the gambling enterprise was Thomas Orlando. The enterprise operated a series of "wirerooms," where bets were accepted on sporting events. The "wirerooms" also were the site of casino gambling nights at which the invitees played blackjack, craps, and poker.

Janis was a gambler and was acquainted with members of the Orlando organization, including Thomas Orlando. In the fall of 1982, Merino, who unknown to Janis was a government informant, rented a house owned by

Janis for one of Merino's friends, Pluta, a bookmaker affiliated with the Orlando organization. Pluta did not immediately move in, and from November 1982 until July 1983 the house was in continuous use as a wireroom, operated first by Merino and then by Pluta. Merino and Pluta were the tenants, although the rent was paid by Michael Gioringo, whom Janis knew to be a member of the Orlando organization. After Pluta moved out of the house, Janis offered the key to the house to Orlando.

Janis claimed that he did not know that the house was being used as a wireroom. The government responded that the rented house was a short distance from the street that Janis drove down on his way to work. "It would have been easy for him to drive by the house from time to time to see what was doing, and if he had done so he might have discovered its use as a wireroom."

Janis claimed that the judge improperly had issued a "willful blindness" instruction to the jury because there was no evidence that he knew or strongly suspected that the house was being used for "shady dealings" or that he took steps to make sure that he did not "acquire full or exact knowledge of the nature and extent of those dealings." Is Janis correct in his claim? See *U.S. v. Giovannetti*, 919 F.2d 1223 (7th Cir. 1991).

You can find the answer at www.sagepub.com/lippmanccl3e.

RECKLESSLY

We all know people who enjoy taking risks and skirting danger and who are confident that they will beat the odds. These reckless individuals engage in obviously risky behavior that they know creates a risk of substantial and unjustifiable harm and yet do not expect that injury or harm will result.

Why does the law consider individuals who are reckless less blameworthy than individuals who act purposely or knowingly?

- Individuals who act purposely deliberately create a harm, and individuals who act knowingly are aware that injury is certain to follow.
- Individuals acting recklessly, in contrast, disregard a strong probability that harm will result.

Recklessness is big, bold, and outrageous. Recklessness involves a conscious disregard of a substantial and unjustifiable risk. This must constitute a gross deviation from the standard of conduct that a law-abiding person would observe in a similar situation. The reckless individual speeds down a street where children usually play, builds and sells to an uninformed buyer a house that is situated on a dangerous chemical waste dump, manufactures an automobile with a gas tank that likely will explode in the event of an accident, or locks the exit doors of a rock club during a performance in which a band ignites fireworks.

The Model Penal Code provides a two-fold test for reckless conduct:

- *A Conscious Disregard of a Substantial and Unjustifiable Risk.* The defendant must be *personally aware* of a severe and serious risk. *Unjustifiable* means that the harm was not created in an effort to serve a greater good, such as speeding down the street in an effort to reach the hospital before a passenger who was in auto accident bleeds to death.
- *A Gross Deviation From the Standard That a Law-Abiding Person Would Observe in the Same Situation.* The defendant must have acted in a fashion that demonstrates a clear lack of judgment and concern for the consequences. This must clearly depart from the behavior that would be expected of other law-abiding individuals. Note this is an *objective test based on the general standard of conduct.*

In *Hranicky v. State,* the next case in the chapter, the court is confronted with the challenge of determining whether the defendant recklessly caused serious bodily injury to his stepdaughter.

Was the defendant aware of the risk posed by the tigers to his daughter?

Hranicky v. State, 13–00–431–CR (Tex. App. 2005). Opinion by: Castillo, J.

Bobby Lee Hranicky appeals his conviction for the second-degree felony offense of recklessly causing serious bodily injury to a child. A jury found him guilty, sentenced him to eight years confinement in the Institutional Division of the Texas Department of Criminal Justice, and assessed a $5,000 fine. On the jury's recommendation, the trial court suspended the sentence and placed Hranicky on community supervision for ten years.

Facts

A newspaper advertisement offering tiger cubs for sale caught the eye of eight-year-old Lauren Villafana. She decided she wanted one. She expressed her wish to her mother, Kelly Dean Hranicky, and to Hranicky, her stepfather. Over the next year, the Hranickys investigated the idea by researching written materials on the subject and consulting with owners of exotic animals. They visited tiger owner and handler Mickey Sapp several times. They decided to buy two rare tiger cubs from him, a male and a female whose breed is endangered in the wild. . . .

Sapp trained Hranicky in how to care for and handle the animals. In particular, he demonstrated the risk adult tigers pose for children. Sapp escorted Hranicky, Kelly Hranicky, and Lauren past Sapp's tiger cages. He told the family to watch the tigers' focus of attention. The tigers' eyes followed Lauren as she walked up and down beside the cages.

The Hranickys raised the cubs inside their home until they were six or eight months old. Then they moved the cubs out of the house, at first to an enclosed porch in the back and ultimately to a cage Hranicky built in the yard. The tigers matured into adolescence. The male reached 250 pounds, the female slightly less. Lauren actively helped Hranicky care for the animals.

By June 6, 1999, the tigers were two years old. Lauren was ten. She stood fifty-seven inches tall and weighed eighty pounds. At dusk that evening, Lauren joined Hranicky in the tiger cage. Suddenly, the male tiger attacked her. It mauled the child's throat, breaking her neck and severing her spinal cord. She died instantly.

The record reflects four different versions of the events that led to Lauren's death. Hranicky told the grand jury that he and Lauren were sitting side-by-side in the cage about 8:00 p.m., petting the female tiger. A neighbor's billy goat cried out. The noise attracted the male tiger's attention. He turned toward the sound. The cry also caught Lauren's attention. She stood and looked at the male tiger. When Lauren turned her head toward the male tiger, "That was too much," Hranicky told the grand jury. The tiger attacked. Hranicky yelled. The tiger grabbed Lauren by the throat and dragged her across the cage into a water trough. Hranicky ran after them. He struck the tiger on the head and held him under the water. The tiger released the child.

Kelly Dean Hranicky testified she was asleep when the incident occurred. She called for emergency assistance. Through testimony developed at trial, she told the dispatcher her daughter had fallen from a fence. She testified she did not remember giving that information to the dispatcher. However, police officer Daniel Torres, who responded to the call, testified he was told that a little girl had cut her neck on a fence.

Hranicky gave Torres a verbal statement that evening. Torres testified Hranicky told him that he had been grooming the female tiger. He asked Lauren to come and get the brush from him. Lauren came into the cage and grabbed the brush. Hranicky thought she had left the cage, because he heard the cage door close. Then, however, Hranicky saw Lauren's hand "come over and start grooming the female, start petting the female cat, and that's when the male cat jumped over." The tiger grabbed the child by the neck and started running through the cage. It dragged her into the water trough. Hranicky began punching the tiger in the head, trying to get the tiger to release Lauren.

Justice of the Peace James Dawson performed an inquest at the scene of the incident. Judge Dawson testified Hranicky gave him an oral statement also. Hranicky told him Lauren went to the cage on a regular basis and groomed only the female tiger. He then corrected himself to say she actually petted the animal. Hranicky was "very clear about the difference between grooming and petting." Hranicky maintained that Lauren never petted or groomed the male tiger. Hranicky told Dawson that Lauren asked permission to enter the cage that evening, saying "Daddy, can I come in?"

Sapp, the exotic animal owner who sold the Hranickys the tigers, testified Hranicky told him yet another version of the events that night. When Sapp asked Hranicky how it happened, Hranicky replied, "Well, Mickey, she just snuck in behind me." Hranicky admitted to Sapp he had allowed Lauren to enter the cage. Hranicky told Sapp he had lied because he did not want Sapp to be angry with him.

Hranicky told the grand jury that Sapp and other knowledgeable sources had said "there was no problem in taking a child in the cage." He did learn children were especially vulnerable, because the tigers would view them as prey. However, Hranicky told the grand jury, he thought the tigers would view Lauren differently than they would an unfamiliar child. He believed the tigers would not attack her, he testified. They would see her as "one of the family." Hranicky also told the grand jury the tigers' veterinarian allowed his young son into the Hranickys' tiger cage.

Several witnesses at trial contradicted Hranicky's assessment of the level of risk the tigers presented, particularly to children. Sapp said he told the Hranickys it was safe for children to play with tiger cubs. However, once the animals reached forty to fifty pounds, they should be confined in a cage and segregated from any children. "That's enough with Lauren, any child, because they play rough, they just play rough." Sapp further testified he told the Hranickys to keep Lauren away from the tigers at that point, because the animals would view the child as prey. He also said he told Lauren directly not to get in the cage with the tigers. Sapp did not distinguish between children who were strangers to the tigers and those who had helped raise the animals. He described any such distinction as "ludicrous." In fact, Sapp testified, his own two children had been around large cats all of their lives. Nonetheless, he did not allow them within six feet of the cages.

The risk is too great, he told the jury. The Hranickys did not tell him that purchasing the tigers was Lauren's idea. Had he known, he testified, "that would have been the end of the conversation. This was not for children." He denied telling Hranicky that it was safe for Lauren to be in the cage with the tigers.

Charles Currer, an animal care inspector for the U.S. Department of Agriculture (USDA), met Hranicky when Hranicky applied for a USDA license to exhibit the tigers. Currer also denied telling Hranicky it was permissible to let a child enter a tiger's cage. He recalled giving his standard speech about the danger big cats pose to children, telling him that they "see children as prey, as things to play with."

On his USDA application form, Hranicky listed several books he had read on animal handling. One book warned that working with exotic cats is very dangerous. It emphasized that adolescent males are particularly volatile as they mature and begin asserting their dominance. Big cat handlers should expect to get jumped, bit, and challenged at every juncture. Another of the listed books pointed out that tigers give little or no warning when they attack. The book cautioned against keeping large cats such as tigers as pets.

Veterinarian Dr. Hampton McAda testified he worked with the Hranickys' tigers from the time they were six weeks old until about a month before the incident. McAda denied ever allowing his son into the tigers' cage. All large animals present some risk, he testified. He recalled telling Hranicky that "wild animals and female menstrual periods . . . could cause a problem down the road" once both the animals and Lauren matured. Hranicky seemed more aware of the male tiger, the veterinarian observed, and was more careful with him than with the female. . . .

James Boller, the chief cruelty investigator for the Houston Society for the Prevention of Cruelty to Animals, testified that tigers, even those raised in captivity, are wild animals that act from instinct. Anyone who enters a cage with a conscious adult tiger should bring a prop to use as a deterrent. Never take one's eyes off the tiger, Evans told the jury. Never make oneself appear weak and vulnerable by diminishing one's size by crouching or sitting. Never bring a child into a tiger cage. The danger increases when the tigers are in adolescence, which begins as early as two years of age for captive tigers. Entering a cage with more than one tiger increases the risk. Entering with more than one person increases the risk further. Entering with a child increases the risk even more. Tigers' activity level depends on the time of day. . . . Boller identified eight o'clock on a summer evening as a high activity time. A child should never enter a tiger cage in the first place, Boller testified. Taking a child into a tiger cage "during a high activity time for the animal is going to increase your risk dramatically."

Dr. Richard Villafana, Lauren's biological father, told the jury he first learned of the tigers when his daughter told him over the phone she had a surprise to show him at their next visit. When he came to pick her up the following weekend, he testified, she took him into the house and showed him the female cub. Villafana described his reaction as "horror and generalized upset and dismay, any negative term you care to choose." He immediately decided to speak to Kelly Hranicky about the situation. He did not do so in front of Lauren, however, in an effort to avoid a "big argument." Villafana testified he later discussed the tigers with Kelly Hranicky, who assured him Lauren was safe. . . . As the tigers matured, no one told Villafana the Hranickys allowed Lauren in the cage with them. Had he known, he "would have talked to Kelly again" and "would have told her that [he] was greatly opposed to it and would have begged and pleaded with her not to allow her in there." He spoke to his daughter about his concerns about the tigers "almost every time" he saw her.

Kelly Hranicky told the jury Lauren was a very obedient child. Villafana agreed. Lauren would not have gone into the tiger cage that evening without Hranicky's permission.

Issue

. . . Did Hranicky act in a reckless fashion?

Reasoning

The record reflects that each of the witnesses who came into contact with Hranicky in connection with the tigers testified they told him that (1) large cats, even those raised in captivity, are dangerous, unpredictable wild animals; and (2) children were particularly at risk from adolescent and adult tigers, especially males. Expert animal handlers whom Hranicky consulted and written materials he claimed to have read warned Hranicky that the risks increased with adolescent male tigers, with more than one person in the cage, with more than one tiger in the cage, at dusk during the animals' heightened activity period, and when diminishing one's size by sitting or crouching on the ground. They each cautioned that tigers attack swiftly, without warning, and are powerful predators.

Further, Hranicky's initial story to Sapp that Lauren had sneaked into the cage evidences Hranicky's awareness of the risk. The jury also could have inferred his awareness of the risk when he concealed from Sapp that the family was purchasing the tigers for Lauren. The jury also could have inferred Hranicky's consciousness of guilt when he gave several different versions of what happened.

On the other hand, the record shows that before buying the tigers, Hranicky researched the subject and conferred with professionals. He received training in handling the animals. Further, Kelly Hranicky testified she also understood the warnings about not allowing children in the tiger cage to apply to strangers, not to Lauren. Hranicky told the grand jury he did not think the warnings applied to children, like Lauren, who had helped raise the animal. He said he had seen other handlers, including Sapp and McAda, permit Lauren and other children to go into tiger cages. He testified Currer told him it was safe to permit children in tiger cages. Further, while the State's witness described zoo policies for handling tigers, those policies were not known to the general public. Finally, none of the significant figures in Lauren's life fully appreciated the danger the tigers posed for Lauren. Hranicky was not alone in not perceiving the risk. . . .

Holding

Hranicky testified to the grand jury he did not view the risk to be substantial, because he thought the tigers were domesticated and had bonded with the family. He claimed not to have any awareness of any risk. The tigers were acting normally. Lauren had entered the cage numerous times to pet the tigers with no incident. Further, he asserted, other than a minor scratch by the male as a cub, the tigers had never harmed anyone. Thus, he argues, he had no knowledge of any risk.

Viewing all the evidence neutrally, favoring neither Hranicky nor the State, we find that proof of Hranicky's guilt of reckless injury to a child is not so obviously weak as to undermine confidence in the jury's determination. Nor do we do not find that the proof of his guilt is greatly outweighed by contrary proof.

Questions for Discussion

1. Did Hranicky's disregard constitute a substantial and unjustifiable risk? Did his actions constitute a gross deviation from the standard of conduct that a law-abiding person would observe in a similar situation?

2. Why does the court consider it to be a close call as to whether Hranicky was aware of the risk posed by the tigers to Lauren?

3. Would the result be the same in the event that the tigers attacked Lauren when they were tiger cubs and were first living in the home?

4. What if Bobby Lee Hranicky had been mauled and killed by the tiger? Would a court convict Kelly Hranicky of recklessly causing Bobby Lee's death?

You Decide

5.3 Norma Suarez left home with her son P. and her daughters N.E. and A.E. in the car. She stopped to visit Michelle Dominguez and then drove to the home of Violanda Corral, P.'s grandmother. Suarez left P. at Corral's home and started toward home. N.E. was in the front passenger seat and A.E. was in the back seat. Suarez arrived home to find that A.E. was not in the auto. It later was learned that A.E. had fallen from the car as the vehicle crossed the Continental Bridge, was struck by another car, and died of head injuries. Suarez was convicted of recklessly endangering A.E., who was three years old at the time, by failing to properly supervise her child. It was a crime in Texas at the time of this incident for the operator of a motor vehicle to fail to secure a child over two and younger than four years of age by a seat belt or child seat.

An investigating police officer testified that A.E. fell out of the front passenger window. The officer also found that the seat belt clips in the back seat were "pushed down . . . along the crease" indicating "non-use." Suarez contended that A.E. put the belt on herself when they left home. Dominguez testified that she later buckled A.E. in the car. Corral stated that she told Suarez to "make sure you buckle up the girls" and testified that she saw Suarez look toward the back seat and then put N.E. in the front seat. Corral indicated that she had no doubt that A.E. was properly secured with a seat belt. There was testimony that A.E. could unbuckle the seat belt herself. Other evidence indicated that Suarez stopped at a red light before driving across the bridge to ensure that A.E. was asleep.

Did Suarez recklessly cause A.E.'s death? See *Suarez v. State*, Tex. App. LEXIS 10799 (2003).

You can find the answer at www.sagepub.com/lippmanccl3e.

NEGLIGENTLY

Recklessness entails creating and disregarding a risk. The reckless individual consciously lives on the edge, walking on a ledge above the street. Negligence, in contrast, involves engaging in harmful and dangerous conduct while being unaware of a risk that a reasonable person would appreciate. The reckless individual would "play around" and push someone off a cliff into a pool of water that he or she knows contains a string of dangerous boulders and rocks. The negligent individual simply does not bother to check whether the water conceals a rock quarry before pushing another person off the cliff. Recklessness involves an awareness of harm that is lacking in negligence, and for that reason is considered to be of greater "moral blameworthiness."

In considering negligence, keep the following in mind:

- *Mental State.* The reckless individual is aware of and disregards the substantial and unjustifiable risk; the negligent individual is not aware of the risk.
- *Objective Standard.* Recklessness and negligence ask juries to decide whether the individual's conduct varies from that expected of the general public. The reckless individual grossly deviates from the standard of care that a law-abiding person would demonstrate in the situation; the negligent individual grossly deviates from the standard of care that a reasonable person would exhibit under a similar set of circumstances.

It is not always easy to determine whether a defendant was unaware of a risk and is guilty of negligence rather than recklessness. In *Tello v. State,* the defendant was convicted of criminally negligent homicide after a trailer that he was pulling came unhitched, jumped a curb, and killed a pedestrian. Tello argued that he had not previously experienced difficulties with the trailer and claimed to have been unaware that safety chains were required or that the hitch was clearly broken and in need of repair. The court convicted Tello of negligent homicide based on the fact that a reasonable person would have been aware that the failure to safely secure the trailer hitch constituted a gross deviation from the standard of care that an ordinary person would have exhibited and posed a substantial risk of death. Is it credible to believe that Tello regularly used the trailer and yet lacked awareness that the trailer was secured so poorly that a bump in the road was able to separate the trailer from the truck?[11]

People v. Baker illustrates the difficulty of distinguishing negligence from recklessness.

Was the babysitter guilty of negligent or reckless homicide?

People v. Baker, 771 N.Y.S.2d 607 (N.Y. App. Div. 2004). Opinion by: Rose, J.

Facts

After a three-year-old child died while defendant was babysitting in the child's home, she was charged with both intentional and depraved indifference murder. At trial, the evidence established that, on a warm summer night, the victim died of hyperthermia as a result of her prolonged exposure to excessive heat in a bedroom of her foster parents' apartment. The excessive heat was caused by the furnace having run constantly for many hours as the result of a short circuit in its wiring. The victim was unable to leave her bedroom, because defendant engaged the hook and eye latch on its door after putting her to bed for the night. Defendant then remained in the apartment watching television while the furnace ran uncontrollably. The victim's foster parents and another tenant testified that when they returned in the early morning hours and found the victim lifeless in her bed, the living room of the apartment where defendant sat waiting for them felt extremely hot, like an oven or a sauna, and the victim's bedroom was even hotter. Temperature readings taken later that morning during a police investigation while the furnace was still running indicated that the apartment's living room was 102 degrees Fahrenheit, the victim's bedroom was 110 degrees Fahrenheit, and the air coming from the vent in the bedroom was more than 130 degrees Fahrenheit.

In characterizing defendant's role in these events, the prosecutor argued that the key issue for the jury was whether or not defendant had intended to kill the victim. The prosecution's proof on this issue consisted primarily of the second of two written statements given by defendant to police during a four-hour interview conducted a few hours after the victim was found. In the first statement, defendant related that she had been aware of the oppressive heat in the victim's bedroom, kept the victim latched in because the foster parents had instructed her to do so, had not looked at or adjusted the thermostat even though the furnace was running on a hot day, heard the victim kicking and screaming to be let out, and felt the adverse effects of the heat on herself. The second statement, which defendant disavowed at trial, described her intent to cause the victim's death by turning up the thermostat to its maximum setting, closing all heating vents except the one in the victim's bedroom, and placing additional clothing on the victim, which she then removed after the victim died. Because these actions differed from those described in the first statement, and each reflects an intent to kill the victim, the jurors' initial task, as proposed by the prosecutor during summation, was to decide which statement they would accept.

After trial, the jury acquitted defendant of intentional murder, thereby rejecting the second statement, and instead convicted her of depraved indifference murder of a child. County court sentenced her to a prison term of fifteen years to life, and she now appeals.

Issue

Could the jury reasonably infer from the evidence a culpable mental state greater than criminal negligence due to the unique combination of events that led to the victim's death, as well as the lack of proof that defendant actually perceived and ignored an obvious and severe risk of serious injury or death?

Reasoning

The jury's finding that defendant was not guilty of intentional murder clearly indicates that it rejected defendant's second statement containing an explicit admission of an intent to kill. Although the excessive heat ultimately proved fatal, and defendant failed to remove the victim from her bedroom and made no effort to reduce the heat, the evidence does not establish that the defendant created dangerous conditions supporting the jury verdict of a wanton indifference to human life or a depravity of the mind.

Is the defendant guilty of reckless or negligent homicide? There is no evidence that defendant knew the actual temperature in any portion of the apartment or subjectively perceived a degree of heat that would have made her aware that serious injury or death from hyperthermia would almost certainly result. Put another way, the risk of serious physical injury or death was not so obvious under the circumstances that it demonstrated defendant's actual awareness. There was only circumstantial evidence on this point, consisting of the subjective perceptions of other persons who later came into the apartment from cooler outside temperatures. Defendant, who had been in the apartment as the heat gradually intensified over many hours, and who was described by others as appearing flushed and acting dazed, could not reasonably be presumed to have had the same perception of oppressive and dangerous heat. Rather, defendant testified that she knew only that the heat made her feel dizzy and uncomfortable, and she denied any awareness of a risk of death. Most significantly, there is no dispute that defendant remained in a room that was nearly as hot as the victim's bedroom for approximately nine hours and checked on the victim several times before the foster parents returned. This evidence of defendant's failure to perceive the risk of serious injury stands unrefuted by the prosecution.

Defendant's ability to appreciate such a risk was further brought into doubt by the prosecution's own expert witness, who described her as having borderline intellectual function, learning disabilities, and a full-scale IQ of only seventy-three. We also note that here, unlike where an unclothed child is shut outside in freezing temperatures, the circumstances are not of a type from which it can be inferred without a doubt that a person of even ordinary intelligence and experience would have perceived a severe risk of serious injury or death. . . .

A person is guilty of manslaughter in the second degree when he or she recklessly causes the death of another person and of criminally negligent homicide when, with criminal negligence, he or she causes the death of another person. Reckless criminal conduct occurs when the actor is aware of and consciously disregards a substantial and unjustifiable risk, and criminal negligence is the failure to perceive such a risk.

As we have noted, there is no support for a finding that defendant perceived and consciously disregarded the risk of death that was created by the combination of the "runaway" furnace and her failure to release the victim from her bedroom. None of defendant's proven conduct reflects such an awareness, and the fact that she subjected herself to the excessive heat is plainly inconsistent with a finding that she perceived a risk of death.

Holding

However, the evidence was sufficient to establish defendant's guilt beyond a reasonable doubt of criminally negligent homicide. A jury could reasonably conclude from the evidence that defendant should have perceived a substantial and unjustifiable risk that the excessive heat, in combination with her inaction, would be likely to lead to the victim's death. . . . Since defendant was the victim's caretaker, this risk was of such a nature that her failure to perceive it constituted a gross deviation from the standard of care that a reasonable person in the same circumstances would observe in such a situation. Thus, defendant's conduct was shown to constitute criminal negligence, and such a finding would not be against the weight of the evidence. Accordingly, we reduce the conviction from depraved indifference murder to criminally negligent homicide and remit the matter to county court for sentencing on the reduced charge.

Questions for Discussion

1. Explain the court's factual basis for determining that the defendant should be held liable for negligent rather than reckless homicide.

2. Should the appellate court overturn the verdict of the jurors who actually observed the trial?

3. As a judge, what would be your ruling in this case?

See more cases on the study site: **Koppersmith v. State,** *www.sagepub.com/lippmanccl3e.*

You Decide

5.4 Ginger McLaughlin was charged with child neglect. Both McLaughlin and her husband were 22 years of age. Ginger had two daughters aged 4 and 6; and two years into her marriage, she and her husband had a child. One year after they were married, Ginger's husband while babysitting spanked one of her daughters with a toy broom handle causing severe bruising on the 6-year-old, and he was charged with assault. He had gone to the bedroom of one of Ginger's daughters on two previous occasions and had spanked her without provocation; and he had hit Ginger on two separate occasions, in one instance causing a black eye. Between October 1977 and March 1978, Ginger's husband left the house; and when he returned, Ginger stated that he seemed to have greater control over his emotions. Between March and June 1978, her husband babysat for the girls without incident and seemed genuinely happy over the birth of their new baby in April 1978. A Children's Services Division caseworker warned Ginger that allowing her husband to be alone with the children created a "high risk" and urged her to enter into counseling with her husband. As a precaution following the birth of the new baby, Ginger's husband periodically stayed with a friend.

On the day of the relevant incident, Ginger and her husband had gotten into an argument and she left for 45 minutes to exchange some bottles for money at the store. "When she returned she found that the baby had been injured by her husband. They immediately took the child to a hospital, but he died a few days later of head injuries received from at least two blows." Ginger's husband was convicted of manslaughter, and Ginger was indicted for child neglect. Is Ginger guilty of negligence? See *State v. McLaughlin*, 600 P.2d 474 (Ore. Ct. App. 1979).

You can find the answer at www.sagepub.com/lippmanccl3e.

STRICT LIABILITY

We all have had the experience of telling another person that "I don't care why you acted in that way; you hurt me and that was wrong." This is similar to a strict liability offense. A **strict liability** offense is a crime that does not require a *mens rea,* and an individual may be convicted based solely on the commission of a criminal act.

Strict liability offenses have their origin in the industrial development of the United States in the middle of the nineteenth century. The U.S. Congress and various state legislatures enacted a number of **public welfare offenses** that were intended to protect society against impure food, defective drugs, pollution, and unsafe working conditions, trucks, and railroads. These *mala prohibita* offenses (an act is wrong because it is prohibited) are distinguished from those crimes that are *mala in se* (inherently wrongful, such as rape, robbery, and murder).

The common law was based on the belief that criminal offenses required a criminal intent; this ensured that offenders were morally blameworthy. The U.S. Supreme Court has pronounced that the requirement of a criminal intent, although not required under the Constitution, is "universal and persistent in mature systems of law."[12] Courts, however, have disregarded the strong policy in favor of requiring a criminal intent in upholding the constitutionality of *mala prohibita* laws. Congress and state legislatures typically indicate that these are strict liability laws by omitting language such as "knowingly" or "purposely" from the text of the law. Courts look to several factors in addition to the textual language in determining whether a statute should be interpreted as providing for strict liability:

- The offense is not a common law crime.
- A single violation poses a danger to a large number of people.
- The risk of the conviction of an "innocent" individual is outweighed by the public interest in preventing harm to society.
- The penalty is relatively minor.
- A conviction does not harm a defendant's reputation.
- The law does not significantly impede the rights of individuals or impose a heavy burden. Examples are the prohibition of acts such as "selling alcohol to minors" or "driving without a license."
- These are acts that most people avoid, and individuals who engage in such acts generally possess a criminal intent.

The argument for strict liability offenses is that these laws deter unqualified people from participating in potentially dangerous activities, such as the production and selling of pharmaceutical drugs, and that those who engage in this type of activity will take extraordinary steps to ensure that they proceed in a cautious and safe fashion. There is also concern that requiring prosecutors to establish a criminal intent in these relatively minor cases will consume time and energy and divert resources from other cases.

There is a trend toward expanding strict liability into the non–public welfare crimes that carry relatively severe punishment. Many of these statutes are criticized for imposing prison terms without providing for the fundamental requirement of a criminal intent. For instance, in *State v. York,* the defendant was sentenced to one year in prison in Ohio after he was convicted of having touched the buttocks of an eleven-year-old girl. The appellate court affirmed his conviction for "gross sexual imposition" and ruled that this was a strict liability offense and that the prosecutor was required to demonstrate only a prohibited contact with an individual under thirteen that could be perceived by the jury as sexually arousing or gratifying to the defendant.[13]

The U.S. Supreme Court indicated in *Staples v. United States* that it may not be willing to continue to accept the growing number of strict liability public welfare offenses. The National Firearms Act was intended to restrict the possession of dangerous weapons and declared it a crime punishable by up to ten years in prison to possess a "machine gun" without legal registration. The defendant was convicted for possession of an AR-15 rifle, which is a semiautomatic weapon that can be modified to fire more than one shot with a single pull of the trigger. The Supreme Court interpreted the statute to require a *mens rea,* explaining that the imposition of a lengthy prison sentence has traditionally required that a defendant possess a criminal intent. The Court noted that gun ownership is widespread in the United States and that a strict liability requirement would result in the imprisonment of individuals who lacked the sophistication to determine whether they purchased or possessed a lawful or unlawful weapon.[14]

The Model Penal Code, in section 1.04(5), accepts the need for strict liability crimes while limiting these crimes to what the code terms "violations." Violations are not subject to imprisonment and are punishable only by a fine, forfeiture, or other civil penalty, and they may not result in the type of legal disability (e.g., result in loss of the right to vote) that flows from a criminal conviction.

In the next case in the chapter, *State v. Walker,* Walker was convicted of knowingly or intentionally delivering cocaine within 1,000 feet of a school. The Indiana Supreme Court was asked to decide whether the trial court was correct in ruling that the prosecution was not required to establish that Walker knew that there was a school nearby, because this is a strict liability offense. The answer was important to Walker, because delivering the cocaine within 1,000 feet of a school enhanced his sentence from ten to twenty years in prison. Pay attention to the majority and to the dissenting opinion, and ask yourself whether this should be a strict liability offense.

Is dealing in cocaine within 1,000 feet of a school a strict liability offense?

State v. Walker, 668 N.E.2d 243 (Ind. 1996). Opinion by: Shephard, C.J.

Issue

Appellant Aaron Walker contends that to sustain a conviction for dealing in cocaine within 1,000 feet of a school, as a class A felony, the State must prove that the defendant had actual knowledge that the sale was occurring within 1,000 feet of a school.

Facts

The State charged Walker with dealing in cocaine after he sold the drug to an undercover police officer, Ernie Witten. Armed with a $20 bill to make a purchase and a microphone taped to his chest, Witten drove to the parking lot of an Indianapolis apartment complex near Public School No. 114. He noticed a group of young men sitting under a shade tree. One of these motioned to Witten, a sign the officer interpreted as asking what the officer wanted. Witten held up one finger, intending to indicate that he wanted one rock of cocaine. The young man made another motion that Witten construed as an instruction to pull around. The officer did so.

Once Witten had parked his truck, Walker approached and asked what he was looking for. Witten replied he wanted "a twenty," which is street slang for $20 worth of crack cocaine. Walker reached into his pocket, took out a plastic bag containing "several rocky hard white substances," and handed one to Witten. Witten gave Walker the marked $20 bill and the transaction was over.

Walker was eventually arrested and charged. A jury found him guilty of dealing in cocaine as a class A felony and determined that he was a habitual offender. The trial judge gave him the presumptive sentence for dealing, thirty years, and added thirty years for the habitual offender finding.

The statute under which Walker was convicted declares that "(a) A person who: (1) Knowingly or intentionally . . . (C) Delivers . . . cocaine . . . commits dealing in cocaine, a Class B felony punishable by ten years in prison." The statute elevates the offense to a class A felony punishable by thirty years in prison if the person "delivered . . . the drug in or on school property or within one thousand (1,000) feet of school property or on a school bus."

Reasoning

Walker does not dispute the evidence offered at trial that the transaction occurred 542 feet from the school. The statute does not contain any express requirement that a defendant know that a transaction is occurring within 1,000 feet of a school, but Walker argues that permitting enhancement of the crime to a class A felony without such proof violates the due process requirement that a conviction rest on proof of each element of the crime. While Walker's argument is difficult to assess in its summary form, we perceive the question to be whether we should interpret the statute as requiring separate proof of scienter with respect to an element for which the legislature has not specifically required proof of knowledge. We have encountered this question in a variety of settings, including statutes we concluded were meant to establish strict liability for so-called white collar crimes. Conversely, Indiana courts have required proof of mental culpability in a number of [other] crimes where statutes did expressly provide that element (e.g., a statute punishing possession of a handgun with an altered serial number requires proof of knowledge of modification of serial number).

Professors LaFave and Scott accurately describe this question as "whether the legislature meant to impose liability without fault or, on the other hand, really meant to require fault, though it failed to spell it out clearly." We noted with approval the seven factors LaFave and Scott have suggested be balanced in deciding this question. One of these factors, the severity of the punishment, suggests that the legislature might have intended to require proof of mental state for the enhancement of dealing in cocaine. Other factors, particularly the great danger of the prohibited conduct

and the great number of expected prosecutions, suggest that the General Assembly likely did intend to create a strict liability enhancement. These factors are

1. the legislative history, title, or context of a criminal statute;

2. similar or related statutes;

3. the severity of punishment (greater penalties favor a culpable mental state requirement);

4. the danger to the public of the prohibited conduct (greater danger disfavors need for culpable mental state requirement);

5. the defendant's opportunity to ascertain the operative facts and avoid the prohibited conduct;

6. the prosecutor's difficulty in proving the defendant's mental state; and

7. the number of expected prosecutions (greater numbers suggest that crime does not require a culpable mental state).

Holding

Our assessment of these factors makes it difficult to conclude that the General Assembly intended to require separate proof the defendant knew that the dealing occurred near a school but failed to articulate its intent. Moreover, we can imagine an altogether rational reason the legislature might decide to write a statute with a strict liability punishment provision. As Judge Staton wrote for the court of appeals, "A dealer's lack of knowledge of his proximity to the schools does not make the illegal drug any less harmful to the youth in whose hands it may eventually come to rest." Accordingly, we hold that the conviction was not deficient for failure to prove that Walker knew he was within 1,000 feet of a school when he committed the crime. Accordingly, we affirm the judgment of the trial court.

Dissenting, *DeBruler, J.*

The pertinent language of the Indiana Dealing in Cocaine statute reads as follows:

(a) A person who knowingly or intentionally . . . delivers . . . cocaine or a narcotic drug, pure or adulterated, classified in schedule I or II . . . : commits dealing in cocaine or a narcotic drug, a Class B felony, except as provided in subsection (b).

Subsection (b) further provides that

the offense is a Class A felony if the amount of the drug involved weighs three (3) grams or more; the person delivered or financed the delivery of the drug to a person under eighteen (18) years of age at least three (3) years junior to the person; or the person delivered or financed the delivery of the drug in or on school property or within one thousand (1,000) feet of school property or on a school bus.

Given this language, we are confronted with the question of which parts of the statute the "knowingly or intentionally" language is supposed to modify. The "knowingly or intentionally" phrase in Indiana's Dealing in Cocaine statute, as well as the lack of any language manifesting a contrary purpose, causes it to be more plausibly read to target the drug trade involving children near schools rather than to create a drug free zone around our state's schools. However, its purpose is rather to target those who would sell to school age children and, worse still, recruit them as distributors of illicit drugs. The intent language actually used in the statute indicates a legislative intent to punish the schoolyard pusher more harshly than those who sell to adults in their apartments and homes that merely happen to be within a zone. The legislature could have reasonably believed that drug dealers who sell to adults are bad enough, but those who lurk in the playgrounds of our nation's school to prey upon school age children

are worse still. By this reading of the statute, the greater harm created by this particular form of drug trafficking and the greater moral culpability of one involved in such trafficking led the legislature to require proof of a greater level of knowledge for the Class A felony conviction than for the Class B conviction under the Dealing in Cocaine statute. I therefore believe that this reading of the statute clearly requires the State to prove that evil intent by showing that appellant knew that he was dealing within 1,000 feet of a school when he was dealing cocaine.

When the State fails to prove all the elements of a criminal statute, the conviction cannot stand. In the present case, the prosecution made no showing at trial that appellant knew his distance from the school. The only proof addressing the "within 1,000 feet of a school" element of the statute was Detective Witten's testimony that he and his colleagues measured the distance from the site of the controlled buy to the front of Public School 114. Therefore, even the evidence most favorable to the verdict and the reasonable inferences therefrom fail to provide probative evidence from which a reasonable trier of fact could infer the requisite scienter beyond a reasonable doubt.

I would remand this case to the trial court for appellant to be sentenced for the Class B felony of dealing in cocaine.

Questions for Discussion

1. What are the facts in *Walker* that resulted in the enhancement of his sentence for the narcotics offense?

2. Discuss the impact on Walker's prison sentence of his having been convicted of selling narcotics within 1,000 feet of a school.

3. Why does the majority opinion conclude that the selling of cocaine within 1,000 feet of a school is a strict liability offense? Does the arrest of Walker fit within the purpose of the statute?

4. Summarize the argument of the dissenting judge.

5. How would you decide this case?

Cases and Comments

An Open Bottle of Intoxicating Liquor. Steven Mark Loge was cited for a violation of a Minnesota statute that declares it a misdemeanor for the owner of a motor vehicle, or the driver when the owner is not present, "to keep or allow to be kept in a motor vehicle when such vehicle is upon the public highway any bottle or receptacle containing intoxicating liquors or 3.2 percent malt liquors which has been opened." This does not extend to the trunk or to other areas not normally occupied by the driver or passengers. Loge borrowed his father's pickup truck and was stopped by two police officers while on his way home from work. One of the officers observed and seized an open beer bottle underneath the passenger's side of the seat and also found one full unopened

can of beer and one empty beer can in the truck. Loge passed all standard field sobriety tests and was issued a citation for a violation of the open bottle statute.

At trial, Loge testified that the bottle was not his, but he nevertheless was convicted based on a determination by the trial and appellate court that this was a strict liability offense. The Minnesota Supreme Court affirmed that the plain language of the statute indicated that the legislature intended this to be a strict liability offense and that a knowledge requirement would make conviction for possession difficult, if not insurmountable. The Supreme Court also observed that drivers who are aware of this statute will carefully check any case of packaged alcohol before driving in order to ensure that each container's seal is not broken. The dissent noted that the language "allow to be kept" clearly indicated a knowledge requirement. Absent a provision for intent, there is a risk that individuals will be convicted "not simply for an act that the person does not know is criminal, but also for an act the person does not even know he is committing."

Does the prevention of "drinking and driving" justify the possible conviction of innocent individuals? See *State v. Loge,* 608 N.W.2d 152 (Minn. 2000).

You Decide

5.5 A juvenile court ordered C.R.M. to attend Anoka County, Minnesota, Juvenile Day School. Students' coats are hung outside the classroom and inspected in the morning for contraband. A folding knife with a four-inch blade was discovered in C.R.M.'s coat. C.R.M. immediately reacted, "Oh man, I forgot to take it out, I was whittling this weekend." The head teacher in the school found C.R.M.'s reaction to be "spontaneous" and "believable." C.R.M.'s mother testified that he was wearing a "double jacket" and, although he stated that he patted himself down, he likely would not have felt the knife. The police were contacted, and C.R.M. was charged with possession of a dangerous weapon on school property, a felony-level offense. The judge concluded that C.R.M. likely was not "whittling" and probably accidentally brought the knife to school. C.R.M. nonetheless was convicted under a statute that makes possession of a dangerous weapon on school property a strict liability offense. C.R.M., appealed his conviction and argued that the prosecution was required to establish that he knowingly possessed a dangerous weapon. The Minnesota statute provides that "[w]hoever possesses, stores, or keeps a dangerous weapon or uses or brandishes a replica firearm or a BB gun on school property is guilty of a felony and may be sentenced to imprisonment for not more than two years or to payment of a fine of not more than $ 5,000, or both." Consider whether there is a clear statement of legislative intent, the severity of the punishment of the offense, whether requiring a demonstration of criminal intent will impede enforcement of the law, and whether the law will result in the arrest of significant number of individuals who unknowingly bring prohibited "weapons" onto the school grounds. See *In re C.R.M.,* 611 N.W.2d 802 (Minn. 2000). Should a strict liability standard be applied to a teacher who claims to have brought a firearm to school accidentally? See *Esteban v. Commonwealth,* 587 S.E.2d 523 (Va. 2003).

You can find the answer at www.sagepub.com/lippmanccl3e.

You Decide

5.6 In July 1995, Ronnie Polk was the passenger in an automobile that was stopped for a moving violation in close proximity to Highland Christian School in Lafayette, Indiana. A police officer's search led to the seizure of crack cocaine and several tablets of diazepam. In Indiana, possession of more than three grams of cocaine within 1,000 feet of a school is enhanced from a Class D felony to a Class A felony punishable by thirty years in prison, and possession of a "Schedule IV" drug without a doctor's prescription within 1,000 feet of a school is enhanced from a Class D to a Class C felony, punishable by four years in prison.

Polk was convicted and sentenced for both offenses, and his two sentences were to run concurrently. Polk also was convicted of being a habitual offender, and his combined sentence for the three convictions totaled fifty years. Polk maintains that the legislature did not intend for the possession of cocaine within 1,000 feet of a school to be a strict liability offense that applied to passengers possessing narcotics in automobiles, because this did not advance Indiana's interest in protecting schoolchildren. Applying the statute to individuals in automobiles would allow the police to wait to stop automobiles suspected of containing narcotics as they approached within 1,000 feet of a school.

How would you decide *Polk v. State* in light of the precedent established in *State v. Walker?* See *Polk v. State,* 683 N.E.2d 567 (Ind. 1997).

CONCURRENCE

We now have covered both *actus reus* and *mens rea.* The next step is to understand that there must be a **concurrence** between a criminal act and a criminal intent. *Chronological concurrence* means that a criminal intent must exist at the same time as a criminal act. An example of chronological concurrence is the requirement that a burglary involves breaking and entering with an intent to commit a felony therein. The classic example is an individual who enters a cabin to escape the cold and after entering decides to steal food and clothing. In this instance, the intent did not coincide with the criminal act and the defendant will not be held liable for burglary.

The principle of concurrence is reflected in section 20 of the California Penal Code, which provides that in "every crime . . . there must exist a union or joint operation of act and intent or criminal negligence." The next case is *State v. Rose.* Can you explain why the defendant's guilt for manslaughter depends on the prosecution's ability to establish a concurrence between the defendant's act and intent?

The Legal Equation

Concurrence = *Mens rea* (in unison with)

+ *actus reus.*

Was there a concurrence between the defendant's criminal act and criminal intent?

State v. Rose, 311 A.2d 281 (R.I. 1973). Opinion by: Roberts, J.

These are two indictments, one charging the defendant, Henry Rose, with leaving the scene of an accident, death resulting, . . . and the other charging the defendant with manslaughter. The defendant was tried on both indictments to a jury in the superior court, and a verdict of guilty was returned in each case. Thereafter the defendant's motions for a new trial were denied. . . .

Facts

These indictments followed the death of David J. McEnery, who was struck by defendant's motor vehicle at the intersection of Broad and Summer Streets in Providence at about 6:30 p.m. on April 1, 1970. According to the testimony of a bus driver, he had been operating his vehicle north on Broad Street and had stopped at a traffic light at the intersection of Summer Street. While the bus was standing there, he observed a pedestrian starting to cross Broad Street, and as the pedestrian reached the middle of the southbound lane he was struck by a "dirty, white station wagon" that was proceeding southerly on Broad Street. The pedestrian's body was thrown up on the hood of the car. The bus driver further testified that the station wagon stopped momentarily, the body of the pedestrian rolled off the hood, and the car immediately drove off along Broad Street in a southerly direction. The bus operator testified that he had alighted from his bus, intending to attempt to assist the victim, but was unable to locate the body.

Subsequently, it appears from the testimony of a police officer, about 6:40 p.m. the police located a white

station wagon on Haskins Street, a distance of some 610 feet from the scene of the accident. The police further testified that a body later identified as that of David J. McEnery was wedged beneath the vehicle when it was found and that the vehicle had been registered to defendant. . . .

Issue

The defendant is contending that if the evidence is susceptible of a finding that McEnery was killed upon impact, he was not alive at the time he was being dragged under defendant's vehicle, and defendant could not be found guilty of manslaughter. An examination of the testimony of the only medical witness makes it clear that, in his opinion, death could have resulted immediately upon impact by reason of a massive fracture of the skull. The medical witness also testified that death could have resulted a few minutes after the impact but conceded that he was not sure when it did occur.

Reasoning

We are inclined to agree with defendant's contention in this respect. Obviously, the evidence is such that death could have occurred after defendant had driven away with McEnery's body lodged under his car and, therefore, be consistent with guilt. On the other hand, the medical testimony is equally consistent with a finding that McEnery could have died instantly upon impact and, therefore, be consistent with a reasonable conclusion other than the guilt of defendant.

Holding

It is clear, then, that, the testimony of the medical examiner lacking any reasonable medical certainty as to the time of the death of McEnery, we are unable to conclude that on such evidence defendant was guilty of manslaughter beyond a reasonable doubt.

Therefore, we conclude . . . that it was error to deny defendant's motion for a directed verdict of acquittal. . . .

We are unable, however, to reach the same conclusion concerning the denial of the motion for a directed verdict of acquittal . . . in which defendant was charged with leaving the scene of an accident. . . .

Questions for Discussion

1. Why is it important to determine whether the victim died on impact with Rose's automobile or whether the victim was alive at the time he was dragged under the defendant's automobile? What is the ruling of the Rhode Island Supreme Court?

2. How does this case illustrate the principle of concurrence?

You Decide

5.7 Jackson administered what he believed was a fatal dose of cocaine to Pearl Bryan in Cincinnati, Ohio. Bryan was pregnant, apparently as a result of her intercourse with Jackson. Jackson and a companion then transported Bryan to Kentucky and cut off her head to prevent identification of the body. Bryan, in fact, was still alive when brought to Kentucky and died as a result of the severing of her head. A state possesses jurisdiction over offenses committed within its territorial boundaries. Can Jackson be prosecuted for the intentional killing of Bryan in Ohio? In Kentucky? See *Jackson v. Commonwealth*, 38 S.W. 422 (Ky. 1896).

You can find the answer at www.sagepub.com/lippmanccl3e.

CAUSATION

You now know that a crime entails a *mens rea* that concurs with an *actus reus*. Certain crimes (termed *crimes of criminal conduct causing a criminal harm*) also require that the criminal act cause a particular harm or result: the death or maiming of a victim, the burning of a house, or damage to property.

Causation is central to criminal law and must be proven beyond a reasonable doubt. The requirement of causality is based on two considerations: [15]

- *Individual Responsibility.* The criminal law is based on individual responsibility. Causality connects a person's acts to the resulting social harm and permits the imposition of the appropriate punishment.
- *Fairness.* Causality limits liability to individuals whose conduct produces a prohibited social harm. A law that declares that all individuals in close proximity to a crime are liable regardless of their involvement would be unfair and penalize people for being in the wrong place at the wrong time. If such a law were enacted, individuals might hesitate to gather in crowds or bars or to attend concerts and sporting events.

Establishing that a defendant's criminal act caused harm to the victim can be more complicated than you might imagine. Should an individual who commits a rape be held responsible for the victim's subsequent suicide? What if the victim attempted suicide a week before the rape and then killed herself following the rape? Would your answer be the same if the stress induced by the rape appears to have contributed to the victim contracting cancer and dying a year later? What if doctors determine that a murder victim who was hospitalized would have died an hour later of natural causes in any event? We can begin to answer these hypothetical situations by reviewing the two types of causes that a prosecutor must establish beyond a reasonable doubt at trial in order to convict a defendant: **cause in fact** and legal or **proximate cause**.

As noted, causality arises in prosecutions for crimes that require a particular result, such as murder, maiming, arson, and damage to property. The prosecution must prove beyond a reasonable doubt that the harm to the victim resulted from the defendant's unlawful act. You will find that most causality cases involve defendants charged with murder who claim that they should not be held responsible for the victim's death.

Cause in Fact

The cause in fact or factual cause simply requires you to ask whether "but for" the defendant's act, would the victim have died? An individual aims a gun at the victim, pulls the trigger, and kills the victim. "But for" the shooter's act, the victim would be alive. In most cases, the defendant's act is the only factual cause of the victim's injury or death and is clearly the direct cause of the harm. This is a simple cause-and-effect question. The legal or proximate cause of the victim's injury or death may not be so easily determined.

A defendant's act must be the cause in fact or factual cause of a harm in order for the defendant to be criminally convicted. This connects the defendant to the result. The cause in fact or factual cause is typically a straightforward question. Note that the defendant's act must also be the legal or proximate cause of the resulting harm.

Legal or Proximate Cause

Just when things seemed simple, we encounter the challenge of determining the legal or proximate cause of the victim's death. Proximate cause analysis requires the jury to determine whether it is fair or just to hold a defendant legally responsible for an injury or death. This is not a scientific question. We must consider questions of fairness and justice. There are few rules to assist us in this analysis.

In most cases, a defendant is clearly both the cause in fact and legal cause of the victim's injury or death. However, consider the following scenarios: You pull the trigger and the victim dies. You point out that it was not your fault, since the victim died from the wound you inflicted in combination with a minor gun wound that she suffered earlier in the day. Should you be held liable? In another scenario an ambulance driver rescues the victim, the ambulance's brakes fail, and the vehicle crashes into a wall, killing the driver and victim. Are you or the driver responsible for the victim's death? You later learn that the victim died after the staff of the hospital emergency room waited five hours to treat the victim and that she would have lived had she received timely assistance. Who is responsible for the death? Would your answer be different in the event that the doctors protest that they could not operate on the victim because of a power outage caused by a hurricane? What if the victim was wounded from the gunshot and, although barely conscious, stumbled into the street and was hit by an automobile or by lightning? In each case, "but for" your act, the victim would not have been placed in the situation that led to his or her death. On the other hand, you might argue that in each of these examples you were not legally liable, because the death resulted from an **intervening cause** or outside factor rather than from the shooting. As you can see from the previous examples, an intervening cause may arise from

- The act of the victim wandering into the street
- An act of nature, such as a hurricane
- The doctors who did not immediately operate
- A wound inflicted by an assailant in combination with a previous injury

Another area that complicates the determination of proximate causes is a victim's preexisting medical condition. This arises when you shoot an individual and the shock from the wound results in the failure of the victim's already seriously weakened heart.

Intervening Cause

Professor Joshua Dressler helps us answer these causation problems by providing two useful categories of intervening acts: **coincidental intervening acts** and **responsive intervening acts**.

Coincidental Intervening Acts

A defendant is not considered legally responsible for a victim's injury or death that results from a coincidental intervening act. (Some texts refer to this as an *independent intervening cause*). The classic case is an individual who runs from a mugger and is hit by and dies from injuries sustained when a tree that has been struck by lightning falls on him. It is true that "but for" the robbery, the victim would not have fled. The defendant nevertheless did not order or compel the victim to run and certainly had nothing to do with the lightning strike that felled the tree. As a result, the perpetrator generally is not held legally liable for a death that results from this unpredictable combination of an attempted robbery, bad weather, and a tree.

Coincidental intervening acts arise when a defendant's act places a victim in a particular place where the victim is harmed by an unforeseeable event.

The Ninth Circuit Court of Appeals offered an example of an unforeseeable event as a hypothetical in the case of *United States v. Main.* The defendant in this example drives in a reckless fashion and crashes his car, pinning the passenger in the automobile. The defendant leaves the scene of the accident to seek assistance, and the semiconscious passenger is eaten by a bear. The Ninth Circuit Court of Appeals observed that reckless driving does not create a foreseeable risk of being eaten by a bear and that this intervening cause is so out of the ordinary that it would be unfair to hold the driver responsible for the victim's death.[16] Another example of an unforeseeable coincidental intervening event involves a victim who is wounded, taken to the hospital for medical treatment, and then killed in the hospital by a knife-wielding mass murderer. Professor Joshua Dressler notes that in this case the unfortunate victim has found himself or herself in the "wrong place at the wrong time."[17]

Defendants will be held responsible for the harm resulting from coincidental causes in those *rare* instances in which the event is "normal and foreseeable" or could have been reasonably predicted. In *Kibbe v. Henderson,* two defendants were held liable for the death of George Stafford, whom they robbed and abandoned on the shoulder of a dark, rural two-lane highway on a cold, windy, and snowy evening. Stafford's trousers were down around his ankles, his shirt was rolled up toward his chest, and the two robbers placed his shoes and jacket on the shoulder of the highway and did not return Stafford's glasses. The near-sighted and drunk Stafford was sitting in the middle of a lane on a dimly lit highway with his hands raised when he was hit and killed by a pickup truck traveling ten miles per hour over the speed limit that coincidentally happened to be passing by at the precise moment that Stafford wandered into the highway.[18]

The defendant generally is legally liable for foreseeable coincidental intervening acts.

Responsive Intervening Acts

The response of a victim to a defendant's criminal act is termed a *responsive intervening act* (some texts refer to this as a *dependent intervening act*). In most instances, the defendant is considered responsible because his or her behavior caused the victim to respond. A defendant is relieved of responsibility only in those instances in which the victim's reaction to the crime is both abnormal and unforeseeable. Consider the case of a victim who jumps into the water to evade an assailant and drowns. The assailant will be charged with the victim's death despite the fact that the victim could not swim and did not realize that the water was dangerously deep. The issue is the *foreseeability* of the victim's response rather than the *reasonableness* of the victim's response. Again, courts generally are not sympathetic to defendants who set a chain of events in motion and generally will hold such defendants criminally liable.

In *People v. Armitage,* David Armitage was convicted of "drunk boating causing [the] death" of Peter Maskovich. Armitage was operating his small aluminum speedboat at a high rate of speed while zigzagging across the river when it flipped over. There were no floatation devices on board, and the intoxicated Armitage and Maskovich clung to the capsized vessel. Maskovich disregarded Armitage's warning and decided to try to swim to shore and drowned. A California appellate court ruled that Maskovich's decision did not break the chain of causation. The "fact that the panic stricken victim recklessly abandoned the boat and tried to swim ashore was not a wholly abnormal reaction to the peril of drowning," and Armitage could not exonerate himself by claiming that the "victim should have reacted differently or more prudently."[19]

Defendants have also been held liable for the response of individuals other than the victim. For instance, in the California case of *People v. Schmies,* defendant Schmies fled on his motorcycle from a traffic stop at speeds of up to ninety miles an hour and disregarded all traffic regulations. During the chase, one of the pursuing patrol cars struck another vehicle, killing the driver and injuring the officer. Schmies was convicted of grossly negligent vehicular manslaughter and of reckless driving. A California court affirmed the defendant's conviction based on the fact that the officer's response and the resulting injury were reasonably foreseeable. The officer's reaction, in other words, was not so extraordinary that it was unforeseeable, unpredictable, and statistically extremely improbable. [20]

Medical negligence has also consistently been viewed as foreseeable and does not break the chain of causation. In *People v. Saavedra-Rodriquez,* the defendant claimed that the negligence of the doctors at the hospital rather than the knife wound he inflicted was the proximate cause of the death and that he should not be held liable for homicide. The Colorado Supreme Court ruled that medical negligence is "too frequent to be considered abnormal" and that the defendant's stabbing of the victim started a chain of events, the natural and probable result of which was the defendant's death. The court added that only the most gross and irresponsible medical negligence is so removed from normal expectations as to be considered unforeseeable.[21]

In *United States v. Hamilton,* the defendant knocked the victim down and jumped on and kicked his face. The victim was rushed to the hospital, where nasal tubes were inserted to enable him to breathe, and his arms were restrained. During the night the nurses changed his bedclothes and negligently failed to reattach the restraints on the victim's arms. Early in the morning the victim went into convulsions, pulled out the nasal tubes, and suffocated to death. The court held that regardless of whether the victim accidentally or intentionally pulled out the tubes, the victim's death was the ordinary and foreseeable consequence of the attack and affirmed the defendant's conviction for manslaughter.[22]

In sum, a defendant who commits a crime is responsible for the natural and probable consequences of his or her actions. A defendant is responsible for foreseeable responsive intervening acts.

The Model Penal Code

The Model Penal Code eliminates legal or proximate causation and requires only "but-for causation." The code merely asks whether the result was consistent with the defendant's intent or knowledge or was within the scope of risk created by the defendant's reckless or negligent act. In other words, under the Model Penal Code, you merely look at the defendant's intent and act and ask whether the result could have been anticipated. In cases of a resulting harm or injury that is "remote" or "accidental" (e.g., a lightning bolt or a doctor who is a serial killer), the Model Penal Code requires that we look to see whether it would be unjust to hold the defendant responsible. [23]

The next case, *People v. Cervantes,* asks whether it is just and fair to hold a defendant liable as the legal or proximate cause of the victim's death.

The Legal Equation		
Causality	=	Cause in fact
	+	legal or proximate cause.
Cause in fact	=	"But for" the defendant's criminal act, the victim would not be injured or dead.
Legal or proximate cause	=	Whether just or fair to hold the defendant criminally responsible.
Intervening acts	=	Coincidental intervening acts limit liability where unforeseeable; responsive intervening acts limit liability where unforeseeable and abnormal.

Should Cervantes be held criminally liable for causing the killing of a rival gang member?

People v. Cervantes, 29 P.3d 225 (Cal. 2001). Opinion by: Baxter J.

Issue

This case presents a question concerning proof of proximate causation in a provocative act murder case. We granted review to decide whether defendant, a member of a street gang, who perpetrated a nonfatal shooting that quickly precipitated a revenge killing by members of an opposing street gang, is guilty of murder on the facts before us.

Facts

Shortly after midnight on October 30, 1994, defendant and fellow Highland Street gang members went to a birthday party in Santa Ana thrown by the Alley Boys gang for one of their members. Joseph Perez, the prosecution's gang expert, testified the Highland Street and Alley Boys gangs were not enemies at the time. Over 100 people were in attendance at the party, many of them gang members.

Outside of the house, defendant approached a woman he knew named Grace. She was heavily intoxicated and declined defendant's invitation to go to another party with him, which prompted him to call her a "ho," leading, in turn, to an exchange of crude insults. Juan Cisneros, a member of the Alley Boys, approached and told defendant not to "disrespect" his "homegirl." Richard Linares, also an Alley Boy, tried to defuse the situation, but Cisneros drew a gun and threatened to "cap [defendant's] ass." Defendant responded by brandishing a handgun of his own, which prompted Linares to intervene once again, pushing or touching defendant on the shoulder in an effort to separate him from Cisneros. In response, defendant stated "nobody touches me" and shot Linares through the arm and chest.

A crowd of some 50 people was watching these events unfold. Someone yelled, "Why did you shoot my home boy?" or "your home boy shot your own homeboy," to which someone responded "Highland [Street] is the one that shot." A melee erupted, and gang challenges were exchanged.

A short time later a group of Alley Boys spotted Hector Cabrera entering his car and driving away. Recognizing him as a member of the Highland Street gang, they fired a volley of shots, killing him. A variety of shell casings recovered from the street evidenced that at least five different shooters had participated in the murder of Cabrera.

Perez testified that although the Highland Street and Alley Boys gangs were not enemies at the time of the shootings, both gangs would be expected to be armed. He opined that the Alley Boys would consider defendant's conduct in shooting Linares to be an act of "major disrespect" to their gang. To avenge the shooting, they would be expected to respond quickly with equal or greater force against defendant or another member of his gang. Therefore, Perez opined, Cabrera's death was a reasonably foreseeable consequence of defendant's actions.

Defendant testified he did not intend to shoot Linares, but was simply trying to protect himself from Cisneros, who drew his weapon first. He was surprised when his gun went off, because he did not feel it fire or see any flash. He testified, "I don't know if I shot [Linares] or somebody else shot [him], but what I do know is that if I [had] attempted to murder anybody, I would have shot [him] while he was on the floor." In the confusion following the shooting of Linares, defendant heard someone say, "[Y]our home boy shot your own home boy," and then he heard someone say "Highland's the one that shot." Realizing he was in danger, defendant ran from the party and sped off with several others. He heard shots being fired as they drove away. He was stopped by police and arrested a short distance away.

Defendant was charged with murdering Cabrera. . . . The jury was . . . instructed that liability for homicide requires a causal connection between an unlawful act and death, namely, that the act's direct, natural and probable consequences must be death. . . . [T]he jury was instructed that a direct, natural and probable consequence must be reasonably foreseeable, measured objectively under a reasonable person test. . . . The jury convicted defendant of the murder of Cabrera, fixed at second degree.

Reasoning

The question before us is, therefore, whether sufficient evidence supports defendant's conviction of murder based on the provocative act murder theory. . . . In particular, the essential element with which we are here concerned is proximate causation in the context of a provocative act murder prosecution.

In homicide cases, a "cause of the death of [the decedent] is an act or omission that sets in motion a chain of events that produces as a direct, natural and probable consequence of the act or mission the death of [the decedent] and without which the death would not occur." In general, "[p]roximate cause is clearly established where the act is directly connected with the resulting injury, with no intervening force operating." In this case there was an intervening force in operation—at least five persons in attendance at the party, presumably all members of the Alley Boys, shot and killed Highland Street gang member Hector Cabrera in a hail of bullets shortly after the melee erupted. . . .

The provocative act murder doctrine has traditionally been invoked in cases in which the perpetrator of the underlying crime instigates a gun battle, either by firing first or by otherwise engaging in severe, life-threatening, and usually gun-wielding conduct, and the police, or a victim of the underlying crime, responds with privileged lethal force by shooting back and killing the perpetrator's accomplice or an innocent bystander. In *People v. Gilbert*, 408 P.2d 365 (Cal. 1965) we stated that:

> When the defendant or his accomplice, with a conscious disregard for life, intentionally commits an act that is likely to cause death, and his victim or a police officer kills in reasonable response to such act, the defendant is guilty of murder. In such a case, the killing is attributable, not merely to the commission of a felony, but to the intentional act of the defendant or his accomplice committed with conscious disregard for life.

We then discussed causation:

> [T]he victim's self-defensive killing or the police officer's killing in the performance of his duty cannot be considered an independent intervening cause for which the defendant is not liable, for it is a reasonable response to the dilemma thrust upon the victim or the policeman by the intentional act of the defendant or his accomplice.

In short, *Gilbert* described provocative act murder liability in traditional terms of proximate causation and . . . reaffirmed the general rule that no criminal liability attaches . . . for an unlawful killing that results from an independent intervening cause (i.e., a superseding cause). In contrast, when the death results from a dependent intervening cause, the chain of causation ordinarily remains unbroken and the initial actor is liable for the unlawful homicide. . . .

In an early Illinois case, *Belk v. The People* (1888) 125 Ill. 584, the defendants were alleged to have negligently allowed their team of horses to break loose on a narrow country lane. The team collided with a wagon in plain sight just ahead, causing that wagon's team of horses to panic and run away and thereby throwing the victim, a passenger, to her death. The Illinois court reversed the resulting manslaughter convictions on other grounds, but reasoned that "Between the acts of omission or commission of the defendants, by which it is alleged the collision occurred, and the injury of the deceased, there was not an interposition of a human will acting independently . . . or any extraordinary natural phenomena, to break the causal connection." We also cited *Madison v. State* (1955) 234 Ind. 517, in which "the court . . . affirmed a conviction of second degree murder when the defendant threw a hand grenade at one Couch who, presumably impulsively [i.e., instinctively and not as an act of will], kicked it to another who was killed. The fact that Couch kicked the grenade did not break the line of causation. . . ." And we recalled *Wright v. State* (Fla. Dist. Ct. App. 1978) 363 So.2d 617, wherein the defendant was convicted of manslaughter for firing from his car into his intended victim's car. The intended victim had "rapidly accelerated his car while 'ducking bullets'" and fatally ran over a pedestrian. We found the significance of the facts and holding in *Wright* to be as follows: "Shots that cause a driver to accelerate impulsively and run over a nearby pedestrian suffice to confer liability; but if the driver, still upset, had proceeded for several miles before killing a pedestrian, at some point the required causal nexus would have become too remote for the [shooter] to be liable for homicide."

The principles derived from these and related authorities have been summarized as follows.

In general, an "independent" intervening cause will absolve a defendant of criminal liability. However, in order to be "independent" the intervening cause must be "unforeseeable . . . an extraordinary and abnormal occurrence, which rises to the level of an exonerating, superseding cause." On the other hand, a "dependent" intervening cause will not relieve the defendant of criminal liability. . . . If an intervening cause is a normal and reasonably foreseeable result of defendant's original act the intervening act is "dependent" and not a superseding cause, and will not relieve defendant of liability.

. . . The precise consequence need not have been foreseen; it is enough that the defendant should have foreseen the possibility of some harm of the kind which might result from his act.

Turning to the facts at hand, we agree with defendant that the evidence introduced below is insufficient as a matter of law to support his conviction of provocative act murder, for it fails to establish the essential element of proximate causation. The facts of this case are distinguishable from the classic provocative act murder case in a number of respects. Defendant was not the initial aggressor in the incident that gave rise to the provocative act. There was no direct evidence that Cabrera's unidentified murderers were even present at the scene of the provocative act, i.e., in a position to actually witness defendant shoot Linares. Defendant himself was not present at the scene where Cabrera was fatally gunned down; the only evidence introduced on the point suggests he was already running away from the party or speeding off in his car when the victim was murdered.

But the critical fact that distinguishes this case from other provocative act murder cases is that here the actual murderers were not responding to defendant's provocative act by shooting back at him or an accomplice, in the course of which someone was killed. They were not in the shoes of police officers . . . who shot back and killed an accomplice as an objectively "reasonable response to the dilemma thrust upon [them]" by the defendant's malicious and life-endangering provocative acts. . . . and were not like the intermediary in *Madison v. State* who instinctively kicked away a live hand grenade thrown at him by defendant Madison, resulting in the death of another.

On the contrary . . . nobody forced the Alley Boys' murderous response in this case, if indeed it was a direct response to defendant's act of shooting Linares. The willful and malicious murder of Cabrera at the hands of others was an independent intervening act on which defendant's liability for the murder could not be based.

The circumstance that the murder occurred a very short time after defendant shot Linares, and the opinion of prosecution gang expert Perez that Cabrera's murder was a foreseeable consequence of defendant's shooting of Linares in the context of a street gang's code of honor mentality, was essentially the only evidence on which the jury was asked to find that Cabrera's murder was "a direct, natural, and probable consequence" of defendant's act of shooting Linares. Given that the murder of Cabrera by other parties was itself felonious, intentional, perpetrated with malice aforethought, and directed at a victim who was not involved in the original altercation between defendant and Linares, the evidence is insufficient . . . to establish the requisite proximate causation to hold defendant liable for murder.

Holding

The judgment of the Court Appeal is reversed to the extent it affirms defendant's conviction of murder. . . .

Questions for Discussion

1. Summarize the facts in *Cervantes*.

2. Explain the court's distinction between dependent intervening acts and independent intervening acts.

3. Why did the court find that the killing of Cabrera was an independent intervening act and that Cervantes was not liable for murder?

4. Do you agree with the prosecution's gang expert that Cabrera's murder was a foreseeable consequences of the shooting of Linares.

5. As a juror would you hold Cervantes liable for murder?

Cases and Comments

1. **Apparent Safety Doctrine.** Preslar kicked and choked his wife and beat her over the head with a thirty-inch-thick piece of wood. He also threatened to kill her with his axe. The victim gathered her children and walked over two miles to her father's home. Reluctant to reveal her bruises and injuries to her family, she spread a quilt on the ground and covered herself with cotton fabric and slept outside. The combination of the exhausting walk, her injuries, and the biting cold led to a weakened condition that resulted in her death. The victim's husband was acquitted by the North Carolina Supreme Court, which ruled that the chain of causation was broken by the victim's failure to seek safety. The court distinguished this case from the situation of a victim who in fleeing is forced to wade through a swamp or jump into a river. Is it relevant that the victim likely feared that her family would force her to return to her marital home and that she would have to face additional physical abuse from her husband? See *State v. Preslar*, 48 N.C. 421 (1856).

2. **Drag Racing.** In *Velazquez v. State*, the defendant Velazquez and the deceased Alvarez agreed to drag race their automobiles over a quarter-mile course on a public highway. Upon completing the race, Alvarez suddenly turned his automobile around and proceeded east toward the starting line. Velazquez also reversed direction. Alvarez was in the lead and attained an estimated speed of 123 mph. He was not wearing a seat belt and had a blood alcohol content of between .11 and .12. Velazquez had not been drinking and was traveling at roughly 90 mph. As both approached the end of the road, they applied their brakes, but Alvarez was unable to stop. He crashed through the guardrail and was propelled over a canal and landed on the far bank. Alvarez was thrown from his car, pinned under the vehicle when it landed, and died. The defendant crashed through the guardrail, landed in the canal, and managed to escape.

A Florida district court of appeal determined that the defendant's reckless operation of his vehicle in the drag race was technically the cause in fact of Alvarez's death under the "but for" test. There was no doubt that "but for" the defendant's participation, the deceased would not have recklessly raced his vehicle and would not have been killed. The court, however, ruled that the defendant's participation was not the proximate cause of the deceased's death because the "deceased, in effect, killed himself by his own volitional reckless driving," and that it "would be unjust to hold the defendant criminally responsible for this death." The race was completed when Alvarez turned his car around and engaged in a "near-suicide mission."

From the point of public policy, would it have been advisable to hold Velazquez liable? Was Alvarez's death foreseeable? See *Velazquez v. State,* 561 So. 2d 347 (Fla. Dist. Ct. App. 1990).

3. **The Year-and-a-Day Rule.** Defendant Wilbert Rogers stabbed James Bowdery in the heart with a butcher knife on May 6, 1994. During an operation to repair Bowdery's heart, he suffered a cardiac arrest. This led to severe brain damage as a result of a loss of oxygen. Bowdery remained in a coma and died on August 7, 1995, from kidney complications resulting from remaining in a vegetative condition for such a lengthy period of time. Rogers was convicted of second-degree murder and appealed on the grounds that the prosecution was barred by the year-and-a-day rule, which prohibits a murder conviction when more than a year has transpired between the defendant's criminal act and the victim's death. The Tennessee Supreme Court observed that the rule was based on the fact that thirteenth-century medical science was incapable of establishing causation beyond a reasonable doubt when a significant amount of time elapsed between the injury to the victim and the victim's death. The rule has also been explained as an effort to moderate the common law's automatic imposition of the death penalty for felonies.

The Tennessee Supreme Court, in abolishing the year-and-a-day rule, noted that almost one-half of the states had now eliminated the rule. The court explained that medical science now possessed the ability to determine the cause of death with greater accuracy and that it no longer made sense to terminate a defendant's liability after a year. In addition, medicine was able to sustain the life of a victim of a criminal act for a lengthy period of time, and the year-and-a-day rule would result in the perpetrators of slow-acting poisons or viruses escaping criminal prosecution and punishment. The court declined to adopt a revised period in which prosecutions for murder must be undertaken and, instead, stressed that prosecutors possessed the burden of establishing causation. The U.S. Supreme Court later ruled that the Tennessee court's abolition of the year-and-a-day rule was not in violation of the *Ex Post Facto* Clause of the U.S. Constitution. See *State v. Rogers,* 992 S.W.3d 393 (Tenn. 1999), *aff'd* 532 U.S. 451 (2001).

Should there be a time limit on prosecutions for homicide? See *Commonwealth v. Casanova,* 708 N.E.2d 86 (Mass. 1999).

You Decide  **5.7** Larry Roberts along with other inmates was convicted of the murders of a fellow inmate, Charles Gardner, and of a correctional officer, Albert Path. Gardner was an inmate at the California Medical Facility in Vacaville and was attacked by Roberts and other inmates and was stabbed eleven times. The knife fell to the ground and was grabbed by Gardner, who pursued one of his assailants up a flight of stairs where Gardner plunged the knife into the chest of a prison guard, Officer Patch. Patch died within an hour at the prison clinic. Gardner died shortly thereafter.

It was established that Gardner was dazed and in shock from the loss of blood as he staggered up the stairs and that Patch was the first individual that he encountered. Several witnesses alleged that Roberts had stabbed Gardner. Gardner was described as a well-behaved inmate who had no motive to intentionally kill Patch. Roberts was charged with killing Gardner and claimed that the cause of Gardner's death was medical negligence.

Will this defense prove successful? Should Roberts be held criminally liable for causing the death of Path? See *People v. Roberts,* 826 P.2d 274 (Cal. 1992).

You can find the answer at www.sagepub.com/lippmanccl3e.

You can find more cases on the study site:
Banks v. Commonwealth, People v. Kern, *www.sagepub.com/lippmanccl3e.*

CRIME IN THE NEWS

On his fifteenth birthday Christopher Hill received his first cell phone, and a year later he celebrated his birthday by purchasing a used red pickup truck. Christopher devoted himself to keeping the truck in peak condition and prided himself on his clean driving record. The day Christopher turned 20, he pulled out of a parking lot and dialed a neighbor on his cell phone to tell her that he had found a dresser in a store that seemed perfect for her bedroom. Christopher clearly was excited and distracted and ran a red light and hit a sports utility vehicle, killing 61-year-old Linda Doyle. He had been going 45 miles per hour, admitted that he had never seen the light, and pled guilty to negligent homicide.

Studies indicate that drivers talking on cell phones possess reduced reaction time and are four times as likely to cause an accident as other drivers. The probability that they will crash is the same as a driver with a .08 percent blood alcohol level, the amount required in most states be considered guilty of drunk driving. Between 35 percent and 46 percent of teenage drivers report that when talking on the phone that they are distracted from driving.

Text messaging and talking on a phone are more than twice as high among drivers 16-to-24 years of age. Over 70 percent of young people state that they text while driving. Texting and driving is even more dangerous than talking on a cell phone. A year-long study by Virginia Tech Transportation Institute placed video cameras in the cabs of long-haul trucks and concluded that when drivers text, the risk of collision increased 23 times as opposed to a driver who is not texting. A University of Utah study of college students using a driving simulator found that drivers who are texting are eight times more likely to crash than drivers who are not texting.

In 2009, an experiment conducted by the editor of a leading automotive magazine found that at 70 miles per hour, a drunk driver's stopping distance increased by 4 feet. Reading an e-mail increased the distance by 36 feet, and sending a text message increased the distance by another 70 feet. The National Transportation Administration compares texting while driving 55 miles per hour to driving the length of a football field with your eyes closed.

Despite these statistics, drivers tend to overestimate their ability to multitask while behind the wheel. Our cars increasingly are being turned into electronic centers for phones, navigation devices, laptops, e-mail, Web surfing, and media. The auto industry is manufacturing what are termed "digital cars," equipped with an array of sophisticated electronic systems. The number of accidents resulting from multitasking likely will continue to rise as the number of wireless subscribers and minutes talked continues to increase.

States in recent years have enacted statutes regulating the use of cell phones and texting while driving.

All Cell Phones. Thirty states and the District of columbia ban cell phone use by "novice drivers," and nineteen states and the District of Columbia prohibit bus drivers from using a cell phone while passengers are in the vehicle.

Hand-Free Cell Phones. Nineteen states and the District of Columbia require hand-free talking on a cell phone while driving.

Text Messaging. Thirty-four states, the District of Columbia, and Guam prohibit text messaging while driving. An additional seven states prohibit texting by novice drivers. Three states ban texting by school bus drivers.

Most drivers, when polled, recognize that texting or e-mailing while driving poses a safety risk. Where should we draw the line on the use of electronic devices while driving? Should a vehicular homicide resulting from the use of a electronic device be punished more severely? Nine countries and several Canadian provinces prohibit talking on a cell phone while driving. Are we overemphasizing the dangers that are posed by the use of cell phones while failing to address other forms of "distracted driving?"

CHAPTER SUMMARY

It is a fundamental principle of criminal law that a criminal offense requires a criminal intent that occurs concurrently with a criminal act. The requirement of a *mens rea,* or the mental element of a criminal act, is based on the concept of "moral blameworthiness." The notion of blameworthiness, in turn, reflects the notion that individuals should be subject to criminal punishment and held accountable only when they consciously choose to commit a crime or to create a high risk of harm or injury.

We cannot penetrate into the human brain and determine whether an individual harbored a criminal intent. In some cases, a defendant may confess to the police or testify as to his or her intent in court. In most instances, prosecutors rely on circumstantial evidence and infer an intent from a defendant's motives and patterns of activity.

The Model Penal Code proposed four levels of *mens rea* or criminal intent. The four in order of severity or culpability are as follows:

- *Purposely.* You aimed and shot the arrow at William Tell with the purpose of killing him rather than with the intent of hitting the apple on his head. (Tell is the national hero of Switzerland who was required to shoot an apple off his son's head.)
- *Knowingly.* You know that you are a poor shot, and when shooting at the apple on William Tell's head, you knew that you were practically certain to kill him.
- *Recklessly.* You clearly appreciated and knew the risk of shooting the arrow at William Tell with your eyes closed. Nevertheless, you proceeded to shoot the arrow despite the fact that this was a gross deviation from the standard of care that a law-abiding person would exhibit.

- **Negligently.** You claim that you honestly believed that you were such an experienced hunter that there was no danger in shooting the apple from William Tell's head. This was a gross deviation from the standard of care that a reasonable person would practice under the circumstances.

Strict liability crimes require only an *actus reus* and do not require proof of a *mens rea*. These offenses typically are public welfare crimes whose creation is meant to protect the safety and security of society by regulating food, drugs, and transportation. These offenses are *mala prohibita* rather than *mala in se* and usually are punishable by a small fine. Strict liability offenses are criticized as inconsistent with the traditional concern with "moral blameworthiness."

A criminal act requires the unison or concurrence of a criminal intent and a criminal act. This means that the intent must dictate the act.

Crimes such as murder, aggravated assault, and arson require the achievement of a particular result. Particularly in the case of homicide, defendants may claim that their act did not cause the victim's death. The prosecution must establish beyond a reasonable doubt that an individual's act was the cause in fact, or "but for" cause, that set the chain of causation in motion. The defendant's act must also be the legal or proximate cause of the death. Normally this is not difficult. Cases involving complex patterns of causation, however, may require judges to make difficult decisions concerning whether it is fair and just to hold an individual responsible for the consequences of intervening acts.

We saw that two types of intervening acts are important in examining the chain of causation:

- A *coincidental intervening act* is unforeseeable and breaks the chain of causation.
- A *responsive intervening act* breaks the chain of causation only when the reaction is both abnormal and unforeseeable.

CHAPTER REVIEW QUESTIONS

1. What is the reason that the law requires a *mens rea?*

2. Why is it difficult to prove *mens rea* beyond a reasonable doubt? Discuss some different ways of proving *mens rea*.

3. Explain the difference between purpose and knowledge. Which is punished more severely? Why?

4. Distinguish recklessness from negligence. Which is punished more severely? Why?

5. What is the difference between a crime requiring a criminal intent and strict liability?

6. Explain the "willful blindness" rule.

7. What is the importance of the principle of concurrence? Provide an example of a lack of concurrence.

8. Disputes over causation typically arise in prosecutions for what types of crimes?

9. Explain the statement that an individual's criminal act must be shown to be both the cause in fact and the legal or proximate cause.

10. What is meant by the statement that legal or proximate cause is based on a judgment of what is just or fair under the circumstances? How does this differ from the determination of a cause in fact or a "but for" analysis?

11. What is the difference between a coincidental intervening act and a responsive intervening act? Provide examples.

12. Discuss the test for determining whether coincidental intervening acts and responsive intervening acts break the chain of causation.

13. Provide concrete examples illustrating a coincidental intervening act and a responsive intervening act that do not break the chain of causation. Now provide examples of coincidental and intervening acts that do break the chain of causation.

14. What is the year-and-a-day rule? Why are states now abandoning this principle?

15. What are the arguments for and against strict liability offenses?

16. What is the approach of the Model Penal Code toward causality? Use some of the cases in the text to illustrate your answer.

17. Are we too concerned with criminal intent? Why not impose the same punishment on criminal acts regardless of an individual's intent? Is the father or mother of a child hit by a car concerned whether the driver was acting intentionally, knowingly, recklessly, or negligently?

LEGAL TERMINOLOGY

causation	intervening cause	responsive intervening act
cause in fact	knowingly	scienter
circumstantial evidence	*mens rea*	specific intent
coincidental intervening act	negligently	strict liability
concurrence	proximate cause	transferred intent
constructive intent	public welfare offense	willful blindness
crimes of cause and result	purposely	year-and-a-day rule
general intent	recklessly	

CRIMINAL LAW ON THE WEB

Log on to the Web-based student study site at www.sagepub.com/lippmanccl3e to assist you in completing the Criminal Law on the Web exercises, as well as for additional features such as podcasts, Web quizzes, and video links.

1. Read about the risks of texting while driving.

2. Explore states laws on the ownership of exotic animals.

3. Learn about state laws on hate crimes.

BIBLIOGRAPHY

American Law Institute, *Model Penal Code and Commentaries*, vol. 1, pt. 1 (Philadelphia: American Law Institute, 1985), pp. 225–266. A detailed examination of intent and causality with proposals for reforms.

Joshua Dressler, *Understanding Criminal Law*, 3rd ed. (New York: Lexis, 2001), pp. 115–150, 179–200. A comprehensive and easily understood examination of intent, concurrence, and causation.

Hyman Gross, *A Theory of Criminal Justice* (New York: Oxford University Press, 1979), pp. 74–113, 232–254, 342–374. A difficult discussion of the theory of intent and causation for the sophisticated reader.

Jerome Hall, *General Principles of Criminal Law*, 2nd ed. (Indianapolis, IN: Bobbs-Merrill, 1960), pp. 70–105. A difficult and challenging discussion of intent.

Wayne R. LaFave, *Criminal Law*, 3rd ed. (St. Paul, MN: West Publishing, 2000), pp. 224–320. A detailed discussion with citations to relevant cases on intent, concurrence, and causation.

Rollin M. Perkins and Ronald N. Boyce, *Criminal Law*, 3rd ed. (Mineola, NY: Foundation Press, 1982), pp. 760–906. A valuable discussion of the historical evolution of criminal intent and causation.

Richard G. Singer and John Q. La Fond, *Criminal Law Examples and Explanations* (New York: Aspen, 2001), pp. 45–78, 99–146. An accessible overview of intent and causality with questions to test your understanding.

PARTIES TO CRIME AND VICARIOUS LIABILITY

Did Gometz intend to kill a prison guard?

In the morning, Silverstein, while being escorted from the shower to his cell, stopped next to Randy Gometz's cell, and while two of the escorting officers were for some reason at a distance from him, reached his handcuffed hands into the cell. The third officer, who was closer to him, heard the click of the handcuffs being released and saw Gometz raise his shirt to reveal a home-made knife ("shank")—which had been fashioned from the iron leg of a bed—protruding from his waistband. Silverstein drew the knife and attacked one of the guards, Clutts, stabbing him 29 times and killing him.

Learning Objectives

1. Know the four categories of parties to a crime under the common law and the two categories of parties to a crime used in contemporary statutes.

2. Describe the *actus reus* and *mens rea* of accomplice liability.

3. Understand the natural and probable consequences doctrine.

4. Know the elements of accessory after the fact.

5. Understand the reasons that the law imposes vicarious liability and list some examples of vicarious liability under the laws of the various states.

INTRODUCTION

Thus far, we have established a number of building blocks of criminal conduct. First, there are *constitutional limits* on the government's ability to declare acts criminal. Second, *actus reus* requires that an individual commit a voluntary act or omission. People are punished for what they do, not for what they think or for who they are. Third, the existence of a criminal intent or *mens rea* means that punishment is limited to morally blameworthy individuals. Last, there must be a *concurrence* between a criminal act and a criminal intent. The criminal act must be established as both the *factual cause* and the *legal* or *proximate cause* of a prohibited harm or injury. We now add another building block to this foundation by observing that more than one individual may be liable for a crime. In this chapter, we will discuss two situations in which multiple parties are held liable for a crime.

- ***Parties to a Crime* or *Complicity*.** Individuals who assist the perpetrator of a crime before, during, or following the crime are held criminally responsible. *In other words, individuals who assist in the commission of a crime are held liable for the criminal conduct of the perpetrator of the offense.*

- ***Vicarious Liability*.** Individuals may be held liable based on their *relationship* with the perpetrator of a crime. The most common instance involves extending guilt to an employer for the acts of an employee or imposing liability on a corporation for the acts of a manager or employee. Two other instances of vicarious liability are reviewed in this chapter. The first involves holding the owner of an automobile liable for traffic tickets issued to the car despite the fact that the auto may have been driven by another individual. The second entails imposing responsibility on parents for the acts of their children.

In reading cases concerning complicity, you should ask yourself whether the appellant intended to assist and assisted a crime. As a matter of social policy, consider why we punish people who assist another to commit a crime. Should the parties to a crime be subject to the same punishment as the perpetrator? As for vicarious liability, consider whether criminal responsibility should be extended to individuals who were neither present nor involved or perhaps even aware of the crime.

PARTIES TO A CRIME

Common law judges appreciated that criminal conduct often involves a range of activities: planning the crime, carrying out the offense, evading arrest, and disposing of the fruits of the crime. The common law divided the participants in a crime into *principals* and **accessories**. Principals were actually present and carried out the crime, while accessories assisted the principals. Holding individuals accountable for intentionally assisting the criminal acts of another is termed *accomplice* or *accessory liability*.

The four categories of **parties to a crime** under the common law are as follows:

- *Principals in the First Degree.* The perpetrator of the crime. For example, the person or persons actually robbing the bank.
- *Principals in the Second Degree.* Individuals assisting the robbers. This includes look-outs, getaway drivers, and those disabling burglar alarms. Principals in the second degree are required to be either physically present at the bank or constructively present, meaning that they directly assist the robbery at a distance by engaging in such activities as serving as a lookout.
- *Accessory Before the Fact.* Individuals who help prepare for the crime. In the case of a bank robbery, accessories before the fact may purchase firearms or masks, plan the crime, or encourage the robbers. Accessories, in contrast to principals, are neither physically nor constructively present.
- *Accessory After the Fact.* Individuals who assist the perpetrators, knowing that a crime has been committed. This includes those who help the bank robbers escape or hide the stolen money.

Both principals and accessories were punishable as felons under the common law. All felonies were subject to the death penalty. Common law judges desired to limit the offenses for which capital punishment might be imposed, and they developed various rules to frustrate the application of the death penalty. Judges, for instance, held that principals and accessories could be prosecuted only following the conviction of the principal in the first degree. This posed a barrier to prosecution in those instances in which the principal in the first degree was acquitted, fled, or died, or in which the principal's conviction was reversed. There were additional requirements that complicated prosecutions, such as the fact that an accessory who assisted a crime while living in another state could be prosecuted only in the state in which the acts of accessoryship occurred. These jurisdictions typically had little interest in prosecuting an accessory for crimes committed outside the state.[1]

Today we are no longer required to overcome these complications. Virtually every jurisdiction has abandoned the common law categories. States typically provide for two parties to a crime:

- *Accomplices.* Individuals involved before and during a crime
- *Accessories.* Individuals involved in assisting an offender following the crime

Returning to our bank robbery example, the perpetrator of the bank robbery and the individuals planning and organizing the robbery as well as the lookout, the driver of the getaway car, and the individuals disabling the bank guard will all be charged with bank robbery. In the event that the accomplices are convicted, they will receive the same sentence as the perpetrator of the crime.

Individuals who assist the perpetrators following the crime will be charged as accessories after the fact. Accessoryship is no longer viewed as being connected to the central crime. It is considered a separate, minor offense involving the frustration of the criminal justice process, and it is punishable as a misdemeanor. Despite these changes to the law of parties, you will find that common law categories are frequently referred to in judicial decisions and in various state statutes.

Holding an individual liable for the conduct of another seems contrary to the American value of personal responsibility. Why should we punish an individual who drives a getaway car in a bank robbery to the same extent as the actual perpetrator of the crime? The law presumes that the individuals who assisted the robber implicitly consented to be bound by the conduct of the principal in the first degree and, in the words of Joshua Dressler, "forfeited their personal identity." Professor Dressler refers to this as **derivative liability**, in which the accessory's guilt flows from the acts of the primary perpetrator of the crime.[2]

In a later chapter we will discuss *conspiracy*, which is an agreement to commit a crime, such as bank robbery. A defendant may be liable both for an agreement to rob a bank and for the bank robbery itself. This is the so-called **Pinkerton rule**, which provides that a conspiracy to commit a crime and the crime itself are separate and distinct crimes. An individual may be charged with one or both of these offenses.[3]

For a deeper look at this topic, visit the study site.

ACTUS REUS OF ACCOMPLICE LIABILITY

Statutes and judicial decisions describe the *actus reus* of accomplice liability using a range of seemingly confusing terms such as *aid, abet, encourage,* and *command.* Whatever the terminology, keep in mind that the *actus reus* of accomplice liability is satisfied by even a relatively insignificant degree of material or psychological assistance. In a well-known English case, a journalist bought a ticket to attend and review a concert by American jazz musician Coleman Hawkins and was convicted of encouraging and supporting Hawkins, who had not received permission to perform from British immigration authorities.[4] Consider the range of conduct in the following examples that courts have considered to constitute aiding and abetting a crime and as accomplice liability.[5]

- Two men attacked and broke Clifton Robertson's leg. The men initially approached Robertson's brother-in-law, Carl Brown, who pointed to Robertson. Following the attack, one of the assailants remarked to Brown, "You can pay me now." Brown was found guilty of aggravated assault for inciting, encouraging, or assisting the perpetrators of the assault.[6]
- Guadalupe Steven Mendez was an incarcerated felon who directed Patricia Morgan over the phone to molest and to take nude photos of her fourteen-year-old granddaughter. Mendez was found to have aided and abetted aggravated rape.[7]
- Delfino Alejandres was convicted of aggravated robbery for having encouraged an offense by either word or deed. Alejandres parked his car so as to prevent Peter Pham, a fifteen-year-old fellow high school student, from pulling out of a driveway. Chris Valoretta exited Alejandres's auto and shot Pham when Pham refused to give Valoretta the keys to Pham's automobile. Alejandres and Valoretta then drove away. The evidence indicated that Alejandres and Valoretta were friends and that Alejandres had earlier commented to a fellow student that he was considering "wasting" Pham.[8]
- Donald Jones directed his son Michael to rob Guy Justice, who owed money to Donald. Michael agreed to rob Justice later in the day. The three subsequently met at Andra Wright's house, and Donald asked Michael whether he still planned to rob Justice. Michael demanded that Justice empty his pockets, and during the struggle Donald shouted at Michael to "shoot . . . it's either you or him, you better shoot." Michael responded by shooting Justice in the chest. Donald was convicted of having aided and abetted Michael through his "conduct, presence and companionship."[9]
- Prentiss Phillips was a high-ranking member of the Gangster Disciples street gang and was convicted of the murder and aggravated kidnapping of Vernon Green, whom he suspected of standing outside of a meeting of the street gang in order to identify members of the Gangster Disciples for the rival Vice Lords. Phillips allegedly ordered gang members to seize Green and watched as Green was dragged upstairs, where he was beaten. Phillips later ordered three gang members to take Green outside and kill him. Phillips was convicted of aiding and abetting the murder and aggravated kidnapping of Green.[10]

It is important to note that although an accomplice must assist in the commission of a crime, there is no requirement that the prosecution demonstrate that the accomplice's contribution was essential to the commission of the crime. This seems to be intended to deter individuals from assisting in the commission of a crime. Such a rule, of course, may prove unfair, because the punishment of an accomplice who makes a small contribution may be much more severe than he or she actually deserves. The Maine Supreme Court ruled that a defendant who offered to make an automobile available was guilty of being an accomplice to murder despite the fact that the perpetrator carried out the crime without the car.[11] How should a court rule in those instances in which the perpetrator is unaware of the accomplice's effort to assist in the crime? The consensus is that as long as an individual acts with the required intent, he or she will be held liable as an accessory.

In the cases discussed previously, the defendants' acts clearly assisted, encouraged, or incited criminal conduct. The **mere presence rule** provides that being present and watching the commission of a crime is not sufficient to satisfy the *actus reus* requirement of accomplice liability. Why is that? A mere presence is ambiguous. On the one hand, it is sometimes the case that an individual's presence encourages and facilitates a defendant's criminal conduct. On the other hand, silence may indicate disapproval.

Courts have struggled with what action is required for an individual who is present to be considered an accomplice. Judges have ruled that a gang member was not guilty of aiding and abetting when he, along with fifteen or twenty others, chanted the name of the gang while two gang members smashed and beat an automobile containing a member of a rival gang.[12] On the other hand, an Illinois court convicted a gang member who joined a group in chasing a truck and watched as the driver was pulled out of a truck and then silently stood over the driver as a codefendant kicked the victim while another man hit the victim with a bat. The defendant then left the crime scene with the assailants.[13] Is there a clear and meaningful difference in the contribution of these two gang members?

The leading case on the mere presence rule is *Bailey v. United States*. Bailey spent most of the day conversing with his partner and then left to shoot craps while his partner stood across the street near the entrance to the Center Market. Bailey later joined his partner. The two watched as an employee followed his regular routine and exited the market carrying money receipts in a bag. As the employee approached, Bailey retreated to the curb ten feet from his partner, who then seized the deposits at gunpoint. Bailey and his partner then fled, and only Bailey was subsequently apprehended. The District of Columbia Court of Appeals determined that Bailey was innocently talking to his partner when his partner suddenly pulled out a gun and seized the bag of money. The fact that Bailey fled did not mean that he was involved with his partner in either planning or in assisting the crime as a lookout. The court stressed that individuals who are entirely innocent sometimes flee based on a fear of being considered guilty or in order to avoid appearing as a witness.[14]

An exception to the mere presence doctrine arises where defendants possess a duty to intervene. In *State v. Walden,* a mother was convicted of aiding and abetting an assault with a deadly weapon when she failed to intervene to prevent an acquaintance from brutally beating her young son. The North Carolina Supreme Court reasoned that a parent's failure to protect his or her child communicates an approval of the criminal conduct.[15]

The mere presence rule is explored in the next case, *State v. Ulvinen*. This decision also raises the liability of individuals who utter words of approval and the difficulty of determining whether these individuals intend to assist the perpetrator.

Was the defendant an accomplice to the murder of her daughter-in-law?

State v. Ulvinen, 313 N.W.2d 425 (Minn. 1981). Opinion by: Otis, J.

Issue

Was the appellant properly convicted of first degree murder pursuant to Minnesota Statutes section 609.05, subdivision 1 (1980), imposing criminal liability on one who "intentionally aids, advises, hires, counsels, or conspires with or otherwise procures" another to commit a crime?

Facts

Carol Hoffman, daughter-in-law of appellant Helen Ulvinen, was murdered late on the evening of August 10 or the very early morning of August 11, 1980, by her husband, David Hoffman. She and David had spent an amicable evening together playing with their children, and when they went to bed David wanted to make love to his wife. However, when she refused him he lost his temper and began choking her. While he was choking her he began to believe he was "doing the right thing" and that to get "the evil out of her" he had to dismember her body.

After his wife was dead, David called down to the basement to wake his mother, asking her to come upstairs to sit on the living room couch. From there she would be able to see the kitchen, bathroom, and bedroom doors and could stop the older child if she awoke and tried to use the bathroom. Appellant didn't respond at first but after being called once, possibly twice, more she came upstairs to lie on the couch. In the meantime David had moved the body to the bathtub. Appellant was aware that while she was in the living room her son was dismembering the body but she turned her head away so that she could not see.

After dismembering the body and putting it in bags, Hoffman cleaned the bathroom, took the body to Weaver Lake, and disposed of it. On returning home he told his mother to wash the cloth covers from the bathroom toilet and tank, which she did. David fabricated a story about Carol leaving the house the previous night after an argument, and Helen agreed to corroborate it. David phoned the police with a missing person report and during the ensuing searches and interviews with the police, he and his mother continued to tell the fabricated story.

On August 19, 1980, David confessed to the police that he had murdered his wife. In his statement he indicated that not only had his mother helped him cover up the crime, but she had known of his intent to kill his wife that night. After hearing Hoffman's statement, the police arrested appellant and questioned her with respect to her part in the cover-up. Police typed up a two-page statement, which she read and signed. The following day a detective questioned her further regarding events surrounding the crime, including her knowledge that it was planned.

Appellant's relationship with her daughter-in-law had been a strained one. She had moved in with the Hoffmans on July 26, 1980—two weeks before the crime—to act as a live-in babysitter for their two children. Carol was unhappy about having her move in and told friends that she hated Helen, but she told both David and his mother that they could try the arrangement to see how it worked.

On the morning of the murder, Helen told her son that she was going to move out of the Hoffman residence, because "Carol had been so nasty to me." In his statement to the police, David reported the conversation that morning as follows:

A: Sunday morning I went downstairs and my mom was in the bedroom reading the newspaper and she had tears in her eyes, and she said in a very frustrated voice, "I've got to find another house." She said, "Carol don't want me here," and she said, "I probably shouldn't have moved in here." And

I said then, "Don't let what Carol said hurt you. It's going to take a little more period of readjustment for her." Then I told mom that I've got to do it tonight so that there can be peace in this house.

Q: What did you tell your mom that you were going to have to do that night?

A: I told my mom I was going to have to put her to sleep.

Q: Dave, will you tell us exactly what you told your mother that morning, to the best of your recollection?

A: I said I'm going to have to choke her tonight and I'll have to dispose of her body so that it will never be found. That's the best of my knowledge.

Q: What did your mother say when you told her that?

A: She just—she looked at me with very sad eyes and just started to weep. I think she said something like, "It will be for the best."

David spent the day fishing with a friend of his. When he got home that afternoon, he had another conversation with his mother. She told him at that time about a phone conversation Carol had had in which she discussed taking the children and leaving home. David told the police that during the conversation with his mother that afternoon, he told her, "Mom, tonight's got to be the night."

Q: When you told your mother, "Tonight's got to be the night," did your mother understand that you were going to kill Carol later that evening?

A: She thought I was just kidding her about doing it. She didn't think I could.

Q: Why didn't your mother think that you could do it?

A: Because for some time I had been telling her I was going to take Carol scuba diving and make it look like an accident.

Q: And she said?

A: And she always said, "Oh, you're just kidding me."

Q: But your mother knew you were going to do it that night?

A: I think my mother sensed that I was really going to do it that night.

Q: Why do you think your mother sensed you were really going to do it that night?

A: Because when I came home and she told me what had happened at the house, and I told her, "Tonight's got to be the night," I think she said, again I'm not certain, "that it would be the best for the kids."

Reasoning

It is well settled in this state that presence, companionship, and conduct before and after the offense are circumstances from which a person's participation in the criminal intent may be inferred. The evidence is undisputed that appellant was asleep when her son choked his wife. She took no active part in the dismembering of the body but came upstairs to intercept the children, should they awake, and prevent them from going into the bathroom. She cooperated with her son by cleaning some items from the bathroom and corroborating David's story to prevent anyone from finding out about the murder. She is insulated by statute from guilt as an accomplice after the fact for such conduct because of her relation as a parent of the offender. The jury might well have considered appellant's conduct in sitting by while her son dismembered his wife so shocking that it deserved punishment. Nonetheless, these subsequent actions do not succeed in transforming her behavior prior to the crime to active instigation and encouragement. Minnesota Statutes section 609.05, subdivision 1 (1980) implies a high level of activity on the part of an aider and abettor in the form of conduct that encourages another to act. Use of terms such as *aids, advises,* and *conspires* requires something more of a person than mere inaction to impose liability as a principal.

Holding

The evidence presented to the jury at best supports a finding that appellant passively acquiesced in her son's plan to kill his wife. The jury might have believed that David told his mother of his intent to kill his wife that night and that she neither actively discouraged him nor told anyone in time to prevent the murder. Her response that "it would be the best for the kids" or "it will be the best" was not, however, active encouragement or instigation. There is no evidence that her remark had any influence on her son's decision to kill his wife. The Minnesota statute imposes liability for actions that affect the principal, encouraging him to take a course of action that he might not otherwise have taken. The State has not proved beyond a reasonable doubt that appellant was guilty of anything but passive approval. However morally reprehensible it may be to fail to warn individuals of their impending death, our statutes do not make such an omission a criminal offense.

David told many people besides appellant of his intent to kill his wife, but no one took him seriously. He told a co-worker approximately three times a week that he was going to murder his wife and confided two different plans for doing so. Another co-worker heard him tell his plan to cut Carol's air hose while she was scuba diving, making her death look accidental, but did not believe him. Two or three weeks before the murder, David told a friend of his that he and Carol were having problems, and he expected Carol "to have an accident sometime." None of these people has a duty imposed by law to warn the victim of impending danger, whatever their moral obligation may be. The State has not proved beyond a reasonable doubt that appellant was guilty of any of the activities enumerated in the statute. Appellant's comment is not sufficient additional activity on her part to constitute planning or conspiring with her son. She did not offer advice on how to kill his wife, nor offer to help him. She did not plan when to accomplish the act or tell her son what to do to avoid being caught. She was told by her son that he intended to kill his wife that night and responded in a way that, while not discouraging him, did not aid, advise, or counsel him to act as he did. Where, as here, the evidence is insufficient to show beyond a reasonable doubt that appellant was guilty of active conduct sufficient to convict her of first degree murder . . . her conviction must be reversed.

Questions for Discussion

1. What facts did the prosecutor likely rely on to establish Ms. Ulvinen's guilt? What facts would the defense attorney rely on in urging her acquittal? Why did the Minnesota Supreme Court reverse Ms. Ulvinen's conviction?

2. Does Ms. Ulvinen's behavior following Carol Hoffman's murder indicate that she approved of Carol's killing? Does Ms. Ulvinen's involvement in covering up the crime indicate that she shared her son's intent to kill Carol Hoffman?

3. Is it significant that Ms. Ulvinen took the initiative in informing David that she was not getting along with Carol and that Carol planned to leave with the children? What of Ms. Ulvinen's response that killing Carol would be beneficial for the children? How do Ms. Ulvinen's comments to David on the day of the killing differ from her earlier responses to David's statements that he intended to kill Carol? Is it significant that David killed Carol only following his conversation with his mother? Was the court correct in characterizing Ms. Ulvinen as passive?

4. Did Ms. Ulvinen have a duty to protect her daughter-in-law, the mother of her grandchildren?

5. Why does the statute exempt a parent from liability for accessory after the fact?

6. Draft your own one-page opinion in this case.

Cases and Comments

The Mere Presence Rule. The four- or five-year-old victim lived with her mother, Holly Swanson-Birabent, for a year while in kindergarten. On one occasion, the victim wandered into her mother's bedroom and observed her mother sexually engaged with her boyfriend, Don Umble. The victim got into the bed with her mother and Umble and fell asleep. At some point, Umble woke up the victim and molested her. The defendant stood to the side of the bed watching silently. Swanson-Birabent then entered the bed, and her daughter returned to her own bedroom. The same series of events was repeated on another occasion. The California appellate court determined that the defendant was aware of what was occurring, because there was light from the street shining through the window, and the sheet was sufficiently thin to reveal what was transpiring.

The appellate court concluded that the defendant engaged in an "act" sufficient to constitute aiding and abetting in that the "defendant was not merely present. Her presence served an important purpose: it encouraged the victim to comply with Umble rather than resist . . . [and] also encouraged Umble to continue molesting the victim." In addition, the court concluded that Swanson-Birabent failed to fulfill her duty as a parent to protect her daughter from molestation. This failure to act both encouraged the "victim to comply with Umble rather than resist, and . . . encouraged Umble to continue molesting the victim." The appellate court affirmed Swanson-Birabent's conviction of two counts of committing a lewd or lascivious act on a child under the age of fourteen. See *People v. Swanson-Birabent*, 7 Cal. Rptr. 3d 744 (Cal. Ct. App. 2003).

You Decide

6.1 A woman entered a bar in New Bedford, Massachusetts, in March 1983, in order to purchase cigarettes. As she started to leave, she was knocked to the ground by two men who tore off her clothing, and for the next seventy-five minutes she was forced to commit various sexual acts, which she resisted. The victim cried for help, but the sixteen men in the bar yelled, laughed, and cheered. No one came to her assistance. Were all fifteen male customers and the male bartender accomplices to the sexual attack? Various procedural issues are discussed in *Commonwealth v. Cordeiro*, 519 N.E.2d 1328 (Mass. 1988), and *Commonwealth v. Vieira*, 519 N.E.2d 1320 (Mass. 1988).

You can find the answer at www.sagepub.com/lippmanccl3e.

MENS REA OF ACCOMPLICE LIABILITY

A conviction for accomplice liability requires that a defendant both assist and intend to assist the commission of a crime. In *People v. Perez*, the defendant, Victor Perez, encountered a group of individuals on the street. Victor was not a member of a gang and was unaware that several of the individuals were affiliated with the Maniac Latin Disciples and that they were confronting Victor's former acquaintance, Pedro Gonzalez. Victor was unaware that the Maniac Latin Disciples alleged that Pedro had taunted them by displaying hand signs of the rival Latin Kings. In an effort to defend himself when they asked Victor whether Pedro was a member of the Latin Kings, Victor responded that he knew that Pedro had been a member of the Kings, but was uncertain whether he still was a member. Pedro was immediately shot and killed. An Illinois appellate court concluded that the defendant lacked the requisite criminal intent to be held liable for first-degree murder. What intent was required? Would it have been sufficient for Victor to know that Pedro might be killed if Victor identified him? Or was it necessary that Victor intended to assist in Pedro's death?[16]

There is a lack of agreement over the required *mens rea* for accomplice liability. In most cases, an accomplice is required to intend or possess the purpose that an individual commit a specific crime. This is described as *dual intents:*

- the intent to assist the primary party, and
- the intent that the primary party commit the offense charged.[17]

The requirement of purposeful conduct was articulated in the decision of Judge Learned Hand in *United States v. Peoni*. Peoni sold counterfeit money to Regno in the Bronx. The money later was sold to Dorsey, who was arrested attempting to pass the money in Brooklyn. Peoni was charged with aiding and abetting Dorsey's possession of counterfeit money.

Judge Hand recognized that Joseph Peoni was aware that Regno would sell the money to another individual and that the money then would be circulated in the economy. Hand, however, ruled that Peoni's "mere knowledge" was not sufficient to hold him liable for aiding and abetting. Peoni was required to "associate himself with the venture, that he participate in it as something that he wishes to bring about, that he seek by his action to make it succeed." Hand stressed that Peoni had to possess a "purposive attitude" that Regno sell the money on the street. Peoni, however, once having sold the money to Regno, had no real interest in whether the money was circulated.[18]

In other words, an individual will not be held liable as an accomplice for knowingly rather than purposely

- selling a gun to an individual who plans to rob a bank.
- renting a room to someone who plans to use the room for prostitution.
- repairing the car of a stranded motorist who intends to use the auto to rob a bank.

Some judges have ruled that the *mens rea* requirement for accomplice liability in the case of serious crimes is satisfied by knowledge of a defendant's criminal plans. In *Backun v. United States,* Max Backun sold stolen silver to Zucker in New York and was aware that Zucker intended to sell the silver in various southern states. Backun's conviction as an accomplice to Zucker's transporting stolen merchandise in interstate commerce was affirmed by a federal court on the grounds that the

seller may not ignore the purpose for which the purchase is made if he is advised of that purpose, or wash his hands of the aid that he has given the perpetrator . . . by the plea that he has merely made a sale of merchandise.[19]

Regardless of whether a "purposive" or "knowledge" standard is employed, an accomplice is subject to the **natural and probable consequences** doctrine. This provides that a person encouraging or facilitating the commission of a crime will be held liable as an accomplice for the crime he or she aided and abetted as well as for crimes that are the natural and probable outcome of the criminal conduct. Two issues arise: What crimes are the natural and probable consequence of a criminal act? Should an accomplice be held liable for crimes he or she did not intend to assist or knowingly assist?

The next case in the chapter, *United States v. Fountain,* explores the best approach to determining the *mens rea* for complicity. Ask yourself whether Judge Richard Posner relied on a "purpose" or "knowledge" standard for aiding and abetting. What approach do you favor?

Model Penal Code

Section 2.06. Liability for Conduct of Another; Complicity

. . .

(1) A person is an accomplice of another person in the commission of an offense if:

   (a) with the purpose of promoting or facilitating the commission of the offense, he

     (i) aids or agrees or attempts to aid such other person in planning or committing it, or

     (ii) having a legal duty to prevent the commission of the offense, fails to make proper effort so to do; or

   (b) his conduct is expressly declared by law to establish his complicity. . . .

. . .

(2) An accomplice may be convicted on proof of the commission of the offense and of his complicity therein, though the person claimed to have committed the offense has not been prosecuted or convicted . . . .

The Legal Equation

Complicity = Aiding, assisting, inciting, counseling, encouraging another to commit a crime; or failure to fulfill a legal duty to intervene

+ purpose to encourage or assist another to commit the crime.

Did Gometz aid and abet the killing of the prison guard?

United States v. Fountain, 768 F.2d 790 (7th Cir. 1985). Opinion by: Posner, J.

We have consolidated the appeals in two closely related cases of murder of prison guards in the control unit of the federal penitentiary at Marion, Illinois—the maximum-security cell block in the nation's maximum-security federal prison—by past masters of prison murder, Clayton Fountain and Thomas Silverstein.

Facts

Shortly before these crimes were committed, Fountain and Silverstein, both of whom were already serving life sentences for murder, had together murdered an inmate in the control unit of Marion, and had again been sentenced to life imprisonment. After that, Silverstein killed another inmate, pleaded guilty to that murder, and received his third life sentence. At this point Fountain and Silverstein had each killed three people. (For one of these killings, however, Fountain had been convicted only of voluntary manslaughter. And Silverstein's first murder conviction was reversed for trial error, and a new trial ordered, after the trial in this case). The prison authorities—belatedly, and as it turned out ineffectually—decided to take additional security measures. Three guards would escort Fountain and Silverstein (separately), handcuffed, every time they left their cells to go to or from the recreation room, the law library, or the shower (prisoners in Marion's control unit are confined, one to a cell, for all but an hour or an hour and a half a day, and are fed in their cells). But the guards would not be armed; nowadays guards do not carry weapons in the presence of prisoners, who might seize the weapons.

The two murders involved in these appeals took place on the same October day in 1983. In the morning, Silverstein, while being escorted from the shower to his cell, stopped next to Randy Gometz's cell, and while two of the escorting officers were for some reason at a distance from him, reached his handcuffed hands into the cell. The third officer, who was closer to him, heard the click of the handcuffs being released and saw Gometz raise his shirt to reveal a home-made knife ("shank")—which had been fashioned from the iron leg of a bed—protruding from his waistband. Silverstein drew the knife and attacked one of the guards, Clutts, stabbing him 29 times and killing him. While pacing the corridor after the killing, Silverstein explained that "this is no cop thing. This is a personal thing between me and Clutts. The man disrespected me and I had to get him for it." Having gotten this off his chest he returned to his cell.

Fountain was less discriminating. While being escorted that evening back to his cell from the recreation room, he stopped alongside the cell of another inmate (who, however, apparently was not prosecuted for his part in the events that followed) and reached his handcuffed hands into the cell, and when he brought them out he was out of the handcuffs and holding a shank. He attacked all three guards, killing one (Hoffman) with multiple stab wounds (some inflicted after the guard had already fallen), injuring another gravely (Ditterline, who survived but is permanently disabled), and inflicting lesser though still serious injuries on the third (Powles). After the wounded guards had been dragged to safety by other guards, Fountain threw up his arms in the boxer's gesture of victory, and laughing walked back to his cell.

A jury convicted Fountain of first-degree murder and of lesser offenses unnecessary to go into here. The judge sentenced him to not less than 50 nor more than 150 years in prison and also ordered him, pursuant to the Victim and Witness Protection Act of 1982 . . . to make restitution of $92,000 to Hoffman's estate, $98,000 to Ditterline, and nearly $300,000 to the Department of Labor. . . .

Silverstein and Gometz were tried together (also before a jury, and before the same judge who presided at Fountain's trial) for the murder of Clutts, and both received the same 50- to 150-year sentences as Fountain and were ordered to pay restitution to Clutts's estate and to the Department of Labor of $68,000 and $2,000 respectively. Fountain and Silverstein are now confined in different federal prisons, in what were described at argument as "personalized" cells. . . .

Issue

Gometz argues that the evidence was insufficient to convict him of aiding and abetting Silverstein in murdering Clutts. This argument requires us to consider the mental element in "aiding and abetting." . . . Under the older cases . . . it was enough that the aider and abettor knew the principal's purpose. Although this is still the test in some states, after the Supreme Court adopted Judge Learned Hand's test—that the aider and abettor "in some sort associate himself with the venture, that he participate in it as in something that he wishes to bring about, that he seek by his action to make it succeed," . . . it came to be generally accepted that the aider and abettor must share the principal's purpose in order to be guilty of violating . . . the federal aider and abettor statute. But . . . there is support for relaxing this requirement when the crime is particularly grave. . . .

Reasoning

In *People v. Lauria*, 59 Cal. Rptr. 628, 634 (1967)—not a federal case, but illustrative of the general point—the court, en route to holding that knowledge of the principal's purpose would not suffice for aiding and abetting of just any crime, said it would suffice for "the seller of gasoline who knew the buyer was using his product to make Molotov cocktails for terroristic use." Compare the following hypothetical cases. In the first, a shopkeeper sells dresses to a woman whom he knows to be a prostitute. The shopkeeper would not be guilty of aiding and abetting prostitution unless the prosecution could establish the elements of Judge Hand's test. Little would be gained by imposing criminal liability in such a case. Prostitution, anyway a minor crime, would be but trivially deterred, since the prostitute could easily get her clothes from a shopkeeper ignorant of her occupation. In the second case, a man buys a gun from a gun dealer after telling the dealer that he wants it in order to kill his mother-in-law, and he does kill her. The dealer would be guilty of aiding and abetting the murder. This liability would help to deter—and perhaps not trivially given public regulation of the sale of guns—a most serious crime. We hold that aiding and abetting murder is established by proof beyond a reasonable doubt that the supplier of the murder weapon knew the purpose for which it would be used. . . .

Holding

Gometz argues that there is insufficient evidence that he knew why Silverstein wanted a knife. We disagree. The circumstances make clear that the drawing of the knife from Gometz's waistband was prearranged. There must have been discussions between Silverstein and Gometz. Gometz must have known through those discussions or others that Silverstein had already killed three people in prison—two in Marion—and while this fact could not be used to convict Silverstein of a fourth murder, it could ground an inference that Gometz knew that Silverstein wanted the knife in order to kill someone. If Silverstein had wanted to conceal it on his person in order to take it back to his cell and keep it there for purposes of intimidation, escape, or self-defense (or carry it around concealed for any or all of these purposes), he would not have asked Gometz to release him from his handcuffs (as the jury could have found he had done), for that ensured that the guards would search him. Since the cuffs were off before Silverstein drew the shank from Gometz's waistband, a reasonable jury could find beyond a reasonable doubt that Gometz knew that Silverstein, given his history of prison murders, could have only one motive in drawing the shank and that was to make a deadly assault. . . .

Questions for Discussion

1. Judge Richard Posner states that it is sufficient that the aider and abettor "knew the principal's purpose" and contrasts this with an intent to "bring about" a crime. Explain the difference between these two approaches to intent using the facts in *Fountain* to illustrate your answer.

2. What was Gometz's intent in providing the shank to Fountain? Would Gometz have been convicted under either a purpose or knowledge standard?

3. Do you agree with Judge Posner that a knowledge standard should be employed for more serious offenses in order to deter criminal activity? Would such a test cause difficulties for businesses selling goods to the public?

You Decide

6.2 Mark Manes, twenty-two, met Eric Harris, a seventeen-year-old student at Columbine High School in Littleton, Colorado, at a gun show. Manes purchased a semiautomatic handgun for Harris and accompanied Harris to a target range. After hitting a target, Harris excitedly proclaimed that this could have been someone's brain. Several months later, Manes sold Harris one hundred rounds of ammunition for $25. The next day, Harris and Dylan

Klebold entered Columbine High School and killed twelve students and a teacher and then took their own lives. Harris and Klebold left a tape recording thanking Manes for his help and urged that he not be arrested, because they would have eventually found someone willing to sell them guns and ammunition.

As a prosecutor, would you charge Manes as an accomplice to the murders? To the suicides? What if Harris and Klebold arrived at the school armed with weapons and ammunition provided by Manes but used other weapons to kill? What if they left the weapons and ammunition provided by Manes at home? See *Joshua Dressler, Cases and Materials on Criminal Law,* 3rd ed. (St. Paul, MN: West Publishing, 2003), p. 886.

You can find the answer at www.sagepub.com/lippmanccl3e.

Natural and Probable Consequences Doctrine

The leading case on the natural and probable consequences doctrine is the Maine case of *State v. Linscott.* Joel Fuller enlisted Linscott in a robbery scheme. The plan was for Linscott and Fuller to enter through the backdoor to prevent Grenier from grabbing a shotgun that he kept in the bedroom. Linscott carried a hunting knife and a switchblade and Fuller was armed with his shotgun. As they approached the house, they saw that the snow blocked the backdoor. They revised their plan. Linscott was to break the living room picture window whereupon Fuller would freeze Grenier with the shotgun while Linscott seized the cash.

Linscott broke the window with his body and Fuller immediately fired a shot through the broken window killing Grenier. Fuller entered the house through the broken window and took $1,300 from Grenier's pocket. Fuller gave Linscott $500.

Linscott later was arrested. He claimed that it was not unusual for Fuller carry a shotgun, that he was unaware that Fuller had a reputation for violence, and that although he may have been negligent, he had no intention of killing Grenier during the course of the robbery. Linscott was convicted of intentionally or knowingly killing Grenier. He was found to possess the intent to commit the crime of robbery and that the murder was a reasonably foreseeable consequence of the Linscott's participation in the robbery. The court recognized that Linscott did not intend to kill Grenier and that he probably would not have participated in the robbery had he believed that Grenier would be killed during the course of the robbery.[20]

You can find State v. Linscott *on the study site www.sagepub.com/lippmanccl3e.*

In reading *State v. Robinson,* ask yourself whether the natural and probable consequence doctrine unfairly punishes defendants.

Should Robinson be held liable for murder?

State v. Robinson, 715 N.W.2d 44 (Mich. 2006). Opinion by: Young, J.

Issue

Defendant and a codefendant, Samuel Pannell, committed an aggravated assault, and Pannell shot and killed the victim, Bernard Thomas. After a bench trial, the trial court convicted defendant of second-degree murder under an aiding and abetting theory. The Court of Appeals reversed the trial court's judgment, because it concluded that there was insufficient evidence that defendant shared or was aware of Pannell's intent to kill. The government appealed the judgment of the appellate court.

Facts

According to the evidence adduced at trial, defendant and Pannell went to the house of the victim, Bernard Thomas, with the stated intent to "f__ him up." Under Pannell's direction, defendant drove himself and Pannell to the victim's house. Pannell knocked on the victim's door. When the victim opened the door, defendant struck him. As the victim fell to the ground, defendant struck the victim again. Pannell began to kick the victim. Defendant told Pannell that

"that was enough," and walked back to the car. When defendant reached his car, he heard a single gunshot.

Following a bench trial, the trial court found defendant guilty of second-degree murder. . . . Specifically, the court found that defendant drove Pannell to the victim's house with the intent to physically attack the victim. The court also found that once at the victim's home, defendant initiated the attack on the victim, and that defendant's attack enabled Pannell to "get the upper-hand" on the victim. The court sentenced defendant to a term of 71 months to 15 years.

The Court of Appeals reversed defendant's murder conviction, holding that there was insufficient evidence to support defendant's second-degree murder conviction. The Court held that the trial court improperly convicted defendant of second-degree murder because there was no evidence establishing that defendant was aware of or shared Pannell's intent to kill the victim.

This Court granted the prosecution's application for leave to appeal. . . .

Reasoning

This case involves liability under our aiding and abetting statute, MCL 767.39, which provides:

> Every person concerned in the commission of an offense, whether he directly commits the act constituting the offense or procures, counsels, aids, or abets in its commission may hereafter be prosecuted, indicted, tried and on conviction shall be punished as if he had directly committed such offense.

Aiding and abetting is simply a theory of prosecution that permits the imposition of vicarious liability for accomplices. This Court recently described the three elements necessary for a conviction under an aiding and abetting theory:

> (1) the crime charged was committed by . . . some other person; (2) the defendant performed acts or gave encouragement that assisted the commission of the crime; and (3) the defendant intended the commission of the crime or had knowledge that the principal intended its commission at the time that [the defendant] gave aid and encouragement.

The primary dispute in this case involves the third element. Under the Court of Appeals analysis, the third element would require the prosecutor to prove beyond a reasonable doubt that a defendant intended to commit the identical offense, here homicide, as the accomplice or, alternatively, that a defendant knew that the accomplice intended to commit

the homicide. We reaffirm that evidence of defendant's specific intent to commit a crime or knowledge of the accomplice's intent constitutes sufficient *mens rea* to convict under our aiding and abetting statute. However, as will be discussed later in this opinion, we disagree that evidence of a shared specific intent to commit the crime of an accomplice is the exclusive way to establish liability under our aiding and abetting statute.

We hold that when the Legislature abolished the distinction between principals and accessories, it intended for all offenders to be convicted of the intended offense, in this case aggravated assault, as well as the natural and probable consequences of that offense, in this case death. The case law that has developed since the Legislature codified these common-law principles provides examples of accomplice liability under both theories.

Under the natural and probable consequences theory, "[t]here can be no criminal responsibility for any thing not fairly within the common enterprise, and which might be expected to happen if the occasion should arise for any one to do it." . . .

The propriety of the trial court's verdict is clear. The victim's death is clearly within the common enterprise the defendant aided because a homicide "might be expected to happen if the occasion should arise" within the common enterprise of committing an aggravated assault. The evidence establishes that the victim threatened Pannell's children in Pannell's presence, enraging Pannell. When defendant woke up at 10:00 that evening, Pannell was still "ranting and raving" in the house. Despite knowing that Pannell was in an agitated state, defendant agreed to drive to the victim's house with the understanding that he and Pannell would "f__ him up." When the pair arrived at the victim's home, defendant initiated the assault by hitting the victim once in the face and once in the neck with the back of his hand. After the victim fell to the ground, Pannell punched him twice and began kicking him. In our judgment, a natural and probable consequence of a plan to assault someone is that one of the actors may well escalate the assault into a murder. . . . Pannell's anger toward the victim escalated during the assault into a murderous rage.

Defendant argues that he should not be held liable for the murder because he left the scene of the assault after telling Pannell, "That's enough." We disagree. Defendant was aware that Pannell was angry with the victim even before the assault. Defendant escalated the situation by driving Pannell to the victim's house, agreeing to join Pannell in assaulting the victim, and initiating the attack. He did nothing to protect Thomas and he did nothing to defuse the situation in which Thomas was ultimately killed by Pannell. A "natural and probable consequence" of leaving the enraged Pannell alone with the victim is that Pannell would ultimately murder the victim. That defendant . . . left the scene of the crime moments before Thomas's murder does not under these circumstances exonerate him from responsibility for the crime.

The fact that Pannell shot the victim, rather than beat him to death, does not alter this conclusion. It cannot be that a defendant can initiate an assault, leave an already infuriated principal alone with the victim, and then escape liability for the murder of that victim simply because the principal shot the victim to death, instead of kicking the victim to death. The defendant is criminally liable as long as the crime is within the natural and probable consequences of the intended assaultive crime. . . .

We hold that a defendant must possess the criminal intent to aid, abet, procure, or counsel the commission of an offense. A defendant is criminally liable for the offenses the defendant specifically intends to aid or abet, or has knowledge of, as well as those crimes that are the natural and probable consequences of the offense he intends to aid or abet. Therefore, the prosecutor must prove beyond a reasonable doubt that the defendant aided or abetted the commission of an offense and that the defendant intended to aid the charged offense . . . or, alternatively, that the charged offense was a natural and probable consequence of the commission of the intended offense.

Holding

We hold that under Michigan law, a defendant who intends to aid, abet, counsel, or procure the commission of a crime, is liable for that crime as well as the natural and probable consequences of that crime. In this case, defendant committed and aided the commission of an aggravated assault. One of the natural and probable consequences of such a crime is death. Therefore, the trial court properly convicted defendant of second-degree murder. We reverse the judgment of the Court of Appeals and reinstate defendant's conviction of second-degree murder.

Dissenting, *Cavanagh, J.*

[I]t cannot fairly be said that this death was the natural and probable consequence of this beating where the trial court found that the victim did not die from injuries inflicted during the beating, defendant did not intend to kill, and defendant did not know Pannell would shoot and kill the victim. Thus, I disagree with the majority's rationale that because under some circumstances a death may result from a beating, defendant's conviction of second-degree murder was proper.

Dissenting, *Kelly, J.*

[The defendant's] conviction of second-degree murder, as an aider and abettor, be reversed.

Robinson went along "only to beat up" the victim. In Robinson's words, "it was understood between us that we were going to f___ him up." As a practical matter, f___ing up someone necessarily entails leaving them alive. In the context of this case, it most likely means to "put [the victim] in an extremely difficult or impossible situation." Offensive as the word is, it is not used to mean "to kill." We have many other slang words that mean "to kill," such as "bump off," "ice," "knock off," "waste," "rub out," and "whack." Applying the trial judge's factual findings, it is clear that Robinson agreed to harm the victim, not to kill him.

The victim's death here was not within Robinson and Pannell's common enterprise; a homicide by gun is not a natural and probable consequence of an intended assault and battery. The majority is mistaken in concluding otherwise. It errs by determining that the unintended result of an intentional act was a "natural and probable consequence" for which a defendant may be held criminally liable.

I would reduce the charge of which Robinson was convicted to assault with intent to do great bodily harm . . . and remand for resentencing on that reduced charge.

Questions for Discussion

1. Describe the facts leading up to Pannell's killing of Thomas.

2. What is the basis for prosecuting and convicting Robinson for second degree murder?

3. Summarize the arguments of the dissenting judges in *Robinson*.

4. As a judge, would you hold Robinson liable for second degree murder or assault with an intent to do great bodily harm?

You Decide

6.3 Leon McCoy conspired with Keith Lamar Bellamy to rob a McDonald's restaurant where McCoy worked. Andre Randall, McCoy's co-worker, was aware of the plan.

C.B. was working the evening shift as the assistant manager of a McDonald's in Wilmington, Delaware. She was assisted by defendant Leon McCoy and Andre Randall. C.B. locked the doors at 10:00 p.m. Randall took out the trash, and, contrary to restaurant policy, failed to notify C.B. who ordinarily opened and locked the door. Randall simply opened the door and rather than closing the door turned the deadbolt so as to prop the door open. Bellamy, who was armed, entered at around 11:30 p.m. as McCoy was mopping the hallway and C.B. was preparing the night deposit. Bellamy put the gun to the side of C.B.'s head and seized the deposit money, and also took C.B.'s personal cash. He demanded a bag for the cash. McCoy, went to the front of the store and got a bag. Although there were several silent alarms in this area, McCoy did not activate any of the alarms.

Once he bagged the money, Bellamy told C.B. to undress. As she was unbuttoning her shirt, he said she was taking too long and he told her to just drop her pants and underwear. He then demanded that she spread her labia apart. He stooped down to inspect her genitals, and used the barrel of his gun to pull her labia further apart. He noticed that she had a tampon inserted, and told her that she was "lucky." Following the robbery, Bellamy fled and McCoy went to the front of the store and hit a silent alarm. There were security cameras in the store recording the robbery. Is McCoy guilty of sexual assault? See *State v. Bellamy,* 617 S.E. 2d 81 (N.C. Ct. App. 2005).

You can find the answer at www.sagepub.com/lippmanccl3e.

You Decide

6.4 The defendant, Jacqueline Annette Williams, and Fedell Caffey had been dating for two years. Williams and Caffey had unsuccessfully attempted to conceive a child. Caffey is described as wanting "a baby boy with light skin so that the baby would resemble him."

Williams's cousin, Lavern Ward, was angry with Debra Evans with whom he had a two-year-old child, Jordan. Williams, Caffey, and Ward met and visited Debra, who was pregnant by Ward with her fourth child, who was to be named Elijah. Williams was aware that Ward and Caffey wanted to talk to Evans about the unborn baby and to "teach her a lesson."

Williams was in the bathroom when she heard a loud noise. She emerged to see Caffey standing over Evans with a small automatic gun while Ward appeared to be stabbing her in the neck. Williams stood next to Caffey as he used poultry shears to cut across Evans's abdomen and to remove a male child from Evans's womb and cut the umbilical cord. Caffey stated that he did not want the baby because he believed that the infant boy was dead. Williams blew into Elijah's nose and mouth, and he began breathing.

Caffey and Ward went into a bedroom where they killed ten-year-old Samantha. Seven-year-old Joshua began crying, and Williams put him in a car with Caffey and Ward, and they dropped him at a friend's house. They learned that Joshua had talked about the murders, and the next day Williams and Caffey drove Joshua to a suburban location where Caffey stabbed him to death. Williams left the body in an alley and discarded the sheet in which Joshua was wrapped. The killing occurred on November 19, and Evans was scheduled to be admitted to the hospital to give birth on November 20. Williams, for several months prior to the murders, represented that she was pregnant and later claimed that Elijah was her son.

The defendant argued that she did not inflict injuries on Evans or Samantha and that she should not be held liable as an accomplice. Williams was convicted and sentenced to death. Do you agree that Williams possessed the required criminal intent to be held liable for Evans's and Samantha's murders? For Joshua's murder? See *People v. Williams,* 739 N.E.2d 455 (Ill. App. 2000).

You can find the answer at www.sagepub.com/lippmanccl3e.

ACCESSORY AFTER THE FACT

The Common Law

For an international perspective on this topic, visit the study site.

Conviction as an accessory after the fact at common law required that a defendant conceal or assist an individual whom he or she knew had committed a felony in order to hinder the perpetrator's arrest, prosecution, or conviction. For example, an individual would be held liable for assisting a friend—whom he or she knew had killed a member of an opposing gang—to flee to a foreign country.

Accessories after the fact at common law were treated as accomplices and were subject to the same punishment as the principal who committed the offense. A wife, however, could not be held liable as an accessory after the fact because it was expected that she assisted her husband.

The Elements of Accessory After the Fact

The elements of accessory after the fact are as follows:

- *Commission of a Felony.* There must be a completed felony. The crime need not have been detected or formal charges filed.
- *Knowledge.* The defendant must possess knowledge that the individual whom he or she is assisting has committed a felony. A reasonable but mistaken belief does not create liability. In *Wilson v. State,* the defendant was speaking with Mendez and Valentine outside a convenience store. Valentine later followed Mendez back to Mendez's trailer and took two gold chains from Mendez's neck. Mendez and three others chased, apprehended, and struggled with Valentine. Wilson was driving by the altercation and intervened to defend Valentine. The Florida District Court of Appeal ruled that there was no evidence that Wilson was aware of the robbery and that he had not intended to assist Valentine after the fact.[21]
- *Affirmative Act.* The accessory must take affirmative steps to hinder the felon's arrest; a refusal or failure to report the crime or to provide information to the authorities is not sufficient. Examples of conduct amounting to accessoryship after the fact are hiding or helping a felon escape, destroying evidence, or providing false information to the police in order to mislead law enforcement officials. In *Melahn v. State,* the defendant was acquitted by a jury that accepted his testimony that he let his roommate drive his automobile, he fell asleep, and as a result, he was unaware that his roommate had burglarized a pet store. The Florida court further ruled that Melahn's refusal to cooperate with the police was not sufficient to establish accessoryship after the fact, which requires a false statement to the police.[22]
- *Criminal Intent.* The defendant must provide assistance with the intent or purpose of hindering the detection, apprehension, prosecution, conviction, or punishment of the individual receiving assistance. In *State v. Jordan,* Kenneth Jordan shot Pendley after seeing Pendley talking to Teresa Jordan, Kenneth Jordan's wife. Teresa was convicted of being an accessory after the fact based on her falsely reporting to friends, nurses, and the police and testifying at trial that Pendley had been shot while attempting to rape her. Kenneth Jordan pled guilty to the voluntary manslaughter of Pendley prior to Teresa's conviction.[23]

The modern view is that because accessories after the fact are involved following the completion of a crime, they should not be treated as harshly as the perpetrator of the crime or accomplices. Accessories after the fact are now held liable for a separate, less serious felony or for a misdemeanor. Some states have abandoned the crime of accessory after the fact and have created the new offense of "hindering prosecution," which punishes individuals frustrating the arrest, prosecution, or conviction of individuals who have committed felonies as well as misdemeanors.

Most states have abandoned the common law requirements that frustrated the conviction of individuals for accessoryship after the fact. A spouse is no longer immune in most states from prosecution for being an accessory after the fact. The common law rule that an individual may be prosecuted only for being an accessory after the fact following the conviction of the principal has also been modified. Most states require only proof of a completed felony.

Modern statutes increasingly follow the Model Penal Code and list the specific types of assistance that are prohibited. Other statutes retain the common law language and look to courts to define the acts constituting the offense of being an accessory after the fact.

The next case, *State v. Chism,* summarizes the distinction between liability as a principal and as an accessory after the fact. The case also asks whether an individual may be held liable for both offenses.

Model Penal Code

Section 242.3. Hindering Apprehension or Prosecution

A person commits an offense if, with purpose to hinder the apprehension, prosecution, conviction or punishment of another for crime, he:

(1) harbors or conceals the other; or

(2) provides or aids in providing a weapon, transportation, disguise or other means of avoiding apprehension or effecting escape; or

(3) conceals or destroys evidence of the crime, or tampers with a witness, informant, document or other source of information, regardless of its admissibility in evidence; or

(4) warns the others of impending discovery or apprehension . . . ; or

(5) volunteers false information to a law enforcement officer. . . .

Analysis

The Model Penal Code views accessory after the fact as obstruction of justice and does not require that the person providing assistance be aware that the alleged offender actually committed a crime. The essence of the crime is interference with the functioning of the legal process. An individual charged with accessory after the fact is thus not treated as a principal. Liability extends to assisting an individual avoid apprehension for a misdemeanor as well as a felony. The Model Penal Code specifies the type of assistance that is prohibited to prevent courts from too narrowly or too broadly interpreting the behavior that is prohibited.

Accessory after the fact is punished as a felony carrying five years' imprisonment if the person aided is charged or is liable to be charged with a felony of the first or second degree. Assisting a felony of the third degree (one to two years in prison) or a misdemeanor is punished as a misdemeanor. The fact that the person who provides assistance is related to the individual confronting a criminal charge is to be considered as a mitigating factor at sentencing rather than being considered as a complete defense.

The Legal Equation

Accessory after the fact **=** Conceals or assists an individual

+ whom he or she knows to have committed a felony

+ in order to hinder the perpetrator's arrest, prosecution, or conviction.

Was Chism an accessory after the fact?

State v. Chism, 436 So. 2d 464 (La. 1983). Opinion by: Dennis, J.

Facts

On the evening of August 26, 1981, in Shreveport, Tony Duke gave the defendant, Brian Chism, a ride in his automobile. Brian Chism was impersonating a female, and Duke was apparently unaware of Chism's disguise. After a brief visit at a friend's house, the two stopped to pick up some beer at the residence of Chism's grandmother. Chism's one-legged uncle, Ira Lloyd, joined them, and the three continued on their way, drinking as Duke drove the automobile. When Duke expressed a desire to have sexual relations with Chism, Lloyd announced that he wanted to find his ex-wife Gloria for the same purpose. Shortly after midnight, the trio arrived at the St. Vincent Avenue Church of Christ and persuaded Gloria Lloyd to come outside. As Ira Lloyd stood outside the car attempting to persuade Gloria to come with them, Chism and Duke hugged and kissed on the front seat as Duke sat behind the steering wheel.

Gloria and Ira Lloyd got into an argument, and Ira stabbed Gloria with a knife several times in the stomach and once in the neck. Gloria's shouts attracted the attention of two neighbors, who unsuccessfully tried to prevent Lloyd from pushing Gloria into the front seat of the car alongside Chism and Duke. Lloyd climbed into the front seat also, and Duke drove off. One of the bystanders testified that she could not be sure but she thought she saw Chism's foot on the accelerator as the car left.

Lloyd ordered Duke to drive to Willow Point, near Cross Lake. When they arrived, Chism and Duke, under Lloyd's direction, removed Gloria from the vehicle and placed her on some high grass on the side of the roadway, near a wood line. Lloyd was unable to help the two, because his wooden leg had come off. Afterwards, as Lloyd requested, the two drove off, leaving Gloria with him.

There was no evidence that Chism or Duke protested, resisted, or attempted to avoid the actions that Lloyd ordered them to take. Although Lloyd was armed with a knife, there was no evidence that he threatened either of his companions with harm.

Duke proceeded to drop Chism off at a friend's house, where Chism changed to male clothing. He placed his blood-stained women's clothes in a trash bin. Afterward, Chism went with his mother to the police station at 1:15 a.m. He gave the police a complete statement and took the officers to the place where Gloria had been left with Ira Lloyd. The police found Gloria's body in some tall grass several feet from that spot. An autopsy indicated that stab wounds had caused her death. Chism's discarded clothing disappeared before the police arrived at the trash bin.

Chism was convicted of being an accessory after the fact and appealed to the Louisiana Supreme Court. Louisiana statute 14.25 provides that an accessory after the fact is an individual who, "after the commission of a felony, shall harbor, conceal, or aid the offender, knowing or having reasonable ground to believe that he has committed the felony, and with the intent that he may avoid or escape from arrest, trial, conviction or punishment." An individual who is convicted under this statute is subject to a fine and imprisonment of up to five years in prison, provided that "in no case shall his punishment be greater than one-half of the maximum provided by law for a principal offender."

Issue

Defendant appealed his conviction and sentence and argued that the evidence was not sufficient to support the judgment. Consequently, we must determine whether, after viewing the evidence in the light most favorable to the prosecution, any rational trier of fact could have found beyond a reasonable doubt that (a) a completed felony had been committed by Ira Lloyd before Brian Chism rendered him the assistance described below, (b) Chism knew or had reasonable grounds to know of the commission of the felony by Lloyd, and (c) Chism gave aid to Lloyd personally under circumstances that indicate either that he actively desired that the felon avoid or escape arrest, trial, conviction, or punishment or that he believed that one of these consequences was substantially certain to result from his assistance.

Reasoning

An accessory after the fact is any person, who, after the commission of a felony, shall harbor, conceal, or aid the offender, knowing or having reasonable ground to believe that the offender has committed the felony and with the intent that the offender may avoid or escape from arrest, trial, conviction, or punishment. . . . We conclude that a person may be punished as an accessory after the fact if he aids an offender personally, knowing or having reasonable ground to believe that the offender has committed the felony, and having a specific or general intent that the offender will avoid or escape from arrest, trial, conviction, or punishment. An accessory after the fact may be tried and convicted notwithstanding the fact that the principal felon may not have been arrested, tried, convicted, or amenable to justice. However, it is still necessary to prove the guilt of the principal beyond a reasonable doubt, and an accessory after the fact cannot be convicted or punished where the principal felon has been acquitted. Furthermore, it is essential to prove that a felony was committed and completed prior to the time the assistance was rendered the felon, although it is not also necessary that the felon have been already charged with the crime. . . .

There was clearly enough evidence to justify the finding that a felony had been completed before any assistance was rendered to Lloyd by the defendant. The record vividly demonstrates that Lloyd fatally stabbed his ex-wife before she was transported to Willow Point and left in the high grass near a wood line. Thus, Lloyd committed the felonies of attempted murder, aggravated battery, and simple kidnapping before Chism aided him in any way. A person cannot be convicted as an accessory after the fact to a murder because of aid given after the murderer's acts but before the victim's death, but under these circumstances the aider may be found to be an accessory after the fact to the felonious assault. In this particular case, it is of no consequence that the defendant was formally charged with accessory after the fact to second degree murder, instead of accessory after the fact to attempted murder, aggravated battery, or simple kidnapping. The defendant was fairly put on notice of the actual acts underlying the offense with which he was charged, and he does not claim or demonstrate in this appeal that he has been prejudiced by the form of the indictment.

The evidence overwhelmingly indicates that Chism had reasonable grounds to believe that Lloyd had committed a felony before any assistance was rendered. In his confessions and his testimony, Chism indicates that the victim was bleeding profusely when Lloyd pushed her into the vehicle, that she was limp and moaned as they drove to Willow Point, and that he knew Lloyd had inflicted her wounds with a knife. The Louisiana offense of accessory after the fact deviates somewhat from the original common law offense in that it does not require that the defendant actually know that a completed felony has occurred. Rather, it incorporates an objective standard by requiring only that the defendant render aid "knowing or having reasonable grounds to believe" that a felony has been committed.

The closest question presented is whether any reasonable trier of fact could have found beyond a reasonable doubt that Chism assisted Lloyd under circumstances that indicate that either Chism actively desired that Lloyd would avoid or escape arrest, trial, conviction, or punishment, or that Chism believed that one of these consequences was substantially certain to result from his assistance. After carefully reviewing the record, we conclude that the prosecution satisfied its burden of producing the required quantity of evidence.

Holding

In this case we conclude that the evidence is sufficient to support an ultimate finding that the reasonable findings and inferences permitted by the evidence exclude every reasonable hypothesis of innocence. Despite evidence supporting some contrary inferences, a trier of fact reasonably could have found that Chism acted with at least a general intent to help Lloyd avoid arrest, because (1) Chism did not protest or attempt to leave the car when his uncle, Lloyd, shoved the mortally wounded victim inside; (2) he did not attempt to persuade Duke, his would-be lover, to exit out the driver's side of the car and flee from his uncle, whom he knew to be one-legged and armed only with a knife; (3) he did not take any of these actions at any point during the considerable ride to Willow Point; (4) at their destination, he docilely complied with Lloyd's directions to remove the victim from the car and leave Lloyd with her, despite the fact that Lloyd made no threats and that his wooden leg had become detached; (5) after leaving Lloyd with the dying victim, he made no immediate effort to report the victim's whereabouts or to obtain emergency medical treatment for her; (6) before going home or reporting the victim's dire condition, he went to a friend's house, changed clothing, and discarded his own in a trash bin from which the police were unable to recover them as evidence; (7) he went home without reporting the victim's condition or location; (8) and he went to the police station to report the crime only after arriving home and discussing the matter with his mother.

The defendant asserted in his statement given to the police and during trial that he helped to remove the victim from the car and to carry her to the edge of the bushes, because he feared that his uncle would use the knife on him. The defense of justification can be claimed in any crime, except murder, when it is committed through the compulsion of threats by another of death or great bodily harm and the offender reasonably believes the person making the threats is present and would immediately carry out the threats if the crime were not committed. However, Chism did not testify that Lloyd threatened him with death, bodily harm, or anything. Moreover, fear as a motivation to help his uncle is inconsistent with some of Chism's actions after he left his uncle. Consequently, we conclude that despite Chism's testimony, the trier of fact could have reasonably found that he acted voluntarily and not out of fear when he aided Lloyd and that he did so under circumstances indicating that he believed that it was substantially certain to follow from his assistance that Lloyd would avoid arrest, trial, conviction, or punishment.

For the foregoing reasons, it is also clear that . . . there is evidence in this record from which a reasonable trier of fact could find a defendant guilty beyond a reasonable doubt. . . . Therefore, we affirm the defendant's conviction.

Dissenting, *Dixon, C.J.*

I respectfully dissent from what appears to be a finding of guilt by association. The majority lists five instances of inaction, or failure to act, by defendant: He (1) did not protest or leave the car, (2) did not attempt to persuade Duke to leave the car, (3) did neither (1) nor (2) on the ride to Willow Point, (5) made no immediate effort to report the crime or get aid for the victim, and (7) failed to report the victim's condition or location after changing clothes. The three instances of defendant's action relied on by the majority for conviction were stated to be (4) complying with Lloyd's direction to remove the victim from the car and leave the victim and Lloyd at Willow Point, (6) changing clothes and discarding bloody garments, and (8) discussing the matter with defendant's mother before going to the police station to report the crime. None of these actions or failures to act tended to prove defendant's intent, specifically or generally, to aid Lloyd to avoid arrest, trial, conviction, or punishment.

You Decide

6.5 In a Mississippi case, Xavier Sherron began having intercourse with his thirteen-year-old stepdaughter Jane in December 2001 or January 2002. Charlotte Sherron, Xavier's wife and Jane's mother, learned that Xavier had been having intercourse with Jane on a regular basis, and Xavier assured her that he would not continue to molest Jane. Charlotte directed Jane to lock her door. Charlotte needed Xavier's monthly disability checks to make ends meet and did not ask him to leave her house. Charlotte also was fearful that in the event the rape was revealed, her three children would be taken from her by state authorities.

In February 2002, Jane told Charlotte that she was pregnant and stated that an abortion was preferable to bearing Xavier's child. Charlotte agreed with other family members that she would not contact the police, and Charlotte arranged for an abortion for Jane in Tuscaloosa, Alabama. Both Charlotte and Xavier drove Jane to Tuscaloosa. In May 2002, Jane's uncle approached the police, and Xavier subsequently was arrested and convicted of statutory rape. Charlotte was arrested as an accessory after the fact to the statutory rape based on her assisting Jane in obtaining an abortion. There was no charge brought against Charlotte for failing to fulfill her obligation as a Mississippi school employee to inform the police of Jane's abuse.

The court analyzed Jane's abortion under the requirements of Mississippi law and noted that Alabama law presumably contained the same provisions. A child who desires to obtain an abortion in Mississippi either must get parental consent or must petition the court for permission to obtain an abortion. The appellate court stated that the question is whether Charlotte's consent to Jane's abortion was based on a desire to support her daughter or based on an intent to conceal her husband's crimes. There was testimony at trial that Charlotte was afraid of Xavier, who in the past had choked Charlotte and had thrown objects at her.

Would you convict Charlotte of being an accessory after the fact to Xavier's statutory rape? See *Sherron v. State,* 959 So. 2d 30 (Miss. Ct. App. 2006).

You can find the answer at www.sagepub.com/lippmanccl3e.

CRIME IN THE NEWS

In December 2010, Mississippi Governor Haley Barbour indefinitely suspended the life prison sentences of sisters Jamie and Gladys Scott. In 1994, The Scott sisters were convicted of armed robbery and had spent over 16 years in a Mississippi prison before their release.

At the sisters' trial, the prosecution claimed that Jamie, then 23, and Gladys, then 19, had lured two men to drive them to a nightclub in Forest, Mississippi. The allegation was that one of the sisters pretended to be sick and, when their car pulled off the road, their two companions were robbed of $11 by three young men between the ages of 14 and 18, one of whom was wielding a shotgun. The two sisters admittedly knew the assailants and jumped into their car as they drove away.

The sister's consistently denied their involvement in the robbery. The young men responsible for the robbery all received reduced sentences on the condition that they implicate the sisters and were released after a few months in jail. One of the young men submitted an affidavit in 1998 claiming that he had been threatened with being sent to a dangerous and violent prison if he did not cooperate and implicate the sisters in the robbery.

Governor Haley Barbour was under intense pressure from civil rights groups to release the sisters. These groups pointed out that the sisters already had served lengthy prison sentences and that there was substantial evidence supporting their claim of innocence. In announcing his decision, Governor Barbour stressed that the decision to release the sisters was supported by the state parole board and that the sisters, who were eligible for parole in 2014, had been rehabilitated and no longer posed a threat to society.

Governor Barbour did not pardon the sisters, which would have relieved them of the stigma of their criminal convictions. Instead he indefinitely suspended their sentences on the condition that Gladys Scott, 36, donate a kidney to her sister Jamie. Some commentators speculated that the sisters had been released because Jamie's dialysis treatment was costing Mississippi as much as $200,000 per year and that State correctional authorities were unable to ensure that the dialysis was conducted in a safe and hygienic fashion.

The sisters, on their release, announced that they were moving to Florida, which relieved Mississippi of the health care costs. Although it was claimed that the sisters suggested the kidney donation, various medical ethicists questioned whether this was a free and voluntary donation. Do you agree with Governor Barbour's decision to make the sisters' release contingent on Gladys donating a kidney to Jamie?

VICARIOUS LIABILITY

We have seen that **strict liability** results in holding a defendant criminally responsible for the commission of a criminal act without a requirement of a criminal intent. An act, in other words, is all that is required. **Vicarious liability** imposes liability on an individual for a criminal act committed by another. They act and you are responsible. **Accomplice liability**, in contrast, holds individuals responsible who affirmatively aid and abet a criminal act with a purposeful intent.

Vicarious liability is employed to hold employers and business executives and corporations (which are considered "legal persons") liable for the criminal acts of employees. Vicarious liability has also been used to hold the owner of an automobile liable for parking violations committed by an individual driving the owner's car. Another example of vicarious liability is imposing liability on parents for crimes committed by their children.

Vicarious liability is contrary to the core principle that individuals should be held responsible and liable for their own conduct. The primary reason for this departure from individual responsibility in the case of corporations is to encourage employers to control and to monitor employees so as to insure that the public is protected from potential dangers, such as poisoned food.[24]

We have distinguished strict and vicarious liability. Keep in mind that statutes that are intended to protect the public health, safety, and welfare typically combine both doctrines. In the California case of *People v. Travers,* Mitchell, a service station employee, misrepresented the quality of motor oil he sold to the public. The defendant Charles Travers was the owner of the station and was prosecuted along with Mitchell under a statute that punished the sale of a misbranded product. Travers objected that he was completely unaware of Mitchell's actions. The court reasoned that the importance of smoothly running motor vehicles and the right of the public to receive what they paid for justified the imposition of vicarious liability on Travers without the necessity of demonstrating that he possessed a criminal intent. The court explained that it was reasonable to expect a service station owner to supervise the sale of motor oil, and requiring the prosecution to establish criminal intent would permit owners to escape punishment by pleading that they were unaware of the quality or contents of the motor oil sold in their service stations.[25]

Is it fair to impose strict liability on Travers for the acts of Mitchell? Would a significant number of guilty people be acquitted in the event that the court required the prosecution to establish a criminal intent? Professor LaFave poses a choice between punishing one hundred people for selling tainted food under a strict liability statute or using an intent standard that would result in the conviction of five of the one hundred. The first alternative would result in some innocent people being convicted; the second alternative would result in some guilty people avoiding a criminal conviction. What is the better approach?[26]

In the next section we will consider the use of vicarious liability to hold corporations criminally liable.

The Legal Equation

Vicarious liability $=$ Voluntary act or omission or possession by another

$+$ status of employer or parent or owner of automobile.

Corporate Liability

The early common law adopted the logical position that corporations are not living and breathing human beings and therefore could not be held criminally liable. There was no doctrine of **corporate liability**, prosecution, and punishment. Prosecution and punishment were limited to corporate officers and employees. Over time, corporations were subject to fines for failing to maintain the repair of public works such as roads and bridges.[27] The increasing power and prominence of large-scale business enterprises resulted in the gradual growth of the idea of corporate criminal liability and the punishment of corporations through the imposition of financial penalties. The U.S. Supreme Court noted in 1909 that acts of an employee "may be controlled in the interests of public policy, by imputing his act to his employer and imposing penalties upon the corporation for which he is acting."[28]

The U.S. Supreme Court, in *United States v. Dotterweich,* affirmed in 1943 that corporations, along with corporate executives and employees, could be held criminally liable under the Food and Drug Act. The Court stressed that holding the president of the corporation and the corporation vicariously liable for the strict liability crimes of employees was intended to ensure that company executives and managers closely monitor the distribution of potentially dangerous drugs to the public.[29] In *United States v. Park,* the Supreme Court upheld the conviction of a large national food store chain, along with the president of the company, for shipping adulterated food in interstate commerce.[30]

Keep in mind that a corporate crime may result in the criminal conviction of the employee committing the offense as well as the extension of vicarious liability to the owner and the corporation. There is nothing mysterious about a corporation. It is a method of organizing a business that provides certain financial benefits in return for complying with various state regulations. Note that most small corporations typically are run by an owner or by several partners, although moderately sized and larger corporations may be organized with boards of directors and outside investors or shareholders. The corporation possesses a life of its own separate and apart from all the executives, managers, and employees and is considered a "person" under the law.

The first step in determining whether a corporation may be criminally liable is to examine whether the legislature intends the criminal statute to apply to corporations. In *United States v. Dotterweich,* the U.S. Supreme Court affirmed the conviction of the defendant and corporation under the Federal Food, Drug, and Cosmetic Act for introducing an "adulterated or misbranded" drug into interstate commerce. The Court stressed that "a person" under the act was defined to include corporations. Courts have ruled in other instances that the term *person* was limited to "natural persons" and did not include "corporate persons."[31]

Once it is determined that a statute encompasses corporations, there are two primary tests for determining whether a corporation should be criminally liable under the statute.

- *Respondeat Superior or the Responsibility of a Superior.* A corporation may be held liable for the conduct of an employee who commits a crime within the scope of his or her employment who possesses the intent to benefit the corporation.
- *Model Penal Code Section 2.07.* Criminal liability is imposed in those instances that the criminal conduct is authorized, requested, commanded, performed, or recklessly tolerated by the board of directors or by a high managerial official acting on behalf of the corporation within the scope of his or her office or employment.

Respondeat superior extends vicarious criminal liability to a corporation for the acts of employees, even when such acts are contrary to corporate policy. It may seem unfair to impose liability on a corporation for the independent criminal acts of an employee, such as Mitchell's selling of motor oil in the *Travers* case. On the other hand, Mitchell's sale of misbranded motor oil increased the company's profits. The Model Penal Code test limits vicarious liability to acts approved or tolerated by high-level corporate officials. Managers, corporate boards, and corporate entities under this approach are liable only for acts that they direct or tolerate. Under this test, decision makers may not possess an incentive to closely monitor employees to ensure that they are not engaging in acts that have not been approved by management. A corporation, for instance, would not be convicted for Mitchell's independent decision to misrepresent the quality of motor oil sold to consumers. Which test do you favor?

Public Policy

There is an increasing trend toward holding corporations criminally liable. The aim is to encourage corporate executives to vigorously prevent and punish illegal activity. Executives know that a criminal conviction may lead to a decline in consumer sales and investment in the firm as well as to criminal fines, and they have a powerful incentive to ensure that the corporation acts in a legal fashion. The prevention of corporate misconduct is important, because we depend on large firms to provide safe and secure health care, transportation, food, and products in the home. Holding a corporation strictly and vicariously liable also makes good sense because business decisions often involve a large number of individuals, and it often is difficult to single out a specific individual or individuals as responsible for designing, manufacturing, marketing, and delivering a defective drug or automobile.[32]

On the other hand, it seems unfair to hold a corporation strictly and vicariously liable and to impose a heavy fine for crimes that may have been committed by low-level employees or managers or secretly approved by a high-level corporate executive. A criminal fine against a corporation is merely paid out of the corporate treasury, and the threat of a financial penalty may not encourage corporate officials to monitor the activities of employees. A fine may also be passed on to consumers, who will be charged a higher price. In the final analysis, the profits to be gained from misrepresenting the effectiveness of a drug may far outweigh any fine that may be imposed. Critics of corporate liability argue that it makes more sense to limit criminal liability to the individuals who committed the offense.[33]

Are there penalties other than fines that might be used against corporations? One federal district court imposed a three-year prison sentence on a corporation that was later suspended. The court observed that this could be carried out by ordering the U.S. marshal to seize corporate assets such as computers, machinery, and trucks. This type of punishment has the advantage of completely shutting down a business. On the other hand, it would likely result in innocent individuals losing their jobs.[34]

Should vicarious liability be extended to corporations for crimes requiring a criminal intent or recklessness? Why should a corporation be immune from a conviction for serious crimes that may result in injury or death? But how can a corporation possess a criminal intent?

In *Commonwealth v. Penn Valley Resorts, Inc.,* the resort and owner were convicted of involuntary manslaughter. The resort was fined $10,000. A Pennsylvania statute, section 307(a)(2), provides that a corporation may be convicted of an offense "authorized, requested, commanded, performed or recklessly tolerated by the board of directors or by a high managerial agent acting in behalf of the corporation within the scope of employment." The court held that the resort owner qualified as a "high managerial agent" under the statute and that the law did not limit the vicarious responsibility of corporations to strict liability health, safety, and welfare offenses.

In *Penn Valley,* Edwin Clancy, the president of the resort, permitted a group of underage students to engage in a drinking binge at the resort. William Frazier, a twenty-year-old, drank excessively for five or six hours. Clancy personally served alcohol to Frazier and seized and later handed Frazier back the keys to Frazier's automobile and encouraged the drunk and hostile student to leave the resort. Frazier was subsequently killed when his car drove off the road and hit a bridge. He was found to possess a blood alcohol content of .23. The Supreme Court of Pennsylvania concluded that the resort, "through its managerial agent, committed involuntary manslaughter and reckless endangerment." How can a corporation act with gross disregard for the safety of customers? On the other hand, Clancy was president, and his acts legally obligated and financially benefited the corporation.[35]

In the next case in the chapter, *Commonwealth v. Koczwara,* the Supreme Court of Pennsylvania examines whether the legislature intended to impose vicarious liability on the owner of a bar for the sale of liquor by his employees. Pay particular attention to the supreme court's reasoning in reviewing the constitutionality of the owner's criminal conviction. Give careful consideration to the views of the dissenting judge.

Can a bar owner be held vicariously liable and sentenced to jail for the sale of liquor to a minor?

Commonwealth v. Koczwara, 155 A.2d 825 (Pa. 1959). Opinion by: Cohen, J.

This is an appeal from the judgment of the Court of Quarter Sessions of Lackawanna County sentencing the defendant to three months in the Lackawanna County Jail and a fine of $5,000 and the costs of prosecution in a case involving violations of the Pennsylvania Liquor Code.

Facts

John Koczwara, the defendant, is the licensee and operator of an establishment on Jackson Street in the City of Scranton known as J.K.'s Tavern. At that place he had a restaurant liquor license issued by the Pennsylvania Liquor Control Board. . . .

At the conclusion of the Commonwealth's evidence, count three of the indictment, charging the sale by the defendant personally to the minors, was removed from the jury's consideration by the trial judge on the ground that there was no evidence that the defendant had personally participated in the sale or was present in the tavern when sales to the minors took place. . . . The case went to the jury, and the jury returned a verdict of guilty as to each of the remaining three counts: two counts of permitting minors to frequent the licensed premises without parental or other supervision and the count of permitting sales to minors. . . . Judge Hoban . . . sentenced the defendant to pay the costs of prosecution, a fine of $500, and to undergo imprisonment in the Lackawanna County Jail for three months. The defendant took an appeal to the Superior Court, which . . . affirmed the judgment and sentence of the lower court.

Judge Hoban found as fact that

in every instance the purchase [by minors] was made from a bartender, not identified by name, and service to the boys was made by the bartender. There was no evidence that the defendant was present on any one of the occasions testified to by these witnesses, nor that he had any personal knowledge of the sales to them or to other persons on the premises.

Issue

We, therefore, must determine the criminal responsibility of a licensee of the Liquor Control Board for acts committed by his employees upon his premises; without his personal knowledge, participation, or presence; which acts violate a valid regulatory statute passed under the commonwealth's police power.

Reasoning

While an employer in almost all cases is not criminally responsible for the unlawful acts of his employees unless he consents to, approves, or participates in such acts, courts all over the nation have struggled for years in applying this rule within the framework of "controlling the sale of intoxicating liquor." . . . At common law, any attempt to invoke the doctrine of respondeat superior in a criminal case would have run afoul of our deeply ingrained notions of criminal jurisprudence that guilt must be personal and individual.

In recent decades, however, many states have enacted detailed regulatory provisions in fields that are essentially noncriminal, e.g., pure food and drug acts; speeding ordinances; building regulations; and child labor, minimum wage, and maximum hour legislation. Such statutes are generally enforceable by light penalties, and although violations are labeled crimes, the considerations applicable to them are totally different from those applicable to true crimes, which involve moral delinquency and which are punishable by imprisonment or another serious penalty. Such so-called statutory crimes are in reality an attempt to utilize the machinery of criminal administration as an enforcing arm for social regulations of a purely civil nature, with the punishment totally unrelated to questions of moral wrongdoing or guilt. It is here that the social interest in the general well-being and security of the populace has been held to outweigh the individual interest of the particular defendant. The penalty is imposed despite the defendant's lack of a criminal intent or *mens rea*.

Not the least of the legitimate police power areas of the legislature is the control of intoxicating liquor. . . . It is abundantly clear that the conduct of the liquor business is lawful only to the extent and manner permitted by statute. Individuals who embark on such an enterprise do so with knowledge of considerable peril, since their actions are rigidly circumscribed by the Liquor Code.

Because of the peculiar nature of this business, one who applies for and receives permission from the commonwealth to carry on the liquor trade assumes the highest degree of responsibility to his fellow citizens. As the licensee of the board, he is under a duty not only to regulate his own personal conduct in a manner consistent with the permit he has received but also to control the acts and conduct of any employee to whom he entrusts the sale of liquor. Such fealty is the price that the commonwealth demands in return for the privilege of entering the highly restricted and, what is more important, the highly dangerous business of selling intoxicating liquor.

The question here raised is whether the legislature intended to impose vicarious criminal liability on the licensee-principal for acts committed on his premises without his presence, participation, or knowledge.

In the Liquor Code, section 493, the legislature has set forth twenty-five specific acts that are condemned as unlawful, and for which penalties are provided in section 494. Subsections (1) and (14) of section 493 contain the two offenses charged here. In neither of these subsections is there any language that would require the prohibited acts to have been done knowingly, willfully, or intentionally, there being a significant absence of such words as *knowingly, willfully,* and so forth. . . . It indicates a legislative intent to eliminate both knowledge and criminal intent as necessary ingredients of such offenses.

As the defendant has pointed out, there is a distinction between the requirement of a *mens rea* and the imposition of vicarious absolute liability for the acts of another. It may be that the courts below, in relying on prior authority, have failed to make such a distinction. In any case, we fully recognize it. Moreover, we find that the intent of the legislature in enacting this code was not only to eliminate the common law requirement of a *mens rea* but also to place a very high degree of responsibility upon the holder of a liquor license to make certain that neither he nor anyone in his employ commit any of the prohibited acts upon the licensed premises. Such a burden of care is imposed upon the licensee in order to protect the public from the potentially noxious effects of an inherently dangerous business. We, of course, express no opinion as to the wisdom of the legislature's imposing vicarious responsibility under certain sections of the Liquor Code. There may or may not be an economic-sociological justification for such liability on a theory of deterrence. Such determination is for the legislature to make, so long as the constitutional requirements are met. . . .

Holding

Defendant, by accepting a liquor license, must bear this financial risk. Because the defendant had a prior conviction for violations of the code, however, the trial judge felt compelled under the mandatory language of the statute, section 494(a), to impose not only . . . a fine of $500 but also a three month sentence of imprisonment. Such sentence of imprisonment in a case where liability is imposed vicariously cannot be sanctioned by this court consistently with . . . clause of section 9, article I of the constitution of the Commonwealth of Pennsylvania . . . which prohibits an individual from being "deprived of his life, liberty or property, unless by the judgment of his peers or the law of the land."

. . . We have found no case in any jurisdiction that has permitted a prison term for a vicarious offense. . . . Our own courts have stepped in time and again to protect a defendant from being held criminally responsible for acts about which he had no knowledge and over which he had little control. . . . We would be utterly remiss were we not to so act under these facts. . . . In holding that the punishment of imprisonment deprives the defendant of due process of law under these facts, we are not declaring that Koczwara must be treated as a first offender under the code. He has clearly violated the law for a second time and must be punished accordingly. Therefore, we are only holding that so much of the judgment as calls for imprisonment is invalid, and we are leaving intact the $500 fine imposed by Judge Hoban under the subsequent offense section.

Dissenting, *Musmanno, J.*

The majority of this court is doing something that can find no justification in all the law books that ornament the libraries and enlighten the judges and lawyers in this commonwealth. It sustains the conviction of a person for acts admittedly not committed by him, not performed in his presence, not accomplished at his direction, and not even done within his knowledge. It is stigmatizing him with a conviction for an act that, in point of personal responsibility, is as far removed from him as if it took place across the seas. The majority's decision is so novel, so unique, and so bizarre that one must put on his spectacles, remove them to wipe the lenses, and then put them on again in order to assure himself that what he reads is a judicial decision proclaimed in Philadelphia, the home of the Liberty Bell, the locale of Independence Hall, and the place where the fathers of our country met to draft the Constitution of the United States, the Magna Carta of the liberties of Americans and the beacon of hope of mankind seeking justice everywhere. The decision handed down in this case throws a shadow over that Constitution, applies an eraser to the Bill of Rights, and muffles the Liberty Bell that many decades ago sang its song of liberation from monarchical domination over man's inalienable right to life, liberty, and the pursuit of happiness. . . .

The majority introduces into its discussion a proposition that is shocking to contemplate. It speaks of "vicarious criminal liability." Such a concept is as alien to American soil . . . [T]here was a time in China when a convicted felon sentenced to death could offer his brother or other close relative in his stead for decapitation. The Chinese law allowed such "vicarious criminal liability." I never thought that Pennsylvania would look with favor on anything approaching so revolting a barbarity.

The majority says that it cannot permit the sentencing of a man to jail "for acts about which he had no knowledge and over which he had little control." It says, "Such sentence of imprisonment in a case where liability is imposed vicariously cannot be sanctioned by this court consistently with the law of the land" . . . but if the Majority cannot sanction the incarceration of a person for acts of which he had no knowledge, how can it sanction the imposition of a fine? How can it sanction a conviction at all? . . .

Questions for Discussion

1. What is the justification for the vicarious liability of John Koczwara? Is there any evidence that Koczwara was aware of the sale of liquor to juveniles? Could he have prevented his employees from selling the liquor? Why does the Supreme Court of Pennsylvania not affirm Koczwara's prison sentence?

2. What are the reasons that Judge Musmanno dissents from Koczwara's conviction? Would following Judge Musmanno's view weaken the law against serving alcohol to minors?

3. This was Koczwara's second offense. Will fining Koczwara encourage him to prevent the sale of alcohol to minors in the future? Under what factual conditions do you believe that the Supreme Court of Pennsylvania would have affirmed Koczwara's prison sentence?

You Decide

6.6 A seventeen-year-old rented sexually oriented videotapes on two occasions from VIP Video in Millville, Ohio. The first time, the seventeen-year-old used his father's driver's license for identification, and the second time he paid in cash and the clerk did not ask him for an identification or proof of age. The owner of the store, Peter Tomaino, did not post a sign in the store indicating that sexually oriented rentals would not be made to juveniles.

Tomaino was absent from the store at the time of the rentals. He was convicted under a statute that provides that "no person, with knowledge of its character or content, shall recklessly . . . sell . . . material . . . that is obscene or harmful to juveniles."

Should Tomaino's conviction be overturned? Could he constitutionally be sentenced to prison? Consider whether this statute differs from the legislative enactment in *Koczwara*. See *State v. Tomaino*, 733 N.E.2d 1191 (Ohio Ct. App. 1999).

You can find the answer at www.sagepub.com/lippmanccl3e.

AUTOMOBILES, PARENTS, AND VICARIOUS LIABILITY

Vicarious liability, as we have seen, is typically applied to extend criminal liability to corporate executives or businesses. Vicarious liability is also used to hold the owners of automobiles liable for traffic tickets issued to their automobiles. In addition, there is a recent trend toward holding parents vicariously liable for the crimes of their children. Ask yourself the reason for applying vicarious liability in these two instances.

Traffic Tickets

You lend your automobile to a friend, who later informs you that she was ticketed for illegal parking. She forgets to pay the ticket as promised, and later you receive a letter reminding you that you owe money to the county. You protest and are told that as the owner of the automobile, you are vicariously liable for tickets issued to the car. Will you be successful in fighting this ticket in court?

Most parking statutes consider the owner of the vehicle *prima facie* responsible for paying the ticket. This means that unless you present evidence that you were not responsible, you are presumed liable for the ticket. In other words, the prosecutor is not required to present any evidence to establish your responsibility; you are presumed responsible unless you appear in court and establish that someone else was driving your car. This approach seems to be based on the desirability of a smooth and efficient method of collecting money that avoids court hearings on the identity of drivers. Why is this unfair? After all, you always can be reimbursed by your friend.

See more cases on the study site: Commonwealth v. Rudinski, Idris v. Chicago; *www.sagepub.com/lippmanccl3e.*

Parents

Susan and Anthony Provenzino of St. Clair Shores, Michigan, were aware that their son Alex was experiencing difficulties. He was arrested in May 1995, and the Provenzinos obtained Alex's release from juvenile custody in the fall of 1995, fearing that he would be mistreated by violent juveniles housed in the facility. Over the course of the next year, Alex was involved in a burglary, excessive drinking, and using and selling marijuana. Alex verbally abused his parents at home and on one occasion attacked his father with a golf club. In May 1996, the Provenzinos were convicted of violating a two-year-old local ordinance that placed an affirmative responsibility on parents to "exercise reasonable control over their children." The jury required only fifteen minutes to find them guilty; each was fined $100 and ordered to pay $1,000 in court fees.[36]

Roughly seventeen states and cities today have similar **parental responsibility laws**. States have a long history of passing laws against parents who abuse, neglect, or abandon their children or fail to ensure that their children attend school. In 1903, Colorado was the first state to punish "contributing to the delinquency of a minor." Similar provisions were subsequently adopted by roughly forty-two states and the District of Columbia. These statutes are not limited to parents and require some affirmative act on the part of an adult that aids, encourages, or causes the child's delinquent behavior.[37]

The first wave of parental responsibility statutes were passed in the late 1980s and early 1990s, when various states and municipalities adopted laws holding parents strictly and vicariously liable for the criminal conduct of their children. It was presumed that parents possess a duty to supervise their offspring and that this type of statute would encourage parents to monitor and to control their kids. These strict and vicarious liability statutes were ruled unconstitutional in Connecticut, Louisiana, Oregon, and Wyoming.

Parental responsibility statutes generally hold parents responsible for the failure to take reasonable steps to prevent their children from engaging in serious or persistent criminal behavior. A New York law, for instance, punishes a parent who "fails or refuses to exercise reasonable diligence in the control of . . . a child to prevent him from becoming . . . a 'juvenile delinquent' or a 'person in need of supervision.' . . ." These statutes, as illustrated by the New York law, generally lack clear and definite standards.[38]

There are a variety of laws that hold adults liable for teenage drinking. **Social host liability laws** hold adults liable for providing liquor in their home to minors in the event that an accident or injury occurs. Variants are so-called **teen party ordinances**, which declare that it is criminal for an adult to host a party for minors at which alcohol is served.

In 1993 in *Williams v. Garcetti,* the Supreme Court of California upheld the constitutionality of a California parental responsibility statute. The law stated that a "parent or legal guardian to any person under the age of 18 years shall have the duty to exercise reasonable care, supervision, protection, and control over their minor child." A parent or guardian whose "act or omission causes or tends to cause or encourage" a child to violate a curfew, be habitually truant or commit a crime" is held liable under the statute. In other words, a parent is held liable who knows (intentionally) or should know (negligently) "that their child is at risk of delinquency and . . . they are able to control the child." A violation of the statute is punished as a misdemeanor although the charges may be dismissed prior to trial against a parent or guardian who completes an education, treatment, or rehabilitation program. The legislature passed the law as part of an effort to combat "violent street gangs whose members threaten, terrorize, and commit a multitude of crimes against the peaceful citizens of . . . neighborhoods." The California Supreme Court stressed that the provision for parental diversion from criminal prosecution in "less serious cases " means that parents will face criminal penalties for a "failure to supervise only in those cases in which the parent's culpability is great and the causal connection correspondingly clear."[39] On the one hand, it seems unfair to hold parents vicariously liable for the criminal acts of their children. On the other hand, parents would certainly seem to have an obligation and responsibility to society to supervise their children. Holding parents liable may lead them to closely monitor their children's activities and may serve to protect society. What is your view?

You can find Williams v. Garcetti *on the study site www.sagepub.com/lippmanccl3e.*

You Decide

6.7 The City of Trenton, New Jersey adopted a "parent responsibility ordinance." The ordinance presumed that a parent is responsible for the misbehavior of a child who twice within one year is adjudged guilty of acts defined as violations of the public peace. The acts so defined include adjudications for delinquency and of the status of being a juvenile in need of supervision). A parent convicted under the ordinance may be fined up to $ 500. Should parents be held responsible for the "misbehavior of a child?" See *Doe v. Trenton*, 362 A.2d 1200 (N.J. Super. Ct. App. Div. 1976).

You can find the answer at www.sagepub.com/lippmanccl3e.

See more cases on the study site: State v. Akers, *www.sagepub.com/lippmanccl3e.*

CHAPTER SUMMARY

We have seen that under the common law there were four parties to a crime. The procedural requirements surrounding the prosecution of parties developed by judges were intended to impede the application of the death penalty. Today there are two parties to a crime:

- *Accomplices.* Individuals participating before and during a crime
- *Accessories.* Individuals involved following a crime

The *actus reus* of accomplice liability is described as "aiding," "abetting," "encouraging," and "commanding" the commission of a crime. This is satisfied by even a small degree of material or psychological assistance. Mere presence is not sufficient. The *mens rea* of accomplice liability is typically described as the intent to assist the primary party to commit the offense with which he or she is charged. Some judges have argued for a knowledge standard, but other courts have recognized liability based on recklessness. The criminality of an accessory after the fact is distinguished from that of accomplices by the fact that the legal guilt of an accessory after the fact is not derived from the primary crime. Instead, accessory after the fact is now considered a separate and minor offense involving an intent and an act undertaken with the purpose of hindering the detection, apprehension, prosecution, conviction, or punishment of the individual receiving assistance.

Strict liability holds an individual liable based on the commission of a criminal act while dispensing with the requirement of a criminal intent. Vicarious liability imposes liability on an individual for the criminal act of another. A corporate officer or corporation may be held vicariously liable under a statute where there is a legislative intent to impose vicarious liability for the act of an employee or corporate agent. This typically arises in the case of strict liability offenses that are punishable by a fine and are intended to protect the societal health, safety, and welfare. Vicarious liability is extended to the owners of automobiles for traffic tickets based on their legal title to the car and the interest in efficiently processing tickets. Parents, under some state statutes, are held vicariously liable for the criminal conduct of their children based on their status relationship.

CHAPTER REVIEW QUESTIONS

1. What were the four categories of common law parties? How does this differ from the modern categorization of parties?

2. Illustrate the definition of common law accomplices and accessories using the example of a bank robbery. Should accomplices be held liable for the same crime as the primary perpetrator of the crime?

3. What *actus reus* is required for an accomplice? Provide some illustrations of acts satisfying the *actus reus* requirement. What is the mere presence rule? Is there an exception to the mere presence rule?

4. Discuss the *mens rea* of accomplice liability. Distinguish this from the minority position that "knowledge" is sufficient. How would these two approaches result in a different outcome in a case? Which approach do you favor?

5. What are the requirements for an individual to be considered an accessory after the fact? Is this considered as serious a criminal violation as that of being an accomplice?

6. Distinguish accomplice liability, strict liability, and vicarious liability.

7. How does the language of a statute determine whether a corporation may be held vicariously liable? What are the two primary tests for determining corporate liability? Discuss some of the arguments for and against the vicarious liability of corporations.

8. What constitutional considerations are involved in holding the owner of an automobile vicariously liable for the traffic tickets issued to the car?

9. Is it constitutional to hold parents strictly and vicariously liable for the criminal acts of their children? In your view, are there any situations in which parents should or should not be held vicariously liable?

10. Write a brief essay summarizing the law of parties.

LEGAL TERMINOLOGY

accessories

accessories after the fact

accessories before the fact

accomplices

corporate liability

derivative liability

mere presence rule

natural and probable consequences

parental responsibility laws

parties to a crime

Pinkerton rule

principals in the first degree

principals in the second degree

social host liability laws

strict liability

teen party ordinances

vicarious liability

 ## CRIMINAL LAW ON THE WEB

Log on to the Web-based student study site at www.sagepub.com/lippmanccl3e to assist you in completing the Criminal Law on the Web exercises, as well as for additional features such as podcasts, Web quizzes, and video links.

1. Read about social host laws and other laws imposing vicarious responsibility for criminal offenses involving alcohol.

2. Learn about civil and criminal vicarious liability for fraternity hazing.

3. Read about the "Texas law of parties" and develop your own view on accomplice liability and the death penalty.

BIBLIOGRAPHY

American Law Institute, *Model Penal Code and Commentaries,* vol. 1, pt. 1 (Philadelphia: American Law Institute, 1985), pp. 295–348. An influential discussion of the appropriate standard of the law of accomplice and corporate liability.

George P. Fletcher, *Rethinking Criminal Law* (New York: Oxford University Press, 2000), pp. 649–682. A review of some of the most difficult issues arising in accomplice liability.

Wayne R. LaFave, *Criminal Law,* 3rd ed. (St. Paul, MN: West Publishing, 2000), pp. 257–283, 614–650. A clear and comprehensive summary of the law of parties and of strict and vicarious liability.

Herbert L. Packer, *The Limits of the Criminal Sanction* (Palo Alto, CA: Stanford University Press, 1968), pp. 121–131. An interesting discussion of the decline in the traditional standards for judging criminal liability and the increasing reliance on strict and vicarious liability.

Rollin M. Perkins and Ronald N. Boyce, *Criminal Law,* 3rd ed. (Mineola, NY: Foundation Press, 1982), pp. 718–769, 911–914. An outline of the history and development of the law of parties and of the criminal liability of corporations.

Richard G. Singer and John Q. La Fond, *Criminal Law Examples and Explanations,* 2nd ed. (New York: Aspen, 2001), pp. 99–119, 329–359. A good summary of the law of parties with accompanying review questions.

7 | ATTEMPT, CONSPIRACY, AND SOLICITATION

Did Dwight Ralph Smallwood intend to kill his victims?

On August 29, 1991, Dwight Ralph Smallwood was diagnosed as being infected with the human immunodeficiency virus (HIV). . . . On September 26, 1993, Smallwood and an accomplice robbed a woman at gunpoint and forced her into a grove of trees where each man alternately placed a gun to her head while the other one raped her. On September 28, 1993, Smallwood and an accomplice robbed a second woman at gunpoint and took her to a secluded location, where Smallwood inserted his penis into her with "slight penetration." On September 30, 1993, Smallwood and an accomplice robbed yet a third woman, also at gunpoint, and took her to a local school where she was forced to perform oral sex on Smallwood and was raped by him. In each of these episodes, Smallwood threatened to kill his victims if they did not cooperate or to return and shoot them if they reported his crimes. Smallwood did not wear a condom during any of these criminal episodes. . . . Based upon his attack on September 28, 1993, Smallwood was charged with, among other crimes, . . . attempted second-degree murder of each of his three victims.

Learning Objectives

1. Understand the *mens rea* of criminal attempts.
2. Know the legal tests for the *actus reus* of criminal attempts.
3. Summarize the law of factual impossibility and the law of legal impossibility.
4. Define and give examples of the defense of abandonment.
5. State the *actus reus* and *mens rea* of a criminal conspiracy and know the difference between chain conspiracies and wheel conspiracies.
6. Understand and *mens rea* and *actus reus* of the crime of solicitation.

INTRODUCTION

We live in fear of a terrorist bombing or hijacking and certainly do not want to wait for an attack to occur before arresting terrorists. On the other hand, at what point can we be confident that individuals are intent on terrorism?

Inchoate or "beginning" crimes provide that individuals can be convicted and punished for an intent to commit a crime when this intent is accompanied by a significant step toward the commission of the offense. At this point, society is confident that the individual

presents a threat and that society is justified in acting to protect itself. There are three **inchoate crimes**:

- *Attempt* punishes an unsuccessful effort to commit a crime.
- *Conspiracy* punishes an agreement to commit a crime and an overt act in furtherance of this agreement.
- *Solicitation* punishes an effort to persuade another individual to commit a crime.

The conviction of an individual for an inchoate crime requires

- a specific intent or purpose to accomplish a criminal offense, and
- an act to carry out the purpose.

Individuals who commit inchoate offenses may be punished less severely than or as severely as they would have been punished if they had completed the crime that was the object of the attempt, conspiracy, or solicitation.

ATTEMPT

Professor George Fletcher notes that attempts are failures. A sniper misses an intended victim, two robbers are apprehended as they enter a store, and a pickpocket finds that the victim's pocket is empty. In this section, we will ask at what point an **attempt** is subject to criminal punishment. Must we wait until a bullet misses its mark and whistles past the head of an intended victim to arrest the shooter? What defenses are available? May a pickpocket plead that the intended victim's pocket was empty?

There are two types of attempts: **complete attempt** (but "imperfect") and **incomplete attempt**. A complete, but imperfect, attempt occurs when an individual takes every act required to commit a crime and yet fails to succeed. An example is an individual firing a weapon and missing the intended victim. In the case of an incomplete attempt, an individual abandons or is prevented from completing a shooting due to the arrival of the police or as a result of some other event outside his or her control. A third category that we should mention is the *impossible attempt*. This arises where the perpetrator makes a mistake, such as aiming and firing the gun only to realize that it is not loaded. You should keep these categories in mind as you read the cases on attempt.[1]

Judges and lawyers, as we mentioned, disagree over how far an individual must progress toward the completion of a crime to be held legally liable for an attempt. It is only when an assailant pulls the trigger that we can be confident that he or she possesses an intent to kill or to seriously wound a potential victim. Yet, the longer we wait to arrest an individual, the greater the risk that he or she may carry out the crime and wound or kill a victim. At what point do you believe the police are entitled to arrest a potential assailant?

In *People v. Miller,* Miller, while slightly inebriated, threatened to kill Albert Jeans. Miller appeared that afternoon on a farm owned by Sheriff Ginochio where Jeans worked. Miller was carrying a .22-caliber rifle and walked toward Ginochio, who was 250 or 300 yards in the distance. Jeans stood roughly thirty yards behind Ginochio. The defendant walked about one hundred yards, stopped, appeared to load his rifle, and then continued to walk toward Ginochio, who seized the weapon from Miller without resistance. Jeans, as soon as he saw Miller, fled on a right angle to Miller's line of approach, but it is not clear whether this occurred before or after Miller crouched and appeared to place a bullet in the rifle. The firearm was loaded with a high-speed cartridge. At no time did Miller raise and aim his rifle. Was this sufficient for an attempt? What was Miller's intent, to frighten Jeans or to kill him? Should we consider Jeans's reaction in determining Miller's guilt? Consider the presence of Ginochio in formulating your view. Would you hold Miller liable for an attempt to kill Jeans? Should an attempt be punished to the same extent as the actual offense? Do you believe that the resources of the legal system should be devoted to prosecuting Miller under the circumstances?[2]

History of Attempt

Scholars and philosophers dating back to the ancient Greeks have wrestled with the appropriate punishment for an attempted crime. After all, an attempt arguably does not result in any harm. In 360 B.C., the famous Greek philosopher Plato argued that an individual who possesses "the purpose and intention to slay another . . . should be regarded as a murderer and tried for murder." Plato, however, also recognized that an attempt does not result in the death of the victim and that banishment rather than the death penalty would be an appropriate penalty.[3]

The early common law did not punish attempts. Henry of Bracton explained this by asking, "What harm did the attempt cause, since the injury took no effect?"[4] English law, rather than relying on the prosecution of attempts to prevent and punish the first steps toward crime, adopted laws against unlawful assemblies, walking at night, and unemployed persons wandering in the countryside, as well as other prohibitions on activities that may result in crime, such as keeping guns or crossbows in the house, lying in wait, or drawing a sword to harm a judge. Gaps in the law were filled by the Court of Star Chamber, which was authorized by the king to maintain order by modifying common law rules where necessary. These were volatile and violent times, and the Star Chamber began to introduce the concept of attempts into the law by punishing threats and verbal confrontations that were likely to escalate into armed confrontations, challenges, and attempts to enter into duels. In 1614, Sir Francis Bacon prosecuted a case before the Star Chamber for dueling in which he argued that acts of preparation for a sword fight should be punished in order to discourage armed confrontations.

The law of attempt was finally recognized by the common law in the important decision of *Rex v. Scofield* in 1784. The defendant was charged with placing a lighted candle and combustible material in a house with the intent of burning down the structure. Lord Mansfield, in convicting the defendant, stressed the importance of intent, writing that "the intent may make an act, innocent in itself, criminal. . . . Nor is the completion of an act, criminal in itself, necessary to constitute criminality."[5] In 1801, the law of attempt was fully accepted in the case of *Rex v. Higgins,* which involved the indictment of an individual for urging a servant to steal his master's goods. The court proclaimed that "all offenses of a public nature, that is, all such acts or attempts as tend to the prejudice of the community, are indictable."[6] This common law rule was subsequently accepted by courts in the United States, which ruled that it was a misdemeanor to attempt to commit any felony or misdemeanor.

Public Policy and Attempt

Why punish an act that does not result in the successful commission of a crime? There are at least three good reasons:

- *Retribution.* An individual who shoots and misses or makes efforts to commit a murder is as morally blameworthy as a successful assailant. Success or failure may depend on unpredictable factors, such as whether the victim moved to the left or to the right or whether the police happened to drive by the crime scene.
- *Utility.* The lesser punishment for attempt provides an incentive for individuals to halt before completing a criminal act in order to avoid being subjected to a harsher punishment.
- *Incapacitation.* The individual has demonstrated that he or she poses a threat to society.

The Elements of Criminal Attempt

Criminal attempt comprises three elements:

- an intent or purpose to commit a crime,
- an act or acts toward the commission of the crime, and
- a failure to complete the crime.

A general attempt statute punishes an attempt to commit any criminal offense. Other statutes may be directed at specific offenses, such as an attempt to commit murder, robbery, or rape.

The Legal Equation

Attempt = Step toward the completion of a crime

+ specific intent or purpose to commit the crime attempted

+ failure to complete the crime.

Mens Rea of Attempt

A criminal attempt involves a dual intent.

- An individual must intentionally perform acts that are proximate to the completion of a crime.
- An individual must possess the specific intent or purpose to achieve a criminal objective.

In the case of an individual accused of attempted murder, the prosecution must demonstrate that (1) the defendant intentionally aimed and engaged in an act toward the shooting of the arrow, and (2) this was undertaken with the intent to kill a hunter walking on the trail. A defendant who did not notice the hunter and lacked the intent to kill would not be held liable for attempted premeditated murder.[7] As noted by an Illinois appellate court, "a finding of specific intent to kill is a necessary element of intent to kill."[8]

The commentary to the Model Penal Code offers the example of an individual who detonates a bomb with the purpose of demolishing a building knowing that people are inside. In the event that the bomb proves defective, the commentary notes that the defendant likely would not be held responsible for attempted murder, because his or her purpose was to destroy the building rather than to kill the individuals inside the structure. The Model Penal Code section 5.01(1)(b) adopts a broad approach to intent and argues that when a defendant knows that death is likely to result from the destruction of the building, it is appropriate to hold him or her liable for attempted murder.[9]

In *Smallwood v. State,* the Maryland Court of Appeals was confronted with the issue of whether a rapist with HIV was guilty of an attempt to murder his victims. In reading this case, consider whether it makes sense to limit the intent for attempt to purpose. Is purpose too difficult a standard to satisfy? Should knowledge also result in criminal liability?

Did Smallwood possess the intent to kill through the transmission of HIV?

Smallwood v. State, *680 A.2D 512 (MD. 1996). Opinion by: Murphy, J.*

Facts

On August 29, 1991, Dwight Ralph Smallwood was diagnosed as being infected with the human immunodeficiency virus (HIV). According to medical records from the Prince George's County Detention Center, he had been informed of his HIV-positive status by September 25, 1991. In February 1992, a social worker made Smallwood aware of the necessity of practicing "safe sex" in order to avoid transmitting the virus to his sexual partners, and in July 1993, Smallwood told health care providers at Children's Hospital that he had only one sexual partner and that they always used condoms. Smallwood again tested positive for HIV in February and March of 1994.

On September 26, 1993, Smallwood and an accomplice robbed a woman at gunpoint, and forced her into a grove of trees where each man alternately placed a gun to her head while the other one raped her. On September 28, 1993, Smallwood and an accomplice robbed a second woman at gunpoint and took her to a secluded location, where Smallwood inserted his penis into her with "slight penetration." On September 30, 1993,

Smallwood and an accomplice robbed yet a third woman, also at gunpoint, and took her to a local school where she was forced to perform oral sex on Smallwood and was raped by him. In each of these episodes, Smallwood threatened to kill his victims if they did not cooperate or to return and shoot them if they reported his crimes. Smallwood did not wear a condom during any of these criminal episodes.

Based upon his attack on September 28, 1993, Smallwood was charged with, among other crimes, attempted first-degree rape, robbery with a deadly weapon, assault with intent to murder, and reckless endangerment. In separate indictments, Smallwood was also charged with the attempted second-degree murder of each of his three victims. On October 11, 1994, Smallwood pled guilty in the Circuit Court for Prince George's County to attempted first-degree rape and robbery with a deadly weapon. The circuit court also convicted Smallwood of assault with intent to murder and reckless endangerment based upon his September 28, 1993, attack, and it convicted Smallwood of all three counts of attempted second-degree murder.

Following his conviction, Smallwood was sentenced to concurrent sentences of life imprisonment for attempted rape, twenty years imprisonment for robbery with a deadly weapon, thirty years imprisonment for assault with intent to murder, and five years imprisonment for reckless endangerment. The circuit court also imposed a concurrent thirty-year sentence for each of the three counts of attempted second-degree murder. The circuit court's judgments were affirmed in part and reversed in part by the Court of Special Appeals, which found that the evidence was sufficient for the trial court to conclude that Smallwood intended to kill his victims and upheld all of his convictions. . . .

Issue

Smallwood asserts that the trial court lacked sufficient evidence to support its conclusion that he intended to kill his three victims. Smallwood argues that the fact that he engaged in unprotected sexual intercourse, even though he knew that he carried HIV, is insufficient to infer an intent to kill. The most that can reasonably be inferred, Smallwood contends, is that he is guilty of recklessly endangering his victims by exposing them to the risk that they would become infected themselves. The State disagrees, arguing that the facts of this case are sufficient to infer an intent to kill. The State likens Smallwood's HIV-positive status to a deadly weapon and argues that engaging in unprotected sex when one is knowingly infected with HIV is equivalent to firing a loaded firearm at that person.

Reasoning

HIV is a retrovirus that attacks the human immune system, weakening it, and ultimately destroying the body's capacity to ward off disease. We also noted that the virus may reside latently in the body for periods as long as ten years or more, during which time the infected person will manifest no symptoms of illness and function normally. HIV typically spreads via genital fluids or blood transmitted from one person to another through sexual contact, the sharing of needles in intravenous drug use, blood transfusions, infiltration into wounds, or from mother to child during pregnancy or birth. . . . AIDS . . . is the condition that eventually results from an immune system gravely impaired by HIV. Medical studies have indicated that most people who carry the virus will progress to AIDS. AIDS patients by definition are profoundly immunocompromised; that is, they are prone to any number of diseases and opportunistic infections that a person with a healthy immune system might otherwise resist. AIDS is thus the acute clinical phase of immune dysfunction. . . . AIDS is invariably fatal. In this case, we must determine what legal inferences may be drawn when an individual infected with the HIV virus knowingly exposes another to the risk of HIV infection and the resulting risk of death by AIDS.

As we have previously stated, the required intent in the crimes of assault with intent to murder and attempted murder are the specific intent to murder, that is, the specific intent to kill under circumstances that would not legally justify or excuse the killing or mitigate it to manslaughter. . . . Smallwood . . . was properly found guilty of attempted murder and assault with intent to murder only if there was sufficient evidence from which the trier of fact could reasonably have concluded that Smallwood possessed a specific intent to kill at the time he assaulted each of the three women. . . .

The State argues that Smallwood similarly knew that HIV infection ultimately leads to death and that he knew that he would be exposing his victims to the risk of HIV transmission by engaging in unprotected sex with them. Therefore, the State argues, a permissible inference can be drawn that Smallwood intended to kill each of his three victims. . . .

Death by AIDS is clearly one natural possible consequence of exposing someone to a risk of HIV infection, even on a single occasion. It is less clear that death by AIDS from that single exposure is a sufficiently probable result to provide the sole support for an inference that the person causing the exposure intended to kill the person who was exposed. While the risk to which Smallwood exposed his victims when he forced them to engage in unprotected sexual activity must not be minimized, the State has presented no evidence from which it can reasonably be concluded that death by AIDS is a probable result of Smallwood's actions to the same extent that death is the probable result of firing a deadly weapon at a vital part of someone's body. Without such evidence, it cannot fairly be concluded that death by AIDS was sufficiently probable to support an inference that Smallwood intended to kill his victims in the absence of other evidence indicative of an intent to kill.

In this case, we find no additional evidence from which to infer an intent to kill. Smallwood's actions are wholly explained by an intent to commit rape and armed robbery, the crimes for which he has already pled guilty. For this reason, his actions fail to provide evidence that he also had an intent to kill. . . . Smallwood's knowledge of his HIV-infected status provides the only evidence in this case supporting a conclusion that he intended anything beyond the rapes and robberies for which he has been convicted. . . .

The evidence in *State v. Haines*, 545 N.E.2d 834 (Ind. App. 1989), contained both statements by the defendant demonstrating intent and actions solely explainable as attempts to spread HIV. There, the defendant's convictions for attempted murder were upheld where the defendant slashed his wrists and sprayed blood from them on a police officer and two paramedics, splashing blood in their faces and eyes. Haines attempted to scratch and bite them and attempted to force blood-soaked objects into their faces. During this altercation, the defendant told the officer that he should be left to die because he had AIDS, that he wanted to "give it to him," and that he would "use his wounds" to spray the officer with blood. Haines also "repeatedly yelled that he had AIDS, that he could not deal with it and that he was going to make [the officer] deal with it." . . .

Holding

We have no trouble concluding that Smallwood intentionally exposed his victims to the risk of HIV infection. The problem before us, however, is whether knowingly exposing someone to a risk of HIV infection is by itself sufficient to infer that Smallwood possessed an intent to kill. . . .

Questions for Discussion

1. Why does the court conclude that Smallwood lacked a specific intent to kill? As a prosecutor, what arguments would you present in support of the contention that Smallwood, in fact, possessed a specific intent to kill? What additional evidence might you present to persuade the court to affirm Smallwood's conviction?

2. As a juror, would you vote to convict or to acquit Smallwood?

You Decide

7.1 Jarmaal Smith challenges the sufficiency of the evidence to support his conviction for two counts of attempted murder. On February 18, 2000, Karen A. drove her boyfriend, Renell T., Sr. (Renell), to a friend's house on Greenholme Lane in Sacramento, California. Renell was seated in the front passenger seat, and their three-month-old baby, Renell T., Jr., was secured in an infant car seat in the back seat directly behind Karen. Karen parked alongside the curb on the street in front of the house, and Renell got out of the car. As Karen waited in the car to ensure that Renell's friend was home, she spotted Smith in her rearview mirror approaching the automobile. Karen had last talked to Smith eight to nine months earlier, and he had told her that the next time he saw her that he would "slap the shit out of [her]." Defendant walked up to the open front passenger window of Karen's car, looked inside and said, "Don't I know you, bitch?" Renell walked back toward the car and defendant responded by lifting his shirt to display a handgun tucked in his waistband. Renell said, "It is cool," and backed away from defendant. As Renell returned to the car, a group of men on the street corner began approaching the car, and the defendant and the other men began hitting him.

As Renell reentered the car, Karen started to pull away from the curb. After moving roughly one car length, she looked in her rearview mirror and saw Smith standing "straight behind" her holding a gun. She heard a single gunshot, and the bullet shattered the rear windshield, narrowly missed Karen and her baby, passed through the driver's head rest, and lodged in the driver's side door. Renell testified that the defendant had a .38-caliber pistol in his possession. After the shooting, a Sacramento County deputy sheriff searched defendant's room at his mother's home and recovered two .38-caliber shell casings. The jury convicted Smith of two counts of attempted murder and he was sentenced to two 27-year prison terms, which were to be served concurrently.

Was the jury justified in convicting Smith of the attempted murder of Karen and of the attempted murder of Karen's baby? See *People v. Smith*, 124 P.3d 730 (Cal. 2005).

You can find the answer at www.sagepub.com/lippmanccl3e.

Actus Reus of Attempt

There are two steps in considering the *actus reus* of attempt. First, determine the legal test to be applied. Second, apply the legal test to the facts.

The often-confusing legal tests reflect two different approaches to attempt. The **objective approach to criminal attempt** requires an act that comes extremely close to the commission of the crime. An arsonist, for instance, must spread kerosene on the ground surrounding a house and then strike a match to ignite the fire. At this point, although the match may be blown out by the wind, we can be confident that the defendant possesses the intent to commit arson and is determined to act on this desire. The fact that the arsonist went so far as to light the match also confirms that the defendant poses a threat to society. This objective approach distinguishes **preparation**, or the planning and purchasing of the materials to commit arson, from acts taken to perpetrate the crime, such as spreading the kerosene and lighting the match.

The **subjective approach to criminal attempt** focuses on an individual's intent rather than on his or her acts. This approach is based on the belief that society should intervene as soon as an individual who possesses the required intent takes an act toward the commission of a crime. The intent may be revealed by an individual's statements or actions. The subjective approach dictates that the police arrest an arsonist as soon as the individual approaches the crime scene with the kerosene and matches. At this point, we can be fairly certain that the individual poses a threat to society. Critics point out that this early intervention fails to provide individuals with the opportunity to change their mind and risks punishing individuals for bad thoughts who may never, in fact, engage in bad acts.

The objective approach stresses the danger posed by a defendant's acts; the subjective approach focuses on the danger to society presented by a defendant who possesses a criminal intent.

In considering these two perspectives, review *People v. Miller* in the introductory section on the law of attempt. Now consider which approach you would adopt in analyzing whether the two twelve-year-old girls in *State v. Reeves* are guilty of attempted murder. Molly and Tracie agreed to poison their teacher, Janice Geiger, and then steal her car. They shared their scheme with an older high school student the night before the planned murder and were unable to persuade him to drive them to the mountains in Geiger's automobile. Later, Molly revealed the plan to a student named Mary while on the bus to school and went so far as to show Mary a packet of rat poison. Mary informed Janice Geiger of the plan, and during homeroom Geiger observed the two girls lean over her desk, giggle, and run back to their seats. Geiger also noticed a purse lying next to her coffee cup on the desk and arranged for Molly to be called to the principal's office, where rat poison was found in the purse.

Were Molly and Tracie liable for an attempt to kill Janice Geiger? At what point did they cross the line from preparing to poison Ms. Geiger to attempting to poison Ms. Geiger? When they entered the school? When they approached the desk? Should we wait until the poison is actually placed in the coffee cup? Is the best approach to emphasize the defendants' acts (objective approach) or intent (subjective approach)?[10]

Three Legal Tests

There are three major legal tests for the *actus reus* of attempt. All three ask whether an individual's actions so clearly indicate an intent to commit a crime that we can confidently charge him or her with an attempt.

- *Physical Proximity to the Commission of a Crime.* The defendant's acts come close to completing the crime. The focus is on the remaining steps required to complete the crime.
- *Unequivocality or Clarity of Purpose to Commit a Crime.* Without any other information, an ordinary person looking at the defendant's acts would conclude without a doubt that the defendant intends to commit the crime. .
- *Model Penal Code or Substantial Step Toward the Commission of a Crime.* The defendant's acts are sufficient to clearly indicate that he or she possesses an intent to commit the crime.

At common law, the first steps toward the commission of a crime were considered mere preparation and did not constitute an attempt. The common law followed the **last step approach** and provided that an attempt occurred only following the completion of the final step required for the commission of a crime. In other words, an attempted arson required that an individual actually ignite a fire before he or she could be arrested for attempt. In an attempted murder, the assailant must pull the trigger only to miss the target or to find that the gun is unloaded.

The modern **physical proximity test** is employed by some state courts and is slightly less demanding than the last step approach. The physical proximity test follows an objective approach and provides that an attempt occurs when an act is "very near" or "dangerously close" to the completion of a crime. An individual must possess the immediate ability to complete the crime or, in the words of the courts, the act must "amount to the commencement of the consummation." Under this test, the arsonist would be required to spread the kerosene on the building and to strike a match. The killer must aim a rifle at the victim and have the victim within the sights of his or her rifle. There is good reason to believe at that point the arsonist or killer intends to take the final steps required to commit the crime.

The **unequivocality test** or clarity test, or ***res ipsa loquitur*** ("the thing speaks for itself"), asks whether an ordinary individual observing the defendant's acts would conclude that the defendant clearly and indisputably intends to commit a crime. The focus is on what an individual has "already done." The defendant's statements to the police or to other persons are not considered in this analysis. The moment the arsonist arrives at the crime scene with the necessary materials, most people would conclude that the defendant possesses a clear intent and is guilty of an attempted arson. This test is criticized for lacking clear guidelines and providing jurors with considerable discretion.

The Model Penal Code **substantial step test** simplifies matters by providing an understandable and easily applied test for attempt. The Model Penal Code states that to constitute an attempt, an act must be a *clear step* toward the commission of a crime. *This step is not required to come close to the completion of the crime itself. The Model Penal Code does state that the act must be "strongly corroborative of the actor's criminal purpose." The focus is on the acts already taken by the defendant toward the commission of the crime.* The code offers a number of factual examples:

- Lying in wait, searching for, or following the contemplated victim of a crime
- Enticing the victim of the crime to go to the place contemplated for its commission
- Surveillance of the site of the contemplated crime
- Unlawful entry of a building or vehicle that is the site of the contemplated crime
- Possession of materials specifically designed for the commission of a crime
- Soliciting an individual to engage in conduct constituting a crime

The Model Penal Code substantial step analysis concentrates on asking whether an individual has taken affirmative acts toward the completion of a crime that, in combination with other evidence, indicate a defendant possesses a criminal intent. These steps are not required to be physically proximate or close to the offense, and there is no firm distinction between preparation and perpetration of a crime. The concern is with detaining dangerous persons rather than with delaying an arrest until an individual comes close to committing a dangerous act.

The Physical Proximity and Substantial Step Tests

You are likely understandably confused by the broad and uncertain nature of the law of attempt. Keep in mind that the fundamental question is whether the law of attempt should be concerned with a defendant's intent or with the proximity of his or her acts to the completion of a crime.

The substantial step test significantly broadens the authority of the police to arrest individuals for an attempt to commit a crime. For instance, the Model Penal Code provides that "lying in wait" and "searching for or following the contemplated victim" satisfies the *actus reus* requirement.

The commentary to the Model Penal Code recognizes that these acts would not satisfy the demanding standard for *actus reus* under the physical proximity test. For example, in *People v. Rizzo,* the New York Court of Appeals rejected that searching for a "contemplated victim" constituted an attempt. In *Rizzo,* four men planned to rob Charles Rao of the payroll that he was scheduled to carry from the bank to his company's offices. The four conspirators, two of whom were armed, drove to the bank and to the firm's various worksites, but failed to find Rao. The four were arrested and subsequently acquitted by the New York Court of Appeals. The court ruled that in searching for Rao, the four men had not progressed beyond preparation and that their acts were not "immediately near" or "dangerously close" to the commission of the crime of robbery, because they had yet to encounter and confront Rao. They clearly could still change their minds. On the other hand, the commentary to the Model Penal Code notes that the defendants' "following, searching," and "lying in wait" with a criminal purpose were sufficiently dangerous to constitute an attempted robbery.[11]

Another illustration of the difference between the substantial step and dangerous proximity tests is *Commonwealth v. Gilliam.* Gilliam was a prisoner at the Dallas State Correctional Institution, and correctional officers discovered that the bars on the window in his cell had been cut and were being held in place by sticks and paper. A search of the cell revealed vise grips concealed inside Gilliam's mattress, and two knotted extension cords attached to a hook were found in a box of clothing. The vise grips were sufficiently strong to cut through the barbed wire along the top of the fence surrounding the prison compound, and the extension cords presumably were to be used to scale the surrounding penitentiary wall. The Pennsylvania Superior Court ruled that Gilliam's sawing through the bars and gathering of tools indicated a clear intent to escape from prison and constituted a substantial step under the Model Penal Code. The court, however, noted that these same acts would not constitute an attempt under a physical proximity test, because a number of additional steps were required to escape from the prison.[12]

A number of states avoid the complexities of attempt by providing that preparation for specific offenses constitutes a crime. For instance, California Penal Code section 466 provides that

> Every person having upon him or her in his or her possession a picklock, crowbar, keybit . . . or other instrument or tool with intent feloniously to break or enter into any building . . . is guilty of misdemeanor.

The next case, *Bolton v. State,* illustrates the difficulty of determining at what point an individual should be held liable for attempt. Early intervention removes a potentially dangerous individual from the street. This, of course, risks arresting and punishing individuals for crimes that they might never have committed. On the other hand, delaying an arrest until an act is proximate to a crime presents the threat that a crime will have been committed before the police are able to intervene. How would you balance these two competing concerns?

Was the defendant guilty of attempted burglary with the intent to commit sexual assault, or were his acts mere preparation?

Bolton v. State, 07–02–0357–CR (Tex. App. 2003). Opinion by: Reavis, J.

Following a plea of not guilty, appellant was convicted by a jury of attempted burglary of a habitation with intent to commit sexual assault, and punishment was assessed by the court at twenty years' confinement. Presenting a sole issue, appellant asserts the evidence is insufficient to support his conviction. Based upon the rationale expressed herein, we . . . reverse and remand for a new trial on punishment.

Facts

At approximately 10:00 p.m. on January 6, 2002, Ramiro Reyna was walking from his home to his mother's home situated one block over when he observed a "suspicious person" walking through complainant's backyard. A street light on the corner lot where complainant's house was located provided some light. Reyna knew that complainant and her mother were the only residents. He observed the person, later identified as appellant, "just peeping" through the kitchen window and later another window. However, Reyna also testified that an eighteen-wheeler was parked on the curb to the side of the house that night, and from where he was standing initially, the truck was between him and appellant, and he was able to observe only the back of appellant's legs. As Reyna proceeded down the block undetected by appellant, he again observed appellant "peeping" through a window, holding his right hand next to his face. Reyna was unable to see appellant's

left hand. Once Reyna reached his mother's house, four houses down from complainant's, he called the police.

Reyna testified he was inside his mother's house for three to five minutes after calling the police. He then waited in his mother's back yard where he could still observe complainant's back yard through other neighbors' gates. Although he could not see appellant, he believed appellant was still standing by a window in complainant's back yard, until he noticed appellant return to complainant's back yard from a business parking lot located across the alley from complainant's house. Reyna's mother waited for the police in her front yard, and upon Officer Jordan's arrival, she alerted him to appellant's location. Reyna observed Jordan's patrol car drive down the alley, where Jordan apprehended appellant behind a neighbor's house adjacent to complainant's.

Complainant testified she turned on her bath water, undressed in the bathroom, and took her clothes to the laundry room. She then walked to her mother's room to answer the telephone before returning to the bathroom to take a bath. According to complainant, she bathed for approximately an hour.

Officer Jordan testified he was dispatched to investigate a burglary in progress. . . . After handcuffing appellant, Jordan conducted a protective frisk and discovered an open jar of petroleum jelly in appellant's pocket. The officer also noticed that appellant was wearing camouflage pants that were unbuttoned and unzipped.

Issue

Appellant's sole contention is that the evidence is legally and factually insufficient to support his conviction for attempted burglary of a habitation with intent to commit sexual assault. We agree.

Reasoning

A person attempts an offense if he commits an act amounting to more than mere preparation that tends but fails to effect the commission of the offense intended. Burglary requires a person to enter a habitation without the effective consent of the owner with the intent to commit a felony, a theft, or an assault. A person commits sexual assault if he intentionally or knowingly

A. causes the penetration of the anus or female sexual organ of another person by any means, without that person's consent;

B. causes the penetration of the mouth of another person by the sexual organ of the actor, without that person's consent; or

C. causes the sexual organ of another person, without that person's consent, to contact or penetrate the mouth, anus, or sexual organ of another person, including the actor. . . .

The State established that appellant was in complainant's backyard without permission and was peeping through more than one window. The arresting officer testified appellant possessed an open jar of petroleum jelly in his pocket and that his camouflage pants were unbuttoned and unzipped. Complainant and her mother testified the window to complainant's room showed new damage. . . . Complainant also testified she had seen appellant before when she and her friends were walking to a nearby grocery store and noticed appellant following them. She explained that when she turned back to look, he stopped, and she did not see him again. However, the State did not establish a time frame between that occurrence and the incident on January 6. . . .

Complainant testified that while she was bathing she did not hear anything outside the window. She further testified she did not see appellant at any of her windows nor elsewhere, nor did she witness him attempt to pull the screen away from the window. According to complainant, she inspected her window every other day when taking out the trash and acknowledged the window was locked on the night of the incident and very difficult to open.

The defense established the window screen was old and dark from rust, and the window frame was broken. No evidence was presented that appellant pulled up the screen or caused any damage to the window. In fact, the only eyewitness, Reyna, testified on direct and redirect examination that when Reyna saw appellant, he was "just peeking through the window."

Referencing the record, in its brief the State notes appellant was "clothed in camouflage pants, which are typically worn when one does not want to be seen." However . . . there is no evidence by Officer Jordan or any other witness establishing that camouflage pants are "worn when one does not want to be seen." Moreover, the State did not argue to the jury, as it does on appeal, that they could reasonably deduce that camouflage clothing is worn by persons who do not want to be seen.

The State also contends it was reasonable for a jury to conclude appellant had a "sexual interest" in complainant, because he had previously followed her while she was walking to a grocery store. However, no evidence was presented of any acts, conduct, or words by appellant to indicate he was sexually interested in complainant.

In support of its final contention that the evidence is sufficient to support appellant's conviction, the State relies on the open jar of petroleum jelly found in appellant's pocket to prove his intent to commit sexual assault. The State cites six cases in which petroleum jelly was used during sexual crimes; however . . . five of the cases relied upon by the State involve sexual acts against minors, and the sixth case was a prosecution for murder resulting from a heinous rape from which the victim sustained fatal injuries. Moreover, no evidence was introduced regarding the amount of petroleum jelly in the jar, if any. The State does not cite any cases, and we have been unable to find any, holding that possession of a jar of petroleum jelly under the circumstances presented here is sufficient to support a conviction for attempted sexual assault, and we decline to so hold.

Additionally, when Officer Jordan first observed appellant, he was leaving complainant's back yard and walking down the alley at a normal pace. . . .

Holding

The record does not establish appellant committed an act amounting to more than mere preparation with the intention to enter complainant's house to commit sexual assault. Proof of a culpable mental state generally relies upon circumstantial evidence and may be inferred from the circumstances under which the prohibited act occurred. . . . However, the circumstances in the underlying case do not establish that through his acts, words, or conduct, appellant had the requisite intent to enter complainant's house to commit sexual assault. . . . Accordingly, we hold the evidence is legally insufficient to support a conviction for attempted burglary of a habitation with intent to commit sexual assault. . . .

Accordingly, the judgment for attempted burglary of a habitation with intent to commit sexual assault is reformed to reflect conviction for the lesser included offense of criminal trespass. . . .

Questions for Discussion

1. What was the legal test for attempt used by the court? List the facts that the prosecutor would rely on to support Bolton's conviction for attempted burglary with the intent to commit sexual assault. What facts would be relied on by the defense? Why did the appellate court reverse the defendant's conviction? How would you rule?

2. Would the court have ruled differently had the prosecution been able to prove that Bolton had stalked the victim earlier? Are there other facts that the prosecution failed to prove that might have led the court to convict the defendant of burglary?

3. As a matter of public policy, do you believe that the criminal law should broadly define attempt and intervene as early as possible to prevent and to punish sexual offenses?

4. Why was the prosecutor not content to charge Bolton with criminal trespass?

You Decide

7.2 Mark Collier was separated from his wife Nancy after seventeen years of marriage. Nancy obtained an order of protection prohibiting Mark from contacting her. A month later Collier visited his neighbor Charles Cameron and stated that he feared losing his home because his wife had filed for divorce. Collier later returned to Cameron's house and proclaimed that "tonight's the night." Collier asked Cameron to take care of the cats and dogs and gave him keys to the house and to his pickup truck.

Collier then went into his bedroom and "started prayin', you know, telling—saying, 'God, forgive me for what I'm gonna do.'" Collier "started kinda cryin'. Then he . . . started kinda chuckling." He then hugged Cameron, and told him, "Tonight's the night. I'm gonna do it." Collier gathered an ice pick, a box cutter, and a pair of binoculars and said, "I'm gonna stab her in the effin' heart twice. I'm gonna cut her effin' throat." Collier also said that he would ram Nancy with his pickup. Collier drove off in his pickup.

Collier next stopped at the house of another of his neighbors, Billy Fansler. Fansler says that Collier told her that "he was going to end it and he had had enough." Collier also told Fansler "to tell [her] husband thank you and tell [her] children God bless them." When Collier was at Fansler's house, "he looked kind of wobbly," "he wasn't quite steady on his feet," "he kinda slurred [his speech] a little," and he was depressed and "maybe a little angry." Both the police and Nancy were alerted that Collier posed a threat.

At approximately 10:40 p.m., Officer Sandy Justice and Officer Robert Cunningham received a radio dispatch and arrived at the hospital and spotted Collier's pickup. The pickup was backed into a parking space in the last row across the street from the emergency area of the hospital. Collier from this location would have been able to see the emergency room exit door. The door was the only exit available to individuals leaving the hospital after 10:00 p.m.

The lights of Collier's automobile were turned off, and Collier was asleep inside the vehicle. The officers ordered Collier to exit the car. Once Collier was out of the vehicle, the officers noticed that he was intoxicated and took him into custody. They then searched the interior of the pickup and found an ice pick, a box cutter, a pair of binoculars, and an open container of beer that was partially full. Indiana has adopted the Model Penal Code provision on attempts. Is Collier guilty of attempted murder? See *Collier v. State*, 846 N.E.2d 340 (Ind. Ct. App. 2006).

You can find the answer at www.sagepub.com/lippmanccl3e.

Impossibility

Consider whether the following defendants should be held liable for an attempted offense.

*   A pickpocket reaches into your pocket only to find that there is no wallet.
*   An individual hands $100 to a seller in an effort to purchase narcotics and is arrested by the police before the seller is able to hand over what is later revealed to be baking powder.
*   A doctor begins to perform an illegal abortion on a woman who is, in fact, a government undercover agent who is not pregnant.
*   In an attempt to kill a romantic rival, an individual enters the rival's bedroom and shoots into the bed, not realizing that it is empty.
*   A male forces himself on a sleeping female with the belief that she did not consent and then discovers that the victim died of a heart attack an hour prior.

In each instance, the defendant possessed an intent to commit a crime that was factually impossible to complete. The perpetrators would have successfully completed the offenses had the facts been as the individuals "believed them to be" (i.e., there is a wallet in the pocket, the seller possessed drugs, the woman was pregnant, the romantic rival is in his or her bed, and the potential victim is alive).

A **factual impossibility** is not a defense to an attempt to commit a crime. This is based on the fact that an offender who possesses a criminal intent and who takes steps to commit an offense should not be free from legal guilt. The factual circumstance that prevents an individual from actually completing the offense is referred to in some state statutes as an **extraneous factor**, or an event outside of an individual's control.

Factual impossibility should be distinguished from **legal impossibility**, which is recognized as a defense. Legal impossibility arises when an individual mistakenly believes that he or she is acting illegally. An example is taking a tax deduction that an individual believes is illegal but that, in fact, is perfectly permissible. A group of eighteen-year-old college freshmen will not be guilty of an attempt in the event that they go to a bar and order beers while mistakenly believing that the drinking age is twenty-one. As Professor Jerome Hall notes, it is not a crime to throw a Kansas steak into the garbage, and an individual who makes an effort to toss the steak into the garbage is not guilty of an attempted offense. The individual attempting to discard the steak possesses a criminal intent to violate a nonexistent law. Our old friend *the principle of legality* prohibits punishing an individual for a crime that is the product of his or her imagination.[13]

The rule is that a mistake concerning the facts is not a defense; a mistake concerning the law is a defense. Ask yourself whether the individual charged with an attempt was mistaken concerning the facts or mistaken concerning the law.

We also should refer to the defense of **inherent impossibility**. This occurs in those rare situations in which a defendant could not possibly achieve the desired result. An English case provides an example, in which an individual who stuck pins in a voodoo doll was acquitted of attempted murder.[14]

The Model Penal Code does not recognize the defense of factual impossibility under any circumstances. The code provides a "safety valve" in section 5.05(2) by providing that an act should be treated as a minor offense in those instances in which neither the offender nor his or her conduct presents a serious threat to the public.

Consider the voodoo doll example. Is there a social benefit in punishing an individual who possesses a criminal intent and who has committed a factually impossible act? Should there be a defense of legal impossibility? A Colorado statute provides that neither factual nor legal impossibility is a defense "if the offense would have [been] committed had the attendant circumstances been as the actor believed them to be. . . ."[15]

People v. Dlugash, decided by the New York Court of Appeals in 1977, is a well-known example of factual impossibility. A fight developed, and Bush shot Geller three times, killing Geller. Dlugash then approached the body and shot Geller five times in the head. The New York court determined that Dlugash believed at the time he fired his pistol that Geller was alive. As a prosecutor, would you charge Dlugash with attempted murder? Should we judge the dangerousness of Dlugash's acts by his intent or by the actual facts? What if Dlugash arrived following Bush's shooting of Geller and believed that Geller was lying wounded in bed and proceeded to shoot into the bed, only to discover that he shot a large toy bear? Would you charge Dlugash with attempted murder?[16]

The next case, *State v. Glass,* involves a sting operation conducted by an Idaho sheriff's department in an effort to investigate and arrest sex offenders. This directly raises the question of whether an individual should be held liable for an attempt to commit a crime involving a "nonexistent" victim.

Is Glass guilty of attempted lewd conduct with a minor under sixteen?

State v. Glass, 87 P.3d 302 (Idaho Ct. App. 2003). Opinion by: Lansing, J.

Jimmy Thomas Glass appeals from the judgment and sentence entered after a jury found him guilty of attempted lewd conduct with a minor under sixteen.

Glass contends that . . . the evidence was insufficient to support a finding that he had taken a substantial step toward the completion of the crime. . . . We affirm.

Facts

The Ada County Sheriff's Office conducted an "online crimes" investigation targeting Internet chat rooms. As part of the investigation, Detective Bart Hamilton created a profile for a fictional fourteen-year-old female with the screen name "boredboisegir114" (BBG14). On November 30, 2000, Detective Hamilton, using this profile, entered a chat room and waited for subjects to contact BBG14 via private instant messages. BBG14 soon received an instant message from Glass, who was using the screen name "s3x_slave_f0r_u." At the start of the online conversation, BBG14 informed Glass that she was fourteen years old. During their conversation, Glass described for BBG14, in graphic detail, the sexual acts that he would like to perform with her. He also asked her about her past sexual experiences and offered to go to her house that day to be her "sex slave." BBG14 said that he could not go to her house because her mom was there, but she told Glass that she would see whether they could use her friend's house at a later date.

Glass contacted BBG14 again one week later. During this online chat, Glass asked BBG14 whether she had found a house that the two could use. BBG14 said that her friend's house would be available the week of December eighteenth. Glass responded that he "[couldn't] wait." On December fifteenth, Glass again contacted BBG14 about meeting during the week of December eighteenth. When BBG14 said that she could arrange an apartment for the next day, Glass agreed to meet then and said that he would bring a box full of condoms. BBG14 also wrote that she would place a picture of herself in a brown paper bag and leave it in a trash can in the parking lot of a local high school swimming pool for him to pick up. Glass said that he would retrieve it and would be driving a black Honda Civic. Then, before ending the conversation, the two agreed to meet at 10 a.m. the next day at the swimming pool, from which they would go to the apartment.

Immediately following this conversation, a police detective drove to the swimming pool and placed in the trash can a paper bag containing a photograph of an anonymous juvenile female. Shortly thereafter, as the detective watched from a distance, a black Honda Civic entered the parking lot of the swimming pool, and the driver retrieved the bag from the garbage can.

The next day, December sixteenth, at approximately 10:20 a.m., police detectives observed the same black Honda enter the parking lot of the swimming pool, turn around, and then go back out. Immediately after the car left the parking lot, the police initiated a stop. Glass, the driver of the car, was arrested. In a search of his automobile, the police officers found a box of condoms. During a subsequent police interview, Glass admitted to logging onto the chat room with the screen name of s3x_slave_f0r_u.

Glass was charged by indictment with attempted lewd conduct with a minor under sixteen. . . . He filed a motion to dismiss the charge, contending that (1) it was legally impossible to commit the crime of attempted lewd conduct with a minor, because there was no minor child involved, and (2) there was insufficient evidence to support the indictment, because his conduct did not constitute an attempt. The district court denied the motion.

At the conclusion of the trial, the jury found Glass guilty. The court imposed a unified sentence of five years with one year determinate, but suspended the sentence and placed Glass on probation for a period of seven years. Glass now appeals the . . . conviction. . . .

Issue

Glass first contends that the district court erred in rejecting his impossibility defense. He argues that it would have been impossible for him to commit the crime of lewd conduct with a minor, because there was no actual minor child involved, and as a result, it was also impossible for him to commit an attempt of that crime.

Reasoning

The same argument was recently rejected by this court. . . . We . . . held that impossibility is not a recognized defense to attempt crimes in the State of Idaho. In determining that Idaho's attempt statute . . . does not allow for an impossibility defense, we stated that the "statute provides no exception for those who intend to commit a crime but fail because they were unaware of some fact that would have prevented them from completing the intended crime."

Holding

Accordingly, we held that factual or legal impossibility for the defendant to commit the intended crime was not relevant to a determination of the defendant's guilt of attempt. It follows that the district court here correctly rejected Glass's proffered impossibility defense.

Issue

Glass next contends that the trial court erred in denying his motion for judgment of acquittal, because there was insufficient evidence to show that he had attempted to commit lewd and lascivious conduct. . . .

Reasoning

Glass argues that his act of driving to the swimming pool was, at most, mere preparation and fell short of a substantial step toward consummation of the crime. What conduct will constitute the requisite substantial step turns upon the facts and circumstances of each case.

Of importance in this analysis is "the proximity of the act, both spatially and temporally, to the completion of the criminal design." . . . It has been said that for a criminal attempt to occur, there "must be a dangerous proximity to success." . . . It is our conclusion in the present case that Glass's acts were sufficient to show an act, beyond mere preparation, toward commission of the attempted crime. After having arranged with BBG14 to meet at a specific time and place for the expressed purpose of sexual activity, Glass arrived at that meeting place at approximately the appointed time with a box of condoms in his vehicle. This conduct goes beyond remote preparatory activity and unequivocally confirms a criminal design. He was unable to proceed further only because no fourteen-year-old girl appeared at the rendezvous point.

Glass contends that his driving through the parking lot without stopping was consistent with his position, presented at trial, that he went to the meeting place out of curiosity to see the girl rather than to pick her up. This defense theory does not explain, however, why Glass had taken care to have a supply of condoms at the ready during this trip to the swimming pool.

The evidence is more than adequate to support a jury finding that Glass drove away merely because there was no girl to stop for. The jury could reasonably find that Glass had not abandoned his effort to commit the crime but had simply been prevented from proceeding further because BBG14 was not there. . . .

Glass further asserts that the trial evidence is insufficient to support a finding that he had the intent necessary to support a guilty verdict. We conclude, to the contrary, that the record includes abundant evidence of his culpable intent. Glass initiated at least three online conversations with BBG14 in which he expressed his desire for a sexual relationship with her. He made arrangements to meet with her for a sexual encounter and arrived at the appointed time and place with a box of condoms in his car.

Holding

This evidence is sufficient to allow a jury to infer that Glass intended to commit lewd and lascivious conduct with a child under the age of sixteen. . . . Accordingly, the judgment of conviction and sentence are affirmed.

Questions for Discussion

1. What are the reasons for holding Glass criminally liable despite the fact that BBG14 was a fictional creation of the police? Should Glass be held liable for attempted lewd conduct with a child who does not exist? What is the social harm?

2. Did Glass go beyond mere preparation? Would you hold Glass guilty in the event that he only communicated with BBG14 over the Internet and never made an effort to meet her? What is the significance of the fact that Glass drove away from the swimming pool? See *Gladish v. United States*, 536 F.3d 646 (7th Cir. 2008).

3. Do you believe that it is appropriate for the police to use this type of investigative strategy?

Cases and Comments

Knowledge of the Facts. Ralph Damms was convicted of attempted murder and was sentenced to a term of imprisonment of not more than ten years. The verdict was affirmed by the Wisconsin Supreme Court. Marjory Damms had initiated a divorce action against Ralph and was also estranged from her mother, Mrs. Laura Grant. Ralph stopped Marjory on her way to work, claimed that Mrs. Grant was dying, and drove Marjory to her mother's home. Marjory then discovered that her mother was perfectly fine. Damms took advantage of this opportunity to attempt a reconciliation with his former wife. After two hours of unproductive conversation, Ralph offered to drive Marjory to work. During the drive, Ralph commented that it was possible for a person to die "quickly" and that "judgment day" may occur without warning. He then removed a cardboard box from under the seat and opened it and took a gun out of a paper bag. He aimed at Marjory and said that this is "to show you that I'm not kidding . . . [or] fooling."

Ralph announced that he was taking Marjory "up north" for a few days. Ralph wanted to eat and drove the car into a restaurant parking lot. Marjory told Ralph that she had a "couple of dollars," and when she refused to let Ralph inspect her checkbook, an argument ensued. Marjory opened the car door and started to run around the restaurant building, screaming for help. Ralph pursued her with a pistol in his hand. Two officers eating lunch rushed out of the restaurant. Marjory slipped and fell, and Ralph crouched down, held the pistol against her head, and pulled the trigger. Ralph exclaimed, "It won't fire. It won't fire." The officers arrested Ralph and discovered that the weapon was unloaded and that the clip containing the cartridges for the gun was in the cardboard box in the car. Damms later stated to two officers that he thought that the gun was loaded. However, Damms testified at trial that he knew that the pistol was not loaded.

The Wisconsin Supreme Court ruled that the jury could have reasonably concluded that Damms believed that the gun actually was loaded and that he therefore was legally liable for attempted murder. The court observed that an

> unequivocal act accompanied by intent should be sufficient to constitute a criminal attempt . . . and he should not escape punishment . . . by reason of some fact unknown to him [that made it] impossible to effectuate the intended result.

The Supreme Court noted that Damms was excited when he grabbed the weapon and pursued his wife and may not have noticed the opening in the end of the butt caused by the absence of an ammunition clip, which would have clearly informed him that the gun was unloaded.

Judge Dietrich, in dissent, noted that Ralph had the gun in his hand several times and that "it would be impossible for him not to be aware or know that the pistol was unloaded. He could feel the hole in the bottom of the butt." Ralph, for example, certainly must have felt the hole in the bottom of the butt caused by the lack of an ammunition clip when he first removed the pistol from the box and then when he chased his wife with the pistol.

Why is the question whether Ralph knew the gun was unloaded significant? Do you agree with the Wisconsin Supreme Court? See *State v. Damms*, 100 N.W.2d 592 (Wis. 1960).

You Decide

7.3 Francisco Martin Duran was a twenty-six-year-old upholsterer from Colorado. On September 13, 1994, Duran bought an assault rifle and roughly 100 rounds of ammunition. Two days later, he purchased a thirty-round clip and equipped the rifle with a folding stock. Thirteen days later, Duran bought a shotgun and, the following day, additional ammunition. On September 30, 1993, Duran left work and, without contacting his family or employer, began a journey to Washington D.C. He purchased another thirty-round clip and a large coat in Virginia. On October tenth, Duran arrived in Washington D.C., and he stayed in various hotels over the next nineteen days.

On October 29, 1994, Duran positioned himself outside the White House fence and observed a group of men in dark suits, one of whom was Dennis Basso, who strongly resembled then-President Bill Clinton. Two eighth-grade students remarked that Basso looked like Bill Clinton. Duran almost immediately began firing twenty rounds at Basso, who managed to take cover. Duran was tackled by a pedestrian when attempting to reload a second clip. The Secret Service searched Duran's automobile and found incriminating evidence, including a map with the phrase "kill the Pres!" and an "X" drawn across a photo of President Clinton. A subsequent search of Duran's home led to the seizure of other incriminating evidence, including a business card on the back of which Duran called for the killing of all government officers and department heads.

Was Duran guilty of an attempt to kill the president of the United States despite the fact that this was impossible given that President Clinton was not on the lawn of the White House? See *United States v. Duran*, 96 F.3d 1495 (D.C. Cir. 1996).

You can find the answer at www.sagepub.com/lippmanccl3e.

Abandonment

An individual who abandons an attempt to commit a crime based on the intervention of outside or extraneous factors remains criminally liable. On the other hand, what about an individual who voluntarily abandons his or her criminal scheme after completing an attempt?

In *People v. Staples,* Staples intentionally rented an office above a bank. He learned that no one was in the building on Saturday and received permission from the owner to move items into his office over the weekend. Staples took advantage of the fact that no one was in the building and drilled several holes partway through the floor, which he then covered with a rug. He placed the drilling tools in the closet and left the key in the office. Later, the landlord discovered the holes and notified the police. Staples was arrested and confessed, explaining that he abandoned his criminal plan after realizing that he could not enjoy life while living off stolen money.

Is the defendant guilty of an attempt? Assuming that the defendant committed an attempted burglary (breaking and entering with an intent to steal), does the defendant's change of heart or abandonment constitute a defense? Would it make a difference if the defendant changed his mind only after hearing voices in the bank?[17]

The Model Penal Code, in section 5.01(4), recognizes the affirmative defense of **abandonment** in those instances in which an individual commits an attempt and "abandoned his effort . . . under circumstances manifesting a complete and voluntary renunciation of criminal purpose." The important point is that an individual can commit an attempt and then relieve himself or herself from liability by voluntarily abandoning the criminal enterprise. A renunciation is not voluntary when motivated by a desire to avoid apprehension, provoked by the realization that the crime is too difficult to accomplish, or where the offender decides to postpone the crime or to focus on another victim. For example, abandonment has not been recognized as a defense where the lock on a bank vault or the door on a cash register proved difficult to open, the police arrived during the commission of a crime, or a victim broke free and fled. Once having completed the commission of a crime, the fact that an offender is full of regret and rushes the victim to the hospital also does not free the assailant from criminal liability. Abandonment, in short, is a defense to attempt when an individual freely and voluntarily undergoes a change of heart and abandons the criminal activity.[18]

In some cases, courts have continued to hold that once an attempt is complete, an individual cannot avoid criminal liability. Why should an attempt be treated differently than any other crime? The vast majority of decisions recognize that there are good reasons for recognizing the defense of abandonment, even in cases where the individual's acts are "dangerously close" to the completion of a crime.[19]

- *Lack of Purpose.* An individual who abandons a criminal enterprise lacks a firm commitment to complete the crime and should be permitted to avoid punishment.
- *Incentive to Renounce Crime.* The defense of abandonment provides an incentive for individuals to renounce their criminal conduct before completing the crime.

In reading the next case, *Ross v. State,* ask yourself whether the defendant was guilty of an attempted rape and whether he voluntarily or involuntarily abandoned his criminal activity. Does reading this case persuade you that the affirmative defense of abandonment is a good or a bad idea?

Did Ross voluntarily abandon his attempt to rape the victim?

Ross v. State, 601 So. 2d 872 (Miss. 1992). Opinion by: Prather, J.

This attempted-rape case arose on the appeal of Sammy Joe Ross from the ten-year sentence imposed on July 7, 1988, by the Circuit Court of Union County. . . .

Facts

Dorothy Henley and her seven-year-old daughter lived in a trailer on a gravel road. Henley was alone at home and answered a knock at the door to find Sammy Joe Ross asking directions. Henley had never seen Ross before. She stepped out of the house and pointed out the house of a neighbor who might be able help him. When she turned back around, Ross pointed a handgun at her. He ordered her into the house, told her to undress, and shoved her onto the couch. Three or four times Ross ordered Henley to undress and once threatened to kill her. Henley described herself as frightened and crying. She attempted to escape from Ross and told him that her daughter would be home from school at any time. She testified that,

I started crying and talking about my daughter, that I was all she had because her

daddy was dead, and he said if I had a little girl he wouldn't do anything, for me just to go outside and turn my back.

As instructed by Ross, Henley walked outside behind her trailer. Ross followed and told her to keep her back to the road until he had departed. She complied.

Henley was able to observe Ross in her sunlit trailer with the door open for at least five minutes. She stated that she had an opportunity to look at him and remember his physical appearance and clothing. Henley also described Ross's pickup truck, including its color, make, and the equipment—that is, a toolbox.

On December 21, 1987, a Union County grand jury indicted Sammy Joe Ross for the attempted rape of Henley, charging that Ross "did unlawfully and feloniously attempt to rape and forcibly ravish" the complaining witness, an adult female. . . . On June 23, 1988, the jury found Ross guilty. On July seventh, the court sentenced Ross to a ten-year term. . . . Ross timely filed a notice of appeal.

Issue

The primary issue here is whether sufficient evidence presents a question of fact as to whether Ross abandoned his attack as a result of outside intervention. Ross claims that the case should have gone to the jury only on a simple assault determination. Ross asserts that "it was not . . . Henley's resistance that prevented her rape nor any independent intervening cause or third person, but the voluntary and independent decision by her assailant to abandon his attack." The State, on the other hand, claims that Ross "panicked" and "drove away hastily."

Reasoning

Henley told Ross that her daughter would soon be home from school. She also testified that Ross stated if Henley had a little girl, he wouldn't do anything to her and to go outside [the house] and turn her back [to him].

The trial court instructed the jury that if it found that Ross did "any overt act with the intent to have unlawful sexual relations with [the complainant] without her consent and against her will," then the jury should find Ross guilty of attempted rape. The court further instructed the jury that before you can return a verdict against the defendant for attempted rape, that you must be convinced from the evidence and beyond a reasonable doubt "that the defendant was prevented from completing the act of rape or failed to complete the act of rape by intervening, extraneous causes. If you find that the act of rape was not completed due to a voluntary stopping short of the act, then you must find the defendant not guilty."

This court has held that lewd (indecent) suggestions coupled with physical force constituted sufficient evidence to establish intent to rape. . . . Attempt consists of "1) an intent to commit a particular crime, 2) a direct ineffectual act done toward its commission, and 3) failure to consummate its commission." The Mississippi attempt statute requires that the third element, failure to consummate, result from extraneous causes. . . . Where the assailant released his throat-hold on the unresisting victim and told her she could go, after which a third party happened on the scene, this court has held that the jury could not have reasonably ruled out abandonment. In comparison, this court has held that where the appellant's rape attempt failed because of the victim's resistance and ability to sound the alarm, the appellant cannot establish an abandonment defense. In another case, the defendant did not voluntarily abandon his attempt but instead fled after the victim, a hospital patient, pressed the nurse's buzzer; a nurse responded and the victim spoke the word "help." The court concluded, "The appellant ceased his actions only after the victim managed to press the buzzer alerting the nurse." . . . In another case, the court properly sent the issue of attempt to the jury where the attacker failed because the victim resisted and freed herself.

Thus, abandonment occurs where, with no physical resistance or external intervention, the perpetrator changes his mind. At the other end of the scale, a perpetrator cannot claim that he abandoned his attempt when, in fact, he ceased his efforts because the victim or a third party intervened or prevented him from furthering the attempt. Somewhere in the middle lies a case . . . where the victim successfully sounded an alarm, presenting no immediate physical obstacle to the perpetrator's continuing the attack, but sufficiently intervening to cause the perpetrator to cease his attack.

The key inquiry is a subjective one: What made Ross leave? According to the undisputed evidence, he left because he responded sympathetically to the victim's statement that she had a little girl. He did not fail in his attack. No one prevented him from completing it. Henley did not sound an alarm. She successfully persuaded Ross, of his own free will, to abandon his attempt. No evidence shows that Ross panicked and hastily drove away, but rather, the record shows that he walked the complainant out to the back of her trailer before he left. Thus . . . this is not to say that Ross committed no criminal act, but "our only inquiry is whether there was sufficient evidence to support a jury finding that [Ross] did not abandon his attempt to rape."

Holding

Ross raises a legitimate issue of error in the sufficiency of the evidence supporting his conviction for attempted rape because he voluntarily abandoned the attempt. This court reverses. . . .

Questions for Discussion

1. What is the subjective legal test for abandonment?

2. The Mississippi Supreme Court rules that Ross voluntarily abandoned the attempted rape. How do you explain this ruling in light of the fact that Ross likely would have raped Henley (not her real name) had she not informed him that her daughter would be returning home? Was this information an extraneous factor?

3. Can you distinguish this rule from the holding of the ruling of the Wisconsin Supreme Court in *Le Barron v. State*, 145 N.W.2d 79 (Wis. 1966)? In *Le Barron*, the Wisconsin court ruled that the defendant did not voluntarily abandon an attempted rape when he failed to complete the rape after determining that the victim was telling the truth in claiming that she was pregnant.

4. How would you rule in *Ross v. State*?

You can find Le Barron v. State *at www.sagepub.com/lippmanccl3e.*

CONSPIRACY

The crime of **conspiracy** comprises an agreement between two or more persons to commit a criminal act. There are several reasons for punishing an agreement:

- *Intervention.* Protecting society by arresting individuals before they commit a dangerous crime.
- *Group Activity.* Crimes committed by groups have a greater potential to cause social harm.
- *Deterrence.* Group pressure makes it unlikely that the conspirators will be deterred from carrying out the agreement.

The common law crime of conspiracy was complete with the agreement to commit a crime. Most modern statutes require an affirmative act, however slight, toward carrying out the conspiracy. The important point is that the law of conspiracy permits law enforcement to arrest individuals at an early stage of criminal planning.

For a deeper look at this topic, visit the study site.

Bear in mind that in common law, the conspiracy did not merge into the criminal act. Today, this continues to be the rule; conspiracy does not merge into the attempted or completed offense that is the object of conspiracy. *As a result, an individual may be convicted of both the substantive offense that is the object of the conspiracy and of a conspiracy. A defendant may be held liable for both armed robbery and for a conspiracy to commit armed robbery.* Remember that under the Pinkerton rule, an individual is guilty of all criminal acts committed by one of the conspirators in furtherance of the conspiracy, regardless of whether the individual aided or abetted or was even aware of the offense.

State statutes differ on the punishment of a conspiracy. Some provide that a conspiracy is a misdemeanor, others that the sentence for conspiracy is the same as the target offense, and a third group provides a different sentence for conspiracies to commit a misdemeanor and for conspiracies to commit a felony. *In general, a conspiracy to commit a felony is a felony; a conspiracy to commit a misdemeanor is a misdemeanor.*[20]

The law of conspiracy is one of the most difficult areas of the criminal law to understand. As former Supreme Court Justice Robert Jackson observed, "The modern crime of conspiracy is so vague that it almost defies definition."[21]

Actus Reus

The *actus reus* of conspiracy consists of

- entering into an agreement to commit a crime, and,
- under some modern statues, an overt act in furtherance of the agreement is required.

The core of a conspiracy charge is an agreement. Individuals do not normally enter into a formal contractual agreement to commit a crime. Prosecutors typically are forced to point to circumstances that strongly indicate that the defendants agreed to commit a crime. In *Commonwealth v. Azim*, Charles Azim pulled his automobile over to the curb, and one of the passengers, Thomas Robinson, called to a nearby Temple University student. The student refused to respond, and Robinson and Mylice James exited the auto, beat and choked him, and took his wallet. The three then drove away from the area. The Pennsylvania Superior Court, in affirming Azim's conviction for conspiracy, pointed to Azim's association with the two assailants, his presence at the crime scene, and Azim's waiting in the automobile with the engine running and lights on as the student was beaten. The case for conspiracy would have been even stronger had this been part of a pattern of criminal activity or if there was evidence that the three divided the money.[22]

United States v. Brown is often cited to illustrate the danger that courts will find a conspiracy based on even the slightest evidence suggesting that the defendants cooperated with one another. An undercover officer approached Valentine on the street seeking to purchase marijuana. Brown joined the conversation and advised Valentine three times that the officer "looks okay to me." Brown told Valentine that there was no reason to distrust the customer or to take precautions and persuaded Valentine to personally hand the drugs over to the undercover agent. The federal court of appeals concluded that the facts indicated that Brown had agreed with Valentine to direct or advise Valentine on drug sales. Judge Oakes observed in dissent that there was not a "shred of evidence" that Brown was involved with Valentine and that when "numerous other inferences could be drawn from the few words of conversation . . . I cannot believe that there is proof of conspiracy . . . beyond a reasonable doubt." Judge Oakes asked, "What conspiracies might we approve tomorrow? The majority opinion will come back to haunt us, I fear."[23]

Critics like to point to a series of trials of anti–Vietnam War activists conducted during the 1960s to demonstrate the potential abuse of conspiracy charges. In the "Chicago Eight" trial, eight activists, most of whom did not even know one another, were prosecuted for conspiring to cross state lines to incite a riot at the 1968 Chicago Democratic Convention; they were ultimately freed.[24]

Overt Act

Under the common law, an agreement was sufficient to satisfy the elements of a conspiracy. Most states and the federal statute now require proof of an **overt act** in furtherance of the conspiracy. The overt act requirement is satisfied by even an insignificant act that is far removed from the commission of a crime. As observed by Justice Oliver Wendell Holmes, "The essence of the conspiracy is being combined for an unlawful purpose—and if an overt act is required, it does not matter how remote the act may be from accomplishing the purpose, if done to effect it. . . ."[25] Attending a meeting of the Communist Party was considered to constitute an overt act in furtherance of a Communist conspiracy to overthrow the United States government,[26] and purchasing large quantities of dynamite satisfied the overt act requirement for a conspiracy to blow up a school building.[27] In other cases, the overt act has been satisfied by observing the movements of an intended kidnapping victim or by purchasing stamps to send poison through the mail.[28]

An overt act by any party to a conspiracy is attributed to every member and provides a sufficient basis for prosecuting all the participants. The requirement of an overt act is intended to limit conspiracy prosecutions to agreements that have progressed beyond the stage of discussion and that therefore present a social danger.[29]

Mens Rea

The *mens rea* of conspiracy is the intent to achieve the object of the agreement. Some judges continue to express uncertainty over whether this requires a purpose to cause the result or whether it is sufficient that an individual knows that a result will occur. Under the knowledge standard, all that is required is that the seller be aware of a buyer's "intended illegal use." A purpose standard requires that the seller possess an intent to further, promote, and cooperate in the buyer's specific illegal objective.

A knowledge standard may deter individuals from providing assistance to individuals whom they are aware or suspect are engaged in illegal activity. On the other hand, limiting liability to individuals with a criminal purpose targets individuals who intend to further criminal conduct.

The Model Penal Code reflects the predominant view that a specific intent to further the object of the conspiracy is required. In *United States v. Falcone,* the U.S. Supreme Court ruled that individuals who provided large quantities of sugar, yeast, and cans to individuals whom they knew were engaged in illegally manufacturing alcohol were not liable for conspiracy. The court held that the government was required to demonstrate that the suppliers intended to promote the illegal enterprise.[30]

In *People v. Lauria,* a California appellate court was confronted with the challenge of determining whether the operator of a telephone message service was merely providing a service to his clients knowing that they were prostitutes or whether he conspired to further acts of prostitution. The court rejected the prosecution's claim that Lauria's knowledge that three of the customers were prostitutes satisfied the mental element required to hold Lauria liable for conspiring to commit prostitution. The court ruled that the prosecution must demonstrate that Lauria possessed an intent to further a criminal enterprise and that there was insufficient evidence that "Lauria took any direct action to further, encourage or direct . . . call-girl activities."

The California court provided some direction to prosecutors in future cases by observing that Lauria's intent to further prostitution might be established by evidence that he promoted and encouraged the prostitutes' pursuit of customers or received substantial financial benefits from their activities. It was significant that only a small portion of Lauria's business was derived from the prostitutes and that he received the same fee regardless of the number of messages left for his prostitute customers.[31]

Parties

A conviction for conspiracy requires that two or more persons intentionally enter into an agreement with the intent to achieve the crime that is the objective of the conspiracy. This is referred to as the **plurality requirement**. As noted by former Supreme Court Justice Benjamin Cardozo, "It is impossible . . . for a man to conspire with himself."[32]

This joint or **bilateral** conception of conspiracy means that a charge of conspiracy against one conspirator will fail in the event that the other party to the conspiracy lacked the required *mens rea*. A conspirator in a two-person conspiracy, for example, would be automatically acquitted in the event that the other party was an undercover police officer or was legally insane and was legally incapable of entering into an agreement. In a joint trial of two conspirators at common law, the acquittal of one alleged conspirator resulted in the dismissal of the charges against the other conspirator. Keep in mind that under the bilateral approach, "There must be at least two guilty conspirators or none."[33]

The bilateral approach is criticized for undermining the enforcement of conspiracy laws. An individual who intends to enter into a conspiracy to commit a crime is a threat to society, and he or she should be subject to punishment regardless of whether the other party turns out to be an undercover police informant who lacks a criminal intent. On the other hand, the bilateral approach is consistent with the view that the law of conspiracy should be directed against group crime.[34]

The Model Penal Code adopts a **unilateral** approach that examines whether a single individual agreed to enter into a conspiracy rather than focusing on whether two or more persons entered into an agreement. This scheme has been incorporated into a number of modern state statutes. Under the unilateral approach, the fact that one party is an undercover police officer or lacks the capacity to enter into a conspiracy does not result in the acquittal of the other conspirator. The commentary to the Model Penal Code notes that under the unilateral approach, it is "immaterial to the guilt of a conspirator . . . that the person or all of the persons with whom he conspired have not been or cannot be convicted."[35]

The unilateral approach has been criticized for permitting the prosecution of individuals for a conspiracy who, in fact, have not actually entered into a criminal agreement. The fear is expressed by civil libertarians that the unilateral approach enables undercover agents to manufacture crime by enticing individuals into unilateral conspiratorial agreements.[36]

The Structure of Conspiracies

The structure of a conspiracy is important. Defendants may be found guilty only of the conspiracy charged at trial and may offer the defense that there were separate agreements to commit different crimes rather than a single conspiracy. A defendant might admit that he or she was involved in a conspiracy to kidnap and hold a corporate executive for ransom and also argue that the other kidnappers entered into a separate conspiracy to kill the executive. Remember, in the event of a single conspiracy, our kidnapper would be held liable for all offenses committed in furtherance of the agreement to kidnap the executive, including the murder.[37]

Most complex conspiracies can be categorized as either a **chain conspiracy** or **wheel conspiracy**.

A chain conspiracy typically arises in the distribution of narcotics and other contraband. This involves communication and cooperation by individuals linked together in a vertical chain to achieve a criminal objective.

The classic case is *United States v. Bruno,* in which eighty-eight defendants were indicted for a conspiracy to import, sell, and possess narcotics. This involved smugglers who brought narcotics into New York and sold them to middlemen who distributed the narcotics to retailers who, in turn, sold narcotics to operatives in Texas and Louisiana for distribution to addicts. The petitioners appealed on the grounds that there were three conspiracies rather than one large conspiracy. The court ruled that this was a single chain conspiracy in which the smugglers knew that the middlemen must sell to retailers for distribution to addicts, and the retailers knew that the middlemen must purchase drugs from smugglers. In the words of the court, the "conspirators at one end of the chain knew that the unlawful business would not and could not, stop with their buyers; and those at the other end knew that it had not begun with their sellers." Each member of the conspiracy knew "that the success of that part with which he was immediately concerned, was dependent upon the success of the whole." Remember that this means that every member of the conspiracy was liable for every illegal transaction carried out by his co-conspirators in Texas and in Louisiana.[38]

A circle or wheel conspiracy involves a single person or group that serves as a *hub,* or common core, connecting various independent individuals or spokes. The *spokes* typically interact with the hub rather than with one another. In the event that the spokes share a common purpose to succeed, there is a single conspiracy. On the other hand, in those instances that each spoke is unconcerned with the success of the other spokes, there are multiple conspiracies.

The most frequently cited case illustrating a wheel conspiracy is *Kotteakos v. United States.* Simon Brown, the hub, assisted thirty-one independent individuals to obtain separate fraudulent loans from the government. The Supreme Court held that although all the defendants were engaged in the same type of illegal activity, there was no common purpose or overall plan, and the defendants were not liable for involvement in a single conspiracy. Each loan "was an end in itself, separate from all others, although all were alike in having similar illegal objects. Except for Brown, the common figure, no conspirator was interested in whether any loan except his own went through." As a result, the Supreme Court found that there were thirty-two separate conspiracies involving Brown rather than one common conspiracy.[39]

State v. McLaughlin is an example of a wheel conspiracy in which the individuals involved share a common purpose. In *McLaughlin,* several gamblers independently agreed to subscribe to an illegal horse racing service that provided racing results and information. The customers realized that the success of this financially expensive venture depended on the willingness of each of the other gamblers to support the service, and the defendants were held liable for involvement in a single, common wheel conspiracy.[40]

Criminal Objectives

The crime of conspiracy traditionally punished agreements to commit a broad range of objectives, many of which would not be criminal if committed by a single individual. The thinking was that these acts assume an added danger when engaged in by a group of individuals.

In 1832, English jurist Lord Denman pronounced that a conspiracy indictment must "charge a conspiracy either to do an unlawful act or a lawful act by unlawful means."[41] English and American courts interpret "unlawful" to include acts that are not punishable under the criminal law. It was "enough if they are corrupt, dishonest, fraudulent, immoral, and in that sense illegal, and it is in the combination to make use of such practices that the dangers of this offense consist."[42]

The U.S. Supreme Court has recognized the danger that broadly defined conspiracy statutes may fail to inform citizens of the acts that are prohibited and may provide the police, prosecutors, and judges with broad discretion in bringing charges. *Musser v. Utah,* 333 U.S. 95 (1948). The doctrine of conspiracy, for instance, was used in a New York court against workers who went on strike in protest against a fellow employee who agreed to work below union wages.[43] The English House of Lords upheld the conviction of an individual for "conspiracy to corrupt public morals" who agreed to publish a directory of prostitutes.[44] In ruling that a Utah conspiracy statute that punished conspiracies to commit acts injurious to the public morals was unconstitutional, the U.S. Supreme Court noted that the statute

> would seem to be a warrant for conviction for an agreement to do almost any act which a judge and jury might find . . . contrary to his or its notions of what was good for health, morals, trade, commerce, justice or order.[45]

Modern statutes generally limit the criminal objectives of conspiracy to agreements to commit crimes. Several jurisdictions, however, continue to enforce broadly drafted statutes. California Penal Code section 182(5) punishes a conspiracy to commit "any act injurious to the public health, to public morals, or to pervert or obstruct justice, or the due administration of justice." Would a religious group's vocal opposition to "safe sex" through the use of condoms potentially result in criminal liability under this statute? The U.S. government's conspiracy statute broadly punishes persons conspiring either to "commit any offense against the United States, or to defraud the United States." A conspiracy to commit a felony under this statute is punishable by up to five years in prison in addition to a fine, while a conspiracy to commit a misdemeanor is subject to the maximum penalty for the target offense.[46]

Pinkerton v. United States established that all criminal acts undertaken in furtherance of a conspiracy or that are "reasonably foreseeable as the necessary or natural consequences of the conspiracy" are attributable to each member by virtue of his or her membership. In *Pinkerton*, Daniel Pinkerton was held liable for conspiring with his brother Walter to avoid federal taxes. Daniel was held criminally responsible for Walter's failure to pay taxes, despite the fact that Daniel was in prison at the time that Walter submitted his fraudulent tax return.[47] The Model Penal Code rejects the Pinkerton rule, because each conspirator may be held liable for "thousands of crimes" of which he or she was "completely unaware" and did not "influence at all." Consider the often-cited example of a woman who refers individuals to a criminal abortionist who may find herself being held liable for unlawful abortions performed on women referred to the abortionist by a co-conspirator whom she has never met.

You also should be aware of **Wharton's rule**. This provides that an agreement by two persons to commit a crime that requires the voluntary and cooperative action of two persons cannot constitute a conspiracy. The classic examples of consensual crimes that require the participation of two individuals and do not permit a charge of conspiracy under Wharton's rule are adultery, bigamy, the sale of contraband, bribery, and dueling. These offenses already punish a cooperative agreement between two individuals to commit a crime, and there is no reason to further punish individuals for entering into a conspiratorial agreement. Wharton's rule does not prevent a conspiracy involving more than the required number of individuals. Three individuals, for example, may conspire for two of them to engage in bribery. Two individuals may also conspire for other individuals to pay and receive bribes.[48]

Another principle is the **Gebardi rule**. This provides that an individual who is in a class of persons that are excluded from criminal liability under a statute may not be charged with a conspiracy to violate the same law. In *Gebardi v. United States*, the U.S. Supreme Court reversed the conspiracy conviction of a man and woman for violation of the Mann Act. This statute prohibited and punished the transportation of a woman from one state to another for immoral purposes. The Court reasoned that the statute was intended to protect women from sexual exploitation and was defined so as to solely punish the individual transporting the women. The Court therefore reasoned that the two defendants could not enter into a criminal conspiracy and reversed their conviction.[49]

Conspiracy Prosecutions

Judge Learned Hand called conspiracy the "darling of the modern prosecutor's nursery."[50] Judge Hand was referring to the fact that conspiracy constitutes a powerful and potential tool for prosecuting and punishing defendants.[51]

- Conspiracies are not typically based on explicit agreements and may be established by demonstrating a commitment to a common goal by individuals sharing a criminal objective.
- Defendants may be prosecuted for both conspiracy and the commission of the crime that was the object of the conspiracy.
- A prosecution may be brought in any jurisdiction in which the defendants entered into a conspiratorial agreement or committed an overt act.
- The defendants all may be joined in a single trial, creating the potential for "guilt by association."
- All conspirators are held responsible for the criminal acts and statements of any co-conspirator in furtherance of the conspiracy. An individual may be held liable who is not present or even aware of a co-conspirator's actions.
- Individuals may abandon the conspiracy and escape liability for future offenses only if this abandonment is communicated to the other conspirators. Some statutes require that individuals persuade the other conspirators to abandon the conspiracy.

The federal law of conspiracy was further expanded in 1970 when Congress passed the Racketeer Influenced and Corrupt Organizations Act (RICO). This law is intended to provide prosecutors with a powerful and potent weapon against organized crime. The RICO law essentially eliminates the need to prove that individuals are part of a single conspiracy, and, instead, holds defendants responsible for all acts of racketeering undertaken as part of an "enterprise." Racketeering includes a range of state and federal offenses typically committed by organized crime including murder, kidnapping, gambling, arson, robbery, bribery, extortion, and dealing in narcotics or obscene material. Critics have voiced concern over the government's power to bring a counterfeiter to trial for murders committed by individuals involved in an unrelated component of a criminal enterprise.

The next case in the text, *United States v. Garcia,* asks you to determine whether Garcia entered into a conspiratorial agreement with members of his gang to commit an assault with a dangerous weapon.

Model Penal Code

Section 5.03. Criminal Conspiracy

(1) . . . A person is guilty of conspiracy with another person or persons to commit a crime if with the purpose of promoting or facilitating its commission he:

    (a) agrees with such other person or persons that they or one or more of them will engage in conduct that constitutes such crime or an attempt or solicitation to commit such crime; or

    (b) agrees to aid such other person or persons in the planning or commission of such crime or of an attempt or solicitation to commit such crime. . . .

(2) . . . If a person guilty of conspiracy . . . knows that a person with whom he conspires to commit a crime has conspired with another person or persons to commit the same crime, he is guilty of conspiring with such other person or persons, whether or not he knows their identity, to commit such crime.

(3) . . . If a person conspires to commit a number of crimes, he is guilty of only one conspiracy so long as such multiple crimes are the object of the same agreement or continuous conspiratorial relationship.

(4) . . .

    (a) . . . two or more persons charged with criminal conspiracy may be prosecuted jointly. . . .

Analysis

1. The Model Penal Code limits conspiracies to crimes and does not extend conspiracy to broad categories of immoral or corrupt behavior.

2. A defendant must possess the purpose of promoting or facilitating the commission of a crime. Knowledge does not satisfy the intent requirement of conspiracy (§5.04(1)).

3. The essence of a conspiracy is an agreement. This determination should be based on clear evidence (§5.04(1)(a)).

4. A unilateral approach to conspiratorial agreements is adopted (§5.04(1)(b)).

5. The Model Penal Code does not use the terminology of wheel or chain conspiracies. The code, instead, examines whether a specific individual has entered into a conspiratorial agreement and whether the individual is aware that the person with whom he or she has conspired has entered into agreements with other individuals. The end result would not differ from the wheel or chain analysis (§5.04(2)).

6. An overt act is required other than in the case of serious felonies.

7. The Pinkerton rule is rejected (§2.06); individuals are responsible only for crimes that they solicited, aided, agreed to aid, or attempted to aid.

8. Conspiracy is punished to the same extent as the most serious offense that is attempted or solicited or is an object of the conspiracy (§5.05(1)).

9. The code does not permit conviction of both a conspiracy and the substantive crime that is the object of the conspiracy (§1.07(1)(b)).

10. The Gebardi rule is incorporated into the Model Penal Code (§5.04(2)).

The Legal Equation

Conspiracy = Agreement (or agreement and overt act in furtherance of agreement)

+ specific intent or purpose to commit a crime. (Some courts employ a knowledge standard.)

Was there a single conspiracy or were there multiple conspiracies to commit arson? Did the members of the Bloods street gang enter into a conspiratorial agreement to assault members of the Crips?

United States v. Garcia, 151 F.3d 1243 (9th Cir. 1998). Opinion by: Reinhardt, J.

Issue

Defendant Leon "Cody" Garcia was convicted in U.S. District Court of conspiracy to assault with a dangerous weapon. He appeals his conviction and sentence of 60 months in prison on the grounds that the evidence did not establish that he entered into a conspiracy. In this case, we consider whether testimony regarding the existence of an implicit, general agreement among gang members to support one another in fights against rival gangs can constitute sufficient evidence to support a conviction of conspiracy to commit assault. . . .

Facts

One evening, a confrontation broke out between rival gangs at a party on the Pasqua Yaqui Indian reservation. The resultant gunfire injured four young people, including appellant Cody Garcia. Two young men involved in the shooting, Garcia and Noah Humo, were charged with conspiracy to assault three named individuals with dangerous weapons. A jury acquitted Humo but convicted Garcia. Because there is no direct evidence of an agreement to commit the criminal act which was the alleged object of the conspiracy, and because the circumstances of the shooting do not support the existence of an agreement, implicit or explicit, the government relied heavily on the gang affiliation of the participants to show the existence of such an agreement.

The party at which the shootings occurred was held in territory controlled by the Crips gang. The participants were apparently mainly young Native Americans. While many of the attendees were associated with the Crips, some members of the Bloods gang were also present. Appellant Cody Garcia arrived at the party in a truck driven by his uncle, waving a red bandanna (the Bloods claim the color red and the Crips the color blue) out the truck window and calling out his gang affiliation: "ESPB Blood!" Upon arrival, Garcia began "talking smack" to (insulting) several Crips members. Prosecution witnesses testified that Garcia's actions suggested that he was looking for trouble and issuing a challenge to fight to the Crips at the party.

Meanwhile, Garcia's fellow Bloods member Julio Baltazar was also "talking smack" to Crips members, and Blood Noah Humo bumped shoulders with one Crips member and called another by a derogatory Spanish term. Neither Baltazar nor Humo had arrived with Garcia, nor is there any indication that they had met before the party to discuss plans or that they were seen talking together during the party.

At some point, shooting broke out. Witnesses saw both Bloods and Crips, including Garcia and Humo, shooting at one another. Baltazar was seen waving a knife or trying to stab a Crip. The testimony at trial does not shed light on what took place immediately prior to the shooting, other than the fact that one witness heard Garcia ask, "Who has the gun?" There is some indication that members of the two gangs may have "squared off" before the shooting began. No testimony establishes whether the shooting followed a provocation or verbal or physical confrontation.

Four individuals were injured by the gunfire: the defendant, Stacy Romero, Gabriel Valenzuela, and Gilbert Baumea. Stacy Romero who at the time was twelve years old was the cousin both of Garcia's co-defendant Humo and his fellow Blood, Baltazar. No evidence presented at trial established that any of the injured persons was shot by Garcia, and he was charged only with conspiracy. The government charged both Garcia and Humo with conspiracy to assault Romero, Valenzuela, and Baumea with dangerous weapons under 18 U.S.C. §§ 371, 113(a)(3) and 1153; Humo alone was charged with two counts of assault resulting in serious bodily injury under 18 U.S.C. §§113(a)(6) and 1153.

After a jury trial, Humo was acquitted on all counts. Garcia was convicted of conspiracy to assault with a dangerous weapon and sentenced to 60 months in prison. He appeals on the ground that there was insufficient evidence to support his conviction.

Reasoning

In order to prove a conspiracy, the government must present sufficient evidence to demonstrate both an overt act and an agreement to engage in the specific criminal activity charged in the indictment. While an implicit agreement may be inferred from circumstantial evidence, proof that an individual engaged in illegal acts with others is not sufficient to demonstrate the existence of a conspiracy. Both the existence of and the individual's connection to the conspiracy must be proven beyond a reasonable doubt.

The government claims that it can establish the agreement to assault in two ways: first, that the concerted provocative and violent acts by Garcia, Humo and Baltazar are sufficient to show the existence of a prior agreement; and second, that by agreeing to become a member of the gang, Garcia implicitly agreed to support his fellow gang members in violent confrontations.

However, no inference of the existence of any agreement could reasonably be drawn from the actions of Garcia and other Bloods members on the night of the shooting. An inference of an agreement is permissible only when the nature of the acts would logically require coordination and planning. An example is two identical trucks traveling down the same highway, both of which contain drugs concealed in identical types of containers.

The government presented no witnesses who could explain the series of events immediately preceding the shooting, so there is nothing to suggest that the violence began in accordance with some prearrangement. The facts establish only that perceived insults escalated tensions between members of rival gangs and that an ongoing gang-related dispute erupted into shooting. Testimony presented at trial suggest more chaos than concert. Such evidence does not establish that parties to a conspiracy "worked together understandingly, with a single design for the accomplishment of a common purpose."

Given that this circumstantial evidence fails to suggest the existence of an agreement, we are left only with gang membership as proof that Garcia conspired with fellow Bloods to shoot the three named individuals. The government points to expert testimony at the trial by a local gang unit detective, who stated that generally gang members have a "basic agreement" to back one another up in fights, an agreement which requires no advance planning or coordination. This testimony, which at most establishes one of the characteristics of gangs but not a specific objective of a particular gang—let alone a specific agreement on the part of its members to accomplish an illegal objective—is insufficient to provide proof of a conspiracy to commit assault or other illegal acts.

Recent authority in this circuit establishes that "membership in a gang cannot serve as proof of intent, or of the facilitation, advice, aid, promotion, encouragement or instigation needed to establish aiding and abetting." *Mitchell v. Prunty,* 107 F.3d 1337 (9th Cir. 1997). In overturning the state conviction of a gang member that rested on the theory that the defendant aided and abetted a murder by "fanning the fires of gang warfare," the *Mitchell* opinion expressed concern that allowing a conviction on this basis would "smack of guilt by association." The same concern is implicated when a conspiracy conviction is based on evidence that an individual is affiliated with a gang which has a general rivalry with other gangs, and that this rivalry sometimes escalates into violent confrontations.

The *Mitchell* court reasoned that the conviction in that case necessarily rested on the faulty assumption that gang members typically act in a concerted fashion. Such an assumption would be particularly inappropriate here. Acts of provocation such as "talking smack" or bumping into rival gang members certainly does not prove a high level of planning or coordination. Rather, it may be fairly typical behavior in a situation in which individuals who belong to rival gangs attend the same events. At most, it indicates that members of a particular gang may be looking for trouble, or ready to fight. It does not demonstrate a coordinated effort with a specific illegal objective in mind. (Conspiracy requires proof of both "an intention and agreement to accomplish a specific illegal objective.") The fact that gang members attend a function armed with weapons may prove that they are prepared for violence, but without other evidence it does not establish that they have made plans to initiate it. And the fact that more than one member of the Bloods was shooting at rival gang members also does not prove a prearrangement—the Crips, too, were able to pull out their guns almost immediately, suggesting that readiness for a gunfight requires no prior agreement. Such readiness may be a sad commentary on the state of mind of many of the nation's youth, but it is not indicative of a criminal conspiracy.

Finally . . . allowing a general agreement among gang members to back each other up to serve as sufficient evidence of a conspiracy would mean that any time more than one gang member was involved in a fight it would constitute an act in furtherance of the conspiracy and all gang members could be held criminally responsible—whether they participated in or had knowledge of the particular criminal act, and whether or not they were present when the act occurred. Indeed, were we to accept "fighting the enemy" as an illegal objective, all gang members would probably be subject to felony prosecutions sooner rather than later, even though they had never personally committed an improper act. This is contrary to fundamental principles of our justice system. "There can be no conviction for guilt by association. . . ."

Because of these concerns, evidence of gang membership cannot itself prove that an individual has entered a criminal agreement to attack members of rival gangs. Moreover, here the conspiracy allegation was even more specific: the state charged Garcia with conspiracy to assault three specific individuals—Romero, Baumea and Valenzuela—with deadly weapons. Even if the testimony presented by the state had sufficed to establish a general conspiracy to assault Crips, it certainly did not even hint at a conspiracy to assault the three individuals listed in the indictment. Of course, a more general indictment would not have solved the state's problems in this case. In some cases, when evidence establishes that a particular gang has a specific illegal objective such as selling drugs, evidence of gang membership may help to link gang members to that objective. However, a general practice of supporting one another in fights, which is one of the ordinary characteristics of gangs, does not constitute the type of illegal objective that can form the predicate for a conspiracy charge.

Holding

We hold that gang membership itself cannot establish guilt of a crime, and a general agreement, implicit or explicit, to support one another in gang fights does not provide substantial proof of the specific agreement required for a conviction of conspiracy to commit assault. The defendant's conviction therefore rests on insufficient evidence, and we reverse.

Because the government introduced no evidence from which a jury could reasonably have found the existence of an agreement to engage in any unlawful conduct, the evidence of conspiracy was insufficient as a matter of law. A contrary result would allow courts to assume an ongoing conspiracy, universal among gangs and gang members, to commit any number of violent acts, rendering gang members automatically guilty of conspiracy for any improper conduct by any member. We therefore reverse Garcia's conviction and remand to the district court to order his immediate release. As a result of this decision, Garcia is not subject to retrial. He has already served over a year in prison.

Questions for Discussion

1. What is act and intent requirement to prove a conspiracy.

2. In *Garcia* why does the government contend that the actions of Garcia, Humo, and Baltazar provide circumstantial evidence of a conspiracy?

3. Explain why the Court of Appeals decides that gang membership cannot serve as evidence that individuals have entered into a conspiracy.

4. Do you agree that "a general practice of supporting one another in fights . . . does not constitute the type of illegal objective that can form the predicate for a conspiracy charge?"

5. Are you persuaded that Garcia and Humo did not enter into a conspiracy to assault Romero, Valenzuela, and Baumea with dangerous weapon?

Cases and Comments

1. *Conspiracy and Agreement.* Jason Escobedo (Jason) went to Will Rogers State Beach with Billy Reyes (Billy), Zonia Petraza (Zonia) and Irene Melendez (Irene). Jason went into a restroom and became involved in an argument with the defendant. Billy ran into the restroom and intervened. Defendant, a member of the "Southside 13" gang, issued a gang challenge and asked Jason "where he was from." Jason replied that he was not a gang member. Defendant told Billy that for $2 the defendant "would forget the whole thing." Billy then hit the defendant in the face and a fight ensued. Zonia arrived on the scene to find defendant on top of Billy. Jason separated the two combatants and apologized to defendant for "intruding into his territory." Defendant responded by punching Jason in the mouth. Immediately thereafter, defendant whistled toward the beach and made a beckoning motion with his hands.

Jason, Billy, Zonia, and Irene then ran for Billy's truck. It appeared to them that the situation had calmed down, and Jason and Billy went down to the beach to retrieve the belongings they had left behind.

While the two girls remained in the truck, another truck pulled up alongside of them containing the defendant and seven other young men. Defendant and two others began bashing the truck with a trash can, a barbell, and their fists. Irene succeeded in starting the truck and began backing it up.

Billy managed to make it back to the truck, but Jason was surrounded and could not escape. Billy, Zonia, and Irene drove to a gas station nearby and telephoned "911." The police arrived in "about 15 minutes." However, by that time Jason lay dead on the other side of the restroom, his head having been severely beaten. There was "graffiti with gang-related monikers at the scene in the vicinity of the decedent's body."

Defendant's fellow gang member Rojas testified that after the bloodied defendant told him what had happened to him in the restroom, one of the other gang members said, "F___ it. Let's go get him." Thereupon, the gang members proceeded to seek out the two individuals whom they believed had intruded into their territory. Was there sufficient evidence of a conspiratorial agreement and a specific intent to accomplish the aims of the conspiracy to hold the defendants liable? See *People v. Quinteros,* 16 Cal. Rptr. 2d 462 (Cal Ct. App. 1993).

2. *Termination of a Conspiracy.* In *United States v. Jimenez Recio,* the U.S. Supreme Court considered whether individuals who joined a conspiracy to distribute illegal narcotics could be held liable for acting in furtherance of the conspiracy despite the fact that the government earlier had seized the drugs that were the object of the conspiracy. In November 1997, the police stopped a truck in Nevada and seized a large quantity of unlawful drugs. The government then arranged for the truck to be driven to the drivers' destination, a mall in Idaho. The original drivers cooperated with a government sting and contacted the individual that they were to meet. Three hours later, defendants Francisco Jimenez Recio and Adrian Lopez-Meza arrived at the mall. Recio took control of the truck and drove away from the mall. Lopez-Meza followed him in a car. The police stopped both individuals and charged them together with others with conspiracy to possess and to distribute unlawful narcotics.

The Ninth Circuit Court of Appeals reversed the defendants' conviction and held that the evidence was not sufficient to demonstrate that the defendants had joined the conspiracy prior to the seizure of drugs in Nevada. In other words, the conspiracy automatically had terminated, because the government, unknown to some of the conspirators, had "defeated the object of the conspiracy." Recio and Lopez-Meza therefore could not be held liable for joining the conspiracy.

The prosecution appealed this reversal to the U.S. Supreme Court, and Justice Stephen Breyer in a near-unanimous decision held that the essence of a conspiracy is the agreement to commit a crime, and the fact that the government has defeated the object of the conspiracy does not prevent individuals from agreeing to participate in criminal activity and being held criminally liable. There are good reasons for this rule. A conspiracy poses a threat to the public over and above the threat of the commission of the object of the conspiracy, because the participants in a conspiracy are likely to commit other crimes. Justice Breyer noted that the general consensus among legal commentators is that impossibility of success is not a defense to a conspiracy charge. A decision that government intervention terminates a conspiracy also would significantly interfere with police sting operations.

Do you agree with the judgment of the U.S. Supreme Court? See *United States v. Jimenez Recio,* 527 U.S. 270 (2003).

You can find more cases on the study site:
United States v. Handlin, www.sagepub.com/lippmanccl3e.

You Decide

7.5 The minors, Angel G., Jose E., Sergio G., Pedro G., and Diego G., appeal from their conviction in juvenile court of various offenses including conspiracy.

Twenty-year-old Daniel Garcia and his girlfriend, Sylvia Villa, were standing outside of Garcia's house. Several cars were parked in the driveway, including Garcia's red maroon rental car. Garcia was wearing a red shirt. The three defendants and several other juveniles were walking down the street after leaving school. Garcia testified that he saw "about 15 guys walking [in] the middle of the street." The defendants gathered in front of Garcia's house and began "throwing 3 signs." gesturing with three fingers extended, and saying "sur." Garcia interpreted these signals as indicating that the juveniles were affiliated with a Sureno gang.

Garcia told the boys to "get out of here." The boys "stood there for a moment" and ran to a neighboring yard, where they picked up rocks. Garcia testified that they "all started running toward me and starting throwing them." Garcia picked up a plastic broomstick and chased one boy across the street. He caught up to the boy and struck him on the back of the leg. Garcia returned to his driveway and was struck by a rock, fell and hit his head on a cement pole, and lost consciousness.

When Garcia was struck by the rock, one of the boys shouted, "I hit him. I hit him. I got him. I got him." Garcia spent a week in the hospital after undergoing surgery on his head. Doctors put three metal plates in his head and used about 30 staples to treat his injury. At the time of trial, Garcia still suffered from poor memory, a stutter, and pain in his head. He could not return to work.

Angel, Jose, Diego, and Sergio had admitted that they were members of the VML (Varrio Mexicanos Locos) gang, while Pedro had admitted he was a member of the Calle Ocho gang; both of these gangs are Suereno gangs. Surenos identify themselves with the color blue and with the number 13. Their main rivals are Nortenos, who identify themselves with the color red and the number 14. Are the defendants guilty of conspiracy? See *People v. Angele C.,* Case No. 21725 (Cal. Ct. App. 2002).

You can find the answer at www.sagepub.com/lippmanccl3e.

CRIME IN THE NEWS

Massachusetts passed one of the nation's most comprehensive bullying prevention laws in response to the suicide of two Massachusetts students: Phoebe Prince, age 15, from South Hadley, Massachusetts, and Carl Joseph Walker-Hoover, age 11, from Springfield.

The Massachusetts law requires teachers and school staff to report bullying to the school principal, provides for mandates training for teachers on bullying, and requires education for students on bullying behavior. Bullying is defined as:

> The severe or repeated use by one or more students of a written, verbal or electronic expression or a physical act or gesture, or any combination thereof, directed at another student that has the effect of: (i) causing physical or emotional harm to the other student or damage to the other student's property; (ii) placing the other student in a reasonable fear of harm to himself or of damage to his property; (iii) creating a hostile environment at school for the other student; (iv) infringing on the rights of the other student at school or (v) materially and substantially disrupting the education process or the orderly operation of a school.

Virtually every state has a bullying statute. Only a handful of these state statutes carry a criminal penalty. Most stipulate that schools adopt anti-bullying policies and provide for the punishment of students who engage in bullying. Subsection (v) of the Massachusetts law reaches conduct outside of school that affects the educational process.

Today traditional forms of bullying are being replaced by cyberbullying. The Pew Internet & American Life Project in a 2007 survey found that one-third of teens report that they have been harassed online, which is defined as "receiving threatening messages; having their private emails or text messages forwarded without consent; having an embarrassing picture posted without permission; or having rumors spread about them online." A 2006 Harris poll found that 43 percent of teens reported having experienced cyberbullying in the past year.

October is National Bullying Prevention Awareness Week. The key for victims of cyberbullying is to understand that they can take action against bullies and can reach out for support. Victims are urged not to return nasty texts or IMs, to make copies of messages, to establish filters to block bullies, and to discuss the matter with a caring adult.

In 2011, six students at South Hadley High School were criminally charged with bullying Phoebe Prince, a 15-year-old ninth grade student from Ireland. Phoebe had a history of emotional problems, of physically "cutting" herself, and of the using prescription anti-depressants. She was a new freshman at South Hadley and was the center of attention of a number of seniors who played on the football team. The girlfriends of these young men and their friends attacked Phoebe and called her an "Irish slut" on Facebook. Three of the students wrote "Irish b__ is a C__" next to her name on the library sign-up sheet; and one of the students yelled "whore" and "close your legs" and "I hate stupid sluts." As Phoebe left school that same day, three of the students encountered Phoebe; and one of the students again called her a whore.

As Phoebe walked home, one of the students drove by and again shouted "whore" and threw an empty can at her.

Phoebe texted a young man whom she was seeing, two hours before she killed herself, and wrote: "I cant do it anymore . . . im literally hme cryn, [the] my scar on my chest [from cutting] is potentially permanent, my bodies fukd up wht mre du they want frm me?" Phoebe at some point that evening hung herself in the stairwell with a black scarf that her sister had given her. The police later discovered a note that asked for forgiveness.

Phoebe's tormentors initially faced serious felony charges. At the request of Phoebe's family, five of the six defendants received probation and community service for harassing Phoebe. Statutory rape charges were dropped against the sixth defendant.

Was there evidence that the students who bullied Phoebe entered into a conspiratorial agreement? What accounts for the sudden identification of bullying as a serious problem that deserves criminal punishment? Is bullying really on the increase or is this "crisis" being manufactured by a generation of overly protective parents and school administrators? Can bullying be addressed through training programs? Is this a matter that schools should address, or should bullying be left to parents?

SOLICITATION

Solicitation is defined as commanding, hiring, or encouraging another person to commit a crime. The crime was largely unknown until the prosecution of the 1801 English case of *Rex v. Higgins,* in which Higgins was convicted of unsuccessfully soliciting a servant to steal his master's goods. A number of states do not have solicitation statutes and continue to apply the common law of solicitation. States with modern statutory schemes have adopted various approaches. Some punish solicitation of all crimes, and others limit solicitation to felonies, particular felonies, or certain classes of felonies. Solicitation generally results in a punishment slightly less severe or equivalent to the punishment that is usual for the crime solicited.[52]

We all read about the greedy spouse who approaches a contract killer to murder his or her partner in order to collect insurance money. The act of proposing the killing of a spouse with the intent that the murder be carried out constitutes solicitation. Solicitation is a form of accomplice liability, and in the event that the spouse is murdered, both the greedy spouse and contract killer are guilty of homicide. In the event the assassination proves unsuccessful, both the greedy spouse and inaccurate assassin are guilty of attempted murder. An agreement between the two that leads to an overt act that is not carried out results in liability for a conspiracy. The contract killer, of course, may refuse to become involved with the greedy spouse. Nevertheless, the spouse is guilty of solicitation; the crime of solicitation is complete when a spouse attempts to hire the killer.

Public Policy

Solicitation remains a controversial crime; this accounts for the fact that some states have not yet enacted solicitation statutes. The thinking is that there is no necessity for the crime of solicitation. Solicitation, it is argued, is not a threat to society until steps are taken to carry out the scheme. At this point, the agreement can be punished as a conspiracy. A solicitor depends on the efforts of others, and simply approaching another person to commit a crime does not present a social danger. There is also the risk that individuals will be convicted based on a false accusation or as a result of a casual remark. Lastly, punishing individuals for solicitation interferes with freedom of speech. As observed by a nineteenth-century court, holding every individual who "nods or winks" to a married person on the sidewalk "indictable for soliciting to adultery . . . would be a dangerous and difficult rule of criminal law to administer."[53]

On the other hand, there are convincing reasons for punishing solicitation:

- *Cooperation Among Criminals.* Individuals typically encourage and support one another, which creates a strong likelihood that the crime will be committed.
- *Social Danger.* An individual who is sufficiently motivated to enlist the efforts of a skilled professional criminal clearly poses a continuing social danger.
- *Intervention.* Solicitation permits the police to intervene before a crime is fully implemented. The police should not be placed in the position of having to wait for an offense to occur before arresting individuals intent on committing a crime.

States typically protect individuals against wrongful convictions by requiring corroboration or additional evidence to support a charge of solicitation. This might involve an e-mail, a voice recording, or witnesses who overheard the conversation. As for the First Amendment, society possesses a substantial interest in prohibiting acts such as the solicitation of adolescents by adults for sexual activity on computer chat rooms that more than justifies any possible interference with individual self-expression.[54]

The Crime of Solicitation

Solicitation involves a written or spoken statement in which an individual intentionally advises, requests, counsels, commands, hires, encourages, or incites another person to commit a crime with the purpose that the other person commit the crime. You are not liable for a comment that is intended as a joke or uttered out of momentary frustration.

The *mens rea* of solicitation requires a specific intent or purpose that another individual commit a crime. You would not be liable in the event that you humorously advise a friend to "blow up" the expensive car of a neighbor who regularly parks in your friend's parking space. On the other hand, you might harbor a long-standing grudge against the neighbor and genuinely intend to persuade your friend to destroy the automobile.

The *actus reus* of solicitation requires an effort to get another person to commit a crime. A variety of terms are used to describe the required act, including *command, encourage,* and *request.* The crime of solicitation occurs the moment an individual urges, asks, or encourages another to commit a crime with the requisite intent. The individual is guilty of solicitation even in those instances in which the other person rejects the offer or accepts the offer and does not commit the crime.

There are three important points on *actus reus:*

1. The crime is complete the moment the statement requesting another to commit a crime is made. This is the case despite the fact that an additional step, such as a phone call or the payment of money, is required to trigger the crime.

2. A statement justifying or hoping that the neighbor's automobile is damaged is not sufficient. There must be an effort to get another person to commit the crime. A solicitation may be direct or indirect. For instance, in cases involving the enticement of children into sexual activity, courts will consider a defendant's use of suggestive and seductive remarks and materials.

3. The Model Penal Code provides that an individual is guilty of solicitation even in instances in which a letter asking others to commit a crime is intercepted by prison authorities and does not reach gang members outside of prison. States that do not follow the Model Penal Code require that the solicitation actually is received by the intended recipient.

For an international perspective on this topic, visit the study site.

In reading the next case, *State v. Cotton,* compare and contrast the New Mexico statute with the Model Code and ask yourself how *Cotton* would be decided under the Model Penal Code. Which of these two approaches to crime of solicitation do you think is better as a matter of public policy?

Model Penal Code

Section 5.02. Criminal Solicitation

(1) . . . A person is guilty of solicitation to commit a crime if with the purpose of promoting or facilitating its commission he commands, encourages or requests another person to engage in specific conduct that would constitute such crime or an attempt to commit such crime or would establish his complicity in its commission or attempted commission.

(2) . . . It is immaterial . . . that the actor fails to communicate with the person he solicits to commit a crime if his conduct was designed to effect such communication.

(3) . . . It is an affirmative defense that the actor, after soliciting another person to commit a crime, persuaded him not to do so or otherwise prevented the commission of the crime, under circumstances manifesting a complete and voluntary renunciation of his criminal purpose.

Analysis

1. Solicitation for a felony or misdemeanor is a crime. This also includes solicitation for an attempt and aiding and abetting.

2. An individual is guilty of solicitation even in those instances that the solicitation is not communicated.

3. The defense of renunciation is recognized in those instances that the other person is persuaded not to commit or prevented from committing the offense.

The Legal Equation

Solicitation **=** Intent or purpose for another person to commit a crime (felony)

+ words, written statements, or actions inviting, requesting, or urging another to commit a crime.

Was Cotton guilty of criminal solicitation despite the fact that the solicitation never was communicated to his wife?

State v. Cotton, 790 P.2d 1050 (N.M. Ct. App. 1990). Opinion by: Donnelly, J.

Facts

Defendant appeals his convictions of two counts of criminal solicitation.

In 1986, defendant, together with his wife Gail, five children, and a fourteen-year-old stepdaughter, moved to New Mexico. A few months later, defendant's wife, children, and stepdaughter returned to Indiana. Shortly thereafter, defendant's stepdaughter moved back to New Mexico to reside with him. In 1987, the New Mexico Department of Human Services investigated allegations of misconduct involving defendant and his stepdaughter. Subsequently the district court issued an order awarding legal and physical custody of the stepdaughter to the Department, and she was placed in a residential treatment facility in Albuquerque.

In May 1987, defendant was arrested and charged with multiple counts of criminal sexual penetration of a minor and criminal sexual contact with a minor. While in the Eddy County Jail awaiting trial on those charges, defendant discussed with his cellmate, James Dobbs, and with Danny Ryan, another inmate, his desire to persuade his stepdaughter not to testify against him. During his incarceration, defendant wrote numerous letters to his wife; in several of his letters he discussed his strategy for defending against the pending criminal charges.

On September 23, 1987, defendant addressed a letter to his wife. In that letter he requested that she assist him in defending against the pending criminal charges by persuading his stepdaughter not to testify at his trial. The letter also urged his wife to contact the stepdaughter and influence her to return to Indiana or give the stepdaughter money to leave the state so that she would be unavailable to testify. After writing this letter, defendant gave it to Dobbs and asked him to obtain a stamp for it so that it could be mailed later. Unknown to defendant, Dobbs removed the letter from the envelope, replaced it with a blank sheet of paper, and returned the sealed stamped envelope to him. Dobbs gave the original letter written by defendant to law enforcement authorities, and it is undisputed that defendant's original letter was never in fact mailed nor received by defendant's wife.

On September 24 and 26, 1987, defendant composed another letter to his wife. He began the letter on September 24 and continued it on September 26, 1987. In this letter defendant wrote that he had revised his plans and that this letter superseded his previous two letters. The letter stated that he was arranging to be released on bond; that his wife should forget about his stepdaughter for a while and not come to New Mexico; that defendant would request that the court permit him to return to Indiana to obtain employment; that his wife should try to arrange for his stepdaughter to visit her in Indiana for Christmas; and that his wife should try to talk the stepdaughter out of testifying or to talk her into testifying favorably for defendant. Defendant also said in the letter that his wife should "warn" his stepdaughter that if she did testify for the State, "it won't be nice and she'll make [New Mexico] news," and that, if the stepdaughter was not available to testify, the prosecutor would have to drop the charges against defendant.

Defendant secured his release on bail on September 28, 1987, but approximately twenty-four hours later was rearrested on charges of criminal solicitation and conspiracy. At the time defendant was rearrested, law enforcement officers discovered, and seized from defendant's car, two personal calendars and other documents written by defendant. It is also undisputed that the second letter was never mailed to defendant's wife.

Following a jury trial, defendant was convicted on two counts of criminal solicitation. The criminal solicitations were alleged to have occurred on or about September 23, 1987. Count 1 of the amended criminal information alleged that defendant committed the offense of criminal solicitation by soliciting another person "to engage in conduct constituting a felony to-wit: Bribery or Intimidation of a Witness (contrary to Sec. 30-24-3, NMSA 1978)." Count 2 alleged that defendant committed the offense of criminal solicitation by soliciting another "to engage in conduct constituting a felony, to-wit: Custodial Interference (contrary to Sec. 30-4-4, NMSA 1978)."

The offense of criminal solicitation, as provided in New Mexico Statutes Annotated (NMSA) 1978 section 30-28-3, is defined in applicable part as follows:

A. Except as to bona fide acts of persons authorized by law to investigate and detect the commission of offenses by others, a person is guilty of criminal solicitation if, with the intent that another person engage in conduct constituting a felony, he solicits, commands, requests, induces, employs or otherwise attempts to promote or facilitate another person to engage in conduct constituting a felony within or without the state.

Issue

Defendant contends that the record fails to contain the requisite evidence to support the charges of criminal solicitation against him, because defendant's wife, the intended solicitee, never received the two letters. In reviewing this position, the focus of our inquiry necessarily turns on whether or not the record contains proper evidence sufficient to establish each element of the alleged offenses of criminal solicitation beyond a reasonable doubt.

The State's brief states that "neither of these letters actually reached Mrs. Cotton, but circumstantial evidence indicates that other similar letters did reach her during this period." The State also argues that under the express language of section 30-28-3(A), where defendant is shown to have the specific intent to commit such offense and "otherwise attempts" its commission, the offense of criminal solicitation is complete. The State reasons that even in the absence of evidence

indicating that the solicitations were actually communicated to or received by the solicitee, under our statute proof of defendant's acts of writing the letters and attempts to mail or forward them, together with proof of his specific intent to solicit the commission of a felony, constitutes sufficient proof to sustain a charge of criminal solicitation. We disagree.

Reasoning

The offense of criminal solicitation, as defined in section 30-28-3 by our legislature, adopts, in part, language defining the crime of solicitation as set out in the Model Penal Code promulgated by the American Law Institute. "As enacted by our legislature, however, Section 30-28-3 significantly omits one section of the Model Penal Code, Section 5.02(2), which pertains to the effect of an uncommunicated criminal solicitation. Under the Model Penal Code, a person is guilty of "solicitation to commit a crime" when

with the purpose of promoting or facilitating its commission he commands, encourages or requests another person to engage in specific conduct which would constitute such crime or an attempt to commit such crime or which would establish his complicity in its commission or attempted commission.

It is immaterial "that the actor fails to communicate with the person he solicits to commit a crime if his conduct was designed to effect such communication."

However, as enacted by our legislature, section 30-28-3 sets out the offense of criminal solicitation in a manner that differs in several material respects from the proposed draft of the Model Penal Code. Among other things, section 30-28-3 specifically omits that portion of the Model Penal Code subsection declaring that an uncommunicated solicitation to commit a crime may constitute the offense of criminal solicitation. This omission, we conclude, indicates an implicit legislative intent that the offense of solicitation requires some form of actual communication from the defendant to either an intermediary or the person intended to be solicited, indicating the subject matter of the solicitation.

Holding

The mere writing and sending of letters by defendant to his wife, without proof of solicitation of a specific felony and proof of defendant's intent to induce another to commit such crime, is insufficient to establish proof of criminal solicitation.

The State contends that under the language of section 30-28-3, where proof is presented that defendant has the requisite intent and has "otherwise attempt[ed] to promote or facilitate another person to engage in conduct constituting a felony within or without the state," the offense of solicitation is complete. This contention must fail, because section 30-28-3 is silent as to any legislative intent to declare that uncommunicated solicitations shall constitute a criminal offense.

Commission of criminal solicitation does not require, however, that defendant directly solicit another; the solicitation may be perpetrated through an intermediary. Thus if A solicits B in turn to solicit C to commit a felony, A would be liable even where he did not directly contact C, because A's solicitation of B itself involves the commission of the offense. Where the intended solicitation is not in fact communicated to an intended intermediary or to the person sought to be solicited, the offense of solicitation is incomplete; although such evidence may support, in proper cases, a charge of attempted criminal solicitation. Defendant's convictions for solicitation are reversed, and the case is remanded with instructions to set aside the convictions for criminal solicitation.

Questions for Discussion

1. What is the holding of the New Mexico court in *Cotton?*

2. Explain why the result likely would be different under the Model Penal Code.

3. As a legislator, would you favor the approach of the Model Penal Code or the State of New Mexico?

4. Could Cotton be held liable for attempted solicitation?

5. Do we need a crime of solicitation?

You Decide

7.6 Casandra informed Lou Tong Saephanh, a California inmate, that she was pregnant with his child. Saephanh was excited to be a father and they talked about the baby every week. Roughly six months later Saephanh wrote a letter to fellow gang member Cheng Saechao: "By the way loc, could you & the homies do me a big favor & take care that white bitch, Cassie for me. ha, ha, ha!! Cuzz, it's too late to have abortion so I think a miss carrage would do just fine. I aint fista pay child sport for this bull-shit loc. You think you can get the homies or home girls do that for me before she have the baby on Aug. '98." Vicki Lawrence, a correctional officer opened and read the letter. The letter was embargoed by prison authorities and was not sent to the addressee. Saephanh later explained that if Cassandra did not let him be a part of the baby's life, he wanted to "get rid of the baby" and did not want to pay child support. Is Saephanh guilty of solicitation? The California solicitation statute reads that "Every person who, with the intent that the crime be committed, solicits another to commit or join in the commission of murder shall be punished by imprisonment." See *People v. Saephanh*, 94 Cal. Rptr. 2d 910 (Cal. Ct. App. 2000).

You can find the answer at www.sagepub.com/lippmanccl3e.

CHAPTER SUMMARY

Attempt, solicitation, and *conspiracy* are inchoate crimes or offenses that punish the beginning steps toward a crime. All require a *mens rea* involving a specific intent or purpose to achieve a crime as well as an *actus reus* that entails an affirmative act toward the commission of a crime. Each of these offenses is subject to the same or a lesser penalty than the crime that is the criminal objective.

An attempt involves three elements:

- an intent or purpose to commit the crime,
- an act toward the commission of the crime, and
- a failure to commit the crime.

A complete (but imperfect) attempt occurs when a defendant takes every act required to complete the offense and fails to succeed in committing the crime. An incomplete attempt arises when an individual abandons or is prevented from completing an attempt.

An individual must possess the intent to achieve a criminal objective. There are two approaches to *actus reus,* the objective and the subjective. The objective centers on the proximity of an individual's acts to the

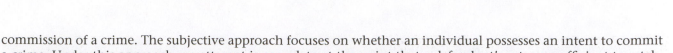

commission of a crime. The subjective approach focuses on whether an individual possesses an intent to commit a crime. Under this approach, an attempt is complete at the point that a defendant's acts are sufficient to establish a criminal intent. Objective approaches include the common law last step analysis as well as the physical proximity test. The unequivocality test is an example of the subjective approach.

The Model Penal Code adopts a substantial step test. This requires that an act must strongly support an individual's criminal purpose. The approach of the Model Penal Code extends attempt to acts that might be considered mere preparation under the objective approach. Among the acts that constitute an attempt under the substantial step test and that are considered "strongly corroborative of an actor's criminal purpose" are

- lying in wait,
- enticing of a victim to go to the place contemplated for the commission of a crime,
- surveying a site contemplated for the commission of a crime,
- unlawful entry of a structure or vehicle in which a crime is contemplated,
- possession of materials to be employed in the commission of a crime, and
- soliciting an individual to engage in conduct constituting an element of a crime.

A factual impossibility does not constitute a defense to an attempt to commit a crime. This is based on the reasoning that the offender has demonstrated a dangerous criminal intent and a determination to commit an offense. The factual circumstance that prevents an individual from actually completing the offense is referred to in some state statutes as an extraneous factor. This should be distinguished from a legal impossibility that is recognized as a defense. Legal impossibility arises in those instances in which an individual wrongly believes that he or she is violating the law. Inherent impossibility arises where an individual undertakes an act that could not possibly result in a crime.

An individual may avoid criminal liability by abandoning a criminal attempt under circumstances manifesting a complete and voluntary renunciation of a criminal purpose. Individuals abandoning a criminal purpose based on the intervention of outside or extraneous factors remain criminally liable.

Conspiracy comprises an agreement between two or more persons to commit a criminal act. Most modern state statutes require an affirmative act in furtherance of this criminal purpose. The common law crime of conspiracy did not merge into the completed criminal act. Today an individual may be convicted of both the substantive offense that is the object of the conspiracy and of the conspiracy itself.

The centerpiece of a charge of conspiracy is an agreement. There is rarely proof of a formal agreement, and an agreement typically must be established by examining the relationship, conduct, and circumstances of the parties. The overt act requirement is satisfied by even an insignificant act in furtherance of a conspiracy. The *mens rea* of conspiracy is the intent or purpose that the object of the agreement is accomplished.

The plurality requirement provides that a conviction for conspiracy requires that at least two persons possess both the intent to agree and the intent to achieve the crime that is the object of the conspiracy. This joint or bilateral approach to conspiracy is distinguished from the Model Penal Code's unilateral approach, which examines whether a single individual agreed to enter into a conspiracy, rather than focusing on whether two or more persons entered into an agreement.

Most complex conspiracies can be categorized as either a wheel or a chain conspiracy. A chain conspiracy entails communication and cooperation by individuals linked together in a vertical relationship to achieve a criminal objective. A wheel conspiracy involves a single person or group that serves as a hub that provides a common core connecting various independent individuals or spokes.

Conspirators are liable for all criminal offenses taken in furtherance of a conspiracy. As a result, defendants will typically attempt to establish that there were multiple conspiracies rather than a single conspiracy. Wharton's rule states that individuals involved in a crime that requires the cooperative action of two persons cannot constitute a conspiracy. The Gebardi rule provides that an individual who is in a class of persons who are excluded from criminal liability under a statute cannot be considered a conspirator.

A conspiracy charge provides prosecutors with various advantages, such as joining the conspirators in a single trial and bringing the charges in any jurisdiction in which an agreement or act in furtherance of the conspiracy is committed.

Solicitation involves a written or spoken statement in which an individual intentionally advises, requests, counsels, commands, hires, encourages, or incites another person to commit a crime with the purpose that the other individual commit the crime. A solicitation is complete the moment the statement requesting another to commit a crime is made. The solicitation need not be actually communicated.

CHAPTER REVIEW QUESTIONS

1. What are the *mens rea* and *actus reus* of inchoate crimes?

2. Distinguish the three categories of inchoate crimes.

3. Provide an example of each crime.

4. Compare the subjective and objective approaches to criminal attempts.

5. How does the Model Penal Code substantial step test differ from the test of physical proximity to attempts? What types of acts satisfy the substantial step test?

6. Discuss and distinguish between legal and factual impossibility.

7. Why is there a defense of abandonment for attempts? What are the legal elements of this defense?

8. What are the reasons for punishing conspiracy?

9. Discuss the *mens rea* and *actus reus* of conspiracy.

10. Why do some states require an overt act for a conspiracy?

11. Is there a difference between the bilateral and unilateral approaches to a conspiratorial agreement?

12. Distinguish between the wheel and chain approaches to conspiracy. Explain why defendants may argue that there are multiple conspiracies rather than a single conspiracy.

13. How does a charge of conspiracy assist a prosecutor in convicting a defendant?

14. Why did Congress adopt the RICO statute?

15. What are the *mens rea* and *actus reus* of solicitation?

16. At what point is the crime of solicitation complete? Is a solicitation required to reach the individual to whom it is directed?

17. How does society benefit by punishing inchoate crimes? Would society suffer in the event that these offenses did not exist?

LEGAL TERMINOLOGY

abandonment

attempt

bilateral

chain conspiracy

complete attempt

conspiracy

criminal attempt

extraneous factor

factual impossibility

Gebardi rule

inchoate crimes

incomplete attempt

inherent impossibility

last step approach

legal impossibility

objective approach to criminal attempt

overt act

physical proximity test

plurality requirement

preparation

res ipsa loquitur

solicitation

subjective approach to criminal intent

substantial step test

unequivocality test

unilateral

Wharton's rule

wheel conspiracy

CRIMINAL LAW ON THE WEB

Log on to the Web-based student study site at www.sagepub.com/lippmanccl3e to assist you in completing the Criminal Law on the Web exercises, as well as for additional features such as podcasts, Web quizzes, and video links.

1. Explore the National Youth Gang survey of 2,500 law enforcement agencies and learn about gangs.

2. Read an account of the death of Phoebe Prince written by journalist Emily Bazelon.

BIBLIOGRAPHY

American Law Institute, *Model Penal Code and Commentaries,* vol. 2, pt. 1 (Philadelphia: American Law Institute, 1985), pp. 293–475. A comprehensive discussion of the major issues in inchoate offenses and suggested reforms.

Joshua Dressler, *Understanding Criminal Law,* 3rd ed. (New York: Lexis, 2001), pp. 373–458. A good introduction to the law of inchoate crimes.

George P. Fletcher, *Rethinking Criminal Law* (New York: Oxford University Press, 2000), pp. 239–328. An in-depth exploration of the philosophical issues in the law of attempt. Fletcher also provides a brief discussion of whether the crime of conspiracy is consistent with the philosophical requirements of Anglo American criminal law.

Wayne R. LaFave, *Criminal Law,* 3rd ed. (St. Paul, MN: West Publishing, 2000), pp. 524–650. A comprehensive discussion of the law of inchoate crimes.

Rollin M. Perkins and Ronald N. Boyce, *Criminal Law,* 3rd ed. (Mineola, NY: Foundation Press, 1982), pp. 611–658, 680–714. A review of the historical development and background of inchoate crimes.

Richard G. Singer and John Q. La Fond, *Criminal Law Examples and Explanations,* 2nd ed. (New York: Aspen, 2001), pp. 131–197, 239–327. A clear and concise summary with a helpful introduction to the Model Penal Code. The authors also provide useful review questions.

JUSTIFICATIONS

Did the defendant have the right to kill her husband in self-defense?

The trial was replete with testimony of forced prostitution, beatings, and threats on the defendant's life. The defendant testified that she believed the decedent would kill her, and the evidence showed that on the occasions when she had made an effort to get away from Norman, he had come after her and beat her. Indeed, within twenty-four hours prior to the shooting, defendant had attempted to escape by trying to take her own life and throughout the day on 12 June 1985 had been subjected to beatings and other physical abuse, verbal abuse, and threats on her life up to the time when decedent went to sleep. Experts testified that in their opinion, defendant believed killing the victim was necessary to avoid being killed. . . .

Learning Objectives

1. Understand the presumption of innocence.
2. Distinguish between justifications and excuses.
3. Differentiate between affirmative defenses and mitigating circumstances.
4. List and explain the elements of self-defense.
5. State the two tests for the defense of others.
6. Know the law on defense of the home.
7. Explain the fleeing felon rule.
8. Distinguish between the American and the European rules for resistance to an unlawful arrest.
9. Summarize the elements of the necessity defense.
10. List the three situations in which the law recognizes consent as a defense to criminal conduct.

INTRODUCTION

The Prosecutor's Burden

The American legal system is based on the **presumption of innocence**. A defendant may not be compelled to testify against himself or herself, and the prosecution is required as a matter of the due process of law to establish every element of a crime beyond a reasonable doubt to establish a defendant's guilt. This heavy prosecutorial burden also reflects the fact that a criminal conviction carries severe consequences and individuals should not be lightly deprived of their liberty. Insisting on a high standard of guilt assures the public that innocents are not being falsely convicted and that individuals need not fear that they will suddenly be snatched off the streets and falsely convicted and incarcerated.[1]

The prosecutor presents his or her witnesses in the **case-in-chief**. These witnesses are then subject to cross-examination by the defense attorney. The defense also has the right to introduce evidence challenging the prosecution's case during the **rebuttal** stage at trial. A defendant, for instance, may raise doubts about whether the prosecution has established that the defendant committed the crime beyond a reasonable doubt by presenting alibi witnesses.

A defendant is to be acquitted if the prosecution fails to establish each element of the offense beyond a reasonable doubt. Judges have been reluctant to reduce the beyond a reasonable doubt standard to a mathematical formula and stress that a "high level of probability"[2] is required and that jurors must reach a "state of near certitude" of guilt.[3] The classic definition of reasonable doubt provides that the evidence "leaves the minds of jurors in that condition that they cannot say they feel an abiding conviction, to a moral certainty of the truth of the charge."[4]

A defendant is entitled to file a motion for judgment of acquittal at the close of the prosecution's case or prior to the submission of the case to the jury. This motion will be granted if the judge determines that the evidence is unable to support any verdict other than acquittal, viewing the evidence as favorably as possible for the prosecution. The judge, in the alternative, may adhere to the standard procedure of submitting the case to the jury following the close of the evidence and instructing the jurors to acquit if they have a reasonable doubt concerning one or more elements of the offense.[5]

Affirmative Defenses

In addition to attempting to demonstrate that the prosecution's case suffers from a failure of proof beyond a reasonable doubt, defendants may present **affirmative defenses**, or defenses in which the defendant typically possesses the **burden of production** as well as the **burden of persuasion**.

Justifications and **excuses** are both affirmative defenses. If the defendant raises an affirmative defense, the defendant possesses the burden of producing "some evidence in support of his defense." In most cases, the defendant then also has the burden of persuasion by a preponderance of the evidence, which is a balance of probabilities, or slightly more than fifty percent. In most jurisdictions, the prosecution retains the burden of persuasion and is responsible for negating the defense by a reasonable doubt.[6]

Assigning the burden of production to the defendant is based on the fact that the prosecution cannot be expected to anticipate and rebut every possible defense that might be raised by a defendant. The burden of rebutting every conceivable defense ranging from insanity and intoxication to self-defense would be overwhelmingly time-consuming and inefficient. Thus, it makes sense to assign responsibility for raising a defense to the defendant. The U.S. Supreme Court has issued a series of rather technical judgments on the allocation of the burden of persuasion. In the last analysis, states are fairly free to place the burden of persuasion on either the defense or prosecution. As noted, in most instances, the prosecution has the burden of persuading the jury beyond a reasonable doubt to reject the defense.

There are two types of affirmative defenses that may result in acquittal:

- *Justifications.* These are defenses to otherwise criminal acts that society approves and encourages under the circumstances. An example is self-defense.
- *Excuses.* These are defenses to acts that deserve condemnation, but for which the defendant is not held criminally liable because of a personal disability such as infancy or insanity.

Professors Singer and La Fond illustrate the difference between these concepts by noting that justification involves illegally parking in front of a hospital in an effort to rush a sick infant into the emergency room, and an excuse entails illegally parking in response to the delusional demand of "Martian invaders."[7] In the words of Professor George Fletcher, "Justification speaks to the rightness of the act; an excuse, to whether the actor is [mentally] accountable for a concededly wrongful act."[8]

In the common law, there were important consequences resulting from a successful plea of justification or excuse. A justification resulted in an acquittal, whereas an excuse provided a defendant with the opportunity to request that the king exempt him or her from the death penalty. Eventually there came to be little practical difference between being acquitted by reason of a justification or an excuse.[9]

Scholars continue to point to differences between categorizing an act as justified as opposed to excused, but these have little practical significance for most defendants.[10] You nevertheless

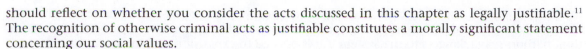

should reflect on whether you consider the acts discussed in this chapter as legally justifiable.[11] The recognition of otherwise criminal acts as justifiable constitutes a morally significant statement concerning our social values.

There are various theories for the defense of justification, none of which fully account for each and every justification defense.[12]

- *Moral Interest*. An individual's act is justified based on the protection of an important moral interest. An example is self-defense and the preservation of an individual's right to life.
- *Superior Interest*. The interests being preserved outweigh the interests of the person who is harmed. The necessity defense authorizes an individual to break the law to preserve a more compelling value. An example might be the captain of a ship in a storm who throws luggage overboard to lighten the load and preserve the lives of those on board.
- *Public Benefit*. An individual's act is justified on the grounds that it is undertaken in service of the public good. This includes a law enforcement officer's use of physical force against a fleeing felon.
- *Moral Forfeiture*. An individual perpetrating a crime has lost the right to claim legal protection. This explains why a dangerous aggressor may justifiably be killed in self-defense.

A defendant who establishes a *perfect defense* is able to satisfy each and every element of a justification defense and is acquitted. An *imperfect defense* arises in those instances in which the requirements of the defense are not fully satisfied. For instance, a defendant may use excessive force in self-defense or possess a genuine, but unreasonable, belief in the need to act in self-defense. A defendant's liability in these cases is typically reduced, for example, in the case of a homicide to manslaughter and to a lower level of guilt in the case of other offenses.[13]

MITIGATING CIRCUMSTANCES

Evidence that is not relevant for justification or excuse may still be relied on during the sentencing stage as a mitigating circumstance that may reduce a defendant's punishment. The jury in death penalty cases is specifically required to consider mitigating as well as aggravating circumstances in determining whether the defendant should be subject to capital punishment or receive a life sentence. An example is *State v. Moore*, in which a nineteen-year-old defendant was convicted of murder during an aggravated robbery and kidnapping. The Ohio Supreme Court affirmed the defendant's death sentence and ruled that the defendant's youth, lack of criminal record, remorse, and religious conversion only modestly mitigated the offense and were clearly outweighed by the aggravating circumstances of the offense. The Ohio Supreme Court also ruled that the defendant's alcohol and drug addiction were not mitigating.[14]

A defendant's *"good motive"* in committing a mercy killing of a severely sick family member may not be considered in determining guilt or innocence and is only considered at sentencing. The law is concerned with *what* crime an individual committed, not *why* he or she committed the crime. During the Vietnam conflict, the Fourth Circuit Court of Appeals ruled that a defendant's **good motive defense** of opposition to what he viewed as a morally reprehensible war could not justify his destruction of draft records. The court of appeals explained that absolving the defendant from guilt based on his "moral certainty" that the war in Vietnam is wrong also would require the acquittal of individuals who might commit breaches of the law to demonstrate their sincere belief in the war. The appellate court stressed that in both cases the defendants "must answer for their acts" to avoid a breakdown in law and order.[15]

At times, lawyers will attempt to indirectly introduce motive by arguing for **jury nullification**. The jury historically has possessed the authority to disregard the law and to acquit sympathetic defendants. This power is based on the jury's historical role as a check on overzealous prosecutors who bring charges that are contrary to prevailing social values. Examples include the acquittal of newspaper publisher Peter Zenger by an eighteenth-century American colonial jury and the acquittal of individuals who assisted fugitive slaves during the nineteenth century. Appellate courts, however, have consistently ruled that trial judges are not obligated to instruct jurors that they possess the power of nullification and that jurors are to be instructed that they are required to strictly apply the law in determining a defendant's guilt. The District of Columbia Circuit Court of Appeals observed that "what is tolerable or even desirable as an informal, self-initiated exception, harbors grave dangers to the system if it is opened to expansion and intensification through incorporation in the judge's instructions." Do you agree?[16]

SELF-DEFENSE

It is commonly observed that the United States is a "government of law rather than men and women." This means that guilt and punishment are to be determined in accordance with fair and objective legal procedures in the judicial suites rather than by brute force in the streets. Accordingly, the law generally discourages individuals from "taking the law into their own hands." This type of "vigilante justice" risks anarchy and mob violence. One sorry example is the lynching of thousands of African Americans by the Ku Klux Kan following the Civil War.

For a deeper look at this topic, visit the study site.

Self-defense is the most obvious exception to this rule and is recognized as a defense in all fifty states. Why does the law concede that an individual may use physical force in self-defense? One federal court judge noted the practical consideration that absent this defense, the innocent victim of a violent attack would be placed in the unacceptable position of choosing between "almost certain death" at the hands of his or her attacker or a "trial and conviction of murder later." More fundamentally, eighteenth-century English jurist William Blackstone wrote that it was "lawful" for an individual who is attacked to "repel force by force." According to Blackstone, this was a recognition of the natural impulse and right of individuals to defend themselves. A failure to recognize this right would inevitably lead to a disregard of the law.[17]

The Central Components of Self-Defense

The common law recognizes that an individual is justified in employing force in self-defense. This may involve deadly or nondeadly force, depending on the nature of the threat. There are a number of points to keep in mind:

- *Reasonable Belief.* An individual must possess a reasonable belief that force is required to defend himself or herself. In other words, the individual must believe and a reasonable person must believe that force is required in self-defense.
- *Necessity.* The defender must reasonably believe that force is required to prevent the imminent and unlawful infliction of death or serious bodily harm.
- *Proportionality.* The force employed must not be excessive or more than is required under the circumstances.
- *Retreat.* A defendant may not resort to deadly force if he or she can safely **retreat**. This generally is not required when the attack occurs in the home or workplace, or if the attacker uses deadly force.
- *Aggressor.* An **aggressor**, or individual who unlawfully initiates force, generally is not entitled to self-defense. An aggressor may claim self-defense only in those instances that an aggressor who is not employing deadly force is himself or herself confronted by deadly force. Some courts require that under these circumstances, the aggressor withdraws from the conflict if at all possible before enjoying the right of self-defense. There are courts willing to recognize that even an aggressor who employs deadly force may regain the right of self-defense by withdrawing following the initial attack. The other party then assumes the role of the aggressor.
- *Mistake.* An individual who is mistaken concerning the necessity for self-defense may rely on the defense so long as his or her belief is reasonable.
- *Imperfect Self-Defense.* An individual who honestly, but unreasonably, believes that he or she confronts a situation calling for self-defense and intentionally kills is held liable in many states for an intentional killing. Other states, however, follow the doctrine of **imperfect self-defense**. This provides that although the defendant may not be acquitted, fairness dictates that he or she should be held liable only for the less serious crime of manslaughter.

The next case in this chapter, *State v. Marshall,* provides a review of the basic principles and application of the law of self-defense and illustrates the core components and challenges of self-defense.

Model Penal Code

Section 3.04. Use of Force in Self-Protection

(1) [T]he use of force upon or toward another person is justifiable when the actor believes that such force is immediately necessary for the purpose of protecting himself against the use of unlawful force by such other person on the present occasion.

(2) Limitations on Justifying Necessity for Use of Force.

   (a) . . .

   (b) The use of deadly force is not justifiable under this Section unless the actor believes that such force is necessary to protect himself against death, serious bodily injury, kidnapping or sexual intercourse compelled by force or threat; nor is it justifiable if:

     (i) the actor, with the purpose of causing death or serious bodily injury, provoked the use of force against himself in the same encounter; or

     (ii) the actor knows that he can avoid the necessity of using such force with complete safety by retreating or by surrendering possession of a thing to a person asserting a claim of right thereto or by complying with a demand that he abstain from any action that he has no duty to take. . . .

Analysis

The Model Penal Code makes some significant modifications to the standard approach to self-defense that will be discussed later in the text. The basic formulation affirms that the use of force in self-protection is justified in those instances in which an individual "employs it in the belief that it is immediately necessary for the purpose of protecting himself against the other's use of unlawful force on the present occasion." The code provides that an aggressor who uses deadly force may "break off the struggle" and retreat and regain the privilege of self-defense against the other party.

The Legal Equation

Self-defense = Reasonable belief

+ immediately necessary

+ to employ proportionate force

+ to protect oneself against unlawful force.

Did the defendant confront an imminent attack?

State v. Marshall, *179 S.E. 427 (N.C. 1935). Opinion by: Stacy, C.J.*

Facts

The homicide occurred in the defendant's filling station. The deceased had been drinking, and, with imbecilic courtesy, undertook to engage the defendant's wife in a whispered conversation. This was repulsed and the deceased ordered to leave the building.

The defendant testified: "I ordered him out two or three times; he would not leave; and the next thing he said you G—d—s—o—b—and b—; pulled off his hat and slammed it on the counter with his right hand and said you haven't got the guts to shoot me, and that he would die like a man; and when he reached to pick up the hammer in the other hand,

I fired. . . . I fired because I thought he was going to kill me with the hammer, or hit me with the hammer and kill me, maybe. He cursed me; I got the pistol and ordered him out, . . . I was scared of the man. No, I was not mad. . . . When I shot him there was the width of the counter between us. We were between 2 1/2 and 3 1/2 feet apart. . . . I did not shoot to kill. . . . I saw him when he grabbed the hammer. I did not say he picked it up, but he grabbed it; he raised the hammer up when he fell back, but he did not have it in a striking position; he was reaching and he grabbed the hammer. I do not say he raised it up in a striking position before I shot. . . . I say he did not draw the hammer back to strike." Defendant's wife testified: "When Rex shot I saw him (deceased) grab for the hammer."

Issue

It appears, therefore, from the defendant's own testimony that he was not in imminent danger of death or great bodily harm when he shot the deceased; nor did he apprehend that he was in such danger. "I did not shoot to kill" is his statement, and it appears from the record that the deceased did not reach for the hammer until after he was shot. The clear inference is that the defendant used excessive force.

Reasoning

The right to kill in self-defense or in defense of one's family or habitation rests upon necessity, real or apparent . . . [O]ne may kill in defense of himself, or his family, when necessary to prevent death or great bodily harm. . . . [O]ne may kill in defense of himself, or his family, when not actually necessary to prevent death or great bodily harm, if he believes it to be necessary and has a reasonable ground for the belief. . . . [T]he jury and not the party charged is to determine the reasonableness of the belief or apprehension upon which he acted. It is also established by the decisions that in the exercise of the right of self-defense, more force must not be used than is reasonably necessary under the circumstances, and if excessive force or unnecessary violence be employed, the party charged will be guilty of manslaughter, at least. . . .

Holding

The defendant's conviction for manslaughter is affirmed.

Questions for Discussion

1. Was Marshall motivated by self-defense or by a desire to punish the deceased? Did Marshall have a reasonable basis for believing that the deceased planned to assault him? At what point would the defendant be justified in using his firearm? Should he wait until the trespasser was clearly ready to strike?

2. What force was Marshall justified in using under the circumstances?

3. Marshall was considered to have engaged in imperfect self-defense and was convicted of manslaughter. Explain the reasoning behind the verdict.

You Decide

8.1 Defendant Roberta Shaffer was separated from her husband and lived with her two children. Her boyfriend, to whom she was engaged, had lived in the house for roughly two years. He had beaten Shaffer on several occasions, and when she asked him to move out, he threatened to kill Shaffer and her children. She claimed that she loved her boyfriend and had urged him to seek psychiatric assistance. Shaffer and her boyfriend argued at breakfast one morning and he allegedly angrily responded that "I'll take care of you right now." The defendant threw a cup of coffee at him and ran to the basement where her children were playing. Shaffer's boyfriend allegedly opened the door at the top of the basement stairs and proclaimed that "If you don't come up these stairs, I'll come down and kill you and the kids." She started to telephone the police and hung up when her boyfriend said that he would leave the house. He soon thereafter reappeared at the top of the stairs, and the defendant, who was fairly experienced in the use of firearms, removed a .22-caliber rifle from the gun rack and loaded the gun. Her boyfriend descended two or three stairs when the defendant shot and killed him with a single shot. Five minutes elapsed from the time she fled to the basement to the firing of the fatal shot. Was this an act of justified self-defense in response to imminent threat? Did Shaffer employ proportionate force? Was this imperfect self-defense? Was Shaffer required to retreat inside her own home? See *Commonwealth v. Shaffer*, 326 N.E.2d 880 (Mass. 1975).

You can find the answer at www.sagepub.com/lippmanccl3e.

Reasonable Belief

The common law and most statutes and modern decisions require that an individual who relies on self-defense must act with a reasonable belief in the imminence of serious bodily harm or death. The Utah statute on self-defense (Utah Code Ann. § 76-2-402) specifies that a person is justified in threatening or using force against another in those instances in which he or she "reasonably believes that force is necessary . . . to prevent death or serious bodily injury." . . . The reasonableness test has two prongs:

- *Subjective.* A defendant must demonstrate an honest belief that he or she confronted an imminent attack.
- *Objective.* A defendant must demonstrate that a reasonable person under the same circumstances would have believed that he or she confronted an imminent attack.

An individual who acts with an honest and reasonable, but mistaken, belief that he or she is subject to an armed attack is entitled to the justification of self-defense. The classic example is the individual who kills an assailant who is about to stab him or her with a knife, a knife that later is revealed to be a realistic-looking rubber replica. As noted by Supreme Court Justice Oliver Wendell Holmes, Jr., "[d]etached reflection cannot be demanded in the presence of an uplifted knife."[18] Absent a reasonableness requirement, it is feared that individuals might act on the basis of suspicion or prejudice or intentionally kill or maim and then later claim self-defense.

The Model Penal Code adopts a subjective approach and only requires that a defendant actually believe in the necessity of self-defense. The subjective approach has been adopted by very few courts. An interesting justification for this approach was articulated by the Colorado Supreme Court, which contended that the reasonable person standard was "misleading and confusing." The right to self-defense, according to the Colorado court, is a "natural right and is based on the natural law of self-preservation. Being so, it is resorted to instinctively in the animal kingdom by those creatures not endowed with intellect and reason, so it is not based on the 'reasonable man' concept."[19]

A number of courts are moving to a limited extent in the direction of the Model Penal Code by providing that a defendant acting in an honest, but unreasonable belief, is entitled to claim *imperfect self-defense* and should be convicted of voluntary manslaughter rather than intentional murder.[20] In *Harshaw v. State*, the defendant and deceased were arguing and the deceased threatened to retrieve his gun. They both retreated to their automobiles and the defendant grabbed his shotgun in time to shoot the deceased as he reached inside his automobile. The deceased was later found to have been unarmed. The Arkansas Supreme Court ruled that the judge should have instructed the jury on manslaughter because the jurors could reasonably have found that Harshaw acted "hastily and without due care" and that he merited a conviction for manslaughter rather than murder.[21]

The New York Court of Appeals wrestled with the meaning of "reasonableness" under the New York statute in the famous "subway murder trial" of Bernard Goetz. Did the law require a subjective standard in which the existence of the threat was "reasonable to the defendant" or an objective standard in which the existence of the threat was "reasonable to a reasonable person"? What is the best test in terms of the interests of society?

Did Goetz reasonably believe that he was threatened with death or great bodily harm?

People v. Goetz, 497 N.E.2d 41 (N.Y. 1986). Opinion by: Wachtler, C.J.

A Grand Jury has indicted defendant on attempted murder, assault, and other charges for having shot and wounded four youths on a New York City subway train after one or two of the youths approached him and asked for $5. The lower courts, concluding that prosecutor's charge to the Grand Jury on the defense of justification was erroneous, have dismissed the attempted murder, assault and weapons possession charges. We now reverse and reinstate all counts of the indictment.

Issue

Is the defense of self-defense based on a subjective standard or a reasonableness under the circumstances, objective standard?

Facts

We have summarized the facts as they appear from the evidence before the Grand Jury. . . .

On Saturday afternoon, December 22, 1984, Troy Canty, Darryl Cabey, James Ramseur, and Barry Allen boarded an IRT express subway train in The Bronx and headed south toward lower Manhattan. The four youths rode together in the rear portion of the seventh car of the train. Two of the four, Ramseur and Cabey, had screwdrivers inside their coats, which they said were to be used to break into the coin boxes of video machines.

Defendant Bernhard Goetz boarded this subway train at 14th Street in Manhattan and sat down on a bench towards the rear section of the same car occupied by the four youths. Goetz was carrying an unlicensed .38 caliber pistol loaded with five rounds of ammunition in a waistband holster. The train left the 14th Street station and headed towards Chambers Street.

It appears from the evidence before the Grand Jury that Canty approached Goetz, possibly with Allen beside him, and stated "give me five dollars." Neither Canty nor any of the other youths displayed a weapon. Goetz responded by standing up, pulling out his handgun and firing four shots in rapid succession. The first shot hit Canty in the chest; the second struck Allen in the back; the third went through Ramseur's arm and into his left side; the fourth was fired at Cabey, who apparently was then standing in the corner of the car, but missed, deflecting instead off of a wall of the conductor's cab. After Goetz briefly surveyed the scene around him, he fired another shot at Cabey, who then was sitting on the end bench of the car. The bullet entered the rear of Cabey's side and severed his spinal cord.

All but two of the other passengers fled the car when, or immediately after, the shots were fired. The conductor, who had been in the next car, heard the shots and instructed the motorman to radio for emergency assistance. The conductor then went into the car where the shooting occurred and saw Goetz sitting on a bench, the injured youths lying on the floor or slumped against a seat, and two women who had apparently taken cover, also lying on the floor. Goetz told the conductor that the four youths had tried to rob him.

While the conductor was aiding the youths, Goetz headed towards the front of the car. The train had stopped just before the Chambers Street station and Goetz went between two of the cars, jumped onto the tracks and fled. Police and ambulance crews arrived at the scene shortly thereafter. Ramseur and Canty, initially listed in critical condition, have fully recovered. Cabey remains paralyzed, and has suffered some degree of brain damage.

On December 31, 1984, Goetz surrendered to police in Concord, New Hampshire, identifying himself as the gunman being sought for the subway shootings in New York nine days earlier. Later that day, after receiving Miranda warnings, he made two lengthy statements, both of which were tape recorded with his permission. In the statements, which are substantially similar, Goetz admitted that he had been illegally carrying a handgun in New York City for three years. He stated that he had first purchased a gun in 1981 after he had been injured in a mugging. Goetz also revealed that twice between 1981 and 1984 he had successfully warded off assailants simply by displaying the pistol.

According to Goetz's statement, the first contact he had with the four youths came when Canty, sitting or lying on the bench across from him, asked "how are you," to which he replied "fine." Shortly thereafter, Canty, followed by one of the other youths, walked over to the defendant and stood to his left, while the other two youths remained to his right, in the corner of the subway car. Canty then said "give me five dollars." Goetz stated that he knew from the smile on Canty's face that they wanted to "play with me." Although he was certain that none of the youths had a gun, he had a fear, based on prior experiences, of being "maimed."

Goetz then established "a pattern of fire," deciding specifically to fire from left to right. His stated intention at that point was to "murder [the four youths], to hurt them, to make them suffer as much as possible." When Canty again requested money, Goetz stood up, drew his weapon, and began firing, aiming for the center of the body of each of the four. Goetz recalled that the first two he shot "tried to run through the crowd [but] they had nowhere to run." Goetz then turned to his right to "go after the other two." One of these two "tried to run through the wall of the train, but . . . he had . . . nowhere to go." The other youth (Cabey) "tried pretending that he wasn't with [the others]" by standing still, holding on to one of the subway hand straps, and not looking at Goetz. Goetz nonetheless fired his fourth shot at him. He then ran back to the first two youths to make sure they had been "taken care of." Seeing that they had both been shot, he spun back to check on the latter two. Goetz noticed that the youth who had been standing still was now sitting on a bench and seemed unhurt. As Goetz told the police, "I said '[y]ou seem to be all right, here's another,'" and he then fired the shot which severed Cabey's spinal cord. Goetz added that "if I was a little more under self-control . . . I would have put the barrel against his forehead and fired." He also admitted that "if I had had more [bullets], I would have shot them again, and again, and again."

After waiving extradition, Goetz was brought back to New York and arraigned on a felony complaint charging him with attempted murder and criminal possession of a weapon. The matter was presented to a Grand Jury in January 1985, with the prosecutor seeking an indictment for attempted . . . murder, assault, reckless endangerment, and criminal possession of a weapon. . . . On January 25, 1985, the Grand Jury indicted defendant on one count of criminal possession of a weapon in the third degree . . . for possessing the gun used in the subway shootings, and two counts of criminal possession of a weapon in the fourth degree . . . for possessing two other guns in his apartment building. It dismissed, however, the attempted murder and other charges stemming from the shootings themselves.

Several weeks after the Grand Jury's action, the People, asserting that they had newly available evidence, moved for an order authorizing them to resubmit the dismissed charges to a second Grand Jury. . . . On March 27, 1985, the second Grand Jury filed a ten-count indictment, containing four charges of attempted murder . . . four charges of assault in the first degree . . . one charge of reckless endangerment in the first degree . . . and one charge of criminal possession of a weapon in the second degree (possession of loaded firearm with intent to use it unlawfully against another). . . .

On October 14, 1985, Goetz moved to dismiss the charges contained in the second indictment alleging, among other things, that the evidence before the second Grand Jury was not legally sufficient to establish the offenses charged . . . and that the prosecutor's instructions to that Grand Jury on the defense of justification were erroneous and prejudicial to the defendant so as to render its proceedings defective. . . .

On November 25, 1985, while the motion to dismiss was pending before Criminal Term, a column appeared in the New York Daily News containing an interview which the columnist had conducted with Darryl Cabey the previous day in Cabey's hospital room. The columnist claimed that Cabey had told him in this interview that the other three youths had all approached Goetz with the intention of robbing him. The day after the column was published, a New York City police officer informed the prosecutor that he had been one of the first police officers to enter the subway car after the shootings, and that Canty had said to him "we were going to rob [Goetz]." . . .

In an order dated January 21, 1986, the Supreme Court . . . granted Goetz's motion to the extent that it dismissed all counts of the second indictment, other than the reckless endangerment charge, with leave to resubmit these charges to a third Grand Jury. The court, after inspection of the Grand Jury minutes . . . held . . . that the prosecutor, in a supplemental charge elaborating upon the justification defense, had erroneously introduced an objective element into this defense by instructing the grand jurors to consider whether Goetz's conduct was that of a "reasonable man in [Goetz's] situation." The court . . . concluded that the statutory test for whether the use of deadly force is justified to protect a person should be wholly subjective, focusing entirely on the defendant's state of mind when he used such force. It concluded that dismissal was required for this error because the justification issue was at the heart of the case.

On appeal by the People, a divided Appellate Division . . . affirmed Criminal Term's dismissal of the charges. The plurality opinion by Justice Kassal, concurred in by Justice Carro, agreed with Criminal Term's reasoning on the justification issue, stating that the grand jurors should have been instructed to consider only the defendant's subjective beliefs as to the need to use deadly force. . . . We agree with the dissenters that neither the prosecutor's charge to the Grand Jury on justification nor the information which came to light while the motion to dismiss was pending required dismissal of any of the charges in the . . . indictment.

Reasoning

New York Penal Law (NYPL) section 35 recognizes the defense of justification, which "permits the use of force under certain circumstances." . . . One such set of circumstances pertains to the use of force in defense of a person, encompassing both self-defense and defense of a third person. New York Penal Law section 35.15(1) sets forth the general principles governing all such uses of force: "[a] person may . . . use physical force upon another person when and to the extent he reasonably believes such to be necessary to defend himself or a third person from what he reasonably believes to be the use or imminent use of unlawful physical force by such other person." Subdivision (1) contains certain exceptions to this general authorization to use force, such as where the actor himself was the initial aggressor. Subdivision (2) sets forth further limitations on these general principles with respect to the use of deadly physical force and provides that a person may not use deadly physical force under circumstances specified in subdivision one unless: "(a) He reasonably believes that such other person is using or about to use deadly physical force . . . or (b) He reasonably believes that such other person is committing or attempting to commit a kidnapping, forcible rape, forcible sodomy or robbery." Section 35.15(2)(a) also provides that even under these circumstances a person ordinarily must retreat "if he knows that he can with complete safety as to himself and others avoid the necessity of [using deadly physical force] by retreating."

Thus, consistent with most justification provisions, NYPL section 35.15 permits the use of deadly physical force only where requirements as to triggering conditions and the necessity of a particular response are met. As to the triggering conditions, the statute requires that the actor "reasonably believes" that another person either is using or about to use deadly physical force or is committing or attempting to commit one of certain enumerated felonies, including robbery. As to the need for the use of deadly physical force as a response, the statute requires that the actor "reasonably believes" that such force is necessary to avert the perceived threat.

Holding

Because the evidence before the second Grand Jury included statements by Goetz that he acted to protect himself from being maimed or to avert a robbery, the prosecutor correctly chose to charge the justification defense in NYPL section 35.15 to the Grand Jury. . . . The prosecutor properly instructed the grand jurors to consider whether the use of deadly physical force was justified to prevent either serious physical injury or a robbery, and, in doing so, to separately analyze the defense with respect to each of the charges. . . .

When the prosecutor had completed his charge, one of the grand jurors asked for clarification of the term "reasonably believes." The prosecutor responded by instructing the grand jurors that they were to consider the circumstances of the incident and determine "whether the defendant's conduct was that of a reasonable man in the defendant's situation." It is this response by the prosecutor—and specifically his use of "a reasonable man"—which is the basis for the dismissal of the charges by the lower courts. As expressed repeatedly in the Appellate Division's plurality opinion, because NYPL section 35.15 uses the term "he reasonably believes," the appropriate test, according to that court, is whether a defendant's beliefs and reactions were "reasonable to him." Under that reading of the statute, a jury which believed a defendant's testimony that he felt that his own actions were warranted and were reasonable would have to acquit him, regardless of what anyone else in defendant's situation might have concluded. Such an interpretation defies the ordinary meaning and significance of the term "reasonably" in a statute, and misconstrues the clear

intent of the Legislature, in enacting NYPL section 35.15, to retain an objective element as part of any provision authorizing the use of deadly physical force.

Penal statutes in New York have long codified the right recognized at common law to use deadly physical force, under appropriate circumstances, in self-defense. . . . These provisions have never required that an actor's belief as to the intention of another person to inflict serious injury be correct in order for the use of deadly force to be justified, but they have uniformly required that the belief comport with an objective notion of reasonableness.

We cannot lightly impute to the Legislature an intent to fundamentally alter the principles of justification to allow the perpetrator of a serious crime to go free simply because that person believed his actions were reasonable and necessary to prevent some perceived harm. To completely exonerate such an individual, no matter how aberrational or bizarre his thought patterns, would allow citizens to set their own standards for the permissible use of force. It would also allow a legally competent defendant suffering from delusions to kill or perform acts of violence with impunity, contrary to fundamental principles of justice and criminal law.

We can only conclude that the Legislature retained a reasonableness requirement to avoid giving a license for such actions. The plurality's interpretation, as the dissenters recognized, excises the impact of the word "reasonably." . . . [W]e have frequently noted that a determination of reasonableness must be based on the "circumstance" facing a defendant or his "situation." . . . Such terms encompass more than the physical movements of the potential assailant. As just discussed, these terms include any relevant knowledge the defendant had about that person. They also necessarily bring in the physical attributes of all persons involved, including the defendant. Furthermore, the defendant's circumstances encompass any prior experiences he had which could provide a reasonable basis for a belief that another person's intentions were to injure or rob him or that the use of deadly force was necessary under the circumstances. . . . Accordingly, the order of the Appellate Division should be reversed, and the dismissed counts of the indictment reinstated.

Questions for Discussion

1. Goetz was acquitted of attempted murder and assault. He was found to have been justified in shooting the four young men in the subway car. Goetz was convicted of unlawful possession of a firearm and was sentenced to one year in prison. He was released after eight months in jail. The jury was composed of eight men and four women, ten whites and two African Americans. Do you agree that Goetz acted in self-defense? In 1996, a six-member civil jury ordered Goetz to pay $18 million in compensatory damages and $25 million in punitive

damages. The Goetz case is discussed in George P. Fletcher's *A Crime of Self-Defense: Bernhard Goetz and the Law on Trial* (New York: The Free Press, 1988).

2. Would problems arise in the event that the law of self-defense was based on a purely subjective test? Are there arguments in support of this approach? In most cases, would it matter whether a jury applied an objective or subjective test?

3. What circumstances should the jury consider in determining the reasonableness of Goetz's actions? Should Goetz's past experiences be considered? The fact that the shooting occurred in the subway? The physical size, age, dress, and behavior of the young males? What about the fact that Ramseur and Cabey had screwdrivers?

4. A number of commentators contend that any explanation of the verdict in the Goetz case must consider the race of the individuals involved. Goetz was Caucasian, while the four young people were African American. Professor George Fletcher raises the issue whether the same verdict would have been returned had Goetz been an African American and his attackers Caucasian juveniles. What is your opinion? Did Goetz stereotype the young men and assume that he was about to be robbed? Was his response based on revenge or self-defense? What would your reaction have been in the event that you found yourself in Goetz's position?

Cases and Comments

The Reasonable Woman. Defendant Yvonne Wanrow was convicted of murder and assault. Her conviction was reversed by the Washington Supreme Court. William Wesler was accused of molestation by Ms. Hooper's children. Hooper's landlord shared that Wesler had earlier attempted to molest a young child who had previously lived in Hooper's house and that Wesler had been committed to a mental asylum; the landlord advised Ms. Hooper that she should arm herself with a baseball bat. Yvonne Wanrow's two children were staying with Hooper at the time, and the two women and several other adults agreed to spend the night together to provide mutual support and security against possible retaliation by Wesler. Two of the men staying with Hooper visited Wesler and persuaded him to accompany them to Hooper's house to discuss the allegations. This led to a noisy and high-pitched verbal exchange. At one point, Wesler provocatively approached a young child sleeping on the couch and Hooper screamed for Wesler to leave her home. Ms. Wanrow, who was five foot four inches in height, had a broken leg, and was using a crutch, had placed a pistol in her purse. She testified that she turned around and found herself confronting the six-foot, two-inch Wesler and that she shot him as a reflex response.

The Washington Supreme Court determined that the judge's self-defense instructions to the jury were deficient, and that although the jury was instructed to consider the relative size and strength of the persons involved, they should also have been instructed to "afford women the right to have their conduct judged in light of the individual physical handicaps which are the product of sex discrimination." The jury had been directed to evaluate the defendant's conduct in accordance with the reactions of a "reasonably and ordinarily cautious and prudent man." The Washington Supreme Court explained that women suffer from a lack of training in the skills required to "effectively repeal a male assailant without resorting to the use of deadly weapons" and that the jury instructions should have directed the jury to consider the defendant's gender. The court also ruled that the trial court had properly declined to permit the defendant to rely on an expert witness to present evidence on the effects of the defendant's Indian culture on her perception and actions. Should juries be instructed to consider the reasonableness of a defendant's actions in light of his or her gender? What else should the jury be instructed to consider? See *State v. Wanrow,* 559 P.2d 548 (Wash. 1977).

You Decide

8.2 A group of five young Latino men was crossing the street when a car speeding around the corner suddenly braked to permit the young men to cross the street. Some of the young men, including the defendant, yelled at the driver to slow down. The driver, Alex Bernal, responded that he was looking for his daughter and that they should move out of the way. Bernal pulled the car to the side of the road and the defendant approached the passenger side of the car. Heated words were exchanged and Bernal exited the auto, removed his shoes, and began kicking into the air without hitting anyone. At some point during the ensuing exchange, Bernal appears to have swung at the defendant and the defendant responded by thrusting a knife into Bernal's heart. Bernal later died; the autopsy indicated that he had been stabbed three times. The defendant testified that he only intended to scare the unarmed Bernal and stab him in the leg, and that he was motivated by a desire to protect his brother from possible injury. The trial court ruled that the testimony of expert witness and sociologist Professor Martin Sanchez Jankowski was irrelevant and inadmissible. Professor Jankowski would have testified that (1) street fighters in the Hispanic culture do not retreat; (2) the Hispanic culture is based on honor; and (3) the defendant's testimony indicates that he was concerned with protecting his younger brother. Should the California Court of Appeal reverse the defendant's conviction because of the failure of the trial court judge to permit the jury to hear Professor Jankowski's testimony? See *People v. Romero*, 81 Cal. Rptr. 2d 823 (Cal. Ct. App. 1999).

You can find the answer at www.sagepub.com/lippmanccl3e.

Imminence

A defendant must reasonably believe that the threatened harm is imminent, meaning that the harm "is about to happen." The requirement that the defendant act out of necessity is based on several considerations:

- *Resolution of Disputes.* The law encourages the peaceful resolution of disputes where possible.
- *Last Resort.* Individuals should only resort to self-help where strictly required.
- *Evidence.* The existence of a clear and measurable threat provides confidence that the defendant is acting out of self-defense rather than out of a desire to punish the assailant or to seek revenge. Also, the existence of a clear threat assists in determining whether there is proportionality between the threatened harm and defensive response.

In *State v. Schroeder,* the nineteen-year-old defendant stabbed a violent cellmate who threatened to make Schroeder his "sex slave" or "punk." Schroeder testified that he felt vulnerable and afraid and woke up at 1:00 a.m. and stabbed his cellmate in the back with a table knife and hit him in the face with a metal ashtray. The Nebraska Supreme Court ruled that the threatened harm was not imminent and that there was a danger in legalizing "preventive assaults."[22]

Some courts have not insisted on a strict imminence standard. An Illinois case, for instance, ruled that a cab driver acted in self-defense in shooting and killing an individual who, along with other gang members, was involved in beating up an elderly man. The assailants threw a brick at the cab in retaliation for the driver's yelling at the gang and allegedly started to move toward the taxi. The Illinois Supreme Court concluded that the attackers possessed the capacity and intent to attack the driver, who was carrying money and was aware that a number of drivers recently had been attacked. On the other hand, consider the fact that the driver could have fled, shot over the heads of the assailants, or waited until the young people presented a more immediate threat.[23]

The Model Penal Code adopts this type of broad approach and provides that force is justifiable when the actor believes that he or she will be attacked on "the present occasion" rather than imminently. The commentary to the Model Penal Code notes that this standard would permit individuals to employ force in self-defense to prevent an individual who poses a threat from summoning reinforcements. The broad Model Penal Code test has found support in the statutes of a number of states, including Delaware, Hawaii, Nebraska, New Jersey, and Pennsylvania. A dissenting judge in *Schroeder* cited the Model Penal Code and argued that the young inmate should have been acquitted on the grounds of self-defense. After all, he could not be expected to remain continuously on guard against an assault by his older cellmate or the cellmate's friends.

The clash between the common law imminence requirement and the Model Penal Code's notion that self-defense may be justified where necessary to prevent an anticipated harm is starkly presented in cases in which defendants invoke the so-called battered spouse defense. In *State v. Norman,* the next case in the chapter, a woman who has been the victim of continual battering by her husband over a number of years kills her abusive spouse while he is sleeping. In reading this case, consider whether we should broadly interpret the imminence standard and, if not, what standard should be adopted.

Did Norman confront an imminent threat from her abusive husband?

State v. Norman, 366 S.E.2d 586 (N.C. Ct. App. 1988). Opinion by: Parker, J.

The primary issue presented on this appeal is whether the trial court erred in failing to instruct on self-defense. We answer in the affirmative and grant a new trial.

Facts

At trial the State presented the testimony of a deputy sheriff of the Rutherford County Sheriff's Department who testified that on 12 June 1985, at approximately 7:30 p.m., he was dispatched to the Norman residence. There, in one of the bedrooms, he found decedent, John Thomas "J.T." Norman (herein decedent or Norman) dead, lying on his left side on a bed. The State presented an autopsy report, stipulated to by both parties, concluding that Norman had died from two gunshot wounds to the head. . . .

Defendant and Norman had been married twenty-five years at the time of Norman's death. Norman was an alcoholic. He had begun to drink and to beat defendant five years after they were married. The couple had five children, four of whom are still living. When defendant was pregnant with her youngest child, Norman beat her and kicked her down a flight of steps, causing the baby to be born prematurely the next day.

Norman, himself, had worked one day a few months prior to his death; but aside from that one day, witnesses could not remember his ever working. Over the years and up to the time of his death, Norman forced defendant to prostitute herself every day in order to support him. If she begged him not to make her go, he slapped her. Norman required defendant to make a minimum of one hundred dollars per day; if she failed to make this minimum, he would beat her.

Norman commonly called defendant "Dog," "Bitch," and "Whore," and referred to her as a dog. Norman beat defendant "most every day," especially when he was drunk and when other people were around, to "show off." He would beat defendant with whatever was handy—his fist, a fly swatter, a baseball bat, his shoe, or a bottle; he put out cigarettes on defendant's skin; he threw food and drink in her face and refused to let her eat for days at a time; and he threw glasses, ashtrays, and beer bottles at her and once smashed a glass in her face. . . . Norman would often make defendant bark like a dog, and if she refused, he would beat her. He often forced defendant to sleep on the concrete floor of their home and on several occasions forced her to eat dog or cat food out of the dog or cat bowl.

Norman often stated both to defendant and to others that he would kill defendant. He also threatened to cut her heart out.

On or about the morning of 10 June 1985, Norman forced defendant to go to a truck stop or rest stop on Interstate 85 in order to prostitute to make some money. Defendant's daughter and defendant's daughter's boy friend accompanied defendant. Some time later that day, Norman went to the truck stop, apparently drunk, and began hitting defendant in the face with his fist and slamming the car door into her. He also threw hot coffee on defendant. . . .

On 11 June 1985, Norman was extremely angry and beat defendant. . . . Defendant testified that during the entire day, when she was near him, her husband slapped her, and when she was away from him, he threw glasses, ashtrays, and beer bottles at her. Norman asked defendant to make him a sandwich; when defendant brought it to him, he threw it on the floor and told her to make him another. Defendant made him a second sandwich and brought it to him; Norman again threw it on the floor, telling her to put something on her hands because he did not want her to touch the bread. Defendant made a third sandwich using a paper towel to handle the bread. Norman took the third sandwich and smeared it in defendant's face.

On the evening of 11 June 1985, at about 8:00 or 8:30, a domestic quarrel was reported at the Norman residence. The officer responding to the call testified that defendant was bruised and crying and that she stated her husband had been beating her all day and she could not take it any longer. The officer advised defendant to take out a warrant on her husband, but defendant responded that if she did so, he would kill her. A short time later, the officer was again dispatched to the Norman residence. There he learned that defendant had taken an overdose of "nerve pills," and that Norman was interfering with emergency personnel who were trying to treat defendant. Norman was drunk and was making statements such as, "If you want to die, you deserve to die. I'll give you more pills," and "Let the bitch die. . . . She ain't nothing but a dog. She don't deserve to live." Norman also threatened to kill defendant, defendant's mother, and defendant's grandmother. The law enforcement officer reached for his flashlight or blackjack and chased Norman into the house. Defendant was taken to Rutherford Hospital. . . .

The next day, 12 June 1985, the day of Norman's death . . . Defendant was driving. During the ride . . . Norman slapped defendant for following a truck too closely and poured a beer on her head. Norman kicked defendant in the side of the head while she was driving and told her he would "cut her breast off and shove it up her rear end."

. . . Witnesses stated that back at the Norman residence, Norman threatened to cut defendant's throat, threatened to kill her, and threatened to cut off her breast. Norman also smashed a doughnut on defendant's face and put out a cigarette on her chest.

In the late afternoon, Norman wanted to take a nap. He lay down on the larger of the two beds in the bedroom. Defendant started to lie down on the smaller bed, but Norman said, "No bitch . . . Dogs don't sleep on beds, they sleep in [sic] the floor." Soon after, one of the Normans' daughters, Phyllis, came into the room and asked if defendant could look after her baby. Norman assented. When the baby began to cry, defendant took the child to her mother's house, fearful that the baby would disturb Norman. At her mother's house, defendant found a gun. She took it back to her home and shot Norman.

Defendant testified that things at home were so bad she could no longer stand it. She explained that she could not leave Norman because he would kill her. She stated that she had left him before on several occasions and that each time he found her, took her home, and beat her. She said that she was afraid to take out a warrant on her husband because he had said that if she ever had him locked up, he would kill her when he got out. She stated she did not have him committed because he told her he would see the authorities coming for him and before they got to him he would cut defendant's throat. Defendant also testified that when he threatened to kill her, she believed he would kill her if he had the chance.

The defense presented the testimony of two expert witnesses in the field of forensic psychology. . . . Dr. Tyson concluded that defendant "fits and exceeds the profile, of an abused or battered spouse." . . . Dr. Tyson stated that defendant could not leave her husband because she had gotten to the point where she had no belief whatsoever in herself and believed in the total invulnerability of her husband. He stated, "Mrs. Norman didn't leave because she believed, fully believed that escape was totally impossible. . . . When asked if it appeared to defendant reasonably necessary to kill her husband, Dr. Tyson responded, "I think Judy Norman felt that she had no choice, both in the protection of herself and her family, but to engage, exhibit deadly force against Mr. Norman, and that in so doing, she was sacrificing herself, both for herself and for her family." . . .

Issue

The State contends that because decedent was asleep at the time of the shooting, defendant's belief in the necessity to kill decedent was, as a matter of law,

unreasonable. The State further contends that even assuming . . . the evidence satisfied the requirement that defendant's belief be reasonable, defendant, being the aggressor, cannot satisfy the third requirement of perfect self-defense. . . . The question then arising on the facts in this case is whether the victim's passiveness at the moment the unlawful act occurred precludes defendant from asserting perfect self-defense.

Reasoning

Applying the criteria of **perfect self-defense** to the facts of this case, we hold that the evidence was sufficient to submit an issue of perfect self-defense to the jury. An examination of the elements of perfect self-defense reveals that both subjective and objective standards are to be applied in making the crucial determinations. The first requirement that it appear to defendant and that defendant believe it necessary to kill the deceased in order to save herself from death or great bodily harm calls for a subjective evaluation. This evaluation inquires as to what the defendant herself perceived at the time of the shooting. The trial was replete with testimony of forced prostitution, beatings, and threats on defendant's life. The defendant testified that she believed the decedent would kill her, and the evidence showed that on the occasions when she had made an effort to get away from Norman, he had come after her and beat her. Indeed, within twenty-four hours prior to the shooting, defendant had attempted to escape by taking her own life and throughout the day on 12 June 1985 had been subjected to beatings and other physical abuse, verbal abuse, and threats on her life up to the time when decedent went to sleep. Experts testified that in their opinion, defendant believed killing the victim was necessary to avoid being killed. . . .

Unlike the first requirement, the second element of self-defense—that defendant's belief be reasonable in that the circumstances as they appeared to defendant would be sufficient to create such a belief in the mind of a person of ordinary firmness—is measured by the objective standard of the person of ordinary firmness under the same circumstances. Again, the record is replete with sufficient evidence to permit but not compel a juror, representing the person of ordinary firmness, to infer that defendant's belief was reasonable under the circumstances in which she found herself. Expert witnesses testified that defendant exhibited severe symptoms of battered spouse syndrome, a condition that develops from repeated cycles of violence by the victim against the defendant. Through this repeated, sometimes constant, abuse, the battered spouse acquires what the psychologists denote as a state of "learned helplessness," defendant's state of mind as described by Drs. Tyson and Rollins. . . . In the instant case, decedent's excessive anger, his constant beating and battering of defendant on 12 June 1985, her fear that the beatings would resume, as well as previous efforts by defendant to extricate herself from this abuse are circumstances to be considered in judging the reasonableness of defendant's belief that she would be seriously injured or killed at the time the criminal act was committed. The evidence discloses that defendant felt helpless to extricate herself from this intolerable, dehumanizing, brutal existence. Just the night before the shooting, defendant had told the sheriff's deputy that she was afraid to swear out a warrant against her husband because he had threatened to kill her when he was released if she did. The inability of a defendant to withdraw from the hostile situation and the vulnerability of a defendant to the victim are factors considered by our Supreme Court in determining the reasonableness of a defendant's belief in the necessity to kill the victim. . . .

To satisfy the third requirement, defendant must not have aggressively and willingly entered into the fight without legal excuse or provocation. By definition, aggression in the context of self-defense is tied to provocation. The existence of battered spouse syndrome, in our view, distinguishes this case from the usual situation involving a single confrontation or affray. The provocation necessary to determine whether defendant was the aggressor must be considered in light of the totality of the circumstances. . . .

Holding

Mindful that the law should never casually permit an otherwise unlawful killing of another human being to be justified or excused, this Court is of the opinion that with the battered spouse there can be, under certain circumstances, an unlawful killing of a passive victim that does not preclude the defense of perfect self-defense. Given the characteristics of battered spouse syndrome, we do not believe that a battered person must wait until a deadly attack occurs or that the victim must in all cases be actually attacking or threatening to attack at the very moment defendant commits the unlawful act for the battered person to act in self-defense. Such a standard, in our view, would ignore the realities of the condition. This position is in accord with other jurisdictions that have addressed the issue. . . .

In this case, decedent, angrier than usual, had beaten defendant almost continuously during the afternoon and had threatened to maim and kill defendant. . . . A jury, in our view, could find that decedent's sleep was but a momentary hiatus in a continuous reign of terror by the decedent, that defendant merely took advantage of her first opportunity to protect herself, and that defendant's act was not without the provocation required for perfect self-defense. The expert testimony considered with the other evidence would permit reasonable minds to infer that defendant did not use more force than reasonably appeared necessary to her under the circumstances to protect herself from death or great bodily harm.

Based on the foregoing analysis, we are of the opinion that, in addition to the instruction on voluntary manslaughter, defendant was entitled to an instruction on perfect self-defense. . . . The jury is to regard evidence of battered spouse syndrome merely as some evidence to be considered along with all other evidence in making its determination whether there is a reasonable doubt as to the unlawfulness of defendant's conduct. . . . New trial.

Questions for Discussion

1. The jury convicted Norman of voluntary manslaughter and, as a result, did not accept that the defendant's killing of her husband was a justified act of perfect self-defense. Summarize the appellate court's reasoning in ruling that the defendant was entitled to have the jury consider her claim of self-defense.

2. What are the dangers of too broad or too narrow a view of the imminence requirement for self-defense? Was the defendant's use of force proportionate to the threat that she confronted from her husband?

3. Does reliance on the "battered spouse syndrome" risk that experts will attribute traits of "helplessness" to the defendant that, in fact, she does not possess?

4. Can men involved in heterosexual or homosexual relationships rely on the "battered spouse syndrome?" Can women rely on it who are involved in homosexual relationships?

Cases and Comments

1. **The North Carolina Supreme Court.** The North Carolina Supreme Court reversed the appellate court and ruled that the trial court properly declined to instruct the jury on self-defense. The supreme court ruled that the evidence did not "tend to show that the defendant reasonably believed that she was confronted by a threat of imminent death or great bodily harm." The court further observed that the "relaxed requirements" for self-defense would "legalize the opportune killing of abusive husbands by their wives solely on the basis of the wives' . . . subjective speculation as the probability of future felonious assaults by their husbands." Do you agree? See *State v. Norman*, 378 S.E.2d 8 (N.C. 1989). An additional case on the right of an abused spouse to rely on self-defense is *State v. Stewart*, 763 P.2d 572 (Kan. 1988).

2. Domestic Violence. A 2005 study by the World Health Organization of women in ten countries concluded that domestic violence is frighteningly common and is accepted as normal within many societies. The percentage of women reported to have been physically or sexually assaulted in their lifetime ranges from fifteen percent in Japan to seventy-one percent in Ethiopia. In the United States, a 2011 Centers for Disease Control and Prevention study reports that that one in four women surveyed had been assaulted by a husband or boyfriend. One in five women had been the victim of a rape or attempted rape at some point in their lives. Most victims had been raped by an intimate partner or by an acquaintance. One-third of women had been victims of a rape, beating, or stalking, or a combination of assaults.

You Decide

8.3 The defendant, seventeen-year-old Andrew Janes, was abandoned by his alcoholic father at age seven. Along with his mother Gale and brother Shawn, Andrew was abused by his mother's lover, Walter Jaloveckas, for roughly ten years. As Walter walked in the door following work on August 30, 1988, Andrew shot and killed him; one 9-millimeter pistol shot went through Walter's right eye and the other through his head. The previous night Walter had yelled at Gale, and Walter later leaned his head into Andrew's room and spoke in low tones that usually were "reserved for threats." Andrew was unable to remember precisely what Walter said. In the morning, Gale mentioned to Andrew that Walter was still mad. After returning from school, Andrew loaded the pistol, drank some whiskey, and smoked marijuana.

Examples of the type of abuse directed against Andrew by Walter included beatings with a belt and wire hanger, hitting Andrew in the mouth with a mop, and punching Andrew in the face for failing to complete a homework assignment. In 1988, Walter hit Andrew with a piece of firewood, knocking him out. Andrew was subject to verbal as well as physical threats, including a threat to nail his hands to a tree, brand his forehead, place Andrew's hands on a hot stove, break Andrew's fingers, and hit him in the head with a hammer.

The "battered child syndrome" results from a pattern of abuse and anxiety. "Battered children" live in a state of constant alert ("hypervigilant"), caution ("hypermonitoring"), and develop a lack of confidence and an inability to seek help ("learned helplessness"). Did Andrew believe and would a reasonable person believe that Andrew confronted an imminent threat of great bodily harm or death? The Washington Supreme Court clarified that imminent means "near at hand . . . hanging threateningly over one's head . . . menacingly near." The trial court refused to instruct the jury to consider whether Andrew was entitled to invoke self-defense. Should the Washington Supreme Court uphold or reverse the decision of the trial court? See *State v. Janes*, 850 P.2d 495 (Wash. 1993).

You can find the answer at www.sagepub.com/lippmanccl3e.

Excessive Force

An individual acting in self-defense is entitled to use the force reasonably believed to be necessary to defend himself or herself. **Deadly force** is force that a reasonable person under the circumstances would be aware will cause or create a substantial risk of death or substantial bodily harm. This may be employed to protect against death or serious bodily harm. The application of excessive rather than proportionate force may result in a defender's being transformed into an aggressor. This is the case where an individual entitled to **nondeadly force** resorts to deadly force. The Model Penal Code limits deadly force to the protection against death, serious bodily injury, kidnapping, or rape. The Wisconsin statute authorizes the application of deadly force against arson, robbery, burglary, and any felony offense that creates a danger of death or serious bodily harm.

In *State v. Pranckus,* the Connecticut Appellate Court ruled that the defendant could not have reasonably believed that the use of a kitchen knife with an eight-inch blade was required to defend himself in a fistfight. The defendant charged at the victims and stabbed each of the victims twice, killing one of them. The Connecticut court noted that both victims suffered wounds on their backs, indicating that they were fleeing and that the defendant was sufficiently confident of his ability to defend himself that he later attempted to continue the fight without a weapon.[24]

The requirement of proportionality is not accepted in various foreign countries that stress the privilege of individuals to respond without limitation to an attack. Should there be a restriction on the right of an innocent individual to respond to an attack?[25]

Retreat

The law of self-defense is based on necessity. An individual may resort to self-protection when he or she reasonably believes it necessary to defend against an immediate attack. The amount of force is limited to that reasonably believed to be necessary. Courts have struggled with how to treat a situation in which an individual may avoid resorting to deadly force by safely retreating or fleeing. The principle of necessity dictates that every alternative should be exhausted before an individual resorts to deadly force and that an individual should be required to **retreat to the wall** (as far as possible). On the other hand, should an individual be required to retreat when confronted with a violent wrongdoer? Should the law promote cowardice and penalize courage?

Virtually every jurisdiction provides that there is no duty or requirement to retreat before resorting to *nondeadly force*. A majority of jurisdictions follow the same **stand your ground rule** in the case of *deadly force,* although a "significant minority" require retreat to the wall. The stand your ground rule is also followed in most former communist countries in Europe and is based on several considerations:[26]

- The promotion of a courageous attitude.
- The refusal to protect a wrongdoer initiating an attack.
- The reluctance of courts to complicate their task by being placed in the position of having to rule on various issues surrounding the duty to retreat.
- The retreat rule may endanger individuals who are required to retreat and encourage wrongdoers who have no reason to fear for their lives.

Most jurisdictions limit the right to "stand your ground" when confronted with nondeadly force to an individual who is without fault, a **true man**. An aggressor employing nondeadly force must clearly abandon the struggle and it must be a **withdrawal in good faith** to regain the right of self-defense. Some courts recognize that even an aggressor using deadly force may withdraw and regain the right of self-defense. In these instances, the right of self-defense will limit the initial aggressor's liability to voluntary manslaughter and will not provide a *perfect self-defense*. A withdrawal in good faith must be distinguished from a **tactical retreat** in which an individual retreats with the intent of continuing the hostilities.

The requirement of retreat is premised on the traditional rule that only necessary force may be employed in self-defense. The provision for retreat is balanced by the consideration that withdrawal is not required when the safety of the defender would be jeopardized. The **castle doctrine** is another generally recognized exception to the rule of retreat and provides that individuals inside the home are justified in "holding their ground."[27]

The Model Penal Code section 3.04(b)(ii) provides that deadly force is not justifiable in those instances in which an individual "knows that he can avoid the necessity of using such force with complete safety by retreating." There is no duty to retreat under the Model Penal Code within the home or place of work unless an individual is an aggressor.

The next case in the chapter, *United States v. Peterson,* explores the right of an "aggressor" to self-defense, the duty to retreat, and the castle doctrine. Do you agree with the federal court's decision that Peterson is not entitled to claim self-defense?

Was Peterson required to retreat?

United States v. Peterson, 483 F.2d 1222 (D.C. Cir. 1973). Opinion by: Robinson, J.

Issues

Peterson was indicted for second-degree murder and was convicted by a jury of manslaughter as a lesser included offense. Bennie L. Peterson . . . complains. . . . that the judge twice erred in the instructions given the jury in relation to his claim that the homicide was committed in self-defense. One error alleged was an instruction that the jury might consider whether Peterson was the aggressor in the altercation that resulted in the homicide. The other was an instruction that a failure by Peterson to retreat, if he could have done so without jeopardizing his safety, might be considered as a circumstance bearing on the question whether he was justified in using the amount of force which he did.

Facts

The events immediately preceding the homicide are not seriously in dispute. The version presented by the Government's evidence follows. Charles Keitt, the deceased, and two friends drove in Keitt's car to the alley in the rear of Peterson's house to remove the windshield wipers from the latter's wrecked car. While Keitt was doing so, Peterson came out of the house. Peterson went back into the house, obtained a pistol, and returned to the yard. In the meantime, Keitt had reseated himself in his car, and he and his companions were about to leave. The car was characterized by some witnesses as "wrecked" and by others as "abandoned." The testimony left it clear that its condition was such that it could not be operated. It was parked on one side of the alley about fifteen feet from the gate in the rear fence which opened into Peterson's back yard. Keitt's car was stopped in the alleyway about four feet behind it. Upon his reappearance in the yard, Peterson paused briefly to load the pistol. "If you move," he shouted to Keitt, "I will shoot." He walked to a point in the yard slightly inside a gate in the rear fence and, pistol in hand, said, "If you come in here I will kill you." Keitt alighted from his car, took a few steps toward Peterson and exclaimed, "What the hell do you think you are going to do with that?" Keitt then made an about-face, walked back to his car and got a lug wrench. With the wrench in a raised position, Keitt advanced toward Peterson, who stood with the pistol pointed toward him. Peterson warned Keitt not to "take another step" and, when Keitt continued onward, shot him in the face from a distance of about ten feet. Death was apparently instantaneous. Shortly thereafter, Peterson left home and was apprehended twenty-odd blocks away. Keitt apparently had been drinking and an autopsy disclosed that he had a .29 percent blood alcohol content. Keitt fell in the alley about seven feet from the gate.

This description of the fatal episode was furnished at Peterson's trial by four witnesses for the Government. Peterson did not testify, but provided a statement to the police in which he related . . . that Keitt had removed objects from his car before, and on the day of the shooting he had told Keitt not to do so. After the initial verbal altercation, Keitt went to his car for the lug wrench, so he, Peterson, went into his house for his pistol. When Keitt was about ten feet away, he pointed the pistol "away of his right shoulder;" adding that Keitt was running toward him, Peterson said he "got scared and fired the gun. He ran right into the bullet." "I did not mean to shoot him," Peterson insisted, "I just wanted to scare him."

At trial, Peterson moved for a judgment of acquittal. The jury returned a verdict finding Peterson guilty of manslaughter. Judgment was entered conformably with the verdict, and this appeal followed.

Reasoning

Peterson's consistent position is that as a matter of law his conviction of manslaughter—alleviated homicide—was wrong and that his act was one of self-preservation—justified homicide. The Government, on the other hand, has contended from the beginning that Keitt's slaying fell outside the bounds of lawful self-defense. The questions remaining for our decision inevitably track back to this basic dispute.

The law of self-defense is a law of necessity . . . the right of self-defense arises only when the necessity begins, and equally ends with the necessity; and never must the necessity be greater than when the force employed defensively is deadly. The "necessity must bear all semblance of reality, and appear to admit of no other alternative, before taking life will be justifiable as excusable." Hinged on the exigencies of self-preservation, the doctrine of homicidal self-defense emerges from the body of the criminal law as a limited though important exception to legal outlawry of the arena of self-help in the settlement of potentially fatal personal conflicts. So it is that necessity is the pervasive theme of the well defined conditions which the law imposes on the right to kill or maim in self-defense. There must have been a threat, actual or apparent, of the use of deadly force against the defender. The threat must have been unlawful and immediate. The defender must have believed that he was in imminent peril of death or serious bodily harm, and that his response was necessary to save himself therefrom. These beliefs must not only have been honestly entertained, but also objectively reasonable in light of the surrounding circumstances. It is clear that no less than a concurrence of these elements will suffice. Here the parties' opposing contentions focus on the roles of two further considerations. One is the provoking of the confrontation by the defender. The other is the defender's failure to utilize a safe route for retreat from the confrontation. The essential inquiry, in final analysis, is whether and to what extent the rule of necessity may translate these considerations into additional factors in the equation. To these questions, in the context of the specific issues raised, we now proceed.

The trial judge's charge authorized the jury, as it might be persuaded, to convict Peterson of second-degree murder or manslaughter, or to acquit by reason of self-defense. On the latter phase of the case, the judge instructed that with evidence of self-defense present, the Government bore the burden of proving beyond a reasonable doubt that Peterson did not act in self-defense; and that if the jury had a reasonable doubt as to whether Peterson acted in self-defense, the verdict must be not guilty. The judge further instructed that the circumstances under which Peterson acted, however, must have been such as to produce a reasonable belief that Keitt was then about to kill him or do him serious bodily harm and that deadly force was necessary to repel him. In determining whether Peterson used excessive force in defending himself, the judge said, the jury could consider all of the circumstances under which he acted.

These features of the charge met Peterson's approval, and we are not summoned to pass on them. There were, however, two other aspects of the charge to which Peterson objected, and which are now the subject of vigorous controversy. The first of Peterson's complaints centers upon an instruction that the right to use deadly force in self-defense is not ordinarily available to one who provokes a conflict or is the aggressor in it. Mere words, the judge explained, do not constitute provocation or aggression; and if Peterson precipitated the altercation but thereafter withdrew from it in good faith and so informed Keitt by words or acts, he was justified in using deadly force to save himself from imminent danger or death or grave bodily harm. And, the judge added, even if Keitt was the aggressor and Peterson was justified in defending himself, he was not entitled to use any greater force than he had reasonable ground to believe and actually believed to be necessary for that purpose. Peterson contends that there was no evidence that he either caused or contributed to the conflict, and that the instructions on that topic could only mislead the jury.

It has long been accepted that one cannot support a claim of self-defense by a self-generated necessity to kill. The right of homicidal self-defense is granted only to those free from fault in the difficulty; it is denied to slayers who incite the fatal attack, encourage the fatal quarrel or otherwise promote the necessitous occasion for taking life. The fact that the deceased struck the first blow, fired the first shot or made the first menacing gesture does not legalize the self-defense claim if in fact the claimant was the actual provoker. In sum, one who is the aggressor in a conflict culminating in death cannot invoke the necessities of self-preservation. Only in the event that he communicates to his adversary his intent to withdraw and in good faith attempts to do so is he restored to his right of self-defense.

This body of doctrine traces its origin to the fundamental principle that a killing in self-defense is excusable only as a matter of genuine necessity. Quite obviously, a defensive killing is unnecessary if the occasion for it could have been averted, and the roots of that consideration run deep with us. A half-century ago, in *Laney v. United States,* this Court declared that, before a person can avail himself of the plea of self-defense against the charge of homicide, he must do everything in his power, consistent with his safety, to avoid the danger and avoid the necessity of taking life. If one has reason to believe that he will be attacked, in a manner which threatens him with bodily injury, he must avoid the attack if it is possible to do so, and the right of self-defense does not arise until he has done everything in his power to prevent its necessity.

In the case at bar, the trial judge's charge fully comported with these governing principles. The remaining question, then, is whether there was evidence to make them applicable to the case. A recapitulation of the proofs shows beyond peradventure that there was.

It was not until Peterson fetched his pistol and returned to his back yard that his confrontation with Keitt took on a deadly cast. Prior to his trip into the house for the gun, there was, by the Government's evidence, no threat, no display of weapons, no combat. There was an exchange of verbal aspersions and a misdemeanor against Peterson's property was in progress but, at this juncture, nothing more. Even if Peterson's postarrest version of the initial encounter were accepted—his claim that Keitt went for the lug wrench before he armed himself—the events which followed bore heavily on the question as to who the real aggressor was.

The evidence is uncontradicted that when Peterson reappeared in the yard with his pistol, Keitt was about to depart the scene. Richard Hilliard testified that after the first argument, Keitt reentered his car and said "Let's go." This statement was verified by Ricky Gray, who testified that Keitt "got in the car and . . . they were getting ready to go;" he, too, heard Keitt give the direction to start the car. The uncontroverted fact that Keitt was leaving shows plainly that so far as he was concerned the confrontation was ended. It demonstrates just as plainly that even if he had previously been the aggressor, he no longer was.

Not so with Peterson, however, as the undisputed evidence made clear. Emerging from the house with the pistol, he paused in the yard to load it, and to command Keitt not to move. He then walked through the yard to the rear gate and, displaying his pistol, dared Keitt to come in, and threatened to kill him if he did. While there appears to be no fixed rule on the subject, the cases hold, and we agree, that an affirmative unlawful act reasonably calculated to produce an affray foreboding injurious or fatal consequences is an aggression which, unless renounced, nullifies the right of homicidal self-defense. We cannot escape the abiding conviction that the jury could readily find Peterson's challenge to be a transgression of that character.

The situation at bar is not unlike that presented in *Laney.* There the accused, chased along the street by a mob threatening his life, managed to escape through an areaway between two houses. In the back yard of one of the houses, he checked a gun he was carrying and then returned to the areaway. The mob beset him again, and during an exchange of shots one of its members was killed by a bullet from the accused's gun. In affirming a conviction of manslaughter, the court reasoned that when defendant escaped from the mob into the back yard . . . he was in a place of comparative safety, from which, if he desired to go home, he could have gone by the back way, as he subsequently did. . . . His appearance on the street at that juncture could mean nothing but trouble for him. Hence, when he adjusted his gun and stepped out into the area-way, he had every reason to believe that his presence there would provoke trouble. We think his conduct in adjusting his revolver and going into the areaway was such as to deprive him of any right to invoke the plea of self-defense.

We think the evidence plainly presented an issue of fact as to whether Peterson's conduct was an invitation to and provocation of the encounter which ended in the fatal shot. We sustain the trial judge's action in remitting that issue for the jury's determination.

The second aspect of the trial judge's charge as to which Peterson asserts error concerned the undisputed fact that at no time did Peterson endeavor to retreat from Keitt's approach with the lug wrench. The judge instructed the jury that if Peterson had reasonable grounds to believe and did believe that he was in imminent danger of death or serious injury, and that deadly force was necessary to repel the danger, he was required neither to retreat nor to consider whether he could safely retreat. Rather, said the judge, Peterson was entitled to stand his ground and use such force as was reasonably necessary under the circumstances to save his life and his person from pernicious bodily harm. But, the judge continued, if Peterson could have safely retreated but did not do so, that failure was a circumstance which the jury might consider, together with all others, in determining whether he went further in repelling the danger, real or apparent, than he was justified in going.

Peterson contends that this imputation of an obligation to retreat was error, even if he could safely have done so. He points out that at the time of the shooting he was standing in his own yard, and argues he was under no duty to move. We are persuaded to the conclusion that in the circumstances presented here, the trial judge did not err in giving the instruction challenged. Within the common law of self-defense there developed the rule of "retreat to the wall," which ordinarily forbade the use of deadly force by one to whom an avenue for safe retreat was open. This doctrine was but an application of the requirement of strict necessity to excuse the taking of human life, and was designed to insure the existence of that necessity. Even the innocent victim of a vicious assault had to elect a safe retreat, if available, rather than resort to defensive force which might kill or seriously injure.

In a majority of American jurisdictions, contrarily to the common law rule, one may stand his ground and use deadly force whenever it seems reasonably necessary to save himself. While the law of the District of Columbia on this point is not entirely clear, it seems allied with the strong minority adhering to the common law. In 1856, the District of Columbia Criminal Court ruled that a participant in an affray "must

endeavor to retreat, . . . that is, he is obliged to retreat, if he can safely." The court added that "[a] man may, to be sure, decline a combat when there is no existing or apparent danger, but the retreat to which the law binds him is that which is the consequence." In a much later era this court, adverting to necessity as the soul of homicidal self-defense, declared that "no necessity for killing an assailant can exist, so long as there is a safe way open to escape the conflict." That is not to say that the retreat rule is without exceptions. Even at common law it was recognized that it was not completely suited to all situations. Today it is the more so that its precept must be adjusted to modern conditions nonexistent during the early development of the common law of self-defense. One restriction on its operation comes to the fore when the circumstances apparently foreclose a withdrawal with safety.

The doctrine of retreat was never intended to enhance the risk to the innocent; its proper application has never required a faultless victim to increase his assailant's safety at the expense of his own. On the contrary, he could stand his ground and use deadly force otherwise appropriate if the alternative were perilous, or if to him it reasonably appeared to be. A slight variant of the same consideration is the principle that there is no duty to retreat from an assault producing an imminent danger of death or grievous bodily harm. "Detached reflection cannot be demanded in the presence of an uplifted knife," nor is it "a condition of immunity that one in that situation should pause to consider whether a reasonable man might not think it possible to fly with safety or to disable his assailant rather than to kill him."

The trial judge's charge to the jury incorporated each of these limitations on the retreat rule. Peterson, however, invokes another—the so-called "castle" doctrine. It is well settled that one who through no fault of his own is attacked in his home is under no duty to retreat therefrom. The oft-repeated expression that "a man's home is his castle" reflected the belief in olden days that there were few if any safer sanctuaries than the home. The "castle" exception, moreover, has been extended by some courts to encompass the occupant's presence within the curtilage outside his dwelling. Peterson reminds us that when he shot to halt Keitt's advance, he was standing in his yard and so, he argues, he had no duty to endeavor to retreat.

Despite the practically universal acceptance of the "castle" doctrine in American jurisdictions wherein the point has been raised, its status in the District of Columbia has never been squarely decided. But whatever the fate of the doctrine in the District law of the future, it is clear that in absolute form it was inapplicable here. The right of self-defense, we have said, cannot be claimed by the aggressor in an affray so long as he retains that unmitigated role. It logically follows that any rule of no-retreat which may protect an innocent victim of the affray would, like other incidents of a forfeited right of self-defense, be unavailable to the party who provokes or stimulates the conflict. Accordingly, the law is well settled that the "castle" doctrine can be invoked only by one who is without fault in bringing the conflict on. That, we think, is the critical consideration here.

Holding

We need not repeat our previous discussion of Peterson's contribution to the altercation which culminated in Keitt's death. It suffices to point out that by no interpretation of the evidence could it be said that Peterson was blameless in the affair. And while, of course, it was for the jury to assess the degree of fault, the evidence well nigh dictated the conclusion that it was substantial.

The only reference in the trial judge's charge intimating an affirmative duty to retreat was the instruction that a failure to do so, when it could have been done safely, was a factor in the totality of the circumstances which the jury might consider in determining whether the force which he employed was excessive.

We cannot believe that any jury was at all likely to view Peterson's conduct as irreproachable. We conclude that for one who, like Peterson, was hardly entitled to fall back on the "castle" doctrine of no retreat, that instruction cannot be just cause for complaint.

As we have stated, Peterson moved for a judgment of acquittal at trial, and in this court renews his contention that the evidence was insufficient to support a conviction of manslaughter. His position is that the evidence, as a matter of law, established a right to use deadly force in self-defense. In considering that contention, we must accept the evidence "in the light most favorable to the Government, making full allowance for the right of the jury to draw justifiable inferences of fact from the evidence adduced at trial and to assess the credibility of the witnesses before it." We have already concluded that the evidence generated factual issues as to the effect, upon Peterson's self-defense claim, of his aggressive conduct and his failure to retreat. The judgment of conviction appealed from is accordingly affirmed.

Questions for Discussion

1. Outline the facts in *United States v. Peterson*.

2. Why does the court of appeals conclude that Peterson was an aggressor who was not entitled to a claim of self-defense?

3. Explain why the court of appeals imposed an obligation on Peterson to retreat before employing deadly force. What type of acts would have fulfilled Peterson's duty to retreat?

4. Why, if Peterson was standing in his own yard, could he not rely on the castle doctrine?

5. Does it make sense to hold Peterson criminally liable for killing a trespasser who was vandalizing Peterson's automobile and whom he had warned not to enter his yard?

6. Do you believe that individuals should be authorized to "hold their ground" under all circumstances rather than retreat?

7. Are the requirements of the law of self-defense too confusing to be understood by most people?

8. Should the area surrounding the home be considered part of a dwelling for purposes of the castle doctrine? What of a porch? In *State v. Blue*, 565 S.E.2d 133 (N.C. 2002), the North Carolina Supreme Court observed that many of the same activities that take place in the home take place on a porch. Should this depend on factors such as the size of the porch, whether the porch is enclosed, and the time of year?

Cases and Comments

The Castle Doctrine and Domestic Violence. John Gartland was found to have abused his wife Ellen for some years and the two had lived in separate bedrooms for ten years. They fought earlier in the evening at a bar and when they returned home, John accused Ellen of hiding the remote control to the television. John later entered her bedroom and threatened to "hurt" her. As he approached Ellen, she grabbed her son's shotgun from the bedroom closet. John vowed to kill her and she shot and killed John as he lunged forward. Ellen was convicted of reckless manslaughter. New Jersey is among the minority of states that impose a duty to retreat on an individual in his or her own home in those instances in which an individual is assaulted by a cohabitant. The New Jersey Supreme

Court determined that Ellen did not have the exclusive right to occupy her bedroom and that John had regular access to the room. As a consequence, Ellen possessed a duty to retreat so long as this could have been safely accomplished prior to resorting to deadly force. See *State v. Gartland*, 694 A.2d 564 (N.J. 1997).

Rhode Island, Massachusetts, and North Dakota follow New Jersey and have statutes imposing a duty to retreat when an individual is attacked by a co-inhabitant of the home.

In 2009, the West Virginia Supreme Court of Appeals in *State v. Harden* reconsidered the "no retreat rule" for victims of domestic violence. On September 5, 2004, Tanya Harden was arrested for shooting and killing her husband, Danuel Harden. Tanya claimed that she acted in self-defense and that the killing was a response to a "night of domestic terror." The evidence indicated that her husband had been drinking heavily (his blood alcohol count at the time of death was 0.22%) and that his violent attack included "brutally beating the defendant with the butt and barrel of a shotgun, brutally beating the defendant with his fists, and sexually assaulting the defendant." The "night of terror" ended when Tanya shot and killed Danuel.

The prosecutor in addition to arguing that there was no reasonable basis for Tanya to believe that there was an imminent threat of serious injury or death contended that Tanya's use of deadly force was not reasonable because she could have retreated to safety. Her husband was "on that couch . . . and she has got control of that shotgun, she . . . could have called the law, and she could have walked out of that trailer. Period. But she didn't."

The West Virginia Supreme Court overturned existing precedent and held that "an occupant who . . . without provocation [is] attacked in his or her home, dwelling or place of temporary abode, by a co-occupant who also has a lawful right to be upon the premises, may invoke the law of self-defense" and has no duty to retreat before exercising the right of self-defense.

The court explained that women who flee the home in many instances are "caught, dragged back inside, and severely beaten again. [Even i]f she manages to escape . . . [w]here will she go if she has no money, no transportation, and if her children are left behind in the care of an enraged man?" The West Virginia Court also reasoned that it was unfair that a woman attacked in the home by a stranger may stand her ground while a woman who is attacked by her husband or partner must retreat. See *State v. Harden*, 679 S.E.2d 628 (W. Va. 2009).

You Decide

8.4 Aiken and the victim had been next-door neighbors in an apartment building in the Bronx, New York, for roughly forty years. The two families had a falling out in 1994 or 1995 when a disagreement arose when the victim accused the defendant of siphoning his family's cable and telephone service. The service providers found no basis for this accusation. In 1997, the victim stabbed Aiken in the back resulting in his hospitalization. The two families continued to live next to one another from 1997 to 1999. This could not have been pleasant because the victim continually "threatened to shoot, stab or otherwise injure defendant. He made these threats to defendant's face, to his father and to neighbors—at one point even brandishing a boxcutter." On December 21, 1999, Aiken and the victim argued through the shared bedroom wall between their apartments. The defendant took a metal pipe and dented his side of the wall. The victim's mother called the police, and the victim left his apartment to go downstairs and open the building's front door for the police. The defendant remained inside his apartment, walked to the front door several times, and then opened the door when he saw the victim standing outside his door with a friend. "Still holding the metal pipe he had earlier used to hit the wall, the defendant then engaged in an angry argument with the victim, who remained in [the defendant's] doorway. . . . [H]e [the defendant] continued standing in the doorway, never going into the hall, when the victim reached into his pocket, came up to defendant's face 'nose to nose,' and said 'he was going to kill' him." The defendant believed that he was about to be stabbed once again and hit the victim in the head with the metal pipe, killing the victim. The trial court instructed the jury that "a person may . . . not use defensive deadly force if he knows he can with complete safety to himself avoid such use of deadly physical force by retreating." The trial court refused the defendant's request to "charge the jury that, if a defendant is in his home and close proximity of a threshold of his home, there is no duty to retreat." Do you agree with these jury instructions? Was Aiken in his home? Did Aiken act in self-defense in killing the victim? See *People v. Aiken*, 828 N.E.2d 74 (N.Y. 2005).

You can find the answer at www.sagepub.com/lippmanccl3e.

DEFENSE OF OTHERS

The common law generally limited the privilege of **intervention in defense of others** to the protection of spouses, family, employees, and employers. This was based on the assumption that an individual would be in a good position to evaluate whether these individuals were aggressors or victims in need of assistance. Some state statutes continue to limit the right to intervene, but this no longer is the prevailing legal rule. The Wisconsin statute provides that a person is justified in "threatening or using force against another when . . . he or she reasonably believes that force is necessary to defend himself or a third person against such other's imminent use of unlawful force."

The early approach in the United States was the **alter ego rule**. This provides that an individual intervening "stands in the shoes" or possesses the "same rights" as the person whom he or she is assisting. The alter ego approach generally has been abandoned in favor of the reasonable person or **objective test for intervention in defense of others** of the Model Penal Code. Section 3.05 provides that an individual is justified in using force to protect another whom he or she reasonably believes (1) is in immediate danger and (2) is entitled under the Model Penal Code to use protective force in self-defense, and (3) such force is necessary for the protection of the other person. An intervener is not criminally liable under this test for a reasonable mistake of fact.

What is the difference between the alter ego rule and the objective test? An individual intervening under the alter ego rule acts at his or her own peril. The person "in whose shoes they stand" may in fact be an aggressor or may not possess the right of self-defense. The objective test, on the other hand, protects individuals who act in a "reasonable," but mistaken belief.

In *People v. Young*, two plainclothes detectives arrested a teenager for blocking traffic. The defendant intervened and hit one of the "two white men" who was "pulling" on the African American teenager. The defendant was convicted under the alter ego rule for intervening to defend an individual who, in fact, did not possess a right to self-defense. The New York Court of Appeals affirmed the conviction, but asked "what public interest is promoted by a principle which would deter one from coming to the aid of a fellow citizen whom he has reasonable grounds to believe is in imminent danger of personal injury at the hands of assailants?"[28] The New York legislature responded by modifying the law to provide that a "person . . . may use physical force upon another person when and to the extent that he reasonably believes such to be necessary to defend himself or a third person from what he reasonably believes to be the use or imminent use of unlawful physical force by such other person. . . ."[29]

Remember, you may intervene to protect another, but you are not required to intervene. George Fletcher notes that the desire to provide protection to those who intervene on behalf of others reflects the belief that an attack against a single individual is a threat to the rule of law that protects us all.[30] Do you agree with the New York Court of Appeals in *Young* that the law should provide protection to individuals who intervene? *State v. Fair* is a well-known case from New Jersey that explains the objective approach.[31]

DEFENSE OF THE HOME

The home has historically been viewed as a place of safety, security, and shelter. The eighteenth-century English jurist Lord Coke wrote that "[a] man's house is his castle—for where shall a man be safe if it be not his own house." Coke's opinion was shaped by the ancient Roman legal scholars who wrote that "one's home is the safety refuge for everyone." The early colonial states adopted the English common law right of individuals to use deadly force in those instances in which they reasonably believe that this force is required to prevent an imminent and unlawful entry. The common law rule is sufficiently broad to permit deadly force against a rapist, burglar, or drunk who mistakenly stumbles into the wrong house on his or her way to a surprise birthday party.[32]

States gradually abandoned this broad standard and adopted statutes that restricted the use of deadly force in defense of the home. There is no uniform approach today, and statutes typically limit deadly force to those situations in which deadly force is reasonably believed to be required to prevent the entry of an intruder who is reasonably believed to intend to commit "a felony" in the dwelling. Other state statutes strictly regulate armed force and only authorize deadly force in those instances in which it is reasonably believed to be required to prevent the entry of an intruder who is reasonably believed to intend to commit a "forcible felony" involving the threat or use of violence against an occupant.[33] The first alternative would permit the use of deadly force against an individual who is intent on stealing a valuable painting, whereas the second approach would require that the art thief threaten violence or display a weapon.

The Model Penal Code balances the right to protect a dwelling from intruders against respect for human life and provides that deadly force is justified in those instances that the intruder is attempting to commit arson, burglary, robbery, other serious theft, or the destruction of property and has demonstrated that he or she poses a threat by employing or threatening to employ deadly force. Deadly force is also permissible under section 3.06(3)(d)(ii)(A)(B) where the employment of nondeadly force would expose the occupant to substantial danger of serious bodily harm.

The most controversial and dominant trend is toward so-called **make my day laws** that authorize the use of "any degree of force" against intruders who "might use any physical force . . . no matter how slight against any occupant." Colorado Revised Statutes section 18-1-704.5 provides:

> [T]he citizens of Colorado have a right to expect absolute safety within their own homes. . . . [A]ny occupant of a dwelling is justified in using any degree of physical force, including deadly physical force, against another person when that other person has made an unlawful entry into the dwelling, and when the occupant has a reasonable belief that such other person has committed a crime in the dwelling in addition to the uninvited entry, or is committing or intends to commit a crime against a person or property in addition to the uninvited entry, and when the occupant reasonably believes that such other person might use any physical force, no matter how slight against any occupant. Any occupant of a dwelling using physical force . . . shall be immune from criminal prosecution for the use of such force . . . [and immune from civil liability] for injuries or death resulting from the use of such force.

Florida Statutes section 776.013 presumes that an intruder who unlawfully enters a home, automobile, or boat intends to commit a forcible felony. An occupant may use nondeadly or deadly force and does not have the burden in court of establishing that the intruder intended to inflict death or great bodily harm.

In *State v. Anderson,* the Oklahoma Court of Criminal Appeals stressed that under the state's Make My Day Law, the occupant possesses unlimited discretion to employ whatever degree of force he or she desires to based "solely upon the occupant's belief that the intruder might use any force against the occupant." In practice, this is a return to the original common law rule because a jury would likely find reasonable justification to believe that almost any intruder poses at least a threat of "slight" physical force against an occupant.[34] The Make My Day Law raises the issue of the proper legal standard for the use of force in defense of the dwelling. Should a homeowner be required to wait until the intruder poses a threat of serious harm?

Professor Joshua Dressler argues that the various legal standards for protection of the dwelling make little difference because in an age marked by fear of "home invasion" and violent crime, a jury will almost always find the use of deadly force is justified against an intruder.[35] The next case, *People v. Ceballos,* discusses whether it is legal to employ a spring gun to protect against illegal entry into the home.

Was Ceballos justified in defending his home with a spring gun?

People v. Ceballos, 526 P.2d 241 (Cal. 1974). Opinion by: Burke, J.

Facts

Defendant lived alone in a home in San Anselmo. The regular living quarters were above the garage, but defendant sometimes slept in the garage and had about $2,500 worth of property there. In March 1970 some tools were stolen from defendant's home. On May 12, 1970, he noticed the lock on his garage doors was bent and pry marks were on one of the doors. The next day he mounted a loaded .22 caliber pistol in the garage. The pistol was aimed at the center of the garage doors and was connected by a wire to one of the doors so that the pistol would discharge if the door was opened several inches.

The damage to defendant's lock had been done by a sixteen-year-old boy named Stephen and a fifteen-year-old boy named Robert. On the afternoon of May 15, 1970, the boys returned to defendant's house while he was

away. Neither boy was armed with a gun or knife. After looking in the windows and seeing no one, Stephen succeeded in removing the lock on the garage doors with a crowbar, and, as he pulled the door outward, he was hit in the face with a bullet from the pistol.

Stephen testified: He intended to go into the garage "[for] musical equipment" because he had a debt to pay to a friend. His "way of paying that debt would be to take [defendant's] property and sell it" and use the proceeds to pay the debt. He "wasn't going to do it [i.e., steal] for sure, necessarily." He was there "to look around," and "getting in, I don't know if I would have actually stolen."

Defendant, testifying in his own behalf, admitted having set up the trap gun. He stated that after noticing the pry marks on his garage door on May 12, he felt he should "set up some kind of a trap, something to keep the burglar out of my home." When asked why he was trying to keep the burglar out, he replied, " . . . Because somebody was trying to steal my property . . . and I don't want to come home some night and have the thief in there . . . usually a thief is pretty desperate . . . and . . . they just pick up a weapon . . . if they don't have one . . . and do the best they can." When asked by the police shortly after the shooting why he assembled the trap gun, defendant stated that "he didn't have much and he wanted to protect what he did have." The jury found defendant guilty of assault with a deadly weapon. . . .

Issue

Defendant contends that had he been present he would have been justified in shooting Stephen since Stephen was attempting to commit burglary . . . that . . . defendant had a right to do indirectly what he could have done directly, and that therefore any attempt by him to commit a violent injury upon Stephen was not "unlawful" and hence not an assault. The People argue that . . . as a matter of law a trap gun constitutes excessive force, and that in any event the circumstances were not in fact such as to warrant the use of deadly force. . . .

Reasoning

In the United States, courts have concluded that a person may be held criminally liable under statutes proscribing homicides and shooting with intent to injure, or civilly liable, if he sets upon his premises a deadly mechanical device and that device kills or injures another. . . . However, an exception to the rule that there may be criminal and civil liability for death or injuries caused by such a device has been recognized where the intrusion is, in fact, such that the person, were he present, would be justified in taking the life or inflicting the bodily harm with his own hands. . . . The phrase "were he present" does not hypothesize the actual presence of the person . . . but is used in setting forth in an indirect manner the principle that a person may do indirectly that which he is privileged to do directly.

Allowing persons, at their own risk, to employ deadly mechanical devices imperils the lives of children, firemen and policemen acting within the scope of their employment, and others. Where the actor is present, there is always the possibility he will realize that deadly force is not necessary, but deadly mechanical devices are without mercy or discretion. Such devices "are silent instrumentalities of death. They deal death and destruction to the innocent as well as the criminal intruder without the slightest warning. The taking of human life [or infliction of great bodily injury] by such means is brutally savage and inhuman."

It seems clear that the use of such devices should not be encouraged. Moreover, whatever may be thought in torts [a civil action for damages], the foregoing rule setting forth an exception to liability for death or injuries inflicted by such devices is inappropriate in penal law for it is obvious that it does not prescribe a workable standard of conduct; liability depends upon fortuitous results. We therefore decline to adopt that rule in criminal cases.

Furthermore, even if that rule were applied here, as we shall see, defendant was not justified in shooting Stephen. California Penal Code (CPC) section 197 provides: "Homicide is . . . justifiable . . . 1. When resisting any attempt to murder any person, or to commit a felony, or to do some great bodily injury upon any person; or, 2. When committed in defense of habitation, property, or person, against one who manifestly intends or endeavors, by violence or surprise, to commit a felony. . . ." Since a homicide is justifiable under the circumstances specified in section 197, it follows that an attempt to commit a violent injury upon another under those circumstances is justifiable.

By its terms, subdivision 1 of CPC section 197 appears to permit killing to prevent any "felony," but in view of the large number of felonies today and the inclusion of many that do not involve a danger of serious bodily harm, a literal reading of the section is undesirable. . . . We must look further into the character of the crime, and the manner of its perpetration. When these do not reasonably create a fear of great bodily harm, as they could not if defendant apprehended only a misdemeanor assault, there is no cause for the exaction of a human life. . . . The term "violence or surprise" in subdivision 2 is found in common law authorities . . . and, whatever may have been the very early common law, the rule developed at common law that killing or use of deadly force to prevent a felony was justified only if the offense was a forcible and atrocious crime.

Examples of forcible and atrocious crimes are murder, mayhem, rape, and robbery. In such crimes "from their atrocity and violence human life [or personal safety from great harm] either is, or is presumed to be, in peril." . . .

Burglary has been included in the list of such crimes. However, in view of the wide scope of burglary . . . it cannot be said that under all circumstances burglary . . . constitutes a forcible and atrocious crime.

Where the character and manner of the burglary do not reasonably create a fear of great bodily harm, there is no cause for exaction of human life or for the use of deadly force. The character and manner of the burglary could not reasonably create such a fear unless the burglary threatened, or was reasonably believed to threaten, death or serious bodily harm.

Holding

In the instant case the asserted burglary did not threaten death or serious bodily harm, since no one but Stephen and Robert was then on the premises. . . . There is ordinarily the possibility that the defendant, were he present, would realize the true state of affairs and recognize the intruder as one whom he would not be justified in killing or wounding. We thus conclude that defendant was not justified under CPC section 197, subdivisions 1 or 2, in shooting Stephen to prevent him from committing burglary. . . .

Questions for Discussion

1. Ceballos contends that the spring gun resulted only in the employment of the same degree of force that he would have been justified in employing had he been present. The California Supreme Court rejects this standard on the ground that the mechanism is "brutally savage and inhumane." What is the basis for this conclusion? Does this suggest that spring guns are prohibited under all circumstances? Do you agree that Ceballos is proposing an "unworkable standard"?

2. Burglary involves breaking and entering with an intent to commit a felony. Why is burglary not considered a forcible and atrocious crime? What types of offenses are considered forcible and atrocious crimes?

3. Is the standard for the use of deadly force against intruders proposed by the court too complicated to be easily understood? Should you have the right to use deadly force against intruders in your house regardless of whether they pose a threat to commit a forcible and atrocious crime?

4. The Model Penal Code section 3.06(5) provides that a "device" such as a spring gun may only be employed in the event that it is not "designed to cause or known to create a substantial risk of causing death or serious bodily injury." Do you agree?

You Decide

8.5 Law is a thirty-two-year-old African American who moved into a white, middle-class neighborhood with his wife. His home was broken into within two weeks and his clothes and personal property were stolen. Law purchased a 12-gauge shotgun and installed double locks on his doors. One week later, a neighbor saw a flickering light in Law's otherwise darkened house at roughly 8:00 p.m., and because the home had previously been burglarized, the neighbor called the police. Officers Adams and Garrison examined whether windows in the house had been tampered with, and they shined their flashlights into the dwelling. Then they entered the back screened porch where they noticed that the windowpanes on the door to the house had been temporarily put into place with a few pieces of molding. They had no way of knowing that Law had placed the windows in the door in this fashion following the burglary. Officer Garrison removed the molding and glass and reached inside to open the door. He determined that it was a deadlock and decided that the door could not have been opened without a key. As the officer removed his hand from the window, he was killed by a shotgun blast. Officer Potts, the next officer to arrive, testified that he saw Officer Adams running to his squad car yelling that he had been shot at from inside the home. A number of officers arrived, and believing there was a burglar in the house, they unleashed a massive attack as indicated by the fact that there were forty bullet holes in the kitchen door alone.

Law was in the bedroom with his wife and testified that he heard noise outside the house, and he went downstairs and armed himself with a shotgun he had purchased two days following the burglary. Law then went to the back door and observed a "fiddling around with the door" and then heard scraping on the windowpane along with a voice saying, "Let's go in." Law could not see the back porch because of curtains covering the window on the door. When Law heard the voice say, "Let's go in," he was admittedly scared and testified that he could have either intentionally or unintentionally pulled the trigger of the shotgun. At one point following his arrest, Law indicated to the police that he believed that the intruders were members of the Ku Klux Klan. The prosecutor conceded in closing argument that Law "probably thought he shot a burglar or whatever that was outside." Did Law act unreasonably and employ excessive force against the "intruders"? Was Law entitled to the justification of defense of habitation? See *Law v. State*, 318 A.2d 859 (Md. Ct. Spec. App. 1974). See also *Law v. State*, 349 A.2d 295 (Md. Ct. Spec. App. 1975).

You can find the answer at www.sagepub.com/lippmanccl3e.

EXECUTION OF PUBLIC DUTIES

The enforcement of criminal law requires that the police detain, arrest, and incarcerate individuals and seize and secure property. This interference with life, liberty, and property would ordinarily constitute a criminal offense. The law, however, provides a defense to individuals executing public duties. This is based on a judgment that the public interest in the enforcement of the law justifies intruding on individual liberty.

There are few areas as controversial as the employment of deadly force by police officers in arresting a fleeing suspect. This, in effect, imposes a fatal punishment without trial. Professor Joshua Dressler writes that until the fourteenth century, law enforcement officers possessed the right to employ deadly force against an individual whom the officer reasonably believed had committed a felony. This was the case even in those circumstances in which a felon could have been apprehended without the use of deadly force. Dressler writes that the authorization of deadly force was based on the notion that felons were "outlaws at war with society" whose lives could be taken to safeguard society. This presumption was strengthened by the fact that felons were subject to capital punishment and to the forfeiture of property. Felons were considered to have forfeited their right to life and the police were merely imposing the punishment that awaited them in any event.[36] The police officer, as noted by the Indiana Supreme Court, is a "minister of justice, and is entitled to the peculiar protection of the law. Without submission to his authority there is no security and anarchy reigns supreme. He must of necessity be the aggressor, and the law affords him special protection."[37] In contrast, only reasonable force could be applied to apprehend a **misdemeanant**. Misdemeanors were punished by a modest fine or brief imprisonment and were not considered to pose a threat to the community. As a consequence, it was considered inhumane for the police to employ deadly force against individuals responsible for minor violations of the law.[38]

The arming of the police and the **fleeing felon rule** were reluctantly embraced by the American public that, although distrustful of governmental power, remained fearful of crime. With a population of three million, Chicago was one of the most violent American cities in the 1920s. A crime survey covering 1926 and 1927 concluded that although most police killings were justified, in other cases "it would seem that the police were hasty and there might be some doubt as to the justification; but in every such instance the coroner's jury returned a verdict of justifiable homicide and no prosecutions resulted. From this we may conclude that the police of the city of Chicago incur no hazard by shooting to kill within their discretion."[39] Some state legislatures attempted to moderate the fleeing felon rule by adopting the standard that a police officer who reasonably believed that deadly force was required to apprehend a suspect would be held criminally liable in the event that he was shown to have been mistaken.[40]

The judiciary began to seriously reconsider the application of the fleeing felon rule in the 1980s. Only a small number of felonies remained punishable by death, and offenses in areas such as white-collar crime posed no direct danger to the public. The rule permitting the employment of deadly force against fleeing felons developed prior to arming the police with firearms in the mid-nineteenth century. As a result, deadly force under the fleeing felon rule was traditionally employed at close range and was rarely invoked to apprehend a felon who escaped an officer's immediate control.[41] An additional problematic aspect of the fleeing felon rule was the authorization for private citizens to employ deadly force, although individuals risked criminal liability in the event that they were proven to have been incorrect.[42]

The growing recognition that criminal suspects retained various constitutional rights also introduced a concern with balancing the interests of suspects against the interests of the police and society.

A number of reasons are offered to justify a limitation on the use of deadly force to apprehend suspects:

1. The shooting of suspects may lead to *community alienation and anger,* particularly in instances in which the evidence indicates that there was no need to employ deadly force or the deceased is revealed to have been unarmed or innocent.

2. *Bystanders* may be harmed or injured by stray bullets.

3. *Substantial monetary damages* may be imposed on a municipality in civil suits alleging that firearms were improperly employed.

4. *Police officers* who employ deadly force can suffer *psychological stress,* strain, and low morale, and may change careers or retire.

The Modern Legal Standard

In 1985, the U.S. Supreme Court reviewed the fleeing felon rule in the next case in the chapter, *Tennessee v. Garner.* The case was brought under a civil rights statute by the family of the deceased who was seeking monetary damages for deprivation of the "rights . . . secured by the Constitution," (42 U.S.C. § 1983). The Supreme Court determined that the police officer violated Garner's Fourth Amendment right to be free from "unreasonable seizures." Although this was a civil rather than criminal decision, the judgment established the standard to be employed in criminal prosecutions against officers charged with the unreasonable utilization of deadly force.

We may question whether it is fair to place the fate of a police officer in the hands of a judge or jury who may not fully appreciate the pressures confronting an officer required to make a split-second decision whether to employ armed force. Others point to the fact that the use of deadly force typically occurs in situations in which there are few witnesses and the judge and jury must rely on the well-rehearsed testimony of the police. What do you think of the standard established in *Garner?*

Model Penal Code

Section 3.07. Use of Force in Law Enforcement

(1) Use of Force Justifiable to Effect an Arrest. . . . The use of force upon or toward the person of another is justifiable when the actor is making or assisting in making an arrest and the actor believes that such force is immediately necessary to effect a lawful arrest.

(2) Limitation on the Use of Force.

　(a) The use of force is not justifiable under this section unless:

　　(i) the actor makes known the purpose of the arrest or believes that it is otherwise known by or cannot reasonably be made known to the person to be arrested; and

　　(ii) when the arrest is made under a warrant, the warrant is valid or believed by the actor to be valid.

　(b) The use of deadly force is not justifiable under this Section unless:

　　(i) the arrest is for a felony; and

　　(ii) the person effecting the arrest is authorized to act as a peace officer or is assisting a person whom he believes to be authorized to act as a peace officer; and

　　(iii) the actor believes that the force employed creates no substantial risk of injury to innocent persons; and

　　(iv) the actor believes that:

　　　(A) the crime for which the arrest is made involved conduct including the use or threatened use of deadly force; or

　　　(B) there is a substantial risk that the person to be arrested will cause death or serious bodily harm if his apprehension is delayed.

Analysis

1. The Model Penal Code substantially restricts the common law on the employment of deadly force against fleeing felons.

2. Deadly force is limited to the police or to individuals assisting an individual believed to be a police officer. This limits the utilization or supervision of the use of deadly force to individuals trained in the employment of firearms.

3. The employment of deadly force is restricted to felonies that the police officer believes involves the use or threatened use of deadly force or to situations in which the police officer believes that a delay in arrest will create a substantial risk that the person to be arrested will cause death or serious bodily harm.

4. The police officer possesses a reasonable belief that there is no substantial risk to innocent individuals.

The Legal Equation

Deadly force

An arrest = Fleeing felon

+ law enforcement officer and civilian acting under officer's direction

+ felony arrest

+ no substantial risk of injury to innocents

+ felony involves use or threatened use of deadly force

+ substantial risk of death or serious injury or death if apprehension delayed

+ warning where feasible.

Was the officer justified in killing the burglar?

Tennessee v. Garner, 471 U.S. 1 (1985). Opinion by: White, J.

Issue

This case requires us to determine the constitutionality of the use of deadly force to prevent the escape of an apparently unarmed suspected felon. We conclude that such force may not be used unless it is necessary to prevent the escape and the officer has probable cause to believe that the suspect poses a significant threat of death or serious physical injury to the officer or others.

Facts

At about 10:45 p.m. on October 3, 1974, Memphis Police Officers Elton Hymon and Leslie Wright were dispatched to answer a "prowler inside call." Upon arriving at the scene they saw a woman standing on her porch and gesturing toward the adjacent house. She told them she had heard glass breaking and that "they" or "someone" was breaking in next door. While Wright radioed the dispatcher to say that they were on the scene, Hymon went behind the house. He heard a door slam and saw someone run across the backyard. The fleeing suspect, who was appellee-respondent's decedent, Edward Garner, stopped at a six-foot-high chain-link fence at the edge of the yard. With the aid of a flashlight, Hymon was able to see Garner's face and hands. He saw no sign of a weapon, and, though not certain, was "reasonably sure" and "figured" that Garner was unarmed. He thought Garner was 17 or 18 years old and about 5'5" or 5'7" tall [In fact, Garner, an eighth grader, was 15. He was 5'4" tall and weighed around 100 or 110 pounds]. While Garner was crouched at the base of the fence, Hymon called out "police, halt" and took a few steps toward him. Garner then began to climb over the fence. Convinced that if Garner made it over the fence he would elude capture, Hymon shot him. The bullet hit Garner in the back of the head. Garner was taken by ambulance to a hospital, where he died on the operating table. Ten dollars and a purse taken from the house were found on his body. . . .

Garner had rummaged through one room in the house, in which, in the words of the owner, "[all] the stuff was out on the floors, all the drawers was pulled out, and stuff was scattered all over." The owner testified that his valuables were untouched but that, in addition to the purse and the 10 dollars, one of his wife's rings was missing. The ring was not recovered.

In using deadly force to prevent the escape, Hymon was acting under the authority of a Tennessee statute and pursuant to Police Department policy. The statute provides that "[if], after notice of the intention to arrest the defendant, he either flee or forcibly resist, the officer may use all the necessary means to effect the arrest." Tenn. Code Ann. § 40-7-108 (1982). The Department policy was slightly more restrictive than the statute, but still allowed the use of deadly force in cases of burglary. Although the statute does not say so explicitly,

Tennessee law forbids the use of deadly force in the arrest of a misdemeanant. The incident was reviewed by the Memphis Police Firearm's Review Board and presented to a grand jury. Neither took any action.

Garner's father then brought this action in the Federal District Court for the Western District of Tennessee, seeking damages under 42 U.S.C. § 1983 for asserted violations of Garner's constitutional rights. . . . After a three-day bench trial, the District Court entered judgment for all defendants. . . . [I]t . . . concluded that Hymon's actions were authorized by the Tennessee statute, which in turn was constitutional. Hymon had employed the only reasonable and practicable means of preventing Garner's escape. Garner had "recklessly and heedlessly attempted to vault over the fence to escape, thereby assuming the risk of being fired upon." The Court of Appeals reversed . . .

Reasoning

Whenever an officer restrains the freedom of a person to walk away, he has seized that person. . . . There can be no question that apprehension by the use of deadly force is a seizure subject to the reasonableness requirement of the Fourth Amendment. A police officer may arrest a person if he has probable cause to believe that person committed a crime. . . . Petitioners and appellant argue that if this requirement is satisfied the Fourth Amendment has nothing to say about how that seizure is made. This submission ignores the many cases in which this Court, by balancing the extent of the intrusion against the need for it, has examined the reasonableness of the manner in which a search or seizure is conducted. . . .

The same balancing process . . . demonstrates that, notwithstanding probable cause to seize a suspect, an officer may not always do so by killing him. The intrusiveness of a seizure by means of deadly force is unmatched. The suspect's fundamental interest in his own life need not be elaborated upon. The use of deadly force also frustrates the interest of the individual, and of society, in judicial determination of guilt and punishment. Against these interests are ranged governmental interests in effective law enforcement. It is argued that overall violence will be reduced by encouraging the peaceful submission of suspects who know that they may be shot if they flee. Effectiveness in making arrests requires the resort to deadly force, or at least the meaningful threat thereof. "Being able to arrest such individuals is a condition precedent to the state's entire system of law enforcement. . . ."

Without in any way disparaging the importance of these goals, we are not convinced that the use of deadly force is a sufficiently productive means of accomplishing them to justify the killing of nonviolent suspects. . . . [W]hile the meaningful threat of deadly force might be thought to lead to the arrest of more live suspects by discouraging escape attempts,

the presently available evidence does not support this thesis. The fact is that a majority of police departments in this country have forbidden the use of deadly force against nonviolent suspects. If those charged with the enforcement of the criminal law have abjured the use of deadly force in arresting nondangerous felons, there is a substantial basis for doubting that the use of such force is an essential attribute of the arrest power in all felony cases. . . . Petitioners and appellant have not persuaded us that shooting nondangerous fleeing suspects is so vital as to outweigh the suspect's interest in his own life [the use of punishment to discourage flight has been largely ignored. The Memphis City Code punishes escape with a $50 fine].

Holding

The use of deadly force to prevent the escape of all felony suspects, whatever the circumstances, is constitutionally unreasonable. It is not better that all felony suspects die than that they escape. Where the suspect poses no immediate threat to the officer and no threat to others, the harm resulting from failing to apprehend him does not justify the use of deadly force to do so. It is no doubt unfortunate when a suspect who is in sight escapes, but the fact that the police arrive a little late or are a little slower afoot does not always justify killing the suspect. A police officer may not seize an unarmed, nondangerous suspect by shooting him dead. The Tennessee statute is unconstitutional insofar as it authorizes the use of deadly force against such fleeing suspects.

It is not, however, unconstitutional on its face. Where the officer has probable cause to believe that the suspect poses a threat of serious physical harm, either to the officer or to others, it is not constitutionally unreasonable to prevent escape by using deadly force. Thus, if the suspect threatens the officer with a weapon or there is probable cause to believe that he has committed a crime involving the infliction or threatened infliction of serious physical harm, deadly force may be used if necessary to prevent escape, and if, where feasible, some warning has been given. As applied in such circumstances, the Tennessee statute would pass constitutional muster.

Officer Hymon could not reasonably have believed that Garner—young, slight, and unarmed—posed any threat. Indeed, Hymon never attempted to justify his actions on any basis other than the need to prevent an escape. . . . The fact that Garner was a suspected burglar could not, without regard to the other circumstances, automatically justify the use of deadly force. Hymon did not have probable cause to believe that Garner, whom he correctly believed to be unarmed, posed any physical danger to himself or others.

The dissent argues that the shooting was justified by the fact that Officer Hymon had probable cause to believe that Garner had committed a nighttime burglary. While we agree that burglary is a serious crime, we cannot agree that it is so dangerous as automatically to justify the use of deadly force. The FBI classifies burglary as a "property" rather than a "violent" crime. Although the armed burglar would present a different situation, the fact that an unarmed suspect has broken into a dwelling at night does not automatically mean he is physically dangerous. This case demonstrates as much. Statistics demonstrate that burglaries only rarely involve physical violence. During the ten-year period from 1973 through 1982, only 3.8 percent of all burglaries involved violent crime. . . .

We hold that the statute is invalid insofar as it purported to give Hymon the authority to act as he did. . . .

Dissenting, *O'Connor, J.,* with whom *Burger, C.J.,* and *Rehnquist, J.,* join.

The public interest involved in the use of deadly force as a last resort to apprehend a fleeing burglary suspect relates primarily to the serious nature of the crime. Household burglaries not only represent the illegal entry into a person's home, but also "[pose] real risk of serious harm to others." According to recent Department of Justice statistics, "[three-fifths] of all rapes in the home, three-fifths of all home robberies, and about a third of home aggravated and simple assaults are committed by burglars." During the period 1973 through 1982, 2.8 million such violent crimes were committed in the course of burglaries. Victims of a forcible intrusion into their home by a nighttime prowler will find little consolation in the majority's confident assertion that "burglaries only rarely involve physical violence." . . .

Admittedly, the events giving rise to this case are in retrospect deeply regrettable. No one can view the death of an unarmed and apparently nonviolent fifteen-year-old without sorrow, much less disapproval. . . . The officer pursued a suspect in the darkened backyard of a house that from all indications had just been burglarized. The police officer was not certain whether the suspect was alone or unarmed; nor did he know what had transpired inside the house. He ordered the suspect to halt, and when the suspect refused to obey and attempted to flee into the night, the officer fired his weapon to prevent escape. The reasonableness of this action for purposes of the Fourth Amendment is not determined by the unfortunate nature of this particular case; instead, the question is whether it is constitutionally impermissible for police officers, as a last resort, to shoot a burglary suspect fleeing the scene of the crime. . . .

I cannot accept the majority's creation of a constitutional right to flight for burglary suspects seeking to avoid capture at the scene of the crime. . . . I respectfully dissent.

Questions for Discussion

1. Did Officer Hymon's shooting of the suspect comply with the Tennessee statute? How does the Tennessee statute differ from the holding in *Garner*? Do you believe that the Supreme Court majority places too much emphasis on protecting the fleeing felon?

2. Justice O'Connor writes at one point in her dissent that the Supreme Court majority offers no guidance on the factors to be considered in determining whether a suspect poses a significant threat of death or serious bodily harm and does not specify the weapons, ranging from guns to knives to baseball bats, that will justify the use of deadly force. Is Justice O'Connor correct that the majority's "silence on critical factors in the decision to use deadly force simply invites second-guessing of difficult police decisions that must be made quickly in the most trying of circumstances"?

3. Summarize the facts that Officer Hymon considered in the "split second" that he decided to fire at the suspect. Was his decision reasonable? What of Justice O'Connor's conclusion that the Supreme Court decision will lead to a large number of cases in which lower courts are forced to "struggle to determine if a police officer's split-second decision to shoot was justified by the danger posed by a particular object and other facts related to the crime."

4. Is Justice O'Connor correct that the Supreme Court majority unduly minimizes the serious threat posed by burglary? Should the Supreme Court be setting standards for police across the country based on the facts in a single case?

Cases and Comments

1. *The Objective Test for Excessive Force Under the Fourth Amendment.* The U.S. Supreme Court clarified the standard for evaluating the use of excessive force by police under the Fourth Amendment in *Graham v. Connor* in 1989. Graham, a diabetic, asked Berry to drive him to a convenience store to purchase orange juice to counteract the onset of an insulin reaction. Graham encountered a long line and hurried out of the store and asked Berry to drive him to a friend's house instead. This aroused the suspicion of a police officer who pulled Berry's automobile over and called for backup officers to assist him in investigating what occurred in the store. The backup officers handcuffed Graham and dismissed Berry's warning

that Graham was suffering from a "sugar reaction." Graham began running around the car, sat down on the curb and briefly collapsed. An officer, concluding that Graham was drunk, cuffed his hands behind his back, placed him face down on the hood, and responded to Graham's pleas for sugar by shoving his face against the car. Four officers grabbed Graham and threw him headfirst into the police car. The police also refused to permit a recently arrived friend of Graham's to give Graham orange juice. The officers then received a report that Graham had done nothing wrong at the convenience store and released him. Graham sustained a broken foot, cuts on his wrists, a bruised forehead, injured shoulder, and claimed to experience a continual ringing in his ear.

The U.S. Supreme Court ruled that claims that the law enforcement officers employed excessive force in the course of an arrest, investigatory stop, or other seizure of a suspect should be analyzed under the Fourth Amendment reasonableness standard. This entails an inquiry into whether the officers' actions are objectively reasonable in light of the facts and circumstances confronting them without regard to their underlying intent or motivation.

The reasonableness of the use of force according to the Supreme Court "must be judged from the perspective of a reasonable officer on the scene, rather than with the 20/20 vision of hindsight. . . . The calculus of reasonableness must embody allowance for the fact that police officers are often forced to make split-second judgments—in circumstances that are tense, uncertain, and rapidly evolving." This analysis should focus on the severity of the crime, whether the suspect poses an immediate threat to the safety of the officers or other individuals, and whether the suspect is actively resisting arrest or evading arrest by flight.

Would you find it difficult as a juror to place yourself in the position of an officer confronting an aggressive and possibly armed or physically imposing suspect? Is it fairer for courts to utilize a "reasonable officer under the circumstances standard" or to use a test that asks whether the degree of force is "understandable under the circumstances?" Are courts "second-guessing" the police? See *Graham v. Connor,* 490 U.S. 386 (1989).

2. ***Hot Pursuit.*** In *Scott v. Harris,* the U.S. Supreme Court confronted the question: May "an officer take actions that place a fleeing motorist at risk of serious injury or death in order to stop the motorist's flight from endangering the lives of innocent bystanders"? In March 2001, a Georgia county deputy clocked Harris's vehicle traveling at seventy-three mph on a road with a fifty-five-mph speed limit. The deputy activated his blue flashing lights indicating that Harris should pull over to the side of the road. He instead sped away, initiating a chase down what is in most portions a two-lane road, at speeds exceeding eighty-five mph. Deputy Timothy Scott heard the radio communication and joined the pursuit along with other officers. "Scott took over as the lead pursuit vehicle. Six minutes and nearly ten miles after the chase had begun, Scott requested permission to terminate the episode by employing a 'Precision Intervention Technique' (PIT) maneuver, which causes the fleeing vehicle to spin to a stop." Scott had received permission to execute this maneuver by his supervisor, who had told him to "go ahead and take him out." Scott concluded that it was safer to apply his push bumper to the rear of respondent's vehicle. As a result, Harris lost control of his vehicle and the automobile left the roadway, ran down an embankment, overturned, and crashed. Harris was badly injured and was rendered a quadriplegic. He filed a civil suit against Deputy Scott and others alleging "violation of his federal constitutional rights, viz. use of excessive force resulting in an unreasonable seizure under the Fourth Amendment." The U.S. Court of Appeals for the Eleventh Circuit "affirmed the District Court's decision to allow respondent's Fourth Amendment claim against Scott to proceed to trial." The Court of Appeals concluded that Scott's actions constituted "deadly force" under *Tennessee v. Garner* and that the use of such force in this context "would violate [respondent's] constitutional right to be free from excessive force during a seizure [and] a reasonable jury could find that Scott violated [respondent's] Fourth Amendment rights."

In 2007, U.S. Supreme Court Justice Antonin Scalia writing for the Supreme Court majority held that Officer Scott had acted in a reasonable fashion. He noted that there was a videotape of the high-speed pursuit that portrays a "Hollywood-style car chase of the most frightening sort, placing police officers and innocent bystanders alike at great risk of serious injury." In evaluating the reasonableness of Officer Scott's actions, Justice Scalia held that the Supreme Court must balance "the risk of bodily harm that Scott's actions posed to Harris" against "the threat to the public that Scott was trying to eliminate." Harris's high-speed flight "posed an actual and imminent threat to the lives of any pedestrians who might have been present, to other civilian motorists, and to the officers involved in the chase." On the other hand, Officer Scott's actions "posed a high likelihood of serious injury or death to Harris—though not the near certainty of death posed by, say, shooting a fleeing felon in the back of the head, or pulling alongside a fleeing motorist's car and shooting the motorist." In this situation, Justice Scalia found that Officer Scott had acted reasonably to protect the innocent members of the public who were placed at risk.

What of abandoning the pursuit? Justice Scalia noted that this would not have insured that Harris would have felt sufficiently free from apprehension by the police to slow down and it would reward a motorist who fled from the police and who placed the public at risk. "The Constitution assuredly does not impose this invitation to impunity-earned-by-reck-lessness. Instead, we lay down a more sensible rule: A police officer's attempt to terminate a dangerous high-speed car chase that threatens the lives of innocent bystanders does not violate the Fourth Amendment, even when it places the fleeing motorist at risk of serious injury or death. . . . The car chase that [Harris] initiated in this case posed a substantial and immediate risk of serious physical injury to others; no reasonable jury could conclude otherwise."

Justice Stevens, in dissent, argued that the reasonable course would have been to abandon the pursuit and proposed the following rule: "When the immediate danger to the public created by the pursuit is greater than the immediate or potential danger to the public should the suspect remain at large, then the pursuit should be discontinued or terminated. . . . Pursuits should usually be discontinued when the violator's identity has been established to the point that later apprehension can be accomplished without danger to the public." As a judge, how would you decide this case? See *Scott v. Harris,* 550 U.S. 374 (2007).

You Decide

8.6 Officer Pfeffer was off duty and spending the day at home. His wife, Sally, noticed a man, later identified as Paul Billingsley, cross the street and attempt to enter their front yard. He was prevented by the bushes from entering the yard and then walked down the sidewalk and entered a neighbor's driveway. Sally called Officer Pfeffer's attention to Billingsley's movements and watched as Billingsley unsuccessfully attempted to enter the locked back door of two homes, before gaining entrance to the home of Gary Machal. Officer Pfeffer asked his wife to call 911, retrieved his service revolver, and confronted Billingsley in Machal's home. Pfeffer drew his revolver and informed Billingsley that he was a police officer and ordered the intruder to halt and to raise his hands. Billingsley had a purse in his left hand and Pfeffer could not observe his right hand. Billingsley ran out the back door onto the deck and jumped some fifteen feet over the railing to the ground. Pfeffer ran to the railing and ordered the suspect to halt. Billingsley landed in a crouched position and then "rotated his left shoulder." Officer Pfeffer fired a shot that struck Billingsley in the lower right back and exited out his groin. Pfeffer did not observe a weapon and the suspect was determined to be unarmed. Billingsley filed an action under 42 U.S.C. § 1983 for a violation of his Fourth Amendment rights. Consider the arguments that might be offered by the prosecution and defense. Was Officer Pfeffer justified in resorting to deadly force? See *Billingsley v. City of Omaha,* 277 F.3d 990 (8th Cir. 2002).

You can find the answer at www.sagepub.com/lippmanccl3e.

CRIME IN THE NEWS

In 2005, Florida passed a so-called "Castle Law," also popularly referred to as the Stand Your Ground law, which expanded the right of self-defense. In the last five years, 31 states have adopted some or all the provisions of the Florida law. These laws are inspired by the common law doctrine that authorizes individuals to employ deadly force without the obligation to retreat against individuals unlawfully entering their home who are reasonably believed to pose a threat to inflict serious bodily harm or death. Individuals under the castle laws possess the right to stand their ground whether they are inside the home or outside the home.

The National Rifle Association (NRA) has been at the forefront of the movement to persuade state legislatures to adopt these "Castle Laws." The NRA argues that it is time for the law to be concerned with the rights of innocent individuals rather than to focus on the rights of offenders. The obligation to retreat before resorting to deadly force according to the NRA restricts the ability of innocent individuals to defend themselves against wrongdoers. The preamble to the Florida law states that "no person . . . should be required to surrender his or her personal safety to a criminal . . . nor . . . be required to needlessly retreat in the face of intrusion or attack." In the words of the spokesperson for the National Association of Criminal Defense Lawyers, "Most people would rather be judged by 12 (a jury) than carried by six (pallbearers)."

The Florida "Castle Law" modified the state's law of self-defense and has three central provisions.

Home. Individuals are presumed to be justified in using deadly force against intruders who forcefully and unlawfully enter their residence or automobile.

Public place. An individual in any location where he or she "has a right to be" is presumed to be justified in the use of deadly force and has no duty to retreat when he or she reasonably believes that such force is required to prevent imminent death or great bodily harm to himself or herself or to another or to prevent a forcible felony.

Immunity. Individuals who are authorized to use deadly force are immune from criminal prosecution and civil liability.

In the past under the Florida law, a jury when confronted with a claim of self-defense by an individual in the home who employed deadly force was asked to decide whether the defendant reasonably believed that an intruder threatened death or serious bodily injury. Under the new Florida law, the issue is whether an intruder forcibly and unlawfully entered the defendant's home.

The immunity provision prevents an individual who possesses a credible claim of self-defense from being brought to trial in criminal or civil court.

The central criticism of the "Castle Doctrine" laws is that the laws create a climate in which people will resort to deadly force in situations in which they previously may have avoided armed violence. This, according to critics, threatens to turn communities into "shooting galleries" reminiscent of the "old West." More than 35 percent of American households contain a firearm, and there are well over a million gun crimes committed in the United States each year. According to critics of the law, the Castle Doctrine laws will increase the number of cases involving claims of self-defense. Claims of self-defense in Florida have more than doubled since the law was passed. Last year there were 278 cases of justified homicide in the United States, the highest total in recent memory.

The early data indicates that self-defense cases in Florida are not being brought to trial The *St. Petersburg Times* studied 93 Florida cases between 2005 and 2010 involving claims of self-defense and found that in well over half of these cases, either individuals claiming self-defense were not charged with a crime or the charges were dropped by prosecutors or dismissed by a judge before trial.

In 2006, Jason Rosenbloom was shot by his neighbor Kenneth Allen in the doorway to Allen's home. Allen had complained about the amount of trash that Rosenbloom was putting out to be picked up by the trash collectors. Rosenbloom knocked on Allen's door and the two engaged in a shouting match. Allen claimed that Rosenbloom prevented Allen from closing the door

to his house with his foot and that Rosenbloom tried to push his way inside the house. Allen shot the unarmed Rosenbloom in the stomach and then in the chest. Allen claimed that he was afraid and that "I have a right . . . to keep my house safe."

The case came down to a "swearing contest" between Rosenbloom and Allen. Allen claimed that the unarmed Rosenbloom "unlawfully" and "forcibly" attempted to enter his home. Rosenbloom's entry created a presumption that Allen acted under reasonable fear of serious injury or death, and the prosecutors did not pursue the case. Under the previous law, the prosecution likely could have established that Allen unlawfully resorted to deadly force because he lacked a reasonable fear that the unarmed Rosenbloom threatened serious injury or death.

In 2009, Billy Kuch wandered in a drunken stupor into the wrong home in his "cookie cutter neighborhood." He unsuccessfully tried to open the door. Gregory Stewart secured his wife and baby inside the home and emerged with his gun. Kuch raised his hands and asked for a light. He testified that he may have stumbled forward and Stewart fired a shot that ripped through the stomach and spleen hospitalizing Kuch for a month. The prosecutor decided that the shooting was undertaken in self-defense. Kuch questioned whether the 6-1, 250-pound Stewart felt threatened by his 5-9, 165-pound drunken, unarmed neighbor. Kuch's parents responded that they were not against gun ownership, but they were against a law that exempts from prosecution a person who kills. In the past, the law in Florida may have required Stewart to lock the door and call 911 rather than to confront Kuch.

The Florida Stand Your Ground law became the topic of intense national debate when George Zimmerman, a neighborhood watch coordinator, claimed that he had killed 17-year-old Treyvon Martin in self-defense. On April 11, 2012, a special prosecutor charged Zimmerman with second degree murder for the killing of the unarmed Treyvon Martin.

Do you favor the adoption of Florida-type Stand Your Ground Laws?

See more cases on the study site:
Hair v. State, www.sagepub
.com/lippmanccl3e.

RESISTING UNLAWFUL ARRESTS

English common law recognized the right to resist an unlawful arrest by reasonable force. The only limitation was that this did not provide a defense to the murder of a police officer. The philosophical basis for the defense of resisting an unlawful arrest is explained in the famous case of *Queen v. Tooley,* in which Chief Justice Holt of the King's Bench pronounced that "if one is imprisoned upon an unlawful authority, it is sufficient provocation to all people out of compassion . . . it is a provocation to all the subjects of England."[43]

The U.S. Supreme Court, in *John Bad Elk v. United States,* in 1900, recognized that this rule had been incorporated into the common law of the United States. The Supreme Court ruled that "[i]f the officer had no right to arrest, the other party might resist the illegal attempt to arrest

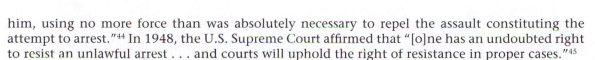

him, using no more force than was absolutely necessary to repel the assault constituting the attempt to arrest."[44] In 1948, the U.S. Supreme Court affirmed that "[o]ne has an undoubted right to resist an unlawful arrest . . . and courts will uphold the right of resistance in proper cases."[45]

The English common law rule was recognized as the law in forty-five states as late as 1963. Today, only twelve states continue to recognize the English rule and have not adopted the **American rule for resistance to an unlawful arrest**. The jurisdictions that retain the rule are generally located in the South, perhaps reflecting the region's historical distrust of government.[46] The abandonment of the recognition of a right to resist by a majority of states is because of the fact that the rule is no longer thought to make much sense. The common law rule reflected the fact that imprisonment, even for brief periods, subjected individuals to a "death trap" characterized by disease, hunger, and violence. However, today,

- Incarcerated individuals are no longer subjected to harsh, inhuman, and disease-ridden prison conditions that result in illness and death.
- An arrest does not necessarily lead to a lengthy period of incarceration. Individuals have access to bail and are represented by hired or appointed attorneys at virtually every stage of the criminal justice process.
- The complexity of the law often makes it difficult to determine whether an arrest is illegal. An officer might, in good faith, engage in what is later determined to be an illegal search, discover drugs, and arrest a suspect. The legality of the search and resulting arrest may not be apparent until an appeals court decides the issue.
- The development of sophisticated weaponry means that confrontations between the police and citizens are likely to rapidly escalate and result in severe harm and injury to citizens and to the police.
- Individuals have access to a sophisticated process of criminal appeal and may bring civil actions for damages.
- The common law rule promotes an unacceptable degree of social conflict and undermines the rule of law.[47]

Individuals continue to retain the right to resist a police officer's application of unnecessary and unlawful force in executing arrest. Judges reason that individuals are not adequately protected against the infliction of death or serious bodily harm by the ability to bring a civil or criminal case charging the officer with the application of excessive force.[48]

The **English rule for resistance to an unlawful arrest**, which provides that an individual may resist an illegal arrest, is still championed by some state courts. The Mississippi Supreme Court noted in *State v. King* that "every person has a right to resist an unlawful arrest; and, in preventing such illegal restraint of his liberty, he may use such force as may be necessary."[49]

Judge Sanders of the Washington Supreme Court dissented from his colleagues' abandonment of the English rule and observed that the police power is "not measured by how hard the officer can wield his baton but rather by the rule of law. Yet by fashioning the rule as it has, the majority legally privileges the aggressor while insulting the victim with a criminal conviction for justifiable resistance."[50] The Maryland Supreme Court observed that law enforcement officers were rarely called to account for illegal arrests by civil or criminal prosecutions and that the right to resist provides an effective deterrent to police illegality.[51]

See more cases on the study site: State v. Hobson, *www.sagepub.com/lippmanccl3e.*

Model Penal Code

Section 3.04. Use of Force in Self-Defense

(1) Use of Force Justifiable for Protection of the Person. . . . The use of force upon or toward another person is justifiable when the actor believes that such force is immediately necessary for the purpose of protecting himself against the use of unlawful force by such other person on the present occasion.

(2) Limitations on Justifying Necessity for Use of Force.

    (a) The use of force is not justifiable under this Section:

        (i) to resist an arrest that the actor knows is being made by a peace officer, although the arrest is unlawful. . . .

Analysis

An individual is not entitled to forcefully resist an unlawful arrest by a law enforcement officer. This restriction does not apply where the aggressor "is not known to the actor to be a peace officer." Self-defense, however, is permitted against a police officer's use of "more force than is necessary" to arrest an individual.

The Legal Equation

A lawful or unlawful arrest   ≠   resistance by physical force.

Excessive force in an arrest   =   proportionate self-defense.

NECESSITY

The **necessity defense** recognizes that conduct that would otherwise be criminal is justified when undertaken to prevent a significant harm. This is commonly called "the **choice of evils**" because individuals are confronted with the unhappy choice between committing a crime or experiencing a harmful event. The harm to be prevented was traditionally required to result from the forces of nature. A classic example is the boat captain caught in a storm who disregards a "no trespassing sign" and docks his or her boat on an unoccupied pier. Necessity is based on the assumption that had the legislature been confronted with this choice, the legislators presumably would have safeguarded the human life of sailors over the property interest of the owner of the dock. As a result, elected officials could not have intended that the trespass statute would be applied against a boat captain confronting this situation.[52]

English common law commentators and judges resisted recognition of necessity. The eighteenth-century English justice Lord Hale objected that recognizing starvation as a justification for theft would lead servants to attack their masters. Roughly one hundred years later, English historian J.F. Stephen offered the often-cited observation that "[s]urely it is at the moment when temptation to crime is the strongest that the law should speak most clearly and emphatically to the contrary."[53]

In 1884, English judges confronted what remains the most challenging and intriguing necessity case in history, *The Queen v. Dudley and Stephens.* The three crew members of the yacht, the *Mignonette,* along with the seventeen-year-old cabin boy, were forced to abandon ship when a wave smashed into the stern. The four managed to launch a thirteen-foot dinghy with only two tins of turnips to sustain them while they drifted sixteen hundred miles from shore. On the fourth day, they managed to catch a turtle that they lived on for a week; they quenched their thirst by drinking their own urine and, at times, by drinking seawater. On the nineteenth day, Captain Thomas Dudley murdered young Richard Parker with the agreement of Edwin Stephens and over the objection of Edmund Brooks. The three survived only by eating Parker's flesh and drinking his blood until rescued four days later. The English court rejected the defense of necessity and the proposition that the members of the crew were justified in taking the life of Parker in order to survive. Lord Coleridge asked, "[b]y what measure is the comparative value of lives to be measured? Is it to be strength, or intellect, or what? . . . It is . . . our duty to declare that the prisoners' act in this case was willful murder. . . . [T]he facts . . . are no legal justification of the homicide. . . ."[54] The defendants were sentenced to death, but released within six months.[55]

The limitation of necessity to actions undertaken in response to the forces of nature has been gradually modified, and most modern cases arise in response to pressures exerted by social conditions and events. *State v. Salin* is representative of this trend. Salin, an emergency medical services technician, was arrested for speeding while responding to a call to assist a two-year-old child who was not breathing. The Delaware court agreed that Salin reasonably assumed that the child was in imminent danger and did not have time to use his cell phone to check on the child's progress. His criminal conviction was reversed on the grounds of necessity. Judge Charles Welch concluded that Salin was confronted by a choice of evils and that his "slightly harmful conduct" was justified in order to "prevent a greater harm."[56]

There are several reasons for the defense of necessity:[57]

- *Respect.* Punishing individuals under these circumstances would lead to disrespect for the legal system.
- *Equity.* Necessity is evaluated on a case-by-case basis and introduces flexibility and fairness into the legal system.

The necessity defense nevertheless remains controversial and subject to criticism:[58]

- *Self-Help.* Individuals should obey the law and should not be encouraged to violate legal rules.
- *Mistakes.* Society suffers when an individual makes the wrong choice in the "choice of evils."
- *Politicalization of the Law.* The defense has been invoked by antiabortion and antinuclear activists and individuals who have broken the law in the name of various political causes.
- *Irrelevancy.* Relatively few cases arise in which the necessity defense is applicable, and too much time is spent debating a fairly insignificant aspect of the criminal law.

Roughly one-half of the states possess necessity statutes and the other jurisdictions rely on the common law defense of necessity. There is agreement on the central elements of the defense.[59]

- *There was an immediate and imminent harm.* In *State v. Green*, the defendant was assaulted and twice sodomized in his cell. He pretended to commit suicide on two occasions so as to be removed from his cell, but was informed that he would have to "fight it out, submit to the assaults or go over the fence." Three months later he was threatened with sexual assaults by five inmates and escaped from prison. The Missouri Supreme Court ruled that the trial court properly denied Green the defense of necessity because "[t]his is not a case where defendant escaped while being closely pursued by those who sought by threat of death or bodily harm to have him submit to sodomy."[60]
- *The defendant also must not have been substantially at fault in creating the emergency.* In *Humphrey v. Commonwealth*, the Virginia Court of Appeals recognized that the defendant, a convicted felon, was justified in violating a gun possession statute in an effort to protect himself from an armed attack. The court stressed that the appellant was "without fault in provoking the altercation."[61]
- *The harm created by the criminal act is less than that caused by the harm confronting the individuals.* Dale Nelson's truck became bogged down in a marshy area roughly 250 feet from the highway. He was fearful that the truck might topple over, and he and two companions unsuccessfully sought to free the vehicle. A passerby drove Nelson to the Highway Department yard where he ignored No Trespassing signs and removed a dump truck that also became stuck. He returned to the heavy equipment yard and took a front-end loader that he used to remove the dump truck. He freed the dump truck, but both the front-end loader and truck suffered substantial damage. Nelson was ultimately convicted of the reckless destruction of personal property and joyriding. The Alaska Supreme Court ruled that "the seriousness of offenses committed by Nelson were disproportionate to the situation he faced." Nelson's "fears about damage to his truck roof were no justification for his appropriation of sophisticated and expensive equipment."[62]
- *An individual reasonably expected a direct causal relationship between his acts and the harm to be averted.* In *United States v. Maxwell*, the First Circuit Court of Appeals dismissed the defendant's contention that he could have reasonably believed that disrupting military exercises at a naval base would cause the U.S. Navy to withdraw nuclear submarines from the coast of Puerto Rico. The court ruled that this was "pure conjecture" and that the defendant "could not reasonably have anticipated that his act of trespass would avert the harm that he professed to fear." Political activists have been equally unsuccessful in contending that they reasonably believed that acts such as splashing blood on walls of the Pentagon or the vandalizing of government property would impede the U.S. production of military weaponry.[63]
- *There were no available legal alternatives to violating the law.* The District of Columbia Court of Appeals confirmed the conviction of a defendant charged with the unlawful possession of

marijuana where the defendant failed to demonstrate that she had tried the dozens of drugs commonly prescribed to alleviate her medical condition.[64] Necessity also was not considered to justify the kidnapping and "deprogramming" of a youthful member of the Unification Church. The Colorado Court of Appeals explained that even assuming that the young woman confronted an imminent harm from a religious cult, her Swedish parents might have pursued legal avenues such as obtaining a court order institutionalizing the twenty-nine-year-old church member as an "incapacitated or incompetent person."[65]

- *The criminal statute that was violated does not preclude the necessity defense.* Courts examine the text or legislative history of a statute to determine whether the legislature has precluded a defendant from invoking the necessity defense. There is typically no clear answer, and judges often ask whether the legislature would have recognized that the statute may be violated on the grounds of necessity under the circumstances. In *United States v. Oakland Cannabis Buyers' Cooperative,* U.S. Supreme Court Justice Clarence Thomas ruled that "a medical necessity exception for marijuana is at odds with the terms of the Controlled Substances Act."[66] In *United States v. Romano,* Romano was bloodied and battered and fleeing the scene of a fight when stopped by a police officer and charged with DWI (driving while intoxicated). A New Jersey superior court ruled that the state legislature did not preclude the necessity defense in those cases in which an intoxicated driver was fleeing a brutal and possibly deadly attack.[67]

The next case, *Commonwealth v. Kendall,* presents the issue whether a defendant arrested for driving under the influence of intoxicating liquor was entitled to a jury instruction on the defense of necessity. The defendant, at the time of his arrest, was rushing his seriously injured girlfriend to the hospital for medical care; he appealed the trial judge's determination that he had failed to exhaust all available legal alternatives. Ask yourself whether you agree with the decision of the Supreme Judicial Court of Massachusetts.

Model Penal Code

Section 3.02. Justification Generally: Choice of Evils

(1) Conduct that the actor believes to be necessary to avoid a harm or evil to himself or to another is justifiable, provided that:

    (a) the harm or evil sought to be avoided by such conduct is greater than that sought to be prevented by the law defining the offense charged; and

    (b) neither the Code nor other law defining the offense provides exceptions or defenses dealing with the specific situation involved; and

    (c) a legislative purpose to exclude the justification claimed does not otherwise plainly appear.

(2) When the actor was reckless or negligent in bringing about the situation requiring a choice of harms or evil or in appraising the necessity for his conduct, the justification afforded by this Section is unavailable in a prosecution for any offense for which recklessness or negligence, as the case may be, suffices to establish culpability.

Analysis

The commentary to the Model Penal Code observes that the letter of the law must be limited in certain circumstances by considerations of justice. The commentary lists some specific examples:

1. Property may be destroyed to prevent the spread of a fire.

2. The speed limit may be exceeded in pursuing a suspected criminal.

3. Mountain climbers lost in a storm may take refuge in a house or seize provisions.

4. Cargo may be thrown overboard or a port entered to save a vessel.

5. An individual may violate curfew to reach an air-raid shelter.

6. A druggist may dispense a drug without a prescription in an emergency.

Several steps are involved under the Model Penal Code:

- *A Belief That Acts Are Necessary to Avoid a Harm.* The actor must "actually believe" the act is necessary or required to avoid a harm or evil to himself or to others. A druggist who sells a drug without a prescription must be aware that this is an act of necessity rather than ordinary law breaking.
- *Comparative Harm or Evils.* The harm or evil to be avoided is greater than that sought to be prevented by the law defining the offense. Human life generally is valued above property. A naval captain may enter a port from which the vessel is prohibited to save the life of a crew member. On the other hand, the possibility of financial ruin does not justify the infliction of physical harm. The question of whether an individual has made the proper choice is determined by the judge or jury rather than by the defendant's subjective belief.
- *Legislative Judgment.* A statute may explicitly preclude necessity; for instance, prohibiting abortions to save the life of the mother.
- *Creation of Harm.* An individual who intentionally sets a fire may not later claim necessity. However, an individual who negligently causes a fire may still invoke necessity to destroy property to control the blaze. He or she may be prosecuted for causing the fire.

The Legal Equation

Necessity = Criminal action believed to be necessary to prevent a harm

+ the harm prevented is greater than will result from the criminal act

+ absence of legal alternatives

+ legislature did not preclude necessity

+ did not intentionally create the harm.

Did the defendant exhaust all reasonable alternatives?

Commonwealth v. Kendall, 883 N.E.2d 269 (Mass. 2008). Opinion by: Spina, J.

Issue

In this case, we consider whether the defendant, Clinton Kendall, was entitled to a jury instruction on the defense of necessity with respect to a charge of operating while under the influence of intoxicating liquor, where the defendant was driving in order to get his seriously injured girl friend to a hospital for medical care. A jury found the defendant guilty of operating a motor vehicle while under the influence of intoxicating liquor (OUI), and he was sentenced to two years of probation, with conditions.

Facts

On the evening of November 25, 2001, the defendant and his girl friend, Heather Maloney, went out to the Little Pub in Marlborough for drinks. They were able to travel there on foot because the establishment was no more than a ten-minute walk from the defendant's trailer home. Over the course of several hours, the defendant and Maloney consumed enough alcohol to become intoxicated. They left the Little Pub around 10 P.M. and walked to a nearby Chinese restaurant to get something to eat. The kitchen was closed, but the bar remained open and they each consumed another drink. Maloney wanted to stay at the restaurant for additional drinks, but the defendant persuaded her that they should return to his home.

After they walked back to the defendant's trailer, he opened the door for Maloney, and she went inside, stopping at the top of the stairs to remove her shoes. As the defendant entered the trailer, he stumbled and bumped into Maloney, causing her to fall forward and hit her head on the corner of a table. The impact opened a wound on her head, and she began to bleed profusely. The defendant was unsuccessful in his efforts to stop the bleeding, so the two decided to seek immediate medical attention.

The trailer did not have a telephone, and neither Maloney nor the defendant had a cellular telephone. Approximately seventy-five to eighty other trailers were located in the mobile home park (each about twenty-five feet apart), at least one nearby neighbor (who lived about forty feet from the defendant) was at home during the time of the incident, and a fire station was located approximately one hundred yards from the neighbor's home. Nonetheless, Maloney and the defendant got into his car, and he drove her to the emergency room of Marlborough Hospital. A breathalyzer test subsequently administered to the defendant at the Marlborough police station, after he had been placed under arrest, showed a blood alcohol level of .23 per cent.

At the close of all the evidence at trial, defense counsel informed the judge that he intended to argue a defense of necessity to the charge of OUI, and he requested an appropriate jury instruction. The judge denied counsel's request for an instruction on necessity, concluding that evidence had not been presented to demonstrate that such a defense was applicable in the circumstances of this case, where the parties were in a highly populated area and the defendant could have availed himself of nearby resources to obtain medical attention for Maloney. As a consequence, during his closing statement, defense counsel did not mention the OUI charge to the jury.

The defendant now contends in this appeal that the judge erred in refusing to allow him to present a defense of necessity during his closing argument and in refusing his request for a jury instruction on such defense. The defendant asserts that, contrary to the judge's conclusion, there were no legal alternatives which would have been effective in abating the danger to Maloney given that her wound was extremely serious and time was a critical factor. Moreover, the defendant continues, by determining that alternative courses of action were available, the judge simply substituted his own judgment, with the benefit of hindsight, for that of the jury. We disagree.

Reasoning

In a prosecution for OUI, the Commonwealth must prove beyond a reasonable doubt that the defendant's consumption of alcohol diminished the defendant's ability to operate a motor vehicle safely. The Commonwealth need not prove that the defendant actually drove in an unsafe or erratic manner, but it must prove a diminished capacity to operate safely. . . .

The defense of necessity, also known as the "competing harms" defense, exonerates one who commits a crime under the "pressure of circumstances" if the harm that would have resulted from compliance with the law . . . exceeds the harm actually resulting from the

defendant's violation of the law. At its root is an appreciation that there may be circumstances where the value protected by the law is, as a matter of public policy, eclipsed by a superseding value. . . . In other words, "[a] necessity defense is sustainable [o]nly when a comparison of the competing harms in specific circumstances clearly favors excusing the defendant's conduct."

The common-law defense of necessity is available in limited circumstances. It can only be raised if each of the following conditions is met: "(1) the defendant is faced with a clear and imminent danger, not one which is debatable or speculative; (2) the defendant can reasonably expect that his action will be effective as the direct cause of abating the danger; (3) there is [no] legal alternative which will be effective in abating the danger; and (4) the Legislature has not acted to preclude the defense by a clear and deliberate choice regarding the values at issue." In those instances where the evidence is sufficient to raise the defense of necessity, the burden is on the Commonwealth to prove the absence of necessity beyond a reasonable doubt.

In considering whether a defendant is entitled to a jury instruction on the defense of necessity, we have stated that a judge shall so instruct the jury only after the defendant has presented some evidence on each of the four underlying conditions of the defense. That is to say, an instruction on necessity is appropriate where there is evidence that supports at least a reasonable doubt whether operating a motor vehicle while under the influence of intoxicating liquor was justified by necessity. Notwithstanding a defendant's argument that the jury should be allowed to decide whether the defendant has established a necessity defense, a judge need not instruct on a hypothesis that is not supported by evidence in the first instance. Thus, if some evidence has been presented on each condition of a defense of necessity, then a defendant is entitled to an appropriate jury instruction.

The only issue here is whether the defendant presented some evidence on the third element of the necessity defense, namely that there were no legal alternatives that would be effective in abating the danger posed to Maloney from her serious head wound. "Where there is an effective alternative available which does not involve a violation of the law, the defendant will not be justified in committing a crime. . . . Moreover, it is up to the defendant to make himself aware of any available lawful alternatives, or show them to be futile in the circumstances."

When viewing the evidence in the light most favorable to the defendant, we conclude that he failed to present any evidence to support a reasonable doubt that his operation of a motor vehicle while under the influence of intoxicating liquor was justified by necessity. There is no question that Maloney's head wound was serious and that time was of the essence in securing medical treatment. Nonetheless, the record is devoid of evidence that the defendant made any effort to seek assistance from anyone prior to driving a motor vehicle while intoxicated. The defendant did not try to contact a nearby neighbor to place a 911 emergency telephone call or, alternatively, to drive Maloney to the hospital. There is also no evidence that the defendant attempted to secure help from the fire station or Chinese restaurant, both in relatively close proximity to the defendant's trailer. This is not a case where, because of location or circumstances, there were no legal alternatives for abating the medical danger to Maloney. Moreover, there has been no showing by the defendant that available alternatives would have been ineffective, leaving him with no option but to drive while intoxicated.

Holding

Because the defendant did not present at least some evidence at trial that there were no effective legal alternatives for abating the medical emergency, we conclude that the judge did not err in refusing to allow counsel to present a defense of necessity and in denying his request for an instruction on such a defense.

Dissenting, *Cowin, J.,* with whom *Marshall, C.J.,* and *Cordy, J.,* join.

The necessity defense recognizes that circumstances may force individuals to choose between competing evils. In particular, it may be reasonable at times for an individual to engage in the "lesser evil" of committing a crime in order to avoid greater harms; when this occurs, the individual should not be punished by the law for his actions.

As the court states, our common law requires a defendant to present some evidence on each of the four elements of the necessity defense before a judge is required to instruct the jury on such defense. Once a judge determines that the evidence, viewed in the light most favorable to the defendant, permits a finding that the defendant reasonably acted out of necessity, the judge must instruct on the defense. The jury then decide what the facts are, and resolve the ultimate question whether the defendant's actions were justified by necessity. . . .

The problem with the court's decision is that it puts unreasonable demands on the defendant to show in every instance that he has tested the legal alternatives. In this case, the court apparently requires the defendant

to have knocked on a neighbor's door, or walked to the fire station or Chinese restaurant. This is too burdensome a threshold. To get to the jury, the defendant need only present evidence that he did not explore the legal alternatives because he reasonably deemed them to have been too high a risk. . . . If it was unreasonable to forgo the lawful alternatives, then the defendant has not made out a case that should go to the jury.

The legal alternatives available to the defendant here carried considerable risk of failure. The defendant had already spent valuable time attempting to stop Maloney's bleeding using towels, but was unable to do so. The first neighbor from whom the defendant might have sought help might not have owned a car, or might have been unable or unwilling to drive Maloney to a hospital; the defendant would then have had to proceed to other neighbors, or to the fire station, where there might not have been anyone available to help; even had there been, it could have meant unacceptable delay in getting a badly injured person to the hospital. In short, any of the alternatives proposed today by the court would have consumed valuable time to no purpose; their exploration raised the real possibility of a chain of events that could have resulted in Maloney's serious injury or death. Given the element of risk associated with the situation and the uncertain likelihood of success with respect to the legal alternatives, a jury could find that it was reasonable for the defendant to reject those alternatives and to select the unlawful solution because of the greater likelihood that it would work. The court's decision, however, punishes a reasonable person for taking the "lesser evil" of the unlawful but more effective alternative. . . .

Of course, a defendant would not be entitled to an instruction on necessity if a reasonable person in his position would have found the legal alternatives to be viable. It would have been proper, for instance, for the judge to deny the defendant's request for an instruction on necessity had there been a hospital within walking distance or a neighbor who offered to drive Maloney to the hospital immediately. In most instances, the unlawful path will not be deemed to be reasonable. On this record, however, the defendant was entitled to make a case to the jury that it was reasonable for him to drive his heavily bleeding girl friend to the hospital to receive treatment without first exploring potentially ineffective alternatives. Although the jury might ultimately reject the defendant's argument, it was for them to decide whether he chose the lesser of two evils. I respectfully dissent.

Questions for Discussion

1. What are the elements of necessity under Massachusetts law?

2. Why does the Massachusetts Supreme Judicial Court uphold the decision of the trial court judge not to issue an instruction on the necessity defense?

3. Summarize the argument of the dissenting judges.

4. Do you agree with the majority decision or with the dissenting opinion?

Cases and Comments

1. ***Imminence.*** Steven Gomez was about to be released from prison in March 1992, when he was approached by a fellow inmate, Imran Mir, who was waiting trial on involvement in an international drug conspiracy. Mir solicited Gomez to murder six witnesses in Mir's case and offered $10,000 or half a kilogram of heroin for each witness Gomez killed. Gomez contacted government authorities and agreed to assist in gathering evidence against Mir. Mir provided Gomez with the names of the individuals to be killed, promised to supply the required weapons, and provided a $1,000 down payment. The government subsequently charged Mir with five counts of solicitation to commit murder. The indictment against Mir revealed Gomez's identity as the informant in the case.

In October 1992, Gomez was stopped by a man with a gun who threatened to kill him, and Gomez later learned that there was a contract out on his life. Gomez unsuccessfully sought assistance from U.S. Customs, which had promised him protection; his parole officer; the Sacramento County Sheriff; and Catholic and Protestant churches. He even resorted to detailing his plight in an interview with a local newspaper.

Gomez was scared and started sleeping in the park, living on the streets, spending the night at the homes of friends, and riding buses all night. At one point, he intentionally violated parole and during his month-long incarceration received a written threat. On February 1, 1993, one of his friends received a death threat meant for Gomez. Gomez reacted by arming himself with a 12-gauge shotgun. On February 4, 1993, two days after Gomez began carrying a weapon, he was arrested by Customs agents and was charged with possession of a firearm by a felon.

The Ninth Circuit Court of Appeals held that the threat against Gomez was more than a vague promise of future harm; Gomez possessed good reason to believe that Mir would seek retribution because his hiring of Gomez substantiated that he was willing to kill witnesses. Gomez confronted an international drug cartel boss who posed a danger that satisfied the "present and immediate" requirement of the necessity defense. The Ninth Circuit also noted that it was the government's filing of an indictment against Mir, rather than Gomez's behavior, that placed Gomez in this precarious position.

Gomez exhausted reasonable alternatives before arming himself and could not leave California and join his wife and son in Texas while on parole. In any event, Mir clearly possessed the resources to track down Gomez in Texas. Gomez also possessed a network of family and friends in California who assisted in hiding him from Mir. In short, there were few alternatives available to Gomez other than arming himself.

There was no indication that Gomez armed himself with a shotgun for any purpose other than self-defense.

Gomez immediately dropped the firearm when confronted with customs agents to demonstrate his cooperation. The Ninth Circuit Court of Appeals concluded that the prosecution of Gomez for "trying to protect himself, when the government refused to protect him from the consequences of its own indiscretion, is not what we would expect from a fair-minded sovereign." How would you rule? See *United States v. Gomez*, 92 F.3d 770 (9th Cir. 1996).

2. ***Property Versus Human Life.*** In *State v. Celli*, defendants Brooks and Celli left Deadwood, South Dakota, in search of employment in Newcastle, Wyoming, a distance of roughly seventy-five miles. They planned to hitchhike in the sunny but chilly weather and dressed warmly. The two defendants failed to secure a ride and by late afternoon had walked roughly twelve miles. Celli slipped in the snow along the road and grabbed Brooks, and the two then tumbled down a steep embankment. In an effort to get back to the road, they were forced to cross a frozen stream and fell through the ice. Their shoes and pants were soaked. The temperature quickly dropped to below freezing and they unsuccessfully attempted to hitch a ride back to Deadwood. Brooks and Celli began the trek back and when they spotted a cabin, they broke the lock on the front door and found matches to start a fire with which to dry their clothes. They spent the night in the bed and, in the morning, shared a can of beans. A neighbor noticed the smoke and notified the police. The South Dakota Supreme Court reversed their conviction on fourth-degree burglary on a technicality and found it unnecessary to reach the issue whether the two were entitled to an instruction on the necessity defense. Were the defendants justified in breaking into the cabin and spending the night? See *State v. Celli*, 263 N.W.2d 145 (S.D. 1978).

3. ***Economic Necessity.*** Jesus Bernardo Fontes was arrested after he presented a false identification card to a convenience store clerk and attempted to cash a forged payroll check in the amount of $454.75. The defendant claimed that his three children, who ranged in age from sixteen months to eleven years, experienced serious health problems. The children had not eaten for more than twenty-four hours, and three different food banks had turned down his request for food. The defendant appealed the trial court's refusal to recognize the defense of necessity. Fontes feared that the lack of food would further complicate his children's health problems and lead to malnutrition and death. "While we are not without sympathy for the downtrodden, the law is clear that economic necessity alone cannot support a choice of crime." Should the judge have issued a jury instruction on (economic) necessity? See *People v. Fontes*, 89 P.3d 484 (Colo. Ct. App. 2003)?

You can find more cases on the study site: **The Queen v. Dudley and Stephens, State v. Caswell, People v. Gray, U.S. v. Schoon,** *www.sagepub.com/lippmanccl3e.*

You Decide

8.7 Butterfield and his friend drank throughout the evening. Butterfield's friend carried the drunk defendant to his a garage apartment at about 1:45 a.m. As Butterfield entered his bedroom "he received a lick on his head which rendered him unconscious." When Butterfield awoke, he found himself lying on the floor in a pool of blood." Butterfield realized he was bleeding from the wound and that he required immediate medical attention. Butterfield lived alone and had no telephone in his apartment. He decided to drive to the hospital, fainted while driving, and wrecked his car. Butterfield was arrested for DWI and was sentenced to thirty days in jail and was fined $50.00. Was Butterfield entitled to the necessity defense? See *Butterfield v. State*, 317 S.W.2d 943 (Tex. Crim. App. 1958).

You can find the answer at www.sagepub.com/lippmanccl3e.

You Decide

8.8 Matthew Ducheneaux was charged with possession of marijuana. He was arrested on a bike path in Sioux Falls, South Dakota, during the city's annual "Jazz Fest" in July 2000. He falsely claimed that he lawfully possessed the two ounces of marijuana as a result of his participation in a federal medical research project. Ducheneaux is thirty-six and was rendered quadriplegic by an automobile accident in 1985. He is almost completely paralyzed other than some movement in his hands. Ducheneaux suffers from spastic paralysis that causes unpredictable spastic tremors and pain throughout his body. He testified that he had not been able to treat the symptoms with traditional drug therapies and these protocols resulted in painful and potentially fatal side effects. One of the prescription drugs for spastic paralysis is Marinol, a synthetic tetrahydrocannabinol (THC). THC is the essential active ingredient of marijuana. Ducheneaux has a prescription for Marinol, but he testified it causes dangerous side effects that are absent from marijuana. The South Dakota legislature has provided that "no person may knowingly possess marijuana" and has declined on two occasions to create a medical necessity exception. Would you convict Ducheneaux of the criminal possession of marijuana? The statute provides that the justification defense is available when a person commits a crime "because of the use or threatened use of unlawful force upon him or upon another person." See *State v. Ducheneaux*, 671 N.W.2d 841 (S.D. 2003).

You can find the answer at www.sagepub.com/lippmanccl3e.

CONSENT

The fact that an individual consents to be the victim of a crime ordinarily does not constitute a defense. For example, the Massachusetts Supreme Judicial Court held that an individual's consensual participation in a sadomasochistic relationship was not a defense to a charge of assault with a small whip. The Massachusetts judges stressed that as a matter of public policy, an individual may not consent to become a victim of an assault and battery with a dangerous weapon.[68]

Professor George Fletcher writes that although an individual is not criminally responsible for self-abuse or for taking his or her own life, those who assist him or her are criminally liable. Why does consent not constitute a justification?[69]

The most common explanation is that the criminal law punishes acts against individuals that harm and threaten society. The fact that an individual may consent to a crime does not mean that society does not have an interest in denouncing and deterring this conduct. The famous eighteenth-century English jurist William Blackstone observed that a criminal offense is a "wrong affecting the general public, at least indirectly, and consequently cannot be licensed by the individual directly harmed."[70]

For an international perspective on this topic, visit the study site.

- Condoning a crime under such circumstances undermines the uniform application of the law and runs the risk that perpetrators will become accustomed and attracted to a life of crime.
- A sane and sensible person would not consent to being a victim of a crime.
- The victim's consent could not possibly constitute a reasonable and rational decision and may be the result of subtle coercion. Society must step in under these circumstances to insure the safety and security of the individual.
- The perpetrator of a crime can easily claim consent, and courts do not want to unravel the facts.

In *State v. Brown*, a New Jersey Superior Court ruled that a wife's instructions to her husband that he was to beat her in the event that she consumed alcoholic beverages did not constitute a justification for the severe beating he administered. Judge Bachman ruled that to "allow an otherwise criminal act to go unpunished because of the victim's consent would not only threaten the security of our society but also might tend to detract from the force of the moral principles underlying the criminal law."[71]

There are three exceptions or situations in which the law recognizes consent as a defense to criminal conduct:

- *Incidental Contact.* Acts that do not cause serious injury or harm customarily are not subject to criminal prosecution and punishment. People, for example, often are bumped and pushed on a crowded bus or at a music club.
- *Sporting Events.* Ordinary physical contact or blows are incident to sports such as football, boxing, or wrestling.
- *Socially Beneficial Activity.* Individuals benefit from activities such as medical procedures and surgery.

Consent must be free and voluntary and may not be the result of duress or coercion. An individual may also limit the scope of consent by, for instance, only authorizing a doctor to operate on three of the five fingers on his or her left hand.

- *Legal Capacity.* Young people below the age of consent, the intoxicated, and those on drugs, as well as individuals suffering from a mental disease or abnormality, are not considered capable of consent.
- *Fraud or Deceit.* Consent is not legally binding in those instances in which it is based on a misrepresentation of the facts.
- *Forgiveness.* The forgiveness of a perpetrator by the victim following a crime does not constitute consent to a criminal act.

A Nassau County court ruled that the defendants went beyond the consent granted by fraternity pledges to "hazing." The judge found that the intentional and severe beating administered exceeded the terms of any consent and observed that "consent obtained through fraud . . . or through incapacity of the party assaulted, is no defense. The consent must be voluntary and intelligent. It must be free of force or fraud. . . . [T]he act should not exceed the extent of the terms of consent."[72]

The next case in the chapter, *State v. Dejarlais,* involves a court order protecting a female against harassment by her former boyfriend. The case asks whether the victim's continuing consensual relationship with her boyfriend following the issuance of the order of protection constitutes a defense.

Model Penal Code

Section 2.11. Consent

(1) In General. The consent of the victim to conduct charged to constitute an offense or to the result thereof is a defense if such consent negatives an element of the offense or precludes the infliction of the harm or evil sought to be prevented by the law defining the offense.

(2) Consent to Bodily Injury. When conduct is charged to constitute an offense because it causes or threatens bodily injury, consent to such conduct or to the infliction of such injury is a defense if:

(a) the bodily injury consented to or threatened by the conduct consented to is not serious; or

(b) the conduct and the injury are reasonably foreseeable hazards of joint participation in a lawful athletic contest or competitive sport or other concerted activity not forbidden by law; or

(c) the consent establishes a justification of the conduct under Article 3 of the Code.

(3) Ineffective Consent. Unless otherwise provided by the Code or by the law defining the offense, assent does not constitute consent if:

(a) it is given by a person who is legally incompetent to authorize the conduct charged to constitute the offense; or

(b) it is given by a person who by reason of youth, mental disease or defect or intoxication is manifestly unable or known by the actor to be unable to make a reasonable judgment as to the nature or harmlessness of the conduct charged to constitute the offense; or

(c) it is given by a person whose improvident consent is sought to be prevented by the law defining the offense; or

(d) it is induced by force, duress or deception of a kind sought to be prevented by the law defining the offense.

Analysis

1. Section 2.11(1) notes that a lack of consent is an essential part of the definition of certain crimes that the prosecution must establish at trial beyond a reasonable doubt. Rape, for instance, requires a male's sexual penetration of a female without her consent.

2. Section 2.11(2)(b) repeats that consent constitutes a defense in the case of an offense causing minor injury or a foreseeable injury that occurs during a lawful sporting event. Section 2.11(2)(c) provides authorization for doctors to undertake emergency medical procedures on patients incapable of consent in those instances in which a reasonable person wishing to safeguard the welfare of the patient would consent.

3. There are four situations in which consent is not a defense under Section 211(3). The first involves an individual who is not entitled to consent, such as a stranger who consents to the "removal of another's property." The second covers a lack of personal capacity to consent. The third addresses an offense, such as the molestation or rape of a minor, in which the law seeks to protect individuals who are considered to be incapable of knowing and intelligently consenting. The last situation addresses consent obtained by fraud. A good example is a patient who consents to a medical procedure and after the administration of an anesthetic is sexually molested.

The Legal Equation

Consent $\neq$ A justification, generally.

Consent $=$ A justification only for

1. minor physical injury;
2. foreseeable injury in legal sporting event; and
3. beneficial medical procedure; where

$+$ consent is voluntarily given by an individual with legal capacity.

May the defendant raise the defense of consent to a violation of an order of protection?

State v. Dejarlais, 969 P. 2d 90 (Wash. 1998). Opinion by: Dolliver, J.

Defendant Steven Dejarlais was convicted in Pierce County Superior Court of violating a domestic violence order for protection. . . . The Court of Appeals affirmed the defendant's convictions, and we granted his petition for review. We now affirm.

Facts

Ms. Shupe met the defendant in 1993 after separating from her husband. She filed for divorce in June 1993 and began seeing the defendant regularly. Their relationship included his frequent overnight stays at her home. Ms. Shupe testified that, during divorce proceedings with her husband, a temporary parenting plan [the judge issued an order providing that Ms. Shupe and her husband were to share custody of the children until the issue of child custody was resolved] was filed, and she feared being found in violation of its terms because of her relationship with the defendant. She further testified her husband gave her $1,500 to help her move, and requested she petition for an order for protection against the defendant to avoid being found in violation of the parenting plan.

On September 9, 1993, Ms. Shupe signed a declaration in support of the request for a protection order, claiming she was a victim of defendant's harassment. She stated: "I met Steve back in February 1993. I'm married but going through a divorce. I decided to stop seeing him because it was becoming too much. He and my husband got into it a few times also. Steve follows me, calls numerous times a day, calls my work, comes to my work. He just don't get the hint it's over."

On September 23, 1993, an Order for Protection from Civil Harassment was entered [an order of protection issued by a Washington court is based on an allegation of domestic violence that includes physical harm or the fear of imminent physical harm or the stalking of one family or household member by another family or household member]. The Order restrained the defendant from contacting or attempting to contact Ms. Shupe in any manner, making any attempts to keep her under surveillance, and going within "100 feet" of her residence and workplace. The order stated it was to remain in effect until September 23, 1994, and that any willful disobedience of its provisions would subject the defendant to criminal penalties as well as contempt proceedings. Police Officer Stephen Mauer served the defendant with the order on November 23, 1993. Ms. Shupe testified her relationship with the defendant continued despite the order.

The defendant went to jail in May 1994, apparently for an offense unrelated to his relationship with Ms. Shupe. During that time, Ms. Shupe discovered he had been seeing another woman. Following his stay in jail, on May 22, 1994, the defendant went to Ms. Shupe's home and let himself in through an unlocked door. Ms. Shupe, who had been asleep on the floor by the couch, confronted the defendant, telling him she knew about the other woman and wanted nothing more to do with him. She did not tell him to leave, fearing he would get "mad and furious," but walked back to her bedroom. The defendant followed her, saying he would "have [her] one more time." . . . He threw her on the bed, and, disregarding her protestations and refusals, had intercourse with her twice.

The defendant was arrested and charged with one count of violation of a protection order and one count of rape in the second degree. At trial, the defendant testified he was aware of the protection order and clearly understood its terms. He testified he did not rape Ms. Shupe but that the two of them had consensual sex.

The trial court declined to give defense counsel's proposed instruction, which stated: "If the person protected by a Protection Order expressly invited or solicited the presence of the defendant, then the defendant is not guilty of Violation of Protection Order." . . . Instead, the trial court instructed the jury as follows: "A person commits the crime of violation of an order for protection when that person knowingly violates the terms of an order for protection."

The jury found the defendant guilty of violation of a protection order and rape in the third degree. The Court of Appeals affirmed his convictions. We granted review and now affirm, holding consent is not a defense to the charge of violating a domestic violence order for protection.

The defendant was convicted of a misdemeanor violation of a protection order under RCW 26.50.110(1) which provides that whenever an order for protection is granted and the respondent or person to be restrained knows of the order, a violation of the restraint provisions or of a provision excluding the person from a residence, workplace, school, or day care is a gross misdemeanor.

Issue

The defendant contends that, where a person protected by an order consents to the presence of the person restrained by the order, the jury should be instructed that consent is a defense to the charge of violating that order. We note at the outset that, even if consent were a defense to the crime of violating a protection order, it is far from clear that the contact in this case was consensual. Contrary to the defendant's proposed instruction, Ms. Shupe does not appear to have invited or solicited the defendant's presence on the night in question. More importantly, the jury found defendant guilty of rape in the third degree. . . . The protection order prohibited any contact; even if Ms. Shupe consented to earlier contacts or to defendant's presence at her home that day, the rape was clearly a nonconsensual contact. We nevertheless reach the issue defendant raises because he seems to suggest that Ms. Shupe's repeated invitations and ongoing acquiescence to defendant's presence constituted a blanket consent or waiver of the order's terms. We disagree.

Reasoning

A domestic violence protection order does not protect merely the "private right" of the person named as petitioner in the order. In fact, the court recognized, the statute reflects the Legislature's belief that the public has an interest in preventing domestic violence. The Legislature has clearly indicated that there is a public interest in domestic violence protection orders. In its statement of intent for RCW 26.50, the Legislature stated that domestic violence, including violations of protective orders, is expressly a public, as well as private, problem, stating that domestic violence is a "problem of immense proportions affecting individuals as well as communities" which is at "the core of other social problems."

We agree. Indeed, the Legislature's intent is clear throughout the statute, and allowing consent as a defense is not only inconsistent with, but would undermine, that intent.

The order served on the defendant warned him "that any willful disobedience of the order's provisions would subject the respondent to criminal penalties and possibly contempt." We are convinced the Legislature did not intend for consent to be a defense to violating a domestic violence protection order.

The statute also requires police to make an arrest when they have probable cause to believe a person has violated a protection order. There is no exception to this mandate for consensual contacts; rather, the obligation to arrest does not even depend upon a complaint being made by the person protected under the order but only on the respondent's awareness of the existence of that order. . . .

Holding

Our reading of the statute is consistent with the Legislature's intent and clear statement of policy. Requests for modification of that policy should be directed to the Legislature not this court. The statute, when read as a whole, makes clear that consent should not be a defense to violating a domestic violence protection order. The defendant is not entitled to an instruction which inaccurately represents the law. We affirm the defendant's convictions.

Questions for Discussion

1. Why did Kimberly Shupe petition for an order of protection against Steven Dejarlais? Was it motivated by a desire to prevent Dejarlais from continuing to abuse or threaten her?

2. Shupe and Dejarlais continued their relationship for roughly six months following the order of protection. Why did Shupe suddenly complain that Dejarlais was violating the order?

3. Was there continuing consent by Kimberly Shupe to engage in a relationship with Dejarlais following the issuance of the order of protection? Should Dejarlais be able to use Shupe's continuing consent as a defense to his violation of the order of protection?

4. Does society have an interest in enforcing the order of protection that takes precedence over Shupe's consent to a continuing relationship with Dejarlais?

Cases and Comments

Sports. In *State v. Shelley*, Jason Shelley and Mario Gonzalez played on opposing teams during an informal basketball game at the University of Washington Intramural Activities Building. These games were not refereed and the players called fouls on opposing players. Gonzalez had a reputation for aggressive play and fouled Shelley several times. At one point, Gonzalez slapped at the ball and scratched Shelley's face and drew blood. Shelley briefly left and returned to the game. Shelley, after returning to the court, hit Gonzalez and broke his jaw in three places, requiring the jaw to be wired shut for six weeks. Shelley was convicted of assault in the second degree. Gonzalez testified that the assault was unprovoked. Shelley, however, contended that Gonzalez continually slapped and scratched him and that Shelley was getting increasingly angry. Shelley explained that the two went for a ball and claimed that Gonzalez raised his hand toward Shelley's face and that Shelley hit Gonzalez as a reflex reaction to protect himself from being scratched.

The Court of Appeals of Washington held that consent is a defense to assaults occurring as part of athletic contests. Absent this rule, most athletic contests would have to be prohibited. The court of appeals rejected the standard proposed by the prosecution that a victim cannot be considered to have consented to conduct that falls outside the rules of an athletic contest, explaining that various "excesses and inconveniences are to be expected beyond the formal rules of the game. . . . However, intentional excesses beyond those reasonably contemplated in the sport are not justified." The court of appeals adopted the Model Penal Code standard that "reasonably foreseeable hazards of joint participation in a lawful athletic contest or competitive sport or other concerted activity are not forbidden by law." The issue is not the injury suffered by the alleged victim, but "whether the conduct of the defendant constituted foreseeable behavior in the play of the game. . . . [T]he injury must have occurred as a by-product of the game itself." The Court of Appeals of Washington affirmed Shelley's conviction and held that there is "nothing in the game of basketball" that would recognize consent as a defense to the conduct engaged in by Shelley. See *State v. Shelley*, 929 P.2d 489 (Wash. Ct. App. 1997).

Compare *Shelley* to *People v. Schacker*. In this New York hockey case, the defendant Robert Schacker struck Andrew Morenberg in the back of the neck after the whistle had blown and play had stopped. Morenberg was standing near the goal net and struck his head on the crossbar of the net, causing a concussion, headaches, blurred vision, and memory loss.

This was a "no-check" hockey league that involved limited physical contact between opposing players. The District Court for Suffolk County dismissed the charges of assault in the third degree against Schacker based on the fact that Morenberg had assumed the risk of injury during the normal course of a hockey game.

Are the differing results in *Shelley* and *Schacker* based on the distinction between basketball and hockey? Is the judge in *Schacker* correct that Morenberg's injury was "connected with the competition"? See *People v. Schacker*, 670 N.Y.S.2d 308 (N.Y. Dist. Ct. 1998).

You Decide

8.9 Givens Miller, an eighteen-year-old, 210-pound football player, had a disagreement with his parents following a high-school football game. Givens's father, George, responded by taking away Givens's cell phone and car keys. Givens repeatedly shouted at his parents telling his father to "take your G.D. money and 'f__' yourself with it." He then baited George, uttering "What the 'f__,' man. I'm going to—you going to hit me, man? Are you going to hit me? What the 'f__,' man."

George responded, "No, I'm not going to hit you," and shoved Givens away from him. Givens kicked and punched George in his side; and, as Givens charged toward him, George punched Givens in the face. George then threw two more punches. Givens testified that at the time of the incident, he "was all jazzed up" from the game and "in an aggressive mood" and "kind of wanted to hit [George]" and he "kind of wanted [George] to hit [him]." Givens "suffered dental fractures and loose teeth. He also received two blows to the head, and testified that he may have lost consciousness for a brief moment." At the close of evidence, George objected to the jury charge because the court did not include an instruction on the defense of consent. Was the judge correct in not issuing an instruction on consent? See *Miller v. State*, 312 S.W.3d 209 (Tex. App. 2010).

You can find the answer at www.sagepub.com/lippmanccl3e.

CHAPTER SUMMARY

Justification defenses provide that acts that ordinarily are criminal are justified or carry no criminal liability under certain circumstances. This is based on the fact that a violation of the law under these conditions promotes important social values, advances the social welfare, and is encouraged by society.

Self-defense, for instance, preserves the right to life and bodily integrity of an individual confronting an imminent threat of death or serious bodily harm. Individuals are also provided with the privilege of intervening to defend others in peril. Defense of the dwelling preserves the safety and security of the home. The execution of public duties justifies the acts of individuals in the criminal justice system that ordinarily would be considered criminal. A police officer, for instance, may use deadly force against a "fleeing felon" who poses an imminent threat to the police or to the public. The right to resist an illegal arrest is still recognized in various states, but has been sharply curtailed based on the fact that the state and federal governments provide effective criminal and civil remedies for the abuse of police powers. Necessity or "choice of evils" justifies illegal acts that alleviate an imminent and greater harm. The defense of consent is recognized in certain isolated instances in which the defendant's criminal conduct advances the social welfare. These include incidental contact, sports, and medical procedures.

The justifiability of a criminal act is ultimately a matter for the finder of fact, either the judge or jury, rather than the defendant. The law generally requires that individuals relying on self-defense or necessity believe their acts are justified and that a reasonable person would find that the act is justified under the circumstances.

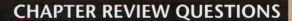

CHAPTER REVIEW QUESTIONS

1. Distinguish the affirmative offenses of justification and excuse.

2. List the elements of self-defense. Explain the significance of reasonable belief, imminence, retreat, withdrawal, the castle doctrine, and defense of others.

3. What are the two approaches to intervention in defense of another? Which test is preferable?

4. What is the law pertaining to the defense of the home? Discuss the policy behind this defense. Compare the laws pertaining to defense of habitation and self-defense.

5. How does the rule regulating police use of deadly force illustrate the defense of execution of public duties? Does this legal standard "handcuff" the police?

6. Why have the overwhelming majority of states abandoned the defense of resistance to an illegal arrest? Distinguish this from the right to resist excessive force.

7. What are the elements of the necessity defense? Provide some examples of the application of the defense.

8. Why do most state legal codes provide that an individual cannot consent to a crime? What are the exceptions to this rule?

9. Write a brief essay outlining justification defenses.

LEGAL TERMINOLOGY

affirmative defenses

aggressor

alter ego rule

American rule for resistance to an unlawful arrest

burden of persuasion

burden of production

case-in-chief

castle doctrine

choice of evils

deadly force

English rule for resistance to an unlawful arrest

excuses

fleeing felon rule

good motive defense

imperfect self-defense

intervention in defense of others

jury nullification

justification

make my day laws

misdemeanant

necessity defense

nondeadly force

objective test for intervention in defense of others

perfect self-defense

presumption of innocence

rebuttal

retreat

retreat to the wall

self-defense

stand your ground rule

tactical retreat

true man

withdrawal in good faith

 # CRIMINAL LAW ON THE WEB

Log on to the Web-based student study site at www.sagepub.com/lippmanccl3e to assist you in completing the Criminal Law on the Web exercises, as well as for additional features such as podcasts, Web quizzes, and video links.

1. Read more about the events surrounding the Bernhard Goetz case and developments following the criminal trial. Would a jury acquit Goetz if he stood trial today?

2. Read about legal liability for violence in hockey.

3. Watch a video of the police hot pursuit in *Scott v. Harris*.

BIBLIOGRAPHY

American Law Institute, *Model Penal Code and Commentaries*, vol. 1, pt. 1 (Philadelphia: American Law Institute, 1985), pp. 31–307. A detailed and technical discussion of self-defense, defense of others, and defense of property.

Joshua Dressler, *Understanding Criminal Law*, 3rd ed. (New York: Lexis, 2001), chaps. 17–20. A clear discussion comparing justifications and excuses and discussing the law of self-defense, defense of others, and defense of property and habitation.

George P. Fletcher, *A Crime of Self-Defense: Bernhard Goetz and the Law on Trial* (New York: The Free Press, 1988). An analysis of the Bernhard Goetz case that explores the history and philosophical basis for self-defense.

Leo Katz, *Bad Acts and Guilty Minds: Conundrums of the Criminal Law* (Chicago: University of Chicago Press, 1987), pp. 8–81.

A thought-provoking discussion of the philosophical bases of the necessity defense.

Wayne R. LaFave, *Criminal Law*, 3rd ed. (St. Paul, MN: West Publishing, 2000), pp. 477–523. A comprehensive explanation of justification defenses with useful citations to the leading cases and law review articles.

Arnold H. Loewy, *Criminal Law in a Nutshell*, 4th ed. (St. Paul, MN: West Publishing, 2003), chap. 6. A short and precise summary of justification defenses.

Herbert L. Packer, *The Limits of the Criminal Sanction* (Palo Alto, CA: Stanford University Press, 1968), pp. 113–121. A brief but insightful discussion of the reasons for justification defenses.

9 EXCUSES

Was Nathaniel Brazill legally responsible for murdering his teacher?

On the last day of the 1999–2000 school year, thirteen-year-old Nathaniel Brazill shot and killed a teacher at his middle school, Barry Grunow. . . . As Grunow attempted to close the classroom door, Brazill pulled the trigger and Grunow fell to the floor, with a gunshot wound between the eyes. A school surveillance videotape of the hallway revealed that Brazill had pointed the gun at Grunow for nine seconds before shooting. Brazill exclaimed: "Oh s__t," and fled.

Learning Objectives

1. Know the four tests for legal insanity.
2. Distinguish between the defense of voluntary and the defense of involuntary intoxication.
3. Know the relationship between age and the capacity to form a criminal intent and the factors that a court will consider in determining whether a juvenile is capable of forming a criminal intent.
4. State the elements of the duress defense.
5. Understand the two tests for entrapment.
6. Describe the "new defenses" and the arguments for recognizing "new defenses."

INTRODUCTION

An act that is ordinarily subject to a criminal penalty is considered to be *justified* and carries no criminal liability when it preserves an important value and benefits society. Self-defense, for instance, protects human life against wrongdoers. The defendant insists, "I broke the law, but I did nothing wrong. Society benefited from my act."

Excuses, in contrast, provide a defense based on the fact that although a defendant committed a criminal act, he or she is not considered responsible. The defendant claims that although "I broke the law and my act was wrong, I am not responsible. I am not morally blameworthy." This is illustrated by legal insanity that excuses criminal liability based on a mental disease or defect. Individuals are also excused due to youth or intoxication or in those instances when they lack a criminal intent as a result of a mental disease

or defect. Defendants are further excused in those instances when they commit a criminal act in response to a threat of imminent harm or a mistake of fact or are manipulated and entrapped into criminal conduct.

Excuses are very different from one another and each requires separate study. The common denominator of excuses is that the defendants are not morally blameworthy and therefore are excused from criminal liability.

THE INSANITY DEFENSE

English common law initially did not consider a mental disturbance or insanity as relevant to an individual's guilt. In the thirteenth century, it was recognized that a murderer of "unsound mind" was deserving of a royal pardon and, as the century drew to a close, "madness" was recognized as a complete defense.[1] This more humanistic approach reflected the regrettable "wild beast" theory that portrayed "madmen" as barely removed from "the brutes who are without reason."[2]

For a deeper look at this topic, visit the study site.

The **insanity defense** is one of the most thoroughly studied and hotly debated issues in criminal law. The debate is not easy to follow because the law's reliance on concepts drawn from mental health makes this a difficult area to understand. Texas residents must have scratched their heads in 2004 when Deanna Laney was acquitted by reason of insanity for crushing the skulls of her three sons with heavy stones. She then proceeded to call the police and informed them that "I just killed my boys." The youngest at the time was a fourteen-month-old who was left brain injured and nearly blind. Two years earlier, another Texas mother, Andrea Yates, received a life sentence for drowning her five children in the bathtub. Yates told the police that the devil had told her to kill her children; and despite Yates's history of mental problems and claim of insanity, the jury found that she was able to distinguish right from wrong. Yates's conviction was overturned on appeal and, in July 2006, a Texas jury ruled Yates not guilty by reason of insanity. Laney reportedly believed that she and Yates had been selected by God to be witnesses to the end of the world.

Defendants who rely on the insanity defense are typically required to provide notice to the prosecution. They are then subject to examination by a state-appointed mental health expert and they will usually hire one or more of their own "defense experts." These experts will interview the defendant and conduct various psychological tests. The prosecution and defense experts will then testify at trial and additional testimony is typically offered on behalf of the defendant by people who are able to attest to his or her mental disturbance. The nature of a defendant's criminal conduct is also important. The prosecutor may argue that a well-planned crime is inconsistent with a claim of insanity. The jury is then asked to either return a verdict of "guilty," "not guilty," or "not guilty by reason of insanity" (NGRI). In some jurisdictions, the jury considers the issue of insanity in a separate hearing in the event that the defendant is found guilty.

A defendant found NGRI in some states is subject to immediate committal to a mental institution until he or she is determined to be sane and no longer poses a threat to society. In most states, a separate **civil commitment** hearing is conducted to determine whether the defendant poses a danger and should be interned in a mental institution. Keep in mind that this period of institutionalization may last longer than a criminal sentence for the crime for which the defendant was convicted.

Why do we have an insanity defense? Experts cite three reasons:

- *Free Will.* The defendant did not make a deliberate decision to violate the law. His or her criminal act resulted from a disability.
- *Theories of Punishment.* A defendant who is unable to distinguish right from wrong or to control his or her conduct cannot be deterred by criminal punishment, and it would be cruel to seek retribution for acts that result from a disability.
- *Humanitarianism.* An individual found not guilty by reason of insanity may pose a continuing danger to society. He or she is best incapacitated and treated by doctors in a noncriminal rather than in a criminal environment.

In the United States, courts and legislators have struggled with balancing the protection of society against the humane treatment of individuals determined to be not guilty by reason of insanity. There have been several tests for insanity:

- M'Naghten (twenty-eight states and the federal government recognize all or a part of this test)
- Irresistible impulse (roughly seventeen states recognize this in conjunction with another test)
- Durham product test (New Hampshire)
- American Law Institute, Model Penal Code standard (roughly fourteen states)

The fundamental difference among these tests is whether the emphasis is placed on a defendant's ability to know right from wrong or whether the stress is placed on a defendant's ability to control his or her behavior. You might gain some appreciation of what is considered an inability to tell right from wrong by considering a young child who has not been taught right from wrong and takes an object from a store without realizing that this is improper. As an example of an inability to control behavior, think about a motorist who suddenly erupts in "road rage" and violently threatens you for driving too slowly.

Keep in mind that an individual who is "mentally challenged" may not necessarily meet the legal standard for insanity. A serial killer, for instance, may be mentally disturbed but still not considered to be so impaired by a mental illness or so retarded as to be considered legally insane. Juries generally find the determination of insanity to be highly complicated and they experience difficulty in following the often technical testimony of experts. As a result, jurors often follow their own judgment in determining whether a defendant should be determined to be not guilty by reason of insanity.

You also should be aware that insanity is distinct from **competence to stand trial**. Due process of law requires that defendants should not be subjected to a criminal trial unless they possess the ability to intelligently assist their attorney and to understand and follow the trial. The prosecution of an individual who is found incompetent is suspended until he or she is found competent.

The Right-Wrong Test

Daniel M'Naghten was an ordinary English citizen who was convinced that British Prime Minister Sir Robert Peel was conspiring to kill him. In 1843, M'Naghten retaliated by attempting to assassinate the British leader and, instead, mistakenly killed Sir Robert's private secretary. The jury acquitted M'Naghten after finding that he "had not the use of his understanding, so as to know he was doing a wrong or wicked act." This verdict sent shock waves of fright through the British royal family and political establishment, and the judges were summoned to defend the verdict before Parliament. The judges articulated a test that continues to be followed by roughly one-half of American states.[3]

- At the time of committing the act, the party accused must have been suffering from such a defect of reason, from a disease of the mind as a result of which:
  - the defendant "did not know what he was doing" (did not know the "nature and quality of his or her act"); or
  - the defendant "did not know he was doing wrong."

The "mental disease or defect" requirement is satisfied by any organic (physical) damage, psychological disorder, or intellectual deficiency (e.g., a low I.Q. or feeblemindedness) that results in an *individual's either not knowing "what he was doing" or not knowing he was "doing wrong."*

The most common mental disorder or defect that results in legal insanity under the **M'Naghten test** for insanity is a psychosis, a psychological disorder that results in an inability to distinguish between reality and fantasy. Another frequent mental disorder is a neurosis, which is simply a compulsive drive to engage in certain behavior. The mental conditions that are generally not considered to fall within the notion of a mental disease or defect are sociopathic or personality disorders that lead individuals to engage in patterns of antisocial or criminal conduct. These people are generally aware of the difference between right and wrong and believe that they are above the law and the rules that apply to other people.[4] A Georgia appellate court offered a fairly straightforward definition of mental disease or defect. The court stated that the "term mentally ill means having a disorder of thought or mood which significantly impairs judgment, behavior, capacity to recognize reality, or ability to cope with the ordinary demands of life. The term mentally ill does not include a mental state shown only by repeated unlawful or antisocial conduct."[5]

The requirement that the defendant did not know the *"nature and quality of the act"* requirement is extremely difficult to satisfy. The common example is that an individual squeezing the victim's neck must be so detached from reality that he or she believes that he or she is squeezing a lemon. Individuals suffering this level of mental disturbance are extremely rare, and the M'Naghten test assumes that these individuals should be detained and receive treatment and that criminal incarceration serves no meaningful purpose and is inhumane.[6]

Courts differ on what it means for a defendant to *"know" that his or her act "was wrong."* Some judges interpret "know" to require an understanding that to kill is prohibited. Others demand a demonstration that the defendant fully comprehends that the reason killing is wrong is that it causes pain, hurt, and harm to the victim and to his or her family. Professor Arnold Loewy explains that "one may 'know' that it is wrong to kill in the sense that he can articulate these words, but lack the capacity to really feel or appreciate the wrongfulness of killing."[7]

There also is an ongoing debate over whether a defendant must know that an act is a "legal wrong" or whether the defendant must know that the act is a "moral wrong."

State v. Crenshaw attempted to resolve this conflict. The defendant Rodney Crenshaw was honeymooning with his wife in Canada and suspected that she was unfaithful. Crenshaw beat his wife senseless, stabbed her twenty-four times, and then decapitated the body with an axe. He then drove to a remote area and disposed of his wife's body and cleaned the hotel room. Crenshaw claimed to be a member of the Moscovite religious faith, a religion that required a man to kill a wife guilty of adultery. He claimed he believed that his act, although illegal, was morally justified. Was Crenshaw insane based on his belief that his act was morally justified? Did he possess the capacity to distinguish between right and wrong?

Crenshaw was convicted and appealed on the grounds that the judge improperly instructed the jury that insanity required a finding that as a result of a mental defect or disease, Crenshaw believed that his act was lawful rather than moral. The Washington Supreme Court, however, concluded that under either a legal or moral wrongfulness test, Crenshaw was legally sane. The court noted that Crenshaw's effort to conceal the crime indicated that he was aware that killing his wife was contrary to society's morals as well as the law. The Washington Supreme Court ruled that in the future, courts should not define "wrongfulness," and that jurors should be left free to apply either a societal morality or legal wrongfulness approach. Some states create an exception and consider an individual legally insane who believes that his or her action resulted from a direct command from God (a "deific decree").[8]

In *Clark v. Arizona,* the U.S. Supreme Court upheld the constitutionality of an Arizona statute that limited the insanity defense to individuals who did not know "what [they were] doing was wrong" (moral incapacity). The statute did not include within the definition of insanity the other part of the M'Naghten test that takes in individuals who do not know the "nature and quality of the act (cognitive or mental incapacity)." The Supreme Court held that Arizona's formulation satisfied the requirements of due process because evidence introduced to establish that Clark did not know the nature and quality of his act also would establish that he did not know that his act was wrong. The trial court heard evidence that Clark shot a police officer whom he believed was an alien from outer space. Clark appealed on the grounds that the trial court should have considered this evidence to determine whether Clark knew the nature and quality of his act. The Supreme Court held that Clark was not denied fundamental justice because this very same evidence was considered by the court in deciding whether Clark knew that his act was wrong.[9]

It's likely you are fairly confused at this point. The right-wrong test is clearly much too difficult to be easily applied by even the most educated and sophisticated juror. In the end, juries tend to follow their commonsense notion of whether the defendant was legally sane or insane.

See more cases on the study site: Lopez v. State, Kirkland v. State, www.sagepub.com/lippmanccl3e.

The Legal Equation

M'Naghten right-wrong test = Defect of reason from a disease of the mind

+ at the time of the act did not know

+ the nature and quality of the act or that the act was wrong.

The Irresistible Impulse Test

The M'Naghten test is criticized for focusing on the mind and failing to consider emotions. Critics point out that an individual may be capable of distinguishing between right and wrong and still may be driven by emotions to steal or to kill. Many of us are aware of the dangers of smoking, drinking, or eating too much and yet continue to indulge in this behavior. Various states responded to this criticism by broadening the M'Naghten standard and adopting the irresistible impulse test. This is often referred to as the "third branch of M'Naghten." The irresistible impulse theory was articulated as far back as 1887 when an Alabama court ruled that Nancy Parsons had been driven to assist in the killing of her husband by the delusion that he had cast an evil spell that caused her to suffer from a prolonged and life-threatening illness.

The **irresistible impulse test** requires the jury to find a defendant not guilty by reason of insanity in the event that the jurors find that the defendant possessed a mental disease that prevented him from curbing his or her criminal conduct. A defendant may be found legally insane under this test despite the fact that he or she is able to tell right from wrong. Individuals are not required to act in an explosive or impulsive manner under the irresistible impulse test and may calculate, plan, and perfect their crime. The central consideration is whether the disease overcame his or her capacity to resist the impulse to kill, rape, maim, or commit any other crime. Most courts also do not require that an individual lack total capacity to control his or her criminal impulses.[10] In 1887, in *Parsons v. State,* the Alabama Supreme Court articulated the irresistible impulse test:[11]

1. At the time of the crime was the defendant afflicted with a "disease of the mind"?

2. If so, did the defendant know right from wrong with respect to the act charged? If not, the law excuses the defendant.

3. If the defendant did have such knowledge, the law will still excuse the defendant if two conditions concur:

   A. if mental disease caused the defendant to so far lose the power to choose between right and wrong and to avoid doing the alleged act that the disease destroyed the defendant's free will, and

   B. if the mental disease was the sole cause of the act.

John Hinckley's acquittal by reason of insanity for the attempted assassination of Ronald Reagan sparked a reconsideration and rejection of the irresistible impulse test. After all, why should Hinckley be ruled legally insane because he attempted to kill President Reagan to fulfill an uncontrollable impulse to attract the attention of Jodie Foster, a young female film star?

There was also a recognition that psychiatrists simply were unable to determine whether an individual experienced an irresistible impulse. The Fifth Circuit Court of Appeals concluded that the lack of knowledge concerning human impulses dictated that all criminal impulses should be considered "resistible." The court reasoned the irresistible impulse test had "cast the insanity defense adrift upon a sea of unfounded scientific speculation."[12]

Critics claimed that defendants were regularly making false claims of an irresistible impulse in an attempt to gain an acquittal. In *State v. Quinet,* the defendant conceded that he was able to distinguish right from wrong, but contended that he was unable to control himself and that he was driven to plan the rape and murder of twenty-seven of his former female classmates and an escape to Australia where he planned to commit suicide. The Connecticut Supreme Court rejected the defendant's claim and called attention to his demonstrated ability to plan and patiently wait to initiate the attacks, his reliance on videos to put himself in the proper mood to carry out the sexual assaults, and the emotional stability that enabled him to enjoy a dinner with friends several days prior to his unsuccessful effort to carry out the first attack.[13]

As a result, several jurisdictions abolished the irresistible impulse defense. The U.S. Congress adopted the so-called John Hinckley Amendment that eliminated the defense in federal trials and adopted a strict M'Naghten standard.[14]

- *Affirmative Defense.* It is an affirmative defense to a prosecution under any federal statute that, at the time of the commission of the acts constituting the offense, the defendant, as a result of a severe mental disease or defect, was unable to appreciate the nature and quality or the wrongfulness of his acts. Mental disease or defect does not otherwise constitute a defense. . . .
- *Burden of Proof.* The defendant has the burden of providing the defense of insanity by clear and convincing evidence.

The Durham Product Test

The **Durham product test** was intended to simplify the determination of legal insanity by eliminating much of the confusing terminology. The "product" test was first formulated by the New Hampshire Supreme Court in *State v. Pike* in 1869.[15] This standard was not accepted or even considered by any other jurisdiction until it was adopted in 1954 by the U.S. Court of Appeals for the District of Columbia.[16]

Durham provided that an accused is "not criminally responsible if his unlawful act was the product of mental disease or mental defect." This permitted expert witnesses to provide a broad range of information concerning a defendant's mental health and simplified the task of the jury, which now was only required to determine whether the defendant acted as a result of a mental disease or defect. Jurors no longer were placed in the position of making the difficult determination whether the defendant knew the difference between right and wrong or acted as a result of an irresistible impulse. The only requirement was to evaluate whether the accused was suffering from a disease or defective mental condition at the time he or she committed the criminal act and whether the criminal act was the product of such mental abnormality. However, the decision left the definition of a mental disease or defect undefined.

The District of Columbia Court of Appeals abandoned this experiment after eighteen years, in 1972, after realizing that the "product test" had resulted in expert witnesses playing an overly important role at trial.[17] In *Blocker v United States,* two experts from St. Elizabeth's Hospital concluded that Blocker suffered from a sociopathic personality disorder and testified that this did not amount to a mental disease or defect. Blocker was granted a new trial after pointing out that less than a month following the verdict in his case, another defendant was ruled legally insane as a result of a decision by the psychiatrists at St. Elizabeth's to change their position and to accept that a sociopathic personality disorder did indeed constitute a mental disease or defect.[18]

The Substantial Capacity Test

Psychiatric experts urged the American Law Institute (ALI) to incorporate the Durham product test into the Model Penal Code. The Institute, instead, adopted a modified version of the M'Naghten and irresistible impulse tests. Section 4.01(1)(2) provides:

> A person is not responsible for criminal conduct if at the time of such conduct as a result of mental disease or defect he lacks substantial capacity either to appreciate the criminality [wrongfulness] of his conduct or to conform his conduct to the requirements of law. . . . The terms "mental disease or defect" do not include an abnormality manifested only by repeated criminal or otherwise antisocial conduct.

The important point is that the ALI **substantial capacity test** significantly broadens the test for legal insanity and increases the number of defendants who may be judged to be legally insane.

- *Appreciate.* The ALI test modifies M'Naghten by providing that a defendant may lack a substantial capacity to appreciate rather than know the criminality of his or her conduct. This is intended to highlight that a defendant may be declared legally insane and still know that an act is wrong because he or she still may not appreciate the full harm and impact of his or her criminal conduct. In other words, a defendant may know that sexual molestation is wrong without appreciating the harm a sexual attack causes to the victim.

- *Substantial Capacity.* The ALI test requires that a defendant lack a substantial capacity to appreciate the criminality of his conduct or to conform his conduct to the requirements of the law. The ALI observes that a test calling for total impairment is not "workable" because it limits the application of legal insanity to individuals suffering from a delusional psychosis or to individuals who have absolutely no capacity to conform their conduct to the requirements of the law.
- *Conform Conduct to the Requirements of the Law.* The ALI standard does not use the word "impulse" in order to avoid the suggestion that individuals who are driven by emotions to break the law must act immediately and spontaneously and may not reflect and brood over and plan their criminal conduct.
- *Wrongfulness.* The ALI defines wrongfulness as an inability to appreciate that the community morally disapproves of an act and explains that in most cases, an individual will be unaware that such an act is also contrary to the criminal law. In order to be considered legally insane, a defendant who believes that his or her criminal conduct is justified must possess an inability to appreciate that the community would view his or her act as immoral. In other words, the accused must believe that the community would endorse his or her murder of an individual who, in fact, is a messenger of "Satan the devil." The test also stresses that mental disease or defect does not include sociopaths or an abnormality that causes repeated criminal or antisocial conduct.

The ALI test was adopted by a majority of states and all but one of the federal circuit courts of appeals. The ALI's more tolerant and broader view of legal insanity was abandoned by all but a handful of states following John Hinckley's attempted assassination of President Ronald Reagan in 1981. The trend is to follow the lead of the U.S. Congress and to return to M'Naghten. These revised statutes typically integrate aspects of the substantial capacity test.

The Legal Equation

Substantial capacity test = Mental disease or defect

+ substantial incapacity

+ to appreciate criminality [wrongfulness] of an act or to conform conduct to requirements of the law.

Burden of Proof

The defendant possesses the initial burden of going forward in every state. The defendant is presumed sane until some evidence is produced challenging this assumption. The defendant's burden varies and ranges from a "reasonable doubt" to "some evidence," "slight evidence," or a "scintilla of evidence." In roughly one-half of the states, the prosecution then possesses the burden of persuasion to establish sanity beyond a reasonable doubt. The defendant possesses the burden of persuasion in other jurisdictions by a preponderance of the evidence. In the federal system and in a small number of states, the defendant has the burden of establishing insanity by "clear and convincing evidence." A defendant must meet this burden in order for the issue of insanity to be presented to the jury.[19]

The Future of the Insanity Defense

Critics contend that the insanity defense undermines the functioning of the criminal justice system.

- *Bias.* Wealthy defendants are able to hire experts and are advantaged over the indigent. The insanity defense may also be exploited by perfectly sane defendants who have the resources to mount a credible insanity defense.

- *Theories of Punishment.* The insanity defense undermines the criminal justice system's concern with deterrence, retribution, and incapacitation by acquitting legally guilty defendants by reason of insanity.
- *Moral Blameworthiness.* The legally insane are not considered morally blameworthy and, as a consequence, are not incarcerated. On the other hand, there typically is a fine line between the legally sane and insane. Yet, the legally insane avoid imprisonment. The insanity defense also results in special treatment for individuals who are psychologically disadvantaged, while the law ignores disabilities such as economic deprivation.
- *Experts.* The insanity defense typically involves a battle of experts who rely on technical language that is difficult for jurors to understand. As a result, decisions on legal insanity may be based more on subjective impressions than on reasoned analysis.

Defenders of the insanity defense point out that critics exaggerate the significance of the insanity defense for the criminal justice system and that only a small number of deserving defendants are generally evaluated as legally insane. Statistics suggest that the defense results in an acquittal by reason of insanity in less than one percent of all criminal trials per year; this translates into an average of thirty-three defendants. These individuals may also spend more time institutionalized in a mental institution than they would have served had they been criminally convicted.[20]

Idaho, Montana, Kansas, and Utah have abolished the insanity defense and, instead, permit defendants to introduce evidence of a mental disease or defect that resulted in a lack of criminal intent. Idaho, for example, provides that a "[m]ental condition shall not be a defense to any charge of criminal conduct." Evidence of state of mind is admissible in Idaho to negate criminal intent, and a judge who finds that a defendant convicted of a crime suffers from a mental condition requiring treatment shall incarcerate the defendant in a facility where he or she will receive treatment.[21] State supreme courts have ruled that the insanity defense is not fundamental to the fairness of a trial and that the alternative of relying on evidence of a mental disease or defect to negate criminal intent is consistent with due process. Defendants under this alternative approach, however, continue to rely on experts and highly technical evidence.[22]

Thirteen states have adopted a verdict of **guilty but mentally ill (GBMI)**. Eleven of these states continue to retain the insanity defense, and in these states jurors may select from among four verdicts: guilty, not guilty, not guilty by reason of insanity (NGRI), and GBMI. A verdict of GBMI applies where the jury determines beyond a reasonable doubt that a defendant was mentally ill, but not legally insane, at the time of his or her criminal act. The defendant receives the standard criminal sentence of confinement and is provided with psychiatric care while interned. The intent is to provide jurors with an alternative to the insanity defense that provides greater protection to the public.

The GBMI verdict has thus far not decreased findings of legal insanity. Nevertheless, advocates of the insanity defense remain fearful that jurors will find the GBMI verdict more attractive than verdicts of not guilty by reason of insanity.[23]

In the last analysis, is it realistic to ask judges and juries to evaluate a defendant's mental stability? The Fifth Circuit Court of Appeals questioned whether we are serving the purpose of protecting society and deterring crime by introducing confusing medical concepts into criminal trials. At present, the law limits legal insanity to individuals who are unable to distinguish "right from wrong" while, in most states, refusing to recognize legal insanity in the case of individuals driven by an irresistible impulse. Is this a proper place to draw the line? Do we need an excuse of legal insanity?[24]

The next case, *Galloway v. State,* illustrates the difficulty of distinguishing between a defendant who is mentally challenged and capable of distinguishing between right and wrong and a defendant who is mentally challenged and unable to distinguish between right and wrong. *Galloway* also raises the issue whether the insanity defense endangers society by running the risk that dangerous individuals will be determined to be rehabilitated and released back into the community after a relatively brief period of institutionalization. Consider whether states should abolish the insanity defense and instead allow juries and judges to return a verdict of GBMI.

Was Galloway legally insane at the time of the killing of his grandmother?

Galloway v. State, 938 N.E.2d 699 (Ind. 2010). Opinion by: Sullivan, J.

Issue

Despite nonconflicting expert and lay opinion testimony that defendant Gregory Galloway was insane, the trial court rejected the insanity defense after concluding that the defendant could continue to be a danger to society because of an inadequate State mental health system.

Facts

The defendant, Gregory Galloway, was found guilty but mentally ill for the October, 2007, murder of his grandmother, Eva B. Groves. The defendant raised the "insanity defense" at his bench trial. The trial court found that although the defendant had a long history of mental illness, he did not meet his burden of proving that he was "insane" at the time of the crime.

The trial court concluded, based on the expert testimony and the numerous medical records introduced into evidence, that the defendant suffers from bipolar disorder, an Axis I psychiatric disorder. This evidence showed that prior to his killing his grandmother, the defendant had had a long history of mental illness, and he had had many "contacts" with the mental health system. He had been diagnosed with bipolar disorder by up to twenty different physicians, often with accompanying psychotic and manic symptoms. He had also been voluntarily and involuntarily detained or committed for short-term treatment more than fifteen times.

The defendant was first diagnosed with an Axis I mental illness in 1989, when he was a senior in high school. By 2000, his mental health became more problematic; he had difficulty holding a steady job, he struggled with substance abuse, and his marriage failed. Despite these difficulties, he had very few encounters with law enforcement that were not traffic or mental illness related. After his divorce, the defendant moved in with his grandmother, who lived next door to his parents. He had a great relationship with his grandmother; "he loved [her] very much and considered [her to be] his best friend."

Since 2001, the defendant experienced psychotic episodes with increased frequency and severity. For instance, in February, 2002, he was involuntarily committed after his parents found him with a gun and looking for ammunition—he planned to kill his grandmother because she was the devil and he was Jesus Christ, and he planned to kill his neighbor because he believed his neighbor was controlling his son. Then in June, 2004, the defendant drove to Dayton, Ohio, after God told him to leave his job; he was hospitalized in Ohio after being found in a stranger's driveway looking for the perfect wife for the son of God. In July, 2005, the defendant was admitted to a hospital after crashing his car during a car chase with his mother; he believed she was the devil and was chasing her so that he could kill her, and he believed he was protected because he was an alien.

In the year leading up to the murder, the defendant had at least twelve contacts with the mental health system. In January, 2007, he pulled over on the side of the interstate near Lafayette, Indiana, got out of his car, and began erratically yelling and talking to himself. Because the air temperature was 27 degrees, concerned bystanders called the police. When the medics arrived, he was sitting in the back of a police car; his skin was cold to the touch, and there were ice particles in the facial hair under his nose. At the emergency room, the defendant was uncooperative, mumbling to himself, acting aggressively toward staff, and reacting to audio and visual hallucinations; he was admitted to a Lafayette hospital for a few days.

In March, 2007, after refusing to eat or sleep for one week because he was fearful of something bad happening to him, the defendant lacerated his stomach while trying to get into his grandmother's house through a window after he was accidentally locked out. At the emergency room, he was . . . having difficulty concentrating, and experiencing auditory hallucinations and paranoid delusions. He was transferred to an Anderson hospital, where he was confused and disoriented, detached from reality, and in a catatonic-like state; he was discharged after a few days. Several days later, the defendant was involuntarily committed after the court found him to be a danger to himself because he did not know who or where he was, he had been staying awake all night, he had been trying to sleep with his parents in their bed because he believed someone was in his room, and he was hearing voices; again, he was released after a few days.

In June, 2007, the defendant was admitted to a hospital in Tennessee after police found him driving a semi-truck full of gasoline, threatening to blow up a gas station; he was confused and disoriented, responding to internal stimuli and laughing inappropriately, experiencing racing thoughts and auditory hallucinations, and had not slept for three days. He was discharged from the Tennessee hospital within days. A few days later, he went to counseling where he was delusional about raping a girl (there was no evidence that any rape had occurred). He did not take medications prescribed for him in Tennessee.

In the days leading up to the murder, the defendant heard voices and thought that his grandmother's trailer was haunted. To abate his fears, he slept on the floor next to his parents' bed while holding his mother's hand. The night before the murder, he drank a pint of whiskey, finishing around 3:00 or 4:00 a.m., and did not sleep.

The defendant reported feeling strange on October, 26, 2007, the morning of the murder. He was supposed to pick up his friend from work, but he refused to do so because he was feeling strange. When the friend called to ask about the ride, the defendant uncharacteristically yelled at him. The defendant also spoke with his father that morning, and during their conversation, his father became concerned because his son was not acting normal and seemed to be in another world. The defendant told the police that during this conversation, his father was telling him through coded verbal messages that he needed to kill his grandmother.

During the early afternoon, the defendant went shopping with his grandmother and his aunt (the victim's daughter). They shopped for only fifteen minutes and then went to lunch, though the defendant did not eat much. While eating lunch, the defendant began thinking that his grandmother was against him and "that life should be more colorful" and that it would be if she were gone—life would be better again once he killed his grandmother. He believed that she was the devil, that she was out to get him, and that he needed to kill her to restore his powers. As they sat there eating, he was hoping that his grandmother would die. After lunch, they stopped at a gas station, where the defendant pumped their gas and purchased cigarettes. They returned home a little more than an hour after they had originally left; there had been no arguments, and nothing unusual had occurred during their outing. On the way home, the defendant's grandmother remarked that it had been a wonderful day.

Once they arrived home, the defendant went next door to his parents' house while his grandmother and aunt sat on a couch inside the grandmother's trailer and talked. While at his parents' house, the defendant began believing that he was reading his father's mind; his father was communicating telepathically, telling the defendant that he needed to kill his grandmother "to feel good again, to see like the bright lights and the flowers and the pretty things."

The defendant then went back to his grandmother's house and sat on the porch swing. Shortly thereafter, the defendant's fifteen-year-old son, Cory, arrived and said "hi" to his dad. Cory had seen his father cycle from normal to psychotic before and could tell that something was not quite right. At the same time, the defendant's father, who had come over from next door, was entering the grandmother's house.

The defendant entered the house at the same time as his father and went to his bedroom, grabbed his knife, and came back down the hallway to the living room, where his aunt and grandmother were sitting on a couch. According to his aunt, the defendant had a "wild look" in his eye that she had seen before—it was the look he gets right before he "lose[s] it." With his father, son, and aunt in the room, and with no plan or motive, the defendant jumped on top of his grandmother, straddled her, and stabbed her in the chest while yelling "you're going to die, I told you, you're the devil." His father yelled, "What have you done!" and the defendant responded that she "was going to kill me."

As soon as everyone started screaming, the defendant realized that he did not feel better like he thought he would, and he hoped that his grandmother would survive. His father was able to commandeer the knife and store it in a safe place until the police arrived. As the defendant's son applied pressure to the wound, the defendant told his grandmother that he loved her and that he did not mean to do it. He pleaded for the paramedics to save his grandmother's life. When the police arrived, he told them that he loved his grandmother and would not hurt her. When the police were getting ready to take him to the police station, he did not understand what was happening and asked where he was going. But he was cooperative during the police interrogation, which occurred two-and-a-half hours later.

Prior to trial, the defendant was examined by three experts: Dr. Parker, a psychiatrist engaged by the defense; Dr. Coons, a court-appointed psychiatrist; and Dr. Davidson, a court-appointed psychologist. All three experts agreed that he suffers from a mental illness, suffers paranoid delusions (a symptom of severe psychosis), and has suffered from intermittent psychosis since 1999. Dr. Parker and Dr. Coons both testified (and submitted in their preliminary reports) that the defendant was legally insane at the time of the murder. They both opined that he was jolted out of his delusion when he realized that he did not feel better and had just harmed someone he loved.

The psychologist, Dr. Davidson, submitted a preliminary opinion to the court that the defendant was sane at the time of the murder. The basis for his opinion was that it was unlikely the defendant would have been insane only for the few moments that it took for him to grab the knife and stab his grandmother. But while testifying, Dr. Davidson withdrew his opinion in light of additional facts that he did not have when he submitted his preliminary opinion. Among other things, Dr. Davidson was unaware that the defendant had been experiencing delusions and responding to internal stimuli in the days leading up to the murder and on the day of the murder. Dr. Davidson also was unaware that eyewitnesses heard the defendant call his grandmother the devil as he stabbed her. After being presented with all of the facts while on the witness stand, Dr. Davidson ultimately testified that he could not give an opinion on the matter.

After the close of trial, but before a verdict was rendered, the defendant stopped taking his medication and deteriorated to the point where he was found incompetent to stand trial. He regained competence after treatment at a state mental hospital.

On May 4, 2009, the trial court found the defendant guilty but mentally ill for murdering his grandmother, rejecting the insanity defense. Finding that none of the experts or lay witnesses testified that the defendant was sane, the trial court based its conclusion on demeanor evidence. Specifically, the court found that the defendant and his grandmother had interacted with each other and other people on the day of the murder, he had committed the offense in front of several family members and made no effort to conceal his crime, he had not attempted to evade police, and he had cooperated with law enforcement. Additionally, the defendant had been alert and oriented throughout the trial proceedings and had been able to assist counsel. The court also found that the defendant's "psychotic episodes increased in duration and frequency" and that he "lacks insight into the need for his prescribed medication." The court then found that the defendant had "repeatedly discontinued medication because of side effect complaints and would self medicate" by abusing alcohol and illicit drugs. Furthermore, there was "no evidence that this pattern of conduct [would] not continue if the Defendant [were] hospitalized and released, posing a danger to himself and others in the community." The court concluded that the defendant "is in need of long term stabilizing treatment in a secure facility."

During the sentencing hearing, on June 2, 2009, the trial court indicated that the preferred route would be to commit the defendant to a mental health facility for the rest of his life but concluded that route was not an option.

There is absolutely no evidence that this mental illness is [feigned], or malingered, or not accurate and there is no dispute as to that. But quite frankly, this is a tragedy that's ripped apart a family and there is very little this Court can do to remedy that. This case is as much a trial of our mental health system as it is of a man. For 20 years, Mr. Galloway's family has sought long-standing permanent treatment for Mr. Galloway, and the fact that there may not be the funds available to pay for the mentally ill in the State of Indiana does not mean that we don't have mentally ill people in the State of Indiana. . . . [T]his is difficult for everyone, and I can pick apart about 20 mental health records that were submitted to this Court where I would have begged a mental health

provider to keep Mr. Galloway long term in a civil commitment, but they have not. Mr. Galloway is able to take his medication when forced to do so in a very structured setting, but we have a 20-year history which shows when he is not in that setting that he will not take his medication, that he will continue to have episodes, and most concerning for this Court is that he will endanger others and himself. One of my options is not to say that he's committed for the rest of his life in a mental health institution. That would have been easy, but that's not one of my choices. . . . I cannot in good conscience allow someone with the severe mental health illness to return to the community, and that is what has made this case so very difficult.

The Court of Appeals affirmed the defendant's conviction. . . .

Reasoning

To sustain a conviction, the State must prove each element of the charged offense beyond a reasonable doubt. Even where the State meets this burden, a defendant in Indiana can avoid criminal responsibility by successfully raising and establishing the "insanity defense." A successful insanity defense results in the defendant being found not responsible by reason of insanity ("NGRI").

The defendant bears the burden of establishing the insanity defense by a preponderance of the evidence. . . . Thus a defendant must convince the trier of fact that, in consideration of all the evidence in the case, he or she was more probably legally insane than legally sane at the time of the crime.

To meet this burden, the defendant must establish both (1) that he or she suffers from a mental illness and (2) that the mental illness rendered him or her unable to appreciate the wrongfulness of his or her conduct at the time of the offense. Thus, mental illness alone is not sufficient to relieve criminal responsibility. Rather, a defendant who is mentally ill but fails to establish that he or she was unable to appreciate the wrongfulness of his or her conduct may be found guilty but mentally ill ("GBMI").

A mental disease or defect is defined as a "severely abnormal mental condition that grossly and demonstrably impairs a person's perception." Whether a defendant appreciated the wrongfulness of his or her conduct at the time of the offense is a question for the trier of fact. Indiana Code section 35-36-2-2 provides for the use of expert testimony to assist the trier of fact in determining the defendant's insanity. Such expert testimony, however, is merely advisory, and even unanimous expert testimony is not conclusive on the issue of sanity. The trier of fact is free to disregard the unanimous testimony of experts and rely on conflicting testimony by lay witnesses. And even if there is no conflicting lay testimony, the trier of fact is free to disregard or discredit the expert testimony.

Because it is the trier of fact's province to weigh the evidence and assess witness credibility, a finding that a defendant was not insane at the time of the offense warrants substantial deference from reviewing courts. . . . [T]his Court has long held that . . . the conviction will be set aside "when the evidence is without conflict and leads only to the conclusion that the defendant was insane when the crime was committed." . . .

The strongest showing of an evidentiary conflict occurs where the experts disagree as to whether the defendant was insane at the time of the offense. Our cases have consistently held that conflicting credible expert testimony is sufficiently probative of sanity. Such a conflict arises where one or several experts opine that the defendant was insane at the time of the offense, while one or several other experts opine that the defendant was sane at the time of the offense.

The expert testimony in this case did not conflict. Although Dr. Davidson submitted a preliminary report opining that the defendant was sane at the time of the murder, he recanted that opinion under cross-examination in light of learning critical facts. The State contends that Dr. Davidson's equivocation illustrates that the expert testimony was in conflict. We disagree.

Where there is no conflict among the expert opinions that the defendant was insane at the time of the offense, there must be other evidence of probative value from which a conflicting inference of sanity can be drawn. Such probative evidence is usually in the form of lay opinion testimony that conflicts with the experts or demeanor evidence that, when considered in light of the other evidence, permits a reasonable inference of sanity to be drawn. . . .

Even where there is no conflict among the experts and the lay witnesses, a finding that a defendant was sane at the time of the crime still may be sustained by probative demeanor evidence from which a conflicting inference of sanity may be drawn. We have recognized the importance of demeanor evidence in insanity cases. Demeanor is useful because a defendant's "behavior before, during, and after a crime may be more indicative of actual mental health at [the] time of the crime than mental exams conducted weeks or months later." . . .

In this case, there was not sufficient evidence of probative value from which an inference of sanity could be drawn sufficient to create a conflict with the expert testimony that the defendant was insane at the time of the offense. First, there was no lay opinion testimony given that conflicted with the experts' opinions that the defendant was insane at the time of the stabbing. The three eyewitnesses to the stabbing called by the State testified that the defendant was showing familiar signs of "losing it." The defendant's aunt, who was sitting on the couch as her mother was stabbed only a few feet away, testified that the defendant had a "wild look" in his eye and that she recognized this as the look he gets right before he loses it. She also heard the defendant call his beloved grandmother the devil as he stabbed her. Two other witnesses—the defendant's mother and the defendant's friend—also testified that the defendant was showing signs of losing it in the days and hours leading up to the murder. Thus . . . there were five lay witnesses in this case whose testimony supports the experts' opinions.

Second, there was not sufficient demeanor evidence of probative value from which an inference of sanity could be drawn. The trial court based its findings on very little evidence. It found as probative of sanity the fact that, over the course of an hour, the defendant shopped, ate, and filled a car with gasoline without incident. It also found as probative the fact that the defendant cooperated with police after the fact. Viewed in isolation, each of these events may indeed represent the normal events of daily life. However, when viewed against the defendant's long history of mental illness with psychotic episodes, the defendant's demeanor during the crime, as testified to by three eyewitnesses, and the absence of any suggestions of feigning or malingering, this demeanor evidence is simply neutral and not probative of sanity. . . .

The trial court expressly found that the defendant deteriorates mentally and experiences psychosis when he does not take his medication. At the time of the stabbing, the defendant was supposed to be taking his medications twice a day. He told police, however, that he had not taken any prescription medication in two days. The trial court found this failure to take medication to be probative of sanity, but we do not, especially in light of the trial court's finding that the defendant became psychotic when not on his medication.

The trial court also relied on the defendant's demeanor during trial, when he was competent to stand trial, as probative of his sanity at the time of the crime. As discussed at length a defendant's demeanor during court proceedings is certainly probative of sanity with regard to his or her competence to stand trial. But the probative value of a defendant's courtroom demeanor during trial as to his or her mental state at the time of the crime is doubtful. The justification for considering a defendant's demeanor before and after the crime is that conduct occurring in temporal proximity to the crime "may be more indicative of actual mental health at [the] time of the crime than mental exams conducted weeks or months later." Trial proceedings, however, often occur many months or even years after the crime. In this case, the two-day bench trial occurred nearly a year after the murder. Thus, we do not find the fact that the defendant "was alert and oriented throughout the proceedings and assisted his counsel and the investigator" to be probative of his sanity at the time of the crime.

There also is no evidence or suggestion that the defendant here feigned his mental illness. The trial court expressly found as much with regard to defendant's long history of mental illness. . . .

The trial court committed an error in this case by entering a verdict of guilty but mentally ill when the evidence presented reasonably led only to a conclusion that the defendant was legally insane at the time of the offense. Underlying the trial court's decision was not a concern of malingering or feigning but a concern about the State's mental health system and the defendant's need for structure and constant supervision. Among the trial court's findings is that the defendant "lacks insight into the need for his prescribed medication" and "is in need of long term stabilizing treatment in a secure facility." The trial court also found that the defendant "repeatedly discontinued medication" and there was "no evidence that this pattern of conduct will not continue if [the defendant] is hospitalized and released, posing a danger to himself and others in the community." . . . The trial court confessed at sentencing that it viewed "[t]his case . . . as much a trial of our mental health system as . . . of a man." The court lamented that it could not simply commit the defendant to a mental health institution for the rest of his life—the "easy" decision. What made the court's decision so difficult was that it could not "in good conscience allow someone with . . . severe mental illness to return to the community." . . . Thus, while we sympathize with the difficulty of the trial court's decision, we cannot sustain it.

Holding

We reverse the judgment of the trial court.

Dissenting, *Shepard, C.J.*, with whom *Dicksin, J.*, joins.

Gregory Galloway is someone who went shopping at a going-out-of-business sale in the morning, had some lunch at a local restaurant with his aunt and grandmother, and stopped off at a gas station to buy fuel and cigarettes. Galloway appeared normal all day; "everybody was happy," one of his companions said.

When Galloway arrived home, he stabbed his grandmother to death, and then immediately announced that he regretted what he had done. The finder of fact in this case, Judge Mary Willis, concluded on the basis of the admitted evidence that Galloway was not insane at the time of the crime, that is to say, that he knew killing his grandmother was wrong.

Of course, all of the testimony by psychiatrists and psychologists necessarily came from witnesses who were not present at the scene of the crime. They offered their observations based on records of Galloway's medical history from moments other than the hour of the killing and on direct observations of Galloway that occurred months or even years after the crime. One of these experts, Dr. Glenn Davidson, appointed by the court, concluded that Galloway was not insane at the time of the crime. Eyewitness evidence about how Galloway acted before and after the crime also supported the trial court's decision.

This was one of those cases where the defense argued that the perpetrator was sane right before the crime and sane right after the crime, but insane for the sixty seconds or so it took to commit it. Dr. Davidson's basic view was that it was unlikely that Galloway qualified as insane on the basis of a "very thin slice of disorganized thinking."

Defense counsel's vigorous cross-examination confronted Dr. Davidson with a host of hypotheticals ("now what if I told you") and asked as to each new proposed fact whether it would affect his diagnosis. It was twenty to thirty pages of the sort of energetic cross-examination tactics to which we lawyers are inured but which often befuddle the uninitiated. It finally left the witness saying, in the face of this onslaught, that he was unsure.

As the majority points out, juries and judicial fact finders are not required to take as completely true all or none of what witnesses say. They are entitled to believe and disbelieve some, all, or none of the testimony of experts and non-experts alike. Indeed, their assignment is to sort out truth from cacophony. It was altogether plausible that Judge Willis could credit Dr. Davidson's opinion that Galloway was sane and treat the doctor's answers under cross as less compelling. She could also, of course, give weight to Galloway's own contemporaneous declaration of regret right after he killed his grandmother.

As the majority does acknowledge, there is risk involved when appellate judges second-guess a jury or trial judge and acquit a criminal offender. If Galloway is declared not guilty by this Court, the prosecutor will initiate a civil commitment process to determine whether Galloway should be confined because his mental illness makes him a danger to himself or to others.

The one thing we know for sure about Mr. Galloway is that he is in actual fact a danger to others. . . . We also know what is likely to occur as a result of this Court setting aside Judge Willis's judgment: sooner or later, probably sooner rather than later, Galloway will be determined safe and turned back into society.

The reason we know that is that the civil commitment process has produced such an outcome over and over again with Mr. Galloway. The majority has recited the long trail of medical treatments and mental commitments. It has not focused much in that recitation on how the exercise of expert medical judgments and the civil commitment processes have combined to turn him back out on the street over and over again.

I count perhaps seventeen identifiable encounters by Galloway. But just to name a few, call it number 5, there was a May 1999 event in which Galloway's wife brought him in because he had been carrying around a gun and threatening to use it on his supervisor at work. This trip produced a prescription for medication and a period of outpatient treatment, then a failure to take his medications and a medical trail gone cold.

During encounter number 7, in April 2001, Galloway was admitted to the hospital because of aggressive and frightening behavior at home. He said he had been receiving messages from the television. This interaction with the system produced several months of monitoring during which Galloway took some of his medicines and not others. And then he was out.

During encounter number 8, Galloway was involuntarily committed because he had threatened to kill his neighbor and his grandmother. He was released from commitment and then admitted again just a month later,

in March 2002. He stayed a few months at Richmond State Hospital before being declared safe for release.

In encounter number 13, not long before Galloway killed his grandmother, Galloway came under care after he stopped taking his medicines and began reporting hallucinations and recurring thoughts of suicide. After being stabilized, he was discharged to live with his grandmother, with a result plain and painful for all to see.

I mention this litany—just salient elements in an even longer story—to suggest that some innocent future victim is placed at risk by this Court's decision to second-guess Judge Willis. A society that responds to such violence with tolerance should well expect that it will experience more violence than it would if it finally said, "This is unacceptable." Not knowing what I would say to the next victim, I choose to stand with Judge Willis and affirm the judgment of guilty but mentally ill.

Questions for Discussion

1. What is the legal standard for insanity used by the Indiana Supreme Court?

2. Summarize the evidence that the court majority relies on in concluding that Galloway suffered form a bipolar disorder and was unable to appreciate the wrongfulness of his conduct.

3. Was Galloway's demeanor unusual on the day of his grandmother's death? Why is the demeanor evidence significant in this case?

4. Explain the reason that the trial judge held that Galloway was GBMI rather than legally insane (NGRI).

5. Why did the Indiana Supreme Court overturn the verdict of GBMI?

6. What is the reason that two judges dissented from the majority decision?

7. Why does it matter whether Galloway is found to GBMI rather than NGRI? Do you agree with the decision of the majority or with the decision of the dissenting judges?

See more cases on the study site:
Moler v. State, *cited above with* Lopez, *www.sagepub.com/lippmanccl3e.*

You Decide

9.1 In January 1992, defendant Freddie Armstrong went to Loche's Mortuary to obtain a copy of his father's death certificate. The police responded to a call from the mortuary and found Armstrong with a bloody butcher's knife standing over the body of Reverend Fred Neal. One officer drew his gun and the defendant severed Reverend Neal's head from the corpse and held it up by the ears. Armstrong then placed the headless body in a chair and walked up the stairs and deposited the head in the toilet. Armstrong placed the knife in his briefcase, put on his cap, and walked out the door as if in a trance. Armstrong displayed no concern or awareness of the police. The medical evidence indicated that Armstrong had been evaluated as an acute paranoid schizophrenic and had been medically discharged from military service during the Vietnam War. He had been admitted to mental institutions in 1969, 1970, 1973, 1980, 1983, 1987, and 1992 and had been released three days

prior to killing Reverend Neal. Armstrong apparently had been commanded by a hallucinatory voice to stab Reverend Neal while another voice told him that his actions were wrong. The legal standard for insanity in Louisiana is the ability to "distinguish between right and wrong." The jury determined that Armstrong was legally sane, a decision that was upheld by a Louisiana appellate court.

Armstrong's diagnosis indicated that he was not curable, but that his mental condition might be controlled through constant medication. Armstrong had delusions about his wife working against him, he heard voices of good and evil about Christ and the Antichrist, and he was not sleeping.

Armstrong called his brother on the day of the killing and announced that he had been nominated for President of the United States. He then missed an appointment for a drug injection to control his schizophrenia and drove to the mortuary and waited for Reverend Neal to enter the building. Armstrong was told by a voice that Reverend Neal was the Antichrist and testified that he severed the Reverend's head from the body in order to prove that Neal would not bleed. The prosecutor found it significant that Armstrong's violent impulse was limited to Reverend Neal, whom he was told by a "voice" was the Antichrist and should be killed and sent to hell. At the jail following his arrest, Armstrong refused to take his medication, and stopped up the urinal, flooding his cell. He walked around naked carrying a Bible and stated that the federal government had been attempting to kill him for several years. The initial psychiatric evaluation indicated that he was aggressive, threatened to kill the staff, and believed that he was being poisoned.

The doctor who had been treating Armstrong for ten years stated that Armstrong became grossly psychotic when he was off his medication and had observed Armstrong in a psychotic state on two previous occasions. Armstrong typically was preoccupied during these periods with religion and the Antichrist. The doctor testified that when Armstrong was on his medication, he was capable of distinguishing right from wrong, but that he was unable to make this distinction when he was in a psychotic state. Another doctor examined Armstrong a month after the killing and stated that the defendant was unstable, delusional, and obsessed with destroying the Antichrist. The second doctor testified that Armstrong had heard a divine voice instructing him not to kill Reverend Neal, but that God had forgiven him. Three other psychiatrists testified, two of whom shared the view that Armstrong was legally insane. Another testified that she believed that Armstrong was able to distinguish between divine and human law and that he was aware that killing Reverend Neal was wrong. Was Armstrong capable of distinguishing "right from wrong"? Would you return a verdict of not guilty by reason of insanity? Would he satisfy the irresistible impulse test? See *State v. Armstrong*, 671 So. 2d 307 (La. 1996).

You can find the answer at www.sagepub.com/lippmanccl3e.

DIMINISHED CAPACITY

Diminished capacity is recognized in roughly fifteen states. This permits the admission of psychiatric testimony to establish that a defendant suffers from a mental disturbance that *diminishes* the defendant's capacity to form the required criminal intent. Diminished capacity merely recognizes that an individual has the right to demonstrate that he or she is incapable of forming the intent required for the offense. This is a compromise between finding an individual either not guilty by reason of insanity or fully liable. Some states confine diminished capacity to intentional murder and provide that an accused may still be convicted of second-degree murder, which does not require premeditation.

This often is referred to as the *Wells-Gorshen* rule based on two California Supreme Court decisions.[25] Gorshen, a dock worker, reacted violently when ordered to "get to work" and then precipitated a fight when he was told that he was drunk and should go home. The defendant later returned to work and shot and killed his foreman, Joseph O'Leary. The California Supreme Court affirmed the trial court's decision to convict Gorshen of second rather than first-degree murder, which requires a premeditated intent to kill. Psychiatric testimony indicated that the defendant suffered from chronic paranoiac schizophrenia, a "disintegration of mind and personality . . . [involving] trances during which he hears voices and experiences visions, particularly of devils in disguise committing abnormal sexual acts, sometimes upon the defendant." According to a psychiatrist, Gorshen believed that O'Leary's remarks demeaned his manliness and sexuality and that this sparked enormous rage and anger. The defendant was reportedly out of control and felt that he was slipping into permanent insanity. He blamed O'Leary and developed an obsession with killing him. The appellate court ruled that Gorshen possessed a driving and overwhelming obsession with murdering O'Leary and that he did not make a reasoned and conscious decision to kill.[26]

The diminished capacity defense has been rejected by some state courts that point out that psychiatric testimony is unreliable and too confusing for jurors and that the "medical model" is contrary to the notion that individuals are responsible for their actions.[27] The far-reaching implications of the diminished capacity defense became apparent when a San Francisco jury convicted city official Dan White of manslaughter for the killings of his colleague Harvey Milk and Mayor George Moscone. The defense argued that White's depressions were exaggerated by junk food, diminishing his capacity to form a specific intent to kill. In reaction to this "Twinkie defense," California voters adopted a statute that provides that the "defense of diminished capacity is hereby abolished" and shall not be admissible "to show or negate capacity to form the . . . intent . . . required for the commission of the crime charged."[28] The Model Penal Code, however, provides that evidence that a defendant suffers from a mental disease or defect is admissible "whenever it is relevant to prove that the defendant did or did not have a state of mind that is an element of the offense." In other words, under this approach a defendant may introduce psychiatric evidence to negate the required intent in a prosecution for any criminal offense.[29]

CRIME IN THE NEWS

On January 8, 2011, Gerald Lee Loughner, a 22-year-old college student left his home and took a taxi to a mall where Tuscon, Arizona Congresswoman Gabby Giffords was holding a "Congress on the Corner" for her constituents. Loughner calmly approached Giffords, pulled out a Glock 19 pistol equipped with a high capacity magazine clip, and shot her in the head at point-blank range. He also shot and killed federal Judge John M. Rolls, who was talking to Giffords about the problem of overcrowding in the courts. Loughner next turned and fired wildly into the crowd, killing five people and wounding 19 others. Among the dead were 9-year-old Christina Taylor Green, a vibrant and impressive young girl who had developed a passion for politics, and Gerald Zimmerman, age 30, an aide to Congresswoman Giffords. Loughner also killed three women in their 70s. Giffords miraculously survived and continues to make progress toward overcoming the physical and emotional damage from the violent criminal attack.

A search of Loughner's house revealed a note with the words "I planned ahead," "My assassination," and "Giffords," Although Loughner had embraced the extreme views of fringe right-wing groups, his actions according to most experts were motivated by a psychological disorder rather than by political ideology. On the day of the killings, Loughner had purchased bullets for the gun and picked up photos of himself posing with a Glock pistol while wearing a red g-string that he posted on the Internet.

Loughner earlier had been suspended from Pima Community College and had been told by the college that he would be readmitted when the mental health professional concluded that he no longer posed a danger. The college reluctantly took this decision after a number of instructors and students complained about Loughner's bizarre and disruptive behavior in class. In one incident, Lougher loudly insisted that he had the right to call the number 6 the number 18; and on another occasion, he reacted to a student's reading of a poem by erupting into a loud rant on "strapping babies to bombs" that included references to abortion, wars, and killing people.

In an Internet posting, Loughner complained that the government was trying to control and brainwash people through the rules of grammar and language usage. His fellow students described Loughner as "dark" and "creepy," and his friends described him as living in a dreamlike and delusional trance.

Loughner was charged in Federal District Court with the attempted assassination of a member of Congress and with killing and the attempted killing of four federal government employees. He waived bail and was found by the judge to be incompetent to stand trial, meaning that he was too emotionally disturbed to participate in his own defense. Loughner's lawyers unsuccessfully challenged the government's right to subject him to a course of drug treatment in an effort to psychologically stabilize him in preparation for trial. Loughner also may face state charges based on his killing and wounding of victims who were not affiliated with the federal government.

In August, 2012, Loughner was found competent to stand trial and pled guilty in return for multiple sentences of life imprisonment. He reportedly agreed to the plea bargain because he wanted to avoid a death sentence, and the prosecutors agreed to the plea bargain because they wanted to avoid a finding of legal insanity.

Gerald Loughner is reminiscent of Seung-Hui Cho, the Virginia Tech student, who also used a 9 millimeter pistol in 2007 to kill 32 students and faculty and to wound 19 individuals. A survey by the American College Counseling Association found that 44 percent of university students seeking counseling suffer from severe psychological disorders and 24 percent are on psychiatric medication. Roughly half of the nation's colleges have "threat assessment teams" that monitor at risk students. Experts debate the appropriate role of college and university administrators in dealing with students who reasonably pose a danger. Was Gerald Loughner legally insane at the time of his armed attack?

INTOXICATION

Alcoholic beverages and drugs are commonly used to relax and to enhance enjoyment. These substances, however, can impede coordination and alertness, distort judgment, and cause impulsive and emotional reactions. It is not surprising that some studies suggest that more than half of those arrested for felonies have been drinking or using drugs. Should the law limit the legal responsibility of individuals who are drunk or are "high" on drugs or treat them more harshly? The law has struggled to find a balance between "conflicting feelings" of concern and condemnation for the "intoxicated offender."[30]

Voluntary Intoxication

Voluntary intoxication was not recognized as a defense under the early common law in England. Lord Hale proclaimed that the intoxicated individual "shall have no privilege by this voluntary contracted madness, but shall have the same judgment as if he were in his right senses." William Blackstone went beyond this neutral stance and urged that intoxication should be viewed "as an aggravation of the offense, rather than as an excuse for any criminal behavior."[31]

The common law rule was incorporated into American law. An 1847 textbook recorded that this was a "long established maxim of judicial policy, from which perhaps a single dissenting voice cannot be found."[32]

The rule that intoxication was not a defense began to be transformed in the nineteenth century. Judges attempted to balance their disapproval toward alcoholism against the fact that inebriated individuals often lacked the mental capacity to formulate a criminal intent. Courts created a distinction between offenses involving a specific intent for which voluntary intoxication was an excuse and offenses involving a general intent for which voluntary intoxication was not recognized as an excuse. An individual charged with a crime requiring a specific intent was able to introduce evidence that the use of alcohol prevented him or her from forming a specific intent to assault an individual with the intent to kill. A defendant who proved successful would be held liable for the lesser offense of simple assault. As noted by the California Supreme Court, the difference between an intent to commit a battery and an intent to commit a battery for the purpose of raping or killing "may be slight, but it is sufficient to justify drawing a line between them and considering evidence of intoxication in the one case and disregarding it in the other."[33]

The Model Penal Code section 2.08(1)(2) accepts the common law's distinction between offenses based on intent and substitutes "knowledge" or "purpose" for a specific intent and "negligence or recklessness" for a general intent. The commentary to the code notes that it would be unfair to punish an individual who, due to inebriation, lacks "knowledge or purpose," even when this results from voluntary intoxication.[34]

Professor Jerome Hall observes that *in practice,* the hostility toward the inebriated defendant has resulted in the voluntary intoxication defense only being recognized in isolated instances, typically involving intentional killing.[35] Courts have placed a heavy burden on defendants seeking to negate a specific intent. Even the consumption of large amounts of alcohol is not sufficient. The New Jersey Supreme Court observed that there must be a showing of such a "great prostration of the faculties that the requisite mental state was totally lacking. . . . [A]n accused must show that he was so intoxicated that he did not have the intent to commit an offense. Such a state of affairs will likely exist in very few cases." This typically requires an evaluation of the quantity and period of time that an intoxicant was consumed, blood alcohol content, and the individual's conduct and ability to recall events.[36]

The contemporary trend is to return to the original common law rule and refuse to recognize a defense based on voluntary alcoholism. The Arizona Criminal Code, section 13-503 of the Arizona Revised Statutes Annotated, provides that "[t]emporary intoxication resulting from the voluntary ingestion . . . of alcohol . . . or other psychoactive substances or the abuse of prescribed medications . . . is not a defense for any criminal act or requisite state of mind." The Texas Penal Code Annotated section 8.04 provides that "[v]oluntary intoxication does not constitute a defense to the commission of crime." The right of states to deny defendants the intoxication defense was affirmed by the U.S. Supreme Court in 1996, in *Montana v. Egelhoff*; Justice Antonin Scalia noted that Montana was merely returning to the law at the time of the drafting of the U.S. Constitution and that this rule served to deter excessive drinking.[37]

Involuntary Intoxication

Involuntary intoxication is a defense to any and all criminal offenses in those instances that the defendant's state of mind satisfies the standard for the insanity defense in the state. The Model Penal Code section 2.08(4) requires that the individual "lacks substantial capacity" to distinguish right from wrong or to conform his or her behavior to the law. The code also recognizes "pathological intoxication." This arises in those instances when an individual voluntarily consumes a substance and experiences an extreme and unanticipated reaction. Involuntary intoxication can occur in any of four ways:[38]

- *Duress.* An individual is coerced into consuming an intoxicant.
- *Mistake.* An individual mistakenly consumes a narcotic rather than his or her prescribed medicine.
- *Fraud.* An individual consumes a narcotic as a result of a fraudulent misrepresentation of the nature of the substance.
- *Medication.* An individual has an extreme and unanticipated reaction to medication prescribed by a doctor.

The proliferation of drugs, medicine, and newly developed therapies promises to lead to involuntary intoxication being increasingly raised as a defense in criminal prosecutions. Consider the next case in the textbook, *Brancaccio v. State*.

Did the defendant's medication cause him to kill?

Brancaccio v. State, 698 So. 2d 597 (Fla. Dist. Ct. App. 1997). Opinion by: Klein, J.

Appellant was found guilty of first-degree murder and kidnapping and sentenced to life in prison. He had admitted the killing, and the only issue for the jury to determine was whether he had the mental capacity to form the intent necessary to commit the crimes. We must reverse for a new trial because the court erred in refusing to instruct the jury . . . that he was involuntarily intoxicated as a result of the medication he was taking pursuant to a prescription.

Facts

Appellant, who was 16, gave a statement to the police explaining that he had had a fight with his mother over what they were having for dinner and went for a walk to cool down. He encountered a stranger who asked him to stop cursing and called him low class. He punched her repeatedly, led her to a vacant lot, and continued to punch her and kick her. When a car came along, he became frightened and ran home.

He returned to the scene the next morning to ask the woman if she needed help, but she did not respond. The medical examiner testified that the woman had suffered at least four severe and potentially fatal blows to her head, as well as massive trauma to her chest, and that she would not have been alive at that point. Appellant then went shopping for car parts, but returned later in the day with newspaper and unsuccessfully attempted to set her body on fire. He then left and returned with spray paint, painting her body red in order to cover up his fingerprints.

Appellant's defense was that the medication he was taking, which was prescribed to him during his recent confinement in a mental hospital, had caused him to lose control. Two months prior to the killing,

appellant had been committed to a mental health center . . . after threatening to kill his parents and himself. His parents then moved him to the Savannas Mental Hospital for treatment, where appellant was placed on Zoloft, a drug used to treat depression. The hospital also diagnosed appellant as suffering from alcohol abuse, attention deficit disorder, and oppositional defiant disorder. The hospital noted a change in appellant's personality after he was placed on Zoloft, in that appellant became more irritable, loud, had increased energy and was given to angry outbursts. Appellant had apparently attempted suicide while at the hospital by holding his breath.

[The condition of Oppositional Defiant Disorder is defined in the Diagnostic and Statistical Manual of Mental Disorders. . . . The essential feature of this disorder is a pattern of negativistic, hostile, and defiant behavior. . . . Children with this disorder commonly are argumentative with adults, frequently lose their temper, swear, and are often angry, resentful, and easily annoyed by others. They frequently actively defy adult requests or rules and deliberately annoy other people. They tend to blame others for their own mistakes or difficulties.]

A psychiatrist testifying for the defense, Dr. Wade Myers, testified that Zoloft may have had a reaction that was opposite to what it was supposed to have had in appellant, causing hypomania [which manifested itself by causing the appellant to have trouble concentrating, difficulty sitting still, an increased energy level, irritability, and anxiety]. The medical warnings for the drug state that the following side effects are infrequent but possible: "aggressive reactions, amnesia, anxiety, delusions, depersonalization, depression, aggravated depression, emotional instability . . . hallucinations, neurosis, paranoid reaction, suicidation and suicide attempts." "Infrequent" is medically defined as occurring between . . . 1 in 1,000 patients.

Dr. Myers examined appellant after his arrest and concluded that he suffered from major depression, possible bipolar disorder, alcohol abuse, a learning disability, and a probable brain injury. Appellant had been born prematurely and spent seven days in intensive care for oxygen deprivation. Also, at the age of two, he had nearly drowned and had to be resuscitated. His IQ is just above retarded.

Dr. Myers concluded that appellant did not have the ability to form the intent to commit first-degree murder based on his mental deficiencies and his involuntary intoxication by the Zoloft.

Another expert who testified for the defense, Dr. Peter R. Breggin, testified that in his opinion, Zoloft could have . . ."pushed him over." . . . Dr. Breggin was able to obtain information reported by doctors and pharmacists regarding reactions to the drug. Dr. Breggin found 22 reports tying the drug to hostile reactions, 57 reports linking the drug to an aggravation reaction, 55 suicide attempts, and 64 reports linking the drug to increased agitation. These reports suggest that the drug can cause a loss of impulse control. He opined that the hospital records from the mental hospital where appellant was confined indicate that he was experiencing a similar reaction to the drug. Dr. Breggin diagnosed Brancaccio with substance induced mood disorder brought on by Zoloft.

The State rebutted appellant's experts with experts of its own, who testified that appellant was capable of forming the intent to commit murder and kidnapping. They did agree that he suffered from major depression, but were of the opinion that he was not involuntarily intoxicated because of the Zoloft.

Issue

The jury found appellant guilty of first-degree murder (felony murder), and kidnapping. His primary argument on appeal is that the trial court erred in refusing to instruct the jury on his theory of defense, involuntary intoxication.

Reasoning

The defense of involuntary intoxication has been explained . . . as follows:

Generally speaking, an accused may be completely relieved of criminal responsibility if, because of involuntary intoxication, he was temporarily rendered legally insane at the time he committed the offense. And again speaking generally, the courts have considered one to be involuntarily intoxicated when he has become intoxicated through the fault of another, by accident, inadvertence, or mistake on his own part, or because of a physiological or psychological condition beyond his control.

The practice of relieving one of criminal responsibility for offenses committed while in a state of involuntary intoxication extends back to the earliest days of the common law. Involuntary intoxication, it appears, was first recognized as that caused by the unskillfulness of a physician or by the contrivance of one's enemies. Today, where the intoxication is induced through the fault of another and without any fault on the part of the accused, it is generally treated as involuntary. Intoxication caused by the force, duress, fraud, or contrivance of another, for whatever purpose, without any fault on the part of the accused, is uniformly recognized as involuntary intoxication. This is often stated in an exclusive form, that is, that the intoxication is involuntary only if induced by the fraud, etc., of another. Although this implies that intoxication from any other cause is not involuntary, the courts making such a statement do, in fact recognize that intoxication caused in other ways can be involuntary.

Holding

The State first argues that the evidence did not show that appellant was taking Zoloft at the time of the murder. In his taped statement to the police, which was taken three days after the killing, appellant was asked whether he was on any medication and replied that he was taking Zoloft. Appellant had also told one of his medical experts that he thought that he had taken his medicine on the day of the killing. In addition, a friend of appellant had testified that on the day of the killing he had observed appellant's mother make him go into the house in order to take his medication.

The only other argument advanced by the State is that there was no evidence that Zoloft had an adverse reaction on appellant; however, that argument ignores the expert testimony, the mental hospital records, and testimony of people who knew the appellant and observed a personality change in him. . . .

Questions for Discussion

1. What facts does the appellant present in support of the claim that he lacked a criminal intent and was temporarily insane? What is the government's counterargument?

2. As a juror, would you find that the appellant acted in a fashion that is characteristic of an individual who is legally insane? Are you persuaded that his criminal conduct was caused by Zoloft? What is the significance of the fact that only a small percentage of people have a negative side effect to Zoloft?

You Decide

9.2 Robert Low was president and general manager of a trucking company in Springfield, Missouri. Low and his fourteen-year-old stepson, Sane Low, arranged a hunting trip with two friends. The group met in Creede, Colorado, and drove to the campsite. Robert became increasingly disoriented and asked his stepson why he was being "tricked." The drivers of two trucks stopped to check on Robert's health. Robert then demanded that they all kneel in prayer. This was unusual because Robert was not particularly religious. During the remainder of the ride, Robert speculated on whether he was alive or dead. They arrived at the campsite and Robert was convinced that he was dead and had gone to hell. He requested that his tent be set up on a knoll and stated that this would provide the foundation for a divine temple. Robert then accused McCowan of being the devil, and the three others realized that he was disturbed and prevented him from loading his rifle. He then stabbed McCowan in the upper back, and McCowan was taken by some hunters to the hospital. Robert then unsuccessfully attempted to stab himself and poured kerosene on the floor of the tent and ignited a fire. The police arrived and arrested Robert.

Low had ingested forty to fifty cough drops a day for the past several months. He initially took the cough drops to combat a cold and then continued to ingest these as a substitute for chewing tobacco and to help him to quit smoking. On his trip to Colorado, Robert consumed roughly one hundred twenty cough drops within a twenty-four-hour period. A psychiatrist testified that the cough drops contain a drug called dextromethorphan hydrobromide. This caused a psychotic disorder known as "organic delusional syndrome" or "toxic psychosis." The symptoms include a distorted perception of reality, paranoia, hallucinations, and delusions. The psychiatrist testified that Low was incapable of knowing right from wrong at the time of the hunting trip and did not have the ability to formulate a specific intent to commit a criminal act. A doctor at the Colorado State Hospital testified that Low tested negative for marijuana, alcohol, cocaine, and most other narcotics.

The cough drops were sold over the counter without prescription. The customary warnings included on the label proclaimed, "Not Habit Forming contains 7.5 milligrams of dextromethorphan HBr per lozenge." Robert was charged with first-degree assault that requires a specific intent to cause serious bodily injury or disfigurement by a dangerous weapon or knowingly engaging in conduct that creates a grave risk of death. Second-degree assault involves recklessly causing serious bodily injury by means of a deadly weapon. Third-degree assault is committed when the accused knowingly or recklessly causes bodily injury to another or with criminal negligence causes bodily injury to another person by means of a deadly weapon. Should the jury have been given an instruction on the defense of involuntary intoxication? Would you convict Low of first-degree assault? See *People v. Low*, 732 P.2d 622 (Colo. 1987).

You can find the answer at www.sagepub.com/lippmanccl3e.
See more cases on the study site: **People v. Holloway, www.sagepub.com/lippmanccl3e.**

AGE

The early common law did not recognize **infancy** as a defense to criminal prosecution. Youthful offenders, however, were typically pardoned. A tenth-century statute softened the failure to recognize infancy as a defense by providing that an individual younger than the age of fifteen was not subject to capital punishment unless he or she made an effort to elude authorities or refused to surrender. A further refinement occurred in the fourteenth century when children younger than seven were declared to be without criminal capacity.

The common law continued to develop and reached its final form by the seventeenth century. Juveniles were divided into three categories based on the capacity of adolescents at various ages to formulate a criminal intent. Individuals were categorized on the basis of their actual rather than their mental age at the time of the offense.[39]

- *Children younger than seven lacked a criminal capacity.* There was an *irrebuttable presumption,* an assumption that cannot be overcome by facts, that children younger than seven lack the ability to formulate a criminal intent.
- *Children older than seven and younger than fourteen* were presumed to be without capacity to form a criminal intent. This was a *rebuttable presumption;* the prosecution could overcome the presumption by evidence that the juvenile knew what he or she was doing was wrong. The older the child and the more atrocious the crime, the easier to overcome the presumption. Factors to be considered include the age of the child, efforts to conceal the crime and to influence witnesses, and the seriousness of the crime.

- *Children fourteen and older* possessed the same criminal capacity as adults. Juveniles capable of forming a criminal intent may be prosecuted as adults rather than remain in the juvenile system. Today, the age when a juvenile may be criminally prosecuted as an adult rather than being brought before a juvenile court is determined by state statute. There is no standard approach. One group of states maintains a conclusive presumption of incapacity for juveniles younger than a particular age (usually fourteen); however, other states provide that juveniles regardless of age may be treated as an adult.[40]
- Roughly twenty-five states continue to follow the tripartite common law scheme while modifying the age categories. These states provide that a presumption of incapacity may be overcome when a juvenile is demonstrated to have known the wrongfulness of his or her actions.
- Others specify an age, typically fourteen, younger than which there is a conclusive presumption that a juvenile cannot form a criminal intent.
- A third group of state statutes provides for exclusive jurisdiction by the juvenile court until a specified age. These states typically provide that cases involving individuals between sixteen and eighteen charged with serious crimes may be transferred to adult court.
- Another set of statutes recognizes that the jurisdiction of the juvenile court is not exclusive and that juveniles charged with serious offenses may be subject to criminal prosecution.
- A fifth group of states merely provides that the jurisdiction of the juvenile court does not prevent the criminal prosecution of juveniles.

The common law presumptions of incapacity are not applicable to proceedings in juvenile court because the purpose of the court is treatment and rehabilitation rather than the adjudication of moral responsibility and punishment.[41]

There is a growing trend for state statutes to permit the criminal prosecution of any juvenile as an adult who is charged with a serious offense. These "transfer statutes" adopt various schemes, vesting "waiver authority" in juvenile judges or prosecutors or providing for automatic transfer for specified crimes.[42] The standard to be applied by judges was articulated by the U.S. Supreme Court in *Kent v. United States*. The factors to be considered in the decision whether to prosecute a juvenile as an adult include the seriousness and violence of the offense, the background and maturity of the juvenile, and the ability of the juvenile justice system to protect the public and rehabilitate the offender.[43] The controversial question of certifying juveniles for trial as adults is explored in the next case, *Brazill v. State*.

Can Florida constitutionally prosecute juveniles as adults?

Brazill v. State, 845 So. 2d 282 (Fla. Dist. Ct. App. 2003). Opinion by: Gross, J.

On the last day of the 1999–2000 school year, thirteen-year-old Nathaniel Brazill shot and killed a teacher at his middle school, Barry Grunow.

The state charged Brazill with first-degree murder and aggravated assault with a firearm. The jury convicted him of second-degree murder and aggravated assault with a firearm. The trial judge sentenced him to concurrent sentences: a mandatory minimum sentence of twenty-eight years in prison on the murder charge and five years in prison, with a three year mandatory minimum, on the assault charge.

We affirm in all respects.

Facts

In the early afternoon of May 26, 2000, Brazill and Michelle Cordovaz were suspended for the remainder of the day as the result of a water balloon fight. School counselor Kevin Hinds escorted the two students off campus. Brazill asked Hinds what time he was going home. Hinds indicated that he was leaving around 4:15 to 4:30 p.m. and asked why Brazill wanted to know. Brazill shrugged and did not respond.

As he was walking away with Cordovaz, Brazill told her that he had a gun and was going to return to shoot Hinds. Cordovaz asked: "You wouldn't do that, Nate, would you?" Brazill answered: "Watch. I'm going to be all over the news."

On the way home, Brazill made several stops. Near his grandmother's house, Brazill spoke to Brandon Spann. He asked if Spann was part of a gang or had a gun. Spann asked him why he needed a gun. Brazill replied that he was "going to fuck up the school" because of the suspension.

At his home, Brazill retrieved a gun from his bedroom. The previous weekend, Brazill was at his grandfather's house and found the gun in a cookie jar in his grandfather's bureau. At that time, he loaded the gun, pulled the slide back, engaged the safety, and placed it in his overnight bag. When Brazill left his grandfather's house, he took the gun home with him; upon returning home he hid the gun in his room.

Taking the gun from his bedroom, Brazill rode his bike back to school. On the way, he stopped by his aunt's house and left a note.

Brazill entered the school grounds near the rear parking lot, a designated teachers' area. School security officer Matt Baxter saw him. Baxter followed him, but found only an abandoned bike. After leaving his bike, Brazill ran to the school building. On the way, he advised a student sitting outside to go home.

Once inside the school, Brazill went directly to Barry Grunow's classroom to speak with two friends, Dinora Rosales and Vonae Ware. He had once dated Ware for a time, and was romantically interested in Rosales. Earlier in the day, Brazill gave Rosales two cards and a bouquet of flowers.

When Brazill knocked on Grunow's door, the students in the class were already standing, because they were about to go outside. Brazill sternly asked to speak to Rosales and Ware, who were standing on either side of Grunow. The teacher did not allow the girls to leave the classroom, but said that Brazill could come inside. Brazill refused to enter the classroom. Three more times he asked to see the girls. Each time Grunow calmly declined and told him to go back to class.

Brazill then pulled out the gun and aimed it at Grunow's head. He was in the hallway, approximately an arm's length from Grunow. He backed up slightly and assumed a shooter's stance with his legs apart.

Grunow told Brazill to stop pointing the gun, but he continued to point the gun at the teacher's head. Brazill appeared to be angry but calm; he was not crying or shaking. Brazill pulled the slide back on the gun.

[A crime scene investigator testified that pulling the slide back on this gun put a bullet in the chamber. If a bullet was already in the chamber when the slide was pulled, then a live round would eject. At the crime scene, the investigator found a live cartridge, along with a discharged shell casing.]

As Grunow attempted to close the classroom door, Brazill pulled the trigger and Grunow fell to the floor, with a gunshot wound between the eyes. A school surveillance videotape of the hallway revealed that Brazill had pointed the gun at Grunow for nine seconds before shooting. Brazill exclaimed: "Oh s__t," and fled.

On the way out, Brazill used both hands to aim the gun at math teacher, John James, who was conducting class next door to Grunow. As Brazill aimed the gun, he told James not to bother him, that he was going to shoot. James immediately turned around and led his students back into his classroom.

Brazill ran out of the building. To one teacher, Brazill did not appear to be visibly upset. He was not sweating. He was not crying. Near the school, Officer Michael Mahoney observed Brazill walk into the street, put his hands on his head, and kneel. When the officer asked what he was doing, Brazill stated that he had shot someone at school and the gun was in his pocket. Brazill was then arrested. He acknowledged that he had shot Grunow. Brazill was taken to the police station, where he gave a videotaped statement.

A firearms expert with the FBI testified that the gun used in the shooting had a safety that functioned normally. The gun had a trigger pull that required five and one-half pounds of pressure to fire. It would not discharge unless the trigger was pulled.

Issue

Brazill argues that Florida Statutes (FS) section 985.225 (1999) is unconstitutional as a violation of due process, equal protection, and separation of powers. In pertinent part, section 985.225 provides:

(1) A child of any age who is charged with a violation of state law punishable by death or by life imprisonment is subject to the jurisdiction of the [juvenile] court . . . unless and until an indictment on the charge is returned by the grand jury. When such indictment is returned, the petition for delinquency, if any, must be dismissed and the child must be tried and handled in every respect as an adult:

   (a) On the offense punishable by death or by life imprisonment; and

   (b) On all other felonies or misdemeanors charged in the indictment which are based on the same act or transaction as the offense punishable by death or by life imprisonment or on one or more acts or transactions connected with the offense punishable by death or by life imprisonment.

(2) . . .

(3) If the child is found to have committed the offense punishable by death or by life imprisonment, the child shall be sentenced as an adult. If the juvenile is not found to have committed the indictable offense but is found to have committed a lesser included offense or any other offense for which he or she was indicted as a part of the criminal episode, the court [also] may sentence [as an adult]. . . .

Reasoning

Brazill first contends that his due process rights were violated because he was denied the "rehabilitative aspect of juvenile court" solely because the state decided to procure an indictment as an adult. However, there is no absolute right conferred by common law, constitution, or otherwise, requiring children to be treated in a special system for juvenile offenders. . . . Under article I, section 15(b) of the Florida Constitution, a "child," as defined by "law," may be charged "with a violation of law as an act of delinquency instead of [a] crime." As our supreme court has explained, this provision means that "a child has the right to be treated as a juvenile delinquent only to the extent provided by our legislature. . . ."

Holding

Florida Statutes section 985.225 is related to the state's interest in crime deterrence and public safety. The statute provides treatment as an adult for those offenses serious enough to be punishable by life imprisonment or death. Such crimes are the most violent or dangerous offenses against persons. It is not unreasonable for the state legislature to treat children who commit serious crimes as adults in order to protect societal goals. The legislature could reasonably have determined that for some crimes the rehabilitative aspect of juvenile court must give way to punishment. . . . [T]he Florida legislature "considered carefully the rise in the number of crimes committed by juveniles as well as the growing recidivist rate among this group. The legislature was entitled to conclude that the . . . juvenile system would not work for certain juveniles, or that society demanded greater protection from these offenders than that provided by that system."

Raising a procedural due process argument, Brazill cites *Kent v. United States*, 383 U.S. 541 (1966), to support his argument that a hearing is required before adult sanctions may be imposed upon a child. He attacks FS section 985.225 because it allows the state to bypass a hearing on the suitability of adult sanctions by securing an indictment. . . .

Florida Statutes section 985.225 does not require a court to hold a hearing to decide whether adult sanctions are appropriate. . . . [T]he Florida Supreme Court discussed *Kent* and found that: "Whatever its constitutional ramifications, we do not believe they extend to the statutory provision under consideration here where discretion to prosecute a juvenile as an adult is vested in the prosecutor rather than in a judge." . . . Brazill was afforded the same procedural rights as anyone else

charged with first-degree murder by indictment. Due process does not require anything more because of his status as a child.

Brazill complains that because the statute contains no criteria "to steer prosecutorial discretion," arbitrariness is injected into the decision-making process. . . . These attacks must fail because of the broad discretion accorded a prosecutor under our legal system. As the Florida Supreme Court has written, "the discretion of a prosecutor in deciding whether and how to prosecute is absolute in our system of criminal justice." Florida Statutes section 985.225(1) applies to "[a] child of any age who is charged with a violation of state law punishable by death or by life imprisonment." . . . It does not differentiate between age groups. The statute equally applies to any child who commits an offense punishable by death or by life imprisonment. Additionally, a child transferred to the criminal court becomes similarly situated with defendants in that court, rather than those still in the juvenile system. . . . [T]he state points out, the statutory "requirement of an indictment is for the protection of the accused juvenile," because the grand jurors must concur in the prosecutor's charging decision. When the grand jury does not return an indictment, a juvenile thirteen and under is not subject to FS section 985.226(2).

Holding

The twenty-eight year mandatory minimum sentence was lawful. The jury's verdict was sufficient to support the mandatory minimum sentence. An enhanced sentence is proper when it is based on a jury verdict that specifically refers to the use of a firearm, either as a separate finding or by including a reference to a firearm when identifying the specific crime. . . .

Florida Statutes section 775.087(2)(a) provides that if during the course of the commission of the felony such person discharged a "firearm" . . . and, as a result of the discharge, death or great bodily harm was inflicted upon any person, the convicted person shall be sentenced to a minimum term of imprisonment of not less than 25 years and not more than a term of imprisonment of life in prison. The indictment in this case charged the crime of first-degree murder and alleged that Brazill "did use and have in his possession a handgun, a firearm as defined in Florida Statutes section 790.001(6)." . . .

Questions for Discussion

1. As a prosecutor, would you have treated the thirteen-year-old Brazill as an adult?

2. What is the basis for the Florida court's conclusion that Brazill does not possess a constitutional right to be treated as a juvenile?

3. Brazill argues that he should be given a hearing to determine whether he should be treated as an adult. The court, however, ruled that this decision was within the discretion of the prosecutor and that Brazill was protected by the fact that a grand jury of ordinary citizens determined that there was sufficient evidence to indict him for murder. Is it preferable to have an individual's adult status determined by a court or by a prosecutor?

The next case, *State v. Ramer,* discusses the factors considered by the Washington Supreme Court in determining whether a juvenile was capable of forming a criminal intent.

Did eleven-year-old Andrew Ramer possess the capacity to know that rape was wrong?

State v. Ramer, 86 P.3d 132 (Wash. 2004). Opinion by: Chambers, J.

Issue

Andrew Ramer, an 11-year-old juvenile defendant, was charged with two counts of first degree rape of a child. The Thurston County Superior Court found Ramer lacked the capacity to commit the crime. The State appealed, and . . . Division Two of the Court of Appeals reversed. The Washington Supreme Court was asked to decide whether Ramer was aware that the rape of a child is wrong.

Facts

Ramer, his nine-year-old sister, Kensie, and his mother, Dina Lawrence, were temporarily living with their friends, the Briscoes. On January 25, 2001, another child in the home told Deanna Briscoe that Ramer was in the bathroom with his arm around her seven year old son, ZPG. Briscoe asked Ramer what happened, and Ramer "basically said 'nothing.'" When ZPG was asked about the incident, he told his mother that Ramer had "rubbed his butt."

Briscoe finished getting the children ready for school and did not immediately pursue the subject further. Later that evening when she resumed her inquiry, ZPG told her that, in addition to "rubbing his butt," Ramer had also "placed his penis inside of [ZPG's] butt." When asked, Ramer freely admitted he had done these things.

The next morning, Briscoe and Lawrence contacted several social service agencies. The mothers were advised to go to the police. At the police station, Lawrence waived Ramer's right to counsel by granting Detective Beverly Reinhold's request to talk with Ramer alone. Before speaking with Reinhold, Ramer was not told that he might be arrested or sent to detention.

While talking with Detective Reinhold, Ramer revealed that for about two weeks he had sexual contact with ZPG approximately twice a week. Ramer also recounted having sexual contact with ZPG several years earlier when his family had previously lived with the Briscoes. Detective Reinhold asked Ramer if he thought what he had done to ZPG was wrong. First, Ramer responded "'kind of sort of wrong.'" Then he added, "'it wasn't wrong because he was into it too.'" When asked to give examples of wrong behavior, Ramer said

it would be wrong "to steal, murder, or poach." Ramer also said that he had sexual contact a "few times" with his sister. At the end of the conversation with Detective Reinhold, Ramer was arrested and charged with two counts of first degree rape of a child.

Because of Ramer's age, a hearing to determine his capacity to commit first degree rape of a child was held before Thurston County Superior Court Commissioner Scott Neilson. Detective Reinhold testified about her conversation with Ramer at the police station. The defense was allowed to call two witnesses out of order: Dr. Brett Trowbridge, Ph.D., J.D., a forensic psychologist, and Peg Cain, M.A., a mental health specialist at St. Peter's Hospital psychiatric unit, who also performs juvenile and adult "safe to be at large" evaluations for Thurston County. Both witnesses for the defense had evaluated Ramer and had prepared written reports.

Dr. Trowbridge based his opinion upon his evaluation of Ramer. He testified that, in his opinion, Ramer did not understand the act of having sexual contact with a much younger child. He also testified that he believed Ramer did not understand that it was wrong, especially if the other child enjoys and voluntarily participates in the act. Specifically, Dr. Trowbridge opined, when questioned:

Q. Based on your evaluation and investigation, it's your conclusion that [Ramer] does not possess sufficient information or ability to come to the understanding of what rape of a child meant in terms of his committing the act in this instance?

A. Yes. Because at that time of the alleged offense I don't think he did have that understanding.

Q. He did not understand if something felt good why it was wrong.

A. [H]e thought if the child consented that that made it not wrong.

Similarly, Cain testified that Ramer had "no concept of how serious the charge was." She also testified that Ramer asked her why sexual contact with ZPG was considered rape when it felt good, ZPG wanted to participate, and ZPG "really liked it." Cain also testified that Ramer's attitude and demeanor led her to believe that he did not understand that sexual contact was inappropriate behavior and that he did not know that what he had been doing with ZPG was wrong.

The State then called Thomas Nore, M.S.W., a juvenile court probation counselor with 26 years of experience. Nore did not evaluate Ramer, nor did he take notes of his conversations with Ramer. He based his opinion on the written reports of Dr. Trowbridge and Cain and his conversations with Ramer while transporting him to and from a psychosexual evaluation by Trudy Howe and on other occasions. Nore testified that he believed Ramer understood that his conduct was wrong. Nore also testified that Ramer had been told by his parents that "sexual contact with each other in the home or with anyone else" was wrong. Nore opined that Ramer "had knowledge and experience far beyond any 11-year-old I'd ever met. In fact far beyond some 16-, 17-year-olds." It was Nore's opinion that Ramer knew the serious consequences of sexual contact and had the capacity to commit the crime.

Nore further testified that Ramer knew his biological father was currently incarcerated for sexually molesting his sister. Although it is possible that Ramer may have been molested by his father as well. Ramer has no recollection of being molested. After his father's incarceration, Ramer's mother remarried. This stepfather committed suicide after finding out that he was going to be prosecuted for molesting Ramer's sister.

The commissioner assigned to Ramer's case concluded that Ramer understood his conduct was wrong and "understood the act of Rape of a Child first degree." Ramer moved for revision in the superior court before the Honorable Christine A. Pomeroy. Judge Pomeroy read the record and heard arguments before finding Ramer lacked capacity to commit the crime. Judge Pomeroy found that Ramer was "a highly sexualized young person, who clearly was confused about appropriate sexual behaviors and could not understand the prohibitions on sexual behavior with other children." The State appealed the superior court's finding. Finding Ramer did have capacity to commit the crime, the Court of Appeals reversed. Ramer sought and we granted discretionary review.

Reasoning

We review the record for evidence sufficient to support the superior court's finding. When the superior court finds capacity, we review the record to determine whether there is substantial evidence establishing that the State met its burden of overcoming the statutory presumption that children under 12 years of age are incapable of committing crime by clear and convincing evidence. When the trial court finds a lack of capacity, we review the record to determine whether a rational trier of fact could find that the State failed to overcome the presumption that the child lacked capacity.

By statute, a child "under 12 years of age is presumed incapable of committing any crime." RCW 9A.04.050 provides, in part:

Children under the age of eight years are incapable of committing crime. Children of eight and under twelve years of age are presumed to be incapable of committing crime, but this presumption may be removed by proof that they have sufficient capacity to understand the act or neglect, and to know that it was wrong.

The statute codifies what is known as "the infancy defense." The purpose of the infancy defense is "to protect from the criminal justice system those individuals of tender years who are less capable than adults of appreciating the wrongfulness of their behavior." In order to overcome the presumption of incapacity, the State must provide clear and convincing evidence that the child had sufficient capacity to understand the act and to know that it was wrong. A capacity determination is fact-specific and must be in reference to the specific act charged. It is not necessary, however, for the child to understand that the act would be punishable under the law. The focus is on "whether the child appreciated the quality of his or her acts at the time the act was committed," rather than whether the child understood the legal consequences of the act.

We have identified seven factors to consider in determining capacity: (1) the nature of the crime, (2) the child's age and maturity, (3) whether the child evidenced a desire for secrecy, (4) whether the child told the victim (if any) not to tell, (5) prior conduct similar to that charged, (6) any consequences that attached to that prior conduct, and (7) whether the child had made an acknowledgment that the behavior is wrong and could lead to detention. Also relevant is testimony from those acquainted with the child and the testimony of experts.

Capacity requires the actor to understand the nature or illegality of his acts. In other words, he must be able to entertain criminal intent. A "sense of moral guilt alone, in the absence of knowledge of legal responsibility, is not sufficient," although actual knowledge of the legal consequences is not necessary.

Washington courts have held that when a juvenile is charged with a sex crime, the State carries a greater burden of proving capacity, and must present a higher degree of proof that the child understood the illegality of the act.

Because the superior court found, consistent with the statutory presumption, that Ramer lacked capacity, we review this record to determine whether there was evidence from which a rational trier of fact could find Ramer incapable of first degree rape of a child. We do not substitute our judgment for that of the superior court's. While reasonable minds might differ over conflicting evidence, we will reverse the superior court only if, based upon the record, no rational trier of fact could reach the conclusion that the State failed to meet its burden.

In *J.P.S.*, this court noted that the nature of the crime is an important factor, but with sexual crimes it is very difficult to tell if a child understands the prohibitions on sexual behavior with other children. *J.P.S.*, 954 P.2d 899 (Wash. 1998). In *[State v.] J.P.S.*, the eleven-year-old, mildly retarded defendant was charged with first degree rape of a three-year-old. The defendant took the younger child to a shed, sent the other children away, pulled down both their pants, and then touched the three year old on the vagina. The defendant lied about what contact had occurred but did not admonish the victim not to tell. After talking with the police, the defendant admitted that the conduct was "bad." This court unanimously concluded that the State had failed to establish by clear and convincing evidence that the defendant understood that his conduct was wrong.

We conclude that there is sufficient evidence in the record to support the superior court's finding in this case that the State failed to establish by clear and convincing evidence that Ramer understood that his conduct was wrong. Four witnesses testified at the capacity hearing. Detective Reinhold did not express an opinion as to Ramer's capacity, and her testimony was equivocal, quoting Ramer as saying his conduct was "kind of sort of wrong." Dr. Trowbridge, who evaluated Ramer for the defense, opined that Ramer did not have the capacity to commit the crime charged. Dr. Trowbridge was further of the opinion that Ramer did not understand that sex with someone who consents and likes the sex is wrong. Cain, who performed Ramer's "safe to be at large" evaluation for Thurston County, also expressed the opinion that Ramer did not have the capacity to commit the

crime charged. Nore, who did not evaluate Ramer, was the only witness to express the opinion that Ramer had the capacity to commit the crime of first degree rape of a child. While Nore's testimony supports the State's contention, we conclude a rational trier of fact could find the State failed to meet its burden based upon Detective Reinhold's testimony and the expert opinions of Cain and Dr. Trowbridge.

Holding

We find that there is sufficient evidence in the record to support the superior court's finding that the State failed to overcome the statutory presumption by clear and convincing evidence. We reverse the Court of Appeals.

Concurring, *Madsen, J.*

Although I concur in the result, I write separately to emphasize that the majority has not announced a new standard for reviewing child capacity determinations but continues to adhere to the standard set forth in *State v. J.P.S.* Under RCW 9A.04.050, the State has the burden to rebut the presumption of incapacity by clear and convincing evidence. The standard on review is whether there was evidence from which a rational trier of fact could find capacity by clear and convincing evidence. If the trial court concludes that a child lacks capacity, the State must point to evidence in the record that establishes capacity by clear and convincing evidence in order to prevail on a claim that the trial court's finding is erroneous.

Questions for Discussion

1. What is the reason for the age defense?

2. What factors does a court consider in evaluating whether Andrew Ramer possessed sufficient capacity to understand the act and to know that it was wrong? Why is there a higher standard to prove capacity in the case of a sexual offense?

3. Can you explain the reason that the Washington Supreme Court concludes that Ramer lacked a capacity to understand that rape is "wrong"?

4. What is the significance of Andrew's family history?

5. Do you agree with the court's decision?

See more cases on the study site: **State v. J.P.S.,** *www.sagepub.com/lippmanccl3e.*

You Decide

9.3 K.R.L., 8 years 2 months old, was playing with a friend behind a building. Catherine Alder heard the boys playing and directed them to leave because the area was dangerous. K.R.L. responded in an angry manner and replied that he would leave "in a minute." Alder, with obvious irritation, told the two boys, "No, not in a minute, now, get out of there now." The boys then ran off.

Three days later K.R.L. entered Alder's home without her permission. He removed a goldfish from a fishbowl, chopped it into several pieces with a steak knife and

"smeared it all over the counter." He then went into Alder's bathroom and "clamped a 'plugged in' hair curling iron onto a towel." K.R.L's mother testified that he admitted to her that entering Alder's home was wrong after she had beaten him "with a belt, black and blue." He told her that the "Devil was making him do bad things."

K.R.L. subsequently was charged with residential burglary. Earlier, he had taken "Easter candy" from a neighbor's home without permission. K.R.L. admitted to the police that he "knew it was wrong and he wouldn't like it if somebody took his candy." The same officer testified that on an earlier occasion, K.R.L. had been caught riding the bicycles of two neighbor children without having their permission. K.R.L. told the police officer that he "knew it was wrong" to ride the bicycles.

The assistant principal of K.R.L.'s elementary school testified that K.R.L. was of "very normal" intelligence. K.R.L.'s first grade teacher said that K.R.L. had "some difficulty" in school and that he would place K.R.L. in a "lower age academically."

In Washington State, children younger than eight are incapable of a criminal intent. Children between eight and twelve years of age are presumed to be incapable of committing crime. This presumption may be overcome by proof that they have "sufficient capacity to understand the act or neglect, and to know that it was wrong." The Washington Supreme Court was asked to decide whether the trial court was correct in concluding that there was clear and convincing evidence that K.R.L. had the capacity to commit residential burglary. What is your opinion? See *State v. K.R.L.*, 840 P.2d 210 (Wash. Ct. App. 1992).

You can find the answer at www.sagepub.com/lippmanccl3e.

DURESS

The common law excused an individual from guilt who committed a crime to avoid a threat of imminent death or bodily harm. In several seventeenth- and eighteenth-century cases involving treason or rebellion against the king, defendants were excused who joined or assisted the rebels in response to a threat of injury or death. The common law courts stressed that individuals were obligated to desert the rebels as soon as the threat of harm was removed.[44]

There are various explanations for the **duress** defense:

- *Realism.* The law cannot expect people to act in a heroic fashion and resist threats of death or serious bodily harm.
- *Criminal Intent.* An individual who commits a crime in response to a severe threat lacks a criminal intent.
- *Criminal Act.* Individuals who commit crimes under duress act in an involuntary rather than voluntary fashion.

Realism may be the most persuasive justification for duress. An English court nicely captured this concern in the observation that in the "calm of the courtroom, measures of fortitude or of heroic behavior are surely not to be demanded when they could not in moments for decision reasonably have been expected even of the resolute and the well-disposed."[45]

The Elements of Duress

The defense of duress involves several central elements.

The defendant's actions are to be judged in accordance with a reasonable person standard. In *State v. Van Dyke,* the defendant Sheryl Van Dyke was a thirty-four-year-old mother of two who had been married for fifteen years. She was convicted of the sexual assault and endangerment of the welfare of J.M., a thirteen-year-old male with whom she had a sexual affair. Van Dyke claimed that she entered into and continued the relationship out of fear resulting from J.M.'s periodic physical abuse and threats to seriously assault her daughter and to choke her son to death. A New Jersey Superior Court ruled that society could reasonably expect that the will of the average member of the community would not be overwhelmed by the type of threats directed at Sheryl Van Dyke and her family.[46]

There must be a threat of death or serious bodily harm that causes an individual to commit a crime. Most states also recognize that a threat directed against a member of the defendant's family or a third party may constitute duress. Psychological pressure or blackmail does not amount to a threat for purposes of duress.

Duress does not excuse the intentional taking of the life of another. In the California case of *People v. Anderson,* the defendant Robert Anderson, along with Ron Kiern, abducted Margaret Armstrong, who was suspected of molesting Kiern's daughter. Anderson testified that when he objected to Kiern's request that Anderson give him a rock with which to beat Armstrong, Kiern responded "give me the rock or I'll beat the s____ out of you." The defendant testified that he gave Kiern the rock because he was "not 'in shape' to fight" and he feared that if he refused, Kiern would "punch me out, break my back, break my neck." The California Supreme Court held that the intentional taking of a life was not excused by duress and that the law "should require people to choose to resist rather than kill an innocent person." The majority opinion noted that California is "tormented by gang violence" and "persons who know they can claim duress will be more likely to follow a gang order to kill instead of resisting."[47]

The threat must be immediate and imminent. Judges have insisted on an imminent threat that compels an individual to "involuntarily" commit a crime; a threat of future harm is not considered to prevent an individual from making a reasoned choice whether to violate the law. In *Commonwealth v. Perl,* the defendant, Dr. Alan Perl, was a physician who provided LaCorte, described as a "longtime criminal," with prescriptions and pills after LaCorte mentioned in August 1993 that he knew where Perl's daughter attended school. This was followed by threats in January, March, and July 1994. On the last occasion, LaCorte told Perl that if he did not provide him with pills, "I will see your daughter." Perl continued to provide LaCorte's pills between August 1993 and October 1994. The Massachusetts Appeals Court denied Perl the defense of duress because LaCorte's threats were of future harm and were not "present, immediate, and impending."[48]

An individual must have exhausted all reasonable and available alternatives to violating the law. A defendant must reasonably believe that the criminal act is the only means of preventing imminent death or great bodily harm. Jon Barreau hit Robert Hansen with a bat, knocking him to the floor. Barreau's bat broke, and he directed Jeffery Keeran to hit Hansen "or I'm hitting you." Keeran testified that he grabbed the bat with two hands and struck Hansen twice and then went outside and vomited and waited roughly forty-five minutes for Keeran to complete the robbery. A Wisconsin appellate court noted that Keeran offered no explanation of why he did not run out the back door, threaten Barreau, refuse to hit Hansen, or object that Hansen already was incapacitated. Judge Stuart Schwartz recognized that Keeran was afraid of Barreau, but nevertheless stressed that the duress defense requires that a criminal act is the "only course" and that duress is not a "license to take the safest course." Keeran reasonably believed that Barreau would hunt him down if he refused to cooperate in the criminal enterprise. This, however, would not support a finding of duress, which requires the prevention of "imminent" death or great bodily harm.[49]

The defendant must not create or assist in creating the circumstances leading to the claim of duress. An individual must not intentionally or recklessly become involved in an enterprise in which it is foreseeable that he or she will be coerced into criminal activity. Drug dealer Luis Rafael Santiago Rodriquez approached Cesar Castro Gomez and insisted that Castro "solve his transportation problem." Castro testified that he felt threatened and drove with Santiago and one of his lieutenants to the dock where they boarded Castro's boat to retrieve the drugs. Castro testified that he managed to frustrate the sale by deliberately steering the boat far from the drop point for the drugs. He was subsequently threatened by Santiago and feared that he would be killed. Several days later, Castro agreed to meet Santiago at a local pizza parlor where Castro testified that he again was intimidated into cooperating with the drug dealers. Castro was apprehended several miles off the coast of Puerto Rico while piloting two other passengers in a ship carrying roughly 762 kilograms of cocaine and was charged with various drug offenses. The First Circuit Court of Appeals concluded that by returning to the pizza parlor, Castro "recklessly placed himself in a situation in which it was probable that he would be subjected to duress" and that he was not entitled to raise the duress defense.[50]

Married women were exempted from liability under the common law. The common law presumed that a woman who committed a crime in the presence of her husband acted under his direction and she was not responsible. This is no longer the law.

Keep in mind that the individual exerting the coercion is liable as a principal in the crime despite the fact that the perpetrator may be excused on the basis of duress.

Duress and Correctional Institutions

The most controversial duress cases involve prison escapes in which inmates threatened with physical assault offer the defense of duress to excuse their escape. In *State v. Unger,* the defendant Francis Unger, a twenty-two-year-old full-blooded Crete Indian, pled guilty to a theft charge and was imprisoned for one to three years in Stateville Penitentiary in Joliet, Illinois. During the first two months of Unger's imprisonment, he was threatened by an inmate wielding a six-inch knife who demanded that the defendant engage in homosexual activity. Unger was transferred to a minimum security honor farm and one week later was beaten and sexually assaulted by a gang of inmates.

Unger was warned against informing authorities and several days later received a phone call informing him that he would be killed in retribution for having allegedly contacted correctional officials. Unger responded by escaping from the dairy farm, and he was apprehended two days later while still wearing his prison clothes. He claimed that he had intended to return to the institution.

The court determined that the correctional system was dominated by gangs that were too powerful to be controlled by prison officials. Unger, under these circumstances, was entitled to a jury instruction on duress because he may have reasonably believed that he had no alternative other than to escape or to be killed or to suffer severe bodily harm. The Illinois appellate court held that it was unrealistic to require that a prisoner wait to escape until the moment that he was being "immediately pursued by armed inmates" and it was sufficient that Unger was threatened that he would be dead before the end of the evening.[51]

Inmates relying on duress must establish that they did not use force or violence toward prison personnel or other innocent individuals in the escape and that they immediately contacted authorities once having reached a position of safety. *Unger* is only one of a number of cases that have recognized that inmates are entitled to rely on duress. Do you believe that the judiciary acted correctly in recognizing Unger's claim of duress? Are you confident that he was "telling the truth"? What about his failure to "turn himself in" to correctional authorities?

The Duress Defense

The defense of duress is not fully embraced by all commentators. The famous nineteenth-century English commentator Sir James Stephen argued that the law should stand firm against human frailty and weakness and insist that individuals follow legal rules, even under conditions of stress and strain. Sir James also contended that duress opens the door to individuals committing crimes and later claiming that they had been threatened and were entitled to the defense of duress. In the last analysis, critics argue that the law should encourage people to resist rather than conform to the demands of violent and forceful individuals. Do you agree? In reading the next case, *United States v. Moreno*, consider whether Moreno was entitled to the duress defense.

Model Penal Code Section 2.09. Duress

(1) It is an affirmative defense that the actor engaged in the conduct charged to constitute an offense because he was coerced to do so by the use of, or a threat to use, unlawful force against his person or the person of another, that a person of reasonable firmness in his situation would have been unable to resist.

(2) The defense . . . is unavailable if the actor recklessly . . . [or negligently] placed himself in such a situation. . . .

(3) It is not a defense that a woman acted on the command of her husband. . . .

Analysis

The Model Penal Code significantly amends the common law standard:

1. The threat need not be limited to death or serious bodily harm. The commentary provides for a threat of unlawful force against the individual or another that would coerce an individual of "reasonable firmness" in the defendant's situation. Only threats to property or reputation are excluded in the commentary.

2. The threat is not required to be imminent or immediate.

3. Duress may be used as an excuse for homicide.

4. The threat may be to harm another person and is not limited to friends or relatives.

The Legal Equation

Duress = Reasonable belief of

+ an imminent threat by another

+ of death or severe bodily harm

+ against the defendant or close friend or relative (not limited in the Model Penal Code)

+ that causes defendant (reasonable person standard) to commit a criminal act

+ defendant did not place himself or herself in the situation

+ defendant did not kill another (not in Model Penal Code).

Did Moreno transport crack cocaine to Hawaii because he feared retribution by a gang against himself and his children?

United States v. Moreno, 102 F.3d 994 (9th Cir. 1994). Opinion by: Alarcon, J.

Issue

Danny Moreno appeals from his conviction for possession with intent to distribute cocaine base (21 U.S.C. §§ 841(a)(1) and 841(b)(1)(A)). He contends that the district court erred by preventing him from presenting a duress defense and testifying regarding his state of mind concerning his fear of violence to himself and his children at the time of the commission of the charged offense.

Facts

Honolulu police arrested Moreno at the Honolulu International Airport on May 4, 1994. Moreno's suspicious behavior following his arrival from Los Angeles had attracted the attention of Honolulu Police Department Officer Thomas Krajewski. Officer Krajewski approached Moreno at an exit to the terminal, identified himself as a narcotics investigator, and asked Moreno if he could talk to him briefly. Moreno consented, but appeared nervous. Moreno gave Officer Krajewski permission to inspect his carry-on bag. Officer Krajewski told Moreno that many smugglers tape drugs to their bodies. When Officer Krajewski asked Moreno "Can I search," Moreno fled before Officer Krajewski completed his request for permission to conduct a further search. Officer Krajewski and four officers gave chase as Moreno ran through traffic. Officer Krajewski grabbed Moreno as he was climbing a fence. Moreno kicked Officer Krajewski in the head

"a couple of times" before Moreno was subdued. A search of Moreno's person revealed several packages of cocaine base or "crack" taped to his abdomen and thighs.

Prior to trial, Moreno informed the Government that he intended to assert a duress defense. On January 3, 1995, the Government filed a motion *in limine* (a motion to prohibit the defense at trial). The Government requested that the court strike the proposed defense on the ground that Moreno could not establish a case of duress. In response, Moreno filed a four page, unsigned typewritten document that contained a narration of facts. It is entitled "Defendant Danny Moreno's Proffer of Evidence."

The narrative sets forth the following factual assertions: A senior member of Moreno's Los Angeles area gang approached Moreno three weeks prior to May 4, 1994, and asked if he knew anyone who could transport crack cocaine to Hawaii. The senior gang member, known to Moreno only as "Joker," told Moreno that Moreno would have to do it himself if he did not find a courier. Moreno was "upset by this conversation" because he was aware of Joker's high status within the gang and his violent reputation. Moreno was aware that Joker had killed a man when a drug deal went "bad."

Approximately two weeks later, Joker approached Moreno outside a convenience store in Wilmington, California. He asked Moreno if he had found a courier. Moreno told Joker he had been unsuccessful. Joker then ordered Moreno to transport the crack cocaine to Hawaii.

Moreno told Joker he could not go to Hawaii because of his parental responsibilities to his two young daughters. Joker replied, "If you don't do this job for me, I'll kill you and have your family killed." Joker stated that Moreno had "pretty little girls" and that he knew they lived with their mother on the "westside." Moreno's daughters and their mother lived on the "westside" of Wilmington.

Joker told Moreno to be ready to go to Hawaii on the morning of May 4, 1994. Joker asked Moreno where he lived so he could take him to the airport. Because Moreno did not want to involve his family, he told Joker he would meet him in the San Pedro YMCA parking lot.

Moreno went to the parking lot on May 4, 1994, intending to talk Joker out of the scheme. He told Joker "I don't want to go." Joker told Moreno that if he did not make this trip, Joker would "take you and your family out." Moreno understood this to mean that he and his family would be killed. Because of this threat, Moreno agreed to follow Joker's instructions. Joker gave Moreno a "multi-colored shirt" so Joker's associates could identify him. Joker instructed Moreno to deliver the drugs to Byron's Restaurant near the Honolulu Airport, and assisted Moreno in taping four packages of crack cocaine to his body. Joker opened his own shirt to reveal a gun tucked into his waistband. He warned Moreno, "Don't try anything funny cause they're going to be watching you." Moreno understood this to mean that Joker's associates would be on the plane or at the airport in Honolulu. Joker told Moreno that Joker's "homeboy" would telephone Joker after the delivery was completed. Joker drove Moreno to the airport and watched him board the plane.

Moreno further alleged that Joker's threats prompted his attempt to escape from the police at the airport. Moreno attempted to flee rather than consent to a pat-down search because he thought that Joker's associates were watching him during his initial encounter with Officer Krajewski. Moreno was fearful that Joker would carry out his threats if he knew that Moreno had "consented to a police search."

Moreno was convicted after a trial by jury of possession with intent to distribute cocaine base. This timely appeal followed.

Reasoning

Moreno argues that the facts presented in his proffer of evidence entitled him to assert the defense of duress. A defendant must establish three elements in order to present this defense: (1) an immediate threat of death or serious bodily injury, (2) a well-grounded fear that the threat will be carried out, and (3) lack of a reasonable opportunity to escape the threatened harm. "Fear alone is not enough to establish a prima facie case of duress; the defendant must establish all three elements." . . .

Moreno failed to offer evidence that would support an inference "that he had no opportunity to avoid violating the law without subjecting himself to further immediate danger." Moreno had a reasonable opportunity to escape Joker's threatened harm at any time between his initial encounter with Joker and his encounter with Officer Krajewski in the Honolulu Airport. During this three week period, Moreno saw Joker on only three occasions. No one else made any threats or appeared to follow Moreno. Joker did not know Moreno's address, or where in the "westside" his daughters could be located. Moreno made no effort to flee or hide. Moreno's failure to present evidence that he lacked a reasonable opportunity to escape the threatened harm precludes his duress defense. See *U.S. v. Bailey* (444 U.S. 394 1980) (holding that a criminal defendant charged with escape "must proffer evidence of a bona fide effort to surrender or return to custody as soon as the claimed duress or necessity had lost its coercive force" in order to present a duress or necessity defense).

Moreno argues that *U.S. v. Contento-Pachon,* 723 F.2d 691 (9th Cir. 1984), controls this case. His reliance is misplaced. In *Contento-Pachon,* the defendant proffered evidence that Colombian drug traffickers had forced him to transport drugs to the United States. We held that there was a triable issue of fact "whether one in Contento-Pachon's position might believe that some of the Bogota police were paid informants for drug traffickers and that reporting the matter to the police did not represent a reasonable opportunity of escape." We also held that a jury should decide whether it was reasonable to expect "Contento-Pachon, along with his wife and three-year old child, . . . to pack his possessions, leave his job, and travel to a place beyond the reaches of the drug traffickers." Unlike the defendant in *Contento-Pachon,* however, Moreno presented no evidence that he could not flee from his gang's reach, or that he could not seek help from local law enforcement agencies because they were corrupt and controlled by gang members.

The fact that Moreno claims he was under constant surveillance by Joker's associates during the flight and at the Honolulu Airport does not justify his failure to submit to authorities. Like Moreno, Contento-Pachon claimed "that he was being watched at all times." Contento-Pachon, however, consented to a stomach x-ray "at the first opportunity to cooperate with authorities without alerting the observer." Moreno failed to avail himself of a similar, reasonable opportunity to escape from the threatened harm. When Officer Krajewski approached Moreno at the Honolulu Airport, Moreno could have explained to the officer that he had been coerced to transport crack cocaine without appearing to betray Joker's alleged instruction not to "try anything funny." The encounter with Officer Krajewski presented a clear opportunity for Moreno to save himself and alert authorities about the threat to his family. Instead, he kicked Officer Krajewski in the head twice in his attempt to escape to complete his illegal delivery.

Holding

Because Moreno has failed to demonstrate that he did not have the opportunity to escape the threatened harm, we need not discuss the other elements of duress. The district court did not err in granting the government's motion to strike the proposed defense of duress.

Moreno contends that, pursuant to the constitutional right to testify, the district court was required to permit him to explain to the jury that he behaved in the manner that he did because he was acting under duress. As discussed above, the district court ruled correctly that Moreno's proffered evidence was insufficient to establish the elements of the defense of duress because he failed to demonstrate that he lacked a reasonable opportunity to escape the threatened harm. . . . In *Contento-Pachon*, this court held that "if evidence is insufficient as a matter of law to support a duress defense, . . . the trial court should exclude that evidence." While the constitutional right to testify permits a defendant to choose whether or not to take the witness stand, it does not authorize a defendant to present irrelevant testimony.

Questions for Discussion

1. Why does Moreno claim that he is entitled to rely on the duress defense?

2. How does the appellate court distinguish *Moreno* from *Contento-Pachon*?

3. Assuming that Moreno had no reasonable opportunity to escape, should he be permitted to have the jury consider the duress defense?

Cases and Comments

1. **Contento-Pachon.** Compare and contrast *Contento-Pachon* and *Moreno*. In *Contento-Pachon*, the defendant was a taxi driver. A passenger, Jorge, told Contento-Pachon that he needed a driver and asked Contento-Pachon to meet with him the next day. At the meeting, Jorge asked Contento-Pachon to transport drugs. Contento-Pachon responded that he would think about the offer. The next week, Contento-Pachon told Jorge that he did not want to transport narcotics. Jorge, a notorious and dangerous drug dealer, threatened Contento-Pachon's wife and three-year-old child if he did not transport the narcotics. The defendant proceeded to swallow 129 balloons of cocaine. Jorge warned Contento-Pachon that he would be under constant surveillance during the trip to the United States. Contento-Pachon testified that he did not contact the police when the Los Angeles-bound plane stopped in Panama because he believed that the Panamanian police were as corrupt as the police in Bogota, Columbia. The defendant was stopped at customs in Los Angeles and submitted to X-rays. The Ninth Circuit Court of Appeals held that although the threat against Contento-Pachon had been issued roughly three weeks before the scheduled flight to the United States, it was unreasonable to expect Contento-Pachon to leave his job and pack his possession and to flee to a location in Columbia beyond the reaches of the drug traffickers.

2. **Contento-Pachon *Precedent and Threats to the United States by Foreign Drug Dealers.*** Pakistani authorities arrested a drug courier with several kilograms of heroin on his way to New York City. The Pakistanis reported to U.S. Drug Enforcement Agency officials that the courier planned to contact the defendant Subhan. Agents arranged to meet Subhan and arrested him after he took possession of a bag that he believed contained drugs. Subhan claimed that his son had been kidnapped in Pakistan and that the kidnappers had written a note, which Subhan had destroyed, stating that his son would be returned after their "business" with Subhan had been concluded. Various friends and family members supported Subhan's story. Subhan claimed that he had been instructed to pick up the drugs and that he did not approach American authorities because he feared that they would contact the Pakistani police. Subhan presented a report from the U.S. Department of State in his defense documenting that the Pakistani police were corrupt and involved with the drug dealers. He testified that he had reason to believe that the Pakistani authorities would contact the kidnappers, who would kill his son. Subhan relied on the precedent of *Contento-Pachon*. A federal district court ruled that this precedent was inapplicable since Subhan was in the United States at the time of the coercion and could have approached American law enforcement whereas Contento-Pachon was in Columbia, where the authorities were "allegedly corrupt." See *United States v. Subhan*, 38 Fed. App'x 89 (2d Cir. 2002).

You can find more cases on the study site:
United States v. Contento-Pachon, *www.sagepub.com/lippmanccl3e.*

You Decide

9.4 Georgia Carradine was held in contempt of court based on her refusal to testify after witnessing a gang-related homicide, explaining that she was in fear for her life and the lives of her children. Carradine was sentenced to six months in the Cook County jail. She persisted in this refusal despite the government's offers to relocate her and her family to other areas in Chicago, Illinois, or the continental United States. Carradine had been separated from her husband for roughly four years and supported her six children aged five to eighteen through payments from her husband and supplemental welfare funds. She explained that she distrusted the State's Attorney and doubted that law enforcement authorities could protect her from the Blackstone Rangers youth gang. Carradine's fear was so great that she was willing to go to jail rather than to testify. The Illinois Supreme Court, in affirming the sentence, stated that criminals could not be brought to the bar of justice "unless citizens stand up to be counted." Do you agree with the decision to deny Carradine the defense of duress? See *State v. Carradine*, 287 N.E.2d 670 (Ill. 1972).

You can find the answer at www.sagepub.com/lippmanccl3e.

MISTAKE OF LAW AND MISTAKE OF FACT

A core principle of the common law is that only "morally blameworthy" individuals should be subject to criminal conviction and punishment. What about the individual who commits an act that he or she does not realize is a crime? Consider a resident of a foreign country who is flying to the United States for a vacation and is asked by a new American acquaintance to bring a vial of expensive heart medicine to his or her parents in the United States. The visitor is searched by American customs officials as he or she enters the United States, and the heart medicine is discovered to be an illegal narcotic. Should the victim be held criminally liable for the knowing possession of narcotics despite this "mistake of fact"? What if the visitor was asked by his American friend to transport cocaine and was assured that there was nothing to worry about because the importation and possession of this narcotic is legal in the United States? How should the law address this "mistake of law"? We will be addressing these questions throughout the textbook, and this section merely provides you with an outline of the central issues.

In the previous two hypothetical examples, the question is whether an individual who mistakenly believes that his or her behavior is legal should be held liable for violating the law. Professor Wayne R. LaFave has observed that no area has created "more confusion" than mistakes of law and fact[52]—a confusion that has caused "ulcers in law students."[53]

Mistake of Law

The conventional wisdom is that ***ignorantia lexis non excusat***: "Ignorance of the law is no excuse." The rule that a **mistake of law** does not constitute a defense is based on several considerations:[54]

- *Knowledge.* People are expected to know the law.
- *Evidence.* Defendants may falsely claim that they were unaware of the law. This claim would be difficult for the prosecution to overcome.
- *Public Policy.* The enforcement of the law insures social stability.
- *Uniformity.* Individuals should not be permitted to define for themselves the legal rules that govern society.

The expectation that individuals know the law may have made sense in early England. Critics contend, however, that people cannot realistically be expected to comprehend the vast number of laws that characterize modern society. An individual who, through a lack of knowledge, violates highly technical statutes regulating taxation or banking can hardly be viewed as "morally blameworthy."[55] Some observers note that courts seem to have taken this criticism seriously and, in several instances, have relaxed the rule that individuals are presumed to "know the law."[56] Three U.S. Supreme Court decisions illustrate this trend:

For an international perspective on this topic, visit the study site.

- *Notice.* In *Lambert v. California,* the defendant was convicted of failure to adhere to a law that required a "felon" resident in Los Angeles to register with the police within five days.

The U.S. Supreme Court found that convicting Lambert would violate due process because the law was unlikely to have come to his attention.[57]

- *Intent.* In *United States v. Cheek,* an airline pilot was counseled by antitax activists and believed that his wages did not constitute income and therefore he did not owe federal tax. He was convicted of willfully attempting to evade or defeat his taxes. The U.S. Supreme Court ruled that Congress required a showing of a willful intent to violate tax laws because the vast number of tax statutes made it likely that the average citizen might innocently fail to remain informed of the provisions of the tax code.[58]
- *Reliance.* In the civil rights–era case of *Cox v. Louisiana,* the defendants were convicted of picketing a courthouse with the intent of interfering, obstructing, or influencing the administration of justice. The U.S. Supreme Court reversed the students' convictions on the grounds that the chief of police had instructed them that they could legally picket at a location 101 feet from the courthouse steps.[59]

The Model Penal Code section 2.04(3) recognizes an "ignorance of the law defense" when the defendant does not know the law and the law has not been published or made reasonably available to the public (notice). This defense also applies where the defendant has relied on an official statement of the law (reliance).

Mistake of Fact

A **mistake of fact** constitutes a defense in those instances when the defendant's mistake results in a lack of criminal intent. The Model Penal Code section 2.04(1) states that "ignorance or mistake is a defense when it negatives the existence of a state of mind that is essential to the commission of an offense. . . ." As a first step, determine the intent required for the offense and then compare this to the defendant's state of mind. A defendant may take an umbrella from a restaurant during a rainstorm believing that this is the umbrella that he or she left at the restaurant two years ago. The accused will be acquitted of theft because he or she lacked the intent to take, carry away, and permanently deprive the owner of the umbrella. Some courts require that a defendant's mistake must be objectively reasonable, meaning that a reasonable person would have made the same mistake. A trial court, for instance, might conclude that it was unreasonable for the defendant to believe after two years that his umbrella was still at the restaurant.[60]

Another aspect of the mistake of fact defense is that an individual may be mistaken but nonetheless will be held criminally liable in the event that the facts as perceived by the defendant still comprise a crime. For example, a defendant may be charged with receiving stolen umbrellas and contend that he or she believed that the package contained stolen raincoats. This would not exonerate the defendant. The charge is based on the receipt of stolen property, not stolen umbrellas.[61]

Let us return to the issue of whether a mistake of fact must be reasonable. Clearly an honest and good-faith mistake of fact, however misguided, negates a criminal intent. Should the law, however, insist that mistakes meet a reasonableness standard?

In the well-known English case of *Regina v. Morgan,* the appellant Morgan invited his three male co-appellants to have sexual relations with his wife. He led them to believe that Mrs. Morgan would consent to this group activity. Morgan cautioned that Mrs. Morgan was "kinky" and that she may "struggle a bit" in order to enhance her sense of excitement. She was held down while each of the three men engaged in intercourse with her in the presence of the other appellants. The three were convicted of rape. The judge instructed the jury that the defendants must have reasonably believed that despite Mrs. Morgan's resistance, she consented to the rape. The British Law Lords (the equivalent of our U.S. Supreme Court) overturned the convictions and ruled that a male who acts on an unreasonably mistaken belief in the female's consent does not possess the required criminal intent and should be acquitted. Lord Hailsham observed that "either the prosecution proves that the accused had the requisite intent, or it does not." This decision resulted in Parliament enacting a statutory amendment providing that the mistake of fact defense requires an honest and reasonable mistake. Could the appellants in *Morgan* credibly contend that they honestly and reasonably believed that Mrs. Morgan consented? Should we convict individuals who make a factual, but unreasonable mistake, and who lack a criminal intent?[62]

The Model Penal Code section 2.04(1)(a)(b) diverges from the English parliamentary rule and accepts that a mistake of fact constitutes a defense so long as it "negatives" the intent required under the statute. In other words, the Model Penal Code would acquit the defendants in *Morgan* if the jury believed that the defendants did not "purposely, knowingly, or recklessly" rape Mrs. Morgan.

The Legal Equation

Mistake of law ═ No excuse (some indication may excuse criminal liability in cases involving notice, intent, reliance).

Mistake of fact ═ Mistake is an excuse if it negates the required criminal intent (may require reasonable mistake).

You Decide

9.5 The defendant and his cousin, knowing that their marriage would be illegal in Nebraska, married in Iowa, where such unions are not prohibited. The county prosecutor informed the defendant that he would be prosecuted for sexual relations without marriage ("fornication") in the event that the couple continued to live in Nebraska because the marriage was not recognized in the state. Three private attorneys confirmed that the Iowa marriage was not valid in Nebraska. The defendant subsequently "separated" from his pregnant cousin and remarried another woman. It later was determined that, in fact, the Iowa marriage was valid in Nebraska, and the defendant was charged with bigamy (simultaneous marriage to more than a single spouse). Is the defendant guilty of bigamy? See *Staley v. State*, 89 Neb. 701 (1911).

You can find the answer at www.sagepub.com/lippmanccl3e.

ENTRAPMENT

American common law did not recognize the defense of **entrapment**. The fact that the government entrapped or induced a defendant to commit a crime was irrelevant in evaluating a defendant's guilt or innocence.

The development of the defense is traced to the U.S. Supreme Court's 1932 decision in *Sorrells v. United States*. In *Sorrells*, an undercover agent posing as a "thirsty tourist" struck up a friendship with Sorrells and was able to overcome Sorrells's resistance and persuaded him to locate some illicitly manufactured alcohol. Sorrells's conviction for illegally selling alcohol was reversed by the U.S. Supreme Court.[63]

The decision in *Sorrells* defined entrapment as the "conception and planning of an offense by an officer, and his procurement of its commission by one who would not have perpetrated it except for the trickery, persuasion, or fraud of the officer." The essence of entrapment is the government's inducement of an otherwise innocent individual to commit a crime. Decisions have clarified that the prohibition on entrapment extends to the activities of undercover government agents, confidential informants, and private citizens acting under the direction of law enforcement personnel. The defense has been raised in cases involving prostitution; the illegal sale of alcohol, cigarettes, firearms, and narcotics; and public corruption. There is some indication that the defense may not be invoked to excuse a crime of severe violence.

There are good reasons for the government to rely on undercover strategies:

- *Crime Detection*. Certain crimes are difficult to investigate and to prevent without informants. This includes narcotics, prostitution, and public corruption.
- *Resources*. Undercover techniques, such as posing as a buyer of stolen goods, can result in a significant number of arrests without expending substantial resources.
- *Deterrence*. Individuals will be deterred from criminal activity by the threat of government involvement in the crime.

Entrapment is also subject to criticism:

- The government may "manufacture crime" by individuals who otherwise may not engage in such activity.
- The government may lose respect by engaging in lawbreaking.
- The informants who infiltrate criminal organizations may be criminals whose own criminal activity often is overlooked in exchange for their assistance.
- Innocent individuals are often approached in order to test their moral virtue by determining whether they will engage in criminal activity. They likely would not have committed a crime had they not been approached.

The Law of Entrapment

In developing a legal test to regulate entrapment, judges and legislators have attempted to balance the need of law enforcement to rely on undercover techniques against the interest in insuring that innocent individuals are not pressured or tricked into illegal activity. As noted by U.S. Supreme Court Chief Justice Earl Warren in 1958, "a line must be drawn between the trap for the unwary innocent and the trap for the unwary criminal."[64]

There are two competing legal tests for entrapment that are nicely articulated in the 1958 U.S. Supreme Court case of *Sherman v. United States*. Sherman's conviction on three counts of selling illegal narcotics was overturned by the Supreme Court; and the facts, in many respects, illustrate the perils of government undercover tactics. Kalchinian, a government informant facing criminal charges, struck up a friendship with defendant Sherman. They regularly talked during their visits to a doctor who was assisting both of them to end their addiction to narcotics. Kalchinian eventually was able to overcome Sherman's resistance and persuaded him to obtain and to split the cost of illegal narcotics.[65]

The U.S. Supreme Court unanimously agreed that Sherman had been entrapped. Five judges supported a *subjective test* for entrapment and four supported an *objective test*. The federal government and a majority of states follow a subjective test, whereas the Model Penal Code and a minority of states rely on the objective test. Keep in mind that the defense of entrapment was developed by judges, and the availability of this defense has not been recognized as part of a defendant's constitutional right to due process of law. Entrapment in many states is an affirmative defense that results in the burden being placed on the defendant to satisfy a preponderance of the evidence standard. Other states require the defendant to produce some evidence, and then they place the burden on the government to rebut the defense beyond a reasonable doubt.[66]

The Subjective Test

The subjective test focuses on the defendant and asks whether the accused possessed the criminal intent or "predisposition" to commit the crime or whether the government "created" the offense. In other words, "but for" the actions of the government, would the accused have broken the law? Was the crime the "product of the creative activity of the government" or the result of the defendant's own criminal design?

The first step is to determine whether the government induced the crime. This requires that the undercover agent or informant persuade or pressure the accused. A simple offer to sell or to purchase drugs is a "mere offer" and does not constitute an "inducement." In contrast, an inducement involves appeals to friendship, compassion, promises of extraordinary economic or material gain, sexual favors, or assistance in carrying out the crime.

The second step is the most important and involves evaluating whether the defendant possessed a "predisposition" or readiness to commit the crime with which he or she is charged. The law assumes that a defendant who is predisposed is ready and willing to engage in criminal conduct in the absence of governmental inducements and, for this reason, is not entitled to rely on the defense of entrapment. In other words, the government must direct its undercover strategy against the unwary criminal rather than the unwary innocent. How is predisposition established? A number of factors are considered:[67]

- the character or reputation of the defendant, including prior criminal arrests and convictions for the type of crime involved;
- whether the accused suggested the criminal activity;

- whether the defendant was already engaged in criminal activity for profit;
- whether the defendant was reluctant to commit the offense; and
- the attractiveness of the inducement.

In *Sherman*, the purchase of the drugs was initiated by the informant, Kalchinian, who overcame Sherman's initial resistance and persuaded him to obtain drugs. Kalchinian, in fact, had instigated two previous arrests and was facing sentencing for a drug offense himself. The two split the costs. There is no indication that Sherman was otherwise involved in the drug trade, and a search failed to find drugs in his home. Sherman's nine-year-old sales conviction and five-year-old possession conviction did not indicate that he was ready and willing to sell narcotics. In other words, before Kalchinian induced Sherman to purchase drugs, he seemed to be genuinely motivated to overcome his dependency on narcotics.

The underlying theory is that the jury, in evaluating whether the defendant was entrapped, is merely carrying out the intent of the legislature. The "fiction" is that the legislature did not intend for otherwise innocent individuals to be punished who were induced to commit crimes by government trickery and pressure. The issue of entrapment under the subjective test is to be decided by the jury.

The Objective Test

The objective test focuses on the conduct of the government rather than on the character of the defendant. Justice Felix Frankfurter, in his dissenting opinion in *Sherman*, explained that the crucial question is "whether police conduct revealed in the particular case falls below standards to which common feelings respond, for the proper use of governmental power." The police, of course, must rely on undercover work, and the test for entrapment is whether the government, by offering inducements, is likely to attract those "ready and willing" to commit crimes "should the occasion arise" or whether the government has relied on tactics and strategies that are likely to attract those who "normally avoid crime and through self-struggle resist ordinary temptations."

The subjective test focuses on the defendant; the objective test focuses on the government's conduct. Under the subjective test, if an informant makes persistent appeals to compassion and friendship and then asks a defendant to sell narcotics, the defendant has no defense if he is predisposed to selling narcotics. Under the objective test, there would be a defense because the conduct of the police, rather than the predisposition of the defendant, is the central consideration.[68]

Justice Frankfurter wrote that public confidence in the integrity and fairness of the government must be preserved and that government power is "abused and directed to an end for which it was not constituted when employed to promote rather than detect crime and to bring about the downfall of those who, left to themselves, might well have obeyed the law."[69] These unacceptable methods lead to a lack of respect for the law and encourage criminality. Frankfurter argued that judges must condemn corrupt and uncivilized methods of law enforcement even if this judgment may result in the acquittal of the accused. Frankfurter criticized the predisposition test for providing protection for "innocent defendants," while permitting the government to employ various unethical strategies and schemes against defendants who are predisposed.

In *Sherman*, Frankfurter condemned Kalchinian's repeated requests that the accused assist him to obtain drugs. He pointed out that Kalchinian took advantage of Sherman's susceptibility to narcotics and manipulated Sherman's sympathetic response to the pain Kalchinian was allegedly suffering in withdrawing from drugs. The *Sherman* and *Sorrells* cases suggest that practices prohibited under the objective test include:

- taking advantage of weaknesses;
- repeated appeals to friendship, sympathy;
- promising substantial economic gain;
- pressure or threats;
- providing the equipment required for carrying out a crime; and
- false representations designed to induce a belief that the conduct is not prohibited.

Critics complain that the objective test has not resulted in clear and definite standards to guide law enforcement. Can you determine at what point Kalchinian crossed the line? Critics also charge that it makes little sense to acquit a defendant who is "predisposed" based on the fact that a "mythical innocent" individual may have been tricked into criminal activity by the government's tactics. However, the objective test was adopted by the Model Penal Code, which follows Justice Frankfurter in assigning the determination of entrapment to judges rather than juries based on the fact that judges are responsible for safeguarding the integrity of the criminal justice process.

Due Process

Various defendants have unsuccessfully argued that entrapment tactics violate the Due Process Clause of the U.S. Constitution. In other words, the contention is that the government's conduct is so unfair and outrageous that it would be unjust to convict the defendants under the U.S. Constitution.

The U.S. Supreme Court rejected this argument in *United States v. Russell*. Joe Shapiro, an undercover agent for the Federal Bureau of Narcotics and Dangerous Drugs, met with Richard Russell and his co-defendants John and Patrick Connolly. Shapiro offered to provide them with the chemical phenyl-2-propanone, an essential element in the manufacture of methamphetamine, in return for one-half of the drug produced. The three provided Shapiro with a sample of their most recent batch and showed him their laboratory, where Shapiro observed an empty bottle of phenyl-2-propanone. The next day Shapiro delivered one hundred grams of phenyl-2-propanone and watched as two of the defendants begin to manufacture methamphetamine. Shapiro later was given one-half the drug and he purchased a portion of the remainder. A warrant was obtained and a search revealed two bottles of phenyl-2-propanone, neither of which had been provided by Shapiro.

The defendants, although certainly predisposed to manufacture and sell narcotics, creatively claimed that the government violated due process by prosecuting them for a crime in which the government had been intimately involved. The U.S. Supreme Court rejected this argument and stressed that although phenyl-2-propanone was "difficult to obtain, it was by no means impossible" as indicated by the fact that the defendants had been manufacturing "speed" without the phenyl-2-proponone provided by Shapiro. The court concluded that "[w]hile we may some day be presented with a situation in which the conduct of law enforcement agents is so outrageous that due process principles would absolutely bar the government from invoking judicial process to obtain a conviction . . . the instant case is distinctly not of that breed." The Supreme Court stressed that the investigation of drug-related offenses often requires infiltration and cooperation with narcotics rings and that the law enforcement tactics employed in *Russell* were neither in violation of "fundamental fairness" nor "shocking to the universal sense of justice."[70]

The Entrapment Defense

We might question whether courts should be involved in evaluating law enforcement tactics and in acquitting individuals who are otherwise clearly guilty of criminal conduct. Can innocent individuals really be pressured into criminal activity? Do we want to limit the ability of the police to use the techniques they believe are required to investigate and punish crime? There also appear to be no clear judicial standards for determining predisposition under the subjective test and for evaluating acceptable law enforcement tactics under the objective test. This leaves the police without a great deal of guidance or direction. On the other hand, we clearly are in need of a legal mechanism for preventing government abuse. The next case, *Miller v. State,* raises the issue of whether there should be limits on governmental approaches to criminal investigation.

The Legal Equation

Subjective test **=** Government inducement

+ defendant is not predisposed to commit the crime.

Objective test **=** Police conduct falls below standards to which common feelings respond

+ induces crime by those who normally avoid criminal activity.

Was Miller entrapped by the undercover police decoy?

Miller v. State, 110 P.3d 53 (Nev. 2005). Per Curiam.

Issue

This appeal arises out of an undercover decoy program initiated by the Las Vegas Metropolitan Police Department (LVMPD). The decoy program was designed to combat an increase in street-level robberies occurring in downtown Las Vegas. A street-level robbery is a person-to-person crime where one person walks up to another and either robs that person or picks his or her pocket.

Facts

As part of the decoy operation, Detective Jason Leavitt disguised himself as an intoxicated vagrant to blend in with transient persons who reside in certain areas of Las Vegas. Detective Leavitt carried twenty one-dollar bills in a pocket and left a small portion of the bills exposed. This allowed someone standing close to him to see the money, but the bills were hidden well enough that they did not attract the attention of every passerby. Detective Leavitt wore a monitoring device that allowed surveillance and arrest teams to hear what Detective Leavitt heard and said. When Detective Leavitt gave a predetermined signal, arrest teams would approach the scene and apprehend the suspect.

On July 29, 2003, Detective Leavitt was dressed in black jeans, a dirty t-shirt, a short-sleeved flannel shirt, and a baseball cap. Twenty one-dollar bills were folded inside the breast pocket of the flannel shirt so that only the tips of the bills were exposed. Detective Leavitt rubbed charcoal on his face to appear dirty and wiped beer on his neck to give off the odor of alcohol. He also walked with a limp and carried a can of beer to appear intoxicated.

Detective Leavitt positioned himself on the 200 block of Main Street across from the Greyhound Bus Station and leaned against a chain link fence. Appellant Richard Miller, who was walking southbound on Main Street, approached Detective Leavitt and asked him for money. When Detective Leavitt told Miller that he would not give him any money, Miller put his arm around Detective Leavitt and invited him to get a drink.

Miller stood to the left of Detective Leavitt with his right arm around Detective Leavitt's shoulders. Miller then pulled Detective Leavitt closer to him, quickly reached his hand into Detective Leavitt's pocket, and took the twenty dollars. Miller then loosened his grip on Detective Leavitt and again asked for money.

Detective Leavitt said that he could not give Miller any money because his money was gone. The undercover arrest team then converged on the location and took Miller into custody.

The State charged Miller, by information, with larceny from the person. After a two-day trial, the jury convicted Miller, and the district court sentenced him to a maximum of 32 months and a minimum of 12 months imprisonment. On appeal, Miller argues that he was entrapped, that the prosecutor impermissibly commented on his decision not to testify, and that the prosecutor committed other misconduct.

Reasoning

Miller argues that police officers entrapped him by improperly tempting him with exposed money and a helpless victim. We disagree.

"The entrapment defense is made available to defendants not to excuse their criminal wrongdoing but as a prophylactic device designed to prevent police misconduct. . . . Entrapment encompasses two elements: (1) an opportunity to commit a crime is presented by the state (2) to a person not predisposed to commit the act. . . . The Government may use undercover agents to enforce the law." Nevertheless, undercover agents "may not originate a criminal design, implant in an innocent person's mind the disposition to commit a criminal act, and then induce commission of the crime so that the Government may prosecute."

In *DePasquale v. State,* 757 P.2d 367 (Nev. 1988), we discussed our prior entrapment jurisprudence where an undercover officer posed as a decoy. We cited three earlier cases that collectively held that the defendant was entrapped where the undercover decoy "was apparently helpless, intoxicated, and feigned unconsciousness with cash hanging from his pocket." Specifically, we noted that the "degree of vulnerability, exemplified in [those prior cases] by the decoy's feigned lack of consciousness, . . . cloaks any suggestion of the defendant's predisposition."

However, in *DePasquale,* we held that the defendant was not entrapped when he stole from a female undercover police officer who was walking along open sidewalks around a casino with money zipped into her purse. Thus, we have drawn a clear line between a realistic decoy who poses as an alternative victim of potential crime and the helpless, intoxicated, and unconscious decoy with money hanging out of a pocket. The former is permissible undercover police work, whereas the latter is entrapment.

The opportunity presented to commit a crime was not improper. The theft in this case occurred across from the Greyhound Bus Station at the 200 block of South Main Street in Las Vegas. Twenty one-dollar bills were folded inside the breast pocket of Detective Leavitt's flannel shirt so that only the tips of the bills were exposed. Miller, who was walking southbound on Main Street, approached Detective Leavitt and asked him for money. When Detective Leavitt told Miller that he would not give him any money, Miller put his arm around Detective Leavitt and invited him to get a drink. Miller stood to the left of Detective Leavitt with his right arm around Detective Leavitt's shoulders. Miller then pulled Detective Leavitt closer to him, quickly reached his hand into Detective Leavitt's pocket, and took the twenty dollars.

The police committed no misconduct in this operation. The opportunity presented was sufficient to lead to a criminal act only by a person predisposed to commit a crime. Though a suspect is entrapped when the decoy officer poses as an unconscious vagrant with exposed money hanging from his pockets, Detective Leavitt did not feign unconsciousness nor was his money readily accessible. Only a portion of the bills were exposed; a passerby could see the edges of currency, but not the denominations. Detective Leavitt did not entice Miller into stealing the money. Rather, Miller approached Detective Leavitt and asked him for money. When Detective Leavitt refused to give him money, Miller picked his pocket.

It is clear that Miller was predisposed to commit larceny from the person. We have recognized five factors that, though not exhaustive, are helpful to determine whether the defendant was predisposed: (1) the defendant's character, (2) who first suggested the criminal activity, (3) whether the defendant engaged in the activity for profit, (4) whether the defendant demonstrated reluctance, and (5) the nature of the government's inducement. "Of these five factors, the most important is whether the defendant demonstrated reluctance which was overcome by the government's inducement."

Miller's character is unclear from the record, but it is clear that Miller initiated the conversation and engaged in the larceny for profit. Furthermore, Miller exhibited no reluctance about his actions. Finally, the critical balance between government inducement and Miller's reluctance weighs in favor of predisposition here. Miller approached Detective Leavitt, initiated a conversation, and asked for money. When Detective Leavitt told Miller he would not give him any money, Miller picked his pocket. These facts demonstrate a predisposition to commit the crime of larceny from the person. Since Miller was predisposed to commit the crime, he was not entrapped.

Holding

We conclude that Miller was not entrapped because he was predisposed to commit the crime of larceny from the person. Furthermore, the State did not improperly comment on Miller's failure to testify, nor did the State commit prosecutorial misconduct by implying that Miller was dangerous and preyed upon vulnerable persons. Accordingly, we affirm the conviction.

Questions for Discussion

1. Summarize the facts in *Miller*.

2. Why does the Nevada Supreme Court conclude that Miller was not entrapped?

3. How does the court distinguish *Miller* and *DePasquale* from earlier cases that held that the defendant was entrapped?

4. Was this entrapment under the objective test?

See more cases on the study site: Farley v. State, *www.sagepub.com/lippmanccl3e.*

You Decide

9.6 Russell Thompson, a Carlisle Pennsylvania policeman, was convicted of unlawful delivery of a small amount of marijuana and one count of criminal conspiracy. Beginning on May 1, 1980, state police Trooper Lucinda Hammond engaged in an undercover investigation of the 46-year-old Thompson. Thompson is an African-American and, at that time, was a ten-year police veteran. He was married, with a mentally challenged daughter. Hammond is described as a "young, blonde, white female," who is "very attractive." According to the Pennsylvania court, she regularly wore mid-thigh cut-off shorts during her encounters with Thompson.

During the course of the investigation, Hammond called Thompson eight to ten times, although she never provided him with her telephone number. She initially approached Thompson on three occasions on the street. On the third occasion, the discussion turned to drugs and to "getting high." Hammond accepted Thompson's invitation to meet him at a bar. During their "date," he shared with her that he kept marijuana in his locker that he had seized during drug busts and that although he was a police officer, he liked to "party" and to "get high."

The fifth direct contact occurred on September 29, after she had called him at the police station. Hammond asked Thompson for marijuana for her "personal use" or "maybe [to] make a little money on the side." He responded by trying to get her to go to the Oliver Plunkett bar with him, telling her he could get some marijuana if she would go with him. She declined and left.

The sixth encounter occurred on December 17, 1980. Hammond had talked to Thompson on two occasions in October and, in another telephone conversation on November 25, had "point blank" asked him if he could get her some drugs. At the December 17 meeting, she again asked him about getting drugs. He repeated that if she would meet him after midnight at Oliver Plunkett, he could obtain some drugs. She refused.

Hammond next met the defendant on January 5, 1981. She stated that he "wasn't coming across" with any drugs. She needled him that he was "all talk" and never had produced marijuana. He stated he would get some for her if she'd meet him the next night. They met in a furniture store parking lot and she got into his car only to find he had no drugs. Hammond stated that she was interested in getting some drugs for later in the evening. He nonetheless made no moves to get any drugs, despite having earlier mentioned that he had a "connection" in the Oliver Plunkett, which was right across the street. She asked about paying for the drugs and he insisted that the drugs would not cost her anything.

By December 1980 or January 1981, they were kissing at the end of their dates or meetings; and Hammond testified that he was romantically interested in her. They were seen together in public in Carlisle, and Thompson would put his arm around her and introduce her to his friends. Hammond never offered and he never requested sexual relations.

The telephone calls and meetings continued. Hammond constantly requesting drugs and Thompson "just as consistently failing to provide them." She continued to push him to get her drugs, and he continued to promise to get her drugs and assured her that no payment would be necessary for the drugs he acquired for her. Thompson continued to ask her to meet him after midnight and on one occasion asked her to spend the night with him, all of which she declined.

Hammond arranged a meeting on the evening of March 23 at a bar in Carlisle. At the bar, Thompson obtained 4.5 grams of marijuana from his friend and gave the drugs to Hammond. The two of them left the bar and went to Thompson's car where he rolled some marijuana cigarettes with papers supplied by Hammond. She offered to pay for the marijuana, but he refused to accept any money.

Hammond called Thompson and arranged to meet on April 11, 1981. She told Thompson that she had no marijuana left and he discussed obtaining more drugs for her. Thompson talked about Hammond moving to Carlisle, offered to pay one-half of her rent, and drove her around Carlisle looking at apartments. Pennsylvania's entrapment statute "focus[es] on the conduct of the police and [is] not . . . concerned with the defendant's prior criminal activity or other indicia of a predisposition to commit crime."

Was Thompson entrapped under the objective test? See *Commonwealth v. Thompson*, 484 A.2d 159 (Pa. Super. Ct. 1984).

You can find the answer at www.sagepub.com/lippmanccl3e.

NEW DEFENSES

The criminal law is based on the notion that individuals are responsible and accountable for their decisions and subject to punishment for choosing to engage in morally blameworthy behavior. We have reviewed a number of circumstances in which the law has traditionally recognized that individuals should be excused and should not be held fully responsible. In the last decades, medicine and the social sciences have expanded our understanding of the various factors that influence human behavior. This has resulted in defendants offering various new defenses that do not easily fit into existing categories. These defenses are not firmly established and have yet to be accepted by judges and juries. Most legal commentators dismiss the defenses as "quackery" or "science" and condemn these initiatives for undermining the principle that individuals are responsible for their actions.

One of the foremost critics is Professor Alan Dershowitz of Harvard Law School, who has pointed to fifty "abuse excuses." Dershowitz defines an **abuse excuse** as a legal defense in which defendants claim that the crimes with which they are charged result from their own victimization and that they should not be held responsible. Examples are the "battered wife" and "battered child syndromes" (discussed in Chapter 8).[71] A related set of defenses are based on the claim that the defendant's biological or genetic heredity caused him or her to commit a crime. George Fletcher has warned that these types of defenses could potentially undermine the assumption that all individuals are equal and should be rewarded or punished based on what they do, not on who they are. On the other hand, proponents of these new defenses argue that the law should evolve to reflect new intellectual insights.[72]

Some New Defenses

Four examples of *biological defenses* are:

- *XYY Chromosome.* This is based on research that indicates that a large percentage of male prison inmates possess an extra Y chromosome that results in enhanced "maleness." (Each fetus has two sex chromosomes, one of which is an X. A female has two X chromosomes; a male a Y and an X chromosome.) A Maryland appeals court dismissed a defendant's claim that his robbery should be excused based on the presence of an extra Y masculine chromosome that allegedly made it impossible for him to control his antisocial and aggressive behavior.[73]
- *Premenstrual Syndrome (PMS).* Many women experience cramps, nausea, and discomfort prior to menstruation. PMS has been invoked by defendants who contend that they suffered from severe pain and distress that drove them to act in a violent fashion. Geraldine Richter was detained by an officer for driving while intoxicated, and she verbally attacked and threatened the officer and kicked the breathalyzer. A Fairfax County, Virginia, judge acquitted Richter of driving while intoxicated and resisting arrest and other charges after an expert testified that her premenstrual condition caused her to absorb alcohol at an abnormally rapid rate.[74]
- *Postpartum Psychosis.* This is caused by a drop in the hormonal level following the birth of a child. The result can be depression, suicide, and in its extreme manifestations delusions, hallucinations, and violence. Stephanie Molina reportedly was a happy and outgoing young woman who suffered severe depression and a paranoid fear of being killed. She subsequently killed her child, attempted suicide, and made an effort to burn her house down. A California appellate court ruled that the jury should have been permitted to consider evidence of Molina's condition in evaluating her guilt for the intentional killing of her child.[75]
- *Environmental Defense.* The Massachusetts Supreme Judicial Court rejected a defendant's effort to excuse a homicide based on the argument that the chemicals he used in lawn care work resulted in involuntary intoxication and led him to violently respond to a customer's complaint.[76]

Brainwashing is an example of a *psychological defense* in which an individual claims to have been placed under the mental control of others and to have lost the capacity to make independent decisions. A well-known example is newspaper heiress Patricia Hearst who, in 1974, was kidnapped by a small terrorist group, the Symbionese Liberation Army (SLA). Several months later, she entered a bank armed with a machine gun and assisted the group in a robbery. Patricia Hearst testified at trial that she had been abused and brainwashed by the SLA and had been programmed to assume the identity of "Tanya the terrorist." The jury dismissed this claim and convicted Patty Hearst.[77] Another example of a psychological defense is *post-traumatic stress disorder*. A Tennessee Court of Appeals ruled that a veteran of the Desert Shield and Desert Storm military campaigns who recently had returned to the United States should be permitted to introduce evidence demonstrating that his wartime experiences led him to react in an emotional and violent fashion to his wife's romantic involvement with the victim.[78]

Defendants relying on *sociological defenses* claim that their life experiences and environment have caused them to commit crimes. These include:

- *Black Rage.* Colin Ferguson, a thirty-five-year-old native of Jamaica, in December 1993, boarded a commuter train in New York City and embarked on a shooting spree against Caucasian and Asian passengers that left six dead and nineteen wounded. The police found notes in which Ferguson expressed a hatred for these groups as well as for "Uncle Tom Negroes." His lawyer announced that Ferguson would offer the defense of extreme racial stress precipitated by the destructive racial treatment of African Americans. Ferguson ultimately represented himself at trial and did not raise this defense, which nonetheless has been the topic of substantial discussion and debate.[79]

- *Urban Survivor.* Daimon Osby, a seventeen-year-old student, shot and killed two unarmed cousins who had been demanding that Osby provide them with the opportunity to win back the money they had lost to him while gambling. At one point, a white pickup apparently belonging to one of the cousins pulled alongside Osby's automobile and a rifle barrel was allegedly pointed out the window. Two weeks later, the same truck approached and Osby shot and killed the occupants, Marcus and Willie Brooks, neither of whom were armed. The defense offered the "urban survivor defense" during Osby's first trial. This resulted in a hung jury. He then was retried and convicted. The defense unsuccessfully appealed the fact that Osby was prohibited from introducing experts supporting his claim of the "urban survivor syndrome" at the second trial. The "urban survivor defense" consists of the contention that young people living in poor and violent urban areas do not receive adequate police protection and develop a heightened awareness and fear of threats.[80]
- *Media Intoxication.* Defendants have claimed that their criminal conduct is caused by "intoxication" from television and pornography. Ronald Ray Howard, nineteen, unsuccessfully argued in mitigation of a death sentence that he had killed a police officer while listening to "gangsta rap."[81]
- *Rotten Social Background.* In *United States v. Alexander,* the defendant shot and killed a white Marine who had uttered a racial epithet. The African American defendant claimed that he had shot as a result of an irresistible impulse that resulted from his socially deprived childhood. Alexander's early years were marked by abandonment, poverty, discrimination, and an absence of love. This "rotten social background" (RSB) allegedly created an irresistible impulse to kill in response to the Marine's remark. The United States Court of Appeals for the District of Columbia Circuit affirmed the trial judge's refusal to issue a jury instruction on RSB. Judge David Bazelon dissented and questioned whether society had a right to sit in judgment over a defendant who had been so thoroughly mistreated.[82]
- *Post-Traumatic Stress Disorder* (PTSD). Defendant Bruce Franklin Jerrett was charged with first-degree murder, breaking and entering, kidnapping, and armed robbery. Jerrett and his mother testified to six or seven incidents following the Vietnam War in which he "blacked out," and on one occasion he attacked his sister. He attributed the incidents to the downward spiral of his health as a result of having been exposed to the chemical Agent Orange. Following his blackouts, Jerrett had no memory of what he had done. Jerrett appealed his conviction; and the North Carolina Supreme Court overturned his conviction on the grounds that the jurors should have received an instruction that if they found that the defendant suffered from PTSD and was unconscious at the time of his crime, he should be acquitted.[83]

You can find State v. Jerrett *on the study site, www.sagepub.com/lippmanccl3e.*

The Cultural Defense

Defendants in several cases have invoked the "cultural defense." This involves arguing that a foreign-born defendant was following his or her culture and was understandably unaware of the requirements of American law. Those in favor of the "cultural defense" argue that it is unrealistic to expect that new immigrants will immediately know or accept American practices in areas as important as the raising and disciplining of children.

The acceptance of diversity, however, may breed a lack of respect for the law among immigrant groups and lead Americans who are required to conform to legal standards to believe that they are being treated unfairly. Judges and juries may also lack the background to determine the authentic customs and traditions of various immigrant groups and may be forced to rely on expert witnesses to understand different cultures.

Multiculturalism may be in conflict with important American values regarding respect for women and children. Various Laotian American tribal groups continue the practice of "marriage by capture," in which a prospective bride is expected to protest sexual advances and the male is required to compel the woman to submit in order to establish his courage and ability to be a strong and suitable husband. In one California case, a young woman who did not accept her parents' Hmong cultural practice alleged that she had been raped by her new husband, who invoked a cultural defense.[84]

The next case in the textbook is *State v. Kargar.* The defendant is originally from Afghanistan and argues that the Maine Supreme Judicial Court should invoke an unusual statute that authorizes courts to dismiss charges against defendants whose acts, although technically in violation of the law, result in minimal social harm.

Should the defendant be excused of gross sexual assault based on the practice of the Afghan culture?

State v. Kargar, 679 A.2d. 81 (Me. 1996). Opinion by: Dana, J.

Facts

Mohammad Kargar, an Afghani refugee, appeals from the judgments entered in the Superior Court (Cumberland County) convicting him of two counts of gross sexual assault. Kargar contends on appeal that the court erred in denying his motion to dismiss pursuant to the de minimis statute, Maine Revised Statutes Annotated title 17-A, section 12 (1983). We agree and vacate the judgments. A person is guilty of gross sexual assault if that person engages in a sexual act with another person and . . . the other person, not the actor's spouse, has not in fact attained the age of fourteen years.

On June 25, 1993, Kargar and his family, refugees since approximately 1990, were babysitting a young neighbor. While the neighbor was there, she witnessed Kargar kissing his eighteen-month-old son's penis. When she was picked up by her mother, the girl told her mother what she had seen. The mother had previously seen a picture of Kargar kissing his son's penis in the Kargar family photo album. After her daughter told her what she had seen, the mother notified the police.

Peter Wentworth, a sergeant with the Portland Police Department, went to Kargar's apartment to execute a search warrant. Wentworth was accompanied by two detectives, two Department of Human Services social workers, and an interpreter. Kargar's family was taken outside by the social workers and the two detectives began searching for a picture or pictures of oral/genital contact. The picture of Kargar kissing his son's penis was found in the photograph album. Kargar admitted that it was he in the photograph and that he was kissing his son's penis. Kargar told Wentworth that kissing a young son's penis is accepted as common practice in his culture. Kargar also said it was very possible that his neighbor had seen him kissing his son's penis. Kargar was arrested and taken to the police station.

Prior to the jury-waived trial Kargar moved for a dismissal of the case pursuant to the de minimis statute. . . . The de minimis hearing consisted of testimony from many Afghani people who were familiar with the Afghani practice and custom of kissing a young son on all parts of his body. Kargar's witnesses, all relatively recent emigrants from Afghanistan, testified that kissing a son's penis is common in Afghanistan, that it is done to show love for the child, and that it is the same whether the penis is kissed or entirely put into the mouth because there are no sexual feelings involved.

The witnesses also testified that pursuant to Islamic law any sexual activity between an adult and a child results in the death penalty for the adult. Kargar also submitted statements from expert witnesses that support the testimony of the live witnesses. The State did not present any witnesses during the de minimis hearing. Following the presentation of witnesses the court denied Kargar's motion and found him guilty of two counts of gross sexual assault.

Kargar testified during the de minimis hearing that the practice was acceptable until the child was three, four, or five years old. He also testified during the de minimis hearing that his culture views the penis of a child as not the holiest or cleanest part of the body because it is from where the child urinates. Kargar testified that kissing his son there shows how much he loves his child precisely because it is not the holiest or cleanest part of the body.

Issue

Maine's de minimis statute, Maine Revised Statutes Annotated title 17-A, section 12, provides, in pertinent part:

1. The court may dismiss a prosecution if, . . . having regard to the nature of the conduct alleged and the nature of the attendant circumstances, it finds the defendant's conduct:

    A. Was within a customary license or tolerance, which was not expressly refused by the person whose interest was infringed and which is not inconsistent with the purpose of the law defining the crime; or

    B. Did not actually cause or threaten the harm sought to be prevented by the law defining the crime or did so only to an extent too trivial to warrant the condemnation of conviction; or

    C. Presents such other extenuations that it cannot reasonably be regarded as envisaged by the Legislature in defining the crime.

The court analyzed Kargar's conduct, as it should have, pursuant to each of the three provisions of section 12(1). The language of the statute itself makes it clear that if a defendant's conduct falls within any one of these provisions, the court may dismiss the prosecution. We agree with the State that trial courts should be given broad discretion in determining the propriety of a de minimis motion. In the instant case, however, Kargar asserts that the court erred as a matter of law because it found culture, lack of harm, and his innocent state of mind irrelevant to its de minimis analysis. We agree.

Reasoning

Maine's de minimis statute's . . . purpose is to "introduce a desirable degree of flexibility in the administration of the law." The language of the statute expressly requires that courts view the defendant's conduct "having regard to the nature of the conduct alleged and the nature of the attendant circumstances." Each de minimis analysis will therefore always be case-specific. The Model Penal Code traces the history of de minimis statutes . . . [and] suggests that courts should have the "power to discharge without conviction, persons who have committed acts which, though amounting in law to crimes, do not under the circumstances involve any moral turpitude."

When making a determination under the de minimis statute, an objective consideration of surrounding circumstances is authorized. . . . Although we have not had occasion to articulate circumstances worthy of cognizance, we agree with the courts of New Jersey and Hawaii that the following factors are appropriate for de minimis analysis: the background, experience, and character of the defendant, which may indicate whether he knew or ought to have known of the illegality; the knowledge of the defendant of the consequences to be incurred upon violation of the statute; the circumstances concerning the offense; the resulting harm or evil, if any, caused or threatened by the infraction; the probable impact of the violation upon the community; the seriousness of the infraction in terms of punishment, bearing in mind that punishment can be suspended; mitigating circumstances as to the offender; possible improper motives of the complainant or prosecutor; and any other data that may reveal the nature and degree of the culpability in the offense committed by the defendant. . . . We thus hold that it is appropriate for courts to analyze a de minimis motion by reviewing the full range of factors discussed in the above quoted language.

Our review of the record in the instant case reveals that the court . . . denied Kargar's motion without considering the full range of relevant factors. The court's interpretation of the subsection, which focused on whether the conduct met the definition of the gross sexual assault statute, operated to nullify the effect of the de minimis analysis called for by the statute. The focus is not on whether the conduct falls within the reach of the statute criminalizing it. If it did not, there would be no need to perform a de minimis analysis.

The focus is not on whether the admittedly criminal conduct was envisioned by the Legislature when it defined the crime. If the Legislature did not intend that there be an individual, case-specific analysis then there would be no point to the de minimis statute. Subsection 1(C) provides a safety valve for circumstances that could not have been envisioned by the Legislature. It is meant to be applied on a case-by-case basis to unanticipated "extenuations," when application of the criminal code would lead to an "ordered but intolerable" result. Because the Legislature did in fact allow for unanticipated "extenuations," the trial court was required to consider the possibility that . . . a conviction in this case could not have been anticipated by the Legislature when it defined the crime of gross sexual assault.

In order to determine whether this defendant's conduct was anticipated by the Legislature when it defined the crime of gross sexual assault, it is instructive to review the not-so-distant history of that crime. Maine Revised Statutes Annotated title 17-A, section 253(1)(B), makes criminal any sexual act with a minor (non-spouse) under the age of fourteen. A sexual act is defined as, among other things, "direct physical contact between the genitals of one and the mouth . . . of the other." Prior to 1985 the definition of this type of sexual act included a sexual gratification element. The Legislature removed the sexual gratification element because, "given the physical contacts described, no concern exists for excluding 'innocent' contacts." . . . Thus, the 1985 amendment to section 251(1)(C) illuminates the fact that an "innocent" touching such as occurred in this case has not forever been recognized as inherently criminal by our own law. The Legislature's inability to comprehend "innocent" genital-mouth contact is highlighted by reference to another type of "sexual act," namely, "any act involving direct physical contact between the genitals . . . of one and an instrument or device manipulated by another." Me. Rev. Stat. Ann. tit. 17-A, § 251(1)(C)(3). The Legislature maintained the requirement that for this type of act to be criminal it must be done for the purpose of either sexual gratification or to cause bodily injury or offensive physical contact. Its stated reason for doing so was that "a legitimate concern exists for excluding 'innocent' contacts, such as for proper medical purposes or other valid reasons." . . .

All of the evidence presented at the de minimis hearing supports the conclusion that there was nothing "sexual" about Kargar's conduct. There is no real dispute that what Kargar did is accepted practice in his culture. The testimony of every witness at the de minimis hearing confirmed that kissing a young son on every part of his body is considered a sign only of love and affection for the child. This is true whether the parent kisses, or as the trial court found, "engulfs" a son's penis. There is nothing sexual about this practice. In fact, the trial justice expressly recognized that if the State were required to prove a purpose of sexual gratification it "wouldn't have been able to have done so."

During its sentencing of Kargar, the court stated: "There is no sexual gratification. There is no victim impact." The court additionally recognized that the conduct for which Kargar was convicted occurred in the open, with his wife present, and noted that the photograph was displayed in the family photo album, available for all to see. The court concluded its sentencing by recognizing that this case is "not at all typical [but instead is] fully the exception. . . . The conduct was unequivocally criminal, but the circumstances of that conduct and the circumstances of this defendant call for leniency." Although the court responded to this call for leniency by imposing an entirely suspended sentence, the two convictions expose Kargar to severe consequences independent of any period of incarceration, including his required registration as a sex offender . . . and the possibility of deportation. . . . These additional consequences emphasize why the factors recognized by the court during the sentencing hearing were also relevant to the de minimis analysis. Kargar's wife, Shamayel, testified during the sentencing hearing that she took the picture to send to Kargar's mother to show her how much he loved his son.

Holding

Although it may be difficult for us as a society to separate Kargar's conduct from our notions of sexual abuse, that difficulty should not result in a felony conviction in this case. The State concedes that dismissing this case pursuant to the de minimis statute would pose little harm to the community. The State is concerned, however, with the potential harm caused by courts using the factors of this case to allow for even more exceptions to the criminal statutes. It argues that exceptions should be made by the Legislature, which can gather data, debate social costs and benefits, and clearly define what conduct constitutes criminal activity. The flaw in the State's position is that the Legislature has already clearly defined what conduct constitutes gross sexual assault. It has also allowed for the adjustment of the criminal statutes by courts in extraordinary cases where, for instance, the conduct cannot reasonably be regarded as envisaged by the Legislature in defining the crime.

As discussed above, the Legislature removed the sexual gratification element previously contained within the definition of a sexual act because it could not envision any possible innocent contacts, "given the physical contacts described." In virtually every case the assumption that a physical touching of the mouth of an adult with the genitals of a child under the age of fourteen is inherently harmful is correct. This case, however, is the exception that proves the rule. Precisely because the Legislature did not envision the extenuating circumstances present in this case, to avoid an injustice the de minimis analysis set forth in section 12(1)(C) requires that Kargar's convictions be vacated.

Application of the de minimis statute does not . . . reflect approval of Kargar's conduct. The conduct remains criminal. Kargar does not argue that he should now be permitted to practice that which is accepted in his culture. The issue is whether his past conduct under all of the circumstances justifies criminal convictions.

Questions for Discussion

1. Are you persuaded that Kargar's conduct was a central part of his culture? Did he realize that his conduct was criminal? Should Kargar's cultural background dictate the court's decision?

2. Might Kargar's son suffer long-term harm? Does the court overlook the issue of the victim's incapacity to consent and need for protection? At what age would the Maine court rule that Kargar's conduct toward his son was in violation of the law? Would the Maine Supreme Judicial Court have dismissed Kargar's conviction if his behavior involved acts considered more sexually intrusive?

3. Do you believe that the Maine court would have reached the same decision if Kargar had been born and raised in the United States? If the "victim" had been a female?

4. What if Kargar continues to engage in this conduct? Will the Maine court reach the same decision in the event that another Afghan immigrant engages in similar conduct?

5. Should the cultural concern be raised at sentencing rather than when considering guilt or innocence? Was the Maine court influenced by the threat that Kargar would be required to register as a sexual offender and risk deportation?

You Decide

9.7 A Laotian refugee was indicted for the intentional killing of his wife of one month. The defendant offered the defense that his wife's continuing affection and receipt of phone calls from a former boyfriend brought shame on the defendant and his family and caused him to lose self-control. Should this cultural consideration result in the defendant being held liable for voluntary manslaughter? See *People v. Aphaylath*, 502 N.E.2d 998 (N.Y. 1986).

You can find the answer at www.sagepub.com/lippmanccl3e.

CHAPTER SUMMARY

Excuses comprise a broad set of defenses in which defendants claim a lack of responsibility for their criminal acts. This lack of "moral blameworthiness" is based on a lack of criminal intent or on the involuntary nature of the defendant's criminal act.

The M'Naghten "right-wrong" formula is the predominant test for *legal insanity*. The criminal justice system has experimented with broader approaches that resulted in a larger number of defendants being considered legally insane.

- *Irresistible Impulse.* Emotions cause loss of control to conform behavior to the law.
- *Durham Product Test.* The criminal act was the product of a mental disease or defect.
- *Substantial Capacity.* The defendant lacks substantial (not total) capacity to distinguish right from wrong or to conform his or her behavior to the law.

The diminished capacity defense permits defendants to introduce evidence of mental defect or disease to negate a required criminal intent. This typically is limited to murder. Other excuses include:

- *Duress.* A defendant is excused who commits a crime under a reasonable belief that he or she is threatened with imminent and unavoidable serious bodily harm. Duress does not excuse homicide.
- *Infancy.* The common law and various state statutes divide age into three distinct periods. Infancy is an excuse (younger than seven at common law). There is a rebuttable presumption that adolescents in the middle period lack the capacity to form a criminal intent (between seven and fourteen at common law). Individuals older than fourteen are considered to have the same capacity as adults.
- *Intoxication.* Voluntary intoxication is recognized as a defense to a criminal charge requiring a specific intent. The trend is for abolition of the excuse of voluntary intoxication. Involuntary intoxication is a defense where, as a result of alcohol or drugs, the individual meets the standard for legal insanity in the jurisdiction.
- *Mistake.* A mistake of law is never a defense; a mistake of fact may be relied on to demonstrate a lack of a specific criminal intent. Some courts require that the mistake of fact is reasonable.
- *Entrapment.* Entrapment asks whether the government "implanted a criminal intent" in an otherwise innocent individual. The subjective approach to entrapment focuses on the defendant. This requires that the government induce an individual who lacks a criminal predisposition to commit a crime. The objective test centers on the government. This asks whether the government's conduct falls below accepted standards and would have induced an otherwise innocent individual to engage in criminal conduct. Courts have been reluctant to find that the Due Process Clause protects individuals against outrageous governmental misconduct.

The new defenses surveyed illustrate the effort to base excuses on new developments in biology, psychology, and sociology. Critics contend that many of these are "abuse excuses," in which defendants manipulate the law by claiming that they are victims. On the other hand, defendants ask why some traits and conditions are considered to excuse criminal activity while factors such as poverty, inequality, or abuse are not recognized as a defense. The general trend is for the law to limit rather than to expand criminal excuses.

CHAPTER REVIEW QUESTIONS

1. Define and distinguish between the four major approaches to legal insanity.
2. Discuss the purpose of the diminished capacity defense. What is the result of the application of the defense to a defendant charged with a crime requiring a specific intent?
3. Distinguish between the defenses of voluntary and involuntary intoxication.
4. Describe the common law defense of infancy. How has this been modified under contemporary statutes?
5. What are the elements of the duress defense?

6. Discuss the difference between the mistake of law and mistake of fact defenses.

7. What are the two tests of entrapment? How do these two tests differ from one another? Explain the relationship between these two tests for entrapment and the due process approach.

8. Provide some examples of the "new defenses." How do these differ from established criminal law defenses? Do you agree that some of these defenses deserve to be criticized as "abuse excuses"?

9. Outline the debate over the insanity defense and various efforts at reform. Would you favor abolishing the insanity defense?

10. Under what conditions should defendants younger than eighteen be prosecuted as "adult offenders"?

11. Write a brief essay summarizing the law of excuses and stress their common characteristics.

LEGAL TERMINOLOGY

abuse excuse	excuses	irresistible impulse test
civil commitment	guilty but mentally ill (GBMI)	mistake of fact
competence to stand trial		mistake of law
diminished capacity	*ignorantia lexis non excusat*	
duress	infancy	M'Naghten test
Durham product test	insanity defense	substantial capacity test
entrapment	involuntary intoxication	voluntary intoxication

CRIMINAL LAW ON THE WEB

Log on to the Web-based student study site at www.sagepub.com/lippmanccl3e to assist you in completing the Criminal Law on the Web exercises, as well as for additional features such as podcasts, Web quizzes, and video links.

1. Read about the trial of Ralph Tortorici on the Public Broadcasting Web site. Was he entitled to the defense of insanity?

2. Listen to a program on National Public Radio about counterterrorism and entrapment.

3. Read about the impact on juveniles of prosecuting them as adults.

BIBLIOGRAPHY

Joshua Dressler, *Understanding Criminal Law,* 3rd ed. (New York: Lexis, 2001), Chap. 17. A clear and understandable summary of excuses.

George P. Fletcher, *Rethinking Criminal Law* (New York: Oxford University Press, 2000), pp. 798–855. A comprehensive survey of the approaches and the theoretical basis of excuses in various countries.

Abraham S. Goldstein, *The Insanity Defense* (New Haven, CT: Yale University Press, 1967). A leading law professor's summary and evaluation of the insanity defense.

Hyman Gross, *A Theory of Criminal Justice* (New York: Oxford University Press, 1979), pp. 317–342. A good review of the literature calling for an abolition of excuses.

Norval Morris, *Madness and the Criminal Law* (Chicago: University of Chicago Press, 1984). One of the foremost criminal law scholars argues for the abolition of the insanity defense.

10 HOMICIDE

Was Karen killed with extreme and outrageous depravity?

Karen Slattery was stabbed or cut eighteen times. She was alive when all the wounds were inflicted. She was in terror. She undoubtedly had a belief of her impending doom. Her fear and heightened level of anxiety occurred over a period of time. Most important, the defendant told Dr. McKinley Cheshire that fear in his victim was necessary. The defendant stated that causing deliberate pain and fear would increase the flow of female bodily fluids which he needed for himself. The puncturing of Karen Slattery's lung caused her to literally drown in her own blood. She experienced air deprivation. Each of the eighteen cuts, slashes, and/or stab wounds caused pain by penetrating nerve endings in Miss Slattery's body. The crime of murdering Miss Slattery evidenced extreme and outrageous depravity.

Learning Objectives

1. Understand the development of the common law of homicide and the historical distinction between murder and manslaughter.

2. Know the differing views on when life begins for purposes of homicide and the legal tests for determining death.

3. Describe the *actus reus* of homicide.

4. Know the elements of first degree premeditated murder.

5. Understand the characteristics of capital and aggravated first degree murder.

6. Know the difference between first and second degree murder.

7. List the elements of depraved heart murder.

8. State the law of felony murder and compare and contrast the agency theory of felony murder with the proximate cause theory of criminal responsibility for felony murder.

9. Understand the circumstances under which a corporation may be held criminally liable.

10. State the elements of voluntary manslaughter and state the elements of involuntary manslaughter.

11. Explain misdemeanor manslaughter.

INTRODUCTION

Why is homicide considered the most serious criminal offense? What is the reason that it is the only crime subject to the death penalty?

Supreme Court Justice William Brennan noted that in a society that "so strongly affirms the sanctity of life," it is not surprising that death is viewed as the "ultimate" harm. Justice Brennan went on to observe that death is "truly awesome" and is "unusual in its pain, in its finality, and in its enormity. . . . Death, in these respects, is in a class by itself. . . . [It is] degrading to human dignity. It is this regard for life that reminds us to respect one another and to treat each individual with dignity and regard."[1] In *Coker v. Georgia*, Supreme Court Justice Byron White, in explaining why the death penalty is imposed for murder while it is not imposed for rape, noted that the "murderer kills; the rapist, if no more than that, does not. Life is over for the victim of the murderer; for the rape victim, life may not be nearly so happy as it was, but it is not over and normally is not beyond repair."[2]

There are also religious grounds for treating murder as the most serious of crimes. The influential eighteenth-century English jurist William Blackstone observed that murder is a denial of human life and that human life is a gift from God. He stressed that a mere mortal has no right to take a life and to disrupt the divine order of the universe. Professor George Fletcher expands on this notion and explains that in the view of the Bible, a killer was thought to acquire control over the blood of the victim. The execution of the killer was the only way that the blood could be returned to God.[3]

At common law, murder was defined as the unlawful killing of another human being with malice aforethought (we will discuss the meaning of malice aforethought in the next section). Initially the common law did not distinguish between types of **criminal homicide**. The taking of a life was treated equally as serious whether committed intentionally, in the heat of passion, recklessly, or negligently.

The development of the modern law of homicide can be traced to fifteenth-century England. Members of the clergy were prosecuted for homicide before ecclesiastical or religious courts that, unlike royal courts, were not authorized to impose the death penalty. Offenders, instead, were subject to imprisonment for a year, the branding of the thumb, and the forfeiture of goods. Judges in the religious courts gradually expanded the *benefit of clergy* to any individual who could read, in order to avoid the harshness of the death penalty. Defendants who could not read typically claimed the benefit of clergy by memorizing passages from the Bible in order to prove that they were literate.

The English monarchy resisted expanding the power of religious courts and enacted a series of statutes that established the jurisdiction of royal courts over the most atrocious homicides. These statutes denied the benefit of clergy and provided for the death penalty. The royal courts began to distinguish between murder, which was committed with malice aforethought and was not eligible for the benefit of clergy, and manslaughter, which was committed without malice aforethought and was eligible for the benefit of clergy. This distinction persisted even after royal courts asserted jurisdiction over all homicides. Murder under the royal courts was subject to the death penalty unless a royal pardon was issued, whereas manslaughter was viewed as a less serious offense that did not result in capital punishment. Most state statutes continue to recognize the distinction between murder and manslaughter. Over time, judges created several other categories of homicide, a process that culminated in modern homicide statutes.

In this chapter, we will review the distinction between these various grades of criminal homicide. Your challenge is to understand the distinctions between these various types of criminal homicide.

TYPES OF CRIMINAL HOMICIDE

By the eighteenth century, the law recognized four types of homicide:

- **Justifiable Homicide.** This includes self-defense, defense of others, defense of the home, and police use of deadly force.
- **Excusable Homicide.** Murder committed by individuals who are considered to be legally insane, by individuals with a diminished capacity, or by infants.
- **Murder.** All homicides that are neither excused nor justified.
- **Manslaughter.** All homicides without malice aforethought that are committed without justification or excuse.

As we have seen, by the end of the fifteenth century criminal homicide had been divided into murder, or the taking of the life of another with malice aforethought, and manslaughter, or the taking of the life of another without malice aforethought. **Malice aforethought** is commonly defined as an intent to kill with an ill will or hatred. Aforethought requires that the intent to kill is undertaken with a design to kill. The classic example of a plan to kill is murder committed while "lying in wait" for the victim.[4]

The commentary to the Model Penal Code notes that judges gradually expanded malice aforethought to include various types of murder that have little relationship to the original definition. As observed by the Royal Commission on Capital Punishment in England, malice aforethought has come to be a general name for "a number of different mental attitudes which have been variously defined at different stages in the development of the law, the presence of any one of which has been held by the courts to render a homicide particularly heinous and therefore to make it murder."[5]

The Model Penal Code notes that as the common law developed, malice aforethought came to be divided into several different mental states, each of which was subject to the penalty of death. The first is intent to kill or murder. A second category of murder entails knowingly causing grievous or serious bodily harm. A third category of murder is termed **depraved heart murder** or killing committed with extreme recklessness or negligence. This involves a "depraved mind" or an "abandoned and malignant heart" and entails a wanton and willful disregard of an unreasonable human risk. A fourth category involves an intent to resist a lawful arrest. There is one additional category of murder committed with malice aforethought. This is murder committed during a felony, which today is termed **felony murder**.

Remember that murder requires a demonstration of malice. The Nevada criminal code states that murder is the "unlawful killing of a human being, with malice aforethought, either express or implied. . . . The unlawful killing may be effected by any of the various means by which death may be occasioned."[6] Individuals who have a deliberate intent to kill possess *express malice*. An *implied malice* exists in those cases that an individual possesses an intent to cause great bodily harm or the intent to commit an act that may be expected to lead to death or great bodily harm. Nevada defines express malice as a "deliberate intention unlawfully to take away the life of a fellow creature which is manifested by external circumstances." The Nevada law goes on to provide that malice may be implied "when all the circumstances of the killing show an abandoned and malignant heart."[7]

What about manslaughter? The common law of manslaughter developed into two separate categories. The first entails an intentional killing committed without malice in the heat of passion upon adequate provocation. Murder was also considered manslaughter when it was committed without malice as a result of conduct that was insufficiently reckless or negligent to be categorized as depraved heart murder. Courts typically describe the first category as **voluntary manslaughter** and the second as **involuntary manslaughter**.

In 1794, Pennsylvania adopted a statute creating separate grades of murder and manslaughter that continue to serve as the foundation for a majority of state statutes today. The Pennsylvania statute divided homicide into two separate categories and limited the death penalty to first-degree murder, the most serious form of homicide. Second-degree murder was punishable by life imprisonment.

> All murder, which shall be perpetrated by means of poison, by lying in wait, or by any other kind of willful, deliberate, or premeditated killing or which shall be committed in perpetration or attempt to perpetrate any arson, rape, robbery, or burglary shall be deemed murder in the first degree; and all other kinds of murder shall be deemed murder in the second degree.[8]

Modern state statutes typically divide murder into first- and second-degree murder, both of which require the prosecutor to establish intent and malice. **First-degree murder** is the most serious form of murder, and the prosecutor has the burden of establishing **premeditation and deliberation**. This involves demonstrating that the defendant reflected for at least a brief period of time before intentionally killing another individual. **Second-degree murder** usually includes all murders not involving premeditation and deliberation. Manslaughter typically comprises an additional grade or grades of homicide. These general categories are described in the following list, starting with the most serious degree of homicide. Keep in mind that state statutes differ widely in their approach to defining homicide.[9]

- *First-Degree Murder.* Premeditation and deliberation and murder committed in the perpetration of various dangerous felonies. Some statutes explicitly include the killing of a police officer and murder committed while lying in wait or as a result of torture or poison.
- *Second-Degree Murder.* Killing with malice and without premeditation. This may include a death resulting from an intent to cause serious bodily harm and reckless, depraved heart murders.
- *Voluntary Manslaughter.* Murder in the heat of passion.
- *Involuntary Manslaughter.* Gross negligence.

Some states also single out vehicular manslaughter as a special form of involuntary manslaughter.

ACTUS REUS AND CRIMINAL HOMICIDE

State statutes define the *actus reus* of criminal homicide as the "unlawful killing of a human being" or "causing the death of a person." This may involve an infinite variety of acts, including shooting, stabbing, choking, poisoning, beating with a bat or axe, and "a thousand other forms of death."[10] Homicides can also be carried out without landing a single blow. A wife, for instance, was found to have engaged in a pattern of constant criticism and threats of violence against her husband, weakening him mentally and worsening his heart condition. He died after she coerced him into walking in the deep snow.[11] In another case, a husband was held criminally liable for the murder of his young wife when he threatened to beat her unless she jumped into a stream that subsequently carried her away in the current.[12]

There are two preliminary issues that we must address before examining the various grades of homicide. The first is at what point does life begin for purposes of homicide? This is important in determining whether an assailant can be held criminally liable for the death of a fetus. The question at the opposite end of the scale is at what point does life end? What if a grieving son and daughter "pull the plug" on a sick parent whose brain no longer functions and who is being kept artificially alive on a life support machine?

The Legal Equation

Criminal homicide **=** Unlawful killing of human being

+ purposely or knowingly or recklessly or negligently.

THE BEGINNING OF HUMAN LIFE

We have seen that murder entails the killing of a human being. At what point does life begin? The common law rule adopted in 1348 provided that a defendant was not criminally responsible for the murder of a child in a mother's womb unless the child was born alive with the capacity for an independent existence. Following the child's birth, the question was whether the defendant's acts were the proximate cause of death. This rule reflected the fact that doctors were unable to determine whether an unborn child was alive in the mother's womb at the time of an attack.

The common law rule has been abandoned in most states in the last few decades in favor of a rule that imposes criminal liability when the prosecution is able to establish beyond a reasonable doubt that the fetus was viable, meaning that it was capable of living separate and apart from the mother. In *Commonwealth v. Cass,* in 1984, the Massachusetts Supreme Judicial Court ruled that the "infliction of prenatal injuries resulting in the death of a viable fetus, before or after it is born, is homicide. . . . We believe that our criminal law should extend its protection to viable fetuses."[13]

Keep in mind that the U.S. Supreme Court recognized in *Roe v. Wade,* in 1973, that women have a right to an abortion as part of their constitutional right to privacy. A state may limit this right during the last phase of pregnancy, other than in those instances where an abortion is necessary to protect the health or life of the mother. The decision to hold an attacker criminally responsible for the death of a viable fetus does not limit the right of a woman to voluntarily consent to an abortion by a licensed physician. The Model Penal Code maintains the common law "born alive" rule in order to avoid a possible conflict between the law of abortion and the criminal law rule concerning the fetus.[14]

In *People v. Davis,* the California Supreme Court considered whether to limit criminal liability for the murder of a fetus to viability or to extend criminal liability to the postembryonic stage of pregnancy, seven or eight weeks following fertilization.

Did Davis murder a nonviable fetus?

People v. Davis, 872 P.2d 591 (Cal. 1994). Opinion by: Lucas, C.J.

California Penal Code (CPC) section 187, subdivision (a), provides that "Murder is the unlawful killing of a human being, or a fetus, with malice aforethought." In this case, we consider and reject the argument that viability of a fetus is an element of fetal murder under the statute.

Facts

On March 1, 1991, Maria Flores, who was between twenty-three and twenty-five weeks pregnant, and her twenty-month-old son, Hector, went to a check-cashing store to cash her welfare check. As Flores left the store, defendant pulled a gun from the waistband of his pants and demanded the money ($378) in her purse. When she refused to hand over the purse, defendant shot her in the chest. Flores dropped Hector as she fell to the floor and defendant fled the scene.

Flores underwent surgery to save her life. Although doctors sutured small holes in the uterine wall to prevent further bleeding, no further obstetrical surgery was undertaken because of the immaturity of the fetus. The next day, the fetus was stillborn as a direct result of its mother's blood loss, low blood pressure and state of shock. Defendant was soon apprehended and charged with assaulting and robbing Flores, as well as murdering her fetus. The prosecution charged a special circumstance of robbery-murder.

At trial, the prosecution's medical experts testified the fetus's statistical chances of survival outside the womb were between seven and forty-seven percent. The defense medical expert testified it was "possible for the fetus to have survived, but its chances were only two or three percent." None of the medical experts testified that survival of the fetus was "probable."

Although CPC section 187, subdivision (a), does not expressly require a fetus be medically viable before the statute's provisions can be applied to a criminal defendant, the trial court followed several Court of Appeal decisions and instructed the jury that it must find the fetus was viable before it could find defendant guilty of murder under the statute. The trial court did not, however, give the standard viability instruction, CALJIC No. 8.10, which states: "A viable human fetus is one who has attained such form and development of organs as to be normally capable of living outside of the uterus." The jury, however, was given an instruction that allowed it to convict defendant of murder if it found the fetus had a possibility of survival: "A fetus is viable when it has achieved the capability for independent existence; that is, when it is possible for it to survive the trauma of birth, although with artificial medical aid."

The jury convicted defendant of murder of a fetus during the course of a robbery, assault with a firearm, and robbery. The jury found that, in the commission of each offense, defendant personally used a firearm. The jury found true the special circumstance allegation. Accordingly, because the prosecutor did not seek the death penalty, defendant was sentenced to life without possibility of parole, plus five years for the firearm use.

Issue

On appeal, defendant contended that the trial court prejudicially erred by not instructing the jury pursuant to CALJIC No. 8.10. . . . [D]efendant claimed, rather than defining viability as a "reasonable possibility of survival," the trial court should have instructed the jury under the higher "probability" threshold described in CALJIC No. 8.10.

The People argued that no viability instruction was necessary because prosecution under CPC section 187, subdivision (a), does not require that the fetus be viable. After reviewing the wording of section 187, subdivision (a), its legislative history, the treatment of the issue in other jurisdictions, and scholarly comment on the subject, the Court of Appeal agreed with the People that contrary to prior California decisions, fetal viability is not a required element of murder under the statute. . . .

As explained below, we agree with the People and the Court of Appeal that viability is not an element of fetal murder under CPC section 187, subdivision (a), and conclude therefore that the statute does not require an instruction on viability as a prerequisite to a murder conviction. In addition, because every prior decision that had addressed the viability issue had determined that viability of the fetus was prerequisite to a murder conviction under section 187, subdivision (a), we also agree . . . that application of our construction of the statute to defendant would violate due process and ex post facto principles.

Reasoning

In 1970, CPC section 187, subdivision (a), provided: "Murder is the unlawful killing of a human being, with malice aforethought." In *Keeler v. Superior Court,* 470 P.2d 617 (Cal. 1970), a majority of the court held that a man who had killed a fetus carried by his estranged wife could not be prosecuted for murder because the

Legislature (consistent with the common law view) probably intended the phrase "human being" to mean a person who had been born alive.

The Legislature reacted to the *Keeler* decision by amending the murder statute . . . CPC subdivision (a), to include within its proscription the killing of a fetus. The amended statute reads: "Murder is the unlawful killing of a human being, or a fetus, with malice aforethought." The amended statute specifically provides that it does not apply to abortions complying with the Therapeutic Abortion Act, performed by a doctor when the death of the mother was substantially certain in the absence of an abortion, or whenever the mother solicited, aided, and otherwise chose to abort the fetus.

The legislative history of the amendment suggests the term "fetus" was deliberately left undefined after the Legislature debated whether to limit the scope of statutory application to a viable fetus. The Legislature was clearly aware that it could have limited the term "fetus" to "viable fetus," for it specifically rejected a proposed amendment that required the fetus be at least twenty weeks in gestation before the statute would apply.

In 1973, the United States Supreme Court issued a decision that balanced a mother's constitutional privacy interest in her body against a state's interest in protecting fetal life, and determined that in the context of a mother's abortion decision, the state had no legitimate interest in protecting a fetus until it reached the point of viability, or when it reached the "capability of meaningful life outside the mother's womb." *Roe v. Wade,* 410 U.S. 113 (1973). The court explained that "[v]iability is usually placed at about seven months (28 weeks) but may occur earlier, even at 24 weeks." At the point of viability, the court determined, the state may restrict abortion.

Thereafter . . . the Court of Appeal construed the term "fetus" in CPC section 187, subdivision (a), to mean a "viable fetus" as defined by *Roe v. Wade.* . . . Defendant asserts that section 187, subdivision (a), has no application to a fetus not meeting *Roe v. Wade*'s definition of viability. Essentially, defendant claims that because the fetus could have been legally aborted under *Roe v. Wade,* at the time it was killed, it did not attain the protection of section 187, subdivision (a) and he therefore cannot be prosecuted. . . .

But *Roe v. Wade* does not hold that the state has no legitimate interest in protecting the fetus until viability. . . . As observed by one commentator: . . . The *Roe* decision, therefore, forbids the state's protection of the unborn's interests only when these interests conflict with the constitutional rights of the prospective parent. The Court did not rule that the unborn's interests could not be recognized in situations where there was no conflict. Thus when the state's interest in protecting the life of a developing fetus is not counterbalanced against a mother's privacy right to an abortion or other equivalent interest, the state's interest should prevail.

We conclude, therefore, that when the mother's privacy interests are not at stake, the Legislature may determine whether, and at what point, it should protect life inside a mother's womb from homicide. Here, the Legislature determined that the offense of murder includes the murder of a fetus with malice aforethought. Legislative history suggests "fetus" was left undefined in the face of divided legislative views about its meaning. Generally, however, a fetus is defined as "the unborn offspring in the postembryonic period, after major structures have been outlined." This period occurs in humans "seven or eight weeks after fertilization" and is a determination to be made by the trier of fact. Thus, we agree with the above cited authority that the Legislature could criminalize murder of the postembryonic product without the imposition of a viability requirement. . . .

As the Court of Appeal below observed, the wording of CALJIC No. 8.10, defining viability as "normally capable of living outside of the uterus," while not a model of clarity, suggests a better than even chance—a probability—that a fetus will survive if born at that particular point in time. By contrast, the instruction given below suggests a "possibility" of survival, and essentially amounts to a finding that a fetus incapable of survival outside the womb for any discernible time would nonetheless be considered "viable" within the meaning of CPC section 187, subdivision (a). Because the instruction given by the trial court substantially lowered the viability threshold as commonly understood and accepted . . . we conclude that the trial court erred in instructing the jury pursuant to a modified version of CALJIC No. 8.10.

The question then is whether it is reasonably probable a result more favorable to defendant would have been reached absent the instructional error. The record shows the weight of the medical testimony was against the probability of the fetus being viable at the point it was killed. Defendant's medical expert opined that it was "possible" for the fetus to have survived the trauma of an early birth, but that its chances for survival were about two or three percent. . . . [N]one of the medical experts who testified at defendant's trial believed that the fetus had a "probable" chance of survival. Accordingly, because the evidence on the issue of viability erroneously supported the concept of the "possibility" of survival, and the jury was then instructed that viability means "possible survival," the jury was misinformed that it could find the fetus was viable before it "attained such form and development of organs as to be normally capable of living outside the uterus." Had the jury been given CALJIC No. 8.10, it is reasonably probable it would have found the fetus not viable. We conclude, therefore, that defendant was prejudiced by the instructional error and the conviction of fetal murder must be reversed.

Holding

We conclude that viability is not an element of fetal homicide under CPC section 187, subdivision (a). The third-party killing of a fetus with malice aforethought is murder under section 187, subdivision (a), as long as the state can show that the fetus has progressed beyond the embryonic stage of seven to eight weeks.

We also conclude that our holding should not apply to defendant and that the trial court committed prejudicial error by instructing the jury pursuant to a modified version of CALJIC No. 8.10. We therefore affirm the judgment of the Court of Appeal.

Dissenting, *Mosk, J.*

I dissent. I believe the Legislature intended the term "fetus" in its 1970 amendment to California Penal Code section 187 to mean a viable fetus. . . . The statutory language in issue here—the 1970 amendment to CPC section 187 extending the crime of murder to the killing of "a fetus"—was itself enacted in direct and vigorous response to a judicial opinion (*Keeler*) with which the Legislature disagreed. If the Legislature had also disagreed a few years later with subsequent judicial opinions limiting the statutory prohibition against killing "a fetus" to the killing of a viable fetus, surely it would have spoken again, and equally vigorously. To this day, however, the Legislature has remained silent and taken no remedial action. In these circumstances its acquiescence is persuasive evidence of its intent. . . . .

Having erroneously concluded that the Legislature had no intent with respect to the meaning of the key word "fetus" in the 1970 amendment to CPC section 187, the lead opinion proceeds to legislate on the subject by supplying the assertedly missing definition: "[A] fetus," says the lead opinion, "is defined as 'the unborn offspring in the postembryonic period, after major structures have been outlined.' This period occurs in humans 'seven or eight weeks after fertilization.' . . ." The lead opinion repeats its new definition in concluding that the malicious killing of a fetus is murder under section 187 as long as the state can show the fetus has progressed "beyond the embryonic stage of seven to eight weeks." . . . [I]t is highly unlikely that such was the Legislature's intent.ssss

Yet that is the least of the problems with the lead opinion's new definition of "fetus" in CPC section 187. Because liability after seven weeks necessarily includes liability after eight weeks, we may fairly assume that prosecutors faced with the lead opinion's imprecise definition will opt for the more inclusive figure and charge murder when the fetal death occurs at seven weeks. Do my colleagues have any idea what a seven-week-old product of conception looks like?

To begin with, it is tiny. At seven weeks its "crown-rump length"—the only dimension that can be accurately measured—is approximately 17 millimeters, or slightly over half an inch. It weighs approximately three grams, or about one-tenth of an ounce. In more familiar terms, it is roughly the size and weight of a peanut.

If this tiny creature is examined under a magnifying glass, moreover, its appearance remains less than human. Its bulbous head takes up almost half of its body and is bent sharply downward; its eye sockets are widely spaced; its pug-like nostrils open forward; its paddle-like hands and feet are still webbed; and it retains a vestigial tail. . . . And as concluded in the Comment relied on by the lead opinion, "A being so alien to what we know to be human beings seems hardly worth being made the subject of murder." . . .

The contrast between such a tiny, alien creature and the fully formed "5-pound, 18-inch, 34-week-old, living, viable child" in *Keeler* is too obvious to be ignored. I can believe that by enacting the 1970 amendment the Legislature intended to make it murder to kill a fully viable fetus like Teresa Keeler's baby. But I cannot believe the Legislature intended to make it murder—indeed, capital murder—to cause the death of an object the size of a peanut.

[U]nder the lead opinion's definition a person may be subject to a conviction of capital murder for causing the death of an object that was literally invisible to everyone, and hence that the person had no reason to know even existed. A woman whose reproductive system contains an immature fetus a fraction of an inch long and weighing a fraction of an ounce does not, of course, appear pregnant. In fact, if she is one of many women with some irregularity in her menstrual cycle, she herself may not know she is pregnant: "quickening" does not occur until two or three months later. Unless such a woman knows she is pregnant and has disclosed that fact to the defendant, the defendant has no way of knowing she is carrying a fetus.

Nor is this problem limited to fetuses that are "seven or eight" weeks old. Although the length of time that a woman can be pregnant without her condition's becoming noticeable varies according to such factors as her height and weight, the size of her fetus, and even the style of her clothing, the case at bar demonstrates that it can extend well into her pregnancy. Here Flores testified that in her opinion her pregnancy "showed" on the date of the shooting, March 1, 1991; but defendant testified to the contrary, and there was persuasive evidence to support him. . . . Thomas Moore, M.D., an experienced perinatologist, testified that in his opinion it is "not likely" that on the date of the shooting a woman of Flores's stature would have showed her pregnancy when clothed and standing upright. . . .

Yet the expert testimony agreed that Flores was between twenty-three and twenty-five weeks'—approximately six months'—pregnant on the date of the shooting. This is the very threshold of viability: an expert witness reported on a recent study showing that at twenty-three weeks the survival rate of the fetus is approximately seven percent, at twenty-four weeks thirty-five percent, and at twenty-five weeks forty-seven percent. The case at bar thus demonstrates how long the risk of liability for fetal murder may run under the lead opinion's view before the actor either knows or has reason to know that the victim of the offense even exists. I cannot believe the Legislature intended such an enlargement of liability for the crime of capital murder. . . .

Finally, the lead opinion's construction of the 1970 amendment will make our murder law unique in the nation in its severity: It appears that in no other state is it a capital offense to cause the death of a viable and invisible fetus that the actor neither knew nor had reason to know existed.

To begin with, in the majority of states the killing of a fetus is not a homicide in any degree: "The majority of jurisdictions which have confronted the issue has followed *Keeler* . . . in holding the term 'fetus' does not fall within the definition of a human being under criminal statutes unless the term is so defined by the legislature." . . . In those jurisdictions a live birth remains a prerequisite to a conviction of homicide.

There are, of course, jurisdictions that have enacted statutes criminalizing the killing of a fetus. The lead opinion cites seven such jurisdictions. My research has turned up at least twenty-two, with two additional jurisdictions so holding as a matter of common law. For convenience I have grouped these jurisdictions into three distinct categories.

First, in at least thirteen jurisdictions the killing of a fetus is not criminal unless the fetus is viable or has reached a gestational age significantly more advanced than the "seven or eight weeks" prescribed by the lead opinion (in seven of these states, the crime is manslaughter rather than murder).

The second category of jurisdictions is composed of those in which the legislature has expressly declared that the killing of a product of conception is criminal regardless of its gestational age; contrary to our CPC section 187, therefore, these statutes purport to apply to both viable and nonviable fetuses and to embryos—even to zygotes. There are at least six states in this category. In three the crime is not murder but a lesser offense. Thus Arizona's statute is modeled on those of the eight states discussed above that criminalize the killing of an unborn child by means of an injury to its mother; unlike those statues, the Arizona measure applies not just to a "quick" unborn child but to an unborn child "at any stage of its development"; like those statutes, however, the Arizona offense is deemed manslaughter. . . . In Arkansas the offense is deemed first-degree battery and is punishable by imprisonment for not less than five nor more than twenty years; in New Mexico the offense is

called "injury to [a] pregnant woman" and is punishable by imprisonment for three years with a possible fine of $5,000. . . .

The third and last category of jurisdictions is composed of those in which the statute neither prescribes a minimum gestational age for a conviction of fetal murder nor expressly declares that it applies regardless of gestational age; rather, the statute is facially silent on the matter. California is such a jurisdiction, and there are at least five others. Of these Utah is the only state that, like California, criminalizes the killing of a fetus under its general murder statutes: It defines homicide as the killing of "another human being, including an unborn child." But unlike California, in Utah the crime is a capital offense only if the actor caused the death of the unborn child "intentionally or knowingly," even in a felony-murder case. If, as in the case at bar, the death occurred in the commission of a listed felony but the actor did not kill "intentionally or knowingly," the crime is non-capital murder punishable by imprisonment for not less than five years.

In the remaining four states the offense is given special treatment and is punished much less severely than in California. In two of these states the offense is deemed "feticide." Thus in Indiana one who "knowingly or intentionally" terminates a pregnancy commits feticide, punishable by imprisonment for four years with a possible fine of not more than $10,000. In Louisiana one who kills an unborn child intentionally or in the commission of a listed felony commits first-degree feticide punishable by imprisonment for not more than fifteen years. In South Dakota one who "intentionally kills a human fetus by causing an injury to its mother" commits a felony punishable by imprisonment for ten years with a possible fine of $10,000. And in New Hampshire one who "[p]urposely or knowingly causes injury to another resulting in miscarriage or stillbirth" commits first-degree assault punishable by imprisonment for not more than fifteen years with a possible fine not to exceed $4,000.

Robert Keeler's act of assaulting his estranged wife for the express purpose of terminating her pregnancy by knowingly and intentionally killing her fully viable fetus would have been a crime in all the jurisdictions discussed above that have abrogated the common law rule. It would certainly be a crime in California today. But I cannot believe that in amending CPC section 187 to make that act a crime, the Legislature also intended to make California the only state in the Union in which it is a capital offense to cause the death of a nonviable and invisible fetus that the actor neither knew nor had reason to know existed. Yet this, again, is where the lead opinion's construction of the 1970 amendment inexorably takes us. I dissent from that construction. . . .

Questions for Discussion

1. Why is the wording of the jury instruction significant in this case? How does it impact on the defendant's guilt or innocence? What is the holding of the California Supreme Court?

2. Courts are criticized for "judicial legislation," making the law rather than interpreting statutes passed by the legislative branch. In interpreting the term "fetus," did the California Supreme Court follow the intent of the California legislature or impose its own view?

3. What is the relationship between *Davis* and *Keeler v. Superior Court?* How does the California Supreme Court distinguish the decision in *Davis* from the U.S. Supreme Court holding in *Roe v. Wade?*

4. Judge Mosk argues that the defendant could not have reasonably been aware that Maria Flores was pregnant. Why is this significant?

5. How does the approach of the California Supreme Court in defining a fetus differ from other states? In your opinion, is the California Supreme Court being directed by science, religion, or politics? Do you agree with the decision of the California Supreme Court?

Cases and Comments

1. **Killing of a Fetus That Would Not Have Survived.** In *People v. Valdez,* defendants Elisio Valdez and Johnnie Ray Peraza were convicted of the murders of Andrea Mestas and her fetus and various other serious felonies. Mestas was shot in the chest at close range. The bullet penetrated Andrea's heart and killed Andrea and her sixteen- to seventeen-week-old male fetus. The prosecutor theorized that the defendants were ordered to kill Mestas and her boyfriend on the orders of a prison gang, Nuestra Familia. The gang allegedly viewed Mestas as a "rat" and a "snitch." The two defendants received multiple life sentences as well as additional prison time. A medical examination of the fetus revealed that there was "chronic inflammation of the implantation site where the placenta attaches to the uterine wall as well as acute inflammation of the membrane surrounding the fetus." As a result, doctors concluded that it was "unlikely that the fetus would have survived." The defendants appealed on the grounds of "survivability," meaning that they should have been acquitted based on the fact that it was unlikely that the fetus would have completed gestation and been born in any event. In other words, the "killing of a non-survivable" fetus is "not comparable to murder; it is a much less serious offense because the non-survivable fetus is not a potential human life." The Court of Appeal of California, Third Appellate District, ruled that "just as the state may penalize an act that unlawfully shortens the existence of a terminally ill human being, it may penalize an act that unlawfully shortens the existence of a fetus which later would have perished before birth due to natural causes." Was the fetus a "potential life"? Do you agree with the court's decision? See *People v. Valdez,* 23 Cal. Rptr. 3d 909 (Cal. Ct. App. 2005).

2. **Alcohol Abuse and the Attempted Murder of a Fetus.** Deborah J.Z. was drinking in a local bar one week before her due date. She believed that she was about to give birth and called her mother, who drove her to the hospital. Deborah was "uncooperative, belligerent at times and very intoxicated." Her blood alcohol concentration exceeded 0.30%. She reportedly told the nurse that "if you don't keep me here, I'm just going to go home and keep drinking and drink myself to death and I'm going to kill this thing because I don't want it anyways." Deborah also expressed anxiety concerning the baby's "race, an abusive relationship she was in, and the pain of giving birth." Deborah consented to a cesarean section and gave birth to a baby girl, M.M.Z. M.M.Z. was extremely small, "she had no subcutaneous fat and her physical features—mild dysmorphic abnormalities—presented fetal alcohol effects." M.M.Z.'s blood alcohol level was 0.119%. The baby recovered after several weeks. Deborah was subsequently charged with attempted first-degree murder and first-degree reckless injury. Deborah appealed the trial court's denial of her motion to dismiss the charges.

A Wisconsin appellate court rejected the prosecution's contention that the murder statute's punishment of an individual who caused the death of a "human being" or who caused great bodily harm to a "human being" included a fetus. The court noted that Wisconsin law defined a human being as "one who has been born alive." A broad interpretation of the statute, noted the court, would risk criminal charges against a woman whose behavior during a pregnancy placed an unborn child at risk. A woman under these circumstances might be reluctant to seek prenatal care, fearing that she may be accused of endangering her unborn child. The court also suggested that prosecuting a woman for the treatment of her unborn fetus threatened to burden a woman's right to seek an abortion.

The Wisconsin court concluded that the decision whether to extend protection to an "unborn child" was a matter to be decided by the state legislature. See *State v. Deborah J.Z.,* 596 N.W.2d 490 (Wis. Ct. App. 1999). The South Carolina Supreme Court, on the other hand, ruled that a woman was properly convicted of child neglect who caused her baby to be born with cocaine metabolites in its system by reason of her ingestion of crack cocaine during the third trimester of her pregnancy. See *Whitner v. State,* 492 S.E.2d 777 (S.C. 1997). The Maryland Court of Appeals determined that a pregnant woman may not be held criminally liable for negligent endangerment of a fetus. The court reasoned that holding a woman responsible would potentially subject her to prosecution for a range of activities, including drinking alcohol or smoking in moderation as well as skiing or riding horses. See *Kilmon v. State,* 905 A.2d 306 (Md. 2006).

You Decide

10.1 Kurr killed her boyfriend, Antonio Pena, with a knife. She argued with him over cocaine use, and Pena punched her twice in the stomach. She warned him not to hit her since she was pregnant with their babies. Pena came toward her once again, and she killed him. Kurr claimed the right to "intervene" to protect the fetuses that were sixteen or seventeen weeks in gestation. Will this defense prove successful? See *People v. Kurr,* 654 N.W.2d 651 (Mich. Ct. App. 2002).

You can find the answer at www.sagepub.com/lippmanccl3e.

THE END OF HUMAN LIFE

The question of when life ends seems like a technical debate that should be the concern of doctors and philosophers rather than lawyers and criminal justice professionals.

The traditional definition of death required the total stoppage of the circulation of the blood and the cessation of vital functions, such as breathing. This definition was complicated by technology that, over the last decades, developed to the point that a "brain dead" individual's breathing and blood flow could be maintained through artificial machines despite the fact that the brain had ceased to function.

In 1970, Kansas became the first state to legislate that death occurs when an individual experiences an irreversible cessation of breathing and heartbeat or there is an absence of brain activity. A majority of state legislatures and courts now have adopted a **brain death test** for death.[15] The circulatory and respiratory and brain death tests are incorporated as alternative approaches in the Uniform Determination of Death Act, a model law developed by the American Bar Association and American Medical Association.

The brain death test has also been adopted by courts in states without a statute defining death. In the Arizona case of *State v. Fierro,* the deceased, Victor Corella, was shot in the chest and head by a rival gang member. Corella was rushed to the hospital where he was operated on and, although his brain had ceased to function, he was placed on a life support system. The doctors, convinced that nothing could be done to save Corella's life, removed him from the life support machine after four days. The defendant argued that the removal of Corella from the life support machine was the proximate cause of death. The Arizona Supreme Court ruled that under Arizona law, death could be shown by either a lack of bodily function or brain death and concluded that the victim was legally dead before being placed on life support.[16]

As we noted in discussing causation, the year-and-a-day rule provides that an individual is criminally responsible only for a death that occurs within one year of his or her criminal act. This common law standard is still followed in several states. Note that in these states, under the traditional vital function test, a defendant could not be held legally responsible for the death of an individual who is maintained on a life support machine for longer than a year.

MENS REA AND CRIMINAL HOMICIDE

The *mens rea* of criminal homicide encompasses all of the mental states that we discussed in Chapter 5. The Utah criminal code provides that an individual commits criminal homicide "if he intentionally, knowingly, recklessly, with criminal negligence" or acting with the "mental state . . . specified in the statute defining the offense, causes the death of another human being, including an unborn child at any stage of development."[17] Two forms of criminal homicide that we will review later in the chapter, felony murder and misdemeanor manslaughter, involve strict liability. The **grading**, or assignment of degrees to homicide, is based on a defendant's criminal intent. As we shall see, an individual who kills as a result of premeditation and deliberation is considered more dangerous and morally blameworthy than an individual who kills as a result of a reckless disregard or negligence.

MURDER

We have seen that murder is the unlawful killing of an individual with malice aforethought. Several types of murder are discussed in this chapter, and your challenge is to learn the difference between each of these categories of homicide:

- First-degree murder
- Capital and aggravated first-degree murder
- Second-degree murder
- Depraved heart murder
- Felony murder
- Corporate murder

First-Degree Murder

First-degree murder is the most serious form of homicide and can result in the death penalty in thirty-three states and the federal government. (Two other states have abolished the death penalty; but eleven inmates remain on death row in Connecticut, and two in New Mexico.)

The *mens rea* of first-degree murder requires deliberation and premeditation as well as malice. Premeditation means the act was thought out prior to committing the crime. Deliberation entails an intent to kill that is carried out in a cool state of mind in furtherance of the design to kill. An intent to kill without deliberation and premeditation is second-degree murder.

Why is first-degree murder treated more seriously than other forms of homicide? First, an individual who is capable of consciously devising a plan to take the life of another obviously poses a threat to society. A harsh punishment is both deserved and may deter others from cold and calculated killings. Some commentators dispute whether a deliberate and premeditated murderer poses a greater threat than the impulsive individual who lacks self-control and may explode at any moment in reaction to the slightest insult. Assuming that you were asked to formulate a sentencing scheme, which of these two killers would you punish most severely?

The general rule is that premeditation may be formed in the few seconds it takes to pull a trigger or deliver a fatal blow. A West Virginia court observed that the "mental process necessary to constitute 'willful, deliberate and premeditated' murder can be accomplished very quickly or even in the proverbial 'twinkling of an eye.'"[18]

A small number of judges continue to resist the trend toward recognizing that a premeditated intent to kill need only exist for an instant. These jurists point out that unless the prosecution is required to produce proof of premeditation and deliberation, it is difficult to tell the difference between first- and second-degree murder. In *State v. Bingham*, the Washington Supreme Court reversed a defendant's conviction for aggravated first-degree murder. The court held that the fact that the assailant choked the female victim for three to five minutes during an act of sexual intercourse did not constitute sufficient evidence that the defendant premeditated and deliberated the victim's death. In the view of the court, manual strangulation alone is not sufficient to support a finding of premeditation. The defendant might have placed his hand around the victim's neck to quiet her and there also is a question whether the defendant had the capacity to deliberate while engaged in sexual activity.[19] What evidence might have established premeditation? In order to establish premeditation and deliberation, judges generally require either evidence of planning or evidence that the defendant possessed a motive to kill and that the killing was undertaken in a fashion that indicates that it was planned, such as "lying in wait" or the use of a bomb or poison.[20]

State v. Forrest raises the issue of whether all killings involving premeditation and deliberation should be harshly punished.

Should the defendant be held responsible for the premeditated and deliberate killing of his father?

State v. Forrest, 362 S.E.2d 252 (N.C. 1987). Opinion by: Meyer, J.

Defendant was convicted of the first-degree murder of his father, Clyde Forrest. The . . . defendant was sentenced . . . to life imprisonment. In his appeal to this Court, defendant brings forward three assignments of error . . . [W]e find no error in defendant's trial. We therefore leave undisturbed defendant's conviction and life sentence.

Facts

The facts of this case are essentially uncontested, and the evidence presented at trial tended to show the following series of events. On 22 December 1985, defendant John Forrest admitted his critically ill father, Clyde Forrest, Sr., to Moore Memorial Hospital. Defendant's father, who had previously been hospitalized, was suffering from numerous serious ailments, including severe heart disease, hypertension, a thoracic aneurysm, numerous pulmonary emboli, and a peptic ulcer. By the morning of 23 December 1985, his medical condition was determined to be untreatable and terminal. Accordingly, he was classified as "No Code," meaning that no extraordinary measures would be used to save his life, and he was moved to a more comfortable room.

On 24 December 1985, defendant went to the hospital to visit his ailing father. No other family members were present in his father's room when he arrived. While one of the nurse's assistants was tending to his father, defendant told her, "There is no need in doing that. He's dying." She responded, "Well, I think he's better." The nurse's assistant noticed that defendant was sniffling as though crying and that he kept his hand in his pocket during their conversation. She subsequently went to get the nurse.

When the nurse's assistant returned with the nurse, defendant once again stated his belief that his father was dying. The nurse tried to comfort defendant, telling him, "I don't think your father is as sick as you think he is." Defendant, very upset, responded, "Go to hell. I've been taking care of him for years. I'll take care of him." Defendant was then left alone in the room with his father.

Alone at his father's bedside, defendant began to cry and to tell his father how much he loved him. His father began to cough, emitting a gurgling and rattling noise. Extremely upset, defendant pulled a small pistol from his pants pocket, put it to his father's temple, and fired. He subsequently fired three more times and walked out into the hospital corridor, dropping the gun to the floor just outside his father's room.

Following the shooting, defendant, who was crying and upset, neither ran nor threatened anyone. Moreover, he never denied shooting his father and talked openly with law enforcement officials. Specifically, defendant made the following oral statements: "You can't do anything to him now. He's out of his suffering." "I killed my daddy." "He won't have to suffer anymore." "I know they can burn me for it, but my dad will not have to suffer anymore." "I know the doctors couldn't do it, but I could." "I promised my dad I wouldn't let him suffer."

Defendant's father was found in his hospital bed, with several raised spots and blood on the right side of his head. Blood and brain tissue were found on the bed, the floor, and the wall. Though defendant's father had been near death as a result of his medical condition, the exact cause of the deceased's death was determined to be the four point-blank bullet wounds to his head. Defendant's pistol was a single-action .22-calibre five-shot revolver. The weapon, which had to be cocked each time it was fired, contained four empty shells and one live round.

At the close of the evidence, defendant's case was submitted to the jury for one of our possible verdicts: first-degree murder, second-degree murder, voluntary manslaughter, or not guilty. After a lengthy deliberation, the jury found defendant guilty of first-degree murder. Judge Cornelius accordingly sentenced defendant to the mandatory life term. . . .

Issue

In his second assignment of error . . . defendant argues that the trial court's submission of the first-degree murder charge was improper because there was insufficient evidence of premeditation and deliberation presented at trial. We do not agree, and we therefore overrule defendant's assignment of error. . . .

Reasoning

First-degree murder is the intentional and unlawful killing of a human being with malice and with premeditation and deliberation. Premeditation means that the act was thought out beforehand for some length of time, however short, but no particular amount of time is necessary for the mental process of premeditation. Deliberation means an intent to kill, carried out in a cool state of blood, in furtherance of a fixed design for revenge or to accomplish an unlawful purpose and not under the influence of a violent passion, suddenly aroused by lawful or just cause or legal provocation. The phrase "cool state of blood" means that the defendant's anger or emotion must not have been such as to overcome his reason.

Premeditation and deliberation relate to mental processes and ordinarily are not readily susceptible to proof by direct evidence. Instead, they usually must be proved by circumstantial evidence. Among other circumstances to be considered in determining whether a killing was with premeditation and deliberation are: (1) want of provocation on the part of the deceased; (2) the conduct and statements of the defendant before and after the killing; (3) threats and declarations of the defendant before and during the course of the occurrence giving rise to the death of the deceased; (4) ill will or previous difficulty between the parties; (5) the dealing of lethal blows after the deceased has been felled and rendered helpless; and (6) evidence that the killing was done in a brutal manner. We have also held that the nature and number of the victim's wounds is a circumstance from which premeditation and deliberation can be inferred.

Here, many of the circumstances that we have held to establish a factual basis for a finding of premeditation and deliberation are present. It is clear, for example, that the seriously ill deceased did nothing to provoke defendant's action. Moreover, the deceased was lying helpless in a hospital bed when defendant shot him four separate times. In addition, defendant's revolver was a five-shot single-action gun which had to be cocked each time before it could be fired. Interestingly, although defendant testified that he always carried the gun in his job as a truck driver, he was not working on the day in question but carried the gun to the hospital nonetheless.

Holding

Most persuasive of all on the issue of premeditation and deliberation, however, are defendant's own statements following the incident. Among other things, defendant stated that he had thought about putting his father out of his misery because he knew he was suffering. He stated further that he had promised his father that he would not let him suffer and that, though he did not think he could do it, he just could not stand to see his father suffer any more. These statements, together with the other circumstances mentioned above, make it clear that the trial court did not err in submitting to the jury the issue of first-degree murder based upon premeditation and deliberation. Accordingly, defendant's . . . assignment of error is overruled. . . .

Dissenting, *Exum, C.J.*

Almost all would agree that someone who kills because of a desire to end a loved one's physical suffering caused by an illness which is both terminal and incurable should not be deemed in law as culpable and deserving of the same punishment as one who kills because of unmitigated spite, hatred, or ill will. Yet the Court's decision in this case essentially says there is no legal distinction between the two kinds of killing. Our law of homicide should not be so roughly hewn as to be incapable of recognizing the difference. I believe there are legal principles which, when properly applied, draw the desirable distinction and that both the trial court and this Court have failed to recognize and apply them. . . .

Questions for Discussion

1. What circumstantial evidence supports the conclusion that Forrest acted with premeditation and deliberation in killing his father?

2. Do you agree with Chief Judge Exum in his dissent that the law should distinguish between a killing that is intended to end the suffering of a loved one and a killing that is motivated by "unmitigated spite, hatred, or ill will"?

3. Does *Forrest* suggest that a murder committed with premeditation and deliberation is not necessarily more deserving of punishment than a crime committed out of passion?

Cases and Comments

Suicide. Suicide at common law was considered the felony of "self murder" because it deprived the King of one of his subjects and therefore was a crime against the Crown and against God. The punishment for suicide entailed forfeiture of the deceased person's estate and loss of the right to a formal burial. In 1961, England abolished the offense of suicide although assisting suicide remains a crime.

In the United States, suicide in most states also no longer is considered a criminal offense. Assisting suicide, however, remains a crime. New York provides that an individual who "intentionally causes or aids another person to commit suicide" is guilty of manslaughter in the second degree. N.Y. Penal Law § 125.15.

In November 1997, the Oregon "death with dignity" law went into effect. The law provides for physician-assisted suicide. Ore. Rev. Stat. §§ 127.800, *et seq.* Washington passed a similar law in 2008. Wash. Rev. Code § 70.245. Roughly four hundred individuals have made use of the Oregon law. In May 2009, a 66-year-old woman suffering from pancreatic cancer became the first person in Washington to make use of the law to end her life.

In 2009, in *Baxter v. State*, 224 P.3d 1211 (Mont. 2009), the Montana Supreme Court held that a doctor is not criminally liable for assisting a mature, aware, and terminally ill patient to take his or her life. The court reasoned that the state public policy respected the end-of-life autonomy of patients and that doctors had an ethical obligation to respect a patient's wishes.

In 1999, the late Dr. Jack Kevorkian was convicted of the second-degree murder of Thomas Youk and was sentenced to serve from ten to twenty-five years in prison. Youk was in the final stages of Lou Gehrig's disease and had signed a consent form authorizing Kevorkian to take his life. See *People v. Kevorkian*, 642 N.W.2d 681 (Mich. 2002).

See more cases on the study site: **People v. Kevorkian,** *www.sagepub.com/Lippmanccl3e.*

Capital and Aggravated First-Degree Murder

Thirty-three states and the federal government authorize the death penalty. In some states, this is called **capital murder**. The statutes in these jurisdictions typically provide for the death penalty or a life sentence in the case of a first-degree murder committed under conditions that make the killing deserving of the punishment of death or life imprisonment. Other states create a category termed *aggravated murder* that is subject to the death penalty or to life imprisonment. Those states that do not possess the death penalty punish aggravated murder by life imprisonment rather than death. Seventeen states and Washington, D.C., reject capital punishment. These seventeen states include Connecticut and New Mexico, both of which have inmates still remaining on death row.

For a deeper look at this topic, visit the study site.

State capital murder or aggravated murder statutes typically reserve this harsh punishment for premeditated killings committed with the presence of various **aggravating factors** or special circumstances. The Virginia capital murder statute, for instance, includes willful, deliberate, and premeditated killing of a police officer, a killing by an inmate, and killing in the commission of or following a rape or sexual penetration, along with other factors.[21] These statutes differ from one another, but typically include the following aggravating circumstances:

- *Victim.* A killing of a police officer, a juvenile thirteen years of age or younger, or the killing of more than one victim.
- *Offender.* An escaped prison inmate or an individual previously convicted of an aggravated murder.
- *Criminal Act.* Terrorism, murder for hire, killing during a prison escape or to prevent a witness from testifying.
- *Felony Murder.* Killing committed during a dangerous felony.

The jury, in order to sentence a defendant to death, must find one or more aggravating circumstances and is required to determine whether these outweigh any **mitigating circumstances** that may be presented by the defense attorney. Some statutes list mitigating circumstances that the jury should consider. The Florida death penalty statute specifies a number of mitigating circumstances, including the fact that the defendant does not possess a significant history of criminal activity or suffered from a substantially impaired mental capacity, the defendant was under the influence of extreme mental or emotional disturbance or acted under duress, the victim participated in the defendant's conduct or consented to the act, or the defendant's participation was relatively minor.[22]

Owen v. State illustrates a murder that is considered to be deserving of the death penalty.

Did the defendant kill in a heinous, atrocious, or cruel fashion?

Owen v. State, 862 So. 2d 687 (Fla. 2003). Per curiam.

Facts

This is the second appearance of Duane Owen before this Court to review a conviction and sentence of death for the murder of fourteen-year-old Karen Slattery. In 1990, we reversed his original conviction and sentence of death and remanded for a retrial. In early 1999, following retrial, Owen was again found guilty by a jury of the offense of first-degree murder, and was further found guilty of attempted sexual battery with a deadly weapon or force likely to cause serious personal injury and burglary of a dwelling while armed. In March 1999, the same jury recommended, by a ten-to-two vote, that Owen should be sentenced to death. The judge followed the jury's recommendation, and on March 23, 1999, Owen was adjudicated guilty and sentenced to death for the murder of Karen Slattery.

In support of the sentence of death, the trial court found that four aggravating circumstances existed to support the death sentence: (1) the defendant had been previously convicted of another capital offense or a felony involving the use of violence to some person; (2) the crime for which the defendant was to be sentenced was committed while he was engaged in the commission of or an attempt to commit or flight after committing or attempting to commit the crime of burglary; (3) the crime for which the defendant was to be sentenced was especially heinous, atrocious, or cruel (HAC); and (4) the crime for which the defendant was to be sentenced was committed in a cold and calculated and premeditated (CCP) manner without any pretense of moral or legal justification. In mitigation, the trial judge considered three statutory mitigating factors: (1) the crime for which the defendant was to be sentenced was committed while he was under the influence of extreme mental or emotional disturbance; (2) the capacity of the defendant to appreciate the criminality of his conduct or to conform his conduct to the requirement of the law was substantially impaired; and (3) the age of the defendant at the time of the crime was twenty-three.

The victim was babysitting for a married couple on the evening of March 24, 1984, in Delray Beach. During the evening, she called home several times and spoke with her mother, the last call taking place

at approximately 10 p.m. When the couple returned home, just after midnight, the lights and the television were off and the babysitter did not meet them at the door as was her practice. The police were summoned and the victim's body was found with multiple stab wounds. There was evidence that the intruder entered by cutting the screen to the bedroom window. He then sexually assaulted the victim. A bloody footprint, presumably left by the murderer, was found at the scene.

The facts surrounding the death of Georgianna Worden were substantially similar to those of the Slattery murder. As this Court detailed, "the body of the victim, Georgianna Worden, was discovered by her children on the morning of May 29, 1984, as they prepared for school. An intruder had forcibly entered the Boca Raton home during the night and bludgeoned Worden with a hammer as she slept, and then sexually assaulted her. This Court affirmed the conviction and sentence of death in that case and, notably, held that there was sufficient evidence to support the trial court's findings that the murder was especially heinous, atrocious, or cruel and that the murder was committed in a cold, calculated, and premeditated murder.

We have on appeal the judgment and sentence entered in the Fifteenth Judicial Circuit Court imposing the death penalty upon Duane Owen. For the reasons stated below, we affirm the judgment and sentence under review.

Issue

Owen next challenges the trial court's application of the aggravating factors of HAC and CCP. The law is well settled regarding this Court's review of a trial court's finding of an aggravating factor. . . . Judge Cohen found the State had proven the heinous, atrocious, or cruel aggravating factor beyond a reasonable doubt and applied great weight to that factor. . . .

Reasoning

Karen Slattery was stabbed or cut eighteen times. She was alive when all the wounds were inflicted. She was in terror. She undoubtedly had a belief of her impending doom. Her fear and heightened level of anxiety occurred over a period of time. Most important, the defendant told Dr. McKinley Cheshire that fear in his victim was necessary. The defendant stated that causing deliberate pain and fear would increase the flow of female bodily fluids which he needed for himself. The puncturing of Karen Slattery's lung caused her to literally drown in her own blood. She experienced air deprivation. Each of the eighteen cuts, slashes, and/or stab wounds caused pain by penetrating nerve endings in Miss Slattery's body. The crime of murdering Miss Slattery evidenced extreme and outrageous depravity. The defendant desired to inflict pain and fear on Miss Slattery "to increase the flow of her female bodily fluids which he needed for himself." The defendant showed an utter indifference to Karen Slattery's suffering. He was conscienceless and pitiless and unnecessarily torturous to Miss Slattery. She had an absolute full knowledge of her impending death with unimaginable fear and anxiety.

This Court has consistently upheld the HAC aggravator where the victim has been repeatedly stabbed. Furthermore, we have reasoned that the HAC aggravator is applicable to murders that "evince extreme and outrageous depravity as exemplified either by the desire to inflict a high degree of pain or utter indifference to or enjoyment of the suffering of another." The HAC aggravator focuses on the means and manner in which death is inflicted. . . .

Here, the medical examiner testified that Slattery suffered eighteen stab wounds—eight to her upper back, four cutting wounds to the front of her throat, and six stab wounds to her neck. Five of the wounds penetrated her lungs, causing them to collapse, making it impossible for Slattery to breathe or speak. She would have experienced "air hunger"—the feeling of needing to breathe but not being able to do so. The doctor estimated that Slattery lost nearly her entire blood volume. The result of severe blood loss is shock, an involuntary and uncontrollable condition that causes high anxiety and terror. The doctor explained that pain is a result of the nerve receptors in the skin being injured, and that people can experience a substantial amount of pain without suffering a lethal injury.

Although Slattery did not appear to have any defensive wounds, seven of the stab wounds were lethal and could have produced death. While the medical examiner could not determine which wounds were inflicted first, he believed they were all inflicted in rapid succession and all while Slattery was alive. The doctor opined that Slattery would have been capable of feeling pain as long as she was conscious, which he estimated would have been for between twenty seconds and two minutes, depending upon which wound was inflicted first. He testified that one minute was a reasonable estimate for how long Slattery remained conscious, as twenty seconds was too short,

but two minutes would have been a "little long." During that time she would have felt pain, experiencing the additional stab wounds, would have felt terror and shock, would have been aware of her impending doom, would have become weaker as a result of blood loss, and would have been unable to cry out. Finally, according to the medical examiner, although she may have been dead prior to the occurrence, Slattery was sexually assaulted, and semen was found on both her internal and external genitalia.

In addition to the evidence presented by the medical examiner, the testimony of Owen's own mental health expert supports the finding of HAC. Dr. Frederick Berlin testified that Owen believed that by having sex with a woman he could obtain her bodily fluids, and that this would assist him in his transformation from a male to a female. Owen believed that if he had sex with a woman who was near death, his penis would act as a hose, and her soul would enter his body and they would "become one." Importantly, Owen believed that the more frightened the victim was, the better. This express need to cause his victim extreme fear clearly evinces an utter indifference to his victim's torture. On the basis of the entire record, Owen's killing of Karen Slattery unquestionably satisfies the requirements of HAC.

Owen's challenge to the finding of the CCP aggravator is likewise misplaced. This Court has established a four-part test to determine whether the CCP aggravating factor is justified: (1) the killing must have been the product of cool and calm reflection and not an act prompted by emotional frenzy, panic, or a fit of rage (cold); (2) the defendant must have had a careful plan or prearranged design to commit murder before the fatal incident (calculated); (3) the defendant must have exhibited heightened premeditation (premeditated); and (4) the defendant must have had no pretense of moral or legal justification.

In 1992, we held that the finding of CCP was properly applied to the murder of Georgianna Worden, a second murder for which Owen was convicted and sentenced to death. Although Worden was bludgeoned to death and not stabbed, the remaining facts of that murder were virtually identical to those of the Slattery homicide. . . .

Owen's confession to the Slattery murder demonstrates the similarities between the two murders. Owen admitted to cutting a screen out of a window to gain access to the home where Slattery was babysitting. . . .

According to Owen, he confronted Slattery near the phone as she was concluding a telephone conversation. He ordered her to return the phone to its cradle, and when she did not, he dropped his hammer, grabbed the phone from her hand, returned it to its base, and immediately began stabbing her. After Owen had stabbed Slattery, he checked on the children to ensure they had not awakened during the attack, and he then proceeded to lock the doors and turn off all the lights and the television. Owen then dragged Slattery by her feet into the bedroom, removed her clothes, and sexually assaulted her. He explained to the officer questioning him that he had only worn a pair of "short-shorts" into the house. After he sexually assaulted Slattery, Owen showered to wash the blood from his body, and then exited the house through a sliding glass door. He then returned to the home where he was staying and turned the clocks back to read 9:00 p.m. According to Owen, he did this to provide an alibi based on time. He admitted that after he turned the clocks back, he purposely asked his roommate the time. Owen bragged to the officers about his plan to turn back the clocks, explaining that he "had to be thinking."

Clearly, as with the Worden murder, the murder of Karen Slattery satisfies the requirements of CCP. The fact that Owen stalked Slattery by entering the house, observing her, leaving, and then returning after the children were asleep demonstrates that this murder was the "product of cool and calm reflection and not an act prompted by emotional frenzy, panic, or a fit of rage." . . . Further, Owen unquestionably had "a careful plan or prearranged design to commit murder," as evidenced by the fact that he removed his clothing prior to entering the house, wore socks and then gloves on his hands, confronted the fourteen-year-old girl with a hammer in one hand and a knife in the other, and, by his own admission, did not hesitate before stabbing Slattery eighteen times.

The third element of CCP, heightened premeditation, is also supported by competent and substantial evidence. We have previously found the heightened premeditation required to sustain this aggravator to exist where a defendant has the opportunity to leave the crime scene and not commit the murder but, instead, commits the murder. When Owen first entered the home and saw the fourteen-year-old babysitter styling the hair of one of her charges, he had the opportunity to leave the home and not commit the murder. While he did exit the home at that time, he did not decide against killing Slattery. Instead, he returned a short time later, armed himself, confronted the young girl, and stabbed her eighteen times. Owen clearly entered the home the second time having already planned to commit murder. Heightened premeditation is supported under these facts.

Finally, the appellant unquestionably had no pretense of moral or legal justification. Notably, Owen never even suggested to the officers who questioned him, and to whom he confessed, in 1984 that a mental illness caused him to kill. He did not attempt to justify his actions, as he does in the after-the-fact manner he advances today, by explaining to the officers that he needed a woman's bodily fluids to assist in his transformation from a male to a female. He did not explain or disclose in any way that the more frightened the woman, the more bodily fluids she would secrete, and the more satisfying it would be for him. In fact, during his interrogation, Owen in no way attempted to justify his actions. Also, there is no indication in either of Owen's previous direct appeals to this Court, first for the Slattery murder and then for the Worden murder, that he has ever raised this justification in the past. Although the trial court determined that the statutory mental health mitigators were proven, the court also held that Owen had no pretense of legal or moral justification to rebut the finding of CCP. The trial court's ruling is supported by competent and substantial evidence.

Owen's claim that his mental illness must negate the CCP aggravator is unpersuasive. We have held: "A defendant can be emotionally and mentally disturbed or suffer from a mental illness but still have the ability to experience cool and calm reflection, make a careful plan or prearranged design to commit murder, and exhibit heightened premeditation." . . . Here, the evidence clearly demonstrates that Owen entered the home where Slattery was babysitting with a definite plan to murder the victim and then sexually abuse the body. CCP was properly applied to the Slattery murder.

Holding

Having determined the legitimacy of the conviction, we turn next to the sentence of death. It is well settled that the purpose of our proportionality review is to "foster uniformity in death-penalty law." Further, the number of aggravating factors cannot simply be compared to the number of mitigating factors; rather there must be "a thoughtful, deliberate proportionality review to consider the totality of the circumstances in a case, and to compare it with other capital cases." When compared to other decisions of this Court, the death sentence entered in this case is proportionate.

Questions for Discussion

1. What is the legal standard for determining whether a killing was committed in a "heinous, atrocious, or cruel" fashion?

2. Summarize the facts that support the court's conclusion that this standard was satisfied.

3. Do you believe that a murder committed in a "heinous, atrocious, or cruel" fashion merits a harsher penalty than other murders?

4. What facts support the allegation that the defendant killed in a "cold, calculated, and premeditated" fashion?

Cases and Comments

Torture Murder. Gary Heidnik was convicted of murder, kidnapping, rape, aggravated assault, and involuntary deviate intercourse in Pennsylvania. He was sentenced to death based on a number of aggravating circumstances, including "the torture aggravator." The torture aggravator requires an intent to cause pain and suffering in addition to the intent to kill. There must be suffering beyond that associated with murder. This involves an examination of the manner in which the murder is committed, including the number and type of wounds, whether the wounds were inflicted in areas of the body that indicate an intent to cause pain rather than to kill, whether the victim was conscious during the killing, and the duration of the episode.

The police found three females in Heidnik's basement, shackled at the ankles. Two of his female captives, Sandra Lindsay and Debra Dudley, had been killed. Heidnik chained, beat, and continually raped the captive women. Women who screamed for help or attempted to escape were suspended by their handcuffed wrists from a hook. Sandra Lindsay was subjected to this punishment for three or four days and was fed only bread and water. After three or four days of this abuse, Lindsay collapsed. Heidnik kicked her body into a hole he had constructed in the basement floor. After determining that Lindsay was dead, he decapitated her body and boiled her head in a large pot on the stove. Other body parts were shredded in a food processor and mixed with dog food which he fed to the other women or sealed in plastic bags and placed in the freezer.

Heidnik placed Ms. Lindsay's head in a pot which he displayed to Debra Johnson Dudley, another of his captives. He told Ms. Dudley that unless she changed her attitude, she also would end up with her head in the pot. A few days later, he placed Ms. Dudley in the cement hole, which he filled with water, and attached an electrical wire to Dudley's metal chain and electrocuted her to death. Was the killing of Lindsay as well as Dudley "committed by means of torture"? See *Commonwealth v. Heidnik,* 587 A.2d 687 (Pa. 1991).

You Decide

10.2 Eighteen-year-old Richard Henyard stole a pistol from a family friend. On January 29, 1993, Henyard told a friend that he planned to go to a nightclub in Orlando and to visit his father in South Florida. While displaying the gun, he confided to his friend that in order to make the trip, he planned to steal a car and to kill the owner. Henyard persuaded a fourteen-year-old friend, Alfonza Smalls, to participate in a robbery. The two young men followed Ms. Lewis as she left the grocery store and watched as she put her daughters, Jasmine, age three, and Jamilya, age seven, into her automobile. Smalls pulled up his shirt to reveal a gun and ordered Ms. Lewis and her daughters into the front seat.

Smalls told Ms. Lewis to "shut the girls up" and when Lewis cried out to Jesus for help, Henyard responded that "this ain't Jesus, this is Satan." Henyard subsequently stopped the car, ordered Ms. Lewis out of the car and raped Ms. Lewis on the trunk of the car while her daughters remained in the back seat. Smalls also raped Ms. Lewis on the trunk of the car. Henyard directed Lewis to sit on the edge of the road and when she hesitated he pushed her to the ground and shot her in the leg. Henyard subsequently shot Lewis three more times, wounding her in the neck, mouth, and the middle of the forehead between her eyes. Henyard and Smalls pushed Ms. Lewis's unconscious body off to the side of the road. Lewis subsequently regained consciousness and was able to alert the police. Henyard and Smalls then reentered the auto and drove away as Jasmine and Jamilya continued to cry and plead for their mother. Henyard stopped the car and led the two young girls to a grassy area where "they were each killed by a single bullet fired into the head. Henyard and Smalls threw the bodies of Jasmine and Jamilya Lewis over a nearby fence into some underbrush." The autopsies of Jasmine and Jamilya Lewis indicated that they both died of gunshot wounds to the head at very close range. The forensic evidence indicated that Jasmine's eye was open when she was shot. Henyard claimed that the heinous, atrocious, and cruel aggravating circumstance was not applicable because he had killed the two girls with a single shot and that they had not been physically harmed prior to their murder. What is your view? See *Henyard v. State,* 689 So. 2d 239 (Fla. 1996).

You can find the answer at www.sagepub.com/lippmanccl3e.

CRIME IN THE NEWS

On July 5th, 2011, twenty-five-year-old Casey Lee Anthony was acquitted of criminal charges of first degree murder, aggravated manslaughter, and aggravated child abuse stemming from the death of her two-year-old daughter, Caylee Anthony. The jury convicted Casey of four misdemeanor charges of "lying" to the police, and she was released from custody on July 17, 2011.

Statements by the jurors following the trial indicated that they concluded that the prosecution failed to establish Casey's guilt beyond a reasonable doubt on the felony counts.

The verdict stunned the millions of Americans who had followed the trial on television and who had been exposed to a daily dose of analysis by "legal talking heads." Polls indicated that 65 percent of Americans believed that Casey was guilty of murdering Caylee.

The evidence at trial was that Casey left the family home with Caylee on June 16, 2008. The grandparents did not see Caylee during the next month. Casey deflected the grandparents' requests to see Caylee by telling them that she was too busy to bring her infant daughter over for a visit and at other times Casey explained that Caylee was with the babysitter at the park or at the beach.

On July 13, 2011, George Anthony picked up Casey's car from an impound lot and was alarmed by the smell of "death" that pervaded the auto. Two days later, Casey's mother, Cindy, called 911 to report that Caylee was missing. Casey got on the phone and stated that her daughter had been missing for 31 days.

Casey at one point during the search for Caylee reported to the police that the baby had been left with her babysitter, Zenalda Gonzalez, whom Casey alleged had kidnapped the child. It later was revealed that Caylee had never been cared for by a babysitter and that Casey had falsely implicated Zenalda Gonzalez. In October, Casey was indicted by a grand jury for Caylee's death; and, six months later, the prosecutors announced that they would seek the death penalty.

On December 11, 2011, human remains were found in a park near the home of Casey's parents. There was duct tape over the skull along with a plastic bag and a laundry bag all of which matched items located in Casey's residence. Caylee's Winnie the Poo blanket also was found with the corpse. Four days later, additional remains were uncovered.

The prosecution's theory at trial was that Casey had placed the duct tape over Caylee's mouth and nose and suffocated Caylee to death and then disposed of the body.

The body had decomposed and it was impossible to accurately determine the cause of death. There also was no DNA or blood or fingerprint evidence linking the body to the killer. As a result, the prosecution was reduced to presenting a circumstantial case. The government introduced controversial scientific evidence that linked the smell in the trunk of Casey's car to the smell of a decomposed body. Experts also testified that bugs in the trunk were the type of bugs associated with the decomposition of a body. There was additional evidence connecting hair fibers found in the trunk of Casey's car to Caylee and a "sniffer dog" linked the smell in the trunk to Caylee. An analysis of the computer in the Anthony home indicated that there were 84 computer searches for chloroform, a chemical used to render an individual unconscious. The defense dismissed most of this evidence as "fantasy forensics" and the computer evidence was called into question following the trial.

During the month that Caylee was missing, photos captured Casey partying and drinking and competing in a wet tee shirt contest. She reportedly spent the day following Caylee's death in bed with her boyfriend. Casey had a history of "lying" and had misled her parents into believing, for over two years, that she worked at Universal Studios. The prosecution argued that Casey wanted to dispose of Caylee because the child restricted Casey's lifestyle.

The government pointed out that Casey's failure to report Caylee's death immediately to the police indicated that she was involved in covering-up a murder. She also misled law enforcement officials and private individuals and groups by claiming that Caylee was missing and silently watched as a massive search was undertaken and hundreds of thousand of dollars were spent to find Caylee.

The defense claimed that Caylee had died as a result of an accident in the family pool and that Caylee's grandfather, George, had disposed of the body. According to the defense, George then blamed Casey for Caylee's death in an effort to cover up his own role.

One positive development stemming from the tragedy surrounding Caylee's death is the introduction of laws in various states that criminalizes the failure to report that a child is missing.

In the aftermath of the trial of Casey Anthony, commentators asked whether the press attention lavished on high profile trials helps to insure that a defendant receives a fair trial or whether the press attention helps to create a "circus-like" atmosphere that interferes with a defendant's receiving a fair trial and undermines public confidence in the legal process.

Second-Degree Murder

State second-degree murder statutes typically punish intentional killings that are committed with malice aforethought that are not premeditated, justified, or excused. Most statutes go beyond this simple statement and provide that killings committed with malice aforethought that are not specifically listed as first-degree murder are considered second-degree murder. For instance, several states include felony murder as second- rather than first-degree murder.

Washington State provides that a person is guilty of murder in the second degree when with "the intent to cause the death of another person but without premeditation he causes the death of such person." The statute also includes as second-degree murder a killing committed in furtherance of or in flight from a felony.[23] Idaho merely provides that all killings that are not explicitly included in the first-degree statute "are of the second degree." This means that an Idaho prosecutor is authorized to charge second-degree murder in all instances in which a murder does not fall within the state's first-degree murder statute.[24]

The Louisiana statute states that second-degree murder is the killing of a human being when the "offender has a specific intent to kill or to inflict great bodily harm." The law also provides that second-degree murder includes:

- A killing that occurs during the perpetration or attempted perpetration of aggravated rape, arson, burglary, kidnapping, escape, a drive-by shooting, armed robbery or robbery, despite the fact that the individual possesses no intent to kill or to inflict great bodily harm.
- A killing that occurs in the perpetration of cruelty to juveniles, despite the fact that an individual has no intent to kill or to inflict great bodily harm.
- A killing that directly results from the unlawful distribution of an illegal narcotic.[25]

The next case, *Midgett v. State,* is based on an Arkansas second-degree murder statute that punishes an individual who "knowingly causes the death of another person under circumstances manifesting extreme indifference to the value of human life." The statute also punishes a killing that is committed when, with the "purpose of causing serious physical injury to another person, he [the perpetrator] causes the death of another person."[26] In reading *Midgett,* remember that malice may be express or implied. Express malice is the deliberate intent to unlawfully take the life of another individual. Implied malice involves a killing that results from an intentional act, the natural consequences of which are dangerous to life.

The Legal Equation

Second-degree murder **=** Intentional act dangerous to the life of another

+ intent to kill without premeditation and deliberation or intent to commit underlying felony for felony murder

+ causing the death of another person.

Is Midgett guilty of first- or second-degree murder?

Midgett v. State, 729 S.W.2d 410 (Ark. 1987). Opinion by: Newbern, J.

Issue

This child abuse case resulted in the appellant's conviction of first-degree murder. The sole issue on appeal is whether the state's evidence was sufficient to sustain the conviction. We hold there was no evidence of the ". . . premeditated and deliberated purpose of causing the death of another person . . ." required for conviction of first-degree murder. However, we find the evidence was sufficient to sustain a conviction of second-degree murder . . . as the appellant was shown to have caused his son's death by delivering a blow to his abdomen or chest ". . . with the purpose of causing serious physical injury." The conviction is thus modified from one of first-degree murder to one of second-degree murder and affirmed.

Facts

The facts of this case are as heartrending as any we are likely to see. The appellant is six feet two inches tall and weighs 300 pounds. His son, Ronnie Midgett, Jr., was eight years old and weighed between thirty-eight and forty-five pounds. The evidence showed that Ronnie Jr. had been abused by brutal beating over a substantial period of time. Typically, as in other child abuse cases, the bruises had been noticed by school personnel, and a school counselor . . . had gone to the Midgett home to inquire. Ronnie Jr. would not say how he had obtained the bruises or why he was so lethargic at school except to blame it all, vaguely, on a rough playing little brother. He did not even complain to his siblings about the treatment he was receiving from the appellant. His mother, the wife of the appellant, was not living in the home. The other children apparently were not being physically abused by the appellant.

Ronnie Jr.'s sister, Sherry, aged ten, testified that on the Saturday preceding the Wednesday of Ronnie Jr.'s death, their father, the appellant, was drinking whiskey (two to three quarts that day) and beating on Ronnie Jr. She testified that the appellant would "bundle up his fist" and hit Ronnie Jr. in the stomach and in the back. On direct examination she said that she had not previously seen the appellant beat Ronnie Jr., but she had seen the appellant choke him for no particular reason on Sunday nights after she and Ronnie Jr. returned from church. On cross-examination, Sherry testified that Ronnie Jr. had lied and her father was, on that Saturday, trying to get him to tell the truth. She said the bruises on Ronnie Jr.'s body noticed over the preceding six months had been caused by the appellant. She said the beating administered on the Saturday in question consisted of four blows, two to the stomach and two to the back.

On the Wednesday Ronnie Jr. died, the appellant appeared at a hospital carrying the body. He told hospital personnel something was wrong with the child. An autopsy was performed, and it showed Ronnie Jr. was a very poorly nourished and underdeveloped eight-year-old. There were recently caused bruises on the lips, center of the chest plate, and forehead as well as on the back part of the lateral chest wall, the soft tissue near the spine, and the buttocks. There was discoloration of the abdominal wall and prominent bruising on the palms of the hands. Older bruises were found on the right temple, under the chin, and on the left mandible. Recent as well as older, healed, rib fractures were found.

The conclusion of the medical examiner who performed the autopsy was that Ronnie Jr. died as the result of intra-abdominal hemorrhage caused by a blunt force trauma consistent with having been delivered by a human fist. The appellant argues that in spite of all this evidence of child abuse, there is no evidence that he killed Ronnie Jr. having premeditated and deliberated causing his death. We must agree. . . .

The evidence in this case supports only the conclusion that the appellant intended not to kill his son but to further abuse him or that his intent, if it was to kill the child, was developed in a drunken, heated, rage while disciplining the child. Neither of those supports a finding of premeditation or deliberation.

Perhaps because they wish to punish more severely child abusers who kill their children, other states' legislatures have created laws permitting them to go beyond second-degree murder. . . . Idaho has made murder by torture a first-degree offense, regardless of intent of the perpetrator to kill the victim, and the offense is punishable by the death penalty. . . .

Holding

All of this goes to show that there remains a difference between first- and second-degree murder, not only under our statute, but generally. Unless our law is changed to permit conviction of first-degree murder for something like child abuse or torture resulting in death, our duty is to give those accused of first-degree murder the benefit of the requirement that they be shown by substantial evidence to have premeditated and deliberated the killing, no matter how heinous the facts may otherwise be. . . .

The dissenting opinion begins by stating the majority concludes that one who starves and beats a child to death cannot be convicted of murder. That is not so, as we are affirming the conviction of murder; we are, however, reducing it to second-degree murder. The dissenting opinion's conclusion that the appellant starved Ronnie Jr. must be based solely on the child's underdeveloped condition which could, presumably, have been caused by any number of physical malfunctions. There is no evidence the appellant starved the child. The dissenting opinion says it is for the jury to determine the degree of murder of which the appellant is guilty. That is true so long as there is substantial evidence to support the jury's choice. The point of this opinion is to note that there was no evidence of premeditation or deliberation, which are required elements of the crime of first-degree murder. . . .

In this case we have no difficulty with reducing the sentence to the maximum for second-degree murder. The jury gave the appellant a sentence of forty years imprisonment which was the maximum for first-degree murder, and we reduce that. . . . [T]he obvious effect the beatings were having on Ronnie Jr. and his emaciated condition when the final beating occurred are circumstances constituting substantial evidence that the appellant's purpose was to cause serious physical injury, and that he caused his death in the process. That is second-degree murder. . . . Therefore, we reduce the appellant's sentence to imprisonment for twenty years.

Dissenting, *Hickman, J.*

Simply put, if a parent deliberately starves and beats a child to death, he cannot be convicted of the child's murder. In reaching this decision, the majority . . . substitutes its judgment for that of the jury. The majority has decided it cannot come to grips with the question of the battered child who dies as a result of deliberate, methodical, intentional and severe abuse. A death caused by such acts is murder by any legal standard, and that fact cannot be changed—not even by the majority. The degree of murder committed is for the jury to decide—not us.

In this case the majority, with clairvoyance, decides that this parent did not intend to kill his child, but rather to keep him alive for further abuse. This is not a child neglect case. The state proved Midgett starved the boy, choked him, and struck him several times in the stomach and back. The jury could easily conclude that such repeated treatment was intended to kill the child. . . .

The facts in this case are substantial to support a first-degree murder conviction. The defendant was in charge of three small children. The victim was eight years old and had been starved; he weighed only 38 pounds at the time of his death. He had multiple bruises and abrasions. The cause of death was an internal hemorrhage due to blunt force trauma. His body was black and blue from repeated blows. The victim's sister testified she saw the defendant, a 30-year-old man, 6'2" tall, weighing 300 pounds, repeatedly strike the victim in the stomach and back with his fist. One time he choked the child.

The majority is saying that as a matter of law a parent cannot be guilty of intentionally killing a child by such deliberate acts. Why not? Is it because it is inconceivable to rational people that a parent would intend to kill his own child? Evidently, this is the majority's conclusion, because they hold the intention of Midgett was to keep him alive for further abuse, not kill him. How does the majority know that? How do we ever know the actual or subliminal intent of a defendant? . . . This parent killed his own child, and the majority cannot accept the fact that he intended to do just that.

Undoubtedly, the majority could accept it if the child were murdered with a bullet or a knife; but they cannot accept the fact, and it is a fact, that this defendant beat and starved his own child to death. His course of conduct could not have been negligent or unintentional. . . . He is guilty of first-degree murder in the eyes of the law. His moral crime as a father is another matter, and it is not for us to speculate why he did it.

Questions for Discussion

1. Why did the Arkansas Supreme Court rule that Midgett is guilty of second- rather than first-degree murder? Summarize the dissenting view that Midgett killed his son in a premeditated and deliberate manner.

2. Midgett was charged and convicted of the death of his son inflicted with the purpose of causing serious physical injury. Why was Midgett not charged with knowingly causing the death of another person under circumstances manifesting extreme indifference to the value of human life?

3. Are you confident that judges and juries are able to clearly determine a defendant's intent from the nature of his criminal acts? Do you agree with the majority or with the dissent?

4. Based on this case, do you question whether first-degree murder is always a more serious offense than second-degree murder?

5. One month following the decision in *Midgett*, the Arkansas legislature amended the state's criminal code to authorize a verdict of first-degree criminal homicide when an individual under "circumstances manifesting extreme indifference to the value of human life . . . knowingly causes the death of a person fourteen years of age or younger at the time the murder was committed." See Ark. Code Ann. § 5-10-101(a)(9). Would Midgett be found guilty of first degree murder under this statute?

You Decide

10.3 Jerry Chambers was the babysitter for Tiffany Bennett's four female children: P.B., age three; P.B.2., age four; A.B.2., age six; and A.B., age ten. Chambers shared his apartment with Bennett's sister, Candace Geiger, and his "godmother." In the fall of 2002, Bennett agreed to pay Chambers $80.00 a week in return for the girls moving in with him. Bennett only saw the girls on holidays and later resisted taking the girls back into her home.

"The children lived in deplorable conditions. . . . [The apartment] was filthy and infested with flies and cockroaches. An overwhelming stench of urine permeated the dwelling. The bedroom had two beds: one, next to the radiator, for the children; the other for Chambers and Geiger. There were urine and feces on the floor. The window was covered with dark plastic, and on the wall were taped a number of rules. Number seven instructed the girls to "[r]espect [appellant] at all times or every body [sic] gets their ass whooped."

Chambers "regularly beat the four girls with extension cords, belts, a metal pole, and a broomstick." The girls were prohibited from leaving the apartment, even to attend school. They were instructed to remain in the bedroom and to cover their faces when visitors came to the apartment. P.B. was punished whenever she wet her bed by being confined in a cold shower. The girls were reprimanded by being locked in the basement with two pit bulls. At times, they were told to eat "dog poop" out of the dogs' food bowls.

On the night of August 16, 2003, Chambers beat P.B. with an extension cord while she was in the shower. He then locked A.B. in the basement. Chambers and Geiger along with P.B., A.B.2, and P.B.2 went to bed. Some time after midnight, Chambers and Geiger were having sex when Chambers accused P.B. of staring at them. He directed P.B. to stop watching, and she allegedly kept staring at them. Geiger called P.B. over to the bed, beat her with an extension cord, and struck her in the face several times. Geiger also beat P.B. Chambers "picked her up by her feet and threw her across the room. P.B. struck her head on the cast-iron radiator and ended up lodged between the bed, radiator, and wall. She remained there, slowly suffocating, until the next day. Appellant did nothing to help the child; further, he instructed the other girls not to help her." Around 1:00 p.m., the police were called.

Chambers relied on *Midgett* and argued that he should not be held liable for first-degree, premeditated murder.

P.B.'s death was due to a combination of factors, including the multiple blunt-force traumas. She also suffered inanition (a condition in which a child literally wastes away due to physical trauma and emotional stress), and asphyxia from lying "crumpled up in a heap" jammed between the bed, the wall, and the radiator. Apply the precedent in *Midgett*. Would you hold Chambers liable for first- or for second-degree murder? *Commonwealth v. Chambers*, 980 A.2d 35 (Pa. 2009).

You can find the answer at www.sagepub.com/lippmanccl.3e.

Depraved Heart Murder

An individual may be held criminally responsible for depraved heart murder in those instances that he or she kills another as a result of the "deliberate perpetration of a knowingly dangerous act with reckless and wanton unconcern and indifference as to whether anyone is harmed or not."[27] A defendant who acts in this fashion is viewed as manifesting an "abandoned and malignant heart" or "depraved indifference to human life."[28] Reckless homicide is based on the belief that acts undertaken without an intent to kill that severely and seriously endanger human life are "just as antisocial and . . . just as truly murderous as the specific intent to kill and to harm." Malice is implied in the case of depraved heart murder, and this is typically punished as second-degree murder. The California Penal Code states that malice is "implied . . . when the circumstances attending the killing show an abandoned and malignant heart."[29]

Depraved heart murder requires:

- *Conduct.* The defendant's act must create a very high degree of risk or serious bodily injury. Keep in mind that the act must be highly dangerous.
- *Intent.* The defendant must be aware of the danger created by his or her conduct. Some courts merely require that a reasonable person would have been aware of the risk.
- *Danger.* The common law appeared to require that a number of individuals were placed in danger; the modern view is that it is sufficient that a single individual is at risk.

There is no mathematical formula for determining whether an act satisfies the highly dangerous standard of depraved heart murder. This is decided based on the facts of each case. Examples of depraved heart murder include:

- A defendant plays a game of "Russian Roulette" in which he loads a revolver with one bullet and six "dummy bullets" and spins the chamber. He places the gun to the victim's head and pulls the trigger three times; the third pull kills the victim.[30]
- A defendant shoots into a passing train, unintentionally killing a passenger.[31]
- Two street gangs engage in a lengthy shoot-out on a street in downtown Baltimore, killing an innocent fifteen-year-old.[32]
- The defendant pours gasoline through the mail slot of the victim's house and sets the gasoline on fire, killing two children.[33]

The next case, *State v. Davidson*, involves a conviction for depraved heart murder. Pay attention to the facts that led the Kansas Supreme Court to affirm the defendant's guilt for the killing of a human being "unintentionally but recklessly under circumstances manifesting extreme indifference to the value of human life."[34]

The Legal Equation

Depraved heart murder = Dangerous act creating a high risk of death

+ knowledge of danger created by act.

Did the defendant manifest an extreme indifference to the value of human life based on her failure to control her dogs?

State v. Davidson, 987 P.2d 335 (Kan. 1999). Opinion by: Allegrucci, J.

Defendant Sabine Davidson was convicted of reckless second-degree murder and endangering a child. She appeals her conviction of reckless second-degree murder. . . . The question we must resolve is whether . . . the State presented evidence sufficient to sustain a conviction of reckless second-degree murder.

Facts

Davidson's challenge to the law and the evidence rests on the same premise: that the State proved only that she failed to confine her dogs. The statute requires the State to prove that her conduct "manifested extreme indifference to the value of human life." At best, she argues that her conduct constituted a negligent omission to confine the dogs and she should have been charged with involuntary manslaughter. Simply stated, she argues that the crime of second-degree murder does not fit her conduct. . . . The state's evidence established numerous earlier incidents involving defendant's dogs. . . .

Fifteen-year-old Margaret Smith, who lived near the Davidsons, testified that sometime during the 1995–1996 school year, two dogs chased her and Jeffrey Wilson away from the school bus stop as they waited there in the morning. She believed that the dogs belonged to the Davidsons. . . . Deputy Shumate had been to the Davidson house in January 1996 when one of the neighbors complained about their dogs running loose. The complainant said that his wife was afraid of the dogs, a German shepherd and a Rottweiler. When Shumate went to the Davidson house on that occasion, Sabine Davidson told him that she would keep the dogs in the fenced enclosure. The fenced enclosure that Shumate saw in January 1996 was still in use in April 1997.

Learie Thompson, who lives near the Davidsons, complained to the sheriff's office in January 1996 because the Davidsons' dogs were in his yard. He recalled three to five times earlier when the dogs were loose and came into his yard. For the most part, the dogs Thompson saw in his yard were German shepherds. Thompson testified that he was afraid of the Davidsons' dogs because they were big and aggressive toward people. When the dogs were in their fenced enclosure, they would rage and growl and try to get out of the fence. . . . [E]arly on the morning that Chris was killed, as Thompson opened his garage door to leave for work, the Davidsons' three Rottweiler dogs rushed into the garage. Thompson jumped up onto his truck. The dogs stood on their hind legs and growled and bared their teeth at him for several minutes. . . .

One incident occurred at the intersection where Chris was killed. On June 14, 1996, Tony Van Buren, who lives directly across the street from the Davidson house, was out in his front yard in the evening when he heard dogs barking. He saw three Rottweilers forming a semicircle around two young children, who were approximately 3 to 5 years old. . . .

In addition to keeping and breeding German shepherds, defendant in 1995 began purchasing Rottweilers. Bernardi testified that defendant bought three from her, five from other kennels, and one from a breeder in Germany. Timothy Himelick testified that defendant bought two Rottweilers from him. Himelick's dogs were two years old, had been raised together as family dogs, and were very friendly. He sold them because he was moving, but several months later "repossessed" the dogs because defendant could not control the female. . . . When defendant got close to Himelick's female dog, the dog "went ballistic." When two children rode by on bicycles, defendant said, "One of these days I'm going [to] get even with them." And when asked by Himelick about a German shepherd that obviously had had a litter of puppies, defendant said her Rottweilers had eaten the pups. . . .

About 7 a.m. on April 24, 1997, Walls opened his front door to let his dog back into the house. Three Rottweilers had his dog backed into a corner of the porch. Walls' dog acted scared. Walls' dog slipped into the house, and the Rottweilers advanced toward Walls. He went back inside for his gun, and when he returned a few minutes later, they were gone. Walls went outside where he could see that the Rottweilers were back inside the Davidsons' fenced enclosure. . . .

Violet Wilson dropped her two younger sons, Chris, eleven, and Tramell, nine, off at the school bus stop shortly after 7:15 a.m. The bus stop was located near the residences of Tony Van Buren and defendant. While waiting for the bus, Tramell noticed that defendant's dogs were digging at the fence like they "really wanted to get out." When the dogs got out of the fence, they ran toward the boys, who climbed up into a tree in Van Buren's yard. The three dogs surrounded the tree and barked at the boys for several minutes before the biggest dog left and the other two followed it.

Chris wanted to get down out of the tree and see where the dogs had gone and what they were doing. Tramell urged him to stay in the tree, but Chris got down and looked around for the dogs.

When the school bus arrived at 7:30, no children were at the stop, but the driver noticed two book bags and a musical instrument had been left there. The driver saw Tramell up in a tree on Van Buren's lot. Tramell got out of the tree, ran to the bus stop, gathered up the bags and instruments, and got on the bus. As he got on, Tramell said something about Chris, which the driver thought sounded like Chris had run the dog home. The driver waited several minutes before repositioning the bus to let Chris know that he needed to hurry up. After moving the bus, the driver could see in the side mirror that there were three large black dogs down in the ravine. The dogs appeared to be fighting over something; they were jumping back and forth and thrashing their heads from side to side. One of the children on the bus said, "It looks like they have a rag doll." Then the driver realized that the dogs had Chris. The only movement of Chris' body was that caused by the thrashing motion of the dogs.

The driver began honking the bus horn in an effort to distract the dogs, and she radioed the dispatcher. There were approximately twenty children on the bus, and the driver then made an effort to divert their attention away from the ravine and to calm them.

When David Morrison, a sheriff's deputy, arrived, the bus driver told him that the boy and the dogs were in the ravine. As Morrison walked toward them, the dogs noticed him and moved quickly and steadily in his direction. Morrison could see that Chris was not moving. Morrison's shouts and gestures did not divert the dogs, which continued to move straight toward him with a large male in the lead. Morrison could see blood on the lead dog's face and forelegs. When the lead dog was approximately fifteen feet away from him, Morrison shot and killed him. One of the children on the bus testified that the lead dog had been the one at Chris' neck.

The other two dogs began to run away. Another officer, Sergeant Mataruso, who had just arrived, fired shots at the fleeing dogs. Morrison ran down into the ravine and found Chris. The boy had no pulse. The grass around his body was torn up, and there was a lot of blood spread around the area. Items of clothing, some ripped up, and shoes were scattered about.

When Deputy Shumate arrived, he saw a dead dog in the road. Mataruso told him that he had shot at, and probably wounded, another dog that had been seen running toward a house. Shumate went to the Davidson house, where he found the wounded dog and killed it. Shumate told Mr. and Mrs. Davidson that a child had been attacked by their dogs; he then arrested them. They asked no questions about the identity or condition of the child. Inside the house, there was a dog confined in a large plastic pet carrier. Shumate testified that "the dog was growling, sort of barking, and moving the pet carrier all over the floor like it was trying to get out of the carrier to get at me." The third dog was shot and killed by a highway patrolman later in the day.

An autopsy revealed that Chris had died as a result of trauma from animal bites. There were many injuries, but those affecting the head and neck were immediately responsible for his death. The boy's esophagus and carotid artery were torn. His neck had been broken, and the bones splintered and crushed. . . . [T]he victim's neck was engulfed by the dog's mouth so that even the dog's back teeth left impressions in the tissue.

. . . The dogs that killed Chris were Rottweilers. The one killed by Morrison was an 80-pound male. The other two were females of 70 pounds and 54 pounds, respectively. . . .

Defendant told police that she let the dogs out about 6:30 a.m. on April 24, 1997. Then she took a sleeping pill and went to sleep on the living room couch. Later, when she was told that her dogs had attacked a boy, she said, "The dead one should be one of the Wilson boys." She said that the Wilson boys teased her dogs whenever they came around her property so that the dogs barked and got aggressive when the boys were in the area. She also said that the boy had been at the bus stop, and that is how she knew who was attacked.

Defendant told police that she and her husband had discussed putting a chain on the gate to the fenced enclosure because the dogs got out of that gate "all the time." There was no chain on the gate, however. At trial, defendant testified that after Bernardi watched one of the Rottweilers open the gate by lifting the latch, she put a padlock through the hole in the latch so that it could not be lifted. Defendant testified that after padlocking the latch, she had no more reports of the dogs getting out.

A videotape made by Deputies Snyder and Popovich shows that the Davidsons' back and side yards are enclosed by a 6-foot-tall chain link fence. The gate fastening device is a typical horseshoe-shaped hinged latch. In a horizontal position the arms of the latch are on either side of an adjacent upright fence post on which the next section of fence is attached. A locked padlock through the hole in the latch on the Davidsons' gate kept the latch in a horizontal position. The fence post along with the latch should have kept the gate closed, but the post was easily moved far enough from the vertical for the latch to slip past. On the videotape, Snyder opened and closed the "padlocked" gate without much effort. . . . The fence was installed in 1993.

Issue

Reckless second-degree murder, also known as depraved heart murder, was defined by the legislature as "the killing of a human being committed . . . unintentionally but recklessly under circumstances manifesting extreme indifference to the value of human life."

Here, defendant argues that . . . the State's evidence establishes only that she failed to secure the dogs. She then argues that her conduct, as a matter of law, is not reckless second-degree murder. . . . Defendant's argument . . . is based on the State's failure to establish a "depraved heart scienter." It is difficult at times to follow her arguments, but she seems to imply that the State has failed to prove foreseeability, that she had knowledge the dogs would attack someone, let alone harm or kill someone. Thus, she argues the State failed to prove that her conduct was inherently dangerous to human life and that she was indifferent to that danger. In her view, her conduct did not rise to that level of reckless conduct manifesting an extreme indifference to the value of human life.

Reasoning

Here, defendant argues that all she did was let the dogs into the fenced area, take a pill, and go to sleep. This argument conveniently ignores significant aspects of her conduct that contributed to the tragic death of Chris. The State presented evidence that she selected powerful dogs with a potential for aggressive behavior and that she owned a number of these dogs in which she fostered aggressive behavior by failing to properly train the dogs. She ignored the advice from experts on how to properly train her dogs and their warnings of the dire results which could occur from improper training. She was told to socialize her dogs and chose not to do so. She ignores the evidence of the dogs getting out on numerous occasions and her failure to properly secure the gate. She ignored the aggressive behavior her dogs displayed toward her neighbors and their children. The State presented evidence that she created a profound risk and ignored foreseeable consequences that her dogs could attack or injure someone. The State is not required to prove that defendant knew her dogs would attack and kill someone. It was sufficient to prove that her dogs killed Chris and that she could have reasonably foreseen that the dogs could attack or injure someone as a result of what she did or failed to do.

Davidson was charged with reckless second-degree murder. The jury was instructed that in order to establish the charge, it would have to be proved that defendant killed Christopher Wilson "unintentionally but recklessly under circumstances showing extreme indifference to the value of human life." "Recklessly" was defined as "conduct done under circumstances that show a realization of the imminence of danger to the person of another and a conscious and unjustifiable disregard of that danger." The jury also was instructed that it could consider the lesser included offense of involuntary manslaughter, either an unintentional killing done recklessly or an unintentional killing done in the commission of the offense of permitting a dangerous animal to be at large. The jury found that defendant's conduct involved an extreme degree of recklessness.

Holding

Here, the evidence, viewed in a light most favorable to the State, showed that defendant created an unreasonable risk and then consciously disregarded it in a manner and to the extent that it reasonably could be inferred that she was extremely indifferent to the value of human life. The evidence was sufficient to enable a rational fact finder to find Davidson guilty of reckless second-degree murder. Thus, the district court properly submitted the charge of reckless second-degree murder to the jury, and the evidence was sufficient to support the jury's verdict. The judgment of the district court is affirmed.

Questions for Discussion

1. What facts support Sabine Davidson's "indifference to the value of human life"?

2. Does the decision require her to know that the dogs would kill or was it sufficient that the killing was foreseeable? Could Davidson know or anticipate that the dogs were capable of killing a human being?

3. Do you believe that the defendant's recklessness created such an extreme threat to human life that the gravity of her crime was equivalent to intentional murder and that she was properly convicted of second-degree murder rather than a less serious category of homicide, such as negligent manslaughter?

You Decide

10.4 Michael Berry was charged with depraved heart murder. The defendant purchased a pit bull, Willy, from a breeder of fighting dogs. Berry trained Willy and entered the dog in "professional fights" as far away as South Carolina. Willy was described as possessing stamina, courage, and a particularly "hard bite." He was tied to the inside of a six-foot unenclosed fence so as to discourage access to the 243 marijuana plants that Berry was illegally growing in an area in the back of his house. Berry's next-door neighbor momentarily left her two-year-old child, James Soto, playing on the patio of her home. James apparently wandered across Berry's yard to the other side of Berry's home, where he encountered Willy and was mauled to death. An animal control officer testified that pit bulls are considered "dangerous unless proved otherwise." Is Berry guilty of killing with an abandoned and malignant heart? See *Berry v. Superior Court*, 256 Cal. Rptr. 344 (Cal. Ct. App. 1989).

You can find the answer at www.sagepub.com/lippmanccl.3e.

You Decide

10.5 Following a party for his softball team at a club where he admitted drinking six beers, John Doub admitted that he struck two parked vehicles with his pickup truck. He immediately drove off because he was concerned that the police would detect that he had been drinking. Doub subsequently drank additional liquor and smoked crack cocaine; roughly two hours later, he collided into the rear of an automobile in which nine-year-old Jamika Smith was riding. The accident investigator determined that Doub's pickup was traveling at a rapid rate and drove "up on top of [the car]," driving it down into the pavement and propelling the automobile off the street and into a tree. Doub once again left the scene of the accident; he later denied involvement and claimed that his pickup had been stolen. Jamika Smith died fifteen hours later as a result of blunt traumatic injuries caused by the collision. Six months following these events, Doub admitted to a former girlfriend that prior to the collisions, he had an argument with his second ex-wife, had been drinking alcohol and smoking crack, and had subsequently caused the collision. Doub was charged and convicted of second-degree depraved heart murder

Kansas Statutes Annotated section 21-3402 (2003 Supp.) defines depraved heart murder as the killing of a human being committed "unintentionally but recklessly under circumstances manifesting extreme indifference to the value of human life." Would you convict Doub? See *State v. Doub*, 95 P.3d 116 (Kan. Ct. App. 2004).

You can find the answer at www.sagepub.com/lippmanccl3e.

Felony Murder

A murder that occurs during the course of a felony is punished as murder. In *People v. Stamp,* Koory and Stamp robbed a store while armed with a gun and a blackjack. The defendants ordered the employees along with the owner, Carl Honeyman, to lie down on the floor so that no one "would get hurt" while they removed money from the cash register. Fifteen or twenty minutes following the robbery, Honeyman collapsed on the floor and was pronounced dead on arrival at the hospital. He was found to suffer from advanced and dangerous hardening of the arteries, but doctors concluded that the fright from the robbery had caused the fatal seizure. A California appellate court affirmed the defendants' conviction for felony murder and sentence of life imprisonment.[35]

This use of the felony murder to hold defendants liable for murder was criticized by another California appellate court, which observed that such a "harsh result destroys the symmetry of the law by equating an accidental killing . . . with premeditated murder."[36] Despite this criticism, the fact remains that "but for" the robbery, Honeyman would not have died. Severely punishing Koory and Stamp deters other individuals contemplating thievery and protects society. As you read this section of the textbook, consider whether the felony-murder rule is a fair and just doctrine.

The felony-murder doctrine, as previously noted, provides that any homicide that occurs during the commission of a felony or attempt to commit a felony is murder. This is true regardless of whether the killing is committed with deliberation and premeditation, intentionally, recklessly, or negligently. The intent to commit the felony is considered to provide the malice for the conviction of murder. The doctrine can be traced back to Lord Coke in the early 1600s and is illustrated by Judge Stephens's example that "if a man shot at a fowl with intent to steal it, and accidentally killed a man, he was to be accounted guilty of murder, because the act was done in the commission of a felony."[37]

This common law rule was not viewed as unduly harsh because all felonies in England were subject to the death penalty, and it made little difference whether an individual was convicted of murder or of the underlying felony. Felony murder, however, came under increasing criticism in England as the number of felonies subject to the death penalty was gradually reduced. English lawmakers came to view felony murder as making little sense and abandoned the doctrine in 1957.

The 1794 Pennsylvania murder statute included, within first-degree murder, killings committed in the perpetration or attempt to perpetrate "any arson, rape, robbery or burglary." Killings committed in furtherance of other felonies under the Pennsylvania law were considered second-degree murder. The federal government along with virtually every state has continued to apply the felony-murder rule; only Ohio, Hawaii, Michigan, and Kentucky resist the rule. Four reasons are offered for the felony-murder rule:

- *Deterrence.* Individuals are deterred from committing felonies knowing that a killing will result in a murder conviction.

- *Protection of Life.* Individuals are deterred from committing felonies in a violent fashion knowing that a killing will result in a murder conviction.
- *Punishment.* Individuals who commit violent felonies that result in death deserve to be harshly punished.
- *Prosecution.* Prosecutors are relieved of the burden of establishing a criminal intent. The fact that a killing occurred during a felony is sufficient to establish first-degree murder. The imposition of liability on all the felons carrying out the crime provides an efficient method for incarcerating dangerous felons.

There is some question whether the felony-murder rule is an important tool in the fight against crime. The U.S. Supreme Court, for example, cites statistics indicating that only one-half of one percent of all robberies result in homicide.[38]

State felony-murder statutes generally classify killings committed in the perpetration or attempt to commit dangerous felonies, such as arson, rape, robbery, or burglary, as first-degree murder deserving life imprisonment or in states with the death penalty as a capital felony punishable with either life imprisonment or the death penalty. Killings committed in furtherance of other less dangerous felonies typically are not explicitly mentioned and are prosecuted under second-degree murder statutes that punish "all other kinds of murder that are not listed as first-degree murder."[39] Several states have statutes that punish as second-degree murder a killing that results from the commission or attempt to commit "any felony."[40]

Statutes that do not list specific felonies present courts with the challenge of determining which felonies are sufficiently serious to provide the foundation for felony murder. This has potentially severe consequences for a defendant. For example, in Pennsylvania, a killing during the "perpetration of a felony" is considered second-degree murder and is punished by life imprisonment. A court, however, may decide that the killing should not be punished as felony murder and that instead the killing should be punished as "other kinds of murder" under the third-degree murder statute, which is punishable by twenty years in prison.[41]

What felonies should serve as the foundation or predicate for felony murder? Judges have generally limited felony murder to "inherently dangerous felonies." One approach is to ask whether a particular felony can be committed "in the abstract" without creating a substantial risk that an individual will be killed. The other method is to examine whether the manner in which a particular felony was committed in the specific case before the court created a high risk of death.

The approach of the California Supreme Court is to ask whether the "underlying felony" can be committed without endangering human life. In *People v. Burroughs,* the defendant, a self-proclaimed healer of illness, treated a patient suffering from terminal leukemia with a special blend of lemonade, colored lights, and massage. As the victim's condition worsened, the defendant assured the victim that he would recover. After almost a month of this treatment, the victim suffered a brain hemorrhage and died. The evidence strongly suggested that the hemorrhage was the result of the massages administered by the defendant. The California Supreme Court ruled that the defendant's felonious unlicensed practice of medicine did not constitute an "inherently dangerous felony" because an unlicensed practitioner may be treating a common cold, sprained finger, or an individual who suffers from the delusion that he or she is President of the United States. The California Supreme Court accordingly reversed the defendant's conviction for felony murder. Clearly, an examination of the defendant's specific conduct would result in the conclusion that his unlicensed practice of medicine constituted a dangerous felony. Note that the prosecutor in *Burroughs* could have avoided the entire "dangerous felony" issue by charging the defendant with implied malice, second-degree murder rather than felony murder. Some courts seek to avoid the "dangerous felony" barrier by employing both approaches in determining whether a felony is "dangerous."[42]

The felony also must have "caused" the victim's death. An arsonist who sets fire to a hotel should anticipate that a firefighter or guest may die as a consequence of the fire. On the other hand, a Virginia court ruled that a felon was not liable for the death of his accomplice whose plane crashed while transporting a cache of illegal drugs. The crash resulted from bad weather and there was no indication that the pilot modified his customary flight plan or altitude to avoid detection.[43] Keep in mind that under the theory of accomplice liability, all the co-felons will be liable for a killing committed in furtherance of the felony that is the natural and probable result of the crime.

Felony murder can become complicated where a nonfelon, such as a police officer or victim, kills one of the felons or a bystander. In *Campbell v. State,* a police officer killed an armed fleeing felon who had robbed a taxicab driver. An unarmed co-felon was later apprehended by another officer and was charged with the first-degree murder of his co-felon. The Maryland Court of Appeals adopted the agency theory of felony murder that limits criminal liability to the acts of felons and co-felons and acquitted the defendant. This theory was first stated by the Massachusetts Supreme Judicial Court in *Commonwealth v. Campbell,* which held that a felon was criminally responsible only for acts "committed by his own hand or by some one acting in concert with him in furtherance of a common object or purpose" and that a felon is not liable for the acts of a "person who is his direct and immediate adversary . . . [who is] actually engaged in opposing and resisting him and his confederates."[44]

Other courts have adopted a proximate cause theory of felony murder that holds felons responsible for foreseeable deaths that are caused by the commission of a dangerous felony. In *Kinchion v. State,* the defendant acted as a lookout while his armed accomplice entered a store. The clerk shot and killed Kinchion's co-conspirator in self-defense, and Kinchion was convicted of first-degree felony murder. The Oklahoma court affirmed the defendant's conviction, explaining that his planning and carrying out the armed robbery "set in motion 'a chain of events so perilous to the sanctity of human life' that the likelihood of death was foreseeable."[45]

As you read about felony murder, consider whether this doctrine makes sense. Oliver Wendell Holmes, Jr., in his famous book, *The Common Law,* argues that if a felon stealing chickens accidentally kills a farmer, the defendant should be punished for reckless homicide rather than felony murder. Why should the prosecutor rely on felony murder rather than establishing the elements of implied malice second-degree murder? Note that Holmes argues that prosecuting the defendant for felony murder serves little purpose because few chicken thieves will be deterred from stealing chickens by a conviction of the thief for felony murder, and it would not occur to most chicken thieves to carry a weapon in any event.[46]

The Model Penal Code shares Holmes's point of view and limits the felony murder to killings that are recklessly committed during the course of certain felonies. Section 210.2 punishes, as a felony of the first degree, killings committed purposely or knowingly as well as killings that result from "circumstances manifesting extreme indifference to the value of human life." Reckless indifference is presumed when a killing is committed during the commission, attempted commission, or flight from a robbery, sexual attack, arson, burglary, kidnapping, or felonious escape. The jury must find beyond a reasonable doubt that the defendant possessed a reckless indifference to human life.

The strength of the felony-murder doctrine is indicated by the fact that the Model Penal Code is only followed by a single state. Would you recognize felony murder if you were drafting a new state criminal code?

The next case in the textbook, *People v. Lowery,* asks whether a felon should be held criminally liable for the killing of an innocent bystander by a victim.

The Legal Equation

Felony murder = Killing of another

+ intent to commit a dangerous felony

+ killing during perpetration of a dangerous felony

+ caused by felon or co-felon as a consequence of the felony.

Should the defendant be held liable for a killing committed by a victim of his felony?

People v. Lowery, 687 N.E.2d 973 (Ill. 1997). Opinion by: Freeman, C.J.

Following a jury trial in the circuit court of Cook County, defendant, Antonio Lowery, was convicted of first-degree murder based on the commission of a felony, attempted armed robbery and two counts of armed robbery. The trial court sentenced defendant to thirty-five years' imprisonment for first-degree murder, twenty years for each of the two armed robberies, and twelve years for attempted armed robbery, to be served concurrently. On appeal, the appellate court reversed defendant's conviction and vacated his sentence for felony murder, holding that there was insufficient evidence to support defendant's conviction. We . . . now reverse the judgment of the appellate court.

Facts

On March 20, 1993, defendant was arrested and charged with two counts of armed robbery and one count of attempted armed robbery of Maurice Moore, Marlon Moore, and Robert Thomas. Defendant was also charged with the murder of Norma Sargent. In his statement to the police officers, defendant explained that he and his companion, "Capone," planned to rob Maurice, Marlon, and Robert. As Maurice, Marlon, and Robert walked along Leland Avenue in Chicago, defendant approached them, pulled out a gun, and forced Maurice into an alley. Capone remained on the sidewalk with Robert and Marlon. Once in the alley, defendant demanded Maurice's money. Maurice grabbed defendant's gun and a struggle ensued. Meanwhile, Capone fled with Robert in pursuit. Marlon ran into the alley and began hitting defendant with his fists. As defendant struggled with Maurice and Marlon, the gun discharged. The three continued to struggle onto Leland Avenue. . . . [D]efendant noticed that Marlon now had the gun. Defendant then ran from the place of the struggle to the corner of Leland and Magnolia Avenues, where he saw two women walking. As he ran, he heard gunshots and one of the women scream.

Defendant continued to run, and in an apparent attempt at disguise, he turned the Bulls jacket which he was wearing inside out. He was subsequently apprehended by the police and transported to the scene of the shooting, where Maurice identified him as the man who had tried to rob him.

At the conclusion of testimony and arguments, the jury found defendant guilty of first-degree murder under the felony-murder doctrine, two counts of armed robbery, and one count of attempted armed robbery. The appellate court reversed, holding that there was insufficient evidence to sustain a conviction for felony murder and remanded the cause for resentencing on defendant's armed robbery and attempted armed robbery convictions.

Reasoning

At issue in this appeal is whether the felony-murder rule applies where the intended victim of an underlying felony, as opposed to the defendant or his accomplice, fired the fatal shot which killed an innocent bystander. To answer this question, it is necessary to discuss the theories of liability upon which a felony-murder conviction may be based.

The two theories of liability are proximate cause and agency. In considering the applicability of the felony-murder rule where the murder is committed by someone resisting the felony, Illinois follows the "proximate cause theory." Under this theory, liability attaches under the felony-murder rule for any death proximately resulting from the unlawful activity—notwithstanding the fact that the killing was by one resisting the crime.

Alternatively, the majority of jurisdictions employ an agency theory of liability. Under this theory, "the doctrine of felony murder does not extend to a killing, although growing out of the commission of the felony, if directly attributable to the act of one other than the defendant or those associated with him in the unlawful enterprise." . . . Thus, under the agency theory, the felony-murder rule is inapplicable where the killing is done by one resisting the felony.

Defendant offers several arguments in an attempt at avoiding application of the proximate cause theory in this case. Initially, defendant urges this court to . . . adopt an agency theory of felony murder. We decline to do so. . . . Causal relation is the universal factor common to all legal liability. In the law of torts, the individual who unlawfully sets in motion a chain of events that in the natural order of things results in damages to another is held to be responsible for it.

It is equally consistent with reason and sound public policy to hold that when a felon's attempt to commit a forcible felony sets in motion a chain of events that were or should have been within his contemplation when the motion was initiated, he should be held responsible for any death that by direct and almost inevitable sequence results from the initial criminal act. Thus, there is no reason why the principle underlying the doctrine of proximate cause should not apply to criminal cases. Moreover, we believe that the intent behind the felony-murder doctrine would be thwarted if we did not hold felons responsible for the foreseeable consequences of their actions. . . .

Defendant next argues that we should abandon the proximate cause theory because Illinois originally followed the agency theory of felony murder. Notwithstanding what the law held originally, . . . a felon is liable for the deaths that are a direct and foreseeable consequence of his actions.

Defendant further argues that the plain and clear language of the Illinois Criminal Code of 1961 requires adoption of the agency theory. . . . Defendant refers to section 9–1(a) of the Code, which states:

(a) A person who kills an individual without lawful justification commits first-degree murder if, in performing the acts which cause the death:

. . .

(3) he is attempting or committing a forcible felony other than second-degree murder.

We fail to see how the plain language of the statute demonstrates legislative intent to follow the agency theory. To the contrary, the intent of the legislature is an adherence to the proximate cause theory. The legislative committee comments to section 9–1(a)(3) state that "it is immaterial whether the killing in such a case is intentional or accidental, or is committed by a confederate without the connivance of the defendant, . . . or even by a third person trying to prevent the commission of the felony."

It is the inherent dangerousness of forcible felonies that differentiates them from non-forcible felonies. As noted in the committee comments of the felony-murder statute, "it is well established in Illinois to the extent of recognizing the forcible felony as so inherently dangerous that a homicide occurring in the course thereof, even though accidentally, should be held without further proof to be within the 'strong probability' classification of murder." This differentiation reflects the legislature's concern for protecting the general populace and deterring criminals from acts of violence. . . .

Based on the plain language of the felony-murder statute, legislative intent, and public policy, we decline to abandon the proximate cause theory of the felony-murder doctrine. . . .

Because we have decided to adhere to the proximate cause theory of the felony-murder rule, we must now decide whether the victim's death in this case was a direct and foreseeable consequence of defendant's armed and attempted armed robberies. The State . . . argues that defendant was liable for decedent's death because it was reasonably foreseeable that Marlon would retaliate against defendant. We agree. A felon is liable for those deaths which occur during a felony and which are the foreseeable consequence of his initial criminal acts.

In the present case, when defendant dropped the gun and realized that Marlon was then in possession of the weapon, he believed that Marlon would retaliate, and, therefore, he ran. If decedent's death resulted from Marlon's firing the gun as defendant attempted to flee, it was, nonetheless, defendant's action that set in motion the events leading to the victim's death. It is unimportant that defendant did not anticipate the precise sequence of events that followed his robbery attempt. We conclude that defendant's unlawful acts precipitated those events, and he is responsible for the consequences. . . ."[T]hose who commit forcible felonies know they may encounter resistance, both to their affirmative actions and to any subsequent escape." . . .

Defendant . . . argues that Marlon's act was an intervening cause because it was not foreseeable that Marlon would act as a vigilante and take the law into his own hands. It is true that an intervening cause completely unrelated to the acts of the defendant does relieve a defendant of criminal liability. However, the converse of this is also true: When criminal acts of the defendant have contributed to a person's death, the defendant may be found guilty of murder. . . . Marlon's resistance was in direct response to defendant's criminal acts and did not break the causal chain between defendant's acts and decedent's death. It would defeat the purpose of the felony-murder doctrine if such resistance—an inherent danger of the forcible felony—could be considered a sufficient intervening circumstance to terminate the underlying felony or attempted felony. . . .

Furthermore, we do not believe that Marlon acted as a vigilante, or that because of our holding, the citizenry will have license to practice vigilantism. A vigilante is defined as a member of "a group extra-legally assuming authority for summary action professedly to keep order and punish crime because of the alleged lack or failure of the usual law-enforcement agencies." Regardless of how unreasonable Marlon's conduct may, in hindsight, be perceived, his response was not based on a deliberate attempt to take the law into his own hands, but on his natural, human instincts to protect himself.

Defendant next argues that decedent's death falls outside the scope of felony murder because it did not occur during the course of defendant's armed robbery and attempted armed robbery. In support, he relies on section 7–4 of the Code, which states that justifiable use of force is not available to a person who provokes the use of force against himself unless he withdraws from physical contact with the assailant and indicates clearly to the assailant that he desires to withdraw and terminate the use of force, but the assailant continues or resumes the use of force.

Defendant maintains that he had "overtly retreated" from physical contact with Marlon and that Marlon's pursuit of defendant constituted a new conflict. The State, on the other hand, argues that defendant's election to flee fell within the commission of the armed and attempted armed robberies.

We must agree with the State. This court has consistently held that when a murder is committed in the course of an escape from a robbery, each of the conspirators is guilty of murder under the felony-murder statute, inasmuch as the conspirators have not won their way to a place of safety. . . .

Defendant asserts that he had reached a place of "legal safety" when he ran. We disagree. Defendant was attempting to escape when Marlon fired the shots at him. Apparently, defendant also did not believe he had "won a place of safety" as evidenced by his own act of turning his coat inside out to avoid detection before the police arrested him. Therefore, decedent's death falls within the scope of the felony-murder doctrine.

Defendant's final contention that Marlon was not legally justified in firing at defendant is misplaced. There is no claim that Marlon shot at defendant in self-defense or in an attempt to arrest him. Moreover, the proper focus of this inquiry is not whether Marlon was justified in his actions, but whether defendant's actions set in motion a chain of events that ultimately caused the death of decedent. We hold that defendant's actions were the proximate cause of decedent's death and the issue of whether Marlon's conduct was justified is not before this court.

Holding

In conclusion, we hold that the evidence was sufficient to prove defendant guilty beyond a reasonable doubt under the felony-murder rule. . . . We therefore reverse the appellate court. . . . Accordingly, we remand the cause to the appellate court for consideration of defendant's remaining issues.

Questions for Discussion

1. Explain the difference between the agency and proximate cause theories of causality. Which theory favors the prosecutor in *Lowery*? Which theory favors the defense? What theory do you think is best to apply in *Lowery*?

2. Did Marlon fire the pistol in justifiable self-defense? Was Marlon's pistol shot an intervening act that should limit Lowery's criminal liability? Is Lowery correct when he argues that Marlon's act was "not foreseeable" in that he could not anticipate that Marlon would act as a "vigilante and take the

law into his own hands"? Did Lowery set the events in motion that led to the victim's death?

3. Lowery argues that he clearly retreated, that the felony had been completed, and therefore he should not be held liable for felony murder. How does the court respond to this argument?

4. Would it be fair to hold Capone liable for felony murder, particularly given Marlon's inaccurate marksmanship? Would you hold Capone liable in the event that Marlon shot and killed Lowery in self-defense?

You Decide

10.6 John Malaske was convicted of second-degree felony murder for providing vodka to his underage, eighth-grade sister, knowing that she planned to share the alcohol with her friends. One of her friends later drank to excess, passed out, and died of alcohol poisoning. The Oklahoma second-degree murder statute provides that a felony "must be inherently or potentially dangerous to human life, inherently dangerous in light of the facts and circumstances surrounding both the felony and the homicide, or potentially dangerous in light of the facts and circumstances surrounding both the felony and the homicide."

The Oklahoma Court of Criminal Appeals ruled that furnishing alcohol to a minor who is under twenty-one is a "potentially dangerous" felony and that Malaske was legally liable for felony murder. The court held that the offense was not completed upon the delivery of the alcohol. It reasoned that minors are in a "protected class" and the prohibition on providing alcohol is intended to prevent juveniles from drinking to excess and endangering themselves and others. The dissent pointed out that alcohol does not pose an inherent threat to human life and is fully available in stores to adults. The dissent concluded that this is the "ultimate version of the extremists' blame game," and that Malaske should not be branded a murderer as a result of the poor judgment of juveniles. Was providing alcohol to the defendant's underage sister a felony that is dangerous to human life? Should felony murder be limited to felonies such as robbery, rape, arson, burglary, and kidnapping? How did the defendant's act cause the death of the young juvenile victim? Do you agree with the dissent that the "imposition of liability for unintended deaths erodes the relationship between criminal liability and moral culpability"? See *Malaske v. State*, 89 P.3d 1116 (Okla. Crim. App. 2004).

You can find the answer at www.sagepub.com/lippmanccl3e.

You Decide

10.7 Sanexay Sophophone and three other individuals broke into a house in Emporia, Kansas. Police officers responded to a call from residents and spotted four individuals leaving the back of the house. They shined a light on the suspects and ordered them to stop. An officer ran down Sophophone, handcuffed him, and placed him in a police car.

Another officer chased Somphone Sysoumphone. Sysoumphone crossed railroad tracks, jumped a fence, and then stopped. The officer approached with his weapon drawn and ordered Sysoumphone to the ground and not to move. Sysoumphone complied with the officer's command but, while lying face down, rose up and fired at the officer, who returned fire and killed him. Sophophone was charged with conspiracy to commit aggravated burglary, obstruction of official duty, and felony murder. . . . The question of law before the Kansas Supreme Court is whether Sophophone can be convicted of felony murder for the "killing of a co-felon not caused by his acts but by the lawful acts of a police officer acting in self-defense in the course and scope of his duties in apprehending the co-felon fleeing from an aggravated burglary." The Kansas Supreme Court held that the "overriding fact . . . is that neither Sophophone nor any of his accomplices 'killed' anyone. . . . We believe that making one criminally responsible for the lawful acts of a law enforcement officer is not the intent of the felony-murder statute." The dissent pointed out that the rationale for felony murder is that it serves as a general deterrent. Potential felons will be hesitant to engage in criminal activity if they realize that they risk being convicted of first-degree murder in the event that a death occurs during the commission of a felony. "Sophophone set in motion acts which would have resulted in the death or serious injury of a law enforcement officer had it not been for the highly alert law enforcement officer." This "could have very easily resulted in the death of a law enforcement officer . . . [and] is exactly the type of case the legislature had in mind when it adopted the felony-murder rule. . . . It does not take much imagination to see a number of situations where a death is going to result from an inherently dangerous felony and the majority's opinion is going to prevent the accused from being charged with felony murder." What is your view? Should the Kansas court use the agency or proximate cause theory? See *State v. Sophophone*, 79 P.3d 70 (Kan. 2001).

You can find the answer at www.sagepub.com/lippmanccl3e.

Corporate Murder

Should a corporation be held liable for murder? In 1980, the Ford Motor Company was prosecuted for reckless homicide stemming from the 1978 death of three Indiana teenagers. The three were burned to death when their 1972 Ford Pinto was hit from behind by a van. Prosecutors charged that Ford was aware that the Pinto's gasoline tanks were in danger of catching fire when impacted by a rear-end collision. Ford was alleged to have decided that fixing the problem or recalling the Pinto would deeply cut into profits and decided that it would be less expensive to pay any damage awards that might result from civil suits filed by consumers. By 1977, the Pinto no longer was able to meet tough federal safety standards and, in late 1978, Ford recalled 1.5 million 1971–1976 Pinto sedans. Unfortunately, this recall was not issued in time to save the lives of the three victims. Ford was acquitted in a jury trial in March 1980.[47]

In 1999, A Florida jury found airline maintenance company SabreTech guilty of contributing to the 1996 crash of ValuJet Flight 593, an accident that resulted in the death of 110 passengers. The company allegedly had been responsible for placing prohibited hazardous materials on the ValuJet flight that exploded during flight. SabreTech was convicted on eight counts of mishandling hazardous materials and one count of failing to properly train employees.

In 2003, Motiva Enterprises pled "no contest," or ***nolo contendere*** (a guilty plea for purposes of a particular prosecution), to one felony count of criminally negligent homicide and six misdemeanor counts of assault in the third degree. This plea arose out of a July 2001 explosion and fire at a company factory that resulted in the death of one employee and injury to six others. Prosecutors alleged that Motiva, a joint venture between Saudi Aramco and Royal Dutch Shell, ignored warnings and continued to operate the plant in order to maximize profits. The company's conviction resulted in a fine of $11,500 on the homicide charge and $5,750 for each of the assault charges for a total of $46,000, the maximum then permitted under Delaware law.

In 2005, the Far West Water and Sewer Company was convicted of the murder of two workers who died from toxic chemicals while working on an underground sewer tank. The company was fined 1.7 million dollars and required to pay restitution to the families of the dead workers.

These four cases illustrate that a corporation may be held liable for **corporate murder** in those cases in which conduct is performed or approved by corporate managers or officials. Of course, individual managers and executives may also be held criminally responsible. The extension of criminal responsibility to corporations is based on an interpretation of the term "person" in homicide statutes to encompass both natural persons and corporate entities.

For an international perspective on this topic, visit the study site.

A corporation clearly cannot be incarcerated and, instead, is punished by the imposition of a fine. It is reasoned that the threat of a fine will motivate corporate officials and individuals owning stock in the firm to ensure that the corporation follows the law. On the other hand, some would argue that criminal responsibility is properly limited to the individuals who commit the crimes. A fine on a business hurts only the workers and stockholders who depend on strong corporate profits and creates a poor business climate that leads corporations to move their factories to other countries.

State v. Richard Knutson, Inc. is a prominent case involving corporate criminal liability. Ask yourself whether it serves any purpose to hold the corporation liable in this case or whether responsibility should be limited to corporate officials.[48]

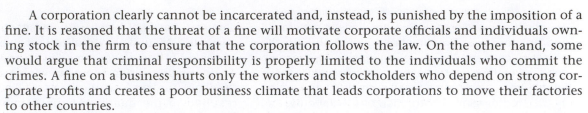

Was the corporation guilty of murder?

State v. Richard Knutson, Inc., 537 N.W.2d 420
(Wis. Ct. App. 1995). Opinion by: Anderson, J.

Facts

In the spring of 1991, Richard Knutson, Inc. (RKI) undertook the construction of a sanitary sewer line for the City of Oconomowoc. On May 20, 1991, while working in an area adjacent to some Wisconsin Electric Power Company power lines, a work crew attempted to place a section of corrugated metal pipe in a trench in order to remove groundwater. The backhoe operator misjudged the distance from the boom of the backhoe to the overhead power lines and did not realize he had moved the stick of the boom into contact with the wires. In attempting to attach a chain to the backhoe's bucket, a member of the crew was instantly electrocuted. . . . The jury found RKI guilty as charged. The trial court entered judgment, concluding that the evidence was sufficient to support the verdict of negligent homicide. The Wisconsin Statute (WS) reads as follows: "§ 940.10 STATS. Homicide by negligent operation of vehicle. Whoever causes the death of another human being by the negligent operation or handling of a vehicle is guilty of a Class E felony."

Issue

RKI raises the same challenges to WS section 940.10, STATS.—homicide by negligent operation of a vehicle statute—as it did in the trial court. The trial court held that section 940.10 covered acts by corporations. . . . The trial court decided that corporate liability was within the spirit of section 940.10, stating, "The purpose of the statute is to protect employees or anyone from the negligent conduct of another which may cause death. It should not matter that the 'another' is a person or corporation as long as the conduct is criminal. . . ."

On appeal, RKI insists that a corporation cannot be held accountable for homicide. RKI argues that "the statute uses the word 'whoever' and the correlative phrase 'another human being.' In the context of this sentence, 'whoever' necessarily refers to a human being. By its own terms, the statute therefore limits culpability for homicide by operation of a vehicle to natural persons." . . . The State argues that when used in the homicide statutes, the word "whoever" refers to natural or corporate persons. The State reasons that either can be liable for taking the life of "another human being."

Here, the statute does not provide a definition of "whoever." It is left to the reader to determine if "whoever" should be read expansively to include natural and artificial persons, or should be read narrowly and have its definition gleaned from its reference to the correlative phrase "another human being." Our task is to ascertain if the legislative intent is to include corporations within the class of perpetrators. This task is made more difficult by the legislature's use of the term "whoever" to identify the perpetrator of a crime and its failure to define that term. Why, when it rewrote the criminal code in 1955, the legislature chose to describe perpetrators with the ambiguous term "whoever" is an enigma. Another mystery is the deletion of any statutory language establishing corporate liability for criminal acts. This was eliminated, upon motion of an advisory committee member who was a house counsel for a large industrial corporation. . . .

Reasoning

Legislative inaction, in the face of repeated Wisconsin Supreme Court pronouncements that corporations can be held liable for criminal acts, convinces us that the legislature concurs in the Supreme Court's decisions. On two separate occasions the legislature significantly revised the homicide statutes; both times it is presumed that the legislature was aware that court decisions have held corporations criminally liable; and on both occasions, the legislature has elected not to undo corporate criminal liability. Our conclusion conforms to the modern trend of the law. A leading treatise on corporations acknowledges that a corporation may be held to answer for its criminal acts, including homicide. . . . The Model Penal Code also has several provisions holding corporations accountable for criminal behavior.

Wayne R. LaFave and Austin W. Scott summarize the persuasive policy considerations supporting corporate criminal liability. Among those considerations is the factor that the corporate business entity has become a way of life in this country and the imposition of criminal liability is an essential part of the regulatory process. Another consideration centers on the premise that it would be unjust to single out one or more persons for criminal punishment when it is the corporate culture that is the origin of the criminal behavior. Also, the size of many corporations makes it impossible to adequately allocate responsibility to individuals. An additional consideration is the "indirect economic benefits that may accrue to the corporation through crimes against the person. To get these economic benefits, corporate management may shortcut expensive safety precautions, respond forcibly to strikes, or engage in criminal anticompetitive behavior." It has also been suggested that the free market system cannot be depended upon to guide corporate decisions in socially acceptable ways, and the threat of imposition of criminal liability is needed to deter inappropriate (criminal) corporate behavior.

RKI insists that Wisconsin has disregarded the modern trend of criminal law to hold corporations liable for criminal acts and that only individuals may be held liable. RKI's argument ignores reality. A corporation acts of necessity through its agents; therefore, the only way a corporation can negligently cause the death of a human is by the act of its agent—another human. Reading the statute to limit its coverage to perpetrators who are human, as suggested by RKI, skirts around the concepts of vicarious and enterprise liability. If a human was operating a vehicle within the scope of his or her employment when the death occurred, RKI's construction would permit the corporation to escape criminal prosecution simply because it is not a human being.

Holding

RKI's attempt to limit the class of perpetrators to natural persons ignores that . . ."finding moral responsibility and criminal liability does not depend on first determining whether an entity is a person." We are satisfied that the history of corporate criminal liability in Wisconsin prescribes the results reached. The construction of WS section 940.10, STATS., to include corporations is consistent with public policy and practice. In reaching this conclusion, we are cognizant of Justice Holmes' observation that, "A word is not a crystal, transparent and unchanged, it is the skin of a living thought and may vary greatly in color and content according to the circumstances and the time in which it is used."

Homicide by negligent use of a vehicle has three elements: "(1) that the defendant cause death (2) by criminal negligence (3) in the operation of a vehicle." . . . Of course, by necessity a corporation can only act through its employees, agents, or officers; therefore, it is the negligence of the employee that must rise to the level of criminal negligence.

Criminal negligence differs from ordinary negligence in two respects. First, the risk is more serious—death or great bodily harm as opposed to simple harm. Second, the risk must be more than an unreasonable risk—it must also be substantial. Criminal negligence involves the same degree of risk as criminal recklessness—an unreasonable and substantial risk of death or great bodily harm. The difference between the two is that recklessness requires that the actor be subjectively aware of the risk, while criminal negligence requires only that the actor should have been aware of the risk—an objective standard. . . .

The evidence permits the reasonable inference that RKI neglected to act with due diligence to ensure the safety of its employees as they installed sewer pipes in the vicinity of overhead electrical lines. RKI's management took no action to have the power lines de-energized or barriers erected; rather, management elected to merely warn employees about the overhead lines. A finder of fact would be justified in reasonably inferring that RKI had ample notice that the existence of overhead power lines would interfere with the job, and unless there was compliance with safety regulations, working in the vicinity of the overhead lines posed a substantial risk to its employees.

The evidence supports the conclusion that if RKI had enforced the written safety regulations issued by the federal government, had abided by its own written safety program, and had complied with the contract requirements for construction on Wisconsin Electric's property, the electrocution death would likely not have happened. The finder of fact was justified in concluding that RKI operated vehicles in close proximity to the overhead power lines without recognizing the potential hazard to its employees in the vicinity of the vehicles. The jury could reasonably find that RKI's failure to take elementary precautions for the safety of its employees was a substantial cause of the electrocution death.

Dissenting, *Brown, J.*

What this debate really comes down to is whether it is desirable that a court avoid the literal meaning of this statute. I acknowledge that there exists a tension between the language of the statute and the announced public policy goal by some of our citizenry that corporations be held to criminal liability for negligent deaths. And I reject the notion that we should never search for the "real" rule lying behind the mere words on a printed page. But when the statute's wording is so clear in its contextual rigidity, the statute has therefore generated an answer which excludes otherwise eligible answers from consideration. Unlike the majority, I take the clear wording of the statute seriously. Since the majority has seen fit to quote Justice Oliver Wendell Holmes, Jr., I too quote from a past justice of the nation's highest court. Justice Robert Jackson wrote: "I should concur in this result more readily if the Court could reach it by analysis of the statute instead of by psychoanalysis of Congress." My sentiments exactly.

Questions for Discussion

1. What facts formed the basis of Richard Knutson, Inc.'s conviction of negligent homicide?

2. List the corporate officials and employees who likely were negligent. Why did the prosecutor choose to prosecute the corporation rather than these individuals? Why did the prosecutor not charge Richard Knutson, Inc., with reckless homicide?

3. Summarize the argument in favor of extending the statute to cover corporations. Are you persuaded by the argument? Should states adopt statutes explicitly punishing corporate murder?

4. What is the purpose of holding Richard Knutson, Inc., criminally liable? Is the death of an employee better addressed through a civil suit seeking monetary compensation to the victim's family?

Cases and Comments

Corporate Criminal Liability for Deaths and Injuries. The charges against Far West Water & Sewer, Inc. arose from an incident that occurred on October 24, 2001, at a sewage collection and treatment facility owned and operated by Far West, an Arizona corporation. Santec Corporation ("Santec") at the time was a subcontractor of Far West. A Far West employee, James Gamble, and a Santec employee, Gary Lanser, died in an underground sewage tank after they were overcome by hydrogen sulfide gas. Another Far West employee, Nathan Garrett, suffered severe injuries when he attempted to rescue Gamble from the tank.

Dr. Daniel Teitelbaum, a physician specializing in occupational medicine and toxicology, and an OSHA expert and consultant, concluded that Gamble and Lanser died from acute hydrogen sulfide poisoning. The Yuma County, Arizona medical examiner found that both of the deceased workers were overcome by inhalation of sewage gas and that the immediate cause of death was asphyxia due to drowning. Although Garrett survived, he suffered life-threatening respiratory distress syndrome and aspiration pneumonia and sustained injuries to his lungs and eyes.

Arizona law § 13-305 holds an "enterprise" criminally liable. Far West was indicted along with Far West's president, Brent Weidman, and other corporate officials. The jury convicted Far West of negligent homicide, aggravated assault and endangerment, and of one count of violating a safety standard or regulation that caused the death of Gamble. Far West President Brent Wiedman was later convicted of two counts of negligent homicide and two counts of endangerment.

An Arizona appeals court concluded that a jury could reasonably conclude that Weidman "consciously disregarded a substantial and unjustifiable risk of death or physical injury by . . . permitting Far West employees to enter dangerous, life-threatening underground tanks without training, equipment, safety measures or rescue capability." See *State v. Far West Water & Sewer, Inc.*, 228 P.3d 909 (Az. Ct. App. 2010).

See more cases on the study site: **State v. Far West Water & Sewer, Inc., People v. O'Neil,** *www.sagepub.com/lippmanccl3e.*

You Decide

10.8 In 1983, an employee at Pyro Science Development Corporation, a fireworks manufacturer, plugged a fan into an electric outlet, generating sparks that caused a fire and explosion, killing one plant employee and injuring several others. Cornellier was director of operations at the plant and was charged and convicted of homicide by reckless conduct. Reckless conduct is defined under Wisconsin law as an act that creates a "situation of unreasonable risk and high probability of death . . . and which demonstrates a conscious disregard for the safety of another and a willingness to take known chances of perpetrating injury." Cornellier was allegedly manufacturing fireworks without a permit, in a factory that did not meet safety requirements.

Three weeks prior to the explosion, Cornellier had been convicted of six violations of safety ordinances at a nearby plant and was advised that there were "high risks" of danger at the plant at which the accident occurred.

Following the explosion, the U.S. Department of Labor found nine separate violations of federal safety standards at the plant, including a lack of adequate precautions against the ignition of flammable vapors, the failure to provide a safe avenue of escape from the building, a mishandling of explosive materials in a fashion hazardous to life, and a failure to protect employees from explosion and fire. The Wisconsin court stressed that gross negligence could result from a failure to fulfill a duty to safeguard workers, as well as from an affirmative act. Judge Eich stressed that Cornellier's failure to provide safe storage of explosive materials and a safe electrical system substantially contributed to producing the employee's death. Can Cornellier be held criminally liable for a failure to act to correct the safety hazards at the plant? See *Cornellier v. Black*, 425 N.W.2d 21 (Wis. Ct. App. 1988).

You can find the answer at www.sagepub.com/lippmanccl3e.

MANSLAUGHTER

Manslaughter comprises a second category of homicide and is defined as an unlawful killing of another human being without malice aforethought.

The common law distinction between voluntary manslaughter and the less severe offense of involuntary manslaughter continues to appear in many state statutes. Other statutes distinguish between degrees of manslaughter, and a third approach provides for a single offense of manslaughter. Voluntary manslaughter is the killing of another human being committed in a sudden heat of passion in response to adequate provocation. Adequate provocation is considered a provocation that would cause a reasonable person to lose self-control. Involuntary manslaughter is the killing of another human being as a result of criminal negligence. Criminal negligence involves a gross deviation from the standard of care that a reasonable person would practice under similar circumstances.[49] Remember, as we discussed in Chapter 8, an unreasonable, but good-faith belief in the necessity of self-defense in many states is recognized as imperfect self-defense and is punished as manslaughter.

Voluntary Manslaughter

One function of criminal law is to remind us that we will be prosecuted and punished in the event that we allow our anger or frustration to boil over and assault individuals or destroy their property. Voluntary manslaughter seemingly is an exception to the expectation that we control our emotions. This offense recognizes that a reasonable person, under certain circumstances, will be provoked to lose control and kill. In such situations, it is only fair that an individual should receive a less serious punishment than an individual who kills in a cool and intentional fashion.

Voluntary manslaughter requires that an individual kill in a sudden and intense **heat of passion** in response to adequate provocation. Heat of passion is commonly described as anger but is sufficiently broad to include fear, jealousy, and panic.

The law of provocation is based on the reaction of the **reasonable person**, a fictional balanced, sober, and fair-minded human being with no physical or mental imperfections. *Adequate provocation* is defined as conduct that is sufficient to excite an intense passion that causes a reasonable person to lose control. The common law restricted adequate provocation to a limited number of situations: aggravated assault or battery, mutual combat defined as a fight voluntarily entered into by the participants, a serious crime committed against a close relative of a defendant, and one spouse observing the adultery of the other spouse. An individual who is overwhelmed by jealousy and anger after observing her boyfriend in an act of sexual interaction with another would not have a claim of heat of passion because the victim is not her husband. Keep in mind that the provocation must cause a reasonable person to lose control (objective component) and the defendant, in fact, must have lost control and killed in a heat of passion (subjective component).[50]

In some cases, courts have instructed jurors that they possess discretion to determine whether an event provides adequate provocation. This certainly seems preferable to limiting provocation to acts that were viewed as provocative by common law judges in England. On the other hand, this

Figure 10.1 Crime on the Streets: Homicide Victimization

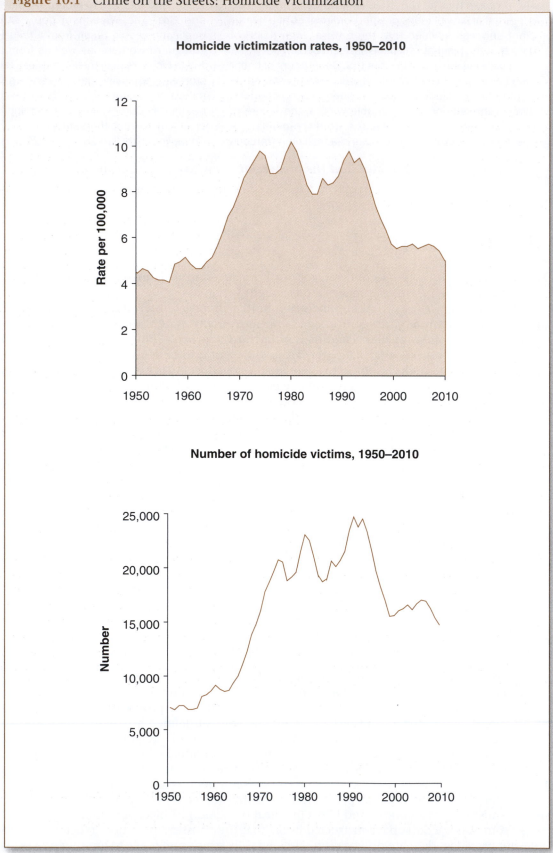

Source: FBI, *Uniform Crime Reports, 1950–2010.*

may lead to controversy concerning questions such as whether a nonviolent homosexual advance constitutes adequate provocation.[51] In a famous English case, the court ruled that an impotent individual was not reasonably provoked when he killed a prostitute who poked fun at his disability. The judges explained that the standard to be used was the hypothetical reasonable person rather than a reasonable person who was impotent.[52]

Jurists increasingly follow the Model Penal Code and are willing to recognize physical traits such as blindness or disease in determining whether an act constitutes reasonable provocation. However, judges are generally reluctant to broaden the traits of a reasonable person to include a defendant's moral views and attitudes. A Pennsylvania court, for instance, rejected a defendant's plea of voluntary manslaughter when he recounted that he was driven to kill a lesbian whom he witnessed making love as a result of his own mother's lesbianism and his anger at having been molested at a young age by a homosexual male.[53] This case illustrates the complexity of considering an individual's views and attitudes. Should we recognize the cultural background of an immigrant parent from the Middle East who kills his daughter as a "matter of honor" after he observes her sexually interacting with her boyfriend?[54]

The defense of sudden heat of passion is unavailable if a reasonable person's passion would have experienced a **cooling of blood** between the time of the provocation and the time of killing. Some common law courts followed an ironclad rule that limited the impact of provocation to twenty-four hours. The modern approach is to view the facts and circumstances of a case and to determine whether a reasonable person's "blood would have cooled" and whether the defendant's "blood had cooled." One court recognized provocation lasting for more than twenty-four hours in the case of a defendant who had been informed that his father-in-law had raped the defendant's wife.[55] In a frequently cited case, the victim was sodomized while unconscious. The perpetrator spread news of the defendant's victimization throughout the community and subjected the victim to what the victim viewed as humiliating comments and embarrassment. The defendant boiled over in rage after two weeks of this harassment and killed the defendant. The court ruled that the cumulative impact of the harassment would not be taken into consideration and that too much time had passed to recognize involuntary manslaughter.[56]

Courts confront the challenge of determining whether an individual killed in a heat of passion as a result of the provocation or killed in a calm and intentional fashion. In *People v. Bridgehouse*, Marylou Bridgehouse informed her husband that while he was working two jobs to support his wife and their two sons, she had been having an affair with William Bahr for the past six or seven months. Bridgehouse later learned that Bahr and Marylou had a joint bank account and that Marylou had used his credit card to purchase a gift for Bahr; and Bridgehouse later discovered Bahr's clothes hanging in his closet. On the morning of the killing, Bridgehouse invited his wife to go skiing; but Marylou stated that she planned to go fishing with her mother and Bahr. Bridgehouse, a former sheriff, placed his service revolver in his belt and went over to his mother-in-law's house to allegedly get a pair of socks for his young son. Bridgehouse testified that he did not realize that Bahr was living there and, when encountering Bahr, lost control and only regained consciousness when he realized that his revolver was clicking on empty and Bahr was dead on the ground. A police officer testified that Bridgehouse was emotionally overwhelmed following the killing and told the officer that he wanted to "tell off" Bahr and that he "didn't want him around my children." Did Bridgehouse go to his mother-in-law's house with the intent to kill Bahr or was he taken by surprise and driven by emotion to kill? Could he have intended to kill Bahr and then found himself overwhelmed by emotion at the time that he arrived at his mother-in-law's home? Is there always a clear line behind intentional and premeditated killing and murder in the heat of passion?[57]

Voluntary Manslaughter Reconsidered

Voluntary manslaughter involves several "hurdles:"[58]

- *Provocation.* An individual must be *reasonably* and *actually* provoked and must *kill in a heat of passion.*
- *Cooling of Blood.* An individual must have *reasonably* and *actually* not "cooled off."

The question remains whether the law should recognize voluntary manslaughter. An individual who loses control and impulsively kills clearly poses a threat to society and might be viewed to be as dangerous as an individual who intentionally and calmly kills. Should we accept that a "reasonable person" can be driven to kill in the heat of passion and therefore should be subject to less severe punishment than other categories of killers?

In the next case, *Girouard v. State,* the court considers whether "words" may be sufficiently provocative to drive a reasonable person to kill. Courts generally have refused to recognize insulting and racist language as adequate provocation. However, there are decisions recognizing that "informational" words may constitute adequate provocation. For instance, in *State v. Flory,* the Wyoming Supreme Court recognized that a husband had been reasonably provoked by his wife's informing him that she had been raped by her father.[59]

The Legal Equation

Voluntary manslaughter **=** Killing another person

+ intent to kill

+ sudden heat of passion based on adequate provocation.

Can words constitute provocation?

Girouard v. State, 583 A.2d 718 (Md. 1991). Opinion by: Cole, J.

In this case we are asked to . . . determine whether words alone are provocation adequate to justify a conviction of manslaughter rather than one of second-degree murder.

Facts

Steven S. Girouard and the deceased, Joyce M. Girouard, had been married for about two months on October 28, 1987, the night of Joyce's death. Both parties, who met while working in the same building, were in the army. They married after having known each other for approximately three months. The evidence at trial indicated that the marriage was often tense and strained, and there was some evidence that after marrying Steven, Joyce had resumed a relationship with her old boyfriend, Wayne.

On the night of Joyce's death, Steven overheard her talking on the telephone to her friend, whereupon she told the friend that she had asked her first sergeant for a hardship discharge because her husband did not love her anymore. Steven went into the living room where Joyce was on the phone and asked her what she meant by her comments; she responded, "[N]othing." Angered by her lack of response, Steven kicked away the plate of food Joyce had in front of her. He then went to lie down in the bedroom.

Joyce followed him into the bedroom, stepped up onto the bed and onto Steven's back, pulled his hair, and said, "What are you going to do, hit me?" She continued to taunt him by saying, "I never did want to marry you and you are a lousy ___ and you remind me of my dad." The barrage of insults continued with her telling Steven that she wanted a divorce, that the marriage had been a mistake, and that she had never wanted to marry him. She also told him she had seen his commanding officer and filed charges against him for abuse. She then asked Steven, "What are you going to do?" Receiving no response, she continued her verbal attack. She added that she had filed charges against him in the Judge Advocate General's Office (JAG) and that he would probably be court-martialed.

There was some testimony presented at trial to the effect that Joyce had never gotten along with her father, at least in part because he had impregnated her when she was fourteen, the result of which was an abortion. Joyce's aunt, however, denied that Joyce's father was the father of Joyce's child. In addition, Joyce lied about filing the charges against her husband.

When she was through, Steven asked her if she had really done all those things, and she responded in the affirmative. He left the bedroom with his pillow in his arms and proceeded to the kitchen where he procured a long handled kitchen knife. He returned to Joyce in the bedroom with the knife behind the pillow. He testified that he was enraged and that he kept waiting for Joyce to say she was kidding, but Joyce continued talking. She said she had learned a lot from the marriage and that it had been a mistake. She also told him she would remain in their apartment after he moved out. When he questioned how she would afford it, she told him she would claim her brain-damaged sister as a dependent and have the sister move in. Joyce reiterated that the marriage was a big mistake, that she did not love him, and that the divorce would be better for her.

After pausing for a moment, Joyce asked what Steven was going to do. What he did was lunge at her with the kitchen knife he had hidden behind the pillow and stab her nineteen times. Realizing what he had done, he dropped the knife and went to the bathroom to shower off Joyce's blood. Feeling like he wanted to die, Steven went back to the kitchen and found two steak knives with which he slit his own wrists. He lay down on the bed waiting to die, but when he realized that he would not die from his self-inflicted wounds, he got up and called the police, telling the dispatcher that he had just murdered his wife.

When the police arrived they found Steven wandering around outside his apartment building. Steven was despondent and tearful and seemed detached, according to police officers who had been at the scene. He was unconcerned about his own wounds, talking only about how much he loved his wife and how he could not believe what he had done. Joyce Girouard was pronounced dead at the scene.

At trial, defense witness, psychologist Dr. William Stejskal, testified that Steven was out of touch with his own capacity to experience anger or express hostility. He stated that the events of October 28, 1987, were entirely consistent with Steven's personality, that Steven had "basically reach[ed] the limit of his ability to swallow his anger, to rationalize his wife's behavior, to tolerate, or actually to remain in a passive mode with that. He essentially went over the limit of his ability to bottle up those strong emotions. What ensued was a very extreme explosion of rage that was intermingled with a great deal of panic." Another defense witness, psychiatrist Thomas Goldman, testified that Joyce had a "compulsive need to provoke jealousy so that she's always asking for love and at the same time destroying and undermining any chance that she really might have to establish any kind of mature love with anybody."

Steven Girouard was convicted, at a court trial in the Circuit Court for Montgomery County, of second-degree murder and was sentenced to twenty-two years incarceration, ten of which were suspended. Upon his release, Petitioner is to be on probation for five years, two years supervised and three years unsupervised. . . . We granted certiorari to determine whether the circumstances of the case presented provocation adequate to mitigate the second-degree murder charge to manslaughter.

Issue

Girouard relies primarily on out-of-state cases to provide support for his argument that the provocation to mitigate murder to manslaughter should not be limited only to the traditional circumstances. . . . Steven argues that the trial judge did find provocation (although he held it inadequate to mitigate murder) and that the categories of provocation adequate to mitigate should be broadened to include factual situations such as this one.

The State counters by stating that although there is no finite list of legally adequate provocations, the common law has developed to a point at which it may be said there are some concededly provocative acts that society is not prepared to recognize as reasonable. Words spoken by the victim, no matter how abusive or taunting, fall into a category society should not accept as adequate provocation. According to the State, if abusive words alone could mitigate murder to manslaughter, nearly every domestic argument ending in the death of one party could be mitigated to manslaughter. This, the State avers, is not an acceptable outcome. Thus, the State argues that the courts below were correct in holding that the taunting words by Joyce Girouard were not provocation adequate to reduce Steven's second-degree murder charge to voluntary manslaughter.

Reasoning

Initially, we note that the difference between murder and manslaughter is the presence or absence of malice. Voluntary manslaughter has been defined as "an intentional homicide, done in a sudden heat of passion, caused by adequate provocation, before there has been a reasonable opportunity for the passion to cool."

There are certain facts that may mitigate what would normally be murder to manslaughter. For example, we have recognized as falling into that group: (1) discovering one's spouse in the act of sexual intercourse with another; (2) mutual combat; (3) assault and battery. There is also authority recognizing injury to one of the defendant's relatives or to a third party, and death resulting from resistance of an illegal arrest as adequate provocation for mitigation to manslaughter. . . . Those acts mitigate homicide to manslaughter because they create passion in the defendant and are not considered the product of free will.

In order to determine whether murder should be mitigated to manslaughter we look to the circumstances surrounding the homicide and try to discover if it was provoked by the victim. Over the facts of the case, we lay the template of the so-called "Rule of Provocation." The courts of this State have repeatedly set forth the requirements of the Rule of Provocation:

1. There must have been adequate provocation;

2. The killing must have been in the heat of passion;

3. It must have been a sudden heat of passion— that is, the killing must have followed the provocation before there had been a reasonable opportunity for the passion to cool; and

4. There must have been a causal connection between the provocation, the passion, and the fatal act.

We shall assume without deciding that the second, third, and fourth of the criteria listed above were met in this case. We focus our attention on an examination of the ultimate issue in this case, that is, whether the provocation of Steven by Joyce was enough in the eyes of the law so that the murder charge against Steven should have been mitigated to voluntary manslaughter. For provocation to be "adequate," it must be "calculated to inflame the passion of a reasonable man and tend to cause him to act for the moment from passion rather than reason." . . . The issue we must resolve, then, is whether the taunting words uttered by Joyce were enough to inflame the passion of a reasonable man so that that man would be sufficiently infuriated so as to strike out in hot-blooded blind passion to kill her. Although we agree with the trial judge that there was needless provocation by Joyce, we also agree with him that the provocation was not adequate to mitigate second-degree murder to voluntary manslaughter.

Before the shooting, the victim had called the appellant "a chump" and "a chicken," dared the appellant to fight, shouted obscenities at him, and shook her fist at him. . . .

[W]ords can constitute adequate provocation if they are accompanied by conduct indicating a present intention and ability to cause the defendant bodily harm. Clearly, no such conduct was exhibited by Joyce in this case. While Joyce did step on Steven's back and pull his hair, he could not reasonably have feared bodily harm at her hands. This, to us, is certain based on Steven's testimony at trial that Joyce was about 5'1" tall and weighed 115 pounds, while he was 6'2" tall, weighing over 200 pounds. Joyce simply did not have the size or strength to cause Steven to fear for his bodily safety. Thus, since there was no ability on the part of Joyce to cause Steven harm, the words she hurled at him could not . . . constitute legally sufficient provocation. . . .

Holding

Thus, with no reservation, we hold that the provocation in this case was not enough to cause a reasonable man to stab his provoker nineteen times. Although a psychologist testified to Steven's mental problems and his need for acceptance and love, we agree with the Court of Special Appeals speaking through Judge Moylan that "there must be not simply provocation in psychological fact, but one of certain fairly well-defined classes of provocation recognized as being adequate as a matter of law." The standard is one of reasonableness; it does not and should not focus on the peculiar frailties of mind of the Petitioner. That standard of reasonableness has not been met here. We cannot in good conscience countenance holding that a verbal domestic argument ending in the death of one spouse can result in a conviction of manslaughter. We agree with the trial judge that social necessity dictates our holding. Domestic arguments easily escalate into furious fights. We perceive no reason for a holding in favor of those who find the easiest way to end a domestic dispute is by killing the offending spouse. . . .

Questions for Discussion

1. Did Joyce's behavior constitute provocation that would cause a reasonable person to lose control or to kill? Was Steven provoked?

2. Do you believe that Steven killed in the "heat of passion" or in a cool and deliberate fashion?

3. Should the jury be permitted to freely determine whether Joyce's words and conduct constituted provocation that should reduce second-degree murder to voluntary manslaughter?

4. Some courts have recognized that "information" may constitute adequate provocation. Would Joyce's alleged filing of abuse charges against Steven qualify as "information" constituting adequate provocation?

Cases and Comments

1. ***Racial Speech and Heat of Passion.*** Rufus Watson, a twenty-year-old African American, was incarcerated for second-degree murder. He was involved in a homosexual relationship for several months with the decedent, Samples, a heavily muscled Caucasian inmate. The two became embroiled in a disagreement in front of a third inmate, Johnny Lee Wilson, shortly before the lights were to be dimmed for the night. Samples was verbally abusing Watson, calling him "N__" and alleging that he was too scared to fight. Samples then made derogatory and obscene comments about Watson's mother. Watson warned Samples, and Samples replied, "Why don't you f_ me up if that's what you want to do. All you're gonna do is tremble, N__." Samples continued to verbally insult Watson, when Watson ran toward Samples' bunk and violently and repeatedly stabbed him with a kitchen paring knife.

The North Carolina Supreme Court ruled that words do not constitute adequate provocation sufficient to reduce the charge to voluntary manslaughter, but noted that a minority of states recognized that an individual who kills in response to a verbal assault might be considered to lack premeditation and be held criminally responsible for second- rather than first-degree murder. The jury seemingly followed this approach on its own accord and convicted Watson of second-degree murder. The Supreme Court affirmed the trial court's refusal to instruct the jury that the informal code of conduct among inmates required Watson to stand up for himself or risk being viewed as weak and easily abused and raped. The Supreme Court pointed out that Watson could have demonstrated his toughness by fighting with his fists rather than with a knife. Is the inmate code of conduct important in determining whether Watson acted in the "heat of passion"? See *State v. Watson*, 214 S.E.2d 85 (N.C. 1975).

2. **Model Penal Code.** The Model Code in Section 210.3 (1)(b) provides that in reducing murder to manslaughter, the question is whether (1) the defendant acted under "extreme mental or emotional disturbance" and (2) the defendant's reaction was reasonable based on the defendant's background, experience, and characteristics. The MPC provision allows the jury to decide for itself whether there is adequate provocation to reduce a defendant's guilt from murder to manslaughter. See *State v. Casassa*, 404 N.E.2d 1219 (N.Y. 1980).

Six states have adopted the Model Penal Code approach to provocation, and several other states have incorporated an aspect of the Model Penal Code provision into their law.

Vidado Dumlao was convicted of killing his mother-in-law, Pacita M. Reyes. He appealed on the grounds that the judge refused to issue an instruction to the jury to consider whether the killing was committed "under the influence of extreme mental or emotional disturbance for which there [was] a reasonable explanation." An expert witness testified that Dumlao suffered from a "paranoid personality disorder" and harbored the belief that other males were involved with his wife throughout his ten-year marriage. He also was easily insulted and reacted with extreme emotion to the smallest slight that other individuals would overlook. He believed that everyone was plotting against him, and he required his wife's family members to speak to her at a distance so that he could monitor what was being said.

Dumlao testified that "because of the way [his brother-in-law] looked at him," he concluded that is wife and brother in-law were having an affair. He thought that the family was talking about him behind his back and, when his brother-in-law allegedly rushed at him with a knife, he tried to scare him by brandishing his gun, which accidentally fired and killed his mother-in-law. See *State v. Dumlao*, 715 P.2d 822 (Haw. Ct. App. 1986).

The Hawaii appellate court found that Dumlao was entitled to have the jury consider his extreme emotional disturbance defense, for which there was a reasonable explanation.

The court clarified that the emotional disturbance defense differs from the defense of voluntary manslaughter: (1) the defense focuses on the defendant's specific background, emotional state, and physical characteristics; (2) there are no limitations on what constitutes adequate provocation; (3) there is no "cooling off period"; and (4) the jury decides whether the defendant's extreme emotional reaction was reasonable based on their particular characteristics. See also *State v. Elliot*, 411 A.2d 3 (Conn. 1977).

Should states replace their statutes on voluntary manslaughter with statutes based on the MPC?

You Decide

10.9 Most courts no longer require a spouse to actually witness adultery. Merely being informed about a spouse's adultery may constitute adequate provocation. George Schnopps shot his wife after fourteen years of marriage. During the previous six months, the two had argued over Schnopps's allegation that his wife was involved with another man who was a "bum," and he threatened her with scissors, a knife, a shotgun, and a plastic pistol. In September 1979, Schnopps's wife informed him that she was moving to her mother's house with the three children. At one point, Schnopps became aware that his wife's alleged boyfriend employed a signal when he telephoned. Schnopps used the signal and his wife answered the phone, "Hello, lover." Schnopps's son shortly thereafter told Schnopps that his wife would not return home.

A few days prior to the killing, Schnopps threatened to make his wife suffer "as she had never suffered before." On October 12, 1979, a co-worker helped Schnopps purchase a gun and ammunition. Schnopps stated that he was "mad enough to kill" and that a "bullet was too good for her, he would choke her to death." On October 13, Schnopps called his wife. For the first time, he stated that he would consider leaving their apartment and arranged for his wife to meet him at the unit. A neighbor agreed to care for the couple's youngest child while the two met. Schnopps's wife refused to reconcile with him and stated that she was going to court and that Schnopps would be left with nothing and pointed to her crotch and said that "[y]ou will never touch this again, because I have got something bigger and better for it." Schnopps stated that these words "cracked" him and he wept that he had "nothing to live for" and that he would never love anyone else. His wife responded that she was "never coming back to you." He then shot her as she began to leave. The evidence indicated that Schnopps fired an additional bullet at his wife while she was on the floor. Schnopps allegedly stated that he wanted to "go with her" and shot himself. Schnopps made a number of statements on his way to the hospital including that he had shot his wife because she was cheating on him and that the "devil made me do it." Marylou died of three gunshot wounds to the heart and lungs fired from within two to four feet. The forensic evidence indicated that the killing occurred within a few minutes of her arrival at the apartment.

Friends and co-workers testified that Schnopps's physical and emotional health deteriorated during the separation from his wife. He began drinking, wept at work, was twice sent home early, and was diagnosed as suffering from a "severe anxiety state."

The jury was instructed that it could convict on manslaughter based on adultery, but returned a verdict of intentional and premeditated murder. The Massachusetts Supreme Court affirmed the verdict. Evidence indicated that following the killing, Schnopps suffered from a major depression in reaction to his wife's death. Can you clearly distinguish between deliberate and premeditated murder and murder in the heat of passion in this case? What verdict would you return? See *Commonwealth v. Schnopps*, 459 N.E.2d 98 (Mass. 1984).

You can find the answer at www.sagepub.com/lippmanccl3e.

Involuntary Manslaughter

We now have reviewed voluntary manslaughter. Involuntary manslaughter is the second branch of manslaughter. Involuntary manslaughter involves the unintentional killing of another without malice and typically includes **negligent manslaughter**, the negligent creation of a risk of serious injury or death of another, as well as **misdemeanor manslaughter** (also referred to as unlawful-act manslaughter), the killing of another during the commission of a criminal act that does not amount to a felony. Some states, such as California, also provide for **vehicular manslaughter**, or the killing of another that results from the grossly negligent operation of an automobile or from driving under the influence of intoxicants.

Negligent Manslaughter

Negligent manslaughter arises when an individual commits an act that he or she is unaware creates a high degree of risk of human injury or death under circumstances in which a reasonable person would have been aware of the threat. Some courts require **recklessness**, meaning that a defendant must have been personally aware that his or her conduct creates a substantial risk of death or serious bodily harm. Other courts do not clearly state whether they require negligence or recklessness.

The Alabama criminal code section 13AA-6-4 provides that a person "commits the crime of criminally negligent homicide if he causes the death of another person by criminal negligence." The Missouri criminal code section 565.024 provides that the crime of involuntary manslaughter involves "recklessly" causing the death of another person. The Model Penal Code uses a negligence standard and holds individuals criminally responsible where they are "grossly insensitive to the interests and claims of other persons in society" and their conduct constitutes a "gross deviation from ordinary standards of conduct."

At this point, it's likely you have correctly concluded that there is not always a clear line separating involuntary manslaughter from depraved heart murder. Depraved heart murder requires recklessness, and involuntary manslaughter in most states requires negligence. Depraved heart murder, as we have seen, involves the commission of an act that exhibits such a gross and obvious indifference to human life that the law implies malice aforethought. This type of recklessness may involve intentionally driving at an excessive and dangerous speed down a crowded street and indifferently killing a pedestrian. Involuntary manslaughter, in contrast, might involve a driver who confidently and negligently takes the wheel without adequate sleep and then falls asleep and kills a pedestrian. Can you explain the difference between the recklessness and negligence? In the next case, *State v. Jones*, the Tennessee Supreme Court is asked to determine whether a mother was criminally negligent in the death of her young son. Pay particular attention to the court's distinction between "ordinary" and "gross" negligence.

Was Jones guilty of the criminal negligent homicide of her two-year-old son?

State v. Jones, 151 S.W.3d 494 (Tenn. 2004). Opinion by: Anderson, J.

Issue

We granted this appeal to determine whether the evidence supported the defendant's conviction for criminally negligent homicide in the death of her two-year-old son. The child was riding on the defendant's lap in the front passenger seat of a rental car and was killed when the passenger-side air bag deployed in a collision.

Facts

On November 9, 1998, defendant Latrece Jones, age eighteen, was riding in the front passenger seat of a rented Chevrolet Cavalier in Chattanooga, Tennessee. Her two-year-old son, Carlon Bowens, Jr., was asleep in her lap. Carlon's aunt, Letitia Abernathy, had rented and was driving the rental car; five children and one adult sat in the backseat. At the intersection of Shallowford Road and Jersey Pike, another car failed to yield the right of way to the rental car, causing a collision ("the accident"). Although the accident was not severe, the passenger-side air bag deployed. The force with which the air bag struck Ms. Jones' son broke his neck, killing him. No one else in the car was seriously injured.

Ms. Abernathy, the driver of the rental car, testified that she operated a day care business out of her home. The five children in the back seat, one of whom was her daughter, were under her care at the time of the accident. The children were aged seven years, six years, five years, four years, and nine months. Ms. Abernathy testified that she normally drove a jeep, but because she had been in an accident earlier in the day she had rented the Chevrolet. She testified that she normally had car seats for the children who required restraints but that she did not use them that day because there was no room in the Chevrolet. It was stipulated that the six passengers in the back seat were unrestrained.

The driver, Ms. Abernathy, was charged separately with criminally negligent homicide and violation of child restraint law. She pled guilty to reckless endangerment and violation of the child restraint law. At the time of the accident, Tennessee's child restraint law required children under four years old to be in a "child passenger restraint system meeting federal motor vehicle standards." The statute only applies to drivers, providing that "any person transporting a child . . . is responsible for . . . properly using a child passenger restraint."

It was undisputed that only Carlon sustained serious injuries. Pediatric surgeon Dr. Joseph Earl Kelley, Jr., testified at trial that he had treated the victim and that the force of the air bag deployment had broken the child's neck. He testified that it was "not the type of injury that would typically be seen from motor vehicle accidents when children are unrestrained and thrown around." Dr. Kelley testified that an injury such as the one Carlon sustained is always fatal.

The prosecution introduced photos showing air bag warnings affixed to the rental automobile's visors and front-passenger seat belt. The visor warning was positioned on the side of the visor facing the passenger and read as follows:

!WARNING
DEATH or SERIOUS INJURY can occur
Children 12 and under can be killed by the air bag.
The BACK SEAT is the SAFEST place for children.
NEVER put a rear-facing child seat in the front.
Sit as far back as possible from the air bag.
ALWAYS use SEAT BELTS and CHILD RESTRAINTS.

The seat belt warning was positioned in the center of the passenger-side seat belt, although it is not clear from the record whether the warning would have been visible if the seat belt were not extended. The seat belt warning read as follows:

!CAUTION
A child in a REAR-FACING CHILD RESTRAINT can be badly injured by the air bag if it inflates. NEVER put a child in a REAR-FACING CHILD RESTRAINT in the front seat of this vehicle. Secure a REAR-FACING CHILD RESTRAINT in the rear seat.
Before Securing a forward-facing child restraint, ALWAYS move the passenger seat as far back as it will go. Or, secure the child restraint in the rear seat.
For more information, see your Owner's Manual and the instructions that came with your child restraint.

The prosecution also introduced other evidence in its effort to establish that Ms. Jones knew her son should have been in a child restraint and should not have been seated in front of an air bag. Lisa McClain, administrator for the Women's and Infant's Services

Division at Erlanger Hospital, testified that when Ms. Jones was discharged after giving birth to Carlon, she would have been given a pamphlet on car seat safety and a videotape including information on car seat safety. The pamphlet included the warnings, "never hold a child in your lap while riding in either the front or back seat" and "be consistent! Always buckle your child in the safety seat." Ms. McClain testified that the hospital's policy was to require parents to place the infant in the car seat themselves upon discharge from the hospital.

Finally, the prosecution offered the testimony of Brooke Pippenger, former assistant director of programs at the Children's Wellness Center, regarding a campaign the Center conducted in 1997 and 1998 aimed at educating the public on preventing injuries to children. Ms. Pippenger testified that the campaign included educating the public about vehicle safety. The State introduced into evidence two television news features and a collection of newspaper articles regarding car seat safety and the dangers of air bags related to the campaign. Ms. Pippenger testified, however, that she had never "found any law that made it illegal to put a child safety seat or a child in front of an air bag."

The jury convicted Ms. Jones of criminally negligent homicide in the death of her son, and the trial judge imposed a sentence of 0.9 years unsupervised probation. Ms. Jones appealed to the Court of Criminal Appeals. She argued that the evidence was insufficient to support the conviction for criminally negligent homicide . . . and that the trial court erred in admitting evidence about the child restraint law and the absence of car seats in the vehicle at the time of the crash. The Court of Criminal Appeals rejected each of Ms. Jones' arguments and affirmed her conviction. We granted review.

Reasoning

First, we address Ms. Jones' argument that the child restraint evidence introduced at trial was irrelevant and prejudicial because, as a passenger, she did not have a statutory (legal) duty to place her child in a car seat and because the medical evidence showed that the child was killed by the air bag. At oral argument, counsel for Ms. Jones asserted that even if Carlon had been in a car seat, he would have been killed by the air bag because he was seated in the front seat. The State counters that the evidence was relevant to show that Ms. Jones ignored well-known risks to the victim. We agree with the State that the evidence was admissible.

The trial court agreed to exclude evidence regarding Tennessee's car seat statute, because it did not apply to Ms. Jones as a passenger. However, the court declined to exclude evidence regarding information in the community regarding car seats.

Ms. Jones was charged with criminal negligence in the death of her son. Therefore the State's burden was to prove that in holding her son on her lap while traveling in a car, she failed to perceive a substantial and unjustifiable risk to her son and that her conduct was a gross deviation from the standard of care. Ms. Jones argues that the evidence was prejudicial because the State was trying to "inflame the jury" by showing Ms. Jones to be irresponsible. However, we agree with the State that the evidence was relevant to show that Ms. Jones was aware, or should have been aware, of the danger to an unrestrained child and that, in any event, its probative value was not outweighed by the danger of unfair prejudice to Ms. Jones. Ms. Jones argues that Carlon was killed by the air bag, not by being unrestrained, but that argument misses the point. The State charged Ms. Jones with criminal negligence for the act of holding her child in her lap in a moving vehicle. Evidence about car seats and car seat safety was directly relevant to show that Ms. Jones' act was negligent. The trial court did not abuse its discretion in admitting the evidence.

We address Ms. Jones' argument that the evidence of criminal negligence was insufficient to sustain her conviction. To establish criminally negligent homicide, the State must prove three elements beyond a reasonable doubt: (1) criminally negligent conduct on the part of the accused; (2) that proximately causes; (3) a person's death. Ms. Jones argues that the evidence is insufficient to support her conviction for criminally negligent homicide because the evidence does not establish that she was grossly negligent. The State argues that the evidence supports the jury's determination that Ms. Jones' failure to perceive the risk to her son was a gross deviation from the standard of care.

When evaluating the sufficiency of the evidence, we must determine whether "any rational trier of fact could have found the essential elements of the crime beyond a reasonable doubt." Criminally negligent conduct that "results in death constitutes criminally negligent homicide." A person acts with criminal negligence with respect to the circumstances surrounding that person's conduct or the result of that conduct when the person ought to be aware of a substantial and unjustifiable risk that the circumstances exist or the result will occur. The risk must be of such a nature and degree that the failure to perceive it constitutes a gross deviation from the standard of care that an ordinary person would exercise under all the circumstances as viewed from the accused person's standpoint.

To be criminally negligent, a defendant must fail to perceive a substantial and unjustifiable risk. Whether the defendant failed to perceive the risk must be determined using a subjective standard; we must view the circumstances "from the accused person's standpoint." ("[The criminally negligent homicide statute] views the situation through the eyes of the [defendant] and whether [s]he could have perceived and then chosen to ignore a 'substantial and unjustifiable risk.'"). The defendant's failure to perceive the risk must be "a gross deviation from the standard of care."

In sum, we must examine the defendant's conduct to determine (1) whether a substantial and unjustifiable risk existed at the time of the conduct or resulting from the conduct; (2) whether, using a subjective standard, the defendant failed at the time of the conduct to perceive the risk; and (3) whether that failure was a gross deviation from the standard of care of an ordinary person under the circumstances.

The Court of Criminal Appeals held that the evidence was sufficient to convict Ms. Jones of criminal negligence, concluding that she "knew that by failing to place her child in a child restraint seat, she was exposing her child to serious bodily harm or death." The court cited Ms. McClain's testimony that Ms. Jones had been verbally informed of child restraint laws and was given written information on vehicle safety. The court also cited the visor warnings and the fact that Ms. Jones and the child's father had a child restraint seat and had used it on other occasions. Finally, the court noted that "during the year prior to the accident, assorted public service announcements in print and television were circulated regarding the importance of using child restraint seats."

We disagree with the Court of Criminal Appeals' conclusion. Viewing the evidence in the light most favorable to the State, as we must, and applying a subjective point of view, as we must, we conclude that the evidence failed to establish that Ms. Jones was criminally negligent. We have little doubt that holding a two-year-old child on one's lap in front of an air bag constitutes a substantial and unjustified risk. However, we must determine whether there was sufficient evidence to permit a rational trier of fact to conclude that Ms. Jones' failure to perceive that risk was a gross deviation from the standard of care. For several reasons, we cannot say that there was.

First, although the State introduced a collection of newspaper articles and two television "spots" to show that information about safely transporting children was available to the community, nothing in the record indicates that Ms. Jones actually saw or read any of that information. Moreover, we disagree with the State that whether the conduct of Ms. Jones was a gross deviation from the standard of care can be established with reference to an advertising campaign, however well-intentioned that campaign might have been.

Second, the articles and television spots entered into evidence illustrate how new the risk of air bags was in 1998. According to one of the articles, 1999 was the first year that all cars had passenger-side air bags.

Third, the very fact that there was a need for a large-scale public information campaign aimed at educating parents about child car safety indicates, sadly, how many people in the community simply were not using child safety restraints at the time of the accident. In fact, one of the newspaper articles in the record, published on October 28, 1998—just twelve days before the accident—noted that a "recent survey" in Hamilton County had found that "only about sixty percent (60%) of youngsters observed riding in cars and trucks were restrained. Some were sitting in laps." If 40% of the children being transported in Ms. Jones' community were being transported without being properly restrained at the time of the accident, it would be difficult for a rational trier of fact to conclude that it was a gross deviation from the standard of care at the time of the accident for Ms. Jones to transport her child improperly.

Finally, the State also introduced evidence that Ms. Jones had received information about car seat safety upon her discharge from the hospital after the birth of her son. That information advised parents to, among other things, always use car seats and to "never hold a child in your lap" in the car. Additionally, the State cited the air bag warning on the passenger-side visor. Although Ms. Jones' failure to heed these warnings and to perceive the danger posed by sitting with her child on her lap in front of an air bag may have been negligent, our cases illustrate that it simply does not rise to the level of gross negligence necessary to uphold a conviction for criminally negligent homicide.

Tennessee courts have sustained convictions for criminally negligent homicide only where a "risk is of such a nature and degree that injury or death is likely and foreseeable." For example, in *State v. Goodwin,* a jury found the defendant guilty of criminally negligent homicide where the defendant had left a cocked shotgun in the woods fifty feet behind a house in a crowded neighborhood. Two children found the gun; it accidentally discharged, killing one child and severely injuring the other. Concluding that the defendant had "exercised extremely poor judgment in his handling of an inherently dangerous weapon," this Court affirmed the judgment. 143 S.W.3d 771, 779 (Tenn. 2004).

We have consistently applied the requirement that death or injury be likely and foreseeable in cases involving automobile accidents. . . . In *Reed v. State,* 110 S.W.2d 308 (Tenn. 1937), a defendant who pulled into "heavy and closely approaching opposing traffic" to pass a truck was guilty of involuntary manslaughter because the resulting collision was "not only a probable result but almost an inevitable result of such negligence as the defendant's." . . . *State v. Ramsey,* 903 S.W.2d 709, 712 (Tenn. Crim. App. 1995) [held that] injury or death was likely and foreseeable where the defendant was "driving fast and carelessly on a hilly, curvy road and [the defendant] consciously disregarded that risk by driving in such a manner."

Likewise, automobile cases in which someone other than the driver was criminally negligent further illustrate the level of negligence required for a finding of a "gross deviation" from the standard of care. In *Flippen v. State,* 365 S.W.2d 895 (Tenn. 1963), this Court held that a passenger was criminally negligent in failing to alert the intoxicated driver that he had struck another car, sending it off the road and into a lake, and for assisting the driver in concealing the car after the accident. In *Freeman v. State*, 362 S.W.2d 251 (Tenn. 1962), this Court held that the defendant was criminally negligent in permitting his intoxicated companion to drive his (the defendant's) car, knowing that the brakes were defective. . . . .

In sum, the above cases demonstrate that something much greater than the want of ordinary care shown by Ms. Jones is necessary to affirm her conviction. There must be a gross deviation from the standard of care. We were unable to find a case anywhere in the country holding a parent criminally liable for a child's death based on the conduct of placing the child in front of a passenger-side air bag. Viewing all the evidence in the light most favorable to the prosecution, the evidence is not sufficient to permit a rational trier of fact to find Ms. Jones criminally negligent beyond a reasonable doubt. We therefore reverse her conviction.

We note that our holding comports with the provisions of the child restraint law in effect at the time of the accident. Although the child restraint law does not apply to Ms. Jones as a passenger rather than a driver,

the child restraint statute in effect at the time of the accident permitted a mother to remove her child from its car seat to nurse the child or to "attend[] to its other physiological needs." The legislative determination that it was permissible for a mother to hold a child for such non-emergency purposes, rather than keeping the child in a restraint also militates against a finding that it was a gross deviation from the standard of care for Ms. Jones to do so in this case.

Holding

The death of a child is a terrible tragedy, particularly when, as here, the death might have been prevented. But however tragic Carlon's death was, the evidence was insufficient to support a conviction for criminally negligent homicide because the defendant's conduct did not constitute a gross deviation from the standard of care. The judgment of the Court of Criminal Appeals is reversed, and the conviction is dismissed. Costs of the appeal are taxed to the State of Tennessee for which execution may issue if necessary.

Questions for Discussion

1. What is the legal test for negligent homicide relied on by the court in *Jones*?

2. Was Letitia Jones aware of the risk of holding her child in her lap in the front seat of the automobile?

3. Did Letitia Jones's holding her child in her lap constitute a gross deviation from the standard of care?

4. Why did the Tennessee Supreme Court reverse the conviction of Letitia Jones?

5. Do you agree with the court decision that Jones did not act in a grossly negligent fashion?

You can find more cases at the study site:
Commonwealth v. Walker, *www.sagepub.com/lippmanccl3e*.

Misdemeanor Manslaughter

An unintentional killing that results from an "unlawful act not amounting to a felony" is termed misdemeanor manslaughter. It may be more accurate to term this *unlawful-act manslaughter* because some states extend this to nonviolent felonies that do not trigger the felony-murder rule, as well as to acts that are not criminal but which are considered "unlawful." Unlawful is broadly interpreted to include "bad" or "immoral conduct." An individual, for instance, was convicted of manslaughter when he accidentally killed his girlfriend while attempting suicide, which is not a crime.[60]

The commission of the unlawful act provides a predicate that the defendant acted in a grossly negligent fashion and is properly held criminally liable for manslaughter. The California Penal Code defines involuntary manslaughter as an unlawful killing of a human being without malice "in the commission of an unlawful act, not amounting to a felony; or in the commission of a lawful act which might produce death, in an unlawful manner, or without due caution and circumspection."[61]

A majority of states retain this rule. Some courts have limited the harshness of the doctrine by requiring a showing of proximate cause. For example, an individual who stole ten dollars from a church collection plate was not held criminally responsible for the death of a congregant who suffered a heart attack while chasing the thief. The Florida court reasoned that this was not the "kind of direct, foreseeable risk of physical harm that would support a conviction of manslaughter."[62] Other courts limit misdemeanor manslaughter to dangerous offenses.

A third approach is to distinguish between a *malum in se* offense, a crime that is "wrong in itself," as opposed to a *malum prohibitum* offense, a social welfare offense regulating areas such as professional licenses, motor vehicles, food service, and housing quality. A *malum prohibitum* offense provides the foundation for manslaughter only in those instances in which a defendant acted negligently. A cosmetologist who had not obtained a professional license was not held criminally liable for death caused by a poisoned facial treatment. The court reasoned that the facial treatment, rather than the failure to possess a license, caused the victim's death.[63]

Various states also provide for the offense of vehicular homicide. California punishes vehicular homicide that results in the death of another stemming from the grossly negligent operation of a motor vehicle, as well as vehicular homicide that results from the operation of a motor vehicle while intoxicated.[64] The Florida vehicular homicide provision encompasses killing caused by the reckless operation of a motor vehicle in a manner "likely to cause the death of, or great bodily harm to, another."[65]

The Model Penal Code rejects the misdemeanor manslaughter rule and merely punishes the reckless killing of another, an approach adopted by various states.

See more cases on the study site: State v. Pray, www.sagepub.com/lippmanccl3e.

The Legal Equation

Misdemeanor manslaughter **=** Misdemeanor

+ (causing) death of another.

CHAPTER SUMMARY

The killing of another human being violates the fundamental right to life and is considered the most serious criminal offense. The common law gradually distinguished between murder (killings committed with malice aforethought), and the less serious crime of manslaughter (killings committed without malice aforethought).

We generally measure the beginning of human life from viability, the point at which a fetus is able to live independently from the mother. Death is measured by the brain death test, or the failure of the brain function.

Malice is an intent to kill with ill will or hatred. Aforethought means a design to kill. Malice aforethought is expressed when there is a deliberate intent to kill or implied where an individual possesses an intent to cause great bodily harm or an intent to commit an act that may lead to death or great bodily harm. Judges gradually expanded the concept of malice aforethought to include various forms of criminal intent.

There is no single approach to defining the law of murder or manslaughter in state statutes. The division of homicide into degrees is intended to divide killings by the "moral blameworthiness of the individual." This division is typically based on factors such as the perpetrator's intent, the nature of the killing, and the surrounding circumstances of the killing.

First-degree murder is the deliberate and premeditated killing of another with malice aforethought. An individual who is capable of devising a plan to take the life of another is considered a serious threat to society. Premeditation may be formed instantaneously and does not require a lengthy period of reflection. Thirty-three states and the federal government recognize the death penalty. Killings viewed as deserving of capital punishment are categorized as capital first-degree murder or aggravated first-degree murder. Conviction results in the death penalty or life imprisonment. In non-death-penalty states, aggravated murder carries life imprisonment. Seventeen states and Washington, D.C., reject capital punishment—including Connecticut and New Mexico, both of which have inmates still remaining on death row. A homicide qualifies as aggravated or capital murder when it is found to have been committed in a heinous or atrocious fashion.

Second-degree murder is comprised of intentional killings with malice aforethought that are not committed in a premeditated and deliberate fashion. Depraved heart murder includes killings resulting from a knowingly dangerous act committed with reckless and wanton disregard as to whether others are harmed. Felony murder entails the death of an individual during the commission of or attempt to commit a felony. This tends to be limited to dangerous felonies, and in various states felony murders are categorized as first-degree murder rather than second-degree murder.

Some state statutes explicitly provide for the criminal liability of corporations for unlawful killings. Absent an explicit provision for corporate liability, the term "person" in criminal statutes generally is interpreted to include corporations, which are legally considered to be "non-natural persons." A corporation is held liable for homicide where the offense is authorized, requested, commanded, or performed by the board of directors or by a high managerial agent acting within the scope of his or her employment. Corporate officers and employees may also be held individually liable.

Manslaughter is comprised of voluntary and involuntary manslaughter. Voluntary manslaughter is the killing of another in a sudden and intense heat of passion in response to an adequate provocation. Adequate provocation is defined as conduct that is sufficient to excite an intense passion that would cause a reasonable person to lose control. Only a limited number of acts are considered to constitute adequate provocation, but some judges have vested the discretion to determine provocation in jurors. The heat of passion is considered to have "cooled" after a reasonable period of time.

Involuntary manslaughter includes negligent manslaughter and misdemeanor manslaughter, also termed unlawful-act manslaughter. Negligent manslaughter involves the creation of a risk of the serious injury or death of another. Courts, in practice, do not clearly distinguish between a negligence and recklessness standard. Misdemeanor manslaughter involves a killing committed during the commission of a misdemeanor. Some states expand misdemeanor manslaughter to include nonviolent felonies and, for this reason, term this offense unlawful-act manslaughter.

CHAPTER REVIEW QUESTIONS

1. Discuss the historical origins and development of criminal homicide into murder and manslaughter. Can you distinguish between murder and manslaughter?

2. Differentiate first-degree murder from first-degree capital or aggravated homicide.

3. What is the difference between first- and second-degree murder?

4. Define depraved heart murder.

5. Why does the law provide for the offense of felony murder? What are the arguments for and against the felony-murder rule?

6. Discuss the legal standards for holding a corporation liable for first-degree murder and for involuntary manslaughter. Is there a social benefit in holding corporations liable for homicide?

7. Define voluntary manslaughter.

8. What acts constitute adequate provocation? What must the defendant prove to establish heat of passion? At what point does a defendant's blood "cool"?

9. Should the law recognize the offense of voluntary manslaughter? Why not?

10. Discuss the difference between negligent homicide and misdemeanor manslaughter. Why is misdemeanor manslaughter termed unlawful-act manslaughter in some states?

11. Discuss the purpose of the various grades of murder and dividing homicide into murder and manslaughter. Why do we make all these technical distinctions between types of homicide?

LEGAL TERMINOLOGY

agency theory of felony murder	capital murder	criminal homicide
aggravating factors	cooling of blood	depraved heart murder
brain death test	corporate murder	excusable homicide

felony murder	manslaughter	proximate cause theory of felony murder
first-degree murder	misdemeanor manslaughter	
grading	mitigating circumstances	reasonable person
heat of passion	murder	recklessness
involuntary manslaughter	negligent manslaughter	second-degree murder
justifiable homicide	*nolo contendere*	vehicular manslaughter
malice aforethought	premeditation and deliberation	voluntary manslaughter

CRIMINAL LAW ON THE WEB

Log on to the Web-based student study site at www.sagepub.com/lippmanccl3e to assist you in completing the Criminal Law on the Web exercises, as well as for additional features such as podcasts, Web quizzes, and video links.

1. Consider how the Wyoming murder of Matthew Shepard, in 1998, illustrates the law of felony murder.

2. Learn about Jack Kevorkian and the right to die.

3. Read about capital murder and the administration of the death penalty.

BIBLIOGRAPHY

American Law Institute, *Model Penal Code and Commentaries,* vol. 1, pt. 11 (Philadelphia: American Law Institute, 1980), pp. 4–107. A discussion and proposal for reform of the law of homicide.

Douglas Birsch and John H. Fielder (Eds.), *The Ford Pinto Case: A Study in Applied Ethics, Business, and Technology* (Albany: SUNY Press, 1994). A compilation of the documents involved in Ford Motor Company's decision not to upgrade the defective design of the Pinto's fuel system.

Joshua Dressler, *Understanding Criminal Law,* 3rd ed. (New York: Lexis, 2001), pp. 499–544. A comprehensive and accessible review of the law of homicide.

Wayne R. LaFave, *Criminal Law,* 3rd ed. (St. Paul, MN: West Publishing, 2000), pp. 651–735. A detailed discussion of homicide with relevant citations.

Rollin M. Perkins and Ronald N. Boyce, *Criminal Law,* 3rd ed. (Mineola, NY: Foundation Press, 1982), pp. 46–150. A comprehensive account of the evolution and contemporary standard for homicide.

11 CRIMINAL SEXUAL CONDUCT, ASSAULT AND BATTERY, KIDNAPPING, AND FALSE IMPRISONMENT

Was C.G. raped by the defendant?

C.G. stated that earlier in the day, M.T.S. had told her three or four times that he "was going to make a surprise visit up in [her] bedroom." She said that she had not taken M.T.S. seriously and considered his comments a joke because he frequently teased her. She testified that M.T.S. had attempted to kiss her on numerous other occasions and at least once had attempted to put his hands inside of her pants, but that she had rejected all of his previous advances. C.G. testified that on May 22, at approximately 1:30 a.m., she awoke to use the bathroom. As she was getting out of bed, she said, she saw M.T.S., fully clothed, standing in her doorway. According to C.G., M.T.S. then said that "he was going to tease [her] a little bit." C.G. testified that she "didn't think anything of it"; she walked past him, used the bathroom, and then returned to bed, falling into a "heavy" sleep within fifteen minutes. The next event C.G. claimed to recall of that morning was waking up with M.T.S. on top of her, her underpants and shorts removed. She said "his penis was into [her] vagina." As soon as C.G. realized what had happened, she said, she immediately slapped M.T.S. once in the face, then "told him to get off [her], and get out." She did not scream or cry out. She testified that M.T.S. complied in less than one minute after being struck; according to C.G., "he jumped right off of [her]." She said she did not know how long M.T.S. had been inside of her before she awoke. C.G. said that after M.T.S. left the room, she "fell asleep crying" because "[she] couldn't believe that he did what he did to [her]." She explained that she did not immediately tell her mother or anyone else in the house of the events of that morning because she was "scared and in shock."

Learning Objectives

1. Know the definition of rape under the common law and the barriers to proving rape under the common law.

2. List some of the changes in the law of rape resulting from rape reform in the 1970s and 1980s.

3. Compare and contrast the extrinsic and intrinsic tests for the *actus reus* of rape.

4. Know the difference between fraud in the *factum* and fraud in inducement.

5. Understand the standard for the *mens rea* of rape.

6. Summarize the law of statutory rape.

7. State the law on the withdrawal of consent.

8. Explain rape shield laws.

9. Describe the difference between a simple and an aggravated battery and between attempted battery assault and the placing of another in fear of a battery assault.

10. Know the elements of the crime of stalking.

11. Describe the intent and act requirement for kidnapping.

12. Distinguish between kidnapping and false imprisonment

INTRODUCTION

This chapter discusses three categories of crimes against the person. The first is freedom from sexual violations (rape and sexual assault), and the second is protection against the threat and infliction of bodily harm (assault and battery). A third category is freedom of movement (kidnapping and false imprisonment).

Next to homicide, sexual offenses are considered the most serious offenses against the person and are punished as felonies, even when the victim is not physically injured. This harsh punishment reflects the fact that nonconsensual acts of sexual intimacy can cause severe physical injury as well as psychological trauma. Assaults and batteries and false imprisonment are typically misdemeanors (other than when committed in an aggravated fashion) that risks or results in physical injury. Kidnapping is considered a serious offense that places people in danger and was subject to the death penalty under the common law.

There clearly is no more fundamental or important interest than protecting the life and bodily integrity of the individual. This is the basic and essential expectation of all members of a community. Imagine a society in which each of us was constantly fearful of an attack to the invasion of our bodily integrity. Are there geographic areas of American society and the world that come close to this type of "lawlessness"?

THE COMMON LAW OF RAPE

The common law treated rape as a capital crime punishable by death.[1] In the United States, only homicide has been historically considered more serious than rape. In 1925, nineteen states and the District of Columbia, as well as the federal government, punished rape with capital punishment. This was particularly controversial because the penalty of death for rape was almost exclusively employed against African Americans, particularly when accused of raping Caucasian women. By 1977, only Georgia provided capital punishment for rape. In that same year, the U.S. Supreme Court declared the death penalty for rape unconstitutional on the grounds that it was disproportionate to the harm caused by the rape.[2] Today, most states divide rape into degrees of seriousness that reflect the circumstances of the offense. Aggravated rape may result in incarceration for a significant period of time and, in some jurisdictions, for life in prison.

The law of rape was rooted in the notion that a man's daughters and wife were his property, and that rape involved a trespass on a male's property rights. In fact, for a brief period of time the common law categorized rape as a trespass subject to imprisonment and fine. William Blackstone recounts that under ancient Hebraic law, the rape of an unmarried woman was punishable by a fine of fifty shekels paid to the woman's father and by forced marriage without the privilege of divorce. The commentary to the Model Penal Code observes that the notion of "the wife as chattel" is illustrated by the fact that as late as 1984, forty states recognized a "marital exemption" that provided that a husband could not be held liable for the rape of his wife.[3] Seventeenth-century English jurist Sir Matthew Hale explained that this exemption was based on the fact that a wife by "matrimonial consent and contract" had forfeited the privilege of refusing sexual favors to her husband.[4]

The law of rape is no longer an expression of property rights and today is designed to punish individuals who violate a victim's bodily integrity, psychological health and welfare, and sexual independence. The stigma and trauma that results from rape contribute to the reluctance of rape victims to report the crime to law enforcement authorities. Another factor contributing to the hesitancy of victims to report a rape is a lack of confidence that the criminal justice system will seriously pursue the prosecution and conviction of offenders.[5]

The criminal justice system is fairly effective in prosecuting what Professor Susan Estrich calls "real rape," or cases in which the victim is attacked by an unknown male. In such instances, the prosecution has little difficulty in demonstrating that the victim was forcibly subjected to sexual molestation. A greater challenge is presented in the area of so-called **acquaintance rape** or date rape. In these cases, although no less serious, the perpetrator typically admits that sexual intercourse occurred and claims that it was consensual, while the victim characterizes the interaction as rape.[6] The reluctance to report rape may be most prevalent in the largely undocumented area of same-sex rape, which is a particularly serious problem in correctional institutions.[7]

Rape has profound, powerful, and long-lasting destructive physical and emotional consequences that often can only be overcome by years of therapy and treatment. The symptoms of **rape trauma syndrome** include:

- *Remembering the Event.* The victim has nightmares and replays the event in his or her mind.
- *Strong Emotions.* The victim may feel anger, guilt, and depression and may be drawn to suicide.

- *Psychological Impact.* Victims can lose self-confidence and develop an apprehension that they will be attacked once again. They often fear trusting others and withdraw from friends and relationships.
- *Physical Symptoms.* In addition to physical injury, victims suffer headaches and stress leading to physical illness. There also is the threat of HIV/AIDS.
- *Self-Medication.* Victims may attempt to find comfort in drug or alcohol abuse.

In reading about rape, keep in mind that jurors and judges are likely to bring a host of prejudices and preconceptions regarding the proper behavior of men and women to their consideration of the facts. What types of issues do you anticipate may arise in rape prosecutions?[8]

The Elements of the Common Law of Rape

The common law defined rape as the forcible carnal knowledge of a woman against her will. Carnal knowledge for purposes of rape is defined as vaginal intercourse by a man with a woman who is not his wife. The vaginal intercourse is required to be carried out by force or threat of severe bodily harm ("by force or fear") without the victim's consent.[9]

The common law of rape reflects a distrust of women, and various requirements were imposed to ensure that the **prosecutrix** (victim) was not engaged in blackmail or in an attempt to conceal a consensual affair or was not suffering from a psychological illness. The fear of an unjust conviction was reflected in Lord Matthew Hale's comment that rape "is an accusation easy to be made, hard to be proved, but harder to be defended by the party accused though innocent." Lord Hale stressed that there was a danger that a judge and jury would be emotionally carried away by the seriousness of the charge and convict a defendant based on false testimony.[10]

The prosecution, as noted, was required to overcome a number of hurdles under the common law in order to convict the defendant:[11]

- *Immediate Complaint.* The absence of a **prompt complaint** by the victim to authorities was evidence that the complaint was not genuine.
- *Corroboration Rule.* The victim's allegation of rape required **corroboration**, evidence such as physical injury or witnesses.
- *Sexual Activity.* The victim's past sexual conduct or reputation for chastity was admissible as evidence of consent or on cross-examination to attack her credibility.
- *Judicial Instruction.* The judge was required to issue a cautionary instruction to the jury that the victim's testimony should be subject to strict scrutiny because rape is a crime easily charged and difficult to prove.

The crucial evidence in a prosecution for rape under the common law was a demonstration that the female "victim" did not consent to the sexual intercourse. Blackstone writes that a "necessary ingredient" in the crime of rape is that it is against the woman's will. He notes that it is a felony to forcibly ravish "even a concubine or harlot because the woman may have forsaken that unlawful course of life."[12]

The victim's lack of consent was demonstrated through outward resistance. In reference to sexual advances, a victim was required to **resist to the utmost** in order to establish a lack of consent. This expectation of combat against an attacker was viewed as reflecting "the natural instinct of every proud female to resist."[13] In *Brown v. State,* a sixteen-year-old woman recovering from the measles was attacked on a path near her family farm. The young woman was a virgin and testified that she tried as "hard as I could to get away" and "screamed as hard I could . . . until I was almost strangled." The Wisconsin Supreme Court ruled that the prosecutrix failed to demonstrate that "she did her utmost" to resist. This not only requires "escape or withdrawal," but involves resistance "by means of hands and limbs and pelvic muscles." The alleged victim also failed to corroborate her complaint by "bruises, scratches and ripped clothing."[14] Courts did rule that resistance to the "utmost" was not required when the woman reasonably believed that she confronted a threat of "great and immediate bodily harm that would impair a reasonable person's will to resist."[15]

In the mid-twentieth century, judges began to relax this harsh resistance requirement. A few courts continued to require a fairly heavy burden of **earnest resistance**. Most, however, adopted the position that a victim was required to engage in **reasonable resistance** under the circumstances. Some judges argued that a more sensible approach would be to require that a female victim engage in that degree of resistance that is necessary to communicate her lack of consent to a reasonable person. This might be satisfied by a verbal protest. After all, it was pointed out that "no means no." Critics of a demanding resistance standard also pointed out that the victims of robbery or burglary are not required to demonstrate resistance. What is the thinking behind the resistance requirement? Which is the best test? How will perpetrators react to "earnest" or "reasonable" resistance?[16]

Resistance is not required under the common law standard when a victim is incapable of understanding the nature of intercourse as a result of intoxication, sleep, or a lack of consciousness.[17] Another exception is so-called **fraud in the *factum*** or "fraud in the nature of the act." In *People v. Minkowski,* a woman consented to treatment for menstrual cramps by a doctor who proceeded to insert a metal instrument. During the procedure he withdrew the metal object and without the consent of the patient inserted his sexual organ. The California court ruled that this was rape because the victim consented to a medical procedure and did not consent to intercourse.[18] This is distinguished from **fraud in inducement** or consent to intercourse that results from a misrepresentation as to the purpose or benefits of the sexual act. In one well-known case, a female consented to intercourse with a donor after being tricked into believing that the donor had been injected with a serum that would cure her alleged disease. The California appellate court ruled that this was not rape because the victim consented to the "thing done" and the fraud related to the underlying purpose rather than to the fact of the sexual interaction.[19]

Rape Reform

During the 1970s and 1980s, a number of states abolished the special procedures surrounding the common law of rape. This included the corroboration and prompt reporting provisions and the judge's cautionary instructions to the jury. Another important development that we will discuss later in the chapter is the adoption of **rape shield laws** prohibiting the introduction of evidence concerning a victim's past sexual activity.

The commentary to the Model Penal Code justifies these reforms on the grounds that rape trials had become focused on the sexual background and resistance of victims rather than on the conduct of defendants. The extraordinary resistance standard placed women in a "no win" situation. Resistance might lead to violent retaliation; a failure to resist might result in the defendant's acquittal.[20]

A number of states adopted new sexual assault statutes that fundamentally changed the law of rape. The statutes treated rape as an assault against the person rather than as an offense against sexual morality. These statutes refer to "criminal sexual conduct" or "sexual assault" rather than rape. The modified statutes widely differ from one another and typically incorporate one or more of the following provisions.[21]

- *Gender Neutral.* A male or a female may be the perpetrator or the victim of rape.
- *Degrees of Rape.* Several degrees of rape are defined that are distinguished from one another based on the seriousness of the offense. These statutes provide that involuntary sexual penetration is more serious than involuntary contact and that the use of force results in a more serious offense than an involuntary contact or penetration that is accomplished without the use of force.
- *Sexual Intercourse.* "Sexual intercourse" is expanded to include a range of forced sexual activity or forced intrusions into a person's body, including oral and anal intercourse and the insertion of an object into the genital or anal opening of another. Some statutes also prohibit "sexual contact" or the intentional and nonconsensual touching of an intimate portion of another individual's body for purposes of sexual gratification.
- *Consent.* Some statutes provide that consent requires free, affirmative, and voluntary cooperation or that resistance may be established by either words or actions. Other statutes provide that physical resistance is not required. Several continue to maintain the traditional standard that rape requires force along with an absence of consent.
- *Coercion.* There is explicit recognition that coercion may be achieved through fraud or psychological pressure as well as through physical force.

- *Marital Exemption.* A husband or wife may be charged with the rape of a spouse. Half of the states continue to recognize the exemption where there is no force or threat of force and some require a prompt reporting requirement. Others only recognize the marital exemption where the spouses are not living separate and apart.[22]

The important point to keep in mind is that these statutes have removed the barriers that have made rape convictions so difficult to obtain. States such as Michigan have adopted far-reaching reforms, whereas states such as Georgia have introduced only modest changes.

The Impact of Rape Reform

Have these statutes fundamentally changed the prosecution of rape? Despite rape reform, prosecutions for "acquaintance rape" inevitably seem to return to considerations of force, resistance, and consent.[23] In the trial of basketball star Kobe Bryant, the public dialogue centered on why the alleged victim accompanied Kobe Bryant to his hotel room and whether there were any indications of resistance. This discussion was complicated by the accusation that the alleged victim had engaged in sexual contact with other men immediately following Kobe Bryant.

We should mention one additional modern innovation in the law of rape. Some courts permit victims to present expert witnesses on the issue of rape trauma syndrome. This expert evidence is intended to support the victim's contention that he or she was raped by pointing out that the victim's psychological and medical condition is characteristic of rape victims. In other instances, experts are employed to educate the jurors that rape victims often have a delayed reaction and that the jurors should not conclude that an individual was not raped because the complainant did not immediately bring charges of rape or was seemingly calm and collected following the alleged attack.[24]

Punishment and Sexual Assault

Sexual assault typically is divided into aggravated rape (or first degree) and simple (or second degree) rape. Aggravated rape in Vermont is punishable by a maximum of life imprisonment, or fine of more than $50,000, or both. This requires serious bodily injury or death, or the use of or threat to use a deadly weapon, or repeated rape, or the perpetration of the rape by more than one individual, or a victim younger than the age of thirteen and a perpetrator who is at least eighteen years of age. In contrast, sexual assault in Vermont is punishable by a maximum of thirty-five years in prison, or by a fine of not more than $25,000, or both. Sexual assault involves the compulsion of another to participate in a sexual act without consent, or through threat or coercion, or by placing the other person in fear of imminent bodily injury, or by the administration of drugs or intoxicants, as well as the sexual assault of a juvenile younger than sixteen by a parent or guardian.[25]

The *Actus Reus* of Modern Rape

The *actus reus* of rape requires the sexual penetration of the body of a rape victim by force. There are three approaches to defining the *actus reus*. Most states adhere to the common law and punish genital copulation. A second group of states that has followed the Model Penal Code expands this to include anal and oral copulation. A third group of states includes digital penetration and penetration with an instrument as well as genital, anal, and oral sex. The Model Penal Code and states such as Utah punish this last form of penetration as a less serious form of sexual assault.[26]

Virtually all jurisdictions have adopted gender-neutral statutes that provide that women as well as men may be the perpetrators or victims of rape. A woman, for instance, may be an accomplice to the rape of a male by restraining a male victim while he is being subjected to homosexual rape. As for sexual penetration, California's statute reflects the majority rule by providing that "sexual penetration, however slight, is sufficient to complete the crime." An emission is not required.[27] An Arizona appellate court noted that it was following the majority rule in affirming the rape conviction of an impotent defendant who was unable to attain an erection and only achieved a penetration of roughly one inch.[28]

The essence of the *actus reus* of the common law crime of rape remains the employment of force to cause another individual to submit to sexual penetration without consent. How much force is required? Professor Wayne LaFave observes that courts have followed two distinct approaches. The first is the **extrinsic force** approach that requires an act of force beyond the physical effort required to accomplish penetration. This ensures that the penetration is without the victim's consent. The **intrinsic force** standard only requires the amount of force required to achieve penetration. The intrinsic force standard is based on the insight that there may be a lack of consent despite the fact that the perpetrator employed little or no force to achieve penetration.

The next cases represent these two approaches to force. *Commonwealth v. Berkowitz* represents the extrinsic force standard. You should contrast this with *In the Interest of M.T.S.,* which illustrates the intrinsic force standard.

Was *Berkowitz* guilty of rape?

Commonwealth v. Berkowitz, 609 A.2d 1338 (Pa. Super Ct. 1992), 641 A.2d 1161 (Pa. 1994)

Facts

Robert Berkowitz . . . was convicted in the Court of Common Pleas, Monroe County, of rape and indecent assault and he appealed. The Superior Court, Philadelphia, reversed the rape conviction. The Commonwealth appealed to the Pennsylvania Supreme Court.

In the spring of 1988, appellant and the victim were both college sophomores at East Stroudsburg State University, ages twenty and nineteen years old, respectively. They had mutual friends and acquaintances. On April 19 of that year, the victim went to appellant's dormitory room. What transpired in that dorm room between appellant and the victim thereafter is the subject of the instant appeal.

During a one day jury trial held on September 14, 1988, the victim gave the following account during direct examination by the Commonwealth. At roughly 2:00 on the afternoon of April 19, 1988, after attending two morning classes, the victim returned to her dormitory room. There, she drank a martini to "loosen up a little bit" before going to meet her boyfriend, with whom she had argued the night before. Roughly ten minutes later she walked to her boyfriend's dormitory lounge to meet him. He had not yet arrived.

Having nothing else to do while she waited for her boyfriend, the victim walked up to appellant's room to look for Earl Hassel, appellant's roommate. She knocked on the door several times but received no answer. She therefore wrote a note to Mr. Hassel, which read, "Hi Earl, I'm drunk. That's not why I came to see you. I haven't seen you in a while. I'll talk to you later, [victim's name]." She did so, although she had not felt any intoxicating effects from the martini, "for a laugh."

After the victim had knocked again, she tried the knob on the appellant's door. Finding it open, she walked in. She saw someone lying on the bed with a pillow over his head, whom she thought to be Earl Hassel. After lifting the pillow from his head, she realized it was appellant. She asked appellant which dresser was his roommate's. He told her, and the victim left the note.

Before the victim could leave appellant's room, however, appellant asked her to stay and "hang out for a while." She complied because she "had time to kill" and because she didn't really know appellant and wanted to give him "a fair chance." Appellant asked her to give him a back rub but she declined, explaining that she did not "trust" him. Appellant then asked her to have a seat on his bed. Instead, she found a seat on the floor, and conversed for a while about a mutual friend. No physical contact between the two had, to this point, taken place. The victim testified on cross-examination that she explained that she was having problems with her boyfriend.

Thereafter, however, appellant moved off the bed and down on the floor, and "kind of pushed [the victim] back with his body. It wasn't a shove, it was just kind of a leaning-type of thing." Next appellant "straddled" and started kissing the victim. The victim responded by saying, "Look, I gotta go. I'm going to meet [my boyfriend]." Then appellant lifted up her shirt and bra and began fondling her. The victim then said "no."

After roughly thirty seconds of kissing and fondling, appellant "undid his pants and he kind of moved his body up a little bit." The victim was still saying "no" but "really couldn't move because [appellant] was shifting at [her] body so he was over [her]." . . . Appellant then tried to put his penis in her mouth. The victim did not physically resist, but rather continued to verbally protest, saying "No, I gotta go, let me go," in a "scolding" manner.

Ten or fifteen more seconds passed before the two rose to their feet. Appellant disregarded the victim's continual complaints that she "had to go," and instead walked two feet away to the door and locked it so that no one from the outside could enter. The victim testified that she realized at the time that the lock was not of a type that could lock people inside the room.

Then, in the victim's words, "[appellant] put me down on the bed. It was kind of like—he didn't throw me on the bed. It's hard to explain. It was kind of like a push but no. . . ." She did not bounce off the bed. "It wasn't slow like a romantic kind of thing, but it wasn't a fast shove either. It was kind of in the middle."

Once the victim was on the bed, appellant began "straddling" her again while he undid the knot in her sweatpants. . . . He then removed her sweatpants and underwear from one of her legs. The victim did not physically resist in any way while on the bed because appellant was on top of her, and she "couldn't like go anywhere." She did not scream out at anytime because, "[i]t was like a dream was happening or something."

Appellant then used one of his hands to "guide" his penis into her vagina. At that point, after appellant was inside her, the victim began saying "no, no to him softly in a moaning kind of way . . . because it was just so scary." After about thirty seconds, appellant pulled out his penis and ejaculated onto the victim's stomach.

Immediately thereafter, appellant got off the victim and said, "Wow, I guess we just got carried away." To this the victim retorted, "No, we didn't get carried away, you got carried away." The victim then quickly dressed, grabbed her school books and raced downstairs to her boyfriend, who was by then waiting for her in the lounge.

Once there, the victim began crying. Her boyfriend and she went up to his dorm room where, after watching the victim clean off appellant's semen from her stomach, he called the police.

Defense counsel's cross-examination elicited more details regarding the contact between appellant and the victim before the incident in question. The victim testified that roughly two weeks prior to the incident, she had attended a school seminar entitled, "Does 'no' sometimes mean 'yes'?" Among other things, the lecturer at this seminar had discussed the average length and circumference of human penises. After the seminar, the victim and several of her friends had discussed the subject matter of the seminar over a speaker-telephone with appellant and his roommate Earl Hassel. The victim testified that during that telephone conversation, she had asked appellant the size of his penis. According to the victim, appellant responded by suggesting that the victim "come over and find out." She declined.

When questioned further regarding her communications with appellant prior to the April 19, 1988 incident, the victim testified that on two other occasions, she had stopped by appellant's room while intoxicated. During one of those times, she had laid down on his bed. When asked whether she had asked appellant again at that time what his penis size was, the victim testified that she did not remember.

Appellant took the stand in his own defense and offered an account of the incident and the events leading up to it that differed only as to the consent involved. According to appellant, the victim had begun communication with him after the school seminar by asking him of the size of his penis and of whether he would show it to her. Appellant had suspected that the victim wanted to pursue a sexual relationship with him because she had stopped by his room twice after the phone call while intoxicated, lying down on his bed with her legs spread and again asking to see his penis. He believed that his suspicions were confirmed when she initiated the April 19, 1988, encounter by stopping by his room (again after drinking) and waking him up.

Appellant testified that, on the day in question, he did initiate the first physical contact, but added that the victim warmly responded to his advances by passionately returning his kisses. He conceded that she was continually "whispering . . . no's," but claimed that she did so while "amorously . . . passionately" moaning. In effect, he took such protests to be thinly veiled acts of encouragement. When asked why he locked the door, he explained that "that's not something you want somebody to just walk in on you [doing]."

According to appellant, the two then lay down on the bed, the victim helped him take her clothing off, and he entered her. He agreed that the victim continued to say "no" while on the bed, but carefully qualified his agreement, explaining that the statements were "moaned passionately." According to appellant, when he saw a "blank look on her face," he immediately withdrew and asked "Is anything wrong, is something the matter, is anything wrong?" He ejaculated on her stomach thereafter because he could no longer "control" himself. Appellant testified that after this, the victim "saw that it was over and then she made her move. She gets right off the bed . . . she just swings her legs over and then she puts her clothes back on." Then, in wholly corroborating an aspect of the victim's account, he testified that he remarked, "Well, I guess we got carried away," to which she rebuked, "No, we didn't get carried, you got carried away.'"

After hearing both accounts, the jury convicted appellant of rape and indecent assault. . . . Appellant was then sentenced to serve a term of imprisonment of one to four years for rape and a concurrent term of six to twelve months for indecent assault. . . .

Pennsylvania Supreme Court

Commonwealth v. Berkowitz, 641 A.2d 1161 (Pa. 1994). Opinion by: Cappy, J.

The Commonwealth appeals from an order of the Superior Court which overturned the conviction by a jury of Appellee, Robert A. Berkowitz, of one count of rape and one count of indecent assault. The Superior Court discharged Appellee as to the charge of rape. . . . For the reasons that follow, we affirm the Superior Court's reversal of the conviction for rape. . . .

Issue

The crime of rape is defined as follows:

18 Pennsylvania Consolidated Statutes Annotated (PCSA) Section 3121. Rape

A person commits a felony of the first degree when he engages in sexual intercourse with another person not one's spouse:

(1) by forcible compulsion;

(2) by threat of forcible compulsion that would prevent resistance by a person of reasonable resolution;

(3) who is unconscious; or

(4) who is so mentally deranged or deficient that such person is incapable of consent.

The victim of a rape need not resist. "The force necessary to support a conviction of rape . . . need only be such as to establish lack of consent and to induce the [victim] to submit without additional resistance. . . . The degree of force required to constitute rape is relative and depends on the facts and particular circumstance of the case." . . .

Was Berkowitz guilty of sexual intercourse with the victim by "forcible compulsion"?

Reasoning

In regard to the critical issue of forcible compulsion, the complainant's testimony is devoid of any statement that clearly or adequately describes the use of force or the threat of force against her. In response to defense counsel's question, "Is it possible that [when Appellee lifted your bra and shirt] you took no physical action to discourage him," the complainant replied, "It's possible." When asked, "Is it possible that [Appellee] was not making any physical contact with you . . . aside from attempting to untie the knot [in the drawstrings of complainant's sweatpants],"

she answered, "It's possible." She testified, "He put me down on the bed. It was kind of like—He didn't throw me on the bed. It's hard to explain. It was kind of like a push but not—I can't explain what I'm trying to say." She concluded that "it wasn't much" in reference to whether she bounced on the bed, and further detailed that their movement to the bed "wasn't slow like a romantic kind of thing, but it wasn't a fast shove either. It was kind of in the middle." She agreed that Appellee's hands were not restraining her in any manner during the actual penetration, and that the weight of his body on top of her was the only force applied. She testified that at no time did Appellee verbally threaten her. The complainant did testify that she sought to leave the room, and said "no" throughout the encounter. As to the complainant's desire to leave the room, the record clearly demonstrates that the door could be unlocked easily from the inside, that she was aware of this fact, but that she never attempted to go to the door or unlock it.

As to the complainant's testimony that she stated "no" throughout the encounter with Appellee, we point out that, while such an allegation of fact would be relevant to the issue of consent, it is not relevant to the issue of force. . . . [W]here there is a lack of consent, but no showing of either physical force, a threat of physical force, or psychological coercion, the "forcible compulsion" requirement . . . is not met. . . .

If the legislature had intended to define rape, a felony of the first degree, as nonconsensual intercourse, it could have done so. It did not do this. It defined rape as sexual intercourse by "forcible compulsion." If the legislature means what it said, then where, as here, no evidence was adduced by the Commonwealth that established either that mental coercion or a threat, or force inherently inconsistent with consensual intercourse was used to complete the act of intercourse, the evidence is insufficient to support a rape conviction. According, we hold that the trial court erred in determining that the evidence adduced by the Commonwealth was sufficient to convict appellant of rape. . . .

Holding

Reviewed in light of the above described standard, the complainant's testimony simply fails to establish that the Appellee forcibly compelled her to engage in sexual intercourse as required under 18 PCSA section 3121. Thus, even if all of the complainant's testimony was believed, the jury, as a matter of law, could not have found Appellee guilty of rape. . . . [T]he crime of indecent assault does not include the element of "forcible compulsion" as does the crime of rape. The evidence described above is clearly sufficient to support the jury's conviction of indecent assault. "Indecent contact" is defined as "any touching of the sexual or other intimate parts of the person for the purpose of arousing or gratifying sexual desire, in either person." 18 PCSA section 3101. Appellee himself testified

to the "indecent contact." The victim testified that she repeatedly said "no" throughout the encounter. Viewing that testimony in the light most favorable to the Commonwealth as verdict winner, the jury reasonably could have inferred that the victim did not consent to the indecent contact. Thus, the evidence was sufficient to support the jury's verdict finding Appellee guilty of indecent assault.

Questions for Discussion

1. Why does the Pennsylvania Supreme Court rule that the evidence was insufficient to hold Berkowitz legally liable for rape? Can you explain why the court rules that Berkowitz was properly convicted of indecent assault?

2. What facts support the court's finding that Berkowitz did not use "forcible compulsion" in his sexual intercourse with the victim? Are there facts supporting the contention that Berkowitz did rely on "forcible compulsion"?

3. Why did the Pennsylvania Supreme Court opinion fail to highlight that in the past, the victim had made sexually provocative remarks to Berkowitz and had visited his dorm room? Is it significant that the victim's boyfriend reported the rape immediately after the victim informed him that she had sexual intercourse with Berkowitz?

Was M.T.S. guilty of rape?

In the Interest of M.T.S., 609 A.2d 1266 (N.J. 1992). Opinion by: Handler, J.

Issue

Under New Jersey law a person who commits an act of sexual penetration using physical force or coercion is guilty of second-degree sexual assault. The sexual assault statute does not define the words "physical force." The question posed by this appeal is whether the element of "physical force" is met simply by an act of nonconsensual penetration involving no more force than necessary to accomplish that result. . . . The factual circumstances of this case expose the complexity and sensitivity of those issues and underscore the analytic difficulty of those seemingly straightforward legal questions.

Facts

On Monday, May 21, 1990, fifteen-year-old C.G. was living with her mother, her three siblings, and several other people, including M.T.S. and his girlfriend. A total of ten people resided in the three-bedroom town-home at the time of the incident. M.T.S., then age seventeen, was temporarily residing at the home with the permission of C.G.'s mother; he slept downstairs on a couch. C.G. had her own room on the second floor. At approximately 11:30 p.m. on May 21, C.G. went upstairs to sleep after having watched television with her mother, M.T.S., and his girlfriend. When C.G. went to bed, she was wearing

underpants, a bra, shorts, and a shirt. At trial, C.G. and M.T.S. offered very different accounts concerning the nature of their relationship and the events that occurred after C.G. had gone upstairs. The trial court did not credit fully either teenager's testimony.

C.G. stated that earlier in the day, M.T.S. had told her three or four times that he "was going to make a surprise visit up in [her] bedroom." She said that she had not taken M.T.S. seriously and considered his comments a joke because he frequently teased her. She testified that M.T.S. had attempted to kiss her on numerous other occasions and at least once had attempted to put his hands inside of her pants, but that she had rejected all of his previous advances. C.G. testified that on May 22, at approximately 1:30 a.m., she awoke to use the bathroom. As she was getting out of bed, she said, she saw M.T.S., fully clothed, standing in her doorway. According to C.G., M.T.S. then said that "he was going to tease [her] a little bit." C.G. testified that she "didn't think anything of it"; she walked past him, used the bathroom, and then returned to bed, falling into a "heavy" sleep within fifteen minutes. The next event C.G. claimed to recall of that morning was waking up with M.T.S. on top of her, her underpants and shorts removed. She said "his penis was into [her] vagina." As soon as C.G. realized what had happened, she said, she immediately slapped M.T.S. once in the face, then "told him to get off [her], and get out." She did not scream or cry out. She testified that M.T.S. complied in less than one minute after being struck; according to C.G., "he jumped right off of [her]." She said she did not know how long M.T.S. had been inside of her before she awoke. C.G. said that after M.T.S. left the room, she "fell asleep crying" because "[she] couldn't believe that he did what he did to [her]." She explained that she did not immediately tell her mother or anyone else in the house of the events of that morning because she was "scared and in shock."

According to C.G., M.T.S. engaged in intercourse with her "without [her] wanting it or telling him to come up [to her bedroom]." By her own account, C.G. was not otherwise harmed by M.T.S. At about 7:00 a.m., C.G. went downstairs and told her mother about her encounter with M.T.S. earlier in the morning and said that they would have to "get [him] out of the house." While M.T.S. was out on an errand, C.G.'s mother gathered his clothes and put them outside in his car; when he returned, he was told that "[he] better not even get near the house." C.G. and her mother then filed a complaint with the police.

According to M.T.S., he and C.G. had been good friends for a long time, and their relationship "kept leading on to more and more." He had been living at C.G.'s home for about five days before the incident occurred; he testified that during the three days preceding the incident they had been "kissing and necking" and had discussed having sexual intercourse. The first time M.T.S. kissed C.G., he said, she "didn't want him to, but she did after that." He said C.G. repeatedly had encouraged him to "make a surprise visit up in her room."

M.T.S. testified that at exactly 1:15 a.m. on May 22, he entered C.G.'s bedroom as she was walking to the bathroom. He said C.G. soon returned from the bathroom, and the two began "kissing and all," eventually moving to the bed. Once they were in bed, he said, they undressed each other and continued to kiss and touch for about five minutes. M.T.S. and C.G. proceeded to engage in sexual intercourse. According to M.T.S., who was on top of C.G., he "stuck it in" and "did it [thrust] three times, and then the fourth time [he] stuck it in, that's when [she] pulled [him] off of her." M.T.S. said that as C.G. pushed him off, she said "stop, get off," and he "hopped off right away."

According to M.T.S., after about one minute, he asked C.G. what was wrong; she replied with a backhand to his face. He recalled asking C.G. what was wrong a second time, and her replying, "[H]ow can you take advantage of me or something like that." M.T.S. said that he proceeded to get dressed and told C.G. to calm down, but that she then told him to get away from her and began to cry. Before leaving the room, he told C.G., "I'm leaving . . . I'm going with my real girlfriend, don't talk to me . . . I don't want nothing to do with you or anything, stay out of my life . . . don't tell anybody about this . . . it would just screw everything up." He then walked downstairs and went to sleep.

On May 23, 1990, M.T.S. was charged with conduct that if engaged in by an adult would constitute second-degree sexual assault of the victim, contrary to New Jersey Statutes Annotated (NJSA) section 2C:14–2c(1). Following a two-day trial on the sexual assault charge, M.T.S. was adjudicated delinquent. After reviewing the testimony, the court concluded that the victim had consented to a session of kissing and heavy petting with M.T.S. The trial court did not find that C.G. had been sleeping at the time of penetration, but nevertheless found that she had not consented to the actual sexual act. Accordingly, the court concluded that the State had proven second-degree sexual assault beyond a reasonable doubt. On appeal, following the imposition of suspended sentences on the sexual assault and the other remaining charges, the Appellate Division determined that the absence of force beyond that involved in the act of sexual penetration precluded a finding of second-degree sexual assault. It therefore reversed the juvenile's adjudication of delinquency for that offense.

Reasoning

The New Jersey Code of Criminal Justice defines "sexual assault" as the commission "of sexual penetration . . . with another person" with the use of "physical force or coercion." An unconstrained reading of the statutory language indicates that both the act of "sexual penetration" and the use of "physical force or coercion" are separate and distinct elements of the offense. The parties offer two alternative understandings of the concept of "physical force" as it is used in the statute. The State would read "physical force" to entail any amount of sexual touching brought about involuntarily. A showing of sexual penetration coupled with a lack of consent would satisfy the elements of the statute. The Public Defender urges an interpretation of "physical force" to mean force "used to overcome lack of consent." That definition equates force with violence and leads to the conclusion that sexual assault requires the application of some amount of force in addition to the act of penetration. The new statutory provisions covering rape were formulated by a coalition of feminist groups assisted by the National Organization of Women (NOW) National Task Force on Rape. Both houses of the Legislature adopted the NOW bill, as it was called, without major changes and the Governor signed it into law on August 10, 1978.

Since the 1978 reform, the Code has referred to the crime that was once known as "rape" as "sexual assault." The crime now requires "penetration," not "sexual intercourse." It requires "force" or "coercion," not "submission" or "resistance." It makes no reference to the victim's state of mind or attitude, or conduct in response to the assault. It eliminates the spousal exception based on implied consent. It emphasizes the assaultive character of the offense by defining sexual penetration to encompass a wide range of sexual contacts, going well beyond traditional "carnal knowledge." Consistent with the assaultive character, as opposed to the traditional sexual character, of the offense, the statute also renders the crime gender-neutral: both males and females can be actors or victims.

The reform statute defines sexual assault as penetration accomplished by the use of "physical force" or "coercion," but it does not define either "physical force" or "coercion" or enumerate examples of evidence that would establish those elements. Some reformers had argued that defining "physical force" too specifically in the sexual offense statute might have the effect of limiting force to the enumerated examples. The task of defining "physical force" therefore was left to the courts.

The New Jersey Code of Criminal Justice does not refer to force in relation to "overcoming the will" of the victim, or to the "physical overpowering" of the victim, or the "submission" of the victim. It does not require the demonstrated nonconsent of the victim. As we have noted, in reforming the rape laws, the Legislature placed primary emphasis on the assaultive nature of the crime, altering its constituent elements so that they focus exclusively on the forceful or assaultive conduct of the defendant.

The Legislature's concept of sexual assault and the role of force were significantly colored by its understanding of the law of assault and battery. As a general matter, criminal battery is defined as "the unlawful application of force to the person of another." The application of force is criminal when it results in either (a) a physical injury or (b) an offensive touching. Thus, by eliminating all references to the victim's state of mind and conduct, and by broadening the definition of penetration to cover not only sexual intercourse between a man and a woman but a range of acts that invade another's body or compel intimate contact, the Legislature emphasized the affinity between sexual assault and other forms of assault and battery. . . .

The understanding of sexual assault as a criminal battery, albeit one with especially serious consequences, follows necessarily from the Legislature's decision to eliminate nonconsent and resistance from the substantive definition of the offense. Under the new law, the victim no longer is required to resist and therefore need not have said or done anything in order for the sexual penetration to be unlawful. The alleged victim is not put on trial, and his or her responsive or defensive behavior is rendered immaterial. We are thus satisfied that an interpretation of the statutory crime of sexual assault to require physical force in addition to that entailed in an act of involuntary or unwanted sexual penetration would be fundamentally inconsistent with the legislative purpose to eliminate any consideration of whether the victim resisted or expressed nonconsent.

We note that the contrary interpretation of force—that the element of force need be extrinsic to the sexual act—would not only reintroduce a resistance requirement into the sexual assault law, but also would immunize many acts of criminal sexual contact short of penetration. The characteristics that make a sexual contact unlawful are the same as those that make a sexual penetration unlawful. An actor is guilty of criminal sexual contact if he or she commits an act of sexual

contact with another using "physical force" or "coercion." NJSA § 2C:14–3(b). That the Legislature would have wanted to decriminalize unauthorized sexual intrusions on the bodily integrity of a victim by requiring a showing of force in addition to that entailed in the sexual contact itself is hardly possible.

Because the statute eschews any reference to the victim's will or resistance, the standard defining the role of force in sexual penetration must prevent the possibility that the establishment of the crime will turn on the alleged victim's state of mind or responsive behavior. We conclude, therefore, that any act of sexual penetration engaged in by the defendant without the affirmative and freely given permission of the victim to the specific act of penetration constitutes the offense of sexual assault. Therefore, physical force in excess of that inherent in the act of sexual penetration is not required for such penetration to be unlawful. The definition of "physical force" is satisfied under NJSA § 2C:14–2c(1) if the defendant applies any amount of force against another person in the absence of what a reasonable person would believe to be affirmative and freely given permission to the act of sexual penetration.

Under the reformed statute, permission to engage in sexual penetration must be affirmative and it must be given freely, but that permission may be inferred either from acts or statements reasonably viewed in light of the surrounding circumstances. Persons need not, of course, expressly announce their consent to engage in intercourse for there to be affirmative permission. Permission to engage in an act of sexual penetration can be and indeed often is indicated through physical actions rather than words. Permission is demonstrated when the evidence, in whatever form, is sufficient to demonstrate that a reasonable person would have believed that the alleged victim had affirmatively and freely given authorization to the act.

Our understanding of the meaning and application of "physical force" under the sexual assault statute indicates that the term's inclusion was neither inadvertent nor redundant. The term "physical force," like its companion term "coercion," acts to qualify the nature and character of the "sexual penetration." Sexual penetration accomplished through the use of force is unauthorized sexual penetration. That functional understanding of "physical force" encompasses the notion of "unpermitted touching" derived from the Legislature's decision to redefine rape as a sexual assault.

As already noted, under assault and battery doctrine, any amount of force that results in either physical injury or offensive touching is sufficient to establish a battery.

Hence, as a description of the method of achieving "sexual penetration," the term "physical force" serves to define and explain the acts that are offensive, unauthorized, and unlawful.

Today the law of sexual assault is indispensable to the system of legal rules that assures each of us the right to decide who may touch our bodies, when, and under what circumstances. The decision to engage in sexual relations with another person is one of the most private and intimate decisions a person can make. Each person has the right not only to decide whether to engage in sexual contact with another, but also to control the circumstances and character of that contact. No one, neither a spouse, nor a friend, nor an acquaintance, nor a stranger, has the right or the privilege to force sexual contact.

We emphasize as well that what is now referred to as "acquaintance rape" is not a new phenomenon. Nor was it a "futuristic" concept in 1978 when the sexual assault law was enacted. Current concern over the prevalence of forced sexual intercourse between persons who know one another reflects both greater awareness of the extent of such behavior and a growing appreciation of its gravity. Notwithstanding the stereotype of rape as a violent attack by a stranger, the vast majority of sexual assaults are perpetrated by someone known to the victim. One respected study indicates that more than half of all rapes are committed by male relatives, current or former husbands, boyfriends, or lovers. Similarly, contrary to common myths, perpetrators generally do not use guns or knives and victims generally do not suffer external bruises or cuts. Although this more realistic and accurate view of rape only recently has achieved widespread public circulation, it was a central concern of the proponents of reform in the 1970s.

The insight into rape as an assaultive crime is consistent with our evolving understanding of the wrong inherent in forced sexual intimacy. It is one that was appreciated by the Legislature when it reformed the rape laws, reflecting an emerging awareness that the definition of rape should correspond fully with the experiences and perspectives of rape victims. Although reformers focused primarily on the problems associated with convicting defendants accused of violent rape, the recognition that forced sexual intercourse often takes place between persons who know each other and often involves little or no violence comports with the understanding of the sexual assault law that was embraced by the Legislature. Any other interpretation of the law, particularly one that defined force in relation to the resistance or protest of the victim, would directly undermine the goals sought to be achieved by its reform.

Holding

In short, in order to convict under the sexual assault statute in cases such as these, the State must prove beyond a reasonable doubt that there was sexual penetration and that it was accomplished without the affirmative and freely given permission of the alleged victim. As we have indicated, such proof can be based on evidence of conduct or words in light of surrounding circumstances and must demonstrate beyond a reasonable doubt that a reasonable person would not have believed that there was affirmative and freely given permission. If there is evidence to suggest that the defendant reasonably believed that such permission had been given, the State must demonstrate either that defendant did not actually believe that affirmative permission had been freely given or that such a belief was unreasonable under all of the circumstances. Thus, the State bears the burden of proof throughout the case. . . .

We acknowledge that cases such as this are inherently fact sensitive and depend on the reasoned judgment and common sense of judges and juries. The trial court concluded that the victim had not expressed consent to the act of intercourse, either through her words or actions. We conclude that the record provides reasonable support for the trial court's disposition.

Accordingly, we reverse the judgment of the Appellate Division and reinstate the disposition of juvenile delinquency for the commission of second-degree sexual assault.

Questions for Discussion

1. Explain why the New Jersey Supreme Court ruled that the definition of "physical force" is satisfied if the defendant applies any amount of force against another person in the absence of what a reasonable person would believe to be the affirmative and freely given permission to the act of sexual penetration.

2. What are the facts the prosecution is required to prove beyond a reasonable doubt to convict a defendant under the legal test established by the New Jersey Supreme Court? List the facts that are relied on by the New Jersey Supreme Court in concluding that the sexual penetration was without the affirmative and freely given permission of the victim.

3. Is there difficulty determining precisely what occurred in this case? As defense attorney, what questions would you ask the victim on cross-examination in order to undermine her direct testimony? What questions would you ask your client, the accused, on direct examination in the event that he testified?

4. As a juror, would you convict the defendant?

5. Do you believe that it is difficult to apply the legal test established by the New Jersey Supreme Court? What problems do you anticipate would confront a defendant in establishing that the alleged victim, in fact, affirmatively and freely engaged in sexual contact?

6. As a lawyer working for a New Jersey university, what points would you include in a guide to sexual conduct to be distributed to students in light of *M.T.S.?*

7. Do you favor the standard in *Berkowitz* or in *M.T.S.?*

Other Approaches to the *Actus Reus* of Modern Rape

The force requirement may also be satisfied by a threat of force. Statutes make this clear by stating that penetration may be accomplished by "force or threat of force" or by "force or coercion" or by "force or fear." In other words, actual force is not required.

There are two requirements that must be satisfied. First, there must be a threat of death or serious personal injury. Second, the victim's fear that the assailant will carry out the threat must be reasonable. The California statute provides that sexual intercourse constitutes rape where accomplished by "threatening to retaliate in the future . . . and there is a reasonable possibility that the perpetrator will execute the threat. . . . Threatening to retaliate means a threat to kidnap or falsely imprison or to inflict extreme pain, serious bodily injury, or death." The California statute also states that the threat requirement is satisfied where a public official threatens to "incarcerate, arrest, or deport the victim or another. . . ."[29] Michigan provides that a person is guilty of criminal sexual conduct where an individual engages in sexual penetration with another under circumstances in which the "actor is armed with a weapon or any article . . . fashioned . . . to lead the victim to reasonably believe it to be a weapon."[30]

Several states have extended the *actus reus* for rape and declare that it is criminal to use a position of trust to cause another person to submit to sexual penetration. Texas, for instance, has provisions covering public servants, therapists, and nurses. Clergy are held liable for causing another person to submit or participate in a sexual act by "exploiting the other person's emotional dependency."[31] Pennsylvania defines "forcible compulsion" to mean "physical, intellectual, moral, emotional or psychological force either express or implied."[32] This was narrowly interpreted by the Pennsylvania Supreme Court in ruling that it was not forcible compulsion for an adult guardian to threaten to send a fourteen-year-old girl back to a juvenile detention home unless she submitted to vaginal intercourse and various deviate sexual acts. The court ruled that although the acts of the sixty-three-year-old guardian were despicable, the young woman was not coerced into the sexual activity because she voluntarily made a deliberate choice to undertake a course of conduct that enabled her to avoid return to the juvenile home.[33]

We have already mentioned that force is not required where penetration is achieved through fraud or when the victim is unconscious, asleep, or insane and is unaware of the nature of the sexual penetration. The California statute provides for rape when the victim is "prevented from resisting by any intoxicating or anesthetic substance, or any controlled substance" or where an individual is "unconscious of the nature of the act. . . ."[34]

In these situations, a victim clearly is incapable of consent and the law finds rape. We shall see when we consider **statutory rape** that force is also not required where the victim is a juvenile.

Mens Rea

Rape at common law required that the male defendant intended to engage in vaginal intercourse with a woman whom he knew was not his wife through force or the threat of force. There was no clear guidance as to whether a defendant was required to be aware that the intercourse was without the female's consent. This issue remained unsettled for a number of years and has been resolved only in the last several decades.

The majority of states accept an "objective test" that recognizes it is a defense to rape that a defendant honestly and reasonably believed that the rape victim consented. This doctrine was first recognized by the California Supreme Court in *People v. Mayberry*. The prosecutrix claimed that she had been kidnapped while shopping and had involuntarily accompanied the kidnapper to an apartment where she was raped by the defendant and another male. The defendant denied the accusations and characterized the victim's story as "inherently improbable."

The California Supreme Court ruled that the defendant was entitled to have the jury receive a mistake of fact instruction, reasoning that the state legislature must have intended that such a defense be available given the seriousness of the charge. The court accordingly held that a defendant who "entertains a reasonable and bona fide belief that a prosecutrix voluntarily consented to accompany him and to engage in sexual intercourse . . . does not possess the wrongful intent that is a prerequisite . . . to a conviction of rape by means of force or threat."[35]

Some commentators argue that given the seriousness of the crime of rape, a defendant who genuinely believes that an individual has consented should be entitled to a mistake of fact defense, no matter how unreasonable his belief. The English House of Lords adopted a subjective approach in *Director of Public Prosecutions v. Morgan.* A married man invited three of his drinking companions over to his house to have intercourse with his wife. He assured his fellow military men that although she might protest, his wife enjoyed "kinky sex" and this was the only way she was able to get "turned on." The three held down the woman while they took turns having sexual intercourse with her.

The House of Lords ruled that rape requires a specific intent to have vaginal intercourse with a woman without her consent and acquitted the defendants based on their honest belief that the officer's wife consented to the sexual intercourse. As Lord Hailsham noted, "either the prosecution proves that the accused had the requisite intent or it does not."[36]

The English test is based on the premise that it is unfair to hold an individual liable who actually believes that an individual has consented. This approach, however, opens the door to defendants claiming this defense in virtually every acquaintance rape case. The subjective approach has been adopted with qualifications by the English Parliament and has been followed by only a small number of American courts.

Several states do not recognize the mistake of fact defense and continue to adhere to the view that a defendant's belief as to a victim's consent should not be considered in determining guilt. In *State v. Plunkett,* a physically imposing twenty-nine-year-old defendant intimidated and forced a seventeen-year-old woman who was a virgin into his house where, in response to his aggressive refusal to let her leave, she fearfully cooperated in oral, anal, and vaginal intercourse. The Kansas Supreme Court ruled that "[w]hether Plunkett thought his victim consented is irrelevant if the State proved that she did not consent and was overcome by fear. . . . Plunkett's argument that there was insufficient evidence to show that he knew his sexual activity was nonconsensual is irrelevant." The difficulty with this approach is that rape is converted into a strict liability defense and that a defendant is liable regardless of the reasonableness of his or her belief concerning the victim's consent.[37]

A majority of states, as noted, have adopted the objective test. This requires *equivocal conduct,* meaning that the victim's nonconsensual reactions were capable of being reasonably, but mistakenly interpreted by the assailant as indicating consent. In the prosecution of heavyweight boxing champion Mike Tyson for the rape of a contestant in a beauty pageant that he was judging, an Indiana Court of Appeals held that Tyson was not entitled to a reasonable mistake of fact defense. Tyson's testimony indicated that the victim freely and fully participated in their sexual relationship. The victim, on the other hand, testified that Tyson forcibly imposed himself on her. The appellate court ruled that "there is no recitation of equivocal conduct by D.W. [the victim] that reasonably could have led Tyson to believe that D.W . . . appeared to consent . . . [N]o gray area existed from which Tyson can logically argue that he misunderstood D.W.'s actions."[38]

Statutory Rape

Sexual relations with a juvenile was not a crime under early English law so long as there was consent. This was modified by a statute that declared that it was a felony to engage in vaginal intercourse with a child younger than the age of ten, regardless of whether there was consent. This so-called statutory rape was incorporated into the common law of the United States. American legislatures gradually raised the age at which a child was protected against sexual intercourse to between eleven and fourteen. Statutory rape is based on several considerations:[39]

- *Understanding.* A minor is considered incapable of understanding the nature and consequences of his or her act.
- *Harmful.* Sexual relations are psychologically damaging to a minor and may lead to pregnancy.
- *Social Values.* This type of conduct is immoral and is contrary to social values.
- *Vulnerability.* The protection of females is based on the fact that males are typically the aggressors and take advantage of the vulnerability of immature females.

The general rule is that statutory rape is a strict liability offense in which a male is guilty of rape by engaging in intercourse with an "underage" female. This rule is intended to ensure that males will take extraordinary steps to ensure that females are of the age of consent.

Commentators have viewed strict liability for statutory rape as unjust in the case of a young woman who is physically mature and misrepresents her age. One reform is to provide that defendants can offer a "promiscuity defense" and document that the victim has had multiple sex partners and can be presumed to have possessed the capacity to appreciate the nature of her act and to have knowingly consented to a sexual relationship. A second approach is to divide the offense into various categories and to provide for more severe penalties for sexual relationships involving younger women. A third approach recognizes that young people will engage in sexual experimentation and that statutory rape should not be a crime where the parties are roughly the same age or should be punished less severely. Forty-five states now recognize statutory rape as a gender-neutral offense; only five states still restrict guilt to males.[40]

The major issue that arises in regard to statutory rape is whether this should remain a strict liability offense in which the mere act of penetration results in criminal liability. Should a reasonable mistake of fact regarding an individual's age constitute a defense?

The Model Penal Code limits strict liability to sexual relations with a woman younger than ten years of age. In such cases, the commentary explains that there is little likelihood of a reasonable mistake regarding the victim's age. The code punishes sexual intercourse with a woman sixteen years of age or younger by a male at least four years older as the less serious offense of corruption of a minor. A defendant charged with corruption of a minor is authorized under the code to offer a reasonable mistake of fact defense as to the age of the victim.[41]

Consider whether you favor imposing strict liability for statutory rape after reading *Garnett v. State*.

Did Garnett know that he was committing statutory rape?

Garnett v. State, 632 A.2d 797 (Md. 1993). Opinion by: Murphy, J.

Facts

Maryland's "statutory rape" law prohibiting sexual intercourse with an underage person is codified in Annotated Code of Maryland (ACM) ... article 27, section 463, which reads in full:

Second-degree rape.

(1) A person is guilty of rape in the second degree if the person engages in vaginal intercourse with another person:

(a) Who is under 14 years of age and the person performing the act is at least four years older than the victim.

(2) Any person violating the provisions of this section is guilty of a felony and upon conviction is subject to imprisonment for a period of not more than 20 years.

Now we consider whether under the present statute, the State must prove that a defendant knew the complaining witness was younger than fourteen and, in a related question, whether it was error at trial to exclude evidence that he had been told, and believed, that she was sixteen years old.

Raymond Lennard Garnett is a young retarded man. At the time of the incident in question he was twenty years old. He has an I.Q. of fifty-two. His guidance counselor from the Montgomery County public school system, Cynthia Parker, described him as a mildly retarded person who read on the third-grade level, did

arithmetic on the fifth-grade level, and interacted with others socially at school at the level of someone eleven or twelve years of age. Ms. Parker added that Raymond attended special education classes and for at least one period of time was educated at home when he was afraid to return to school due to his classmates' taunting. Because he could not understand the duties of the jobs given him, he failed to complete vocational assignments; he sometimes lost his way to work. As Raymond was unable to pass any of the State's functional tests required for graduation, he received only a certificate of attendance rather than a high-school diploma.

In November or December 1990, a friend introduced Raymond to Erica Frazier, then aged thirteen; the two subsequently talked occasionally by telephone. On February 28, 1991, Raymond, apparently wishing to call for a ride home, approached the girl's house at about nine o'clock in the evening. Erica opened her bedroom window, through which Raymond entered; he testified that "she just told me to get a ladder and climb up her window." The two talked, and later engaged in sexual intercourse. Raymond left at about 4:30 the following morning. On November 19, 1991, Erica gave birth to a baby, of whom Raymond is the biological father.

Raymond was tried before the Circuit Court for Montgomery County on one count of second degree rape under § 463(a)(3) proscribing sexual intercourse between a person under fourteen and another at least four years older than the complainant.

The court found Raymond guilty. It sentenced him to a term of five years in prison, suspended the sentence and imposed five years of probation, and ordered that he pay restitution to Erica and the Frazier family.

Issue

Raymond asserts that the events of this case were inconsistent with the criminal sexual exploitation of a minor by an adult. As earlier observed, Raymond entered Erica's bedroom at the girl's invitation; she directed him to use a ladder to reach her window. They engaged voluntarily in sexual intercourse. They remained together in the room for more than seven hours before Raymond departed at dawn. With an I.Q. of fifty-two, Raymond functioned at approximately the same level as the thirteen-year-old Erica; he was mentally an adolescent in an adult's body. Arguably, had Raymond's chronological age, twenty, matched his sociointellectual age, about twelve, he and Erica would have fallen well within the four-year age difference obviating a violation of the statute, and Raymond would not have been charged with any crime at all.

The precise legal issue here rests on Raymond's unsuccessful efforts to introduce into evidence testimony that Erica and her friends had told him she was sixteen years old, the age of consent to sexual relations, and that he believed them. Thus the trial court did not permit him to raise a defense of reasonable mistake of Erica's age, by which defense Raymond would have asserted that he acted innocently without a criminal design.

Reasoning

At common law, a crime occurred only upon the concurrence of an individual's act and his guilty state of mind. In this regard, it is well understood that generally there are two components of every crime, the *actus reus* or guilty act and the *mens rea* or the guilty mind or mental state accompanying a forbidden act. The requirement that an accused acted with a culpable mental state is an axiom of criminal jurisprudence.

To be sure, legislative bodies since the mid-nineteenth century have created strict liability criminal offenses requiring no *mens rea*. Almost all such statutes responded to the demands of public health and welfare arising from the complexities of society after the Industrial Revolution. Typically misdemeanors involving only fines or other light penalties, these strict liability laws regulated food, milk, liquor, medicines and drugs, securities, motor vehicles and traffic, the labeling of goods for sale, and the like. Statutory rape, carrying the stigma of felony as well as a potential sentence of twenty years in prison, contrasts markedly with the other strict liability regulatory offenses and their light penalties.

Modern scholars generally reject the concept of strict criminal liability.

The consensus is that punishing conduct without reference to an actor's state of mind is unjust since a person is subjected to a criminal conviction without being morally blameworthy. An individual who acts without a criminal intent neither needs to be punished in order to be deterred from future criminal activity nor incapacitated or reformed in order to remove a threat to society.

Conscious of the disfavor in which strict criminal liability resides, the Model Penal Code states generally as a minimum requirement of culpability that a person is not guilty of a criminal offense unless he acts purposely, knowingly, recklessly, or negligently. The Code allows generally for a defense of ignorance or mistake of fact negating criminal intent.

The commentators similarly disapprove of statutory rape as a strict liability crime. In addition to the arguments discussed above, they observe that statutory rape prosecutions often proceed even when the defendant's judgment as to the age of the complainant is warranted by her appearance, her sexual sophistication, her verbal misrepresentations, and the defendant's careful attempts to ascertain her true age. Voluntary intercourse with a sexually mature teenager lacks the features of psychic abnormality, exploitation, or physical danger that accompanies such conduct with children.

Statutory rape laws are often justified on the "lesser legal wrong" theory or the "moral wrong" theory; by such reasoning, the defendant acting without *mens rea* nonetheless deserves punishment for having committed a lesser crime, fornication, or for having violated moral teachings that prohibit sex outside of marriage. We acknowledge here that it is uncertain to what extent Raymond's intellectual and social retardation may have impaired his ability to comprehend imperatives of sexual morality in any case.

The legislatures of seventeen states have enacted laws permitting a mistake of age defense in some form in cases of sexual offenses with underage persons. In addition, the highest appellate courts of four states have determined that statutory rape laws by implication required an element of *mens rea* as to the complainant's age. In the landmark case of *People v. Hernandez*, 393 P.2d 673 (Cal. 1964), the California Supreme Court held that, absent a legislative directive to the contrary, a charge of statutory rape was defensible wherein a criminal intent was lacking; it reversed the trial court's refusal to permit the defendant to present evidence of his good faith, reasonable belief that the complaining witness had reached the age of consent. In so doing, the court first questioned the assumption that age alone confers a sophistication sufficient to create legitimate consent to sexual relations, observing that "the sexually experienced fifteen-year-old may be far more acutely aware of the implications of sexual intercourse than her sheltered cousin who is beyond the age of consent." The court then asked whether it could be considered fair to punish an individual who participates in a mutual act of sexual intercourse while reasonably believing his partner to be beyond the age of consent. . . .

We think it sufficiently clear, however, that Maryland's second-degree rape statute defines a strict liability offense that does not require the State to prove *mens rea*; it makes no allowance for a mistake-of-age defense. The plain language of ACM section 463, viewed in its entirety, and the legislative history of its creation lead to this conclusion. Section 463(a)(3) prohibiting sexual intercourse with underage persons makes no reference to the actor's knowledge, belief, or other state of mind. As we see it, this silence as to *mens rea* results from legislative design. . . . Second, an examination of the drafting history of section 463 during the 1976 revision of Maryland's sexual offense laws reveals that the statute was viewed as one of strict liability from its inception and throughout the amendment process. . . . [T]he Legislature explicitly raised, considered, and then explicitly jettisoned any notion of a *mens rea* element with respect to the complainant's age in enacting the law that formed the basis of current section 463(a)(3).

Holding

In the light of such legislative action, we must inevitably conclude that the current law imposes strict liability on its violators. This interpretation is consistent with the traditional view of statutory rape as a strict liability crime designed to protect young persons from the dangers of sexual exploitation by adults, loss of chastity, physical injury, and, in the case of girls, pregnancy. The majority of states retain statutes which impose strict liability for sexual acts with underage complainants. We observe again, as earlier, that even among those states providing for a mistake-of-age defense in some instances, the defense often is not available where the sex partner is fourteen years old or less; the complaining witness in the instant case was only thirteen.

Maryland's second degree rape statute is by nature a creature of legislation. Any new provision introducing an element of *mens rea,* or permitting a defense of reasonable mistake of age, with respect to the offense of sexual intercourse with a person less than fourteen, should properly result from an act of the Legislature itself, rather than judicial fiat. Until then, defendants in extraordinary cases, like Raymond, will rely upon the tempering discretion of the trial court at sentencing.

Dissenting, *Bell, J.*

To hold, as a matter of law, that ACM section 463(a)(3) does not require the State to prove that a defendant possessed the necessary mental state to commit the crime, i.e., knowingly engaged in sexual relations with a female under fourteen, or that the defendant may not litigate that issue in defense, "offends a principle of justice so rooted in the traditions of conscience of our people as to be ranked as fundamental" and is, therefore, inconsistent with due process.

In this case, according to the defendant, he intended to have sex with a sixteen-, not a thirteen-year-old girl. This mistake of fact was prompted, he said, by the prosecutrix herself; she and her friends told him that she was sixteen years old. Because he was mistaken as to the prosecutrix's age, he submits, he is certainly less culpable than the person who knows that the minor is thirteen years old, but nonetheless engages in sexual relations with her. Notwithstanding, the majority has construed ACM section 463(a)(3) to exclude any proof of knowledge or intent. But for that construction, the proffered defense would be viable. I would hold that the State is not relieved of its burden to prove the defendant's intent or knowledge in a statutory rape case and, therefore, that the defendant may defend on the basis that he was mistaken as to the age of the prosecutrix.

The contention that an injury can amount to a crime only when inflicted by intention is no provincial or transient notion. It is as universal and persistent in mature systems of law as belief in freedom of the human will and a consequent ability and duty of the normal individual to choose between good and evil. A relation between some mental element and punishment for a harmful act is almost as instinctive as the child's familiar exculpatory "But I didn't mean to," and has afforded the rational basis for a tardy and unfinished substitution of deterrence and reformation in place of retaliation and vengeance as the motivation for public prosecution. Unqualified acceptance of this doctrine by English common law in the eighteenth century was indicated by Blackstone's sweeping statement that to constitute any crime there must first be a "vicious will." . . . "[G]iven the tremendous difference between individuals, both in appearance and in mental capacity, there can be no . . . rational relationship between the proof of the victim's age and the defendant's knowledge of that fact."

Questions for Discussion

1. Summarize the argument of the majority and of the dissenting opinions.

2. Should vaginal intercourse with an individual younger than the age of fourteen be a strict liability offense? Would you provide for the defense of a reasonable mistake of fact as to the age of the victim?

3. Why did Raymond receive a "suspended sentence" rather than being sentenced to prison? In answering this question, discuss justifications for punishment.

4. Consider this quote from a judgment of the California Supreme Court recognizing a mistake of age defense: "The sexually experienced 15-year-old may be far more acutely aware of the implications of sexual intercourse than her sheltered cousin who is beyond the age of consent. . . . [T]he [older] male is deemed criminally responsible for the act, although himself young and naïve and responding to advances which may have been made to him." Would you modify or abolish the crime of statutory rape? See *People v. Hernandez,* 293 P.2d 673 (Cal. 1964).

Withdrawal of Consent

In 2003, Illinois became the first state to pass a law on the withdrawal of consent. This legislation provides that a person "who initially consents to sexual penetration or sexual conduct is not deemed to have consented to any sexual penetration or sexual conduct that occurs after he or she withdraws consent during the course of that sexual penetration or conduct" (720 Illinois Compiled Statutes (ILCS) 5/12-14 (a)(2)).

The Illinois law was passed in reaction to disagreement among courts in California as to whether an individual who continues sexual intercourse following the other party's withdrawal of consent is guilty of rape. The California Supreme Court resolved this conflict by ruling in the next case in the text, *People v. John Z.* The decision in *John Z.* has been followed by courts in Alaska, Connecticut, Kansas, Maine, Maryland, Minnesota, and South Dakota. Do you believe that a male who continues sexual relations under such circumstances is guilty of rape?

Should the law recognize Laura's withdrawal of consent to sexual intercourse?

People v. John Z., 60 P.3d 183 (Cal. 2003). Opinion by: Chin, J.

The juvenile court . . . found that John Z. committed forcible rape . . . [and] committed him to Crystal Creek Boys Ranch. On appeal, defendant contends the evidence is insufficient to sustain the finding that he committed forcible rape. We disagree.

Facts

During the afternoon of March 23, 2000, seventeen-year-old Laura T. was working at Safeway when she received a call from Juan G., whom she had met about two weeks earlier. Juan wanted Laura to take him to a party at defendant's home and then return about 8:30 p.m. to pick him up. Laura agreed to take Juan to the party, but since she planned to attend a church group meeting that evening she told him she would be unable to pick him up.

Sometime after 6:00 p.m., Laura drove Juan to defendant's residence. Defendant and Justin L. were present. After arranging to have Justin L.'s stepbrother, P. W., buy them alcohol, Laura picked up P. W. and drove him to the store where he bought beer. Laura told Juan she would stay until 8:00 or 8:30 p.m. Although defendant and Juan drank the beer, Laura did not.

During the evening, Laura and Juan went into defendant's parents' bedroom. Juan indicated he wanted to have sex but Laura told him she was not ready for that kind of activity. Juan became upset and went into the bathroom. Laura left the bedroom and both defendant and Justin asked her why she "wouldn't do stuff." Laura told them that she was not ready.

About 8:10 p.m., Laura was ready to leave when defendant asked her to come into his bedroom to talk. She complied. Defendant told her that Juan had said he (Juan) did not care for her; defendant then suggested that Laura become his girlfriend. Juan entered the bedroom and defendant left to take a phone call.

When defendant returned to the bedroom, he and Juan asked Laura if it was her fantasy to have two guys, and Laura said it was not. Juan and defendant began kissing Laura and removing her clothes, although she kept telling them not to. At some point, the boys removed Laura's pants and underwear and began "fingering" her, "playing with [her] boobs" and continued to kiss her. Laura enjoyed this activity in the beginning, but objected when Juan removed his pants and told defendant to keep fingering her while he put on a condom. Once the condom was in place, defendant left the room and Juan got on top of Laura. She tried to resist and told him she did not want to have intercourse, but he was too strong and forced his penis into her vagina. The rape terminated when, due to Laura's struggling, the condom fell off. Laura told Juan that "maybe it's a sign we shouldn't be doing this," and he said "fine" and left the room. (Although Juan G. was originally a codefendant, at the close of the victim's testimony he pled guilty to charges of sexual battery and unlawful sexual intercourse, a misdemeanor.)

Laura rolled over on the bed and began trying to find her clothes; however, because the room was dark she was unable to do so. Defendant, who had removed his clothing, then entered the bedroom and walked to where Laura was sitting on the bed and "he like rolled over [her] so [she] was pushed back down to the bed." Laura did not say anything and defendant began kissing her and telling her that she had "a really beautiful body." Defendant got on top of Laura, put his penis into her vagina "and rolled [her] over so [she] was sitting on top of him." Laura testified she "kept . . . pulling up, trying to sit up to get it out . . . [a]nd he grabbed my hips and pushed me back down and then he rolled me back over so I was on my back . . . and . . . kept saying, will you be my girlfriend." Laura "kept like trying to pull away" and told him that "if he really did care about me, he wouldn't be doing this to me and if he did want a relationship, he should wait and respect that I don't want to do this." After about 10 minutes, defendant got off Laura, and helped her dress and find her keys. She then drove home.

On cross-examination, Laura testified that when defendant entered the room unclothed, he lay down on the bed behind her and touched her shoulder with just enough pressure to make her move, a nudge. He asked her to lie down and she did. He began kissing her and she kissed him back. He rolled on top of her, inserted his penis in her and, although she resisted, he rolled her back over, pulling her on top of him. She was on top of him for four or five minutes, during which time she tried to get off, but he grabbed her waist and pulled her back down. He rolled her over and continued the sexual intercourse. Laura told him that she needed to go home, but he would not stop. He said, "[J]ust give me a minute," and she said, "[N]o, I need to get home." He replied, "[G]ive me some time" and she repeated, "[N]o, I have to go home." Defendant did not stop, "[h]e just stayed inside of me and kept like basically forcing it on me." After about a "minute, minute and [a] half," defendant got off Laura.

Defendant testified, admitting that he and Juan were kissing and fondling Laura in the bedroom, but claimed it was with her consent. He also admitted having sexual intercourse with Laura, again claiming it was consensual. He claimed he discontinued the act as soon as Laura told him that she had to go home.

Reasoning

Although the evidence of Laura's initial consent to intercourse with John Z. was hardly conclusive, we will assume for purposes of argument that Laura impliedly consented to the act, or at least tacitly refrained from objecting to it, until defendant had achieved penetration. As will appear, we conclude that the offense of forcible rape occurs when, during apparently consensual intercourse, the victim expresses an objection and attempts to stop the act and the defendant forcibly continues despite the objection.

People v. Vela, 218 Cal. Rptr. 161 (Cal. Ct. App. 1985), reasoned that "the essence of the crime of rape is the outrage to the person and feelings of the female resulting from the nonconsensual violation of her womanhood. When a female willingly consents to an act of sexual intercourse, the penetration by the male cannot constitute a violation of her womanhood nor cause outrage to her person and feelings. If she withdraws consent during the act of sexual intercourse and the male forcibly continues the act without interruption, the female may certainly feel outrage because of the force applied or because the male ignores her wishes, but the sense of outrage to her person and feelings could

hardly be of the same magnitude as that resulting from an initial nonconsensual violation of her womanhood. It would seem, therefore, that the essential guilt of rape as stated in . . . section 263 is lacking in the withdrawn consent scenario."

As the Court of Appeal in this case stated, while outrage of the victim may be the cause for criminalizing and severely punishing forcible rape, outrage by the victim is not an element of forcible rape, "forcible rape occurs when the act of sexual intercourse is accomplished against the will of the victim by force or threat of bodily injury and it is immaterial at what point the victim withdraws her consent, so long as that withdrawal is communicated to the male and he thereafter ignores it."

In the present case, assuming arguendo that Laura initially consented to, or appeared to consent to, intercourse with defendant, substantial evidence shows that she withdrew her consent and, through her actions and words, communicated that fact to defendant. Despite the dissent's doubt in the matter, no reasonable person in defendant's position would have believed that Laura continued to consent to the act. As the Court of Appeal below observed, "Given [Laura's testimony], credited by the court, there was nothing equivocal about her withdrawal of any initially assumed consent."

Vela appears to assume that, to constitute rape, the victim's objections must be raised, or a defendant's use of force must be applied, before intercourse commences, but that argument is clearly flawed. One can readily imagine situations in which the defendant is able to obtain penetration before the victim can express an objection or attempt to resist. Surely, if the defendant thereafter ignores the victim's objections and forcibly continues the act, he has committed "an act of sexual intercourse accomplished . . . against a person's will by means of force. . . ."

Issue

Defendant, candidly acknowledging *Vela*'s flawed reasoning, contends that, in cases involving an initial consent to intercourse, the male should be permitted a "reasonable amount of time" in which to withdraw, once the female raises an objection to further intercourse. As defendant argues, "By essence of the act of sexual intercourse, a male's primal urge to reproduce is aroused. It is therefore unreasonable for a female and the law to expect a male to cease having sexual intercourse immediately upon her withdrawal of consent. It is only natural, fair and just that a male be given a reasonable amount of time in which to quell his primal urge. . . ."

Holding

We disagree with defendant's argument. Aside from the apparent lack of supporting authority for defendant's "primal urge" theory, the principal problem with his argument is that . . . there is no support for the proposition that the defendant is entitled to persist in intercourse once his victim withdraws her consent.

In any event, even were we to accept defendant's "reasonable time" argument, in the present case he clearly was given ample time to withdraw but refused to do so despite Laura's resistance and objections. Although defendant testified he withdrew as soon as Laura objected, for purposes of appeal we need not accept this testimony as true in light of Laura's contrary testimony. As noted above, Laura testified that she struggled to get away when she was on top of defendant, but that he grabbed her waist and pushed her down onto him. At this point, Laura told defendant that if he really cared about her, he would respect her wishes and stop. Thereafter, she told defendant three times that she needed to go home and that she did not accept his protestations he just needed a "minute." Defendant continued the sex act for at least four or five minutes after Laura first told him she had to go home. According to Laura, after the third time she asked to leave, defendant continued to insist that he needed more time and "just stayed inside of me and kept like basically forcing it on me," for about a "minute, minute and [a] half."

The judgment of the Court of Appeal is affirmed.

Dissenting, *Brown, J.*

The majority finds Laura's "actions and words" clearly communicated withdrawal of consent in a fashion "no reasonable person in defendant's position" could have mistaken. But, Laura's silent and ineffectual movements could easily be misinterpreted. . . . When asked if she had made it clear to John that she didn't want to have sex, Laura says "I thought I had," but she acknowledges she "never officially told him" she did not want to have sexual intercourse. When asked by the prosecutor on redirect why she told John "I got to go home," Laura answers: "Because I had to get home so my mom wouldn't suspect anything."

Furthermore, even if we assume that Laura's statements evidenced a clear intent to withdraw consent, sexual intercourse is not transformed into rape merely because a woman changes her mind. . . . Under the facts of this case, however, it is not clear that Laura was forcibly compelled to continue. All we know is that John Z. did not instantly respond to her statement that she needed to go home. He requested additional time. He did not demand it. Nor did he threaten any consequences if Laura did not comply.

The majority relies heavily on John Z.'s failure to desist immediately. But, it does not tell us how soon would have been soon enough. Ten seconds? Thirty? A minute? Is persistence the same thing as force? . . . And even if we conclude persistence should be criminalized in this situation, should the penalty be the same as for forcible rape? Such questions seem inextricably tied to the question of whether a reasonable person would know that the statement "I need to go home" should be interpreted as a demand to stop. Under these circumstances, can the withdrawal of consent serve as a proxy for both compulsion and wrongful intent?

Questions for Discussion

1. What facts support the conclusion that Laura clearly withdrew her consent? Could John Z. have reasonably believed that Laura desired to continue to engage in sexual intercourse?

2. Did the majority apply the extrinsic or intrinsic force test? Consider California Penal Code section 261.6, which defines consent to "mean positive cooperation in act or attitude pursuant to and exercise of free will. The person must act freely and voluntarily and have knowledge of the nature of the act or transaction involved."

3. How did the court distinguish *Vela* from *John Z.?*

4. Do you agree with dissenting Judge Brown that Laura was not forcibly compelled to continue the intercourse with John Z.? What of Judge Brown's argument that the majority of the California Supreme Court is equating "persistence" with "force"? Should a male be provided a "reasonable amount of time" to respond to a withdrawal of consent? This was the decision of the Kansas Supreme Court in *State v. Bunyard*, 133 P.3d 14 (Kan. 2006).

5. Would you hold John Z. guilty of rape? Is the continuation of sexual intercourse despite a withdrawal of consent the same crime as the forcible attack on an individual?

Rape Shield Laws

The common law permitted the defense to introduce evidence concerning a victim's prior sexual relations with the accused, prior sexual relations with individuals other than the accused, and evidence concerning the alleged victim's reputation for chastity. Would you find this type of evidence valuable in determining a defendant's guilt or innocence?

The law continues to permit the introduction of evidence relating to sexual activity between the accused and victim. The assumption is that an individual who voluntarily entered into a relationship with a defendant in the past is more than likely to have again consented to enter into a relationship with the accused. The thinking is that the defendant is entitled to have the jury consider and determine the weight (importance) to attach to this evidence in determining guilt.

Rape shield laws prohibit the defense from asking the victim about or introducing evidence concerning sexual relations with individuals other than the accused or introducing evidence concerning the victim's reputation for chastity. The common law assumed that such evidence was relevant in that an individual who has "already started on the road of [sexual unchastity] would be less reluctant to pursue her way, than another who yet remains at her home of innocence and looks upon such a [pursuit] with horror."

The other reason for this evidence was the belief that the jury should be fully informed concerning the background of the alleged victim in order to determine whether her testimony was truthful or was the product of perjury or of a desire for revenge.[42]

Rape shield laws prohibiting evidence relating to a victim's general sexual activity are based on several reasons:

- *Harassment.* To prevent the defense attorney from harassing the victim.
- *Relevance.* The evidence has no relationship to whether the victim consented to sexual relations with the defendant and diverts the attention of the jury from the facts of the case.
- *Prejudice.* The evidence biases the jury against the accused.
- *Complaints.* Victims are not likely to report rapes if they are confronted at trial with evidence of their prior sexual activity.

Rape shield laws do not prohibit the introduction of an accused's past sexual activity in every instance. The Sixth Amendment to the U.S. Constitution guarantees individuals a fair trial and provides that individuals have the right to confront the witnesses against them. Courts have permitted the introduction of a victim's past activity with others in those instances when it is relevant to the source of injury or semen or reveals a pattern of activity or a motive to fabricate. For instance, the fact that a victim had a sexual relationship with a man other than the accused before going to the hospital may be relevant for the source of injury or semen.

Consider the issue that confronted the trial court in *State v. Colbath.* The defendant and victim were in a bar. The victim made sexually provocative remarks to the defendant and permitted him to feel her breast and buttocks and rubbed his sexual organ. The two went to the defendant's trailer where they had sexual intercourse. The defendant's significant other arrived and assaulted the woman, who defended her behavior by contending that she had been raped by the defendant. The trial judge rejected the defendant's effort to introduce evidence of the alleged victim's public sexual displays with other men in the bar and evidence that the victim had left the bar with other men prior to her approaching the defendant. The New Hampshire Supreme Court, however, held that despite the rape shield law, the defendant's Sixth Amendment right to confront witnesses against him required admission of evidence of the victim's conduct in the bar because it might indicate that at the time that the victim met the defendant, she possessed a "receptiveness to sexual advances."[43]

On the other hand, in *People v. Wilhelm,* a Michigan appellate court upheld a ruling excluding evidence that the victim had exposed her breasts to two men who were sitting at her table in a bar and that she permitted one of them to fondle her breasts. The court ruled that the victim's conduct in the bar did not indicate that she would voluntarily engage in sexual intercourse with the defendant.[44]

You Decide

11.1 Cottie Brown had been involved with the defendant Alston for roughly six months. They shared an apartment and when they fought she would return to live with her mother until he called to ask her to return. Brown testified that at times that she had sex with the defendant to "accommodate him." On these occasions, Brown would "stand still and remain entirely passive while the defendant undressed and had sexual intercourse with her." She testified that Alston beat her on occasion and that she left the apartment and ended the relationship after he struck her when she refused to give him money.

Alston appeared at Brown's technical school a month later and prevented her from entering the building and grabbed her arm, saying that she was going with him. Brown agreed to walk with Alston and he let go of her arm. Alston stated that Brown was going to miss class that day and threatened to "fix her face" to keep Brown's mother from continuing to interfere in their relations. Brown told him that their relationship was over and Alston replied that he had a right to make love to her again since "everyone could see her but him." Brown agreed to give Alston her address in an unsuccessful effort to persuade him to permit her to return to school. They passed a group of Alston's friends and finally arrived at the home of Lawrence Taylor. Alston briefly went to the back of the house and when he returned asked whether Brown was "ready" and she replied that she "wasn't going to bed with him." Brown complied with Alston's order to lie down on the bed and the defendant pushed apart Brown's legs and had sexual intercourse with her. Brown testified that she cried, but did not attempt to push Alston off her. They then talked. At some point following this incident, Brown let Alston into her apartment after he threatened to kick down the door. He spent the night and the two made love several times. The defendant testified that she did not resist because she "enjoyed it." Was Alston guilty of the rape of Brown? See *State v. Alston,* 312 S.E.2d 470 (N.C. 1984).

You can find the answer at www.sagepub.com/lippmanccl3e.

You can find more cases on the study site: **State v. Rusk, *www.sagepub.com/Lippmanccl3e.***

You Decide

11.2 Stephen F. (Child) appeals his convictions for two counts of criminal sexual penetration and argues that the trial court improperly excluded evidence of the alleged victim's past sexual activities. Child claimed that this evidence would have demonstrated her motive to fabricate. Under sections 30-9-11 through 30-9-15 of New Mexico Statutes, evidence of the victim's past sexual conduct and opinion evidence of the victim's past sexual conduct or of reputation for past sexual conduct shall not be admitted unless, and only to the extent, the court finds that the evidence is material to the case and that its inflammatory or prejudicial nature does not outweigh its probative value.

Child (age fifteen) and the alleged victim (B.G., age sixteen) engaged in sexual intercourse. Child, B.G., and B.G.'s brother had been watching movies in B.G.'s bedroom. Child had been a friend of B.G.'s brother and family for nine years and usually slept on the couch in the living room when he spent the night. B.G. testified that after Child had headed for bed in the living room, he returned to her room and forced her to engage in sexual conduct, including oral, vaginal, and anal intercourse. The morning after the incident, B.G. told her mother that Child had raped her. Child was convicted of two counts of criminal sexual penetration. Child contended that the intercourse was consensual and claimed that B.G. lied because she feared that she would be punished by her religious parents. B.G.

previously had been punished by her parents after having had consensual sexual relations with her then boyfriend. B.G. reportedly had told Child that her mother "was really upset . . . [about my having engaged in sex with my boyfriend;] she said that it was going to take her a long time to trust me again, . . . about three or four months[,] . . . and I wasn't allowed to go out on dates with guys." Child's theory was that B.G. was motivated to fabricate the claim of rape because she feared the punishment and disapproval of her parents, devout Christians who "don't believe in sex before marriage." The State of New Mexico opposed Child's motion to permit the cross-examination of the complaining witness in regard to her prior sexual conduct with her boyfriend on the grounds that this was intended to portray the complaining witness as an individual who is likely to engage in sexual activity outside of marriage. According to the appellate court, there are five areas to consider in making a decision on this issue: (1) whether there is a clear showing that complainant committed the prior acts; (2) whether the circumstances of the prior acts closely resemble those of the present case; (3) whether the prior acts are clearly relevant to a material issue, such as identity, intent, or bias; (4) whether the evidence is necessary to the defendant's case; [and] (5) whether the probative value of the evidence outweighs its prejudicial effect.

As a judge, would you permit Child to cross-examine the complaining witness in regard to her sexual conduct with her boyfriend? See *State v. Stephen F.,* 152 P.3d 842 (N.M. Ct. App. 2007).

You can find the answer at www.sagepub.com/lippmanccl3e.

ASSAULT AND BATTERY

Assault and battery, although often referred to as a single crime, in fact are separate offenses. A battery is the application of force to another person. An assault may be committed either by attempting to commit a battery or by intentionally placing another in fear of a battery. Notice that an assault does not involve physical contact. An assault is the first step toward a battery, and the law takes the position that it would be unfair to hold an individual liable for both an assault and battery. As a result, the assault "merges" into the battery, and an individual only is held responsible for the battery. A Georgia statute provides that an individual "may not be convicted of both the assault and completed crime."[45]

State statutes typically include assault and battery under a single "assault statute." Both offenses are considered misdemeanors. Serious assaults and batteries are punished as aggravated misdemeanors and aggravated batteries that are categorized as felonies.

The Elements of Battery

For an international perspective on this topic, visit the study site.

Modern battery statutes require physical contact that results in bodily injury or an offensive touching, a contact that is likely to be regarded as offensive by a reasonable person. Assault and battery is satisfied under the Model Penal Code by an intentional, purposeful, reckless, or negligent intent. The code punishes an individual who "purposely, knowingly or recklessly causes bodily injury to another or negligently causes bodily injury to another with a deadly weapon."[46]

Most state statutes narrowly limit the required intent. Illinois punishes the intentional or knowing causing of bodily harm to an individual or physical contact with an individual of an insulting or provoking nature.[47] Georgia limits battery to the intentional causing of "substantial physical harm or visible bodily harm to another." This includes, but is not limited to, substantially blackened eyes, substantially swollen lips or substantial bruises.[48]

In thinking about battery, you should be aware that a battery is not confined to the direct application of force by an individual. It can include causing substantial bodily harm by poisoning, bombing, a motor vehicle, illegal narcotics, or an animal. Minnesota, for instance, punishes causing "great or substantial harm" by intentionally or negligently failing to keep a dog properly confined.[49] In states with statutes punishing offensive physical contact, an uninvited kiss or sexual fondling may be considered a battery. A Washington court held that assault and battery includes spitting.[50]

You should also keep in mind that not every physical contact is a battery. We impliedly consent to physical contact in sports, medical operations, while walking in a crowd, or when a friend greets us with a hug or kiss. The law accepts that police officers and parents are justified in employing reasonable force. Reasonable force may also be used in self-defense or in defense of others.

Simple and Aggravated Battery

We earlier mentioned that a battery is a misdemeanor. Aggravated batteries are felonies and typically require:

- serious injury,
- the use of a dangerous or deadly weapon, or
- the intent to kill, rape, or seriously harm.

The Georgia battery statute punishes a second conviction for a simple battery with imprisonment of between ten days and twelve months with the possibility of a fine of not more than one thousand dollars.[51] An aggravated battery in Georgia requires an attack that renders a "member" of the victim's body "useless" or "seriously disfigured" and is punishable by between one and twenty years in prison. The penalty is enhanced to between ten and twenty years when knowingly directed at a police or correctional officer and is punished by between five and twenty years when directed at an individual over sixty-five, is committed in a public transit vehicle or station, or is directed at a student or teacher. An aggravated battery is punished by between three and twenty years in prison when directed at a family member.[52]

California considers a battery as aggravated when committed with a deadly weapon, a caustic or flammable chemical, a taser or stun gun, or when the battery results in grievous bodily harm.[53] Illinois lists as an aggravated battery inserting a substance that may cause death or serious bodily harm in food, drugs, or cosmetics.[54] South Dakota considers the serious physical injury on an unborn child to be an aggravated assault.[55] A Minnesota statute punishes as battery the selling or provision of illegal narcotics that "causes great bodily harm" by imprisonment for not more than ten years and by payment of a fine of not more than $20,000.[56]

The federal government and sixteen states provide criminal penalties for female genital mutilation, a practice in which the sexual organs of young women are bound to prevent premarital sexual relations. The federal statute punishes "whoever knowingly circumcises, excises, or infibulates the whole or any part of the labia majora or labia minora or clitoris of another person." This procedure is punished with a fine and imprisonment of not more than five years when applied against a person younger than eighteen.[57]

The common law crime of **mayhem** is included in the criminal codes of several states, including California. California defines mayhem when one deprives a human being of a "member of his or her body, or disables, disfigures, or renders it useless, or cuts or disables the tongue, or puts out an eye, or slits the nose, ear, or lip."[58] A well-known case of "malicious wounding" involved Lorena Bobbit who, while her husband John was asleep, dismembered his sexual organ and then left the house and tossed it out the car window onto the highway. Lorena claimed that John had raped her and she was subsequently found not guilty by reason of insanity and was committed to a mental institution for observation.

In summary, a battery involves:

- *Act.* The application of force that results in bodily injury or offensive contact. Aggravated battery statutes require a serious bodily injury. The contact must be regarded as offensive by a reasonable person.
- *Intent.* The intentional, knowing, reckless, or negligent application of force.
- *Consent.* An implied or explicit consent may constitute a defense under certain circumstances.

Assault

An assault may be committed by an attempt to commit a battery or by placing an individual in fear of a battery. Georgia defines an assault as an attempt to "commit a violent injury to the person of another; or . . . an act which places another in reasonable apprehension of immediately receiving a violent injury."[59]

California limits assault to an "an unlawful attempt, coupled with a present ability, to commit a violent injury on the person of another."[60] Illinois, on the other hand, provides that a person commits an assault when, "without lawful authority, he engages in conduct that places another in reasonable apprehension of receiving a battery."[61] An assault is a misdemeanor, punishable in California by imprisonment by up to six months in the county jail and by a possible fine of $2,000.[62] Ohio is among the states whose criminal code uses the term "menacing" rather than assault.[63]

A small number of states, including New York, recognize the offense of an attempted assault. The overwhelming majority of jurisdictions reject that an individual may be prosecuted for an "attempt to attempt a battery" on the grounds that this risks the conviction of individuals who have yet to take clear steps toward an assault.[64] A Georgia court in the nineteenth century also pointed out that prosecuting an individual for an attempt to commit a battery "is simply absurd. As soon as any act is done towards committing a violent injury on the person of another, the party doing the act is guilty of an assault, and he is not guilty until he has done the act . . . [a]n attempt to act is too [confused] for practical use." Do you agree that an attempt to commit an assault is "absurd"?[65]

Aggravated Assault

Aggravated assault is a felony and is generally based on factors similar to those constituting an aggravated battery.

Georgia provides three forms of aggravated assault that are punishable by between five and twenty years in prison: assault with intent to murder, rape, or rob; assault with a deadly weapon; and discharge of a firearm from within an automobile. The statute also punishes as an aggravated assault: an assault on a police or correctional officer, a teacher, or an individual sixty-five years of age or older, or an assault committed during the theft of a vehicle engaged in public or commercial transport.[66]

Illinois lists as aggravated assaults an assault committed while "hooded, robed or masked," and an assault committed with "a deadly weapon or any device manufactured and designed to be substantially similar in appearance to a firearm." An individual also commits an aggravated assault in Illinois when he or she knowingly and without lawful justification "shines or flashes a laser gunsight or other laser device that is attached or affixed to a firearm . . . so that the laser beam strikes near or in the immediate vicinity of any person."[67] Illinois also punishes under a separate statute "vehicular endangerment," the dropping of an item off a bridge with the intent to strike a motor vehicle.[68]

The crime of **stalking** is recognized in every state. Some of you may recall the stalking and killing of both young television actress Rebecca Schaeffer and rock star and former member of the Beatles John Lennon, and the attack on tennis star Monica Seles. A recent study indicates that 3.4 million Americans have been the victim of stalking, eleven percent of whom have been stalked for five years or more. The Illinois statute provides that a person commits stalking when he or she "on at least 2 separate occasions" follows another person or places the person under surveillance or any combination of these two acts. This must be combined with the transmittal of a threat of immediate or future bodily harm or the placing of a person in reasonable apprehension of immediate or future bodily harm or the creation of a reasonable apprehension that a family member will be placed in immediate or future bodily harm.[69] These acts, when combined with the causing of bodily harm, restraining the victim, or the violation of a judicial order prohibiting such conduct, constitute aggravated stalking.[70] Illinois, along with a number of other states and the federal government, has also passed laws to combat the new crime of **cyberstalking**. This involves transmitting a threat through an electronic device of immediate or future bodily harm, sexual assault, confinement, or restraint against an individual or family member of that person. The threat must create a "reasonable apprehension of immediate or future bodily harm, sexual assault, confinement, or restraint."[71]

The next case, *Carter v. Commonwealth,* discusses some interesting issues involved in the legal standard for a threatened battery.

The Elements of Assault

In considering an attempted-battery assault, keep in mind:

- *Intent.* An attempt in most states to commit a battery requires an intent (purpose) to commit a battery.
- *Act.* An individual is required to take significant steps toward the commission of the battery.
- *Present Ability.* Some states require the present ability to commit the battery. In these jurisdictions, an individual would not be held liable for an assault where the assailant is unaware that a gun is unloaded. South Dakota, on the other hand, provides for a battery "with or without the actual ability to seriously harm the other person."[72]
- *Victim.* The victim need not be aware of the attempted battery.

The *assault of placing another in fear of a battery* requires:

- *Intent.* The intent (or purpose) to cause a fear of immediate bodily harm.
- *Act.* An act that would cause a reasonable person to fear immediate bodily harm. Words ordinarily are not sufficient and typically must be accompanied by a physical gesture that, in combination with the words, creates a reasonable fear of imminent bodily harm.
- *Victim.* The victim must be aware of the assailant's act and possess a reasonable fear of imminent bodily harm. A threat may be conditioned on the victim's meeting the demands of the assailant.

Model Penal Code Section 211.1. Assault

(1) Simple Assault. A person is guilty of assault if he:

    (a) attempts to cause or purposely, knowingly, or recklessly causes bodily injury to another; or

    (b) negligently causes bodily injury to another with a deadly weapon; or

    (c) attempts by physical menace to put another in fear of imminent serious bodily injury.

Simple assault is a misdemeanor unless committed in a fight or scuffle entered into by mutual consent in which case it is a petty misdemeanor.

(2) Aggravated Assault. A person is guilty of aggravated assault if he:

    (a) attempts to cause serious bodily injury to another, or causes such injury purposely, knowingly, or recklessly under circumstances manifesting extreme indifference to the value of human life; or

    (b) attempts to cause or purposely or knowingly causes bodily injury to another with a deadly weapon.

Aggravated assault under paragraph (a) is a felony of the second degree; aggravated assault under paragraph (b) is a felony of the third degree.

Analysis

1. The Model Penal Code eliminates the common law categories and integrates assaults and batteries into a single assault statute.

2. Assaults are graded into categories based on the gravity of the harm intended or actually caused.

3. Grading is not based on the identity of the victim.

4. Actual or threatened bodily injury or serious bodily injury is required. Offensive contact is excluded.

5. A deadly weapon includes poisons, explosives, caustic chemicals, handguns, knifes, and automobiles.

The Legal Equation

Battery **=** Unlawful bodily injury or offensive contact with another

    + purposeful, knowing, reckless, negligent bodily injury, or offensive contact with another.

Attempted battery **=** Intent to injure

    + an act undertaken to commit a battery.

Threatened battery **=** Intent to place in immediate fear of a battery

    + act undertaken that places a reasonable person in fear of battery.

Did Carter place Officer O'Donnell in imminent fear of a battery?

Carter v. Commonwealth, 594 S.E.2d 284
(Va. Ct. App. 2004). Opinion by: Clements, J.

Michael Anthony Carter was convicted in a bench trial of assaulting a police officer. . . . On appeal, he contends the evidence presented at trial was insufficient to support his conviction because the Commonwealth did not prove he had the present ability to inflict actual violence upon the officer. . . . [W]e affirm Carter's conviction.

Facts

The evidence presented to the trial court established that, on December 29, 1998, around 11:00 p.m., Officer B.N. O'Donnell of the City of Charlottesville Police Department observed a speeding car and, activating his vehicle's overhead flashing blue emergency lights, initiated a traffic stop. O'Donnell, who was on routine patrol at the time in a high crime area of the city, was driving a marked police vehicle and wearing his police uniform and badge. After the car pulled over, O'Donnell shone his vehicle's "take down" lights and spotlight onto the car and approached it on foot.

Two people were inside the car, the driver and Carter, who was seated in the front passenger seat. O'Donnell initiated a conversation with the driver, asking for his driver's license and registration and informing him why he had been stopped. The driver responded to O'Donnell in a "hostile" tone of voice. While conversing with the driver, O'Donnell used his flashlight to conduct a "plain view search" of the car to make sure there were no visible weapons or drugs in it. O'Donnell noticed that Carter had his right hand out of sight "down by his right leg." Carter then suddenly brought his right hand up and across his body. Extending the index finger on his right hand straight out and the thumb straight up, he pointed his index finger at the officer and said, "Pow." Thinking Carter "had a weapon and was going to shoot" him, O'Donnell "began to move backwards" and went for his weapon. A "split second" later, O'Donnell realized "it was only [Carter's] finger." O'Donnell testified: "The first thing I thought was that I was going to get shot. I—it's a terrifying experience, and if I could have gotten my weapon, I would have shot him." Immediately after the incident, O'Donnell, who was "visibly shaken," asked Carter "if he thought it was funny," and Carter responded, "Yes, I think it is funny."

Carter moved to strike the evidence, arguing the Commonwealth's evidence was insufficient to prove assault because it failed to prove Carter had the present ability to inflict actual violence upon the officer. The Commonwealth responded that proof of such ability was unnecessary as long as the evidence proved the officer reasonably believed Carter had the present ability to inflict actual bodily harm upon him.

The trial court agreed with the Commonwealth. Finding Carter's "act of pointing what the officer believed at the time to be a weapon at him" did, "in fact, place Officer O'Donnell in reasonable apprehension or fear," the trial court found the evidence sufficient to prove beyond a reasonable doubt that Carter was guilty of assault. Thus, the trial court denied Carter's motion to strike the evidence and subsequently convicted him of assaulting a police officer. . . . At sentencing, the court imposed a sentence of three years, suspending two years and six months.

Issue

Virginia Code Annotated section 18.2-57I provides, in pertinent part, that "any person [who] commits an assault . . . against . . . a law enforcement officer . . .

engaged in the performance of his public duties as such . . . shall be guilty of a . . . felony."

On appeal, Carter asserts the Commonwealth failed to prove his conduct constituted an assault of a law enforcement officer because, in pointing his finger at the officer and saying "Pow," he did not have the present ability to inflict harm upon the officer, as required under the common law definition of assault. Thus, he contends, the trial court erred, as a matter of law, in finding the evidence sufficient to sustain a conviction for assault.

In response, the Commonwealth contends that, under long-established Virginia case law, a defendant need not have had the present ability to inflict harm at the time of the offense to be guilty of assault. It is enough, the Commonwealth argues, that, as in this case, the defendant's conduct created in the mind of the victim a reasonable fear or apprehension of bodily harm. Accordingly, the Commonwealth concludes, the trial court properly found the evidence sufficient to convict Carter of assaulting a police officer. . . .

Reasoning

While statutorily proscribed and regulated, the offense of assault is defined by common law in Virginia. (In this jurisdiction, we adhere to the common law definition of assault, there having been no statutory change to the crime.) . . . Assault has . . . long been defined at common law "as being (1) an attempt to commit a battery or (2) an intentional placing of another in [reasonable] apprehension of receiving an immediate battery." Today, most jurisdictions include both of these separate types of assault, attempted battery and putting the victim in reasonable apprehension, within the scope of criminal assault. In Virginia, our Supreme Court has long recognized the existence of both concepts of assault in the criminal law context. . . .

The instruction under consideration . . . presents the question on which there is a sharp and irreconcilable conflict in the authorities on the subject; diametrically opposed positions being taken by the authorities. . . . We think that, both in reason and in accordance with the great weight of modern authority, . . . a present ability to inflict bodily harm upon the victim is not an essential element of criminal assault in all cases. Indeed, under those cases, to be guilty of . . . criminal assault, a defendant need have only an apparent present ability to inflict harm.

Holding

As previously discussed, the two types of criminal assault recognized at common law—attempted assault and putting the victim in reasonable apprehension of bodily harm—are separate and distinct forms of the same offense. They have different elements and are, thus, defined differently and applied under different circumstances. For these reasons, we hold that, under the common law definition of assault, one need not, in cases such as this, have a present ability to inflict imminent bodily harm at the time of the alleged offense to be guilty of assault. It is enough that one's conduct created at the time of the alleged offense a reasonable apprehension of bodily harm in the mind of the victim. Thus, an apparent present ability to inflict imminent bodily harm is sufficient to support a conviction for assault.

In this case, the trial court found that Carter's "act of pointing what the officer believed at the time to be a weapon at him" did, "in fact, place Officer O'Donnell in reasonable apprehension or fear." The evidence in the record abundantly supports this finding, and the finding is not plainly wrong. . . . O'Donnell testified that he thought he was "going to get shot." It was, he said, "a terrifying experience, and if I could have gotten my weapon, I would have shot him."

The trial court could reasonably conclude from these facts that the officer was terrified and thought he was about to be shot. That the officer's terror was brief does not alter the fact, as found by the trial court, that the officer believed for a moment that Carter had the intention and present ability to kill him. Moreover, under the circumstances surrounding the incident, we cannot say, as a matter of law, that such a belief was unreasonable. Thus, although Carter did not have a weapon, the trial court could properly conclude from the evidence presented that Carter had an apparent present ability to inflict imminent bodily harm and that his conduct placed Officer O'Donnell in reasonable apprehension of such harm.

Hence, the trial court did not err, as a matter of law, in finding the evidence sufficient to convict Carter of assault. . . . Accordingly, we affirm Carter's conviction for assault . . .

Dissenting, *Benton, J.,* with whom *Fitzpatrick, C.J.,* joins.

The police officer testified that the "first thing I thought was that I was going to get shot. I—it's a terrifying experience, and if I could have gotten my weapon, I would have shot him. But it's—it happens . . . [in] a split second." The officer testified that Carter then "started laughing."

The common law definition of "assault" . . . does not encompass this type of intentional conduct, which is intended to startle but is performed without a present ability to produce the end if carried out. . . . I disagree with the majority opinion's holding that a conviction for criminal assault can be sustained in Virginia even though the evidence failed to prove the accused had a present ability to harm the officer. I would hold that Carter committed an "act accompanied with circumstances denoting an intention" to menace but it was not "coupled with a present ability . . . to use actual violence" or "calculated to produce the end if carried into execution."

Because Virginia continues to be guided by the common law rule concerning assault, I would hold that the conviction is not supported because the evidence failed to prove Carter acted "by means calculated to produce the end if carried into execution." Accordingly, I would reverse the conviction for assault.

Questions for Discussion

1. Explain the distinction between attempted battery and the assault of placing a victim in reasonable apprehension of bodily harm.

2. Compare and contrast the majority and dissenting opinions. Which is more persuasive?

3. Did Officer O'Donnell reasonably believe that he would be subjected to an immediate battery? Would Carter be criminally liable in the event that he pointed a pistol at O'Donnell that Carter knew was unloaded?

4. Would you convict Carter of assault?

Cases and Comments

Indirect Application of Force. Defendant Joseph Sherer placed random phone calls to over thirty women in Bozeman, Montana, in which he impersonated a doctor. Sherer claimed that he was treating their mother or daughter for a hereditary urinary disease for which they also might be at risk. Sherer asked various highly personal questions and requested that the women perform a test on themselves that involved a razor blade, knife, or fingernail polish remover. Three of the women harmed themselves

in response to Sherer's directions and one sliced off her left nipple. Sherer claimed that his directions did not constitute an aggravated assault because he only indirectly caused the injuries. The Montana Supreme Court ruled that aggravated assault does not require that the defendant personally direct force toward a victim and that the resulting injury was precisely what Sherer intended to accomplish. Sherer's communications were held to be the cause of the victims' personal physical abuse. See *State v. Sherer,* 60 P.3d 1010 (Mont. 2002).

The next case, *Waldon v. State,* discusses "stalking." Consider whether the prosecution demonstrated beyond a reasonable doubt that the defendant knowingly or intentionally engaged in a course of conduct involving continuous or repeated harassment that would cause a reasonable person to feel terrorized, frightened, intimidated, or threatened.

Figure 11.1 Crime on the Streets: Intimate Partner Violence

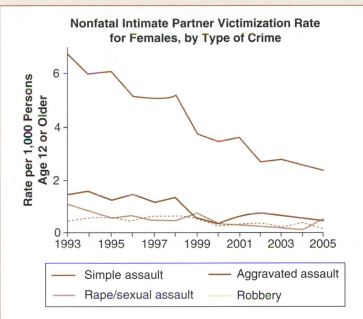

For nonfatal intimate partner violence, as for violent crime in general, simple assault is the most common type of violent crime.

The long-term trend for female victims of nonfatal intimate partner violence shows that between 1993 and 2005,

- the rate of simple assault declined by about two-thirds.
- the rate of aggravated assault declined by two-thirds.

Nonfatal intimate partner violence is more likely to occur between the hours of 6 P.M. and 6 A.M.

- Females and males experienced nonfatal intimate partner victimization at similar times during the day and night.

Source: The National Crime Victimization Survey, U.S. Department of Justice.

On average between 2001 and 2005, females experienced higher rates of nonfatal intimate partner violence than males in each type of crime.

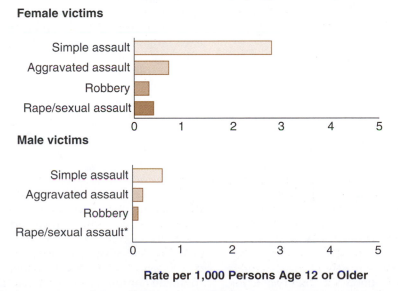

Source: The National Crime Victimization Survey, U.S. Department of Justice.

*Information about rape victimization of males is not provided because the small number of cases is insufficient for reliable estimates.

Did Waldon stalk his former wife?

Waldon v. State, 684 N.E.2d 206 (Ind. Ct. App. 1997). Opinion by: Sharpnack, C.J.

Facts

In April of 1993, Waldon married Val Majors, and they had a son. During the marriage, Majors worked as a nurse and taught classes at a dance studio. In November of 1994, Waldon and Majors divorced, stipulating to a mutual restraining order.

On the morning of December 5, 1994, Majors was driving home from the dance studio when she encountered Waldon. Waldon was walking down the street about two blocks away from the studio. Majors was fearful and alarmed to find Waldon so close to the studio.

On the morning of August 16, 1995, Majors again saw Waldon as she was leaving the parking lot of the studio. Waldon was walking at a "hurried" rate toward the entrance of the parking lot. Majors was intimidated and threatened by Waldon's actions.

On evening of August 20, 1995, Majors was leaving work from the hospital when she saw Waldon standing near the parking lot. He stood about twenty-five feet away from Majors and stared at her. Although Waldon did not approach her, Majors was frightened by him.

A few days later, Majors again saw Waldon near her studio. As Majors was getting into her car, Waldon walked slowly by her. Majors grew more fearful of her encounters with Waldon.

On the morning of November 1, 1995, Majors was getting into her car in the studio parking lot when she noticed Waldon staring at her through a fence about four feet away. Majors became very frightened and ran back to the studio. She then called the police and filed a report.

On the evening of November 7, 1995, Majors was teaching a dance class when she saw Waldon outside of the studio. Waldon was riding his bike around a dumpster in the parking lot. When the class ended, Majors was afraid to walk alone to her car, so she had a friend escort her. As Majors was leaving the parking lot, Waldon rode his bike in front of her car. Majors felt threatened by Waldon's actions.

On December 14, 1995, the State charged Waldon with stalking. After a trial on March 7, 1996, the jury found Waldon guilty as charged. The trial court later sentenced Waldon to six months. Waldon now appeals his conviction.

Issue

The first issue raised for our review is whether there was sufficient evidence to support Waldon's conviction for stalking. . . . Waldon was convicted under Indiana Code section 35-45-10-5(a), which provides that a "person who stalks another person commits stalking, a . . . misdemeanor." Stalking is defined as: "[A] knowing or an intentional course of conduct involving repeated or continuing harassment of another person that would cause a reasonable person to feel terrorized, frightened, intimidated, or threatened and that actually causes the victim to feel terrorized, frightened, intimidated, or threatened."

Waldon argues that the State failed to present sufficient evidence that his encounters with Majors caused her emotional distress. Waldon asserts that based on Majors' testimony, "the jury could not have reasonably inferred that Ms. Majors actually felt terrorized, intimidated or threatened by Mr. Waldon during these six encounters."

Reasoning

The record belies Waldon's assertion. During the trial, Majors testified that when she first saw Waldon near her studio, she became "fearful" and "alarmed." When Majors again saw Waldon near her studio, she felt intimidated and threatened. Next, when Majors encountered Waldon in the parking lot of the hospital, she was "afraid."

A few days later, when Majors saw Waldon near her studio, she suffered emotional distress. Majors described her distress as follows: "It makes you feel like you don't have a life, like you're violated, like you have to go, you know, hurry and lock your doors, look around, scan the parking lot. . . . It's frustrating. It's a violation and makes you fearful about what the intent is."

Majors further testified that "the most frightening" encounter with Waldon occurred outside of the studio. When Majors was getting into her car, she saw Waldon peering at her through a fence. Majors stated that she was "frightened because [she] was so vulnerable" and that the experience "was very upsetting." A week later, Majors had her last encounter with Waldon near the studio. Again, Majors felt threatened and intimidated by Waldon's actions.

Holding

Contrary to Waldon's assertion, Majors testified that each of her encounters with Waldon caused her emotional distress. Based on her testimony, we find evidence of probative value from which the jury could conclude that Waldon caused her "to feel terrorized, frightened, intimidated, or threatened." Therefore, the evidence was sufficient to support Waldon's conviction for stalking. . . .

Dissenting, *Barteau, J.*

In support of the stalking charge, the State presented evidence that six times during a one-year time span, Majors observed Waldon in various degrees of proximity to herself. . . . In all but two instances, no eye contact between Majors and Waldon occurred. Given the lack of eye contact, it is possible that Waldon did not see Majors even though she saw him. Four out of the six times Majors observed Waldon, Majors was driving her car. In describing three of those instances, Majors testified that she felt safe in her car. In none of the instances did Waldon make a verbal threat, a physical threat, or even an obscene gesture.

"Stalking" is defined by statute as: a knowing or an intentional course of conduct involving repeated or continuing harassment of another person that would cause a reasonable person to feel terrorized, frightened, intimidated, or threatened and that actually causes the victim to feel terrorized, frightened, intimidated, or threatened. Specifically excluded from the definition of stalking is statutorily or constitutionally protected activity.

"Harassment" is defined as: conduct directed toward a victim that includes but is not limited to repeated or continuing impermissible contact that would cause a reasonable person to suffer emotional distress and that actually causes the victim to suffer emotional distress. Harassment does not include statutorily or constitutionally protected activity such as lawful picketing pursuant to labor disputes or lawful employer-related activities pursuant to labor disputes.

"Impermissible contact" includes, but is not limited to, knowingly or intentionally following or pursuing the victim.

In this case, there was no evidence to support a finding that Waldon intended to harass Majors. Six sightings of an ex-husband in public places during a one-year period do not amount to harassment. There is nothing in the record which would indicate that the six times Majors observed Waldon were anything but chance or coincidental encounters. In three of those encounters, there is no evidence that Waldon even realized he was involved in an encounter. While an individual might dislike a former spouse or might be afraid to be in the vicinity of a former spouse, that person's subjective reasons for wishing that the former spouse be elsewhere cannot transform the former spouse's innocent activities into behavior sufficient to support a criminal conviction. . . .

Here, the encounters between Majors and Waldon were personally upsetting to Majors; they made her upset and fearful. However, there is nothing about the encounters which could make a reasonable person feel harassed, absent some preconceived idea concerning Waldon's intentions. While Waldon's presence in places Majors frequented may have been frightening to her, Majors' fear cannot form the basis for a determination that Waldon entertained the intent to harass Majors.

Indiana cases interpreting the current stalking statute emphasize this point. In *Burton v. State,* 665 N.E.2d 924 (Ind. Ct. App. 1996), Burton made several telephone calls to the victim, leaving messages such as: "Hi whore," . . ."I hope you have the f—ing windows secured," and "I am coming in the morning." Burton was also seen outside the victim's house. From this behavior a reasonable person could conclude that Burton entertained the intent to harass the victim.

In *Hendricks v. State,* 649 N.E.2d 1050 (Ind. Ct. App. 1995), Hendricks was enamored with a thirteen-year-old girl who did not share his passion. Hendricks threatened the girl, attempted to pay others to assault her, repeatedly telephoned her home and would stand outside her home with a baseball bat and stare. This type of activity by Hendricks could lead a reasonable person to conclude that Hendricks entertained the intent to harass the victim.

In *Johnson v. State,* 648 N.E.2d 666 (Ind. Ct. App. 1995), Johnson came to Indianapolis searching for the victim, who had fled from Johnson several times previously and was secluded at a shelter to avoid Johnson. He harassed people outside the shelter on two occasions while attempting to gather information on the victim. He also appeared beside her when she went to the prosecutor's office for help, and even though she indicated she did not wish to speak with him he continued to whisper to her turned back. A reasonable person examining Johnson's conduct could find that he entertained the intent to harass the victim.

In all of these cases, an examination of the defendant's actions alone, without consideration of the victim's subjective perceptions, could lead a reasonable person to the conclusion that the defendant entertained the intent to harass. That is not true in the case before us. Waldon's actions evince no intent to harass when viewed without considering Majors' subjective perception. For this reason, the evidence is insufficient to sustain Waldon's stalking conviction. I would reverse with instructions to enter a judgment of not guilty.

Questions for Discussion

1. Describe Waldon's behavior. What was Majors's reaction? Was this a reasonable reaction?

2. What is the basis for Judge Barteau's disagreement with the majority? Consider how he differs in his discussion of Waldon's behavior and Majors's reaction.

3. Was Waldon's conduct protected under the First Amendment? Would you hold Waldon liable under the stalking statute?

You Decide

11.3 Norman was charged and convicted of one count of stalking of famed film director and producer, Steven Spielberg. He appealed on the grounds that the victim's fear must be "contemporaneous with the stalker's threats and harassment."

On June 23, 1997, Steven Spielberg left his residence and flew to Ireland, where he was joined by his wife and children. Six days later, Norman pushed the intercom button at the gate to Spielberg's home. When a security guard came to the gate, Norman alleged that he worked for Spielberg's partner, David Geffen, and demanded to see Spielberg. The guard told Norman to leave. Several days later, Norman drove a friend by Spielberg's home, opened his "day planner," and showed his friend a photograph of Spielberg's head affixed to a photo of a naked male body. Norman later told his friend that he was going to climb over the wall at Spielberg's residence and rape him.

Roughly two weeks later, a security guard noticed Norman sitting behind the wheel of an automobile parked across the street from Spielberg's residence and ordered Norman to leave. A few hours later, security guards responded to reports of a man in the backyards of homes adjacent to Spielberg's residence. Norman fled and was apprehended. Norman reported that he was Spielberg's "newly-adopted son" and later identified himself as "David Spielberg." He was released from police custody.

Norman returned to Spielberg's residence on July 11th and backed his car into the driveway and "up to the gate as though he was trying to push it open, then drove away." The police later found and detained Norman two blocks from Spielberg's home. They searched him and found "handcuffs, a box cutter and duct tape on his person and, in his briefcase, two pairs of handcuffs, a day planner with photos of Spielberg, and. . . . razor blades."

On July 17, Norman was released from custody, and Spielberg's lawyer contacted Spielberg in Ireland. The lawyer told Spielberg that Norman had the names of Spielberg's wife and children and that Norman had a record of prior assaultive conduct and had suspicious material in his possession. Spielberg expressed concern and fear for himself and for his family. He authorized additional security measures at his house as well as at his mother's Los Angeles residence, and for himself and his family in Ireland, and later in England.

Norman confessed to the police that he had gone to Spielberg's home on the night of July 11, intending to rape Spielberg and planned to tie up Spielberg's wife and to force her to witness the rape. The police found that Norman possessed a notebook in which Norman wrote about "pursuing Mr. Spielberg to become one of his gay lovers," a map of movie stars' homes with Spielberg's name and address marked in ink, articles about Spielberg, and documents listing the names and other information about various members of the Spielberg family. There was an article chronicling the stalking conduct of John Lennon's killer. Is Norman guilty of stalking Steven Spielberg? See *People v. Norman,* 89 Cal. Rptr. 2d 806 (Cal. Ct. App. 1999).

You can find the answer at www.sagepub.com/lippmanccl3e.

CRIME IN THE NEWS

In late November 2004, Eagle County, Colorado, prosecutor Mark Hurlbert charged Los Angeles Lakers superstar basketball player Kobe Bryant with sexually assaulting a nineteen-year-old woman. Fifteen months later, the same prosecutor stood before the trial court and was granted a motion by the judge to dismiss the case. The prosecutor explained that the victim no longer desired to testify against Kobe Bryant.

The Eagle County prosecutor's office had spent $230,000 on this unsuccessful prosecution, ten percent of its entire budget. Basketball fans of the Lakers expressed relief that the twenty-four-year-old Bryant, who if convicted faced a possible prison sentence of up to life, would be available to play for the Lakers.

Bryant entered the National Basketball Association right out of high school at age eighteen and had been selected to play in the NBA All-Star game five times and already had played on three championship teams. Experts predicted that he would rank among the best players in league history by the time he retired.

Bryant insisted that the intercourse had been consensual. Observers, nevertheless, predicted that Bryant would never recover from the public humiliation. His squeaky clean public image had been tarnished by the public admission that he had committed adultery against his wife of two years, who had given birth to the couple's first child in January. In fact, he admitted during a police interview following his arrest that he earlier had been involved with yet another woman. The transcript of Bryant's conversation with the police indicates that he was primarily concerned with whether his public image would be damaged and with the reaction of his wife. He initially denied intercourse with the complainant and changed his mind after the investigators informed him that the physical evidence indicated that the two had engaged in intercourse.

Several days following Bryant's arrest, he purchased a four-million-dollar diamond ring for his wife, who supported her husband throughout his legal ordeal. It later was revealed that Bryant had contemplated divorce several months earlier and that as a result, his wife had been hospitalized and placed on life support. The mere bringing of the sexual assault charge against Bryant resulted in his losing multimillion-dollar endorsements with leading American corporations.

The alleged victim worked at the front desk at a spa located in small mountain town one hundred miles west of Denver where Bryant was staying while preparing for knee surgery at a nearby clinic. Bryant invited her to his room. She initially alleged that he immediately assaulted her and then later indicated that Bryant unexpectedly escalated a consensual encounter and that she had cried and at least twice said "no" as he pushed her over a chair and entered her from the rear.

The media quickly shifted the focus from Kobe Bryant to the unnamed victim, whose name later was mistakenly posted on an Eagle County court Web site. Press reports indicated that she was a former high school cheerleader who reportedly had auditioned for the show "American Idol" and was described by some acquaintances as desperate for attention and notoriety. Other sources told of two suicide attempts during the past year and her hospitalization by the University of Northern Colorado campus police, who feared that she posed a "danger to herself." The media interviewed every resident of the victim's hometown who was willing to speak, leading to a blizzard of supportive as well as skeptical comments. Two Lakers fans were sufficiently upset to issue death threats against the complainant, resulting in their imprisonment.

Documents and testimony from closed court hearings were leaked to the press. This information indicated that the complainant had sexual intercourse with at least one other individual after she left Bryant and before she contacted the police. The judge would later rule that the defense was entitled to introduce evidence of the complainant's sexual activity during the three days prior to her hospital examination on July 1, 2003, on the grounds that this was relevant to the cause of her injuries, source of DNA, and her credibility as a witness. Information was also leaked documenting that she received a substantial amount of money from a state victim compensation fund to address mental health concerns stemming from her alleged sexual assault by Kobe Bryant.

The news stories on Bryant marveled over the fact that he was able to attend court hearings and then fly hundreds of miles to Lakers games and still perform at a high level. Speculation focused on whether Bryant would continue to play with the Lakers when his contact expired. This question was answered several months prior to the dismissal of the charges, when Bryant signed a seven-year contact for more than $136.4 million.

Following the dismissal of the charges against Bryant, fans expressed relief and the Lakers issued a statement of support. Bryant circulated a press release in which he recognized that although he viewed the affair as consensual, "she did not," and he issued an apology to the complainant and her family for his behavior. Advocates for victims of sexual assault were quoted as bemoaning that the Bryant case would have the impact of discouraging women from bringing complaints of rape to prosecutors who, in turn, would be reluctant to file charges.

Colorado seemed to be a center of controversy concerning sexual abuse. A report issued by an independent investigative commission in May 2004 indicated that at least nine women had alleged that they had been sexually assaulted by University of Colorado football players or recruits since 1997. The most infamous incident involved a female kicker on the football team who alleged that she had been subjected to sexual assaults and rape. Her father, an army surgeon stationed in Iraq, stated that the university had been completely unresponsive to his complaints over the treatment of his daughter. Colorado football coach Gary Barnett was suspended for several months during an investigation after telling the press that Katie was an "awful" player who could not kick the ball through the uprights. Do athletes receive preferential treatment when accused of sexual abuse? On the other hand, are they more often than not the target of false allegations?

KIDNAPPING

Kidnapping at common law was the forcible abduction or stealing away of a person from his or her own country and sending him into another country.[73]

This misdemeanor was intended to punish the taking of an individual to an isolated location where the victim was outside of the protection law. Imagine the fear you would experience in the event that you were locked in a basement under the complete control of an abusive individual.

Kidnapping became of concern in 1932 with the kidnapping of the twenty-month-old son of Charles Lindbergh, the aviation hero who piloted the "Spirit of St. Louis" in the first flight across the Atlantic. Lindbergh paid the $50,000 ransom demanded for the return of Charles Jr., who later was found dead in the woods five miles from the Lindbergh home. A German immigrant, Bruno Hauptmann, was prosecuted for felony murder and executed. The question whether Hauptmann was the perpetrator continues to be a topic of intense debate.

For a deeper look at this topic, visit the study site.

The Lindbergh kidnapping resulted in the adoption of the Federal Kidnapping Act, known as the "Lindbergh Law." This law prohibits the kidnapping and carrying of an individual across state lines for the purpose of obtaining a ransom or reward. Six states shortly thereafter adopted new statutes significantly increasing the penalty for kidnapping. This trend continued and, by 1952, all but four states punished kidnapping by death or life imprisonment.

The Lindbergh Law excluded parents from coverage. In 1981, Congress addressed the 150,000 abductions of children by a parent involved in a custody dispute in the Parental Kidnapping Prevention Act. The statute provides for FBI jurisdiction when a kidnapped child is transported across state lines. The abduction of children by a parent or relative involved in a custody dispute is also the subject of specific state statutes punishing a "relative of a child" who "takes or entices" a child younger than eighteen from his or her "lawful custodian."[74]

The last decades have been marked by a string of high-profile kidnappings of wealthy corporate executives and members of their families. In 1974, Patricia Hearst, heiress to the Hearst newspaper fortune, was kidnapped from her campus apartment at the University of California at Berkeley. The abduction was carried out by the Symbionese Liberation Army (SLA), a self-proclaimed revolutionary group. The case took on a bizarre character when Hearst participated in bank robberies intended to finance the activities of the group. Patricia Hearst was later apprehended and convicted at trial despite her claim of "brainwashing." The major figures in the SLA were subsequently killed in a shoot-out with the Los Angeles police.

More recently we have seen the 1993 abduction and killing of twelve-year-old Polly Klaas and the kidnapping, in 2002, of twelve-year-old Elizabeth Smart. Elizabeth was later found to have been forced into a "marriage" with her abductor, a handyman in the Smart home.

Following American intervention into Iraq, foreigners were regularly kidnapped by criminal gangs who "sold" the victims to political groups opposed to the presence of the United States. The terrorists typically threatened to kill the hostage unless the company for which the prisoner worked or their country of nationality agreed to withdraw from Iraq. In a related incident, *Wall Street Journal* reporter Daniel Pearl was kidnapped and beheaded by an extremist group in Pakistan. The terrorist group's British-born leader, Ahmad Omar Saeed Sheikh, was later prosecuted and sentenced to death in Pakistan. The taking of hostages is recognized as a crime under the International Convention Against the Taking of Hostages, which requires countries signing the treaty to either prosecute offenders or to send them to a nation claiming the right to prosecute the offender.

The U.S. Congress has also enacted the Victims of Trafficking and Violence Protection Act of 2000 and the PROTECT Act of 2003 that together combat the international sexual trafficking industry. Roughly one million children, most of whom are girls, have been persuaded to leave or are forcibly abducted from their mostly rural villages in poor countries and forced into sexual slavery or low-wage industrial labor.

Criminal Intent

Kidnapping statutes vary widely in their requirements. The California Penal Code section 202 provides:

> Every person who forcibly, or by any other means of instilling fear, steals or takes, or holds, detains, or arrests any person in this state, and carries the person into another country, state, or county, or into another part of the same county, is guilty of kidnapping.

The *mens rea* of kidnapping, although subject to dispute, is commonly thought to be an intent to move or to confine the victim without his or her consent. The Wisconsin Statute section 940.31 provides that kidnapping is the carrying of another from one place to another without his or her consent and with the "intent to cause him or her to be secretly confined or imprisoned or to be carried out of this state or to be held to service against his or her will." Some statutes require holding an individual for a specific purpose such as detaining a person for ransom or as a hostage. Florida requires a specific intent, whether to hold an individual for ransom or reward, to serve as a shield or hostage, to inflict bodily harm, to terrorize the victim, or to interfere with the performance of any governmental or political function.[75] Texas considers the intentional or knowing abduction of another person with the intent to commit these acts as "aggravated kidnapping."[76]

Criminal Act

The essence of kidnapping is the *actus reus* of the forcible movement of a person as provided under the North Carolina statute "from one place to another."[77] The central issue is the extent of the movement required. The traditional rule in American law is that any movement, no matter how limited, is sufficient. In the well-known California case of *People v. Chessman,* Caryl Chessman was convicted of kidnapping when he forced a rape victim to move twenty-two feet into his car. The California Supreme Court noted that it is "the fact, not the distance of forcible removal which constitutes kidnapping in this State."[78]

Most courts abandoned this approach after realizing that almost any rape, battery, or robbery involves some movement of a victim. This led to prosecutors charging defendants with the primary crime as well as kidnapping and resulted in life imprisonment for crimes that otherwise did not merit this harsh penalty. In 1969, the California Supreme Court rejected the *Chessman* standard in *People v. Daniels*. The defendants entered the victims' apartments and forced them at knifepoint to move a few feet into another room where they were robbed and raped. The court reversed the kidnapping convictions on the grounds that the victims' movements were a central step in the rape or robbery and should not be considered as constituting the independent offense of kidnapping.[79]

Courts now generally limit the application of kidnapping statutes to "true kidnapping situations and [do] not . . . apply it to crimes which are essentially robbery, rape or assault . . . in which some confinement or asportation occurs as a subsidiary incident."[80] In other words, kidnapping statutes are no longer thought to include unlawful confinements or movements incidental to the commission of other felonies. Under this standard, courts require that for a movement to be considered kidnapping, it "must be more than slight, inconsequential, or an incident inherent in the crime." Judges have ruled that for the movement to constitute kidnapping, it must meaningfully contribute to the commission of the primary crime by preventing the victim from calling for help, reducing the defendant's risk of detection, facilitating escape, or increasing the danger to the victim.[81]

The movement or detention of the victim must be unlawful, meaning without the victim's consent by force or threat of force. The Wisconsin statute requires that the movement of an individual must be undertaken "without his or her consent" and by "force or threat of imminent force."[82] This excludes movements undertaken as a result of a lawful arrest, court order, or consent. An Arkansas court held that where the victim voluntarily accepted the defendant's offer of a ride to a friend's house, the victim revoked her consent when the perpetrator prevented her from leaving the automobile by displaying and threatening her with a firearm, ordering her to place her hands under her thighs, and taking her to his home where she was raped.[83]

Courts are divided over whether an individual is "forcibly" moved where the defendant fraudulently misrepresents his or her intended destination. The California Supreme Court held that a victim's movement was "accomplished by force or any other means of instilling fear" where a rapist falsely represented that he was a police officer and informed the eighteen-year-old victim that she faced arrest unless she accompanied him to a store from which she was suspected of having stolen merchandise. The court concluded that this "kind of compulsion is qualitatively different than if defendant had offered to give Alesandria [the victim] a ride, or sought her assistance in locating a lost puppy, or any other circumstance suggesting voluntariness on the part of the victim."[84]

Some states follow the Model Penal Code in reducing the seriousness of the kidnapping where the victim was "voluntarily released . . . in a safe place." This provides an incentive for a defendant to limit the harm inflicted on victims.[85]

In the next case, *People v. Aguilar*, the California court must decide whether the defendant's movement is sufficient to constitute kidnapping.

Model Penal Code Section 212.1. Kidnapping

A person is guilty of kidnapping if he unlawfully removes another from his place of residence or business, or a substantial distance from the vicinity where he is found, or if he unlawfully confines another for a substantial period in a place of isolation, with any of the following purposes:

(a) to hold for ransom or reward, or as a shield or hostage; or

(b) to facilitate commission of any felony or flight thereafter; or

(c) to inflict bodily injury on or to terrorize the victim or another; or

(d) to interfere with the performance of any governmental or political function.

Kidnapping is a felony of the first degree unless the actor voluntarily releases the victim alive and in a safe place prior to the trial, in which case it is a felony of the second degree (ten years).

A removal or confinement is unlawful . . . if it is accomplished by force, threat, or deception, or, in the case of a person who is under the age of 14 or incompetent, if it is accomplished without the consent of a parent, guardian, or other person responsible for general supervision of his welfare.

Analysis

1. Kidnapping is defined to include any of three acts. First, removing a victim from the protection of the home or business is intended to punish the taking of individuals from the safety and security of a home or business and placing them in danger. Second, removing individuals "a substantial distance from the vicinity" from where they are located. This is intended to ensure that punishment is not imposed for "trivial changes of location." Third, confining an individual for a substantial period of time in an isolated location with a specified intent. This is intended to punish the "frightening and dangerous" removal of a victim from a safe environment to an "isolated" location where he or she is outside the protection of the law. No movement or asportation is required in regard to confining an individual. The requirement that the detention is for a "substantial period of time" in a "place of isolation" is intended to avoid punishing a defendant for a detention that is merely incidental to a rape or other crime of violence.

2. Kidnapping is defined to require one of four specified purposes that commonly appear in kidnapping statutes.

3. An unlawful removal must be accomplished through force, threat, or deception or in the case of an "incompetent" juvenile under the age of fourteen, without the consent of a parent or guardian.

4. The reduction in punishment for kidnapping to a second-degree felony where the perpetrator releases the victim "alive and in a safe place" provides an incentive for the kidnapper to abandon the criminal enterprise. The defendant would remain liable for any battery or sexual assault.

The Legal Equation

Kidnapping = Intent to detain and move or to detain and hide without consent

+ act of detention and moving or detention and hiding

+ through the unlawful use of force or threat of force.

Did Aguilar's movement of the victim constitute kidnapping?

People v. Aguilar, 16 Cal. Rptr. 3d 231 (Cal. Ct. App. 2004). Opinion by: Gilbert, J.

Sergio Barrera Aguilar appeals a judgment after conviction of kidnapping to commit rape and sexual penetration, with a finding that he used a deadly weapon among other things.

The jury made additional special findings of fact that Aguilar personally used a knife, that he inflicted great bodily injury, and that "the movement of the victim in the course of the kidnapping substantially increased the risk of harm to her."

We conclude, among other things, substantial evidence supports: 1) the conviction for aggravated kidnapping, and 2) the finding that Aguilar substantially increased the victim's risk of harm by moving her.

Facts

Aguilar followed Nancy C., age 16, as she walked her dog down a residential street at night. He grabbed her and said "he was going to take [her] somewhere and rape [her]." He inserted his fingers in her vagina and she screamed. He then removed his hands from her vagina and pulled her 133 feet down the sidewalk past a house with a lit porch light to an area in front of a house with no light. He pushed her face down onto the hood of a car, "put his hands down [her] pants" and inserted his fingers in her vagina.

Police Officer James Ella testified that the area to which Nancy C. was moved was "extremely dark." Trees blocked "most of the illumination" coming from the light down the street. In a videotaped confession, Aguilar admitted he had grabbed Nancy C., was aroused, and put his fingers in her vagina. He said he moved "to a place where nobody could see [them]" to have intercourse with her. He admitted that what he did was "wrong" and that Nancy C. did not consent to have sex with him. He said he had a knife with him, but he "didn't pull the knife out."

Martin Molina, a nearby resident, testified that his porch light was the "only light on the street" between the area where Aguilar first grabbed Nancy C. and the location to which she was ultimately dragged. He said the first area was lighter because trees and bushes "[funnel] the light" from his porch light to that area. They deflect light away from the area where the attack ended.

Anthony Ventura Castillo was at home when he heard a woman screaming "help, help" and "save me." He testified it was so dark he had to turn on the porch light to see what was happening. He saw Aguilar throw Nancy C. to the ground and grab her by the neck. Aguilar was holding a knife "12 or 13 inches from her neck." Castillo told him to release her. Aguilar "got up and ran." Castillo and his brother chased Aguilar and apprehended him.

Issue

Aguilar contends the evidence is insufficient to support the aggravated kidnapping conviction.

"Kidnapping to commit rape involves two prongs. First, the defendant must move the victim and this asportation must not be 'merely incidental to the [rape].' . . . Second, movement must increase 'the risk of harm to the victim over and above that necessarily present in the [rape].'" . . . For aggravated kidnapping . . . "'there is no minimum number of feet a defendant must move a victim in order to satisfy the first prong.' . . . Where movement changes the victim's environment, it does not have to be great in distance to be substantial. . . . [W]here a defendant moves a victim from a public area to a place out of public view, the risk of harm is increased even if the distance is short." . . .

Aguilar contends that . . . he did not move Nancy C. into "a hidden location" such as a bathroom or a

back room of a store. He says, "[t]he movement was down the sidewalk," an open area. But this distinction is not dispositive. Courts have held that moving a victim to a more isolated open area that is less visible to public view is sufficient. . . .

Reasoning

Here Aguilar forcibly moved Nancy C. 133 feet down a sidewalk at night, from an area illuminated by a porch light to an "extremely dark" area. The "risk to [Nancy C.] in the dark . . . increased significantly. . . ." Aguilar admitted his goal was to move her "to a place where nobody could see [them]." The movement "decreased [Aguilar's] likelihood of detection. . . ."

A reasonable trier of fact could infer this increased the risk to Nancy C. by making it harder for her to escape and "enhanced [Aguilar's] opportunity to commit additional crimes. . . . An increased risk of harm was manifested by appellant's demonstrated willingness to be violent. . . ." He pulled Nancy C. down the sidewalk, threw her to the ground, grabbed her neck, choked her, bit her, slammed her onto a car hood, held her face down and held a knife near her neck. He told Nancy C. that he was moving her to rape her which, when coupled with his violent acts, "pose[d] a substantial increase in the risk of psychological trauma . . . beyond that to be expected from a stationary" sexual attack.

Aguilar contends that he did not complete his goal because Castillo rescued Nancy C. But that "'does not . . . mean that the risk of harm was not increased [by the movement].'" We conclude the evidence was sufficient.

Holding

Aguilar creates a subjective "apparent purpose" test to determine whether his moving the victim was incidental. He argues his "apparent purpose" in moving the victim 133 feet was to commit the rape, and therefore the movement was "incidental" to the crime. The standard, however, is whether "the jury could reasonably have concluded that [the victim's] movement . . . was not merely incidental" from the "totality of the circumstances. . . . [T]he defendant's intent to commit kidnapping as . . . a necessary component of the target offenses is not determinative of whether the movement is incidental." . . . The interpretation of "incidental" depends on the facts of the particular case. . . .

Questions for Discussion

1. Discuss the legal test for kidnapping relied on by the court.

2. Why does Aguilar contend that this was not kidnapping? Is it significant that the victim was moved only a short distance? What does the court conclude?

3. Do you agree with the court's conclusion that the victim was kidnapped?

You Decide

11.4 Daryl Tindall was charged with (a) Aggravated Kidnapping, (b) Lewd and Lascivious Molestation-Offender under 18, Victim under 12, and (c) Sexual Battery on a Child under 12 by Perpetrator under 18.

J.T. testified that when she was six years old, she visited her friend, Stewart, who is the defendant's nephew. Tindall, who at the time was sixteen years old, answered the door and pulled J.T. by her hair inside the home and into his bedroom. He locked the door, placed J.T. on the bed, and took off her clothes and sexually molested her. The defendant "told her not to tell anybody and that she did not tell anyone afterwards because she was scared." Several days later, Tindall again sexually attacked J.T.

Another child, E.K., testified that Tindall gave E.K. a bear hug, picked her up, took her to his bedroom, and locked the door, took off her clothes and molested her. Two weeks later, he molested her once again and told her not tell anyone.

Tindall argues that he was entitled to a judgment of acquittal on the aggravated kidnapping charges because the acts which constituted kidnapping were incidental to the underlying alleged sexual misconduct. Would you convict Tindall of kidnapping? See *Tindall v. State,* 45 So. 3d 799 (Fla. Dist. Ct. App. 2010).

You can find the answer at www.sagepub.com/lippmanccl3e.

FALSE IMPRISONMENT

Within the common law and in state statutes, **false imprisonment** is defined as the intentional and unlawful confinement or restraint of another person. The Idaho statute simply states that false imprisonment is the "unlawful violation of the personal liberty of another."[86] False imprisonment is generally considered a misdemeanor, punishable in Idaho by a fine not exceeding $5,000 or by imprisonment in the county jail for no more than one year or both.[87]

As with kidnapping, false imprisonment is a crime that punishes interference with the freedom and liberty of the individual.

False imprisonment requires an intent to restrain the victim. Arkansas and other states follow the Model Penal Code and provide that false imprisonment may be committed by an individual who "without consent and without lawful authority . . . knowingly restrains another person so as to interfere with his liberty."[88] The detention must be unlawful and without the victim's consent. A restraint by an officer acting in accordance with the law or by a parent disciplining his or her child does not constitute false imprisonment. The consent of the victim constitutes a defense to false imprisonment. A farmer who secured his wife with a chain while he went to town was not held liable for false imprisonment where the evidence indicated that she requested him to manacle her to the bed.[89]

The *actus reus* is typically described as compelling the victim to "remain where he did not want to remain or go where he did not want to go."[90] The confinement may be accomplished by physical restraint or by a threat of force of which the victim is aware. Confinement may also be achieved without force or the threat of force when, for instance, the perpetrator locks a door. Professors Perkins and Boyce point out that an individual is not confined because he is prevented from moving in one or in several directions so long as he may proceed in another direction. An individual may also be confined in a moving bus or in a hijacked airplane.[91]

False imprisonment may overlap with kidnapping or with assault and battery or robbery, and several states have eliminated the crime. In these states, people who are unlawfully detained are able to seek damages in civil court. The major difference between the crimes of kidnapping and false imprisonment is that false imprisonment does not require an asportation (movement). In addition, kidnapping statutes that punish the confinement as well as the movement of a victim often provide that the victim must be "secretly confined" or "held in isolation" for a specific purpose (e.g., to obtain a ransom). The Model Penal Code provides that the confinement for kidnapping must be for a "substantial period."[92] False imprisonment, in contrast, requires a detention that may take place in public or in the privacy of a home and may be for a brief or lengthy period. States such as Alabama provide for the punishment of aggravated false imprisonment where an individual "restrains another person under circumstances which expose the latter to a risk of serious physical injury."[93]

Juries, at times, are asked to determine whether a defendant should be convicted of false imprisonment or kidnapping. In the Georgia case of *Shue v. State,* James Shue attacked and threatened to kill Elizabeth Guthrie unless she got into her car. Guthrie testified that as Shue pushed her into the car, she tightly grabbed the steering wheel and honked the horn. Shue cursed and walked away. The police report indicated that Guthrie voluntarily entered her car after Shue fled. Shue subsequently was acquitted of kidnapping, but convicted of false imprisonment.

The Georgia statute states that false imprisonment is committed by an "arrest, confinement or detention of the person, without legal authority, which violates the person's personal liberty." A kidnapping occurs when an individual "abducts or steals away any person without lawful authority or warrant and holds such person against his will." The Georgia appellate court noted that the "only difference between false imprisonment and kidnapping is that kidnapping requires asportation." The court ruled that the "jury could have believed Guthrie's testimony that Shue held her against her will but discounted her testimony that he pushed her into the car. Thus, the jury could have found that Guthrie was detained without having been carried away and therefore falsely imprisoned but not kidnapped."[94]

In another example, a North Carolina appellate court affirmed the conviction of Richard Overton for second-degree kidnapping of Elsie Fennell for the purpose of terrorizing her, a purpose that is prohibited under the North Carolina kidnapping statute. Overton beat and punched Fennell, with whom he was living, after she told him that he would have to move out if he did not start paying the bills. Overton "caught and restrained" Fennell from leaving the house and only gave her the keys to the car after she persuaded him that she would not call the police. Overton appealed on the grounds that he should have been convicted of false imprisonment rather than for kidnapping because he had confined and restrained Fennell for the purpose of avoiding detection by the police rather than for the purpose of terrorizing her. The North Carolina court recognized that "if the purpose of the restraint was to accomplish one of the purposes . . . in the kidnapping statute . . . the offense is kidnapping. In the absence of one of the statutorily specified purposes, the unlawful restraint is false imprisonment." The court affirmed Overton's conviction based on the fact that "it is immaterial that defendant's purpose may have changed during the course of the restraint."[95]

A contemporary application of false imprisonment is the United States Victims of Trafficking and Violence Protection Act of 2000 that punishes acts of slavery and forced labor. In *United States v. Bradley,* two New Hampshire defendants were convicted of misrepresenting the working conditions of their tree removal company to workers recruited from Jamaica who were threatened, forcibly detained, and abused when they attempted to return home.[96]

You can find more cases on the study site: Cole v. State, *www.sagepub.com/lippmanccl3e.*

Model Penal Code Section 212.3. False Imprisonment

A person commits a misdemeanor if he knowingly restrains another unlawfully so as to interfere substantially with his liberty.

Section 212.2. Felonious Restraint

A person commits a felony of the third degree if he knowingly:

(1) restrains another unlawfully in circumstances exposing him to risk of serious bodily injury; or

(2) holds another in a condition of involuntary servitude.

Analysis

1. The Model Penal Code requires a substantial interference with an individual's liberty. This eliminates prosecutions for relatively modest interference with an individual's liberty.

2. An individual must knowingly restrain another person. As a result, it is sufficient that a defendant is aware that his or her conduct will result in the detention of another individual.

3. Felonious restraint is intended to punish as felonies restraints that create a risk of serious bodily injury or involuntary servitude.

The Legal Equation

False imprisonment $=$ Restrain by

$+$ force, threat of force, or constructive force.

CHAPTER SUMMARY

There are four categories of offenses against the person: sexual offenses, bodily injury and interference, freedom of movement, and homicide.

The most serious sexual offense is rape. Rape at common law was punishable by death, and only homicide was historically considered a more severe crime. Common law rape required the intentional vaginal intercourse by a man of a woman who was not his wife by force or threat of serious bodily injury against her will. The fear of false conviction led to the imposition of various barriers that the victim was required to overcome. This included immediate complaint, corroboration, the admissibility of evidence pertaining to a victim's past sexual activity, and a cautionary judicial instruction. The focus was on the victim, who was expected to demonstrate her lack of consent by the utmost resistance.

The law of sexual offenses was substantially modified by the legal reforms of the 1970s and 1980s that treated rape as an assault against the person rather than as an offense against sexual morality. Sexual intercourse was expanded to include the forced intrusion into any part of another person's body, including the insertion of an object into the genital or anal opening. Rape was defined in gender-neutral terms, the marital exemption was abolished, and rape was divided into simple and aggravated rape based on the type of the penetration, use of force, or resulting physical injury. Various statutes no longer employ the term rape and consider vaginal intercourse merely to be another form of sexual assault.

The intent of rape reform is to shift attention from resistance by the victim to the force exerted by the perpetrator. There are two approaches to analyzing force. The extrinsic force standard requires an act of force beyond the physical effort required to achieve sexual penetration. The intrinsic force standard requires only that amount of force required to achieve penetration. Rape may also be accomplished by threat of force, through penetration obtained by fraud, or when the victim is incapable of consent stemming from unconsciousness, sleep, or insanity.

The *mens rea* of rape at common law was the intent to engage in vaginal intercourse with a woman whom the defendant knew was not his wife through force or the threat of force against her will. There was no clear guidance as to whether a defendant was required to know that the intercourse was without the female's consent. A majority of states now accept an objective test that recognizes the defense that a defendant honestly and reasonably believed that the victim consented. This requires equivocal conduct by the victim that is capable of being reasonably, but mistakenly, interpreted by the assailant as indicating consent. The English House of Lords rule, in contrast, adopts a subjective approach and examines whether a defendant "knows" whether the victim consented. The third approach is the imposition of strict liability that holds defendants guilty whatever their personal belief or the objective reasonableness of their belief.

Recent legal and statutory developments recognize that individuals may withdraw their consent. The continuation of sexual relations under such circumstances constitutes rape.

Statutory rape is a strict liability offense that holds a defendant guilty of rape based on his intercourse with an "underage" female. Several states permit the defense of a reasonable mistake of age. Rape shield laws prohibit the prosecution from asking the victim about or introducing evidence concerning sexual relations with individuals other than the accused or introducing evidence relevant to the victim's reputation for chastity.

Assault and battery, although often referred to as a single crime, are separate offenses. A battery is the application of force to another person. An assault may be committed by attempting to commit a battery or by intentionally placing another in imminent fear of a battery. Aggravated assaults and batteries are felonies.

Kidnapping is the unlawful and forcible seizure and asportation (carrying away) of another without his or her consent. This requires the specific intent to move the victim without his or her consent. The *actus reus* is the moving or detention of the victim. Courts differ on the extent of this movement, variously requiring slight or, under the Model Code standard, substantial movement. The majority rule is that the movement must not be incidental to the commission of another felony and must contribute to the primary crime by preventing the victim from calling for help, reducing the defendant's risk of detection, facilitating escape, or increasing the danger to the accused.

False imprisonment is the intentional and unlawful confinement or restraint of another person. The restraint may be achieved by force, threat of force, or by other means and does not require the victim's asportation or secret confinement. False imprisonment is a misdemeanor other than when committed in an aggravated fashion.

CHAPTER REVIEW QUESTIONS

1. What was the original justification for the crime of rape?

2. Why does "date rape" present a challenge to prosecutors?

3. How did the common law define rape? List some of the barriers to establishing rape under the common law.

4. Describe the changes in the law of rape introduced by the reform statutes of the 1970s and 1980s.

5. What elements distinguish a simple or second-degree rape from aggravated or first-degree rape?

6. Distinguish the standard of extrinsic force from the intrinsic force standard in the law of rape.

7. In addition to force, what other means might a perpetrator employ to satisfy the *actus reus* requirement in the law of rape?

8. What are the three approaches to the defense of mistake of fact in the *mens rea* of rape?

9. Is a defendant's belief that another individual is above the age of lawful consent a defense to statutory rape? What of the female's past sexual experience?

10. May an individual who withdraws consent claim to be the victim of rape?

11. What is the purpose of rape shield laws? Are there exceptions to rape shield laws?

12. Distinguish an assault from a battery.

13. What are the requirements of a battery? Describe the difference between a simple and an aggravated battery.

14. Discuss the two ways to commit an assault.

15. What is the relationship between the crime of stalking and a criminal assault?

16. What is the definition of kidnapping? What are the various approaches to the asportation requirement?

17. Distinguish between false imprisonment and kidnapping.

LEGAL TERMINOLOGY

acquaintance rape

aggravated rape

assault and battery

corroboration

cyberstalking

earnest resistance

extrinsic force

false imprisonment

fraud in inducement

fraud in the *factum*

intrinsic force

kidnapping

mayhem

prompt complaint

prosecutrix

rape shield laws

rape trauma syndrome

reasonable resistance

resist to the utmost

stalking

statutory rape

withdrawal of consent

CRIMINAL LAW ON THE WEB

Log on to the Web-based student study site at www.sagepub.com/lippmanccl3e to assist you in completing the Criminal Law on the Web exercises, as well as for additional features such as podcasts, Web quizzes, and video links.

1. Read about rape in male prisons.

2. Learn more about stalking.

3. Examine human sexual trafficking and sexual slavery in the United States.

4. Read more about the Kobe Bryant rape prosecution.

BIBLIOGRAPHY

American Law Institute, *Model Penal Code and Commentaries,* vol. 1, pt. 2 (Philadelphia: American Law Institute, 1980), pp. 172–356. A comprehensive discussion of the conceptual development of and the legal and policy considerations underlying the crimes of rape, assault and battery, and kidnapping.

Joshua Dressler, *Understanding Criminal Law,* 3rd ed. (New York: Lexis, 2001), pp. 569–598. A detailed account of the history and reform of rape laws.

Susan Estrich, *Real Rape* (Cambridge, MA: Harvard University Press, 1987). An influential analysis of rape law in the United States with a stress on the difficulties of prosecuting acquaintance rape.

Wayne R. LaFave, *Criminal Law,* 3rd ed. (St. Paul, MN: West Publishing, 2000), pp. 736–789. A comprehensive legal discussion of the laws of assault, battery, and rape with detailed citations.

Rollin M. Perkins and Ronald N. Boyce, *Criminal Law,* 3rd ed. (Mineola, NY: Foundation Press, 1982), pp. 224–236. The best discussion of the history and law of kidnapping and false imprisonment.

12 BURGLARY, TRESPASS, ARSON, AND MISCHIEF

Was the defendant guilty of vandalism?

The evidence : . . proved that in the early morning hours of February 11, 2000, Nicholas Y. wrote on a glass window of a projection booth at an AMC theater with a Sharpie marker. . . . Police saw "approximately 30 incidents" in red magic marker throughout the theater, including the one on the glass. Appellant said the initials stood for "The Right to Crime." At the close of the prosecution's case, appellant's counsel argued that no defacing of or damage to property had been proved, stating: "It's a piece of glass with a marker on it. You take a rag and wipe it off. End of case. It's ridiculous."

Learning Objectives

1. Know the common law of burglary and the changes to the law of burglary introduced by burglary statutes.
2. Understand the law of trespass and the difference between burglary and trespass.
3. Know the *mens rea* and *actus reus* of arson.
4. State the law of criminal mischief and the three categories of acts that constitute criminal mischief under the Model Penal Code.

INTRODUCTION

The notion that the home is an individual's castle is deeply ingrained in the American character. After all, the United States is an immigrant society to which people flocked in order to seek a better life. A fundamental aspect of this dream was the ownership of a home.

The importance of the home was apparent from the early days of the country. The warrantless intrusion and quartering of British soldiers in the homes of the colonists was a central cause of the American Revolution. The Fourth Amendment to the Constitution was specifically adopted to insure that individuals were free from arbitrary searches of their homes and seizures of their property. This amendment reminds us that a home is more than bricks and mortar and is valued as more than an economic investment. It is a safe and secure shelter where we are free to express our personalities and interests without fear of uninvited intrusions. What are your feelings when you think about your home?

In this chapter, we look at the common law offenses developed to protect an individual's dwelling and at the incorporation of these common law crimes in state statutes that cover a broad range of structures and vehicles. The criminal offenses covered in the chapter are:

- *Burglary.* Breaking and entering into a dwelling or other structure.
- *Trespass.* An uninvited intrusion onto an individual's property and dwelling.
- *Arson.* The burning of a dwelling.
- *Malicious Mischief.* The destruction of property or the home.

You should pay particular attention to how statutes have changed the common law of burglary and arson.

BURGLARY

Burglary at common law was defined as the breaking and entering of the dwelling house of another at night with the intention to commit a felony. Burglary was punished by the death penalty, reflecting the fact that a nighttime invasion of a dwelling poses a threat to the home, which is "each man's castle . . . and the place of security for his family, as well as his most cherished possessions."[1] Blackstone observed that burglary is a "heinous offense" that causes "abundant terror," which constitutes a violation of the "right of habitation" and which provides the inhabitant of a dwelling with the "natural right of killing the aggressor."[2] The crime of burglary protects several interests:

- *Home.* The right to peaceful enjoyment of the home.
- *Safety.* The protection of individuals against violent attack and fright within the home.
- *Escalation.* The prevention of a dangerous confrontation that may escalate into a fatal conflict.

In 1990, the U.S. Supreme Court noted that state statutes no longer closely followed common law burglary and that these statutes, in turn, did not agree on a common definition of burglary. This means that in thinking about burglary, you should pay particular attention to the definition of burglary in the relevant state statute. As you read this section, analyze how burglary has been modified by state statutes. In addition, consider whether we continue to need the crime of burglary. What does burglary contribute that is not provided by other offenses?[3]

Breaking

Common law burglary requires a "breaking" to enter the home by a trespasser, an individual who enters without the consent of the owner. A breaking requires an act that penetrates the structure, such as breaking a window or pushing open an unlocked door. Permanent damage is not required; the slightest amount of force is sufficient. Why did the common law require a breaking? Most commentators conclude that this requirement was intended to encourage homeowners to take precautions against intruders by closing doors and windows. In addition, an individual who resorts to breaking also typically lacks permission to enter, and the breaking is evidence of an "unlawful" or "uninvited" entry. A breaking may also occur through constructive force. This entails entry by fraud, misrepresentation, or threat of force; entry by an accomplice; or entry through a chimney.

Most statutes no longer require a breaking. Burglary is typically defined as an unlawful or uninvited entry (e.g., lacking permission to enter). Note that it is not burglary under this definition for an individual to enter a store that is open to the general public. Some courts have interpreted statutes to cover the entry into a store by arguing that an individual who enters a store while concealing that he or she plans to commit a crime has committed a fraud and therefore has entered unlawfully. Courts have ruled that breaking into an ATM machine or other structure "too small for a human being to live in or do business in is not a 'building' or 'structure' for the purpose of burglary."[4]

Entry

The next step after the breaking is "entry." This requires only that a portion of an individual's body enters the dwelling; a hand, foot, or finger is sufficient. Courts also find burglary when there is entry by an instrument that is used to carry out the burglary. This might involve reaching into a window with a straightened coat hanger to pull out a wallet or reaching an arm through a window to pour inflammable liquid into a home. In a recent case, an individual launched an aggressive verbal attack on his lover's husband while reaching his arm threateningly through the open door of the husband's motor home. A Washington appellate court determined that this was sufficient to constitute an intent to assault and a conviction for burglary.[5] Note the general rule is that it is not a burglary when an instrument is used solely to break the structure, such as tossing a brick through a window.[6]

Another point to keep in mind is that the breaking must be the means of entering the dwelling. You might break a window and then realize that the front door is open and walk in and steal a television. There is no burglary because the breaking is not connected to the entry. A burglary may also be accomplished constructively by helping a small, thin co-conspirator enter a home through a narrow basement window.

Some statutes provide that a burglary may be committed by "knowingly . . . remaining unlawfully in a building" with the required intent or "surreptitiously (secretly) remaining on the premises with the intent to commit a crime." In *Dixon v. State,* for example, the defendant entered a church during Sunday services, wandered into the church sanctuary, and stole money from the collection plate in the pastor's office. A Florida appellate court ruled that the defendant illegally entered the sanctuary and surreptitiously remained in the structure when he closed the door to the pastor's office during the robbery.[7]

The important point to keep in mind is that the entry for a burglary must be trespassory, meaning without consent. Note that stealing a computer from your own dorm room would not be a burglary. The essence of burglary is the unlawful interference with the right to habitation of another. In *Stowell v. People,* the Colorado Supreme Court observed that there is no burglary "if the person entering has a right to do so, although he may intend to commit and may actually commit a felony." Otherwise, a schoolteacher "using the key furnished to her . . . to re-open the schoolhouse door immediately after locking it in the evening, for the purpose of taking (but not finding) a pencil belonging to one of her pupils, could be sent to the penitentiary."[8]

Dwelling House

The common law limited burglary to a "dwelling house," a structure regularly used as a place to sleep. A structure may be used for other purposes and still constitute a dwelling so long as the building is used for sleep. The fact that the residents are temporarily absent from a summer cottage does not result in the building's losing its status as a dwelling. However, a structure that is under construction and not yet occupied or a dwelling that has been permanently abandoned is not considered a dwelling. The Illinois burglary statute provides that a dwelling is a "house, apartment, mobile home, trailer, or other living quarters in which at the time of the alleged offense the owners or occupants actually reside or in their absence intend within a reasonable period of time to reside."[9]

A dwelling at common law included the **curtilage**, or the land and buildings surrounding the dwelling, including the garage, tool shed, and barn. A recently decided Washington case held a defendant liable for a burglary when, with the intent to assault his former wife, he jumped over a six-foot wooden fence in the backyard of the house she shared with her current lover.[10]

Most statutes no longer limit burglaries to dwelling houses and typically categorize the burglary of a dwelling as an aggravated burglary. The California statute extends protection to "any house, room, apartment, . . . shop, warehouse, store, mill, barn, stable, outhouse, . . . tent, vessel . . . floating home, railroad car, . . . inhabited camper, . . . aircraft, . . . or mine."[11] Other statutes are less precise and provide that a burglary involves a "building or occupied structure, or separately secure or occupied portion thereof."[12]

Dwelling of Another

The essence of burglary under the common law is interference with an individual's sense of safety and security within the home. In determining whether the home is "of another," you need to examine who resides in the dwelling rather than who owns the dwelling. For example, a husband who separates from his wife and moves out of the home that he owns with his spouse may be liable for the burglary of his former home. Also, an individual generally cannot burglarize a dwelling that he or she shares with another. A burglary of this dwelling is possible only when the individual enters into portions of the home under the exclusive control of his or her roommate with the requisite criminal intent.

The requirement that an entry be of the dwelling of another is not explicitly stated in most statutes. Despite the failure to include this language, you still cannot burglarize your own home because an entry must be unlawful, meaning without a legal right, and you clearly are entitled to enter your own home.

Nighttime

A central requirement of burglary at common law is that the crime be committed at night. The nighttime hours are the time when a dwelling is likely to be occupied and when individuals are most apt to be resting or asleep and vulnerable to fright and to attack. Perpetrators are also less likely to be easily identified during the nighttime hours. The common law determined whether it was nighttime by asking whether the identity of an individual could be identified in "natural light."

State statutes no longer require that a burglary be committed at night. However, a breaking and entering during the evening is considered an aggravated form of burglary and is punished more severely. States typically follow the rule that night extends between sunset and sunrise or from thirty minutes past sunset to thirty minutes before sunrise. English law defines nighttime as extending from six at night until nine in the morning.[13]

Intent

The common law required that individuals possess an intent to commit a felony within the dwelling at the time that they enter the building. An individual is guilty of a burglary when he or she enters the dwelling, regardless of whether he or she actually commits the crime or abandons his or her criminal purpose. The intent must be concurrent with the entry; it is not a burglary when the felonious intent is formed following entry.

Some judges recognize that individuals who enter a building are guilty of a burglary in the event that they develop a felonious intent after entering into a building and unlawfully break into a secured space, such as an office or dorm room.

Statutes have adopted various approaches to modifying the common law intent standard. Pennsylvania requires an intent to commit a crime.[14] California broadens the intent to include any felony or any misdemeanor theft.[15] The expansion of the intent standard is justified on the grounds that an intrusion into the home is threatening to the occupants regardless of whether the intruder's intent is to commit a felony or misdemeanor.

Aggravated Burglary

Burglary is typically divided into degrees. Aggravated first-degree burglary statutes generally list various circumstances as deserving enhanced punishment, including the nighttime burglary of a dwelling, the possession of a dangerous weapon, or the infliction of injury to others. Second-degree burglary may include the burglary of a dwelling, store, automobile, truck, or railroad car. The least serious grade of burglary typically involves entry with the intent to commit a misdemeanor or nonviolent felony.

Arizona punishes as first-degree burglary the entering of or remaining in a residential or nonresidential structure with the intent to commit a felony or theft while knowingly possessing explosives or a deadly weapon or dangerous instrument. The burglary of a residential structure is a second-degree burglary, and the least serious form of burglary involves a nonresidential structure or fenced-in commercial or residential yard.[16]

Most states also prohibit possession of burglar tools. Idaho punishes as a misdemeanor the possession of a "picklock, crow, key, bit, or other instrument or tool with intent feloniously to break or enter into any building or who shall . . . knowingly make or alter any key . . . [to] fit or open the lock of a building, without being requested so to do by some person having the right to open the same. . . ."[17]

Burglary is a distinct offense and does not merge into the underlying offense. An individual, as a result, may be sentenced for both a burglary and assault and battery or for both a burglary and larceny. Pennsylvania, however, provides that a burglary merges into the offense "which it was his intent to commit after the burglarious entry" unless the additional offense was a serious felony.[18]

Do We Need the Crime of Burglary?

Do we really need burglary statutes? Why not just severely punish a crime committed inside a dwelling or other building?

The commentary to the Model Penal Code points out that punishment for burglary can lead to illogical results. An individual entering a store with the intent to steal an inexpensive item under some statutes would be liable for both burglary and shoplifting. On the other hand, an individual who developed an intent to steal only after having entered the store would only be liable for shoplifting. Does this make sense?

In *State v. Stinton,* Stinton violated an order of protection issued by a judge that prohibited Stinton from harassing his former lover, Tyna McNeill. Stinton nevertheless entered and attempted to remove his personal property from the home the two formerly shared. He was held liable for the misdemeanor of violating the order of protection in addition to the felony of burglary for entering a dwelling with the intent to commit a crime. Stinton unsuccessfully argued that this unfairly transformed his violation of an order of protection into a burglary. Had he confronted McNeill on the street, Stinton would be held liable only for a misdemeanor. Do you agree with Stinton's contention?[19]

On the other hand, burglary statutes recognize that there clearly is a difference in the degree of fear, terror, and potential for violence resulting from an assault or theft in the home as opposed to an assault and theft on the street. Do burglary statutes require reform? Should we return to the common law definition of burglary? The Model Penal Code provides a reformed version of the law of burglary.

The next case, *Bruce v. Commonwealth,* challenges you to creatively apply the breaking and entering requirement to an unusual set of circumstances.

Model Penal Code Section 221.1. Burglary

(1) A person is guilty of burglary if he enters a building or occupied structure, or separately secured or occupied portions thereof, with purpose to commit a crime therein, unless the premises are at the time open to the public or the actor is licensed or privileged to enter. It is an affirmative defense to prosecution for burglary that the building or structure was abandoned.

(2) Burglary is a felony of the second degree (maximum sentence of ten years) if it is perpetrated in the dwelling of another at night, or if, in the course of committing the offense, the actor:

    (a) purposely, knowingly or recklessly inflicts or attempts to inflict bodily injury on anyone; or

    (b) is armed with explosives or a deadly weapon.

Otherwise, burglary is a felony of the third degree (a maximum sentence of five years). An act shall be deemed "in the course of committing" an offense if it occurs in an attempt to commit the offense or in flight after the attempt or commission.

(3) A person may not be convicted both for burglary and for the offense which it was his purpose to commit after the burglarious entry or for an attempt to commit that offense, unless the additional offense constitutes a felony of the first or second degree.

Analysis

- A burglary is limited to an occupied building or structure. The building or structure need not be occupied at the precise moment of the burglary; the important point is that the structure is "normally occupied." There is no breaking and entering requirement. The Model Penal Code does not punish remaining unlawfully on the premises as burglary.
- The Model Penal Code does not include stores open to the public or motor vehicles or railcars.
- A burglary may be committed in a separate portion or unit of a building.
- A burglary involves an intent to commit a "crime" and is not limited to a felony. The burglary is aggravated when perpetrated at night or when it involves the infliction or attempted infliction of bodily harm or in those instances that the perpetrator is armed with explosives or a deadly weapon.
- Most burglaries are punished as felonies in the third degree. The burglary merges into the completed crime unless the underlying offense is a felony in the second degree (maximum sentence of ten years) or serious felony such as rape, violent robbery, or murder (maximum imprisonment for life).

The Legal Equation

Burglary = Breaking and entering or unlawfully remaining or unlawful entry

+ specific intent to commit a felony or crime

+ inside a dwelling or other structure at night and other aggravating factors.

Did the appellant break into the trailer?

Bruce v. Commonwealth, *469 S.E.2d 64 (Va. Ct. App. 1996). Opinion by: Elder, J.*

Facts

Appellant and Deborah Bruce (Deborah), although married, lived in separate residences during late 1993. Deborah lived with the couple's son, Donnie Bruce, Jr. (Donnie), and Donnie's girlfriend at Greenfield Trailer Park in Albemarle County, Virginia. Although appellant stayed with Deborah at the residence during a period of time in September or October of 1993, his name was not on the lease, he was not given a key to the residence, and he did not have permission to enter the residence at the time of the alleged offense.

On December 5, 1993, at approximately 2:00 p.m., Deborah, Donnie, and Donnie's girlfriend left their residence. Earlier that morning, Donnie told appellant that Deborah would not be home that afternoon. Upon departing, Donnie and Deborah left the front door and front screen door closed but unlocked. The front door lacked a knob but had a handle, which allowed the door to be pulled shut or pushed open.

After Deborah, Donnie, and Donnie's girlfriend left their residence, a witness observed appellant drive his truck into the front yard of the residence and enter through the front door without knocking. Appellant testified, however, that he parked his truck in the lot of a nearby supermarket and never parked in front of the residence. Appellant stated that the front screen door was open and that the front door was open three to four inches when he arrived. Appellant testified that he gently pushed the front door open to gain access and entered the residence to look for Donnie.

While preparing to leave the residence, appellant answered a telephone call from a man with whom Deborah was having an affair. The conversation angered appellant, and he threw Deborah's telephone to the floor, breaking it. Appellant stated that he then exited through the residence's back door, leaving the door "standing open," and retrieved a .32 automatic gun from his truck, which was parked in the nearby supermarket parking lot. Appellant returned to the residence through the open back door. Appellant, who testified that he intended to shoot himself with the gun, went to Deborah's bedroom, lay on her bed, and drank liquor.

When Deborah, Donnie, and Donnie's girlfriend returned to their residence, appellant's truck was not parked in the front yard. Upon entering the residence, Donnie saw that someone was in the bathroom, with the door closed and the light on. When police arrived soon thereafter, they found appellant passed out on Deborah's bed and arrested him.

On May 24, 1994, a jury in the Circuit Court of Albemarle County convicted appellant of breaking and entering a residence, while armed with a deadly weapon, with the intent to commit assault. Appellant appealed to this court.

Issue

In order to convict appellant of the crime charged, the Commonwealth had to prove that appellant broke and entered into his wife's residence with the intent to assault her with a deadly weapon. Under the facts of this case, the Commonwealth satisfied this burden.

Reasoning

Breaking, as an element of the crime of burglary, may be either actual or constructive. . . . Actual breaking involves the application of some force, slight though it may be, whereby the entrance is effected. Merely pushing open a door, turning the key, lifting the latch, or resorting to other slight physical force is sufficient to constitute this element of the crime. "Where entry is gained by threats, fraud or conspiracy, a constructive breaking is deemed to have occurred." . . . "[A] breaking, either actual or constructive, to support a conviction of burglary, must have resulted in an entrance contrary to the will of the occupier of the house."

Appellant's initial entry into Deborah's residence constituted an actual breaking and entering. Sufficient credible evidence proved that appellant applied at least slight force to push open the front door and that he did so contrary to his wife's will. However, as the Commonwealth concedes on brief, appellant did not possess the intent to assault his wife with a deadly weapon at this time. . . . The Commonwealth therefore had to prove appellant intended to assault his wife when he reentered the residence with his gun.

We hold that the Commonwealth presented sufficient credible evidence to prove the crime charged. On the issue of intent, the jury reasonably could have inferred that the phone call from Deborah's boyfriend angered appellant, resulting in his destruction of the telephone and the formation of an intent to commit an assault with a deadly weapon upon Deborah. Viewed in the light most favorable to the Commonwealth, credible evidence proved that appellant exited the back door of the residence, leaving the door open, moved his truck to a nearby parking lot, and reentered the residence carrying a gun with the intent to assault Deborah.

Well-established principles guide our analysis of whether appellant's exit and reentry into the residence constituted an actual or constructive breaking. As we stated above, an "actual breaking involves the application of some force, slight though it may be, whereby the entrance is effected." . . . "In the criminal law as to housebreaking and burglary, [breaking] means the tearing away or removal of any part of a house or of the locks, latches, or other fastenings intended to secure it, or otherwise exerting force to gain an entrance, with criminal intent. . . ." Virginia, like most of our sister states, follows the view that "breaking out of a building after the commission of a crime therein is not burglary in the absence of a statute so declaring." . . . In this case, appellant exited the back door of the residence on his way to retrieve the

gun from his truck. In doing so, the appellant did not break for the purpose of escaping or leaving. Rather, by opening the closed door, he broke in order to facilitate his reentry. At the time he committed the breaking, he did so with the intention of reentering after retrieving his firearm. Although appellant used no force to effect his reentry into the residence, he used the force necessary to constitute a breaking by opening the closed door on his way out. . . .

Holding

Sound reasoning supports the conclusion that a breaking from within in order to facilitate an entry for the purpose of committing a crime is sufficient to prove the breaking element of burglary. The gravamen of the offense is breaking the close or the sanctity of the residence, which can be accomplished from within or without. A breaking occurs when an accomplice opens a locked door from within to enable his cohorts to enter to commit a theft or by leaving a door or window open from within to facilitate a later entry to commit a crime. . . . Accordingly, a breaking occurred when appellant opened the back door of the victim's residence, even though the breaking was accomplished from within. Thus, because the evidence was sufficient to prove an intent to commit assault at the time of the breaking and the entering, the Commonwealth proved the elements of the offense. Thus, we affirm appellant's conviction.

Questions for Discussion

1. The Virginia burglary statute, section 18.2-91, punishes any person who, in the daytime, breaks and enters a dwelling house with the intent to commit larceny, assault and battery, or enters with the intent to commit any felony other than murder, rape, or robbery. Is the decision of the Virginia appellate court clearly dictated by the language of this statute? Is the court's judgment consistent with the public policy underlying the crime of burglary?

2. Why did the appellant's initial entry not constitute a burglary?

3. Would the case have turned out differently in the event that the back door was wide open when Bruce left and reentered the trailer?

4. How would you decide this case?

Hitt v. Commonwealth raises the issue of the definition of a dwelling under burglary statutes.

Was *Hitt's* breaking and entry into the bedroom a burglary?

Hitt v. Commonwealth, 598 S.E.2d 783
(Va. Ct. App. 2004). Opinion by: Humphreys, C.J.

Andy Dale Hitt appeals his conviction . . . for statutory burglary. . . . Hitt contends that the trial court erred in finding the evidence sufficient, as a matter of law, to support the conviction because the

Commonwealth failed to establish that he broke and entered a dwelling house, with the intent to commit larceny. We agree and reverse Hitt's conviction for statutory burglary.

Facts

On the evening of May 22, 2002, Hitt spent the night at a friend's home. Hitt's friend, Keith, lived at the home with his father, John Burner, as well as his sister, Cara, and her minor son. Burner consented to Hitt's spending the night at the home. Hitt spent the evening, as he had on prior occasions, in the guest bedroom, a converted carport on the first floor of the home. Burner and the others slept in their bedrooms on the second floor of the home.

On the morning of May 23, 2002, Burner had approximately $3,000 in cash, on top of his bedroom dresser. For that reason, Burner locked his bedroom door, by means of an outside lock, when he left for work that morning. Before he left the home, however, Burner went to the guest room and woke up Hitt. Burner asked Hitt if he was going to work that morning, and Hitt replied, "In a little bit." Burner told Hitt not to "oversleep" and left for work. Keith had already left for work.

Sometime after Burner left the home, Cara asked Hitt to take her son to the child's grandmother's house. Hitt did so, then returned to the Burner home. At that time, Cara was still there. Hitt fell asleep "on the couch" in the guest bedroom for about "a half an hour." When he woke up, Cara had already left for work and Hitt was alone in the home.

Hitt then went upstairs to Burner's bedroom and tried to open the door. When the door would not open, Hitt used his body weight to force the door open. Hitt used enough force to open the locked door and to "knock" "a little piece of paneling" "out of place." Hitt found the money on Burner's dresser, took it, and left the home.

On June 5, 2002, Page County Sheriff's Department Investigator Rebecca Hilliard questioned Hitt about the burglary. Hitt admitted to taking the money. . . . On March 19, 2003, Hitt pled guilty to grand larceny but proceeded to . . . trial on the burglary charge.

During the trial, Hitt moved to strike the Commonwealth's evidence, arguing that the Commonwealth's own evidence proved that he had consent to be in the residence that morning and that the Commonwealth failed to establish he had broken into a "separate residence" by breaking into Burner's locked bedroom. Hitt also argued the Commonwealth failed to produce sufficient evidence that he broke into Burner's locked bedroom with the intent to commit larceny.

The trial court denied Hitt's motions, finding: "I think, by analogy . . . to the cases of secreting one's person, I think that, under the common law, an area, even though it may be on the interior of a dwelling house, which is clearly marked and delineated as being off bounds to a guest in the home, would be a sufficient breaking and entering of a dwelling house to sustain a conviction." . . . The court thus found Hitt guilty of burglary on May 7, 2003 and sentenced Hitt to a total of ten years in prison, with nine years suspended upon certain conditions.

Reasoning

Section 18.2-90 provides:

> If any person in the nighttime enters without breaking or in the daytime breaks and enters or enters and conceals himself in a dwelling house or an adjoining, occupied outhouse or in the nighttime enters without breaking or at any time breaks and enters or enters and conceals himself in any office, shop, manufactured home, storehouse, warehouse, banking house, church . . . or other house . . . with intent to commit murder, rape, robbery or arson in violation of . . . he shall be deemed guilty of statutory burglary, which offense shall be a Class 3 felony. However, if such person was armed with a deadly weapon at the time of such entry, he shall be guilty of a Class 2 felony.

Section 18.2-91 provides as follows:

> If any person commits any of the acts mentioned in § 18.2-90 with intent to commit larceny, or any felony other than murder, rape, robbery or arson . . . or if any person commits any of the acts . . . with intent to commit assault and battery, he shall be guilty of statutory burglary, punishable by confinement in a state correctional facility for not less than one or more than twenty years or, in the discretion of the jury or the court trying the case without a jury, be confined in jail for a period not exceeding twelve months or fined not more than $2,500, either or both. However, if the person was armed with a deadly weapon at the time of such entry, he shall be guilty of a Class 2 felony.

> To sustain a conviction for statutory burglary . . . the Commonwealth must [thus] prove: (1) the accused . . . broke and entered the dwelling house in the daytime; and (2) the accused entered with the intent to commit any felony [other than murder, rape, robbery, or arson].

Issue

On appeal, Hitt contends that the Commonwealth failed to present sufficient evidence to establish that he "unlawfully" "broke and entered [a] dwelling house in the daytime." Specifically, Hitt contends that he had permission to be in Burner's residence on the morning of May 23, 2002, and that such permission necessarily extended to Burner's locked bedroom. Consistent with this argument, Hitt contends that a bedroom within a dwelling cannot constitute a separate "dwelling house." . . . As an alternative argument, Hitt contends the Commonwealth failed to produce sufficient evidence to show that he "entered" Burner's locked bedroom with the intent to commit larceny. Because we find that the Commonwealth's evidence failed to establish that Hitt unlawfully broke and entered a "dwelling house," . . . we do not reach the second issue.

Reasoning

"At common law, [burglary was] primarily an offense against the security of the habitation, and that is still the general conception of it." . . . ."Burglary laws are based primarily upon a recognition of the dangers to personal safety created by the usual burglary situation—the danger that the intruder will harm the occupants in attempting to perpetrate the intended crime or to escape and the danger that the occupants will in anger or panic react violently to the invasion, thereby inviting more violence." . . .

The Virginia Code expands notions of traditional common law burglary to include, among other things, entry by "breaking in the daytime of any dwelling house." . . . It is well settled that "[a] breaking, . . . may be either actual or constructive. An actual breaking involves the application of physical force, however slight, to effectuate the entry." . . . In the case at bar, there can be no question that Hitt's conduct, when considered in a "vacuum," constituted a "breaking" into Burner's locked bedroom. Indeed, Hitt conceded that he had to apply some amount of force to the door in order to open it and enter the room.

Specifically, we must determine whether Hitt's conduct, when considered under the totality of the circumstances presented here, could constitute a "breaking" of a "dwelling house," when there is no dispute that Hitt was on the premises of Burner's home with either Burner's consent or Cara's consent. In fact, no evidence was presented that suggested that Hitt "broke" into the home after he returned from delivering Cara's son to the child's grandmother's home.

We have held . . . that "in enacting Code § 18.2-89, the legislature intended to preserve the crime of common law burglary as an offense against habitation." . . . Accordingly, we found that "the term 'dwelling house' . . . means a place which human beings regularly use for sleeping." We further held that

a "house remains a dwelling house so long as the occupant intends to return [to it for that purpose]." . . . Consistent with this definition, we find it clear that the place of habitation on the facts presented here was Burner's home as a whole, not his bedroom within his home.

We have recognized that all "dwelling houses must have an 'occupant' in order to satisfy the definition of 'dwelling house'" and that "all dwelling houses are necessarily 'occupied' in the sense that they are regular residences." . . . It does not contemplate individual rooms or compartments within such a "residence" that are not "dwelling houses" in and of themselves (such as a rented room within a larger dwelling, intended to be the place of habitation/residence for the individual residing therein). . . .

It is of no moment that Burner's bedroom was a place that he regularly used for sleeping. That is but one of the indicia utilized to determine whether a given structure is actually a "dwelling house" or a "regular residence" in which human beings habitate. . . . Indeed, while habitation or occupancy necessarily includes sleeping, it clearly also includes other "dwelling-related" activities, such as preparing and consuming meals, bathing, and other day-to-day activities traditionally associated with "habitation."

This conclusion also comports with the common law application of burglary in similar contexts. Specifically, breaking and entering of private "dwelling houses" used as residences. . . . Although the legislature may, and often has, extended the traditional common law notion of burglary of a "dwelling house," it has not chosen to extend this definition to rooms or compartments within a private "dwelling house," which do not constitute separate residences in and of themselves.

Holding

We thus decline the Commonwealth's invitation to extend the definition in this manner by judicial fiat. Accordingly, because we find that there is no evidence in the record suggesting that Hitt "broke" and entered Burner's home on the morning of May 23, 2002, we reverse Hitt's conviction for statutory burglary and dismiss the indictment.

Questions for Discussion

1. Why does Hitt argue that he is not guilty of burglary? What is the prosecution's argument? Explain the appellate court's reasoning in acquitting Hitt. Did Hitt possess the required criminal intent?

2. Can you interpret the language of the statute to support the trial court's conviction of Hitt? How would you amend or change the statute to strengthen the prosecution's case?

3. What if, after returning from taking Cara's son to the child's grandmother's house, Hitt found the house locked and broke and climbed in a window and thereafter entered Burner's bedroom? Would this be a burglary?

4. List several hypothetical examples of situations that will not constitute burglary under the appellate court's decision.

5. How would you rule as a judge in this case?

You Decide **12.1** Defendant John Martin Sandoval is an alcoholic. He started drinking while watching a football game, and after a twelve-pack of beer, he walked to a bar and drank an additional beer. He does not remember leaving the bar or any other location until he awoke in jail later in the day for first-degree burglary. Sandoval was found to have walked to a stranger's home at 3:20 a.m. and kicked in the front door of the home of Christiansen, a reserve deputy with the police for twelve years. Christiansen demanded to know what Sandoval was "doing in my house." Sandoval responded, "Who are you?" and shoved Christensen in the chest, knocking him back a few feet. Christensen then punched Sandoval in the head, wrestled him to the floor, and held him down until the police arrived. The Washington burglary statute declares that it is burglary in the first degree if, "with the intent to commit a crime against a person or property therein," an individual "enters or remains unlawfully in a building" and if, while entering, "[t]he actor . . . assaults any person." Sandoval and Christiansen had never met. Sandoval did not have burglary tools or take any of Christensen's property. A Washington appellate court reversed Sandoval's conviction. How would you rule? See *State v. Sandoval*, 94 P.3d 323 (Wash. Ct. App. 2004).

You can find the answer at www.sagepub.com/lippmanccl3e.

You Decide **12.2** A neighbor observed Valencia remove a window screen from a bathroom window of the Floreas' home and unsuccessfully attempt to open the window. The defendant a few minutes earlier had pulled a window screen off a bedroom window and unsuccessfully attempted to open that window.

The neighbor watched as Valencia unsuccessfully tried to open the front door. The defendant "banged on the wall in frustration and then sat down for a few minutes." Valencia subsequently was convicted of burglary and appealed his conviction. Under California law, the crime of burglary is committed when a person "enters any . . . building," including a "house," "with intent to commit . . . larceny or any felony." As a judge how would you decide this case? See *People v. Valencia*, 46 P.3d 920 (Cal. 2002).

TRESPASS

Criminal trespass is the unauthorized entry or remaining on the land or premises of another. The *actus reus* is entering or remaining on another person's property without his or her permission. An example is disregarding a "no trespassing" sign and climbing over a fence in order to swim at a private beach. You also may commit a trespass when you swim with the owner's permission and then disregard his or her request to leave.

A **defiant trespass** occurs when an individual knowingly enters or remains on a premises after receiving a clear notice that he or she is trespassing. Keep in mind that the police, firefighters, and emergency personnel are privileged to enter any land or premises.

Criminal trespass entails an unauthorized entry, and unlike burglary, there is no requirement that the intruder intend to commit a felony. Another important point is that statutes punish a trespass on a broad range of private property. The Texas statute provides that an individual commits a trespass who "knowingly and unlawfully" enters or remains in the dwelling "of another" as well as in a motor vehicle, hotel, motel, condominium, or apartment building or on agricultural land. The federal and many state governments also have special statutes that punish trespass in schools, military facilities, and medical facilities.

Most statutes require that an owner provide notice that an entry is prohibited or that individuals specifically receive and ignore a notice to leave premises. The Texas statute states that notice requires oral or written communication, fencing, or other enclosures designed to exclude intruders or to contain livestock; posted signs forbidding entry that are reasonably likely to come to the attention of intruders; special identifying purple paint marks on trees or posts; or the visible cultivation of crops fit for human consumption.[20]

Statutes typically divide trespass into various degrees. First-degree criminal trespass typically entails entering or remaining in the dwelling of another and is punished as a minor felony,

Figure 12.1 Crime on the Streets: Burglary Rates

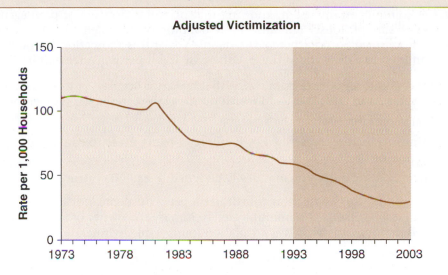

- Burglary accounts for roughly 18% of the FBI Crime Index offenses and 20% of all property crimes and, after a steep decline, has recently experienced a small increase.
- Forcible entry burglaries comprise over 60% of all burglaries. Unlawful entry burglary accounts for roughly 30% of burglaries.
- Residential burglaries are roughly 65% of all burglaries and approximately 34% are nonresidential, including stores and offices.
- Over 60% of residential burglaries occur during daytime hours. A majority of nonresidential burglaries are perpetrated at night.
- The total financial loss due to burglary is estimated at $3.3 billion per year. Residential burglaries result in an average loss of $1,482 per offense; nonresidential burglaries average $1,678 per offense.
- Individuals under 18 account for roughly 17% of all arrests for burglary.

Source: The National Crime Victimization Survey, U.S. Department of Justice.

carrying a jail sentence of between a number of months and several years along with a fine. Second-degree criminal trespass involves entering or remaining in enclosed buildings or fenced-in property and is typically punished as a misdemeanor resulting in imprisonment for up to several months as well as by a fine of several thousand dollars. Third-degree criminal trespass is usually categorized as a petty misdemeanor and entails entering in or remaining on unenclosed land and is punished with a fine. Some states punish all criminal trespass as a misdemeanor.[21]

A recent development in the law of trespass is the felony of *computer trespassing.* New York's law punishes an individual who "intentionally and without authorization" accesses a computer, computer system, or network with the intent to delete, damage, destroy, or disrupt a computer, computer system, or computer network.[22] Colorado does not require an intent to tamper with a computer system and merely requires a knowing and unauthorized use of a computer or intruding without authorization onto a computer system or computer network.[23] In *Fugarino v. State,* the defendant acted angrily after learning that he had been fired from his job as a computer programmer. Fugarino announced that he was going to make certain that the company would "never get to make any money" from his design work. He immediately used a company computer to delete data from the firm's computer system and added layers of password protection to prevent his employer from gaining access to the program he designed. Fugarino was convicted of having "used a computer, owned by his employer, with knowledge that such use was without authority and with the intention of removing programs or data from the computer."[24] Can you think of other ways of applying the law of trespass to modern technology?

Model Penal Code Section 221.2. Criminal Trespass

(1) A person commits an offense if, knowing that he is not licensed or privileged to do so, he enters or surreptitiously remains in any building or occupied structure, or separately secured or occupied portion thereof. An offense under this Subsection is a misdemeanor if it is committed in a dwelling at night. Otherwise it is a petty misdemeanor.

(2) A person commits an offense if, knowing that he is not licensed or privileged to do so, he enters or remains in any place as to which notice against trespasser is given by:

(a) actual communication to the actor; or

(b) posting in a manner prescribed by law or reasonably likely to come to the attention of intruders; or

(c) fencing or other enclosure manifestly designed to exclude intruders.

An offense under this Subsection constitutes a petty misdemeanor if the offender defies an order to leave personally communicated to him by the owner of the premises or other authorized person. Otherwise it is a violation (punishable by fine).

(3) It is an affirmative defense to prosecution under this Section that:

(a) a building or occupied structure involved in an offense under Subsection (1) was abandoned; or

(b) the premises were at the time open to members of the public and the actor complied with all lawful conditions imposed on access to or remaining in the premises; or

(c) the actor reasonably believed that the owner of the premises or other persons empowered to license access thereto, would have licensed him to enter or remain.

Analysis

- An accused is guilty of trespass and a petty misdemeanor in the event that the accused knows that he or she lacks permission to enter and nevertheless enters or surreptitiously (hiding) remains in any building or occupied structure. This is a misdemeanor if committed in a dwelling at night and is a petty misdemeanor if committed during the daytime.
- It is a violation (fine) to enter any other "place" without authorization in which a notice against trespass is posted or in which a prohibition against trespass is clear from the enclosure surrounding the area. This is a petty misdemeanor where the trespasser defies an order personally communicated to him.
- The code requires knowledge of trespass. An individual who accidentally enters on property or mistakenly believes that he or she possesses authorization to enter or remain upon property is not guilty of a trespass.
- There are three affirmative defenses to trespass.

The Legal Equation

Criminal trespass = Entry or remaining on the property of another without authorization

+ purposely, knowingly, strict liability.

You Decide

12.3 J.L. stole a go-cart, a four-wheeler, and a skateboard from the victim's yard. "These items were leaning against the side of the victim's residence." The victim's mother testified that although the yard was not completely enclosed, there was a fence "in the back" and a fence "between . . . my house and my neighbor's house."

In Florida, in order to prove burglary, the State is required to establish that J.L. entered a "dwelling" with the intent to commit an offense. A "dwelling" is defined in Florida to include the "curtilage," which is the area immediately surrounding the home. J.L. claims that he was guilty of trespass rather than burglary. Do you agree? See *Child v. State*, 57 So. 3d 924 (Fla. Dist. Ct. App. 2011).

You can find the answer at www.sagepub.com/lippmanccl3e.

You can find more cases on the study site: State v. Chatelaine, *www.sagepub.com/lippmanccl3e.*

ARSON

Common law **arson** is defined as the willful and malicious burning of the dwelling house of another. The purpose is to protect the home along with the occupants and their possessions. Blackstone justifies treating arson as a felony punishable by death on the grounds that murder "seldom extends beyond the felonious act designed," but "fire too frequently involves in the common calamity persons unknown to the incendiary and not intended to be hurt by him . . . friends as well as enemies."[25] Common law arson has been substantially modified by state statutes.

Burning

The common law requires a burning. This is commonly defined as the "consuming of the material" of the house or the "burning of any part of the house." The burning is not required to destroy the structure or seriously damage the home. The burning is required to affect only a small portion of the dwelling, no matter how insignificant or difficult to detect. Even a small "spot" on the floor is sufficient.[26] Professor Perkins quotes a decision that establishes that the test for burning is whether "the fiber of the wood or other combustible material is charred, and thus destroyed" by fire.[27]

For a deeper look at this topic, visit the study site.

The burning need not involve an actual flame and need merely result in a "charring" of the structure. This does not include "soot," "smudging," "blackening or discoloration or shriveling from heat" or "smoke damage." The common law did not consider an explosion as arson unless the combustion resulted in a fire.[28]

The trend is for state statutes and courts to broadly interpret arson statutes and to find that even smoke damage and soot are sufficient.[29] A New Jersey statute defines third-degree arson without requiring damage and provides that an individual commits arson when he or she "purposely starts a fire and recklessly places a person in danger of death or bodily injury or recklessly places a building or structure in danger of damage or destruction."[30] Florida explicitly punishes a fire or explosion that damages or causes to be damaged an occupied structure as well as other similar structures.[31]

Dwelling

Arson at common law must be committed against a dwelling. This is defined by the familiar formula as a place regularly used for sleeping. The definition reflects the fact that criminal laws against arson are designed to protect individuals and their right to the peace and security in the home. The occupants may be absent at the time of the arson, so long as the structure is regularly used for sleep. The burning of a building under construction or of an abandoned building does not constitute arson. The definition of dwelling extends to all structures within the curtilage, the area immediately surrounding the home. This includes a barn, garage, or tool shed.

Statutes no longer limit arson to a dwelling. Illinois, in addition to prohibiting residential arson, punishes damage to real property (buildings and land) and to personal property (e.g., personal belongings). Aggravated arson is directed against injury to individuals resulting from the arson of "any building or structure, including any adjacent building . . . including . . . a house trailer, watercraft, motor vehicle or railroad car." Statutes that include personal property extend arson to the burning of furniture in a house regardless of whether the fire damages the dwelling.[32]

This broadening of the coverage of arson statutes means that laws against arson no longer protect only the home. Arson now protects buildings and vehicles in which individuals are likely to spend a portion of the day and is intended to fully combat the danger posed by fire to human life and property.

Dwelling of Another

The common law required that the burned dwelling was occupied by another individual. As with burglary, the central issue is occupancy rather than ownership. A tenant would not be guilty of arson for burning his or her rented apartment that is owned by the landlord; the landlord would be guilty of arson for burning the house that he or she owns and rents to the tenant. A husband would not be guilty of arson for burning the home he shares with his wife.

Modern statutes have eliminated the requirement that the arson must be directed at the dwelling "of another." Florida holds an individual liable for arson in the first degree "whether the property [is] of himself or herself or another."[33] Courts have reasoned that a fire poses a threat to firefighters, as well as to the neighbors, and have held that it is not an unreasonable limitation on property rights to hold a defendant liable for burning his or her own property. Do you agree?

For an international perspective on this topic, visit the study site.

Willful and Malicious

The *mens rea* of common law arson is malice. This does not require dislike or hatred. Malice in arson entails either a purpose to burn or a knowledge that the structure would burn or the creation of an obvious fire hazard that, without justification or excuse, damages a dwelling. An "obvious fire hazard" is created when an individual recklessly burns a large pile of dry leaves on a windy day and in the process creates an unreasonable hazard that burns a neighbor's house. A negligent or involuntary burning does not satisfy the requirement for common law arson.

State statutes typically retain the common law intent standard and include language such as "willfully and maliciously." Separate statutes often punish a reckless burning. A number of states also punish a burning committed by an individual with the specific intent to defraud an insurance company.

Grading

State statutes are typically divided into arson and aggravated arson. Some states provide for additional categories. Washington provides for knowing and malicious arson and aggravated arson, as well as for reckless burning. The Washington statute categorizes arson as aggravated based on various factors, including causing a fire or explosion that damages a dwelling or that is dangerous to human life. Aggravated arson also includes causing a fire or explosion on property valued at $10,000 or more with the intent to collect insurance.[34] Washington state punishes aggravated arson by life imprisonment, along with a possible fine of up to $50,000, while arson is punishable by ten years, by a fine of up to $20,000, or by both confinement and a fine.[35] California enhances the punishment of "willful and malicious" burning and of "reckless" burning when the perpetrator has been previously convicted of either offense, a police officer or firefighter is injured, more than one victim suffers great bodily injury, multiple structures are burned, or the defendant employed a device designed to accelerate the fire.[36]

The next case, *In re Jesse L.*, discusses the type of fire damage to a building required to convict a defendant of arson.

Model Penal Code

Section 220.1. Arson and Related Offenses

(1) Arson. A person is guilty of arson, a felony of the second degree, if he starts a fire or causes an explosion with the purpose of:

   (a) destroying a building or occupied structure of another; or

   (b) destroying or damaging any property, whether his own or another's, to collect insurance for such loss. It shall be an affirmative defense . . . that the actor's conduct did not recklessly endanger any building or occupied structure of another or place any other person in danger of death or bodily injury.

(2) Reckless Burning or Exploding. A person commits a felony of the third degree if he purposely starts a fire or causes an explosion whether on his own property or another's, and thereby recklessly:

    (a) places another person in danger of death or bodily injury; or

    (b) places a building or occupied structure of another in danger of damage or destruction.

(3) Failure to Control or Report Dangerous Fire. A person who knows that a fire is endangering a life or a substantial amount of property of another and fails to take reasonable measures to put out or control the fire, when he can do so without substantial risk to himself, or to give a prompt fire alarm, commits a misdemeanor if:

    (a) he knows that he is under an official, contractual, or other legal duty to prevent or combat the fire;

    (b) the fire was started . . . lawfully, by him or with his assent, or on property in his custody or control.

(4) Definitions. "Occupied structure" means any structure, vehicle or place adapted for overnight accommodation of persons, or for carrying on business therein, whether or not a person is actually present. Property is that of another, for the purposes of this section, if anyone other than the actor has a possessory or proprietary interest therein. If a building or structure is divided into separately occupied units, any unit not occupied by the actor is an occupied structure of another.

Analysis

- The Model Penal Code in section 220.1(1)(a) defines arson in terms of starting a fire or causing an explosion with the purpose of destroying a building or occupied structure of another.
- Directing punishment at individuals who start or cause a fire or explosion results in their being held liable for arson despite the fact that the fire is extinguished before damage results.
- Arson is punishable by a maximum of ten years in prison under the Model Penal Code. This would be in addition to punishment for any resulting injury to individuals.
- The requirement of a purpose to destroy a building or occupied structure or to destroy or damage property means that a specific intent is required for arson.
- The commentary states that the terms *building* and *occupied structure* are intended to refer to structures that are capable of occupancy. This restricts arson to fires or explosions dangerous to the life of inhabitants and firefighters. An individual need not be actually present in the dwelling.
- Arson to defraud in section 220.1(1)(b) includes property owned by the defendant as well as another. There must be an intent to defraud an insurance company, and this provision includes all types of property.
- An individual is not liable for arson where the property of another or other persons is not endangered. This is designed to avoid the harsh penalties for arson when another person or his or her property is not recklessly endangered.
- Reckless burning or exploding is punishable by five years in prison. There is no requirement of a purpose to destroy a structure.
- The duty to undertake affirmative action to prevent and control fires is imposed.

The Legal Equation

Arson $=$ Setting fire to dwelling (other structures under state statutes) or structures in cartilage

    $+$ intent to burn, knowing will burn, reckless creation of risk of burning

    $+$ burning of dwelling (smoke damage is sufficient under state statutes).

Did the defendant commit arson when he started a fire that caused minor charring of the structure of the high school as well as damage to a light fixture?

In re Jesse L., 270 Cal. Rptr. 389 (Cal. Ct. App. 1990). Opinion by: Martin, J.

Issue

On May 9, 1989, a petition was filed in the Fresno County Superior Court alleging appellant, Jesse L., a minor, came within Welfare and Institutions Code section 602 in that the minor had committed two acts which, if committed by an adult, would constitute felonies. It was alleged in count I that on or about May 3, 1989, appellant committed arson by the burning of a structure known as Roosevelt High School in violation of Penal Code section 451, subdivision (c), causing in excess of $100,000 in damage as provided in section 12022.6, subdivision (b). Count II alleged appellant had participated in a criminal street gang on that date in violation of section 186.22.

On July 10, 1989, appellant was adjudged a ward of the court and committed to the California Youth Authority (CYA). The maximum sentence, if sentenced as an adult, was determined to be eight years. Appellant filed a timely notice appealing his arson conviction.

Facts

On Wednesday, May 3, 1989, at approximately 3:45 a.m., the fire alarm at Roosevelt High School in Fresno was activated. Firefighters were in the process of extinguishing the two-alarm fire, which involved the administration building, when John Salveson arrived to investigate for the fire department. He later testified that entry into the administration building had been obtained by breaking a window next to a door, reaching around, and pressing the panic bar on the locked door, allowing it to open. Appellant's fingerprints and palm prints were found in three different places on the window. Salveson determined that three fires had been ignited, two on the tops of desks and one on the floor, by heaping files and papers, dousing them with a flammable liquid and igniting them.

The assistant principal at Roosevelt High School, Frank M. Hernandez, testified that it had been estimated that the cost of the "equipment and supplies that were destroyed as well as the damages to the structure itself" was approximately $250,000.

Between 7:45 and 8 a.m. on that same day, May 3, 1989, appellant approached a group of students waiting for classes to begin at the school and claimed responsibility for the fire. He also indicated that his brother, a friend by the name of "Snoopy," and a girl by the name of Latitia were involved.

Several of these students testified against appellant. At least one in the group thought appellant was joking.

About an hour and a half after school began, the fire was announced over the loudspeaker, as was a $5,000 reward offer for information leading to the arrest of the arsonist. . . .

Reasoning

The relevant statute in this case reads:

> Subdivision (c) of Section 451. A person is guilty of arson when he or she willfully and maliciously sets fire to or burns or causes to be burned or who aids, counsels or procures the burning of, any structure, forest land or property. . . .
>
> (c) Arson of a structure or forest land is a felony punishable by imprisonment in the state prison for two, four, or six years. . . .
>
> (d) Arson of property is a felony punishable by imprisonment in the state prison for 16 months, two, or three years. . . .

Acknowledging that the evidence supports the burning of property in violation of section 451, subdivision (d), appellant claims there was insufficient evidence presented to establish that a structure as required by subdivision (c) was burned. It is well established that the setting of a fire which does not burn the structure itself does not violate this subsection of the statute.

In *People v. Haggerty* (1873) 46 Cal. 354, the California Supreme Court stated: Upon the question of what is a sufficient burning to constitute the crime, Mr. Bishop states the rule thus:

> The word "burn" . . . means to consume by fire. If the wood is blackened, but no fibers are wasted, there is no burning; yet the wood need not be in a blaze. And the burning of any part, however small, completes the offense, the same as of the whole. Thus, if the floor of the house is charred in a single place, so as to destroy any of the fibers of the wood, this is a sufficient burning in a case of arson.

In *Haggerty,* the Supreme Court concluded that there was evidence tending to show that a spot was on the floor and that the defendant accordingly was guilty of arson: "[The spot was c]harred, so as to destroy the fibers of the wood by the fire set by the defendant . . . To some of the witnesses, it is true, the spot only appeared to be blackened, and not charred. But we cannot say that the verdict was so contrary to the evidence as to justify us in reversing the judgment on that account."

People v. *Simpson* (1875) 50 Cal. 304 also held that evidence of a wooden partition inside a building, and annexed to it, charred by fire and burned through in one place, was sufficient evidence of burning to constitute arson. . . .

Almost all of the evidence of the nature and extent of the fire in this case was given by fire investigator John Salveson. He testified he had not observed any direct flame damage on the exterior of the building. Inside the building he observed fire damage on two desk tops, computers, typewriters, telephones, and other items of property and paper had been damaged by the fire. Plastic light covers had melted and the "formica covered wood" of a counter top had been charred by fire. "[B]urn patterns" were discovered on the floor on the very bottom edge of the counter and on the face of the doors on that wall to the offices.

In our view, Salveson's uncontradicted testimony of "burn patterns" on the floor of the building and the bottom edge of the counter and on the face of the doors was sufficient to establish that an area on the floor and a door were charred so as to destroy the fibers of the wood by the fire set by the appellant. Mr. Salveson's testimony clearly distinguished between smoke damage and these so-called "burn patterns." To illustrate, he used the term "burn pattern" to describe the burning of both the desks and the counter which

he specifically testified were charred as opposed to "smoke stains" around the entrance doors.

Moreover, Salveson also testified the "plastic covers for the lights had melted and some of them had fallen on the ground, others were hanging from the metal framework that they were originally positioned in." Thus, a further question is presented whether the destruction by fire of a light fixture is sufficient evidence of structural fire damage to support a conviction of arson.

Section 660 of the California Civil Code defines a fixture as "[a] thing is deemed to be affixed to land when it is attached to it by roots, as in the case of trees, vines, or shrubs; or embedded in it, as in the case of walls; or permanently resting upon it, as in the case of buildings; or permanently attached to what is thus permanent, as by means of cement, plaster, nails, bolts or screws. . . ." Thus, a fixture is a thing, originally personal property, but later affixed or annexed to realty, so that it is considered real property. . . .

Holding

Our Civil Code . . . supports the conclusion that a fixture (i.e., personal property affixed to the realty so that it becomes an integral part of the structure) becomes part of the structure to the extent that a burning or charring or destruction by fire is all that is required to constitute a burning sufficient to support a conviction of arson under section 451, subdivision (c). Thus, we conclude that the evidence of "burn patterns" which indicates at least minor charring of the structure together with the damage to the light fixtures is ample evidence of a violation of section 451, subdivision (c).

Questions for Discussion

1. What type of damage to a structure is required for arson?

2. Define a "fixture." Why is the destruction of a fixture considered arson?

3. Explain how the law of arson differentiates between burning, charring, scorching, and smoke damage. Does it make sense to distinguish between these different categories of fire-related damage?

You Decide

12.4 Arbie Jo Buckley poured kerosene onto her husband, George House, as he lay on the sofa. The kerosene soaked into his clothes and onto the sofa and floor of the trailer home. Arbie then threw a match onto George. He awoke to find that he was on fire, and he later died. The fire subsequently spread to the sofa and trailer home. Arbie was convicted of felony murder, carrying the death penalty. The Mississippi statute required a finding that George had been killed during the course of a felony, one of which is arson. Was George killed during the course of an arson? Would you affirm or reverse Arbie's conviction? See *Buckley v. State*, 875 So. 2d 1110 (Miss. Ct. App. 2004).

You can find the answer at www.sagepub.com/lippmanccl3e.

You Decide **12.5** A group of homeless persons erected a structure for overnight sleeping. The sidewalls consisted of unmovable fences. The remaining two walls consisted of carpets draped over clotheslines that extended between the two fences. The entrance was covered by shower curtains and blankets and the shelter was covered by a 30-by-50-foot tarp. Occupants slept in sleeping bags or on mattresses. Electricity was supplied by an extension cord that was connected to a light socket at a nearby subway station. Does a fire initiated by Fox against the structure constitute arson against a building under New York law? *See People v. Fox,* 771 N.Y.S.2D 156 (N.Y. APP. DIV. 2004).

You can find the answer at www.sagepub.com/lippmanccl3e.

CRIMINAL MISCHIEF

The common law misdemeanor of malicious mischief is defined as the destruction of, or damage to, the personal property (physical belongings) of another. The Model Penal Code refers to this offense as **criminal mischief**, and under modern statutes, criminal mischief includes damage to both personal and real (land and structures) **tangible property** (physical property as opposed to ownership of intangible property, such as ownership of a song or the movie rights to a book). The offense is directed against interference with the property of another and punishes injury and destruction to an individual home or personal possessions.

Malicious mischief under most statutes is a minor felony and the punishment is reduced or increased based on the dollar amount of the damage. A sentence may also be increased when the damage is directed against a residence or interferes with the delivery of essential services, such as phone, water, or utilities.

Actus Reus

The Model Penal Code specifies that criminal mischief is composed of three types of acts:

1. *Destruction or Damage to Tangible Property.* Injury to property, including damage by a fire, explosion, flood, or other harmful force.

2. *Tampering With Tangible Property so as to Endanger a Person or Property.* Interference with property that creates a danger, for example, the removal of a stop sign or one-way road sign.

3. *Deception or Threat Causing Financial Loss.* A trick that dupes an individual into spending money. An example is sending a telegram falsely informing an individual that his or her mother is dying in a distant city, causing the individual to spend several hundred dollars on an unnecessary plane flight.

CRIME IN THE NEWS

On January 10, 2011, thousands of Auburn University football fans gathered at historic Toomer's Corner to celebrate the team's national college championship. Toomer's Corner has been called the Times Square or center of Auburn University, the site where students, alumni, and fans traditionally have gathered to mark major football victories. The corner is dotted with historic 130-year-old oak trees, which students roll in toilet paper as a traditional part of their celebrations.

On January 28, Auburn officials discovered that a herbicide had been applied in "lethal amounts" to the area surrounding two trees. The poisoning was discovered following a call to a Birmingham radio station from "Al from Dadeville" who claimed that he used "Spike 80DF" to poison the trees and that the trees "definitely will die." "Al" proclaimed that he was a dyed-in-the-wool University of Alabama Crimson Tide fan and that he had poisoned the trees following the annual Iron Bowl, in which Auburn scored 28 straight points and overcame what seemed like an insurmountable 24-point Alabama lead. Al signed off by exclaiming "Roll Damn Tide."

Jay Gogue, the President of Auburn, responded to the poisoning by vowing: "We will take every step we can to save the Toomer's oaks, which have been the home of countless celebrations and a symbol of the Auburn spirit for generations of students, fans, alumni, and the community."

In Alabama, football is an obsession, and the Alabama–Auburn rivalry is all-consuming. There had been isolated acts of arson and violence over the years, although none of these incidents rivaled the poisoning of trees in Toomer's Corner.

A police investigation led to the arrest of Harvey Almorn Updyke, 62, a resident of Dadeville, Alabama, who is a retired Texas State Trooper. Updyke explained that he believed that Auburn was paying outstanding athletes "under the table" to play football and that he was enraged by the gloating of Auburn fans on the radio. He also alleged that he had seen photos of Auburn fans celebrating following the death of beloved Alabama coach Paul "Bear" Bryant.

Updyke initially characterized the tree poisoning as the type of prank that is a traditional part of college football rivalries. A grand jury charged Updyke with six criminal counts, including two counts of the felony of criminal mischief; and he may face additional federal charges for environmental crimes. Updyke's initial defiance and defense of his actions gradually gave way to a sense of remorse and regret. He stated that he had "done a lot of good things" and he did not want to go to his grave with "Harvey the tree poisoner" as his legacy. He stated that as a Texas trooper, he had arrested a record number of drunk drivers and that he also had been responsible for a significant number of drug busts.

Updyke explained that his entire life people had told him that he cared too much about Alabama football and that he "just had too much 'Bama in me." He admitted that he was a "very unhealthy Alabama fan. . . . I live it, I breathe it. I think about Alabama football 18 hours a day." Updyke explained that his father had died when he was a youngster and that he had been drawn to Alabama's legendary coach Paul "Bear" Bryant as a father substitute. He named his daughter Crimson Tyde and his son Bear and called his dogs Bama and Nicky (after Coach Nick Saban). Updyke owned 46 Alabama hats and had bought out the complete supply of Alabama football championship shirts at a local store. He planned to be buried in a crimson casket. Should Updyke be prosecuted and punished to the full extent of the law?

Mens Rea

The Model Penal Code requires that these acts be committed purposely or recklessly. Damage to property by "catastrophic means," such as an explosion or flood, may be committed negligently. The punishment of criminal mischief under the Model Penal Code is based on the monetary damage of the harm. Keep in mind that property damage resulting from a fire or explosion that purposely endangers the person or property of another may be punished as arson.

The Legal Equation

Criminal mischief = Destruction or damage or tampering with tangible property or deception or threat causing financial loss

+ purposely, knowingly, recklessly or negligently.

Did Nicholas Y. vandalize property when he wrote with a marker on the glass of a projection booth?

People v. Nicholas Y., *102 Cal. Rptr. 2d 511 (Cal. Ct. App. 2000). Opinion by: Epstein, J.*

Issue

Appellant Nicholas Y. appeals from orders of the juvenile court finding that he vandalized property belonging to the AMC theater, declaring him a ward of the court pursuant to Welfare and Institutions Code section 602, and placing him home on probation. He contends the evidence was insufficient to prove he violated Penal Code section 594 (vandalism). . . .

Facts

The evidence . . . proved that in the early morning hours of February 11, 2000, Nicholas Y. wrote on a glass window of a projection booth at an AMC theater with a Sharpie marker. After his arrest, appellant admitted to police that he had written "RTK" on the window. Police saw "approximately 30 incidents" in red magic marker throughout the theater, including the one on the glass. Appellant said the initials stood for "The Right to Crime."

At the close of the prosecution's case, appellant's counsel argued that no defacing of or damage to property had been proved, stating: "[i]t's a piece of glass with a marker on it. You take a rag and wipe it off. End of case. It's ridiculous." The prosecutor countered that appellant trespassed and left fresh marks on the window, thus defacing the window with graffiti. The court found that appellant violated Penal Code section 594, subdivision (a), a misdemeanor.

Reasoning

Penal Code section 594 provides, in relevant part:

(a) Every person who maliciously commits any of the following acts with respect to any real or personal property not his or her own . . . is guilty of vandalism:

(1) Defaces with graffiti or other inscribed material.

(2) Damages.

(3) Destroys. . . .

(4) (A) If the amount of defacement, damage, or destruction is less than four hundred dollars ($400), vandalism is punishable by imprisonment in a county jail for not more than six months, or by a fine of not more than one thousand dollars ($1,000) or by both that fine and imprisonment. . . .

(e) As used in this section the term "graffiti or other inscribed material" includes any unauthorized inscription, word, figure, mark, or design that is written, marked, etched, scratched, drawn, or painted on real or personal property.

Appellant contends he did not violate the statute because the word "deface" contemplates a "permanent alteration" of the surface of an object rather than the easily removed marking he placed on the window. He compares the facts of this case to chalk writing on a sidewalk held not to constitute vandalism in violation of Penal Code section 594 in *MacKinney v. Nielsen*, 69 F.3d 1002 (9th Cir. 1995). As appellant acknowledges, however, the statutory language interpreted in that case was different, making it "illegal to (1) deface 'with paint or any other liquid,' (2) damage or (3) destroy any real or personal property that is not one's own." The Ninth Circuit Court of Appeals reasoned that chalk is not a liquid and did not damage the sidewalk. Therefore, it found the defendant did not violate the statute. The Legislature subsequently amended the statute to delete the phrase "defaces with paint or any other liquid" and substitute in its place the phrase "defaces with graffiti or other inscribed material." Accordingly, the *MacKinney* case is of no assistance to appellant's cause. . . . Certainly the case does not support appellant's contention that there is no defacement unless the act makes a material alteration or permanent change to the surface of the defaced object.

Graffiti may be, and regularly is, created with marker pens. It would be irrational to hold that use of a marker pen on, for example, a painted or stucco surface constitutes vandalism in violation of Penal Code section 594, subdivision (a)(1) while use of a marker pen on glass is not. Each mars the surface with graffiti which must be removed in order to restore the original condition. This pragmatic fact is consistent with the primary meaning of the word "deface" as defined in the Oxford English Dictionary: "To mar the face, features, or appearance of; to spoil or ruin the figure, form, or beauty of; to disfigure." This definition does not incorporate an element of permanence. Thus, it appears that a marring of the surface is no less a defacement because it is more easily removed.

Holding

It follows that appellant was properly found to have violated Penal Code section 594, subdivision (a)(1).

Questions for Discussion

1. Summarize the facts in Nicholas Y.

2. Why does Nicholas Y. claim that the facts do not support his criminal conviction for vandalism?

3. Should a conviction for defacing property require that the defendant cause permanent damage to the property or destroy the property?

4. Do you agree with Judge Epstein that if the use of a marker pen on a "painted or stucco surface" constitutes vandalism, the "use of a marker pen on glass" should constitute vandalism?

5. Would a colorful and attractive mural painted on the wall of a highway underpass violate the California statute?

You Decide

12.6 On July 17, 2007, the defendant painted a wooden fence belonging to his neighbors without permission. He painted along the side of his yard and the back of his yard. The defendant said that the fence facing his property was "in need of repair and painting" and he proceeded to paint the sides facing his property. His intent was to improve the appearance of the fence. The defendant was charged under a New York statute that punished an individual for "intentionally damaging property of another person . . . having no right to do so nor any reasonable ground to believe that he has such right." Would you convict the defendant of "criminal mischief to property"? See *People v. Stockwell*, 18 Misc. 3d 1145A (N.Y. City Ct. 2008). How does this differ from spray painting gang insignia or graffiti on the wall? Compare with *People v. Nicholas Y.*

You can find the answer at www.sagepub.com/lippmanccl3e.

You Decide

12.7. Anthony LeRoy Wallace's wife of two months, Arlissa Pointer Wallace, caught him smoking crack cocaine, called him a crack head, and told him to leave the house she had bought six or seven years before the marriage. The property inside the home originally belonged to Arlissa. Arlissa kept the house in her name, although Wallace apparently acquired some interest in the house based on his contribution to the mortgage payments following their marriage. Anthony reacted angrily and began tearing up the house and property within the home and caused over $9,000 of damage to the house and to over $6,000 of damage to the furniture and furnishings. Anthony claims that he cannot be held criminally liable for vandalizing his own marital property. See *People v. Wallace*, 19 Cal. Rptr. 3d 790 (Cal. Ct. App. 2004).

You can find the answer at www.sagepub.com/lippmanccl3e.

CHAPTER SUMMARY

Crimes against habitation protect individuals' interest in safe and secure homes free from uninvited intrusions. Burglary and arson are the cornerstones of the criminal law's protection of dwellings. Modern statutes have significantly expanded the structures protected by burglary and arson.

Burglary at common law is defined as the breaking and entering of the dwelling house of another at night with the intention to commit a felony. State statutes have significantly modified the common law and differ in their approach to defining the felony of burglary. In general, a breaking no longer is required, and burglary has been expanded to include a range of structures and vehicles. Statutes provide that a burglary may involve entering as well as remaining in a variety of structures with the requisite purposeful intent. The intent standard has been broadened under various statutes to include "any offense" or a "felony or misdemeanor theft." Also, burglary is no longer required to be committed at night.

Criminal trespass is the unauthorized entry or remaining on the land or premises of another. Trespass is distinguished from burglary in that it does not require an intent to commit a felony or other offense and extends to property beyond the curtilage, including agricultural land. Statutes provide that a trespass may be committed knowingly or purposefully, and Missouri defines trespass as a strict liability offense.

Arson at common law is defined as the willful and malicious burning of the dwelling house of another. This is treated as a felony based on the danger posed to inhabitants and neighbors. Statutes no longer require a burning; even smoke damage or soot is sufficient. Arson also extends to a broad range of structures and is no longer limited to the dwelling of another. Arson requires either a purpose to burn or knowledge that a structure will burn. It may also be committed recklessly by creating an unreasonable hazard on an individual's own property that burns a neighbor's dwelling.

Criminal mischief under modern statutes punishes the damage, destruction, or tampering with personal and real tangible property or may involve a deception causing financial loss. Criminal mischief is generally punished as a misdemeanor and may be committed purposefully or recklessly.

Keep in mind that statutes typically aggravate or enhance the penalty for crimes against habitation. Aggravating circumstances include a crime involving a dwelling or an act that endangers human life or safety, is carried out with a weapon, is committed at night, or causes significant financial loss.

CHAPTER REVIEW QUESTIONS

1. What is the definition of burglary? How have the elements of the common law crime of burglary been modified by modern statutes?

2. What is the difference between burglary and trespass?

3. Define *arson*. How have modern statutes modified the common law crime of arson?

4. What are the three types of acts that satisfy the *actus reus* of criminal mischief?

5. Compare and contrast arson and criminal mischief.

6. What are some factors that aggravate burglary, arson, trespass, and criminal mischief?

7. Discuss the justifications for crimes against habitation. Is it accurate to continue to categorize burglary and arson as crimes against habitation?

LEGAL TERMINOLOGY

arson

burglary

criminal mischief

criminal trespass

curtilage

defiant trespass

tangible property

CRIMINAL LAW ON THE WEB

Log on to the Web-based student study site at www.sagepub.com/lippmanccl3e to assist you in completing the Criminal Law on the Web exercises, as well as for additional features such as podcasts, Web quizzes, and video links.

1. Read about the arson fire of a dorm at Seton Hall University, which resulted in the deaths of three students.

2. Learn about graffiti and crime.

3. Examine the conviction of O. J. Simpson for burglary and other offenses.

BIBLIOGRAPHY

American Law Institute, *Model Penal Code and Commentaries*, vol. 2, pt. 11 (Philadelphia: American Law Institute, 1980), pp. 3–94. A review of the laws of arson, criminal mischief, burglary, and trespass with proposed reforms.

John C. Klotter, *Criminal Law*, 7th ed. (New York: Lexis, 2004), pp. 175–214. A review of statutes on crimes against habitation and coverage of relevant federal laws.

Wayne R. LaFave, *Criminal Law*, 3rd ed. (St. Paul, MN: West Publishing, 2000), pp. 883–893. An understandable review of the law of burglary.

Rollin M. Perkins and Ronald N. Boyce, *Criminal Law*, 3rd ed. (Mineola, NY: Foundation Press, 1982), pp. 247–288, 405–456. An analysis of the history and statutory provisions on arson, burglary, and malicious mischief.

13 CRIMES AGAINST PROPERTY

Did the defendant take the purse through force or intimidation?

On December 14, 1996, Elaine Barker was in the parking lot of a K-Mart store unloading items from a shopping cart into the trunk of her car. She left her purse in the shopping cart, and while she was transferring the items she had purchased, the defendant came over, grabbed the purse, and ran away. Barker chased the defendant on foot and caught up with him, but by that time, he had gotten into his car and closed the door. Barker then sat on the hood of the defendant's car, thinking that would prevent him from driving away. Instead, the defendant started and stopped the car several times while Barker held on to a windshield wiper to keep from falling off. The defendant turned the car sharply causing Barker to fall to the ground. As a result of the fall, Barker suffered a broken foot and lacerations that required stitches.

Learning Objectives

1. Know the elements of larceny.
2. Understand embezzlement and the difference between larceny and embezzlement.
3. State the elements of false pretenses and the distinction between false pretenses and larceny by trick.
4. Explain the purpose of theft statutes.
5. Know the elements of identity theft.
6. Understand why it was necessary for states to adopt statutes aimed at combating computer theft and know the content of computer crime statutes.
7. List the elements of receiving stolen property and the purpose of making it a crime to receive stolen property.
8. Define forgery and uttering.
9. Know the elements of robbery and the difference between robbery and larceny.
10. State the elements of carjacking and the difference between carjacking and robbery.
11. Understand the difference between extortion, blackmail, and bribery.

INTRODUCTION

Seventeenth-century English philosopher John Locke asserted in his influential *Second Treatise on Government* that the protection of private property is the primary obligation of government. Locke argued that people originally existed in a "state of nature" in which they were subject to the survival of the fittest. These isolated individuals, according to Locke, came together and agreed to create and to maintain loyalty to a government that, in return, pledged to protect individuals and to safeguard their private property. Locke, as noted, viewed the protection of private property as the most important duty of government and as the bedrock of democracy.

Locke's views are reflected in the Fifth Amendment to the United States Constitution, which prohibits the taking of property without due process of law. Even today, those individuals who may not completely agree with John Locke recognize that the ownership of private property is a right that provides us with a source of personal enjoyment, pride, profit, and motivation and serves as a measure of self-worth.

A complex of crimes was developed by common law judges and legislators to protect and punish the wrongful taking of private property. As we shall see, each of these crimes was created at a different point in time to fill a gap in the existing law. The development of these various offenses was necessary because prosecutors ran the risk that a defendant would be acquitted in the event that the proof at trial did not meet the technical requirements of a criminal charge. Today, roughly thirty states have simplified the law by consolidating these various property crimes into a single theft statute.

In this chapter, we will review the main property crimes. These include:

- *Larceny.* A pickpocket takes the wallet from your purse and walks away.
- *Embezzlement.* A bank official steals money from your account.
- *False Pretenses.* You sell a car to a friend who lies and falsely promises that he or she will pay you in the morning.
- *Receiving Stolen Property.* You buy a car knowing that it is stolen.
- *Forgery and Uttering.* A friend takes one of your checks, makes it payable to himself or herself, signs your name, and cashes the check at your bank.
- *Robbery.* You are told to hand over your wallet by an assailant who points a loaded gun at you.
- *Extortion.* You are told to pay protection to a gang leader who states that otherwise you will suffer retaliatory attacks in the coming months.

We also will look at the newly developing areas of identity theft, computer crimes, and carjacking. In thinking about the property crimes discussed in this chapter, consider that they all involve the seizure of the property of another individual through either a wrongful taking, fraud, or force.

LARCENY

Early English law punished the taking of property by force, a crime that evolved into the offense of robbery. It became apparent that additional protections for property were required to meet the needs of the expanding British economy. Goods now were being produced, transported, bought, and sold. Robbery did not cover acts such as the taking of property under the cover of darkness from a loading dock. Robbery also did not punish employees who stole cash from their employers or a commercial shipper who removed goods from a container that was being transported to the market. Accordingly, the law of larceny was gradually developed to prohibit and punish the nonviolent taking of the property of another without his or her consent.

Common law **larceny** is the trespassory taking and carrying away of the personal property of another with the intent to permanently deprive that individual of possession of the property. You should be certain that you understand each element of larceny:

- Trespassory
- taking and
- carrying away of the
- personal property of another with the
- intent to permanently deprive that individual of
- possession of the property.

Actus Reus: Trespassory Taking

The "trespassory taking" in larceny is different from the trespass against land that we earlier discussed when reviewing crimes against habitation. The term *trespassory taking* in larceny is derived from an ancient Latin legal term and refers to a wrongdoer who removes goods (chattel) or money from the possession of another without consent. **Possession** means physical control over property with the ability to freely use and enjoy the property. Consent obtained by force, fraud, or threat is not valid. An Illinois court observes that there "can be no larceny without a trespass, and there can be no trespass unless the property was in the possession of the one from whom it is charged to have been stolen." The common law, as we shall soon see, established various rules to distinguish possession from **custody**, or the temporary and limited right to control property.[1]

As we observed earlier, the law of larceny developed in response to evolving economic conditions in Great Britain. In the fifteenth century, England changed from an agricultural country into a manufacturing center. Industry depended on carriers to transport goods to markets and stores. These carriers commonly would open and remove goods from shipping containers and sell them for personal profit. An English tribunal ruled in the *Carrier's Case,* in 1473, that the carrier was a bailee (an individual trusted with property and charged with a duty of care) who had possession of the container and custody over the contents. The carrier was ruled to be liable for larceny by "break[ing] bulk" and committing a trespassory taking of the contents. How did the judge reach this conclusion? The judge in the *Carrier's Case* reasoned that possession of the goods inside the crate continued to belong to the owner and that the shipper was liable for a trespassory taking when he broke open the container and removed the contents. Of course, the carrier might avoid a charge of larceny by stealing an unopened container.[2]

The law of larceny was also extended to employees. An employee is considered to have custody rather than possession over materials provided by an employer, such as construction tools or a delivery truck. The employer is said to enjoy constructive possession over this property, and an employee who walks away with the tools or truck with the intent to deprive the owner of possession of the tools or truck is guilty of larceny. A college student who drives a pizza delivery truck for a local business has custody over the truck. In the event that the student steals the truck, he or she has violated the owner's possessory right and has engaged in a "trespassory taking."[3]

Larceny is not limited to business and commerce. Professor Perkins notes that when eating in a restaurant, you have custody over the silverware, and removing it from the possession of the restaurant by walking out the door with a knife in your pocket constitutes larceny. Another illustration is a pickpocket who, by removing a wallet from your pocket, has taken the wallet from your possession.[4] Professor Joshua Dressler offers the example of an individual who test drives an automobile at a dealership, returns to the lot, and drives off with the car when the auto dealer climbs out of the vehicle. The test driver would be guilty of larceny for dispossessing the car dealer of the auto.[5] One last, fairly complicated point: do not confuse ownership with possession. When you wait as a jeweler repairs your watch, you retain possession as well as ownership. The jeweler has custody. In the event that you leave the watch with the jeweler for repair, the law says that although you own the watch, the jeweler has possession over the watch until it is returned to you. The jeweler would not be guilty of larceny in the event that he or she refuses to return the watch unless you pay an additional $50.[6]

Asportation

There must be a taking (caption) and movement (asportation) of the property. The movement of the object provides proof that an individual has asserted control and intends to steal the object. You might notice that an individual waiting in line in front of you has dropped his or her wallet. Your intent to steal the wallet is apparent when you place the wallet in your pocket or seize the wallet and walk away at a brisk pace in the opposite direction.

A taking requires asserting "dominion and control" over the property, however briefly. The property then must be moved, and even "a hair's breath" is sufficient.[7] A pickpocket who manages to move a wallet only a few inches inside the victim's pocket may be convicted of larceny.[8]

Larceny may be accomplished through an innocent party. A defendant was convicted of larceny by falsely reporting to a neighbor that the defendant owned the cattle that wandered onto the neighbor's property. The neighbor followed the defendant's instructions to sell the cattle and then turned the proceeds over to the defendant. The defendant was held responsible for the neighbor's caption (taking) and asportation (carrying away), and the defendant was convicted of larceny.[9] A defendant was also found guilty of larceny for unlawfully selling a neighbor's bicycle to an innocent purchaser who rode off with the bike.[10]

Every portion of the property must be moved. An individual who only succeeds in turning a barrel on its side when he was apprehended was not determined to have "carried away" the property.[11]

Various modern state statutes have followed the Model Penal Code in abandoning the requirement of asportation and provide that a person is guilty of theft if he or she "unlawfully takes, or exercises unlawful control" over property. The Texas statute provides that an individual is guilty of larceny where there is an "appropriation of property" without the owner's consent. Under these types of statutes, a pickpocket who reaches into an individual's purse and seizes a wallet would be guilty of larceny; there is no requirement that the pickpocket move or carry away the wallet.[12]

Property of Another

At common law, only tangible personal property was the subject of larceny. Tangible property includes items over which an individual is able to exercise physical control, such as jewelry, paintings, crops and trees removed from the land, and certain domesticated animals. Property not subject to larceny at common law included services (e.g., painting a house), real estate, crops attached to the land, and intangible property (e.g., property that represents something of value such as checks, money orders, credit card numbers, car titles, and deeds demonstrating ownership of property). Crops were subject to larceny only when severed from the land. For instance, an apple hanging on a tree that was removed by a trespasser was part of the land and was not subject to larceny. An apple that fell from the tree and hit the ground, however, was considered to have entered into the possession of the landowner and was subject to larceny. In another example, wild animals that were killed or tamed were transformed into property subject to larceny. Domestic animals, such as horses and cattle, were subject to larceny, but dogs were considered to possess a "base nature" and were not subject to larceny. These categories are generally no longer significant, and all varieties of property are subject to larceny under modern statutes.

The property must be "of another." Larceny is a crime against possession and is concerned with the taking of property from an individual who has a superior right to possess the object. A landlord who removes the furniture from a furnished apartment that he or she rents to a tenant is guilty of larceny.

Modern statutes, as noted, have expanded larceny to cover every conceivable variety of property. For example, the California statute protects personal property, animals, real estate, cars, money, checks, money orders, traveler's checks, phone service, tickets, and computer data.[13]

Mens Rea

The *mens rea* of larceny is the intent to permanently deprive another of the property. There must be a concurrence between the intent and the act. The intent to borrow your neighbor's car is not larceny. It is also not larceny if, after borrowing your neighbor's car, you find that it is so much fun to drive that you decide to steal it. In this example, the criminal intent and the criminal act did not coincide with one another. Several states have so-called joyriding statutes that make it a crime to take an automobile with an intent to use it and then return it to the owner.

Professor LaFave points out that in addition to a specific intent to steal, there are cases holding that larceny is committed when an individual has an intent to deprive an individual of possession for an unreasonable length of time or has an intent to act in a fashion that will probably dispossess a person of the property. For instance, you may drive your neighbor's car from New York to Alaska with the intent of going on a vacation. The trip takes two months and deprives your neighbor of possession for an unreasonable period, which constitutes larceny. After arriving in Alaska, you park and leave the car in a remote area. You did not intend to steal the car, but it is unlikely that the car will be returned to the owner, and you could also be held liable for larceny based on your having acted in a fashion that will likely dispossess the owner of the car. Lost property or property that is misplaced by the owner is subject to larceny when a defendant harbors the intent to steal at the moment that he or she seizes the property. The defendant is held guilty of larceny, however, only when he or she knows who the owner is or knows that the owner can be located through reasonable efforts. Property is lost when its owner is involuntarily deprived of an object and has no idea where to find or recover it; property is misplaced when the owner forgets where he or she intentionally placed an object. Property that is abandoned has no owner and is not subject to larceny, because it is not the "property of another individual." Property is abandoned when its owner no longer claims ownership. Property delivered to the wrong address is subject to larceny when the recipient realizes the mistake at the moment of delivery and forms an intent to steal the property.

An individual may also claim property "as a matter of right" without committing larceny. This occurs when an individual seizes property that he or she reasonably believes has been taken from his or her possession or seizes money of equal value to the money owed to him or her. In these cases, the defendant believes that he or she has a legal right to the property and does not possess the intent to "take the property of another."

You can see that intent is central to larceny. What if you go to the store to buy some groceries and discover that you left your wallet in the car? You decide to walk out of the store with the groceries and intend to get your wallet and return to the store and pay for the groceries. Is this larceny?

Grades of Larceny

The common law distinguished between **grand larceny** and **petit larceny**. Grand larceny was the stealing of goods worth more than twelvepence, the price of a single sheep. The death penalty applied only to grand larceny.

State statutes continue to differentiate between grand larceny and petit larceny. The theft of property worth more than a specific dollar amount is the felony of grand larceny and is punishable by a year or more in prison. Property worth less than this designated amount is a misdemeanor and is punished by less than a year in prison. States differ on the dollar amount separating petit and grand larceny. South Carolina punishes the theft of an article valued at less than $1,000 as a misdemeanor. Stealing an object worth more than $1,000 but less than $5,000 is subject to five years in prison. Theft of an article valued at $5,000 or more is punishable by ten years in prison.[14] Texas uses the figure of $1,500 to distinguish a theft punishable as a misdemeanor from a theft punishable as a felony. Harsher punishment is imposed as the value of the stolen property increases.[15]

How is property valued? In the case of the theft of money or of a check made out for a specific amount, this is easily calculated. What about the theft of an automobile? Should this be measured by what the thief believes is the value of the property? The Pennsylvania statute is typical and provides that "value means the market value of the property at the time and place of the crime." Courts often describe this as the price at which the mind of a willing buyer and a willing seller would meet.

The application of this test means that judges will hear evidence concerning how much it would cost to purchase a replacement for a stolen car. What about a basketball jersey worn and autographed by megastar Michael Jordan? You cannot go to the store and purchase this item. The Pennsylvania statute states that if the market value "cannot be satisfactorily ascertained, [the value of the property is] the cost of replacement of the property within a reasonable time after the crime." In other words, the court will hear evidence concerning the value of a Michael Jordan autographed jersey in the same condition as the jersey that was stolen. One difficulty is that courts consider the absolute dollar value of items and do not evaluate the long-term investment or sentimental value of the property.[16] The Pennsylvania statute also provides that when multiple items are stolen as part of a single plan or through repeated acts of theft, the value of the items may be aggregated or combined. This means that the money taken in a series of street robberies will be combined and that the perpetrator will be prosecuted for a felony rather than a series of misdemeanors.[17]

The value of property is not the only basis for distinguishing between grand and petit larceny. California uses the figure of $400 to distinguish between grand and petit larceny and categorizes as grand larceny the theft of "domestic fowls, avocados, olives, citrus or deciduous fruits . . . vegetables, nuts, artichokes, or other farm crops" of a value of more than $100. California also considers grand larceny to include the theft of "fish, shellfish, mollusks, crustaceans, kelp, algae . . . taken from a commercial or research operation."[18]

Theft of a firearm, theft of an item from the "person of another," and theft from a home all pose a danger to other individuals and are typically treated as grand larceny.[19] The penalty for stealing property may be increased where the stolen items belong to an "elderly individual" or to the government.[20]

The next case, *People v. Gasparik,* discusses the point at which an individual has exercised sufficient control over property in a store to be convicted of shoplifting. This difficult issue illustrates the challenge of adjusting the law of larceny to meet new developments.

The Legal Equation

Larceny = Unlawful taking and carrying away

+ intent to permanently deprive another of property.

Is a defendant who never leaves the store guilty of shoplifting?

People v. Gasparik, 420 N.E.2d 40 (N.Y. 1978). Opinion by: Cook, J.

Facts

Defendant was in a department store trying on a leather jacket. Two store detectives observed him tear off the price tag and remove a "sensormatic" device designed to set off an alarm if the jacket were carried through a detection machine. There was at least one such machine at the exit of each floor. Defendant placed the tag and the device in the pocket of another jacket on the merchandise rack. He took his own jacket, which he had been carrying with him, and placed it on a table. Leaving his own jacket, defendant put on the leather jacket and walked through the store, still on the same floor, bypassing several cash registers. When he headed for the exit from that floor, in the direction of the main floor, he was apprehended by security personnel. At trial, defendant denied removing the price tag and the sensormatic device from the jacket and testified that he was looking for a cashier without a long line when he was stopped. The court, sitting without a jury, convicted defendant of petit larceny. Appellate Term affirmed.

Issue

The primary issue in this case is whether the evidence, viewed in the light most favorable to the prosecution, was sufficient to establish the elements of larceny as defined by the Penal Law. To resolve this common question, the development of the common law crime of larceny and its evolution into modern statutory form must be briefly traced.

Reasoning

Larceny at common law was defined as a trespassory taking and carrying away of the property of another with intent to steal it. The early common law courts apparently viewed larceny as defending society against breach of the peace, rather than protecting individual property rights, and therefore placed heavy emphasis upon the requirement of a trespassory taking. Thus, a person . . . who had rightfully obtained possession of property from its owner could not be guilty of larceny. The result was that the crime of larceny was quite narrow in scope. One popular explanation for the limited nature of larceny is the "unwillingness on the part of the judges to enlarge the limits of a capital offense." The accuracy of this view is subject to some doubt.

Gradually, the courts began to expand the reach of the offense, initially by subtle alterations in the common law concept of possession. Thus, for instance, it became a general rule that goods entrusted to an employee were not deemed to be in his possession but were only considered to be in his custody, so long as he remained on the employer's premises. And . . . it was held that a shop owner retained legal possession of merchandise being examined by a prospective customer until the actual sale was made. In these situations, the employee and the customer would not have been guilty of larceny if they had first obtained lawful possession of the property from the owner. By holding that they had not acquired possession, but merely custody, the court was able to sustain a larceny conviction.

As the reach of larceny expanded, the intent element of the crime became of increasing importance, while the requirement of a trespassory taking became less significant. As a result, the bar against convicting a person who had initially obtained lawful possession of property faded. In *King v. Pear* (168 Eng. Rep. 208), for instance, a defendant who had lied about his address and ultimate destination when renting a horse was found guilty of larceny for later converting the horse. Because of the fraudulent misrepresentation, the court reasoned, the defendant had never obtained legal possession. Thus, "larceny by trick" was born. . . .

Section 155.05 of the New York Penal Law defines larceny: (1) A person steals property and commits larceny when, with intent to deprive another of property or to appropriate the same to himself or to a third person, he wrongfully takes, obtains, or withholds such property from an owner thereof. (2) Larceny includes a wrongful taking, obtaining, or withholding of another's property, with the intent prescribed in subdivision one of this section, committed in any of the following ways: (a) By conduct heretofore defined or known as common law larceny by trespassory taking, common law larceny by trick, embezzlement, or obtaining property by false pretenses.

This evolution is particularly relevant to thefts occurring in modern self-service stores. In stores of that type, customers are impliedly invited to examine, try on, and carry about the merchandise on display. Thus, in a sense, the owner has consented to the customer's possession of the goods for a limited purpose. That the owner has consented to that possession does not, however, preclude a conviction for larceny. If the customer exercises dominion and control wholly inconsistent with the continued rights of the owner, and the other elements of the crime are present, a larceny has occurred. Such conduct on the part of a customer satisfies the "taking" element of the crime.

Also required, of course, is the intent prescribed by subdivision 1 of section 155.05 of the Penal Law and some movement when property other than an automobile is involved. As a practical matter, in shoplifting cases, the same evidence that proves the taking will usually involve movement.

The movement, or asportation, requirement has traditionally been satisfied by a slight moving of the property. This accords with the purpose of the asportation element, which is to show that the thief had indeed gained possession and control of the property. It is this element that forms the core of the controversy in these cases. The defendants argue, in essence, that the crime is not established, as a matter of law, unless there is evidence that the customer departed the shop without paying for the merchandise.

Although this court has not addressed the issue, case law from other jurisdictions seems unanimous in holding that a shoplifter need not leave the store to be guilty of larceny. This is because a shopper may treat merchandise in a manner inconsistent with the owner's continued rights—and in a manner not in accord with that of a prospective purchaser—without actually walking out of the store. Indeed, depending upon the circumstances of each case, a variety of conduct may be sufficient to allow the trier of fact to find a taking. It would be well-nigh impossible, and unwise, to attempt to delineate all the situations that would establish a taking. But it is possible to identify some of the factors used in determining whether the evidence is sufficient to be submitted to the fact finder.

In many cases, it will be particularly relevant that the defendant concealed the goods under clothing or in a container. Such conduct is not generally expected in a self-service store and may in a proper case be deemed an exercise of dominion and control inconsistent with the store's continued rights. Other furtive or unusual behavior on the part of the defendant should also be weighed. Thus, if the defendant surveys the area while secreting the merchandise or abandons his or her own property in exchange for the concealed goods, this may evince larcenous rather than innocent behavior. Relevant too is the customer's proximity to or movement towards one of the store's exits. Certainly, it is highly probative of guilt that the customer was in possession of secreted goods just a few short steps from the door or moving in that direction. Finally, possession of a known shoplifting device actually used to conceal merchandise, such as a specially designed outer garment or a false-bottomed carrying case, would be all but decisive.

Of course, in a particular case, any one or any combination of these factors may take on special significance. And there may be other considerations, not now identified, which should be examined. So long as it bears upon the principal issue—whether the shopper exercised control wholly inconsistent with the owner's continued rights—any attending circumstance is relevant and may be taken into account.

Under these principles, there was ample evidence . . . to raise a factual question as to the defendant's guilt. . . . As discussed, the same evidence that establishes dominion and control in these circumstances will often establish movement of the property. And the requisite intent generally may be inferred from all the surrounding circumstances. It would be the rare case indeed in which the evidence establishes all the other elements of the crime but would be insufficient to give rise to an inference of intent.

The defendant removed the price tag and sensor device from a jacket, abandoned his own garment, put the jacket on, and ultimately, headed for the main floor of the store. Removal of the price tag and sensor device, and careful concealment of those items, is highly unusual and suspicious conduct for a shopper. Coupled with defendant's abandonment of his own coat and his attempt to leave the floor, those factors were sufficient to make out a prima facie case of a taking.

Holding

In sum, in view of the modern definition of the crime of larceny, and its purpose of protecting individual property rights, a taking of property in the self-service store context can be established by evidence that a customer exercised control over merchandise wholly inconsistent with the store's continued rights. Quite simply, a customer who crosses the line between the limited right he or she has to deal with merchandise and the store owner's rights may be subject to prosecution for larceny. Such a rule should foster the legitimate interests and continued operation of self-service shops, a convenience that most members of the society enjoy. Accordingly, in each case, the verdict at trial should be affirmed.

You Decide

13.1 Carter entered a paint store and placed four 5-gallon buckets of paint, valued at $398.92, in a shopping cart. Browning waited for Carter by the "return desk" where customers take items they previously had purchased and wish to return for a refund. As planned, Browning falsely stated that the paint had been purchased from the store and requested a refund for the paint. A store manager recognized Browning as someone she had been told to look for and contacted an employee of the store who summoned the police. Was Carter guilty of larceny? See *Carter v. Commonwealth*, 694 S.E.2d 590 (Va. 2010).

You can find the answer at www.sagepub.com/lippmanccl3e.

EMBEZZLEMENT

We have seen that larceny requires a taking of property from the possession of another person with the intent to permanently deprive the person of the property. In the English case of *Rex v. Bazeley*, in 1799, a bank teller dutifully recorded a customer's deposit and then placed the money in his pocket. The court ruled that the teller had taken possession of the note and that he therefore could not be held convicted of larceny and ordered his release from custody. The English Parliament responded by almost immediately passing a law that held servants, clerks, and employees criminally liable for the fraudulent misdemeanor of embezzlement of property. Today, embezzlement is a misdemeanor or felony depending on the value of the property.[21]

The law of **embezzlement** has slowly evolved, and although there is no uniform definition of embezzlement, the core of the crime is the fraudulent conversion of the property of another by an individual in lawful possession of the property. The following elements are central to the definition of the crime:

Fraudulent (deceitful) *conversion of* (the serious interference with the owner's rights) *the property* (statutes generally follow the law of larceny in specifying the property subject to embezzlement) *of another* (you cannot embezzle your own property) *by an individual in lawful possession of the property* (the essence of embezzlement is wrongful conversion by an individual in possession).

The distinction between larceny and embezzlement rests on the fact that in embezzlement, the perpetrator lawfully takes possession and then fraudulently converts the property. In contrast, larceny involves the unlawful trespassory taking of property from the possession of another. Larceny requires an intent to deprive an individual of possession at the time that the perpetrator "takes" the property. The intent to fraudulently convert property for purposes of embezzlement, however, may arise at any time after the perpetrator takes possession of the property.

Typically, embezzlement is committed by an individual to whom you entrust your property. Examples would be a bank clerk who steals money from the cash drawer, a computer repair technician who sells the machine that you left to be repaired, or a construction contractor who takes a deposit and then fails to pave your driveway or repair your roof. Embezzlement statutes are often expressed in terms of "property which may be the subject of larceny." In some states, embezzlement is defined to explicitly cover personal as well as real property (e.g., land).

The next case, *Thomas v. State,* illustrates the factual analysis involved in determining whether a defendant is guilty of embezzlement. Ask yourself whether a defendant who did not "convert property to his own use" should be held liable if he or she substantially interferes with the property rights of another.

Model Penal Code

Section 223.2. Theft by Unlawful Taking or Disposition

(1) Movable Property. A person is guilty of theft if he unlawfully takes or exercises unlawful control over movable property of another with purpose to deprive him thereof.

(2) Immovable Property. A person is guilty of theft if he unlawfully transfers immovable property of another or any interest therein with purpose to benefit himself or another not entitled thereto.

Analysis

- The Model Penal Code consolidates larceny and embezzlement.
- The phrase "unlawfully takes" is directed at larceny, while the exercise of "unlawful control" over the property of another is directed at embezzlement. In both instances, the defendant must be shown to possess an intent to "deprive" another of the property. This includes an intent to permanently deprive the other individual of the property as well as treating property in a manner that deprives another of its use and enjoyment.
- Property is broadly defined to include "anything of value," including personal property, land, services, and real estate.
- Asportation is not required for larceny.
- Combining larceny and embezzlement means that the prosecution is able to avoid the confusing issues of custody and possession. The "critical inquiry" is "whether the actor had control of the property, no matter how he got it, and whether the actor's acquisition or use of the property was authorized."

The Legal Equation

Embezzlement $=$ Conversion of property of another

$+$ intent to permanently deprive another of property.

Was Thomas guilty of the embezzlement of a customer's van?

Thomas v. State, *707 S.E.2d 547 (Ga. Ct. App. 2011). Opinion By: Blackwell, J.*

Issue

The record in this case clearly shows that Otis Patrick Thomas, an automotive mechanic in Dougherty County, did not apply best business practices in the operation of his automotive repair business, did not keep the promises that he made to a customer, and may have lied in civil proceedings commenced by that customer. But the question presented here is whether the State offered sufficient evidence at his criminal trial from which a rational trier of fact might conclude beyond a reasonable doubt that Thomas converted a van, which he had promised to repair but apparently never did, in violation of OCGA § 16-8-4 (a).

Facts

The record shows that the customer paid $1,675 to Thomas for the replacement of the engine in her van. Thomas initially promised the customer that the work would be complete by February 8, 2007, but Thomas later told her that the work would not be complete until February 16. On February 16, Thomas again contacted the customer and told her that the work would not be done until February 20, ostensibly because the crew that helped Thomas with his repair work had quit. The customer then informed Thomas that she was renting another vehicle to drive while Thomas was repairing her van, and Thomas told her that he would reimburse the costs of the rental car.

On March 5, the customer called Thomas several times to inquire about her van, but Thomas did not return these calls until March 12, at which time he told her that he was exhausted and would not be done with the work on the van until the following day. On March 13, Thomas went to the customer's residence and showed her a video recording of her van. According to the customer, Thomas attempted to explain all of the work that was required to replace the engine in the van, but she did not understand what he was saying. Thomas told her that he could finish the work by the next day.

When the customer did not hear from Thomas on March 14, she went to the shop at which he worked and saw her van "raised very high on the rack," although Thomas was working at the time on another vehicle. On March 17, the customer and a companion went to the shop, where they confronted Thomas about the van. Her companion told Thomas that he must finish his work on the van before March 23, and Thomas assured them that he would do so "long before" that date. The customer admitted that Thomas had done at least some work on her van by the time they visited the shop on March 17, inasmuch as she observed during that visit that the old engine had been removed and that a crate was sitting next to the old engine in the shop, which, she believes, contained the new engine to be installed in the van. Between March 19 and March 23, the customer attempted to contact Thomas each day, and she saw her van lifted on the rack whenever she drove by the shop. Thomas did not respond, however, to her additional attempts to contact him.

On March 26, the customer filed a civil complaint in magistrate court against Thomas. At a hearing before the magistrate court on May 9, Thomas claimed that the work on the van was complete, and the trial court ordered him to deliver it to the customer on the next morning. The magistrate court also ordered Thomas to pay damages of approximately $3,000 to the customer. Thomas never delivered the van, however, and the customer testified at Thomas's criminal trial about her belief that Thomas no longer worked at the shop where her van had been stored and that the shop's owner had arranged for someone to tow her van to a junkyard in May 2007. The customer apparently never recovered her van.

The district attorney charged Thomas with one count of theft by conversion. The accusation alleged that Thomas, "having lawfully obtained a 1996 Ford Windstar, property belonging to [the customer], under an agreement to make a specified disposition of such property, did knowingly convert such property to his own use in violation of the agreement." The accusation notably did not accuse Thomas of having converted the money that the customer had paid to him for the repair of her van.

Thomas claims that the evidence presented at his bench trial is insufficient to sustain his conviction for theft by conversion. On appeal, Thomas contends that the evidence shows merely that he failed to fulfill his obligations to repair and return the van, not that he converted it.

Reasoning

Thomas was charged with, and convicted of, violating OCGA § 16-8-4 (a), which provides:

> A person commits the offense of theft by conversion when, having lawfully obtained funds or other property of another including, but not limited to, leased or rented personal property, under an agreement or other known legal obligation to make a specified application of such funds or a specified disposition of such property, he knowingly converts the funds or property to his own use in violation of the agreement or legal obligation. . . .

This statute is intended to punish the fraudulent conversion of property, not mere breaches of contract or broken promises. So, evidence sufficient to show that someone breached a contract and broke his promises may not be sufficient to prove that he committed criminal conversion. To prove criminal conversion, the State must prove something more, that the defendant misappropriated the property at issue to his own use with fraudulent intent. In this case, we think the State failed to do so.

Here, the evidence shows clearly that Thomas abandoned his work on the van, that he apparently abandoned the van at the shop at which he had worked, and that he never delivered the van to the customer. He did these things despite his promises to complete the work, his repeated assurances that the work would soon be complete, his statement to the magistrate court that he had, in fact, completed the work, and the direction of the magistrate court to deliver the van to the customer. But there is no evidence that Thomas drove the van, that he cannibalized it for spare parts, or that he used it for any other purpose, except to perform work upon it. There is no evidence that Thomas did anything to conceal the whereabouts of the van from the customer or to keep her from recovering possession of it. And although it appears from the record that the van ultimately was taken from the shop to a junkyard, nothing in the record suggests that Thomas had anything to do with the disposal of the van. Indeed, the customer admitted her belief that the owner of the shop made the decision to tow away her van after Thomas quit working at the shop.

The State notes that the evidence would authorize a trier of fact to conclude that Thomas, on more than one occasion, made misrepresentations about the repair of the van and ignored the customer when she attempted to contact him, and the State contends that this evidence, when combined with the undisputed failure of Thomas to return the van to the possession of the customer, is sufficient to prove that Thomas converted the van with fraudulent intent. In *Terrell v. State,* 621 S.E.2d 515 (2005) we found sufficient evidence to sustain a conviction for criminal conversion of a wood chipper where the defendant made misrepresentations to the company from which he rented it, ignored repeated attempts to contact him, and never returned the chipper to the possession of the company. But *Terrell* differs from this case in several important respects.

First, the misrepresentations in *Terrell* were misrepresentations about where the defendant and the wood chipper might be found, and the defendant ignored repeated attempts to contact him at a time when the owner was searching for, but could not locate, the chipper. Here, on the other hand, the evidence shows that Thomas's misrepresentations concerned the status

of his work on the van and the time of its completion, not its whereabouts. And at the times Thomas failed to respond to attempts by his customer to contact him, the record suggests that the van was at the shop, precisely where the customer knew it to be. Unlike *Terrell,* the evidence in this case does not authorize the inference that Thomas engaged in an effort to conceal the whereabouts of the van.

Second, our opinion in *Terrell* does not suggest that the record there contained any evidence of what really happened to the wood chipper. Here, on the other hand, the record indicates that the van was towed away to a junkyard, not by Thomas, but by the arrangement of the shop owner.

Finally, in *Terrell,* the defendant fled to Mexico after failing to return the wood chipper, where he lived under an alias for some time. Here, there is no evidence that Thomas fled beyond the reach of the courts. To the contrary, Thomas appeared at a hearing in the magistrate court after his customer filed a lawsuit against him.

Here . . . the evidence simply does not prove that Thomas misappropriated the van to his own use and did so with fraudulent intent.

Holding

Thomas's treatment of his customer was contemptible and reprehensible. But the evidence is insufficient to prove beyond a reasonable doubt that it amounted to a crime. We, therefore, reverse Thomas's conviction for theft by conversion.

Questions for Discussion

1. Summarize the facts in *Thomas* and the relevant statute.

2. Can you speculate why the defendant was convicted at trial? What are the reasons that the Georgia appellate court reversed Thomas's conviction?

3. What of the argument that Thomas fraudulently and substantially interfered with the customer's property rights? Various courts hold that an individual converts property to his or her own use when he or she treats the property as "his or her own." See *State v. Lough,* 899 A.2d 468 (R.I. 2006).

You Decide

13.2 Clifford B purchased a new truck. He did not have the down payment with him. Clifford B drove the truck home. A salesman, Jorge Jose Casas, followed Clifford B home in the truck that Clifford had traded in to the dealership. The next day, Clifford B realized that he had left something in the trade-in vehicle. When he returned to the dealership, the dealer realized that the trade-in truck was missing. Casas, instead of returning the trunk to the dealership, had driven 400 miles in search of narcotics and did not return the truck to the dealership for two days. Casas used the $500 down payment to purchase drugs. The court noted that judges were divided over the required intent for embezzlement. Was Clifford B guilty of the embezzlement of both the truck and of the $500 down payment? See *People v. Casas,* 109 Cal. Rptr. 3d 811 (Cal. Dist. Ct. App. 2010).

You can find the answer at www.sagepub.com/lippmanccl3e.

You can find more cases on the study site: State v. Robinson, *www.sagepub.com/lippmanccl3e.*

FALSE PRETENSES

Larceny punishes individuals who "take and carry away" property from the possession of another with the intent to permanently deprive the individual of the property. Obtaining possession through misrepresentation or deceit is termed **larceny by trick**. In both larceny and larceny by trick, the wrongdoer unlawfully seizes and takes your property.

Embezzlement punishes individuals who fraudulently "convert" to their own use the property of another that the embezzler has in his or her lawful possession. In other words, you trusted the wrong person with the possession of your property.

Common law judges confronted a crisis when they realized that there was no criminal remedy against individuals who tricked another into transferring title or ownership of personal property or land. Consider the case of an individual who trades a fake diamond ring that he or she falsely represents to be extremely valuable in return for a title to farmland.

The English Parliament responded, in 1757, by adopting a statute punishing an individual who "knowingly and designedly" by false pretense shall "obtain from any person or persons money, goods, wares or merchandise with intent to cheat or defraud any person or persons of the same." American states followed the English example and adopted similar statutes.

State statutes slightly differ from one another in their definitions of false pretenses. The essence of the offense is that a defendant is guilty of **false pretenses** who

- obtains title and possession of property of another by
- a knowingly false representation of
- a present or past material fact with
- an intent to defraud that
- causes an individual to pass title to his or her property.

Actus Reus

The *actus reus* of false pretenses is a false representation of a fact. The expression of an opinion or an exaggeration ("puffing"), such as the statement that this is a "fantastic buy," does not constitute false pretenses. The most important point to remember is that the false representation must be of a past (this was George Washington's house) or present (this is a diamond ring) fact. A future promise does not constitute false pretenses ("I will pay you the remaining money in a year"). Why? The explanation is that it is difficult to determine whether an individual has made a false promise, whether an individual later decided not to fulfill a promise, or whether outside events prevented the performance of the promise. Prosecuting individuals for failing to fulfill a future promise would open the door to individuals' being prosecuted for failing to pay back money they borrowed or might result in business executives being held criminally liable for failing to fulfill the terms of a contract to deliver consumer goods to a store. The misrepresentation must be material (central to the transaction; the brand of the tires on a car is not essential to a sale of a car) and must cause an individual to transfer title. It would not be false pretenses where a buyer knows that the seller's claim that a home has a new roof is untrue or where the condition of the roof is irrelevant to the buyer.

Silence does not constitute false pretenses. A failure to disclose that a watch that appears to be a rare antique is in reality a piece of costume jewelry is not false pretenses. The seller, however, must disclose this fact in response to a buyer's inquiry as to whether the jewelry is an authentic antique.

Mens Rea

The *mens rea* of false pretenses requires that the false representation of an existing or past fact be made "knowingly and designedly" with the "intent to defraud." This means that an individual knows that a statement is false and makes the statement with the intent to steal. A defendant who sells a painting for an exorbitant price that he or she mistakenly or reasonably believes was painted by Elvis Presley is not guilty of false pretenses.

"Recklessness" or representations made without information, however, are typically sufficient for false pretenses. Representing that a painting was made by Elvis Presley when you are uncertain or are aware that you have no firm basis for such a representation would likely be sufficient for false pretenses.

In other words, the intent requirement is satisfied by *knowledge* that a representation is untrue, an *uncertainty* whether a representation is true or untrue, or an *awareness* that one lacks sufficient knowledge to determine whether a representation is true or false.

Defendants are not considered to possess an intent to defraud a victim of property when they reasonably believe that they actually own or are entitled to own the property. You cannot steal what you believe you are entitled to own.

Keep in mind that when possession passes to an individual and the owner retains the title, the defendant is guilty of larceny by trick rather than false pretenses. An individual who obtains the permission of an auto dealer to take a car for a drive and intends to and, in fact, does steal the car is guilty of larceny by trick. Obtaining the title to the car with a check that the buyer knows will "bounce" constitutes false pretenses. Another important difference is that larceny requires a taking and carrying away of the property. False pretenses requires only a transfer of title and possession.

False pretenses generally has been interpreted to include a wider variety of property than larceny and includes the acquisition of lodging, labor (washing your car), and services (telephone). The next case, *State v. Henry,* asks whether you may be convicted of obtaining sexual gratification through false pretenses.

Model Penal Code

Section 223.3. Theft by Deception

A person is guilty of theft if he purposely obtains property of another by deception. A person deceives if he purposely:

(1) creates or reinforces a false impression, including false impressions as to law, value, intention or other state of mind; but deception as to a person's intention to perform a promise shall not be inferred from the fact alone that he did not subsequently perform the promise;

(2) prevents another from acquiring information which would affect his judgment of a transaction;

(3) fails to correct a false impression which the deceiver previously created or reinforced, or which the deceiver knows to be influencing another to whom he stands in a fiduciary or confidential relationship;

(4) fails to disclose a lien, adverse claim or other legal impediment to the enjoyment of property which he transfers or encumbers in consideration for the property obtained, whether such impediment is or is not valid, or is or is not a matter of official record.

The term "deceive" does not, however, include falsity as to matters having no pecuniary significance or puffing by statements unlikely to deceive ordinary persons in the group addressed.

Analysis

- The term *deception* is substituted for "false pretense or misrepresentation."
- The defendant must possess the intent to defraud. This entails a purpose to obtain the property of another and the purpose to deceive the other person.
- The act requirement is satisfied by a misrepresentation as well as by reinforcing faulty information. False future promises are considered to constitute false pretenses.
- There is no duty of disclosure other than when the defendant contributes to the creation of a false impression or where the defendant has a duty of care toward the victim (fiduciary or confidential relationship) or there is a legal claim on the property. A seller also may not interfere (destroy or hide) information.
- False pretenses does not include a misrepresentation that has no significance in terms of the value of the property (e.g., the political or religious affiliation of a salesperson) or puffing or nondisclosure.

The Legal Equation

False pretenses = Misrepresentation or deceit

+ intent to steal property

+ victim relies on misrepresentation and conveys title and possession.

Did the defendant use false pretenses to obtain sexual gratification?

State v. Henry, 68 P.3d 455 (Ariz. Ct. App. 2003). Opinion by: Howard, J.

Tyrone Henry was convicted of fraudulent scheme and artifice and sentenced to prison. He argues the trial court erred in denying his motion for judgment of acquittal. Finding no abuse of discretion or other reversible error, we affirm.

Facts

In June 2000, Henry approached the victims, fifteen-year-old K. and sixteen-year-old C., at a shopping mall. He claimed to be marketing a new face cream, asked the victims whether they used face creams, and showed them photographs of females with "clumpy," white cream on their faces. Henry said he was conducting a survey of the face cream, using females ages twelve to twenty-five, and appeared to write the victims' responses to questions he asked them about lotions they used. He asked the victims if they would like to further participate in the survey by having facials, offering them $10 each to do so. The victims made an appointment to have facials the next day. Henry telephoned the victims the next day and gave them directions to his apartment. The victims took a male friend along to the apartment, but Henry requested that the friend remain outside during the facials, claiming Henry and the victims "had to talk about secret traits that were in the facial cream." After the friend agreed, the victims entered the living room of Henry's small apartment, and he asked K. to lie on a bed and C. to lie on a couch near the bed.

Wearing cotton shorts and a T-shirt, Henry placed small caps and a bandanna over K.'s eyes and told her she would go blind if any of the face cream got in her eyes. K. felt him brush a substance on her face and then heard him clicking the mouse on his computer. With her eyes still covered, K. then heard heavy breathing and heard Henry telling C., "it's coming soon," and after that, "it spilt." K. then saw camera flashes after Henry said he was going to take photographs. K. heard Henry walking behind her where C. was lying, and then felt him place a thick, warm substance on K.'s face

with his hands. Henry had told K. he would warm the treatment cream in a microwave oven, but she never heard a microwave oven activated. Shortly thereafter, Henry removed the bandanna and caps and gave K. and C. towels to wipe their faces. When K. sat up to wipe her face, she saw "white stuff" on C.'s chin that was "real thick . . . [and] clumped up."

Henry had not covered C.'s eyes but had told her to keep them closed, claiming the applications to her face would burn her eyes. Henry had taken a "before" photograph and had applied two substances to C.'s face with his hands and had taken more photographs. He then had told C. to "hold on because the thick treatment was going to come in just a second." Without feeling Henry's hands, C. had then felt "something . . . warm . . . just [go] all over [her] face" and shirt and had then noticed camera flashes.

Before the victims left the apartment, Henry asked them "how did it feel," giving them a $20 bill. He also asked if they wanted to make another appointment. The victims made another appointment and left with their friend. The victims thereafter discussed what had happened and, based on their suspicions that Henry had ejaculated on C.'s face, contacted the police.

Police officers interviewed the victims and collected C.'s T-shirt. After receiving crime laboratory test results showing the possible presence of semen on the shirt, which deoxyribonucleic acid (DNA) testing later confirmed as Henry's, police searched his apartment. The search did not produce any indication that Henry had been conducting legitimate face cream testing, but police found a day planner with the victims' names in it, along with the names of numerous other females, and sections marked "site" and "White Dew Facials." Officers seized a computer, a scanner, and 300 to 500 photographs, many of them depicting females involved in situations similar to that the victims had described. Officers also found one photograph of C. on an undeveloped roll of film resembling one of the earlier photographs Henry had taken during the incident. Police discovered that Henry was operating a pornographic Internet website titled, "White Dew Original Facials," on which he would charge visitors between $10 and $90 to view images of females with semen on their faces.

The state charged Henry with two counts of kidnapping and one count of fraudulent scheme and artifice. At trial, in addition to the victims, M., whose name had been found in Henry's day planner and photographs of whom had been recovered from Henry's apartment, testified that about two years earlier, she had responded to an advertisement in which Henry had offered money for females to participate in a face cream experiment. She testified that she had made an appointment with Henry and had gone to his apartment. She said Henry had covered her eyes, telling her that the cream would burn her eyes, had surreptitiously ejaculated on her face, and had taken photographs. Tests conducted on a stain from a sweater M. had worn during the incident produced results consistent with Henry's semen.

In his defense, Henry called several females, who testified they had gone to Henry's apartment and had willingly posed for photographs with Henry's semen on their faces, which they had understood would be used on Henry's website. They testified that Henry had paid them as much as $100 per hour for posing for the photographs. During closing argument, Henry suggested that he, in fact, had been engaged in legitimate skin cream testing, that the semen found on C.'s shirt could have been transferred there from the towel she had used to clean her face at Henry's apartment, and that M. had shown up for a face treatment, had flirted with Henry, and had wanted to "play around with some other things."

The trial court granted Henry's motion for judgment of acquittal on the kidnapping charges but denied it on the fraudulent scheme count. The jury found Henry guilty, and the court imposed a presumptive, five-year prison sentence, which the court enhanced by two years after Henry admitted having committed the offense while on release for an unrelated offense.

Issue

Henry argues the trial court erred in denying his motion for judgment of acquittal. . . . The jury found Henry guilty of fraudulent scheme and artifice, in violation of A.R.S. section 13-2310(A), which prohibits a person from, "pursuant to a scheme or artifice to defraud, knowingly obtaining any benefit by means of false or fraudulent pretenses, representations, promises or material omissions." For purposes of section 13-2310(A), "a 'scheme or artifice' is some 'plan, device, or trick' to perpetrate a fraud." "The scheme need not be fraudulent on its face but 'must involve some sort of fraudulent misrepresentations or omissions reasonably calculated to deceive persons of ordinary prudence and comprehension.'" The term "defraud" as used in

the statute is not measured by any technical standard but, rather, by a "broad view." A "benefit" under the statute is "anything of value or advantage, present or prospective."

Henry argues no substantial evidence supports the conviction because sexual gratification does not qualify as a requisite "benefit" under section 13-2310, contrary to the state's argument to the jury. Henry urges that "benefit" as found in section 13-2310 applies only to property and pecuniary gains, not to anything as intangible as sexual gratification. We agree with the state that in this case, sexual gratification does qualify as a benefit under section 13-2310.

Reasoning

The legislative history and language of section 13-2310 support our result. As originally enacted, the language of section 13-2310 . . . stated that a person was guilty of fraudulent schemes and artifices by fraudulently obtaining "money, property or any other thing of value." However, two years later, the legislature aimed for a broader scope, amending section 13-2310 by replacing the clause "property or any other thing of value" with the more inclusive "any benefit." And the legislature did not define "benefit" in section 13-2301. . . . Rather, "benefit" is defined among the definitions for the entire criminal code in section 13-105 and includes "anything of value or advantage," not merely pecuniary gain. Accordingly, we believe the legislature intended "benefit" as used in section 13-2310 to have a broad definition and did not intend to exclude sexual gratification.

Pertinent case law also supports our decision. Although no state case addresses whether sexual gratification qualifies as a benefit under section 13-2310, we find illuminating our supreme court's decision in *Haas*. There, the court addressed the scope of the language in section 13-2310 pertaining to fraud. Although it did not address what qualifies as a benefit, the court concluded, in overriding the defendant's attempt to limit the scope of section 13-2310, that "if the legislature had intended section 13-2310 to be broad enough to cover all of the varieties made possible by boundless human ingenuity," then it did not intend to confine the definition of "benefit" to include only pecuniary gain. . . .

Nevertheless, if we accept the State's position, Henry foresees a "slippery slope" where a broad range of noncommercial and intangible benefits may become the bases for criminal prosecution. He likens the State's position to criminalizing a "practical joke gone awry" and the "'benefit' of laughter" attending it. But the context of Henry's scheme was commercial, a face cream survey and sample application, and not the type of noncommercial activity that troubles Henry. . . . Henry has not pointed out any statute criminalizing practical jokes. We need not decide the maximum reach of section 13-2310 today but only that Henry's conduct falls within the legislative intent. . . .

Holding

Having concluded that sexual gratification qualifies as a benefit under section 13-2310, we return to Henry's claim that substantial evidence did not support his conviction. Henry operated a pornographic website on which he charged visitors money to view photographs of females with semen on their faces. He paid some females up to $100 per hour to willingly pose for such photographs. Henry approached the victims claiming to be conducting a face cream survey and offered them $10 each to have "facials." The victims testified that they had scheduled an appointment with Henry believing they were participating in a legitimate face cream survey and would, in fact, receive a facial. The evidence supports the conclusion that with the victims' eyes covered or closed, Henry ejaculated on C.'s face

and took photographs, in a manner resembling the subterfuge he had employed with M. two years before. No evidence showed that Henry was associated with any legitimate face cream enterprise.

Construed in a light supporting the conviction, the evidence supporting the conviction is substantial. The jury could infer beyond a reasonable doubt that pursuant to a scheme to defraud and through false pretenses, Henry had obtained a benefit through sexual gratification or by an intent to post photographs of C. on his website for profit while paying her substantially less than he did consenting models. . . . Henry points out there was no evidence C.'s photograph was ever posted on his website. But the benefit need be prospective only. And, although Henry argues that the jury was required to find the value of the benefit obtained by the fraudulent scheme, the portion of section 13-2310 under which Henry was convicted contains no such requirement. . . . We affirm Henry's conviction and sentence.

Questions for Discussion

1. Would Henry's scheme be punishable as false pretenses had the Arizona statute on false pretenses not been amended? Why is Henry's scheme punishable under the text of the current statute? Did Henry obtain a "benefit"?

2. What is Henry's defense? What did he mean when he contended that a practical joke under the court's interpretation

could be interpreted as a crime? Are the arguments of the court or of the defendant more persuasive?

3. As a prosecutor, would you have charged Henry with false pretenses?

Cases and Comments

Stolen Valor Act. In *United States v. Alvarez*, the Ninth Circuit Court of Appeals considered the constitutionality of the federal Stolen Valor Act of 2005, 18 U.S.C. § 704(b). The law provides:

> Whoever falsely represents himself or herself, verbally or in writing, to have been awarded any decoration or medal authorized by Congress for the Armed Forces of the United States, any of the service medals or badges awarded to the members of such forces, the ribbon, button, or rosette of any such badge, decoration, or medal, or any colorable imitation of such item shall be fined under this title, imprisoned not more than six months, or both.

The prescribed prison term is enhanced to one year if the decoration involved is the Congressional Medal of Honor, a distinguished-service cross, a Navy cross, an Air Force cross, a silver star, or a Purple Heart. Id. § 704(c), (d).

Xavier Alvarez won a seat on the Three Valley Water District Board of Directors in 2007. At a joint meeting

with a neighboring water district board, Alvarez introduced himself and noted that "I'm a retired marine of 25 years. I retired in the year 2001. Back in 1987, I was awarded the Congressional Medal of Honor. I got wounded many times by the same guy. I'm still around." Alvarez had neither served in the military nor had been awarded the Congressional Medal of Honor. In the past, Alvarez had falsely claimed to have rescued the American Ambassador during the Iranian hostage crisis and had claimed to have been a helicopter pilot during the Vietnam War. Other misrepresentations included playing hockey for the Detroit Red Wings, working as a police officer, and having been secretly married to a Mexican movie star.

Alvarez pled guilty of one count of falsely claiming that he was awarded the Congressional Medal of Honor and was sentenced to pay a $100 special assessment, a $5,000 fine, to serve three years of probation, and to perform 416 hours of community service.

The Court of Appeals held the Stolen Valor Act was unconstitutional reasoning that "the right to speak and write whatever one chooses—including, to some degree, worthless, offensive, and demonstrable untruths—without cowering in fear of a powerful government is, in our view, an essential component of the protection afforded by the First Amendment."

The appellate court also observed that Stolen Valor Act did not constitute a false pretenses fraud statute. The government was not required to establish that Alvarez's "statement was material, intended to mislead, or most critically, did mislead the listener. . . . [I]f anything, Alvarez has no credibility whatsoever and . . . no one detrimentally relied on his false statement."

What of the argument that "Congress certainly has an interest, even a compelling interest, in preserving the integrity of its system of honoring our military men and women for their service and, at times, their sacrifice." See *United States v. Alvarez*, 617 F.3d 1198 (9th Cir. 2010).

The U.S. Supreme Court upheld the appellate court decision. Justice Anthony Kennedy stated that the government may validly limit "false" speech that is used to fraudulently obtain a material benefit such as money or employment. However, "[w]ere the Court to hold that the interest in truthful discourse [on military medals] alone is sufficient to maintain a ban on speech, absent any evidence that the speech was used to gain material advantage, it would give government a broad censorial power unprecedented in the Court's cases or in our constitutional tradition." *United States v. Alvarez*, 617 F.3d 1198 (9th Cir. 2010), *affirmed* ___ U.S. ___ (2012).

You Decide

13.3 Ronald Nellon inherited $142,409. He is described as possessing an obviously "subnormal intellectual capacity." During a period of roughly five weeks in 1992, Nellon purchased six trucks from the general manager of Quirk Chevrolet in Braintree, Massachusetts. Nellon would purchase a truck and soon thereafter "trade it in" for a more expensive truck, receiving far less than he paid in the trade-in. For instance, on July 9, 1992, Nellon received a trade-in allowance of $5,876 on a truck that he bought the previous day for $14,625. The odometer read 26 miles. The dealership's profit margin was $4,313. The next day, he received a trade-in allowance of $5,530 on a truck that he purchased the day before for $13,818. The odometer read 20 miles, and the profit came to $5,085.

Reske, the general manager of the auto dealership, was charged with false pretenses. The court ruled that a false statement may be inferred from the "inordinate profit margin . . . the manifestly unrealistic trade-in allowances, and from the inflation over sticker prices," all of which were contrary to customary practice. Nellon overpaid roughly $23,651 on the six transactions. Was Reske guilty of false pretenses? Was he guilty of larceny by trick? Note that a Florida statute makes it a crime to obtain or use the funds of an elderly or disabled person if the accused "knows or reasonably should know the elderly person or disabled adult lacks the capacity to consent." See *Commonwealth v. Reske*, 684 N.E.2d 631 (Mass. App. Ct. 1997).

You can find the answer at www.sagepub.com/lippmanccl3e.

THEFT

A number of states have consolidated larceny, embezzlement, and false pretenses into a single **theft statute**. The Model Penal Code and several state provisions also include within their theft statutes the property offenses of receiving stolen property, blackmail or extortion, the taking of lost or mistakenly delivered property, theft of services, and the unauthorized use of a vehicle.

Larceny, embezzlement, and false pretenses are all directed against wrongdoers who unlawfully interfere with the property interests of others, whether through "taking and asportation," "converting," or "stealing." The commentary to the Model Penal Code explains that each of these property offenses involves the "involuntary" transfer of property, and in each instance, the perpetrator "appropriates property of the victim without his consent or with a consent that is obtained by fraud or coercion."[22]

How do these consolidated theft statutes make it easier for prosecutors to charge and convict defendants of a property offense? A prosecutor under these consolidated theft statutes may charge a defendant with "theft" and, in most jurisdictions, is not required to indicate the specific form of theft with which the defendant is charged. The defendant will be convicted in the event that the evidence establishes beyond a reasonable doubt either larceny, embezzlement, or false pretenses. A prosecutor under the traditional approach would be required to charge a defendant with the separate offenses of larceny, embezzlement, or false pretenses. The defendant would be acquitted in the event that he or she was charged with deceitfully obtaining possession and title (false pretenses) but the evidence established that the defendant only obtained possession (larceny).[23]

The classic example of how separately defining property offenses impedes prosecution is a case involving an English woman who wrote movie star Clark Gable alleging that he was the father of her child. The letter was based on a misrepresentation of fact because Gable had not been in England at the time that the child was conceived. The writer was charged with the use of the mail to defraud Gable, a form of false pretenses. A federal appeals court reversed her conviction, finding that she intended to intimidate Gable into giving her money based on the unstated threat to publicize her allegations and used the mail to "extort" money rather than to "defraud" Gable.[24]

The Pennsylvania consolidated theft statute, section 3902, provides that conduct considered theft "constitutes a single offense. An accusation of theft may be supported by evidence that it was committed in any manner that would be theft under this chapter, notwithstanding the specification of a different manner in the complaint or indictment."

Consolidated statutes typically grade the severity of larceny, embezzlement, and false pretenses in a uniform fashion based on the value of the property, whether the stolen property is a firearm or motor vehicle, and on factors such as whether the offense took place during the looting of a disaster area.

Model Penal Code

Section 223.1. Consolidation of Theft Offenses

(1) Conduct denominated theft in this Article constitutes a single offense. An accusation of theft may be supported by evidence that it was committed in any manner that would be theft under the Article, notwithstanding the specification of a different manner in the indictment or information, subject only to the power of the Court to ensure fair trial by granting a continuance or other appropriate relief where the conduct of the defense would be prejudiced by lack of fair notice or by surprise.

(2) Grading of Theft Offenses:

(a) Theft constitutes a felony of the third degree if the amount involved exceeds $5,000, or if the property stolen is a firearm, automobile, airplane, motorcycle, motorboat, or other motor-propelled vehicle, or in the case of theft by receiving stolen property, if the receiver is in the business of buying or selling stolen property.

(b) Theft not within the preceding paragraph constitutes a misdemeanor, except that if the property was not taken from the person or by threat, or in breach of a fiduciary obligation, and the author proves by a preponderance of the evidence that the amount involved was less than $50, the offense constitutes a petty misdemeanor.

(c) The amount involved in a theft shall be deemed to be the highest value, by any reasonable standard, of the property or services which the actor stole or attempted to steal. Amounts involved in thefts committed pursuant to one scheme or course of conduct . . . may be aggregated in determining the grade of the offense.

(3) It is an affirmative defense to prosecution for theft that the actor:

(a) was unaware that the property or service was that of another;

(b) acted under an honest claim of right to the property or service involved or that he had a right to acquire or dispose of it as he did; or

(c) took property exposed for sale, intending to purchase and pay for it promptly, or reasonably believing that the owner, if present, would have consented.

(4) It is no defense that theft was from the actor's spouse except that misappropriation of household and personal effects, or other property normally accessible to both spouses, is theft only if it occurs after the parties have ceased living together.

Two new forms of theft that were not anticipated when the consolidated theft statutes were drafted are identity theft and computer crimes.

IDENTITY THEFT

William Shakespeare wrote that stealing "my good name" enriches the thief, while making the victim "poor indeed." Today **identity theft**, the theft of your name and identifying information, can lead to economic damage and has been called the crime of the twenty-first century. The stealing of your social security number, bank account information, credit card number, and other identifying data enables thieves to borrow money and make expensive purchases in your name. The end result is the ruining of your credit and creation of financial hardship, forcing you to spend months restoring your "good name." The Department of Justice points to a case in which a thief accumulated $100,000 in credit card debt, obtained a federal home loan and bought a house and motorcycle, and filed for bankruptcy in the victim's name.

In the past, thieves threatened to "take and carry away" tangible property. Today, the theft of intangible property, such as a credit card or social security number, may lead to even greater harm because the thief can employ the number to make repeated purchases, to borrow money, or to establish phone service or cable access. You might not even be aware that the information was taken until you apply for credit and are rejected.

Thieves collect data by examining receipts you abandon in the trash, observing the numbers you enter at an ATM, intercepting mail from credit card companies, or enticing you to surrender information to what appears to be a reputable e-mail inquiry. A lost or stolen wallet or burglary of a home or automobile can result in valuable numbers and documents falling into the hands of organized identity theft gangs. Information can also be obtained by breaking into a company database. Individuals falsely portraying themselves as legitimate business executives, for instance, gained access to ChoicePoint in 2005 and stole roughly 145,000 credit files. Shortly thereafter, computer hackers copied the files of 30,000 individuals contained on the database of LexisNexis. The Bank of America reported the disappearance of a computer tape containing the files of more than a million customers.

Even strict protections over your personal information may not be effective. A study by the Federal Trade Commission, the federal agency concerned with consumer protection, determined that roughly fifteen percent of identity thefts are committed by a victim's family members, friends, neighbors, or co-workers. The perpetrators of identity theft often transfer the information to sophisticated gangs of identity thieves in return for drugs, cellular phones, guns, and money.

The Federal Trade Commission reports that 3.2 million Americans are victimized by identity theft each year and find themselves with unwarranted financial obligations or may even be charged with crimes as a result of another person assuming their identity. The commission estimates that an identity theft occurs every ten seconds. The Congressional Research Service estimates that more than ten million Americans are the victims of identity theft each year. The states with the highest rates of identity theft per 100,000 people in 2011 were Florida (178.7), Georgia (120.0), California (103.6), Arizona (98.5), Texas (96.1), New York (92.3), Nevada (89.9), New Jersey (86.4), Maryland (86.3), and Delaware (83.5).

Consider how easy it is to intercept your mail and to obtain a preapproved credit card. Think about how many times in an average week you are vulnerable to identity theft resulting from sensitive information in a letter, using your credit card, discarding a receipt, or providing someone with your social security number.

In 1998, the U.S. Congress passed the Identity Theft and Assumption Deterrence Act. This legislation created the new offense of identity theft and prohibits the knowing transfer or use without lawful authority of the "means of identification of another person with the intent to commit, or to aid or abet, any unlawful activity" that constitutes a violation of federal law or a felony under state or local law.[25] A "means of identification" includes an individual's name, date of birth, social security number, driver's license number, passport number, bank account or credit card number, fingerprints, voiceprint, or eye image. Note that merely obtaining another individual's personal documents can result in a year or more in prison. Sentences for violation of the Identity Theft Act can be as severe as fifteen years in prison and a significant fine in those instances in which the perpetrator obtains items valued at $1,000 or more over a one-year period. The perpetrators of identity theft typically are also in violation of statutes punishing credit card fraud, mail fraud, or wire fraud.

The Utah statute on identity theft punishes an individual who knowingly or intentionally obtains "personal identifying information of another person" and uses or attempts to use this information "with fraudulent intent, including to obtain, or attempt to obtain credit, goods, services . . . or medical information in the name of another person." Obtaining items valued at more than $1,000 is a felony under the Utah law.[26]

In 2007, in *State v. Green,* a Kansas Court of Appeals ruled that each time a thief uses a stolen credit card constitutes a separate offense. The defendant opened credit accounts at three stores using another individual's identity and subsequently was convicted of three counts of identity theft. The court explained that a thief who steals money harms the victim only once, whether or not he later spends the money, while each use of a stolen credit card is a "blow to the body of credit established by an innocent person." Every use of the innocent's identity "takes something away from that person in this modern age of credit history and instantaneous commercial transportation."[27] In *City of Liberal, Kansas v. Vargas,* a Kansas appellate court was asked to decide whether an undocumented immigrant who used a false identity to obtain employment committed identity theft.

Did Vargas commit identity theft?

City of Liberal, Kansas v. Vargas,
24 P.3d 155 (Kan. App. 2001). Opinion by: Marquardt, J.

The City of Liberal (City) appeals the district court's ruling that Juan Vargas did not commit the crime of identity theft. We affirm.

Facts

In February 2000, Officer Rogers was on duty and noticed a car being driven without a tag light. Officer Rogers stopped the vehicle and asked the driver for his driver's license and proof of insurance. The driver produced Missouri and National Beef Packing Company (National Beef) identification cards bearing the name of Guillermo Hernandez. Officer Rogers ran the name through a regional system. The system produced five identical matches in other states.

The driver admitted that his name was Juan Vargas. He later confessed that he was not authorized to work in the United States and that he had bought papers identifying himself as Guillermo Hernandez so that he could obtain employment at National Beef. Vargas pled guilty in municipal court to one count of identity theft, one count of no seat belt, one count of no tag light, and one count of no driver's license. Vargas appealed the municipal court's decision to the district court.

The district court convicted Vargas of one count of no tag light, one count of no seat belt, and one count of no driver's license. Vargas was acquitted on the count of identity theft. The district court held that the City failed to meet its burden to show fraud in the use of the false identity card. The City appeals. . . .

Issue

The City asks this court to determine whether Vargas's use of a false identification to secure employment and receive the economic benefit of a salary is tantamount to defrauding another person.

Reasoning

"Identity theft is knowingly and with intent to defraud for economic benefit, obtaining, possessing, transferring, using or attempting to obtain, possess, transfer or use, one or more identification documents or personal identification number of another person other than that issued lawfully for the use of the possessor." K.S.A. 2000 Supp. 21-4018(a).

"'Intent to defraud' means an intention to deceive another person, and to induce such other person, in reliance upon such deception, to assume, create, transfer, alter or terminate a right, obligation or power with reference to property." K.S.A. 21-3110(9).

It is a fundamental rule of statutory construction that the intent of the legislature governs if that intent can be ascertained. The legislature is presumed to have expressed its intent through the language of the statutory scheme it enacted. Without state or federal cases to guide our inquiry, we turn to the legislative history of K.S.A. 2000 Supp. 21-4018.

The crime of identity theft was created by the 1998 Kansas Legislature. Representative Bonnie Sharp, a proponent of the legislation, believed it was necessary to criminalize identity theft. She cited the example of a person's social security number being used by another to obtain an illegal checking account and/or a credit card. Representative Sharp stated that the citizens of Kansas should be protected from "this potentially devastating crime." . . .

Kyle Smith, an assistant attorney general for the Kansas Bureau of Investigation, testified on behalf of the bill. Smith was concerned that the "surreptitious acquisition of information done with the intent to defraud" was not illegal at the time the bill was proposed. Smith cited the examples of a motel clerk selling a credit card number or a "trasher" obtaining a social security number, which would allow an individual to access other information, leading to theft. . . .

The committee also heard testimony from Dave Schroeder, a special agent from the Kansas Bureau of Investigation. Agent Schroeder defined identity theft as "acquiring someone's personal identifying information in an effort to impersonate them or commit various criminal acts in that person's name." Agent Schroeder went on to state that individuals who are armed with a stolen identity can commit numerous forms of fraud. Agent Schroeder was specifically concerned about the theft of personal information such as social security numbers, birth certificates, passports, driver's licenses, dates of birth, addresses, telephone numbers, family history information, credit or bank card numbers, and personal identification numbers.

K.S.A. 2000 Supp. 21-4018 requires that a defendant obtain, possess, transfer, use, or attempt to obtain the identification documents or personal identification numbers of another. This would occur, for example, when a defendant "took" another person's social security number and used that number when applying for a credit card or bank account.

In the case currently before the court, Vargas admitted that he bought the identification. The record on appeal reflects that Vargas obtained a Missouri identification card as well as a social security card. However, there is no evidence that Guillermo Hernandez is a real person who had his identity "stolen."

It appears from the committee minutes that the legislature passed K.S.A. 2000 Supp. 21-4018 in order to protect individuals who have their identity stolen. The testimony is replete with references to individuals who have been defrauded by perpetrators who misappropriate personal information such as a social security or bank account number. There was no mention of any intent by the legislature to protect a third party from identity theft. We do not believe that the legislature intended to criminalize the act of which Vargas is now accused. Vargas lied to his employer and obtained employment through false means. There are other appropriate remedies for a situation such as this. Vargas was terminated from his employment. There may also be federal immigration laws which apply. However, we do not believe that K.S.A. 2000 Supp. 21-4018 is applicable to the situation currently before the court.

However, even if K.S.A. 2000 Supp. 21-4018 did apply to the facts of this case, we do not believe that Vargas could be found guilty. The statute requires an "intent to defraud." Vargas admitted that he intended to use a false identity only to work at National Beef. There is no evidence in the record on appeal that Vargas intended to defraud National Beef by stealing money or by being compensated for services not actually rendered.

In addition, we fail to see how Vargas received the economic benefit mentioned in K.S.A. 2000 Supp. 21-4018. Vargas was paid for the time he worked. Vargas's work product was rated as satisfactory. In fact, the record on appeal indicates that National Beef would rehire Vargas in the future.

The City's main argument is that National Beef relied upon Vargas's statement that he was Guillermo Hernandez when it hired him. Essentially, the City claims that National Beef would not have hired Vargas had it known of his status as an illegal alien. This fact is not borne out in the record on appeal. The personnel director for National Beef testified that he had a "responsibility" to refrain from hiring undocumented workers. However, there is nothing in the record on appeal that would indicate that National Beef was in any way defrauded by Vargas's actions.

Holding

Vargas did not steal Guillermo Hernandez's identity in order to commit a theft. He bought identification under a false name so that he could work in this state. There is no evidence that Vargas intentionally defrauded anyone in order to receive a monetary benefit. In addition, Vargas was appropriately paid for services rendered. The district court correctly interpreted K.S.A. 2000 Supp. 21-4018 when it reversed Vargas's conviction of identity theft.

Questions for Discussion

1. Explain the basis of the Kansas appellate court's reversal of Vargas's criminal conviction for identity theft.

2. Construct an argument that affirms Vargas's conviction using the language of the Kansas identity theft statute.

3. Would this case have been decided differently had Vargas been using a social security number assigned to another individual?

4. Did the need for low-wage labor in the beef-packing industry play a role in the appellate court's decision?

Cases and Comments

Identity Theft. In *Flores-Figueroa v. United States,* Ignacio Flores-Figueroa, a Mexican citizen, provided his employer with a false social security number and a counterfeit alien registration number in 2000. In 2006, Flores-Figueroa presented his employer with new counterfeit Social Security and alien registration cards. These cards (unlike Flores-Figueroa's old alien registration card) used his real name. The numbers on both cards were in fact numbers assigned to other individuals. The U.S. Supreme Court held that Congress was concerned with classic cases of identity theft in which an individual hacked a computer to obtain bank account numbers or in which an individual went through the trash to find credit card numbers. The Supreme Court held that the federal identity theft statute, 18 U.S.C. §1028A(a)(1), requires that a defendant know that the "means of identification" he or she transfers, possesses, or uses belongs to "another person." Flores-Figueroa, as a result, could not be held liable for identity theft under federal law. Does requiring knowledge that the numbers belong to "another person" make enforcement of the statute difficult? See *Flores-Figueroa v. United States,* ___ U.S. ___, 173 L. Ed. 2d 853 (2009).

See more cases on the study site: State v. Ramirez, *www.sagepub.com/lippmanccl3e.*

You Decide

13.4 Alfredo Ramirez was an illegal resident of the United States and did not possess a social security number. In September 1997, he obtained a job at Trek Bicycle in Walworth County, Wisconsin. In June 1999, the firm discovered that Ramirez's social security number belonged to Benjamin Wulfenstein of Elko, Nevada. Ramirez admitted that he used Wulfenstein's social security number without permission to obtain employment and to continue to be paid over the course of almost two years of work. Ramirez was terminated on July 6, 1999. The Wisconsin statute makes it a crime to intentionally use the personal identifying information or document of another for purposes of obtaining "credit, money, goods, services or anything else of value" without the consent of the other person and by representing that the actor is the other person or is acting with the consent or authorization of such person. Did Ramirez obtain "anything of value"? How would you respond to Ramirez's argument that this is an *ex post facto* prosecution based on the fact that the Wisconsin statute did not take effect until April 27, 1999? See *State v. Ramirez,* 644 N.W.2d 656 (Wis. Ct. App. 2001).

You can find the answer at www.sagepub.com/lippmanccl3e.

COMPUTER CRIME

Computer crime poses a challenge for criminal law. Larceny historically has protected tangible (material) property. Courts have experienced difficulty applying the traditional law of larceny to individuals who gain access to intangible property without authorization (nonmaterial property that you cannot hold in your hands). The primary property offenses committed in cyberspace include unauthorized computer access to programs and databases and unlawfully obtaining personal information through deceit and trickery. Can you take and carry away access to a computer program?

In *Lund v. Commonwealth,* the defendant, a graduate student in statistics at Virginia Polytechnic Institute, was charged with the larceny and the fraudulent use of "computer operation and services" valued at $100 or more. The customary procedure at the university was for departments to receive computer dollar credits. These dollar credits were deducted from the departmental account as faculty and students made use of the university's central computer. This was a bookkeeping procedure, and no funds actually changed hands. Lund's adviser failed to arrange for his use of the university computer, and Lund proceeded to gain access to the university computer without authorization and spent as much as $26,384.16 in unauthorized computer time. The Virginia Supreme Court ruled that computer time and services could not be the subject of either false pretenses or larceny "because neither time nor services may be taken and carried away. . . . It [the Virginia statute] refers to a taking and carrying away of a certain concrete article of personal property."[28]

State legislatures and the federal government responded to *Lund* by passing statutes addressing computer theft and crime. These statutes punish various types of activity including unauthorized access to a computer or to a computer network or program; the modification, removal, or disabling of computer data, programs, or software; causing a computer to malfunction; copying computer data, programs, or software without authorization; and falsifying e-mail transmissions in connection with the sending of unsolicited bulk e-mail. The question remains whether law enforcement possesses the expertise and resources to track sophisticated cybercriminals.

In *State v. Schwartz,* an Oregon appellate court applies the state's computer crime statute to a defendant who is charged with the theft of passwords from a computer. In reading this case, ask yourself the difference between stealing tangible property and stealing passwords. It is a sign of our technological world that a thief can often cause greater harm by stealing your personal data than by breaking into your house.

Was the defendant guilty of theft when he gained unauthorized access to passwords?

State v. Schwartz, 21 P.3d 1128 (Ore. Ct. App. 1999). Opinion by: Deits, J.

Facts

Defendant worked as an independent contractor for Intel Corporation beginning in the late 1980s. Defendant's tasks included programming, system maintenance, installing new systems and software, and resolving problems for computer users. In late 1991 or early 1992, defendant began working in Intel's Supercomputer Systems Division (SSD). SSD creates large computer systems that can cost millions of dollars and are used for applications such as nuclear weapons safety. Intel considers the information stored on its SSD computers to be secret and valuable. Each person using SSD computers must use a unique password in order to gain access to electronic information stored there. Passwords are stored in computer files in an encrypted or coded fashion.

In the spring of 1992, defendant and Poelitz, an Intel systems administrator, had a disagreement about how defendant had handled a problem with SSD's e-mail system. The problem was ultimately resolved in an alternative manner suggested by Poelitz, which upset defendant and made him believe that any future decisions he made would be overridden. Accordingly, defendant decided to terminate his SSD contract with Intel. As defendant himself put it, he "hadn't left SSD on the best of terms." At that time, his personal passwords onto all but one SSD computer were disabled so that defendant would no longer have access to SSD computers. His password onto one SSD computer, Brillig, was inadvertently not disabled.

After defendant stopped working with SSD, he continued to work as an independent contractor with a different division of Intel. In March 1993, Brandewie, an Intel network programmer and systems administrator, noticed that defendant was running a "gate" program on an Intel computer called Mink, which allowed access to Mink from computers outside of Intel. "Gate"

programs like the one defendant was running violate Intel security policy, because they breach the "firewall" that Intel has established to prevent access to Intel computers by anyone outside the company. Defendant was using the gate program to access his e-mail account with his publisher and to get access to his Intel e-mail when he was on the road. When Brandewie talked to defendant about his gate program, defendant acknowledged that he knew that allowing external access to Intel computers violated company policy. Even though defendant believed that precautions he had taken made his gate program secure, he agreed to alter his program.

In July 1993, Brandewie noticed that defendant was running another gate program on Mink. This program was similar to the earlier gate program and had the same effect of allowing external access to Intel computers. Defendant protested that changes he had made to the program made it secure, but Brandewie insisted that the program violated company policy. At that point, defendant decided that Mink was useless to him without a gate program, so he asked that his account on that computer be closed. Defendant then moved his gate program onto an Intel computer called Hermeis. Because that computer was too slow for him, defendant finally moved his gate program onto the SSD computer Brillig.

In the fall of 1993, defendant downloaded from the Internet a program called "Crack," which is a sophisticated password guessing program. Defendant began to run the Crack program on password files on various Intel computers. When defendant ran the Crack program on Brillig, he learned the password for "Ron B.," one of Brillig's authorized users. Although he knew he did not have the authority to do so, defendant then used Ron B.'s password to log onto Brillig. From Brillig, he copied the entire SSD password file onto another Intel computer, Wyeth. Once the SSD password file was on Wyeth, defendant ran the Crack program on that file and learned the passwords of more than 35 SSD users, including that of the general manager of SSD. Apparently, defendant believed that if he could show that SSD's security had gone downhill since he had left, he could reestablish the respect he had lost when he left SSD. Once he had cracked the SSD passwords, however, defendant realized that although he had obtained information that would be useful to SSD, he had done so surreptitiously and had "stepped out of my bounds." Instead of reporting what he had found to anyone at SSD, defendant did nothing and simply stored the information while he went to teach a class in California.

After he returned from California, defendant decided to run the Crack program again on the SSD password file, this time using a new, faster computer called "Snoopy." Defendant thought that by running the Crack program on the SSD password file using Snoopy, he would have "the most interesting figures" to report to SSD security personnel. On October 28, 1993, Mark Morrissey, an Intel systems administrator, noticed that defendant was running the Crack program on Snoopy. At that point, Morrissey contacted Richard Cower, an Intel network security specialist, for advice about how to proceed. In investigating defendant's actions, Morrissey realized that defendant had been running a gate program on the SSD computer Brillig, even though defendant's access should have been canceled. On October 29, 1993, Cower, Morrissey, and others at Intel decided to contact police.

Defendant was charged with three counts of computer crime, ORS 164.377, and was convicted of all three counts by a jury.

ORS 164.377 includes three separately defined crimes:

> (2) Any person commits computer crime who knowingly accesses, attempts to access or uses, or attempts to use, any computer, computer system, computer network or any part thereof for the purpose of:
>
> > (a) Devising or executing any scheme or artifice to defraud;
> >
> > (b) Obtaining money, property or services by means of false or fraudulent pretenses, representations or promises; or
> >
> > (c) Committing theft, including, but not limited to, theft of proprietary information.
>
> (3) Any person who knowingly and without authorization alters, damages or destroys any computer, computer system, computer network, or any computer software, program, documentation or data contained in such computer, computer system or computer network, commits computer crime.
>
> (4) Any person who knowingly and without authorization uses, accesses or attempts to access any computer, computer system, computer network, or any computer software, program, documentation or data contained in such computer, computer system or computer network, commits computer crime.

Issue

Defendant argues, the State concedes, and we agree that the indictment alleged, and the State attempted to prove at trial, only that defendant violated ORS 164.377(2)(c). The parties do not dispute that the State proved that defendant "knowingly accessed . . . used . . . any computer, computer system, computer network or any part thereof" as required by ORS 164.377(2)(c). The parties dispute only whether the evidence was sufficient to establish that defendant did so "for the purpose of . . . committing theft, including, but not limited to, theft of proprietary information." . . .

ORS 164.377 does not define "theft." However, the legislature has defined "theft" in a related statute, ORS 164.015. ORS 164.015 provides, in part, that a person commits theft when, with intent to deprive another of property or to appropriate property to the person or to a third person, the person: "Takes, appropriates, obtains or withholds such property from an owner thereof." "Property" entails "any . . . thing of value." The parties do not dispute that the password file and individual passwords have value, and there is evidence in the record to support that proposition. The parties dispute, however, whether defendant "took, appropriated, obtained or withheld" the password file and individual passwords.

Defendant argues that he could not have "taken, appropriated, obtained or withheld" the password file and individual passwords because, even though he moved them to another computer and took them in the sense that he now had them on his computer, the file and passwords remained on Intel's computers after he ran the Crack program. The individual users whose passwords defendant had obtained could still use their passwords just as they had before. Intel continued to "have" everything it did before defendant ran the Crack program, and consequently, defendant reasons, he cannot be said to have "taken" anything away from Intel.

The State responds that by copying the passwords, defendant stripped them of their value. The State contends that like proprietary manufacturing formulas, passwords have value only so long as no one else knows what they are. Once defendant had copied them, the passwords were useless for their only purpose, protecting access to information in the SSD computers. The loss of exclusive possession of the passwords, according to the State, is sufficient to constitute theft.

Reasoning

Under ORS 164.015, theft occurs, among other ways, when a person "takes" the property of another. "Take" is a broad term with an extensive dictionary entry. Some of the dictionary definitions undermine defendant's argument. The first definition of "take" is "to get into one's hands or into one's possession, power, or control by force or stratagem. . . ." Another definition provides "to adopt or lay hold of for oneself or as one's own. . . ." Still another source defines "take" to include "to obtain possession or control. . . ." These definitions indicate that the term "take" might include more than just the transfer of exclusive possession that defendant proposes. For example, "take" could include obtaining control of property, as defendant did with respect to the passwords and password file by copying them.

Turning back to the text of the statute under which defendant was charged, we note that the legislature contemplated that "theft" as used in ORS 164.377(2)(c) could be exercised upon, among other things, "proprietary information." "Proprietary information" includes "scientific, technical or commercial information . . . that is known only to limited individuals within an organization. . . ." Proprietary information, like the passwords and password files at issue here, is not susceptible to exclusive possession; it is information that by definition, can be known by more than one person. Nevertheless, the legislature indicated that it could be subject to "theft."

Holding

We conclude that the State presented sufficient evidence to prove that by copying the passwords and password file, defendant took property of another, namely, Intel, and that his actions, therefore, were for the purpose of theft. The trial court did not err in denying defendant's motion for judgment of acquittal on counts two and three.

Questions for Discussion

1. Explain in clear and understandable fashion the basic facts that form the basis of the charge brought against Schwartz.

2. Why does the defendant contend that he is not guilty? How does Oregon respond to this argument? What is the ruling of the court?

3. Do you agree with the court that a password may be the subject of theft? What is the difference between the theft of a watch and the theft of a password? Why was a special statute required to prosecute the defendant for stealing passwords?

4. At what point did the defendant commit a computer crime? Was it when he looked at a password without authorization or when he transferred the passwords to his own computer? Do passwords retain their value indefinitely?

You Decide

13.5 Lori Drew, forty-nine, created a MySpace account in 2007 under the name of Josh Evans, a fictitious sixteen-year-old male. "Josh" started corresponding with thirteen-year-old Megan Meir. "Josh" told Megan that he recently had moved into a nearby town and they corresponded for several weeks. Josh's tone changed at some point during the correspondence and he wrote in an instant message that "the world would be a much better place without you." Megan responded to Josh that he was the kind of boy that "a girl would kill herself over," and shortly thereafter, on October 16, 2008, Megan committed suicide. Lori was angry over Megan's alleged gossip about her daughter Sarah and knew that Megan suffered from depression and harbored thoughts of suicide. Lori was prosecuted by federal authorities under the Computer Fraud and Abuse Act (CFAA). The jury found that Lori intentionally had violated the terms of service of MySpace and that as a result, she was guilty of "accessing a computer involved in interstate or foreign communication without authorization or in excess of authorization to obtain information. . . ." Is the intentional breach of an Internet Web site's terms of service sufficient to constitute a criminal violation of the CFAA? See *United States v. Lori Drew*, 259 F.R.D. 449 (C.D. Cal. 2009).

CRIME IN THE NEWS

The gains in student test scores achieved by Atlanta students over the last twelve years has been called into question by findings of a report issued by Georgia Governor Nathan Deal. The report discovered a pattern of cheating on standardized tests in 44 schools of the 56 schools that were studied, involving 178 teachers, 82 of whom admitted to changing the answers of students. The scandal in Atlanta has been called the worst educational cheating scandal in U.S. history.

The report on Atlanta public schools found a widespread, far-reaching conspiracy by teachers, administrators, and principals to manipulate the results of the

Criterion Referenced Competing Test (CREPT). Brenda Muhammad, Chair of the Atlanta School Board, called the report "devastating" and lamented that "You just don't cheat children."

Superintendent Beverly Hall's success in allegedly raising the test scores of Atlanta's 55,000 public school students led to her being named as "Superintendent of the Year" in 2009. Atlanta was praised as a model for the rest of the nation, and the district received funding from the Bill and Melinda Gates Foundation and from the Broad Foundation.

The Georgia report documents unethical behavior at every level of the Atlanta school system. Teachers who questioned the accuracy of standardized test scores were threatened, reprimanded, and criticized as troublemakers. One of the most successful principals was Christopher Waller of Park Middle School, who was paid over $100,000 a year. In a single year under Principal Waller's leadership, the eighth graders at Park who met or exceeded standards in reading increased from 50 to 81 percent; and the percentage of eighth graders who met or exceeded the standards in math increased from 24 percent to 86 percent. This bright picture was tarnished by the disclosure that Waller encouraged the teachers in the school to erase incorrect answers and to change them to correct answers.

A grand jury has been convened in Georgia to determine whether criminal charges should be brought against various teachers and administrators. A conviction for making false statements to investigators, altering public documents, or knowingly submitting false test scores could carry a prison term of between 5 and 10 years.

Atlanta schools are not alone in being enveloped in a cloud of suspicion. Investigations are currently underway in at least six states. The cheating scandal has spread to the District of Columbia (D.C.). The former D.C. school superintendent, Michelle Rhee, was praised for her tough-minded leadership of the school system between 2009 and 2010 and for introducing revolutionary reforms to public education. A study by *USA Today* found a high rate of erasures and unusually high test scores in 41 schools during Rhee's tenure as superintendent of the D.C. system. At Noyes Elementary school in 2008, 84 percent of fourth graders were competent in math, an increase of 62 percent in one year. Ninety-seven percent of the erasures on answer sheets at Noyes were "from wrong to right."

A number of states make student test scores the single most important factor in faculty retention and bonuses. Many states have combated the epidemic of cheating by introducing sophisticated computer systems to audit student tests. Arrests in wealthy Nassau County, Long Island, in New York have revealed that a number of students paid between $500 and $3,600 to individuals to take their SAT. Roughly 3,000 SAT scores are canceled each year because of cheating.

Should teachers and administrators who "cheat" on tests be prosecuted criminally for fraud?

RECEIVING STOLEN PROPERTY

There was no offense of **receiving stolen property** under the common law. An English court, in 1602, condemned a defendant who knowingly purchased a stolen pig and cow as an "arrant knave" and complained that there was "no separate crime of receiving stolen property."[29] In the late seventeenth century, the English Parliament passed a law providing that an individual who knowingly bought or received stolen property was liable as an accessory after the fact to theft. In 1827, Parliament passed an additional statute declaring that receiving stolen property was a criminal offense. This law was later incorporated into the criminal codes of American states and, today, is punished as a misdemeanor or felony, depending on the value of the property.

The offense of receiving stolen property requires that an individual

- receive property,
- knowing the property to be stolen,
- with the intent to permanently deprive the owner of the property.

Why do we punish receiving stolen property as a separate offense? Thieves typically sell stolen property to "fences," individuals who earn a living by buying and then selling stolen property. The offense of receiving stolen property is intended to deter "fencing." Generally, an individual may not be charged with both stealing and receiving stolen property.

Actus Reus

The *actus reus* of receiving stolen property requires that an individual control the stolen property, however briefly. An individual receiving the stolen items may take either actual possession of the property or constructive possession of the property by arranging for the property to be delivered to a specific location or to another individual.

Receiving stolen property traditionally was limited to goods that were taken and carried away in an act of larceny. The trend is to follow the approach of the Model Penal Code and to punish the receipt of stolen property, whether taken through larceny, embezzlement, false pretenses, or another illegal method.

Most state statutes on receiving stolen property cover both personal and real property. The Model Penal Code limits the statute to personal property on the grounds that this property is disposed of through fences and that this is not the case with real estate.

The property must actually be stolen. A defendant was acquitted of receiving stolen property in a case in which thieves were arrested in the course of breaking and entering into a railroad car and then cooperated with the police in arranging to "sell" the tires to a fence. The Sixth Circuit Federal Court of Appeals ruled that the defendant was not guilty, because the defendant had not purchased "stolen property."[30]

Mens Rea

State statutes typically require the *mens rea* of actual knowledge that the goods are stolen. Other statutes broaden this standard by providing that it is sufficient for an individual to believe that the goods are stolen. A court would likely conclude that a jeweler believed that a valuable watch was stolen that he or she inexpensively purchased from a known dealer in stolen merchandise. A third group of statutes applies a recklessness or negligence standard to the owners of junkyards, pawnshops, and other businesses where they neglect to investigate the circumstances under which the seller obtained the property. Consider the case of the owner of an art gallery specializing in global art who regularly buys rare and valuable Asian and African artwork that is thousands of years old and who, in one instance, buys a piece for next to nothing from individuals who wander into the shop. These statutes would hold the buyer guilty of receiving stolen property for failing to investigate how the seller obtained the property.

How can we determine whether an individual knows or honestly believes that property is stolen? Courts generally hold that it is sufficient if a reasonable person would have possessed this awareness. In most cases, this is inferred from the price, the seller, whether the type of property is frequently the subject of theft, the circumstances of the sale, and whether the recipient purchased stolen merchandise from the same individual in the past.

For an international perspective on this topic, visit the study site.

The recipient of stolen property must also have the *mens rea* to permanently deprive the owner of possession. A defendant does not possess the required intent who believes that he or she is the actual owner of the property, because there is no intent to deprive another of possession. The required intent is also lacking where the recipient intends to return the property to the rightful owner.

The required intent to permanently deprive an individual of possession must concur with the receipt of the property. The Model Penal Code, however, provides that the required intent may arise when an individual receives and only later decides to deprive the owner of possession.

Hurston v. State, the next case in the textbook, challenges you to determine whether the defendant satisfied the standard for receiving possession of stolen property.

Model Penal Code

Section 223.6. Receiving Stolen Property

(1) A person is guilty of theft if he purposely receives, retains, or disposes of movable property of another knowing that it has been stolen, or believing that it has probably been stolen, unless the property is received, retained, or disposed with purpose to restore it to the owner. "Receiving" means acquiring possession, control or title, or lending on the security of the property.

(2) The requisite knowledge or belief is presumed in the case of a dealer who:

  (a) is found in possession or control of property stolen from two or more persons on separate occasions;

  (b) has received stolen property in another transaction within the year preceding the transaction charged; or

  (c) being a dealer in property of the sort received, acquires it for a consideration which he knows is far below its reasonable value.

"Dealer" means a person in the business of buying or selling goods, including a pawnbroker.

Analysis

- Receiving stolen property is limited to property that can be moved and does not include real estate.
- There is no requirement that the purchaser know that the property is in fact stolen; it is sufficient that an individual believe that the property probably has been stolen.
- A defendant must know or believe that the property probably has been stolen. The intent to restore the property to the owner is a defense.
- The required intent may arise after the property is in the possession of the defendant.
- Knowledge is assumed under certain circumstances, including the fact that an individual is a "dealer."
- The receiver is liable regardless of the method employed by the thief, whether larceny, embezzlement, false pretenses, or other form of theft.

The Legal Equation

Receiving stolen property **=** Control over stolen property

+ purposely, knowing (recklessly, negligently) that property is stolen

+ intent to permanently deprive individual of property.

Was Hurston guilty of receiving stolen property?

Hurston v. State, 414 S.E.2d 303 (Ga. App. 1991). Opinion by: Andrews, J.

Illya Hurston was tried jointly with Demetrious Reese and convicted of theft by receiving stolen property. Hurston appeals from the denial of his motion for new trial.

Issue

The issue to be decided is whether there was sufficient evidence for a jury to find Hurston guilty of receiving stolen property.

Facts

A silver 1986 Pontiac Fiero belonging to Stella Burns was stolen from a parking lot at Underground Atlanta on June 11, 1989, between 10:35 and 11:05 p.m. Two Rockdale County sheriff's deputies observed a silver Fiero at a convenience store later that night at approximately 1:20 a.m. Hurston's co-defendant, Reese, was driving the car, and appellant was slumped in the passenger seat. The deputies became suspicious because of the late hour, the cautious manner in which Reese was walking after exiting the car, and the fact that Hurston appeared to be hiding; and the deputies decided to follow the Fiero. When Reese drove away from the store with the deputies following in their marked car, he crossed the centerline of the highway.

The deputies, who by this time had ascertained from computer records that the car was stolen, turned on the blue lights and siren of their automobile. Reese refused to stop, drove away from the deputies at a speed in excess of 100 miles per hour, and attempted at one point to run the police vehicle into a wall. The deputies pursued the Fiero until Reese lost control and wrecked in a field. Reese ran from the scene and was pursued and apprehended by one deputy. Another officer apprehended Hurston, who had gotten out of the car immediately after the accident and appeared to be ready to run.

At trial, Hurston testified that he spent the day at his former girlfriend's home watching television with her and a friend of hers. Later in the day, the friend called her boyfriend, Reese, whom Hurston testified he had never met, to join them. Hurston testified Reese came to the house and stayed for awhile, left for several hours, and then returned and invited Hurston to ride in the Fiero with him to a relative's home. Hurston recalled that he was suspicious about the ownership of the vehicle because Reese, a teenager, seemed too young to own such a nice car but that in response to his inquiry, Reese stated that the car belonged to his cousin. Hurston testified that after they left the convenience store and Reese saw the deputies in pursuit, he began to speed and admitted to Hurston for the first time that the car was stolen. Hurston's trial testimony differed somewhat from an earlier statement he gave regarding the evening's events.

Burns, the vehicle's owner, testified that the vehicle was driven without keys and that the steering wheel had been damaged, which was consistent with it having been stolen. She testified that various papers, including the car registration and business cards bearing the owner's name and address, had been removed from the glove compartment and were on the floor of the car; that grass, mud, food, drink, and cigarettes were scattered in the car; and that a picture of her daughter was displayed on a visor. Hurston denied noticing the personal items or the damaged steering wheel.

Reasoning

OCGA section 1687 provides that "a person commits the offense of theft by receiving stolen property when he receives, disposes of, or retains stolen property which he knows or should know was stolen unless the property is received, disposed of, or retained with intent to restore it to the owner. Receiving means acquiring possession or control."

Unexplained possession of recently stolen property, alone, is not sufficient to support a conviction for receiving stolen property, but guilt may be inferred from possession in conjunction with other evidence of knowledge. Guilty knowledge may be inferred from circumstances which would excite suspicion in the mind of an ordinary prudent man. "Possession as we know it, is the right to exercise power over a corporeal thing. . . ." Furthermore, "[i]f there is any evidence of guilt, it is for the jury to decide whether that evidence, circumstantial though it may be, is sufficient to warrant a conviction."

First, there was sufficient evidence for a jury to find that Hurston knew, or should have known, that the vehicle was stolen. At trial, Hurston admitted that he doubted that the vehicle belonged to Reese. There was evidence from which the jury could reasonably have concluded that Hurston was aware during the two hours that he spent in the small vehicle that it was stolen, in that the vehicle was being driven without keys, the steering wheel was damaged, and the interior was disorderly, which was inconsistent both with Reese's ownership of

the vehicle and with his explanation that he borrowed it from a relative. Hurston's suspicious behavior at the convenience store and his attempt to flee also indicated that he knew the vehicle was stolen.

There was also evidence from which the jury could conclude that Hurston possessed, controlled, or retained the vehicle. Although Hurston was only a passenger in the vehicle, the inquiry does not end here, for in some circumstances, a passenger may possess, control, or retain a vehicle for purposes of OCGA section 1687. Here, there was sufficient evidence that Hurston exerted the requisite control over the vehicle in that Reese left Hurston alone in the car with the vehicle running when he went into the convenience store.

Dissenting, *Sognier, C.J.*

I respectfully dissent, for I find the evidence was insufficient to establish the essential element of "receiving" beyond a reasonable doubt. A person commits the offense of theft by receiving stolen property when he "receives, disposes of, or retains stolen property which he knows or should know was stolen. . . . 'Receiving' means acquiring possession or control . . . of the property." OCGA section 1687(a). Here, the record is devoid of evidence that appellant exercised or intended to exercise any dominion or control over the car or that he ever acquired possession of it. The "mere presence" of a defendant in the vicinity of stolen goods "furnishes only a bare suspicion" of guilt and thus is insufficient to establish possession of stolen property. Evidence that a defendant was present as a passenger in a stolen automobile, without more, is insufficient to establish possession or control. I disagree with the majority that the circumstantial evidence that appellant, the automobile passenger, was observed to be "slumped" in the seat while Reese parked the car and entered a store was sufficient to constitute the type of "other incriminating circumstances" that would authorize a rejection of the general principle that "the driver of the [stolen] automobile [is] held prima facie in exclusive possession thereof."

The only evidence offered by the State to connect appellant to the stolen car was that he was a passenger in the car several hours after it was stolen.

Questions for Discussion

1. What facts does the court rely on to establish that Hurston knew or should have known that the automobile was stolen? Do these facts establish a criminal intent beyond a reasonable doubt?

2. What facts does the court rely on in finding that Hurston possessed or controlled the automobile? On what basis does the dissent dispute the court's determination?

3. What single fact is crucial in the court's finding of guilt? Are there additional facts that are not in the court's opinion that you would find helpful in determining if Hurston is guilty or innocent?

4. As a judge, how would you rule in this case?

You can find more cases on the study site: People v. Land, *www.sagepub.com/Lippmanccl3e.*

You Decide

13.6 John L. Clough discovered various items missing from his music club. These items included four amplifier speakers, which were used by bands that played at the club. An employee, Gaylord Burton, worked at the club for several months and disappeared at the same time that the speakers were discovered to be missing. An employee reportedly had seen Burton taking the speakers on the morning of November 2, 1989. The equipment later was discovered at a pawnshop. An employee of the pawnshop, Anthony Smith, testified that two men had tried to pawn the speakers. Smith refused to accept the speakers without identification. The two men returned later with Olga Lee Sonnier, who presented a driver's license and pawned the speakers for $225. The four speakers were worth at least $350. Sonnier appealed her conviction for "theft by receiving." There are three elements of this offense. First, a theft by another person. Second, the defendant received the stolen property. Third, the defendant received the stolen property knowing that it was stolen. Was Sonnier in possession of the speakers? Sonnier also claimed that she lacked actual knowledge that the speakers were stolen. The speakers were pawned for a reasonable amount of money, and a reasonable person would have no notion of the monetary value of the speakers. Should the Texas appellate court affirm Sonnier's conviction? See *Sonnier v. State,* 849 S.W.2d 828 (Tex. App. 1992).

You can find the answer at www.sagepub.com/lippmanccl3e.

FORGERY AND UTTERING

The law of forgery originated in the punishment of individuals who used or copied the king's seal without authorization. The seal was customarily affixed to documents that bestowed various rights and privileges on individuals, and employing this stamp without authorization was viewed as an attack on royal power and prerogative. The law of forgery was gradually expanded to include private as well as public documents.

Forgery is defined as creating a false legal document or the material modification of an existing legal document with the intent to deceive or to defraud others. The crime of forgery is complete upon the drafting of the document regardless of whether it is actually used to defraud others. **Uttering** is a separate and distinct offense that involves the *actus reus* of circulating or using a forged document.

Forgery and uttering are typically limited to documents that possess "legal significance." This means that the document, if genuine, would carry some legal importance, such as conveying property or authorizing an individual to drive. A falsified document is not a forgery when it merely impacts an individual's reputation or professional advancement, such as a fabricated newspaper account of a political candidate's evasion of military service.

The Model Penal Code extends forgery and uttering to all varieties of documents. This would include the attempt several years ago to sell a book that was alleged to be Adolf Hitler's diary that, in truth, was a skillfully produced fraud. Fraudulent documents that may be punished as a forgery under state statutes include checks, currency, passports, driver's licenses, deeds, diplomas, tickets, credit cards, immigration visas, and residency and work permits.

There are several elements to establish forgery; each must be proven beyond a reasonable doubt:

- A false document or material modification of an existing document that is
- written with intent to defraud and,
- if genuine, would have legal significance.

The elements of uttering are

- offering a
- forged instrument that is
- known to be false and is
- presented as authentic
- with the intent to defraud or deceive.

Forgery is similar to other property crimes in that the forger is unlawfully obtaining a benefit from another individual. The larger public policy behind criminalizing forgery is to insure that people are able to rely on the authenticity or truth of documents. You want to be confident that when you buy a car, the title you receive is genuine and that the automobile has not been stolen. Combating the forgery of passports and visas has taken on particular importance in securing the borders of the United States against the entry of terrorists.

Actus Reus

The important point to remember is that forgery is falsely making or materially altering an existing document. This may entail creating a false document or materially (fundamentally) changing an existing document without authorization. A material modification is a change or addition that has legal significance.

A forgery may involve manufacturing a "false identification" for a friend who is too young to drink, creating false passports for individuals seeking to illegally enter the United States, or fabricating tickets to a sold-out rock concert. Forgery may also involve materially or fundamentally altering or modifying an existing document. Stealing a check, signing the name of the owner of the account without authorization, and making the check payable to yourself for $100 is forgery. In this example, although the check itself is genuine, the details do not reflect the intent of the owner of the check. On the other hand, merely filling in the date on an undated check would ordinarily not constitute a material alteration because this change typically has no legal significance.

In other words, the question in forgery is whether a document is a "false writing." The document itself may be false, or the material statements in the document may be materially false.

Mens Rea

Forgery requires an intent to defraud; this need not be directed against a specific individual.

Uttering

Uttering is offering a document as genuine that is known to be false with the intent to deceive. This is a different offense from forgery, although the two are often included in a single statute. Merely presenting a forged check to a bank teller for payment knowing that it is inauthentic completes the crime of uttering. The teller need not accept the forged check as genuine.

Simulation

Several states follow the Model Penal Code in providing for the crime of **simulation**. This punishes the creation of false objects with the purpose to defraud, such as antique furniture, paintings, and jewelry. Simulation requires proof of a purpose to defraud or proof that an individual knows that he or she is "facilitating a fraud."

Model Penal Code

Section 224.1. Forgery

    (1) A person is guilty of forgery if, with purpose to defraud or injure anyone, or with knowledge that he is facilitating a fraud or injury to be perpetrated by anyone, the actor:

        (a) alters any writing of another without his authority; or

        (b) makes, completes, executes, authenticates, issues or transfers any writing so that it purports to be the act of another who did not authorize that act, or to have been executed at a time or place or in a numbered sequence other than was in fact the case, or to be a copy of an original when no such original existed; or

        (c) utters any writing which he knows to be forged in a manner specified in paragraphs (a) or (b).

        "Writing" includes printing or any other method of recording information, money, coins, tokens, stamps, seals, credit cards, badges, trade-marks, and other symbols of value, right, privilege, or identification.

(2) Forgery is a felony of the second degree if the writing is or purports to be part of an issue of money, securities, postage or revenue stamps, or other instruments issued by the government, or part of an issue of stock, bonds or other instruments representing interests in or claims against any property or enterprise. Forgery is a felony of the third degree if the writing is or purports to be a will, deed, contract, release, commercial instrument, or other document evidencing, creating, transferring, altering, terminating, or otherwise affecting legal relations. Otherwise forgery is a misdemeanor.

Analysis

- This section applies to "any writing" and to "any other method of recording information, money, coins, credit cards and trade-marks and other symbols." Forgery is not limited to documents having legal significance. As a result, documents such as medical prescriptions, diplomas, and trademarks are encompassed within this provision. The section is not limited to economic harm and may include circulating a false document that injures an individual's reputation.
- Serious forgeries that have the most widespread and serious impact are second-degree felonies, carrying a maximum penalty of ten years. Other forgeries are punishable as felonies of a third degree, carrying a maximum of five years. Forgeries of documents that do not have legal significance, such as diplomas, are misdemeanors.
- Counterfeiting of currency is included in this section rather than being made a separate offense.
- Section 1(c) punishes uttering.

The Legal Equation

Forgery = Creation of false document (of legal significance) or material alteration of an existing document

+ fraudulent intent.

Uttering = Passing of false document (of legal significance)

+ purposely or knowingly deceitful.

Was the defendant guilty of forgery when he signed his name to the corporate check?

United States v. Cunningham,
813 N.E.2d 891 (N.Y. 2004). Opinion by: Rosenblatt, J.

Defendant was convicted of forgery in the second degree for signing his own name to a corporate check, in excess of his authority. Because defendant's conduct does not constitute forgery under our statute, we reverse his conviction.

Facts

As the owner of a logging operation, Peter Morat planned to open a sawmill business in Madison County, under the name Herkimer Precut, Inc.

He engaged defendant as a consultant to arrange for financing and related activities. In exchange for his services, defendant was to receive a 20% interest in the new venture. As the project progressed, Morat turned over various financial aspects of the business to defendant, entrusting him with control over the corporate checkbook. Because defendant was responsible for paying bills, Morat would sometimes provide defendant with blank, signed checks. At no time, however, did Morat authorize defendant to sign any checks.

After Morat discovered that corporate bills were not being paid, he examined the company's bank records and found unauthorized payments, some on checks he had signed in blank and others bearing a signature he did not recognize. Morat alleged that by improperly signing or issuing checks, defendant stole thousands of dollars from Herkimer Precut.

The court was found to lack jurisdiction over nineteen of the charges. A single count survived the trial: the forgery conviction before us, stemming from a $195.50 Herkimer Precut check defendant wrote to Nancy Herrick for work performed by Northeast Woodcraft. The defendant signed his own name to that check, telling Herrick that he owned Herkimer Precut. Herrick was acquainted with defendant personally and professionally and knew that he was affiliated with Herkimer Precut. She did not know, however, that Morat owned the company and that defendant lacked authority to sign checks. The check was for defendant's personal expenses.

Reasoning

In *People v. Levitan,* 399 N.E.2d 1199 (N.Y. 1980), Levitan signed her name to deeds purporting to convey real property she did not own. In reversing her forgery conviction, we noted that "no pretense was ever made that the signatory was anyone other than defendant." We also observed that "under our present Penal Law, as under prior statutes and the common law, a distinction must be drawn between an instrument which is falsely made, altered or completed, and an instrument which contains misrepresentations not relevant to the identity of the maker or drawer of the instrument."

Although the Legislature has updated the statute to cover credit cards and certain other technological advances, it has not abrogated *Levitan*'s classic approach to forgery. In defining forgery, Penal Law section 170.00(4) provides, in pertinent part, that "a person falsely makes a written instrument when he makes or draws a complete written instrument . . . which purports to be an authentic creation of its ostensible maker or drawer, but which is not such . . . because the ostensible maker or drawer . . . did not authorize the making or drawing thereof."

Issue

The terms "authentic creation" and "ostensible maker" are pivotal. In most prosecutions, the forger, acting without authority, signs someone else's name. Thus, in a typical case, the forger, John Doe, wrongfully signs Richard Roe's name, (mis)leading the payee into believing that the check is the authentic creation of Richard Roe, its ostensible maker. Roe, of course, has not granted Doe any such authority and, in most such instances, has never even met Doe. In this simple formulation, the ostensible maker (Roe) and the actual maker (Doe) are two different people. If, however, the

ostensible maker and the actual maker are one and the same, there can be no forgery under the statute. . . .

The People contend that Herkimer Precut is the ostensible maker because its name appears on the check as owner of the account. Further, they argue that because defendant lacked authority to sign company checks, the check in question was not the authentic creation of the company, and a forgery is made out. Defendant counters that the check was an authentic creation of its ostensible maker and that because he signed his own name, he cannot be guilty of forgery: as the ostensible maker, he did not pretend to be anyone other than himself—the actual maker. Moreover, defendant argues that even if Herkimer Precut was the "ostensible maker" of the check, defendant's relationship with Herkimer Precut was sufficient to make the check the "authentic creation" of the company. We have observed that "when an individual signs a name to an instrument and acknowledges it as his own, that person is the 'ostensible maker.'"

Forgery is a crime because of the need to protect signatures and make negotiable instruments commercially feasible. In its common law roots, forgery had little to do with abstract questions of authority. At Queen's Bench, Chief Justice Cockburn wrote that forgery "by universal acceptation . . . is understood to mean the making or altering of a writing so as to make the writing or alteration purport to be the act of some other person, which it is not." As one treatise explains, "it is not forgery for a person to sign his own name to an instrument, and falsely and fraudulently represent that he has authority to bind another by doing so" and "the signer is guilty of false pretenses only." Although statutes vary, most jurisdictions in this country have tended to follow this approach to forgery.

As Blackstone wrote, "forgery" is "the fraudulent making or alteration of a writing to the prejudice of another man's right."

Holding

We conclude that authority and authenticity are not the same thing. Defendant did not commit forgery merely by exceeding the scope of authority delegated by the corporation. Our interpretation leaves no gap in the Penal Law. Although embezzlers who use their own names to sign checks beyond their authority are not guilty of forgery in New York, their conduct would ordinarily fall within our larceny statutes.

Moreover, a contrary ruling would create vexing problems in adjudging forgery cases. If, for example, a corporate officer authorized to sign corporate checks does so for a personal purchase, is that forgery? Would an officer authorized to sign checks up to $20,000 who signs a check for $25,000 be guilty of forgery? While the prosecution argues that we should read our statute to justify convictions in those instances, it has not identified any New York decision interpreting the statute so expansively.

You Decide

13.7 McGovern owed $1,800 to Scull. McGovern purchased $2,400 in traveler's checks from Citibank for purposes of repaying Scull. The checks may be redeemed for money at most banks or stores when signed by the individual to whom the check is issued. Scull and McGovern entered into a corrupt arrangement designed to reimburse Scull. Scull practiced McGovern's signature and took McGovern's driver's license and the traveler's checks and proceeded to cash the checks at two banks and collected $2,400. McGovern then reported to the police that the checks had been stolen from his car, and in accordance with the highly advertised policy concerning traveler's checks, McGovern was provided with replacement checks in the amount of $2,400 by Citibank. Did Scull's impersonation of McGovern constitute forgery? See *United States v. McGovern*, 661 F.2d 27 (3rd Cir. 1981).

You can find the answer at www.sagepub.com/lippmanccl3e.

ROBBERY

Robbery is typically described as aggravated larceny. You should think of robbery as larceny from an individual with the use of violence or intimidation. Professor Perkins observes that in ancient law, the thief who stole quietly and secretly was viewed as deserving harsher punishment than the robber who openly employed violence. The common law reversed this point of view and categorized robbery as among the most serious of felonies, which should be treated as a separate offense deserving of harsher punishment than larceny.[31]

Robbery is the trespassory taking and carrying away of the personal property of another with intent to steal. Robbery is distinguished from larceny by additional requirements:

- The personal property must be taken from the victim's person or presence.
- The taking of the personal property must be achieved by violence or intimidation.

The California criminal code defines robbery as the "felonious taking of personal property in the possession of another, from his person or immediate presence, and against his will, accomplished by means of force or fear." In this chapter, robbery is treated as a property crime, although the FBI categorizes robbery as a violent crime against the person.[32]

Actus Reus

The *property must be taken from the person or presence of the victim.* Property is considered to be on the person of the victim if it is in his or her hands or pockets or is attached to his or her body (an earring) or clothing (a key chain).

The requirement that an object must be taken from the "presence of the victim" is much more difficult to apply. The rule is that the property must be within the proximity and control of the victim. What does this mean? The prosecution is required to demonstrate that had the victim not been subjected to violence or intimidation, he or she could have prevented the taking of the property.

In one frequently cited case, the defendants forced the manager of a drugstore to open a safe at gunpoint. The defendants then locked the manager in an adjoining room and removed the money from the safe. An Illinois court found that the money was under the victim's personal control and protection and that he could have prevented the theft had he not been subjected to an armed threat.[33] Professor LaFave illustrates the requirement that property be taken from the presence of the victim by noting that it would not be robbery to immobilize a property owner at one location while a confederate takes the owner's property from a location several miles away, because the owner could not have prevented the theft.[34]

The *property must be taken by violence or intimidation.* The Florida statute provides that robbery involves a taking through "the use of force, violence, assault, or putting in fear."[35] Keep in mind that it is the use of violence or intimidation that distinguishes robbery from larceny. The line between robbery and larceny, however, is not always clear. In general, any degree of force is sufficient for robbery. You are walking down the street loosely carrying your backpack when a thief snatches the backpack out of your grasp. You are so surprised that you fail to resist. Is this robbery? The consensus is that the incident is not a robbery. This would qualify as robbery in the event that you are pushed, shoved, or struggled to hold on to the backpack. It is also robbery where force is applied to remove an item attached to your clothing or body, such as an earring or necklace. Does it make sense to distinguish between robbery and larceny based on whether the perpetrator employed a small amount of force?

The Model Penal Code attempts to avoid this type of technical analysis and provides that robbery requires "serious bodily injury." This approach has been rejected by most states on the grounds that it excludes street crimes in which victims are pushed to the ground or receive minor injuries. Before we leave this topic, we should note that it is a robbery when an assailant steals your personal items by rendering you helpless through liquor or drugs.

Property may also be seized as a result of intimidation or the fear of immediate infliction of violence. The threat of immediate harm must place the victim in fear, meaning in apprehension or in anticipation of injury.

The threat may be directed against members of the victim's family or relatives, and some courts have extended this to anyone present as well as to the destruction of the home. The threat must also be shown to have caused the victim to hand over the property.

Again, a threat may be "implied." This might involve a large and imposing panhandler who follows an elderly pedestrian down a dark and isolated street and angrily and repeatedly demands that the pedestrian "give up the money in his or her pocket." The threat must place the victim in apprehension of harm and cause him or her to hand over the property. The jury is required to find that the victim was actually frightened into handing over his or her property. Some courts require that a reasonable person would have acted in a similar fashion.

Mens Rea

The assailant must possess the intent to permanently deprive an individual of his or her property. The defendant may rely on the familiar defense that he or she intended only to borrow the property or was playing a practical joke. Courts are divided over whether it is a defense that the thief acted under a "claim of right," that the thief acted under an honest belief that the victim owed him money, or that the defendant reasonably believed that he or she owned the property. Some courts hold that even a claim of right does not justify the resort to force or intimidation to reclaim property.

Concurrence

The traditional view is that the intent to steal and the application of force or intimidation must coincide. The violence or intimidation must be employed for the purpose of the taking. This means that the threat or application of force must occur at the time of the taking. An individual does not commit a robbery who seizes property and then employs force or intimidation. A pickpocket who removes a victim's wallet and resorts to force only in response to the victim's accusation of theft is not guilty of robbery.

A number of states have followed the Model Penal Code in adopting language that provides that force or intimidation may occur "in the course of committing a theft." This is interpreted to mean that force or threat occurs "in an attempt to commit theft or in flight after the attempt or commission." The commentary explains that a thief's use of force against individuals in an effort to escape indicates that the thief would have employed force "to effect the theft had the need arisen." Even under this more liberal approach, an assailant who knocks the victim unconscious and then forms an intent to steal would not be guilty of robbery.[36] The Florida robbery statute defines robbery to include force or intimidation "in the course of the taking" of money or other property. This includes force or threats "prior to, contemporaneous with, or subsequent to the taking of the property . . . if it and the act of taking constitute a continuous series of acts or events."[37]

Grading Robbery

At common law, the theft of property that terrorized the victim resulted in the death penalty. Today, robbery statutes generally distinguish between simple and aggravated robbery. This is based on the degree of dangerousness caused by the defendant's act and the fear and apprehension experienced by the victim, rather than the value of the property. The factors that aggravate robbery include

- the robber was armed with a dangerous or deadly weapon or warned the victim that the robber possessed a firearm;
- the robber used a dangerous instrumentality, such as a knife, hammer, axe, or aggressive animal;
- the robber inflicted serious bodily injury; and
- the robber carried out the theft with an accomplice.

You might question whether we need the crime of robbery. Is there any justification for the crime of robbery other than historical tradition? Why not simplify matters and merely charge a defendant with larceny along with assault and battery? The next case, *Messina v. State*, raises the issue of whether the application or threat of violence requirement in robbery should include force used to flee the crime scene.

Model Penal Code

Section 222.1. Robbery

(1) A person is guilty of robbery if, in the course of committing a theft, he:

    (a) inflicts serious bodily injury upon another; or

    (b) threatens another with or purposely puts him in fear of immediate serious bodily injury; or

    (c) commits or threatens immediately to commit any felony of the first or second degree.

    An act shall be deemed "in the course of committing a theft" if it occurs in an attempt to commit theft or in flight after the attempt or commission.

(2) Robbery is a felony of the second degree, except that it is a felony of the first degree if in the course of committing the theft the actor attempts to kill anyone, or purposely inflicts or attempts to inflict serious bodily harm.

Analysis

- The infliction or threat of harm is limited to "serious bodily injury." The inclusion of the commission or threat to commit a felony of the first or second degree as an element of robbery is intended to encompass the threat or commission of serious injury to an individual other than the victim as well as the threat to destroy or the destruction of property.
- The harm may be inflicted or threatened "in the course of committing the theft." This includes violence or the threat of violence to obtain or retain property and to prevent pursuit or to escape.
- The commentary explains that the same punishment is imposed for both robbery and attempted robbery. It is immaterial whether the assailant actually succeeds in the taking of property. This reflects the view that the essence of robbery is the placing of individuals in danger rather than the deprivation of property.
- The infliction or threat of harm must be immediate.
- The taking is not required to be from the person or in the presence of the victim. An offender might threaten the victim in order to extract ransom from an individual who is not present.
- Robbery is generally punished as a felony of the second degree, subject to ten years' imprisonment. Life imprisonment is viewed as an extreme penalty that is reserved for violent offenders.

The Legal Equation

Robbery = Taking of the property of another from the person or presence of the person

+ by violence or threat of immediate violence placing another in fear

+ intent to permanently deprive another individual of property.

Did Messina take the property through force and/or violence?

Messina v. State, 728 So. 2d 818
(Fla. Dist. Ct. App. 1999). Opinion by: Padovano, J.

Issue

The defendant, Karl C. Messina, appeals his conviction for the crime of robbery. He contends that the evidence is sufficient to support only the lesser crime of petit theft because there is no proof that he used force against the victim in taking her property. We conclude that the evidence is sufficient to support the main charge of robbery because the record shows that the defendant used force to retain the victim's property once he had taken it from her. Therefore, we affirm.

Facts

On December 14, 1996, Elaine Barker was in the parking lot of a K-Mart store unloading items from a shopping cart into the trunk of her car. She left her purse in the shopping cart, and while she was transferring the items she had purchased, the defendant came over, grabbed the purse, and ran away. Barker chased the defendant on foot and caught up with him, but by that time, he had gotten into his car and closed the door. Barker then sat on the hood of the defendant's car, thinking that would prevent him from driving away. Instead, the defendant started and stopped the car several times while Barker held on to a windshield wiper to keep from falling off. The defendant turned the car sharply causing Barker to fall to the ground. As a result of the fall, Barker suffered a broken foot and lacerations that required stitches.

Based on these facts, the State charged the defendant with the crime of robbery. At the close of the State's case in chief, the defendant moved for a judgment of acquittal, contending that the evidence was sufficient to sustain only the lesser included charge of petit theft. The trial court denied the motion and sent the case to the jury on the charge of robbery. The jury found the defendant guilty as charged, and he was convicted and sentenced for the crime of robbery.

Reasoning

The defendant contends that his conviction must be reversed because there is no evidence that he took the victim's purse by force. It is true, as the defendant argues, that a purse snatching is not a robbery if no force was used other than that necessary to take the victim's purse. In the present case, however, the charge of robbery was not based on the force used to remove the property from the shopping cart but rather on the force subsequently used against the victim once she tried to regain possession of her property. The question is not whether force was used but when it was used in relation to the taking.

A conviction for the crime of robbery requires proof that money or other property was taken from the victim and that the offender used force or violence "in the course of the taking." The temporal relationship between the use of force and the taking of the property is addressed in section 812.13(3)(b), which provides that "an act shall be deemed 'in the course of the taking' if it occurs either prior to, contemporaneous with, or subsequent to the taking of the property and if it and the act of taking constitute a continuous series of acts or events." As this definition reveals, the statute is not limited to situations in which the defendant has used force at the precise time the property is taken.

On the contrary, section 812.13 . . . incorporates the modern view that a robbery can be proven by evidence of force used to elude the victim or to retain the victim's property once it has been taken. The rationale for this view is that the force used in the flight after the taking of property is no different from that used to effect the taking. As explained in the Comments to the Model Penal Code, "the thief's willingness to use force against those who would restrain him in flight suggests that he would have employed force to effect the theft had the need arisen." . . . Florida courts have held that the crime of robbery can be proven by evidence that the defendant used force against the victim after the taking has been completed. . . . The common feature of these cases is that in each case, there was no break in the chain of events between the taking and the use of force.

Holding

In the present case, the defendant used force against the victim immediately after he had taken her property and while she was attempting to get it back. The force was used as a part of a continuous set of events beginning with the removal of the victim's purse from the shopping cart and ending with the victim's fall from the hood of the defendant's car. There was no interruption that would lead us to conclude that the subsequent battery on the victim was a new and separate offense. . . . Here, the taking and the use of force were part of the same offense.

The defendant suggests that the evidence is not sufficient to sustain his conviction for robbery because the injury to the victim was not foreseeable. He argues that it was unreasonable to expect that the victim would place herself in danger by sitting on the hood of his car. The short answer to this point is that the defendant was not obligated to drive the car away. In any event, we decline to engraft concepts of tort law onto the statutory elements of robbery. Whether the defendant could have anticipated the victim's reaction is irrelevant. Likewise, whether the victim would have been wiser to allow the defendant to drive away with her property is irrelevant. The robbery statute merely requires proof that the force and the act of the taking were part of a "continuous series of acts or events." That was proven here.

Questions for Discussion

1. Why was Messina not charged with robbery based solely on his snatching of the victim's purse? Was the defendant's use of force part of a "continuous series of acts or events"?

2. Did the victim place herself at risk by her behavior?

3. Should courts limit the use of force for purposes of robbery to the time of the taking? Would it make more sense to punish the force used by Messina as a battery?

You Decide

13.8 At roughly 5 a.m., Alfonso Gomez broke into an Anaheim, California, restaurant. He covered two surveillance cameras with duct tape and broke open and took money from an ATM in the lobby. Gomez then went to the second floor and searched the manager's office for money. As Gomez went downstairs, he heard the manager, Ramon Baltazar, unlock the front door. Gomez removed a handgun from his backpack and went to the restaurant kitchen. Baltazar noted that the alarm had been deactivated and the ATM damaged, and he heard noise in the kitchen and saw the glow of a flashlight. He went outside to his truck and rang 911. As he spoke to the police dispatcher, Baltazar spotted Gomez exit a side door and walk away. Baltazar shadowed Gomez in his car at a distance of 100 to 150 feet. Gomez fired two shots at Baltazar, explaining that he wanted to scare Baltazar. Gomez drove away and was arrested shortly thereafter with money in his backpack. He was convicted of robbery and burglary. Robbery under the California statute is defined as the "felonious taking of personal property in the possession of another, from his person or immediate presence, and against his will, accomplished by means of force or fear." Taking has two aspects: (1) possession of property or caption and (2) carrying the property away or asportation. The asportation continues until the offender reaches a place of safety. Gomez contended on appeal that he did not take "property" from the "person or immediate person" of the defendant through "force or fear." Is Gomez guilty of robbery? See *People v. Gomez*, 179 P.3d 917 (Cal. 2008).

You Decide

13.9 Ronald Williams, and the victim were drinking wine with Frank Morrow and an unidentified male. The victim passed out. Williams rolled the victim onto his side and removed his wallet from his pants pocket. Williams appealed his robbery conviction. The Pennsylvania appellate court was asked to determine whether robbery may be committed against a voluntarily intoxicated, unconscious victim. Did Williams take the victim's wallet through the "use of force however slight." See *Commonwealth v. Williams*, 550 A.2d 579 (Pa. Super. Ct. 1988).

You can find the answer at www.sagepub.com/lippmanccl3e.

Figure 13.1 Crime on the Streets: Property Crimes

Property crimes fell 22.7% between 1996 and 2006. In recent years, there has been a slight increase in property crimes.

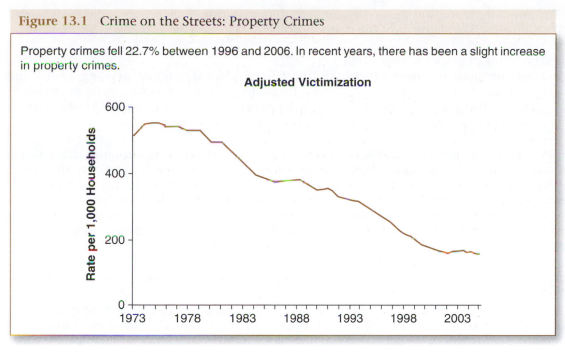

Source: The National Crime Victimization Survey, U.S. Department of Justice.

Note: Property crimes include burglary, theft, and motor vehicle theft.

CARJACKING

Carjacking is a newly recognized form of property crime that is punished under both federal and state statutes. California is typical in defining carjacking as a form of robbery and punishes the taking of a motor vehicle "in the possession of another, from his or her person or immediate presence . . . against his or her will." This must be accomplished by "force or fear." The perpetrator is not required to intend to permanently steal the automobile. The California statute is satisfied by an intent either to "permanently or temporarily deprive the victim of possession of the car."[38]

Several state statutes provide that force must be directed against an occupant of the car. The New Jersey statute requires that while committing the unlawful taking of the automobile, there must be the infliction or use of force against an occupant or person in possession or control of the motor vehicle.[39] Virginia stipulates that the taking be carried out by violence to the person, by assault, or by otherwise putting a person in fear of serious bodily injury.[40]

The trend is to find a defendant guilty of carjacking when an automobile is seized and not to require that the perpetrator move the automobile. A carjacking may be directed against an occupant of the car or against an individual outside the car who is in possession of the keys and is sufficiently close to control of the vehicle.

The punishment of carjacking is based on the degree of harm and apprehension caused by the offense. New Jersey punishes carjacking by between ten and twenty years in prison and a fine of up to $200,000. The Florida statute punishes carjacking with life imprisonment when committed with a firearm or other deadly weapon.[41]

You can find more cases on the study site: People v. Montero, *www.sagepub.com/lippmanccl3e.*

EXTORTION

The common law misdemeanor of **extortion** punished the unlawful collection of money by a government official. William Blackstone defined extortion as "an abuse of public justice, which consists in any officer's unlawful taking, by color of his office, from any man, any money or thing of value that is not due to him, or more than is due, or before it is due."[42]

The law of extortion was gradually expanded to punish threats by private individuals as well as public officials. The elements of the statutory crime of extortion are as follows:

- The taking of property from another by
- a present threat of future violence or threat to circulate secret, embarrassing, or harmful information; threat of criminal charges; threat to take or withhold official government action; or threat to inflict economic harm and other harms listed in the state statute, with
- a specific intent to deprive a lawful possessor of money or property.

Note that while robbery involves a threat of immediate violence, extortion entails a threat of future violence or other harms. The threat to disclose secret or embarrassing information is commonly referred to as the crime of **blackmail**. Robbery must be committed in the presence of the victim, while extortionate threats may be communicated over the phone or in a letter.

The majority of state statutes provide that the crime of extortion is complete when the threat is made. The prosecution must demonstrate that the victim believed that there was a definite threat and believed that this threat would be carried out. A Michigan statute punishes "any person who shall . . . maliciously threaten to accuse another of any crime . . . or . . . maliciously threaten any injury to . . . [a] person or property . . . with intent to thereby . . . extort money or any pecuniary advantage . . . or . . . to compel the person to do . . . any act against his will."[43]

Other statutes require the handing over of money, property, or valuable items in response to the threat. The prosecution must establish a causal relationship between the threat and the conveying of the money or property. The "handing over" requirement is illustrated by the language of the New York statute, which provides that an individual is guilty of extortion when he or she "compels . . . another person to deliver . . . property to himself or to a third person by means of instill[ing] . . . fear."[44]

The object of extortion may be money, property, or "anything of value," including labor or services. The Iowa Supreme Court ruled that a college student who attempted to extort a date from a female acquaintance had attempted to extort "something of value for himself" and that value should be broadly interpreted to include "relative worth, utility, or importance" rather than "monetary worth."[45]

Harrington, a divorce lawyer, represented a female in a divorce action who had been the victim of severe physical abuse by her husband. Harrington arranged for a female to seduce the husband, and while the two were in a romantic embrace in bed, Harrington entered and took photographs. Harrington subsequently threatened to disclose the husband's adultery unless he paid his wife a divorce settlement of $175,000. The Vermont Supreme Court ruled that Harrington "acted maliciously and without just cause . . . with the intent to extort a substantial fee . . . to [Harrington's] personal advantage."[46]

Several commentators contrast extortion with *bribery*. Extortion involves taking money, property, or anything of value from another through threat of violence or harm. In bribery, money or a valuable benefit is offered or provided to a public official in return for an official's action or inaction. This act may involve a legislator voting in favor of or against a law, a judge acquitting or convicting a defendant, or a clerk giving priority to an applicant for a driver's license or passport; the inaction entails a failure to act, such as a building inspector overlooking safety violations in a music club. There must be an intent to corruptly influence an official in the conduct of his or her office. Individuals are held guilty of bribery for offering as well as accepting a bribe. We will discuss bribery in greater detail in Chapter 14 on white-collar crime.

CHAPTER SUMMARY

The common law initially punished only the violent taking of property. This soon proved insufficient. Individuals accumulated farm animals, crops, and consumer goods that were easily stolen by stealth and under the cover of darkness. Larceny developed to protect individuals against the wrongful taking and carrying away of their personal property by individuals harboring the intent to deprive the owner of possession. The economic development of society resulted in clear shortcomings in the coverage of the law that led to the development of embezzlement, false pretenses, and receiving stolen property.

Embezzlement was introduced by the English Parliament to fill a gap in the law of larceny. Embezzlement involves the fraudulent conversion of the property of another by an individual who is in lawful possession of the property. Some statutes extend embezzlement to real (real estate) as well as personal property. The English Parliament also introduced false pretenses to punish individuals who obtain title and possession of the property of another by a false representation of a present or past material fact with the intent to defraud that causes an individual to pass title to his or her property.

A number of states have consolidated larceny, embezzlement, and false pretenses into a single theft statute. These statutes provide a uniform grading of offenses and, in some states, serve to prevent a defendant from being acquitted based on the prosecutor's failure to satisfy the technical factual requirements of the property crime with which the defendant is charged. The grading of larceny, embezzlement, and false pretenses is generally based on the monetary value of the property. Modern theft statutes also provide protection to all varieties of personal property and do not distinguish between tangible (physical objects) and intangible (legal documents) personal property. As noted, various states also extend protection to real property (real estate).

Two new forms of criminal conduct provide a challenge to the law. Identity theft is the fastest growing crime and involves the stealing of individuals' personal information and the use of this information to make purchases or to borrow money. States have responded to computer theft and cybercrime by passing statutes punishing a range of computer offenses, including the unauthorized access to computers, computer networks, and programs and causing computers to malfunction.

Forgery involves the creation of a false legal document or the material modification of an existing legal document with the intent to deceive or to defraud others. The crime of forgery is complete upon the drafting and modification of the document with the intent to defraud others, regardless of whether the document is actually used to commit a fraud. Uttering is the separate offense of circulating or using a forged document.

Robbery is a crime that threatens both the property and the safety and security of the individual. It involves the taking of personal property from the victim's person or presence through violence or intimidation. The grading of robbery depends on the harm inflicted or threatened. The use of a firearm is subject to particularly severe punishment. Carjacking is an increasingly prevalent offense that involves the use of force to unlawfully gain control and possession over a motor vehicle. Extortion is the taking of property from another through the threat of future violence; the threat to circulate secret, embarrassing, or harmful information; the threat to bring criminal charges; the threat to take or withhold official government action; or the threat of economic harm.

All the offenses in this chapter involve the seizure of the property of another individual through unlawful taking, fraud, or force.

CHAPTER REVIEW QUESTIONS

1. Provide an example of how the common law of larceny developed in response to the growth of business and commerce.

2. Distinguish between the requirements of larceny, embezzlement, and false pretenses.

3. Why did various states adopt consolidated theft statutes?

4. Describe the harm to society caused by identity theft. What is the reason that states found it necessary to adopt new criminal statutes punishing identity theft?

5. Why have states passed legislation to address computer crimes rather than using existing statutes punishing crimes against property? What types of acts are prohibited under these statutes?

6. What is a prosecutor required to prove beyond a reasonable doubt in order to establish the crime of receiving stolen property? How does the punishment of this offense deter theft?

7. What is the difference between forgery and uttering?

8. How does robbery differ from larceny?

9. Distinguish robbery from extortion.

10. What are the elements of carjacking?

11. How do courts interpret the "taking and carrying away" element of larceny? Why did the common law require a "carrying away" of property?

12. Discuss the use or threat of harm requirement in regard to robbery.

13. What are some of the factors that aggravate a property offense?

14. Write an essay briefly summarizing property crimes. Address whether we still require the various property offenses that were developed by the common law.

LEGAL TERMINOLOGY

blackmail	forgery	receiving stolen property
carjacking	grand larceny	robbery
computer crime	identity theft	simulation
custody	larceny	theft statutes
embezzlement	larceny by trick	uttering
extortion	petit larceny	
false pretenses	possession	

CRIMINAL LAW ON THE WEB

Log on to the Web-based student study site at www.sagepub.com/lippmanccl3e to assist you in completing the Criminal Law on the Web exercises, as well as for additional features such as podcasts, Web quizzes, and video links.

1. Explore computer crime on the Department of Justice Web site.

2. Read about the notorious forgery of the Hitler diaries. How were the experts fooled?

3. Learn about identity theft.

4. Examine state shoplifting law.

BIBLIOGRAPHY

American Law Institute, *Model Penal Code and Commentaries,* vol. 2, pt. 2 (Philadelphia: American Law Institute, 1980), pp. 96–362. A comprehensive discussion of the history of common law property offenses and a description of various state statutes along with proposed reforms.

Joshua Dressler, *Understanding Criminal Law,* 3rd ed. (New York: Lexis, 2001), pp. 545–568. A clear discussion of larceny, embezzlement, false pretenses, and the consolidation of theft offenses.

Jerome Hall, *Theft, Law, and Society,* 2nd ed. (Indianapolis, IN: Bobbs-Merrill, 1952). The definitive analysis of the social and economic conditions in England that led to the development of various property offenses.

Wayne R. LaFave, *Criminal Law,* 3rd ed. (St. Paul, MN: West Publishing, 2000), pp. 789–885. A well-researched, in-depth discussion of property offenses with useful citations to the leading cases.

Rollin M. Perkins and Ronald N. Boyce, *Criminal Law,* 3rd ed. (Mineola, NY: Foundation Press, 1982), pp. 292–452. A helpful overview of the historical development and common law of property offenses and a discussion of proposed reforms.

14 WHITE-COLLAR CRIME

Did Reverend Davis engage in money laundering?

Reverend Davis became the preacher at the 15th Street Baptist Church in the mid-1980s. Shortly thereafter he began to sell drugs, and by mid-1987 was actively selling crack from two houses. . . . Davis deposited some of the cash he collected from the houses in bank accounts maintained in the name of the 15th Street Baptist Church Development Corporation . . . and the 15th Street Baptist Church . . . and at Illini Federal, a local savings and loan. . . . Davis could write checks on these accounts. Some of these checks were made out to cash, which Davis diverted to his personal use. Others were made out to local vendors who provided services such as beepers and mobile telephones. Still others were made out to the landlord who owned the Swansea, Illinois, residence where Davis lived. Davis also purchased numerous cars, spending over $79,000 on a variety of vehicles for personal and church use. . . .

Learning Objectives

1. Know the different approaches to defining white-collar crime.

2. List the types of acts prohibited by environmental statutes.

3. Know the purpose of the Occupational Safety and Health Act.

4. Understand securities fraud and insider trading.

5. Know the elements of mail fraud and wire fraud.

6. Explain the Travel Act.

7. Outline the type of acts involved in health care fraud.

8. Explain money laundering.

9. Understand the purpose of the Sherman Antitrust Act.

10. Know the difference between bribery of a public official and graft.

INTRODUCTION

In 1949, sociologist Edwin H. Sutherland published his pioneering study, *White Collar Crime*. This volume called attention to the largely overlooked criminal behavior of business managers, executives, and professional groups, which Sutherland labeled **white-collar crime**. Sutherland defined white-collar crime as an offense committed by a "person of respectability and high social status in the course of his [or her] occupation." This definition stresses the social background of offenders and focuses on nonviolent offenses committed in the course of employment. Sutherland's central thesis is that theories that explain crime based on poverty, low social class, and lack of education fail to account for "crimes in the suites." The focus on the poor and disenfranchised diverts our attention from the fact that the financial cost of white-collar crime is several times greater than the economic consequences of common crimes. A second point raised by Sutherland is that despite the social harm caused by the crimes of the powerful, these offenses are typically punished by fines and less severe penalties than the offenses committed by average individuals.[1]

The U.S. Justice Department's definition of white-collar crime focuses on the nature of the criminal activity as well as on the job of the offender. This definition also does not limit white-collar crime to employment-related offenses. White-collar crime is defined as follows:

- Illegal acts that employ deceit and concealment rather than the application of force;
- to obtain money, property, or service;
- to avoid the payment or loss of money or to secure a business or professional advantage.
- White-collar criminals occupy positions of responsibility and trust in government, industry, the professions, and civil organizations.

A third approach defines white-collar crime in terms of the type of criminal activity involved. This has the advantage of drawing attention to the fact that tax and consumer fraud and other offenses characteristic of white-collar crime are committed by individuals of various socioeconomic backgrounds.

You might want to review our previous discussions of corporate criminality. In previous chapters, we discussed the vicarious and strict liability of business owners for regulatory (social welfare) offenses (Chapter 5) and considered the merits of holding corporations as well as individual executives liable for homicide (Chapter 10). You also should keep in mind that many of the property offenses we reviewed (Chapter 13) often are committed by corporate criminals in the course of carrying out fraudulent schemes. This includes larceny, false pretenses, embezzlement, extortion, and bribery.

The focus of the present chapter differs from our previous discussions in that most white-collar crime prosecutions are brought by the U.S. government rather than by state and local officials. You may recall that we discussed the division between federal and state powers in Chapter 1. In this chapter, we primarily examine the federal statutes that most frequently are used to combat white-collar crime, which include the following:

- *Environmental Crimes.* Offenses harming and polluting the environment.
- *Occupational Health and Safety.* Injury and harm to workers.
- *Securities Fraud.* Manipulation of stocks and bonds.
- *Mail and Wire Fraud.* The use of the mail and telephone to commit a fraud.
- *The Travel Act.* Committing certain offenses through the use of interstate travel or the mail.
- *Health Fraud.* Obtaining reimbursement or payment for unwarranted and undelivered medical treatments.
- *Money Laundering.* Transactions involving money derived from illegal activities.
- *Antitrust Violations.* Interference with the competitive marketplace.
- *Public Corruption.* Betrayals of the public trust by government officials.

White-collar crime offenses are typically committed in the regular course of business in an effort to make or save money. These offenses generally involve a betrayal of the trust that we place in business and government. Let me caution that this chapter does not cover the entire field of white-collar crime. Some areas, such as tax evasion and accounting fraud, are not discussed.

Despite the fact that white-collar crime is one of the most active areas of federal prosecution, textbooks generally do not devote significant attention to the subject. This is partially based on the belief that white-collar crime is not a distinct category of crime. It is argued that there is little difference between the theft of money by a corporate executive and the theft of money by a waitress or the theft of tools by a construction worker. As you read this chapter, consider whether the concept of white-collar crime is useful. Should we pay special attention to "crimes in the suites"? Do you believe that the government should devote additional resources to the prosecution and punishment of corporate misconduct? Another question concerns the appropriate form of punishment for white-collar offenders. Should respectable business executives be punished like any other criminals? We begin the chapter with environmental crimes to illustrate the harm that can result from white-collar crime.

ENVIRONMENTAL CRIMES

At times, the drive for corporate profit may lead business executives to disregard their legal obligation to protect the natural environment. There are considerable costs involved in environmental safety and cleanup that can absorb a significant percentage of corporate revenues. The FBI notes that **environmental crimes** threaten the health and natural resources of the United States and that such crimes range from air and water pollution to the illegal transportation and disposal of hazardous waste.

Americans were exposed to the potential danger that illegal business practices pose to the environment when a public health emergency was declared at Love Canal in Niagara Falls, New York. In 1978, a local paper reported that in 1953, Hooker Electro-Chemical Company had buried more than 21,000 tons of toxic waste on land that the company and city government knew was now the site of a housing development and school. Studies revealed that women living nearby experienced an excessive rate of miscarriages and that children suffered high rates of birth defects and disorders of the nervous system. The state and federal government ultimately evacuated the area at a cost of over $42 million, and the area would not be reclaimed for housing until 1990. A second well-known case in Woburn, Massachusetts, in 1979, involved the pollution of the water supply by industrial waste. The industrial firms responsible for the pollution ultimately agreed to a cleanup that cost more than $70 million.

In 1980, Pennsylvania authorities discovered that Hudson Oil Refining Corporation of New Jersey had been dumping waste down an old mine shaft. The waste accumulated and, in July 1979, began pouring out of the mine tunnel into the Susquehanna River. Millions of gallons of toxic waste linked to cancer and birth defects formed an oil slick that threatened the water supply of Danville, Pennsylvania. The company was fined $750,000, and the president of Hudson Oil, the first corporate official imprisoned for illegal environmental dumping in American history, was sentenced to one year in prison. In the mid-1980s, Pennsylvania convicted a corporate executive of illegally dumping 10,000 drums of waste in a Scranton, Pennsylvania, landfill.

In 1989, Rockwell International, a company that had managed the Rocky Flats nuclear weapons plant since 1975, pled guilty to ten federal counts and paid $18.5 million in fines stemming from its mismanagement of the 6,500-acre site fifteen miles northwest of Denver, Colorado. The plant was described as being littered with over 12.9 metric tons of dangerous plutonium, asbestos, lead, and other toxic chemicals.

On March 24, 1989, the oil tanker *Exxon Valdez* ran aground in Alaska, spilling eleven million gallons of oil into Prince William Sound and polluting roughly 1,300 miles of Alaska shoreline. Exxon agreed to pay a $150 million criminal fine. In 2008, the U.S. Supreme Court reduced the civil monetary judgment imposed on Exxon by a jury. The court did affirm the jury's judgment that Exxon was responsible for the actions of the ship's captain, finding that the jury could have reasonably concluded that Exxon "knowingly allowed a relapsed alcoholic repeatedly to pilot a vessel filled with millions of gallons of oil" and that "it was only a matter of time before a crash or spill . . . occurred."

Today, the increased concern with environmental crimes has led the federal and various state governments to establish special prosecution units. The federal government now highlights the seriousness of these offenses through an annual National Environmental Crime Prevention Week. The national dedication to combating environmental crime is illustrated by a recent federal case in which Department of Justice prosecutors obtained the conviction of two individuals for violating the Clean Air Act and the Toxic Substances Control Act. This resulted in the longest federal jail sentences for environmental crimes in history. Alexander Salvagno received twenty-five years in prison and was ordered to forfeit more than $2 million in illegal proceeds and to provide more than $23 million in restitution to the victims. His father, Raul Salvagno, was sentenced to nineteen years in prison and was required to forfeit close to $2 million in illegal proceeds and to pay more than $22 million in restitution. The two falsely represented to clients that they had completely removed dangerous toxic asbestos from homes and schools and directed their young workers to enter into asbestos "hot zones" without adequate protection, exposing more than 500 of their employees to the risk of cancer.

The FBI reports that at any given time, there are roughly 450 environmental criminal cases pending, roughly half of which involve violations of the Clean Water Act. The FBI's investigative priorities are protecting workers against hazardous wastes and pollutants, preventing large-scale environmental damage that threatens entire communities, pursuing organized crime interests that illegally dump solid waste, and monitoring businesses with a history of damaging the environment. The FBI reminds us that a single instance of dumping can poison a river and cost the public millions of dollars in cleanup costs. In Tampa, Florida, Durex Industries repeatedly disregarded warnings to safely dispose of hazardous materials used in the manufacture of aluminum cans. In 1993, two nine-year-old boys playing in a dumpster died when they were overcome by fumes from materials that Durex illegally discarded. The company was ordered to pay a $1.5 million fine, and several Durex officials were criminally convicted.

Most prosecutions for environmental crimes are undertaken by the federal government. Criminal provisions and penalties are typically incorporated into civil statutes regulating the environment. Investigations in this area, for the most part, are carried out by the Environmental Protection Agency (EPA), which refers matters to the Department of Justice for criminal prosecution. The central environmental laws include:

- *The Refuse and Harbors Appropriations Act (1899).* Imposes criminal penalties for improper discharge of refuse (foreign substances and pollutants) into navigable or tributary waters of the United States.
- *Water Pollution Control Act (1972).* Imposes criminal penalties for the discharge of certain pollutants beyond an authorized limit into navigable waterways and a prohibition on unauthorized dredging, the filling of wetlands, and the failure to clean up oil and other hazardous substances.
- *Resource Conservation and Recovery Act (1976).* Punishes knowingly storing, making use of, or disposing of hazardous wastes without a permit. Severe penalties are imposed for placing individuals in danger.
- *Clean Air Act (1990).* Imposes criminal penalties for the knowing violation of emission standards and other related requirements.
- *Safe Drinking Water Act (1974).* Prohibits contamination of the public water system.
- *Toxic Substance Control Act (1976).* Imposes criminal penalties for the failure to follow standards for use of toxic chemicals in manufacturing and industry.
- *Fungicide and Rodenticide Act (1996).* Imposes criminal penalties for the failure to follow standards for the manufacture, registration, transportation, and sale of toxic pesticides.

The *mens rea* for these statutes is generally knowingly committing the prohibited act. A defendant is not required to have knowledge that the act is contrary to a federal statute or that the act poses a health hazard.[2]

You can find more cases on the study site: People v. Thoro Products Company & Newman, *www.sage pub.com/lippmanccl3e.*

OCCUPATIONAL HEALTH AND SAFETY

In 1970, Congress responded to the increasingly high number of job-related deaths and injuries by passing the **Occupational Safety and Health Act** (OSHA) and establishing an agency within the Department of Labor, also known as OSHA, to enforce the act. The act declared that workplace injuries and deaths were resulting in lost production and wages and in preventable medical expenses and disability compensation payments. The act also stated that every working person should be guaranteed safe and healthful working conditions.[3]

For a deeper look at this topic, visit the study site.

OSHA primarily relies on the civil process and financial penalties to insure compliance. A criminal misdemeanor carrying a fine of not more than $10,000 or a prison sentence of up to six months, or both, are provided in the case of a willful violation of the law that results in the death of an employee. A second conviction carries a fine of not more than $20,000 or a prison sentence of up to a year, or both. False statements in any document submitted or required to be maintained under the act may also result in a fine of not more than $10,000 or imprisonment for not more than six months, or both.[4]

OSHA refers cases of intentional, knowing, or reckless violations that result in death for prosecution by state authorities and, in recent years, to the EPA. The most recent data compiled by OSHA found that 4,551 workers died on the job in 2009 and 4,690 workers died on the job in 2010. The most dangerous industry sectors in 2010 were farming, fishing, forestry, transportation, and construction. The overall fatality rate in 2010 was 3.6 workers per 100,000. Between 2007 and 2010, OSHA referred 49 cases for criminal prosecution. One comprehensive study examined data between 1982 and 2002 and found that corporations generally have not been criminally prosecuted either by the federal government or by the roughly two to five states with their own forms of OSHA. OSHA and state agencies initiated 1,798 workplace death investigations and sent a total of 196 cases to federal or state authorities for prosecution. This, in turn, led to 104 prosecutions, 81 convictions, and 16 jail sentences totaling thirty years. In 1992, in a widely discussed North Carolina case, a fire in a poultry plant, which was not equipped with either a fire alarm or sprinkler system, resulted in the death of twenty-five employees and injury to thirty-six workers. The owner pled guilty to twenty-five counts of involuntary manslaughter and was sentenced to twenty years in prison.[5]

SECURITIES FRAUD

Stock market fraud emerged as a subject of intense public interest when it was announced, in June 2002, that domestic diva Martha Stewart was the subject of a criminal investigation for lying to investigators about the sale of stock.

On December 27, 2001, Martha Stewart sold 3,928 shares of stock in the biotech company ImClone, one day before the Federal Drug Administration (FDA) announced that it would not approve the company's new cancer drug, Erbitux. Stewart made roughly $228,000 by selling the stock. Following the FDA announcement, ImClone's stock rapidly fell in value, and had Stewart waited to sell, she would have lost an estimated $45,000. It later was revealed that Stewart lied to federal authorities when she denied having been informed by her stockbroker, Peter Bacanovic, and his assistant, Douglas Faneuil, that the head of ImClone, Sam Waksal, was selling his family's shares at a profit of $7.3 million after learning of the test results.

Martha Stewart and Peter Bacanovic were each sentenced to five months in prison and five months of home detention as a result of their convictions for lying to investigators. Waksal was sentenced to seven years and three months in prison and was ordered to pay more than $4 million in fines and taxes stemming from a variety of criminal offenses, including insider trading. Douglas Faneuil, in return for cooperating with authorities, was fined $2,000. Following Stewart's release from prison, she was confined to home detention on her $16 million estate while being permitted to receive her $900,000 salary and leave her home for up to forty-eight hours a week to work or run errands. Stewart reportedly devoted herself to running her company, Martha Stewart Living Omnimedia, Inc.; writing a magazine column; and preparing for two television shows.

For a deeper look at this topic, visit the study site.

Critics contended that Stewart had been targeted because she was a woman and that the government had wasted valuable resources prosecuting her for the minor offense of lying to authorities about the fact that she had relied on inside information concerning the test results on ImClone's cancer drug. Sixty Wall Street traders were convicted of insider trading between 2009 and 2012. Why does the law punish individuals for buying or selling stocks based on information that is not available to the public at large?

Insider Trading

The stock market rather than banks is increasingly where Americans deposit and look to grow their savings. The average individual has twice as much money in the stock market as in banks. As a result, the federal government has become increasingly concerned with insuring that the stock market functions in a fair fashion and has aggressively brought criminal charges against individuals for stock market fraud.

A corporation that wants to raise money to build new plants, hire workers, or manufacture innovative products typically sells stocks to the public. Individuals purchase stock in hopes that as corporate profits rise, the stock will increase in value, and they eventually will be able to sell it at a substantial profit. This investment in stocks is an important source of money for businesses and provides individuals with the opportunity to invest their money and to save for a house or retirement. Corporate executives and corporate boards of directors possess a **fiduciary relationship** (a high duty of care) to safeguard and to protect the investments of stockholders.

The federal Securities and Exchange Commission (SEC) is charged with insuring that corporate officials comply with the requirements of the Securities Exchange Act of 1934 in the offering and selling of stocks. The act, for instance, requires corporations to provide accurate information on their economic performance in order to enable the public to make informed investment decisions. The SEC typically seeks civil law financial penalties against corporations that violate the law and refers allegations of fraud to the Department of Justice for prosecution. In 2002, Congress passed the **Sarbanes-Oxley Act**. This is a corporate criminal fraud statute that requires the heads of corporations to certify that their firms' financial reports are accurate. A violation of this act is punishable by up to twenty-five years in prison.[6]

In the past decade, the Department of Justice has focused its white-collar crime investigations on **insider trading** in violation of section 10(b) and Rule 10b-5 of the 1934 Act. The enforcement of these provisions is intended to insure that the stock market functions in a fair and open fashion.

Let us return to the ImClone example. Imagine that you are an executive in ImClone and are informed that the company has invented a cure for cancer that has been approved by the federal government. You know that once the information is made public, everyone will be looking to buy ImClone stock and that this will mean that the price of the stock will increase. You tell your relatives the good news and ask them to buy the stock in their name before the announcement and then to sell the stock and to divide the profits with you. When the information is announced, you and your grateful family find that you have made a substantial profit. You believe that this is a just reward for your dedication to the company. The government unfortunately indicts you (the tipper) and your relatives (the tippees) for insider trading. Why is this illegal? Because most people would not put their money in the stock market if a small number of people exploited information that was not available to the public at large to make a profit. This would reduce the money available to businesses and would harm the economy.

Several business law textbooks illustrate insider trading by *Diamond v. Oreamuno*. In this case, several executives of Management Assistance Inc., a computer firm, sold 56,500 shares of company stock for $28 a share at a time when they were aware that the firm's profits were rapidly falling. Then, they publicly announced the company's poor economic performance and the stock declined to $11 a share. The defendants, by selling the stock prior to their announcement, made $800,000 more than they would have earned had they waited to sell the stock. The New York court ruled that "there can be no justification for permitting officers and directors . . . to retain . . . profits which . . . they derived solely from exploiting information gained by virtue of their inside position as corporate officials."[7]

There are two theories of insider trading, both of which prohibit the use of information that is not available to the public to buy or sell a stock. Both theories impose criminal liability on **tippers** (individuals who transmit information) and **tippees** (individuals who receive the information). The **disclose or abstain doctrine** states that corporate officials must publicly reveal information to the public relating to the economic condition of a corporation before they buy or sell the company's stock. The **misappropriation doctrine** as we shall see in the next case, *United States v. Carpenter*, expands the law beyond individuals who work for a corporation and criminally punishes all individuals who take and use inside corporate information that is in the possession of their employer. An example is the U.S. Supreme Court case of *United States v. O'Hagan*, in which a lawyer was convicted of using information that his law firm obtained from a corporate client to make a profit of $4.3 million.[8]

Insider trading is difficult to establish. Investigators must look at who purchases or sells stock and determine whether these individuals relied on inside information in purchasing the securities. The prosecution also must prove a fraudulent intent. In other words, the government must establish that a defendant intentionally purchased or sold the stock knowing that the transaction was in violation of the law. What type of facts would you use to establish a case of insider trading by a corporate executive? Why were the defendants in *United States v. Carpenter* held liable for insider trading?

Did a journalist illegally provide investment advice?

United States v. Carpenter, 791 F.2d 1024 (2d Cir. 1986). Opinion by: Pierce, J.

Facts

Defendants Kenneth P. Felis and R. Foster Winans appeal from judgments of conviction for federal securities fraud in violation of section 10(b) of the 1934 Act and Rule 10b-5, mail fraud, and wire fraud . . . all in connection with certain securities trades conducted on the basis of material, nonpublic information regarding the subject of securities contained in certain articles to be published in the *Wall Street Journal.* Since March 1981, Winans was a *Wall Street Journal* reporter and one of the writers of the "Heard on the Street" column (the "Heard" column), a widely read and influential column in the *Journal.* Carpenter worked as a news clerk at the *Journal* from December 1981 through May 1983. Felis, who was a stockbroker at the brokerage house of Kidder Peabody, had been brought to that firm by another Kidder Peabody stockbroker, Peter Brant ("Brant"), Felis's longtime friend who later became the government's key witness in this case.

Since February 2, 1981, it was the practice of Dow Jones, the parent company of the *Wall Street Journal,* to distribute to all new employees "The Insider Story," a forty-page manual with seven pages devoted to the company's conflicts of interest policy. The district judge found that both Winans and Carpenter knew that company policy deemed all news material gleaned by an employee during the course of employment to be company property and that company policy required employees to treat nonpublic information learned on the job as confidential.

Notwithstanding company policy, Winans participated in a scheme with Brant and later Felis and Carpenter in which Winans agreed to provide the two stockbrokers (Brant and Felis) with securities-related information that was scheduled to appear in "Heard" columns; based on this advance information, the two brokers would buy or sell the subject securities. Carpenter, who was involved in a private, personal, nonbusiness relationship with Winans, served primarily as a messenger between the conspirators. Trading accounts were established in their names. . . . During 1983 and early 1984, defendants made prepublication trades on the basis of their advance knowledge of approximately twenty-seven *Wall Street Journal* "Heard" columns, although not all of those columns were written by Winans. Generally, Winans or Carpenter would inform Brant of the subject of an article the day before its scheduled publication. Winans usually made his calls to Brant from a pay phone and often used a fictitious name. The net profits from the scheme approached $690,000. The district court found that this scheme did not affect the subject matter or quality of Winans's columns, since "maintaining the journalistic purity of the column was actually consistent with the goals of the conspirators," given that the predictability of the columns' market impact depended in large part on the perceived quality and integrity of the columns. . . .

Issue

The fairness and integrity of conduct within the securities markets are a concern of utmost significance for the proper functioning of our securities laws. In broadly proscribing "deceptive" practices in connection with the purchase or sale of securities pursuant to section 10(b) of the Securities Exchange Act of 1934, Congress left to the courts the difficult task of interpreting legislatively defined but broadly stated principles insofar as they apply in particular cases. This case requires us to decide principally whether a newspaper reporter, a former newspaper clerk, and a stockholder, acting in concert, criminally violated federal securities laws by misappropriating material, nonpublic information in the form of the timing and content of the *Wall Street Journal*'s confidential schedule of columns of acknowledged influence in the securities market, in contravention of the established policy of the newspaper, for their own profit in connection with the purchase and sale of securities. It is clear that defendant Winans, as an employee of the *Wall Street Journal,* breached a duty of confidentiality to his employer by misappropriating from the *Journal* confidential prepublication information, regarding the timing and content of certain newspaper columns, about which he learned in the course of his employment. We are presented with the question of whether that unlawful conduct may serve as the predicate for the securities fraud charges herein.

Reasoning

Section 10(b), 15 U.S.C. § 78j(b), prohibits the use in connection with the "purchase or sale of any security . . . [of] any manipulative or deceptive device or contrivance in contravention of such rules and regulations as the Commission may prescribe. . . ."

Rule 10b-5, 17 C.F.R. § 240.10b-5, states:

It shall be unlawful for any person, directly or indirectly, by use of any means or instrumentality of interstate commerce, or of the mails or of any facility of any national securities exchange,

    (a) To employ any device, scheme, or artifice to defraud;

    (b) To make any untrue statement of a material fact or to omit to state a material fact necessary in order to make the statements made, in the light of the circumstances under which they were made, not misleading; or

    (c) To engage in any act, practice, or course of business which operates or would operate as a fraud or deceit upon any person, in connection with the purchase or sale of any security.

The core of appellants' argument is . . . the misappropriation theory may be applied only where the information is misappropriated by corporate insiders or so-called quasi-insiders who owe to the corporation and its shareholders a fiduciary duty of abstention or disclosure. Thus, appellants would have us hold that it was not enough that Winans breached a duty of confidentiality to his employer, the *Wall Street Journal,* in misappropriating and trading on material nonpublic information; he would have to have breached a duty to the corporations or shareholders thereof whose stock appellants purchased or sold on the basis of that information.

We do not say that merely using information not available or accessible to others gives rise to a violation of Rule 10b-5. There are disparities in knowledge and the availability thereof at many levels of market functioning that the law does not presume to address. However, the critical issue is found in the district judge's careful distinction between "information" and "conduct." Whatever may be the legal significance of merely using one's privileged or unique position to obtain material, nonpublic information, here we address specifically whether an employee's use of such information in breach of a duty of confidentiality to an employer serves as an adequate predicate for a securities violation. Obviously, one may gain a competitive advantage in the marketplace through conduct constituting skill, foresight, industry, and the like. But one may not gain such advantage by conduct constituting secreting, stealing, purloining, or otherwise misappropriating material, nonpublic information in breach of an employer-imposed fiduciary duty of confidentiality. Such conduct constitutes chicanery, not competition; foul play, not fair play. Indeed, underlying section 10(b) and the major securities laws generally is the fundamental promotion of "'the highest ethical standards' . . . in every facet of the securities industry." . . . We think the broad language and important objectives of section 10(b) and Rule 10b-5 render appellants' conduct herein unlawful. . . .

The information misappropriated here was the *Journal*'s own confidential schedule of forthcoming publications. It was the advance knowledge of the timing and content of these publications, upon which appellants, acting secretively, reasonably expected to and did realize profits in securities transactions. Since section 10(b) has been found to proscribe fraudulent trading by insiders or outsiders, such conduct constituted fraud and deceit, as it would had Winans stolen material, nonpublic information from traditional corporate insiders or quasi-insiders. The district court found that between October 1983 and the end of February 1984, twenty-seven "Heard" columns were leaked in advance. If an occasional investment plan faltered due to nonpublication of the anticipated corollary "Heard" column, the record nonetheless amply demonstrates that the majority of the securities traded resulted in profits reflecting the predictable price change due to the publication anticipated. This was true, for example, of trades in American Surgery Centers, Institutional Investors, and TIE/Communications, Inc., to mention just a few of the securities traded. In any event, a fraudulent scheme need not be foolproof to constitute a violation of Rule 10b-5. It is enough that appellants reasonably expected to and generally did reap profits by trading on the basis of material, nonpublic information misappropriated from the *Journal* by an employee who owed a duty of confidentiality to the *Journal.*

Nor is there any doubt that this "fraud and deceit" was perpetrated "upon any person" under section 10(b) and Rule 10b-5. It is sufficient that the fraud was committed upon Winans's employer. . . . Appellants Winans and Felis and Carpenter by their complicity perpetrated their fraud "upon" the *Wall Street Journal,* sullying its reputation and thereby defrauding it "as surely as if they took [its] money."

As to the "in connection with" standard, the use of the misappropriated information for the financial benefit of the defendants and to the financial detriment of those investors with whom appellants traded supports the conclusion that appellants' fraud was "in connection with" the purchase or sale of securities under section 10(b) and Rule 10b-5. We can deduce reasonably that those who purchased or sold securities without the misappropriated information would not have purchased or sold, at least at the transaction prices, had they had the benefit of that information. . . . Further, investors are endangered equally by fraud by noninside misappropriators as by fraud by insiders.

Appellants argue that it is anomalous to hold an employee liable for acts that his employer could lawfully commit. . . . In the present case, the *Wall Street Journal* or its parent, Dow Jones Company, might perhaps lawfully disregard its own confidentiality policy by trading in the stock of companies to be discussed in forthcoming articles. But a reputable newspaper, even if it could lawfully do so, would be unlikely to undermine its own valued asset, its reputation, which it surely would do by trading on the basis of its knowledge of forthcoming publications. Although the employer may perhaps lawfully destroy its own reputation, its employees should be and are barred from destroying their employer's reputation by misappropriating their employer's informational property. Appellants' argument that this distinction would be unfair to employees illogically casts the thief and the victim in the same shoes. . . . Here, appellants, constrained by the employer's confidentiality policy, could not lawfully trade by fraudulently violating that policy, even if the *Journal*, the employer imposing the policy, might not be said to defraud itself should it make its own trades.

Holding

Thus, because of his duty of confidentiality to the *Journal*, defendant Winans—and Felis and Carpenter, who knowingly participated with him—had a corollary duty, which they breached, under section 10(b) and Rule 10b-5, to abstain from trading in securities on the basis of the misappropriated information or to do so only upon making adequate disclosure to those with whom they traded.

Questions for Discussion

1. How could Winans violate the law when he did not work for the corporations whose stock he advised Brant and Felis to buy or sell?

2. Winans's columns contained data concerning the economic performance of companies that was available to any sophisticated member of the public who was skilled at corporate analysis. Is this insider information and trading?

3. Explain how the defendants benefited by investing in companies before the publication of Winans's columns.

4. Is it significant that Winans was never certain that a column would appear in a forthcoming issue of the *Wall Street Journal*?

5. Could the *Wall Street Journal* have traded in these securities without violating the law? What about a private individual who did not work for the *Wall Street Journal* who based his or her own analysis on the same information that was provided by Winans to Brant and Felis?

MAIL AND WIRE FRAUD

The U.S. government has relied on the mail and wire fraud statutes to prosecute a variety of corrupt schemes that are not specifically prohibited under federal laws. **Fraud** may be broadly defined as an intentional and knowing misrepresentation of a material (important) fact intended to induce another person to hand over money or property. Mail and wire fraud prosecutions range from fraudulent misrepresentations of the value of land and the quality of jewelry to offering and selling nonexistent merchandise. The common element in these schemes that permits the assertion of federal jurisdiction is the use of the U.S. mails or wires across state lines (phone, radio, television). The federal **mail fraud** statute reads as follows:[9]

> Whoever, having devised or intending to devise any scheme or artifice to defraud, or for obtaining money or property by means of false or fraudulent pretenses . . . for the purpose of executing such scheme . . . places in any . . . authorized depository for mail . . . any matter or thing whatever to be sent or delivered by the Postal Service, . . . or takes or receives therefrom, any such matter or thing, . . . shall be fined under this title or imprisoned not more than 20 years, or both. . . .

A conviction for mail fraud requires the prosecution to demonstrate:

- *Scheme*. Knowing participation in a scheme or artifice to defraud.
- *Falsehood*. Intentional false statement or promise.
- *Money or Property*. The intent to obtain money or property.
- *Reliance*. Statement or promise was of a kind that would reasonably influence a person to part with money or property.
- *Mail*. The mails were used for the purpose of executing the scheme. This includes private mail delivery services. The mail is required to be only incidental to an essential part of the criminal design.

The requirements of mail and **wire fraud** are similar, with the exception that the wire communication must cross interstate or foreign boundaries.[10] A conspiracy to commit mail or wire fraud is also prohibited under federal statutes. A prosecutor under the conspiracy statute is required to demonstrate that the use of the mail or wires would naturally occur in the course of the scheme or that the use of the mail or wires was reasonably foreseeable, although not actually intended.[11]

The next case, *United States v. Duff,* raises interesting issues of the interpretation of fraud under the mail fraud statute. It also clarifies whether the use of the mails or wires must be a central part of a scheme to defraud or may be a minor aspect of the criminal plan.

Did the defendants use the mails to commit mail fraud against the City of Chicago?

United States v. Duff, 336 F. Supp. 2d 852 (N.D. Ill. 2004). Opinion by: Bucklo, J.

Facts

Defendants James M. Duff, William E. Stratton, Patricia Green Duff, and Terrence Dolan move to dismiss the indictment.

The indictment charges defendants with . . . mail fraud . . . money laundering. . . . The indictment charges that defendants conspired among themselves and with others, both named and unnamed, to defraud the City of Chicago (the City) by falsely representing that certain entities, which were in fact owned and managed by Mr. Duff, were qualified as Minority-Owned Businesses (MBEs) or Women-Owned Businesses (WBEs) under . . . the amended Municipal Code of the City of Chicago. The Municipal Code's provisions are designed to provide set-asides for MBEs and WBEs in connection with large contracts let by the City for competitive bidding. In order to qualify for the set-asides, businesses must be at least 51% owned and controlled by one or more minorities or women.

The charging allegations are numerous and specific. Using the charges that involve just two of the Duff businesses as examples, the allegations may be summarized as follows. Windy City Maintenance, Inc. was certified as a WBE in 1991 on the basis of a sworn affidavit and certain other statements made by Patricia Green Duff to the effect that she was the real owner and controlled the operations of Windy City Maintenance. In fact, Ms. Green Duff, who is the mother of James M. Duff, was not the real owner of the business; it was owned and controlled by Mr. Duff. In 1994, Remedial Environmental Manpower, Inc. (REM), was qualified by the City as an MBE on the basis of a sworn affidavit and other statements of Mr. Stratton, an African American, who claimed that he was the real owner and controlled the operations of REM. In fact, Mr. Duff, not Mr. Stratton, owned and controlled REM. Similar allegations are made with respect to the other Duff-owned businesses.

The indictment charges the pattern of deceit did not end with the initial qualification of the entities;

ongoing compliance requirements of the Municipal Code were flouted by similar deceptions made in subsequent years. As a result, the entities specifically mentioned above, together with other Duff-owned and -controlled businesses, obtained direct contracts and subcontracts worth more than $100 million and generated payments and distributions for the benefit of the named defendants and other relatives and associates of Mr. Duff aggregating more than $9 million.

Reasoning

Mail fraud is established as a federal crime by 18 U.S.C. § 1341. In the present case, the defendants' conduct deprived the City of the power to control how its money should be spent. Under the minority set-asides established in the Ordinance, the City made it unmistakably clear how it wanted its money spent, and defendants' conduct, if true, thwarted the City's legislative intent. By allegedly falsely representing the identity of owners and management in Duff family businesses, defendants are accused of causing over $100 million in City money to go to businesses that were neither MBEs nor WBEs.

Defendants also argue that the City suffered no actual loss, because the indictment does not charge that the Duff-owned companies gave less than full value in providing their services. . . . The assertion that the businesses owned and controlled by Mr. Duff performed satisfactorily under the contracts obtained by them through fraud is not a defense . . . using innocent third parties to effect a scheme to defraud also does not shield the perpetrator from criminal penalties. . . .

Issue

Defendants ask that the mail fraud counts be dismissed on the ground that all of the mailings of checks from the City described in those counts were to third-party general contractors such as Waste Management and that in no instance were the mails used to further a scheme; and that the mailings would have occurred regardless of defendants' conduct.

Holding

The defendants' alleged scheme was a continuing one that lasted through the 1990s. The core element of the scheme was the masquerade of Duff-owned and -controlled businesses as MBEs and WBEs, which allowed those entities to gain, fraudulently, large fees as subcontractors of the general contractors to which the checks were mailed. Under § 1341, a mailing will be considered in furtherance of a scheme to defraud if it is incidental to an essential part of the scheme. If the City had not mailed checks to the prime contractors, defendants' entities would have gained nothing from their alleged fraudulently obtained status as qualified subcontractors. . . .

Questions for Discussion

1. Explain how Chicago was the victim of the mail fraud of property or money. Did Chicago receive what it paid for?

2. What about the defendants' argument that the mails were not used to further the alleged fraud?

3. James Duff received a ten-year prison sentence. William Stratton, an African American appointed by Duff to head Remedial Environmental Manpower, received six years in prison. Duff was also ordered to pay more than $22 million in fines. Patricia Green Duff was evaluated as unfit to stand trial. Prosecutors had asked for a sentence of twenty-five years for James Duff. Defense lawyers characterized Duff as a loving husband and father of six, and Duff promised that he would "never be in court under these situations again." Were these sentences too lenient or too harsh? Can you think of a creative punishment for Duff that would have both punished him and assisted society? Duff's conviction and sentence later were affirmed by the Seventh Circuit Court of Appeals. See *United States v. Leahy,* 464 F.3d 773 (7th Cir. 2006). You may be interested in comparing *Duff* to a case involving a sports agent and college athletes in which the Seventh Circuit held that the fraudulent scheme did not fall within the mail fraud statute. See *United States v. Walters,* 997 F.2d 1219 (7th Cir. 1993).

See more cases on the study site: **United States v. Brown,** *www.sagepub.com/lippmanccl3e.*

THE TRAVEL ACT

The **Travel Act** of 1961 was intended to assist state and local governments to combat organized crime. The Travel Act, 18 U.S.C. § 1952, authorizes the federal government to prosecute what are ordinarily considered the state criminal offenses of gambling, the illegal shipment and sale of alcohol and controlled substances, extortion, bribery, arson, prostitution, and money laundering. Federal jurisdiction is based on the fact that the crimes have been committed following travel in interstate or foreign commerce or through the use of the U.S. mails or any other facility in interstate or foreign commerce.

In *United States v. Jenkins,* the Second Circuit Court of Appeals stated that a conviction under the Travel Act requires (1) travel or the use of the mails or some other facility (e.g., wires) in interstate or foreign commerce, (2) with the intent to commit a criminal offense listed in the Travel Act or crime of violence or to distribute the proceeds of an illegal activity, and (3) the commission of a crime or attempt to commit a crime.[12] Performing or attempting to perform an act of violence is punishable by not more than twenty years in prison or a fine or both. Other offenses are punishable by not more than five years in prison or a fine or both.

In *United States v. Goodman,* Goodman promoted records by contacting and persuading radio stations to place records from the companies he represented on their "playlists." Goodman, however, went beyond mere persuasion and was determined to have illegally paid as much as $182,615 a year in cash through the mails to program directors and disc jockeys in return for placing records on their playlists. Goodman was convicted under the Travel Act of the use of interstate mail to commit bribery. The federal appellate court rejected the argument that the payments occurred after the records were added to the playlists and that the payments therefore did not constitute a bribe to induce station managers and disc jockeys to play specific records. The court noted that the receipt of the mailed money was intended to both "reward a past transgression and to influence or promote a future one."[13]

Goodman was also convicted under 47 U.S.C. § 508, the Payola Act, which prohibits the payment of money to radio station employees for the inclusion of material as part of a program unless the payment is disclosed to the recipient's employer. The federal appellate court ruled that the offense is complete upon the payment of money and that the records need not actually be played.

HEALTH CARE FRAUD

Roughly one-fifth of the federal budget is devoted to health care, most of which involves reimbursing doctors and health care workers for services provided under various federal and state programs to the elderly, children, the physically and mentally challenged, and economically disadvantaged individuals. The difficulty of administering programs of this size and complexity creates an opportunity for doctors and other health care providers to submit fraudulent claims for the reimbursement of services that, in fact, were never provided or to seek payment for unnecessary procedures. In 1996, Congress acted to prevent this type of fraud when it adopted a statute on health care fraud that punishes individuals who knowingly and willfully execute or attempt to execute a scheme or artifice

> to defraud any health care benefit program; or to obtain by means of false or fraudulent pretenses, representations, or promises, any of the money or property owned by, or under the custody or control of any health care benefit program.[14]

Health care fraud is punishable by a fine and imprisonment of up to ten years or both. Fraudulent acts that cause serious injury are punishable by a term of imprisonment of up to twenty years, while fraudulent acts that result in death are punishable by up to life in prison.

It is estimated that health care fraud costs the taxpayers $68 billion annually. In *United States v. Baldwin,* the defendants were convicted of submitting a false claim of $275,000 for four dental chairs.[15] The health care fraud statute was interpreted to cover individuals outside the medical profession in *United States v. Lucien.* The defendants paid individuals to cause collisions with other vehicles and to claim that they suffered serious injuries. The defendants referred these alleged "victims" to various medical clinics in return for a fee. The clinics sought reimbursement from New York for medical procedures that, in fact, had not been provided. The "victims" then sued the drivers of the other vehicles in hopes of obtaining a settlement from the drivers' insurance companies.[16]

The type of extreme and grossly fraudulent abuse of the health care system that can take place is illustrated by *United States v. Miles.*[17]

Affiliated Professional Home Health (APRO) was formed in 1993 in Houston, Texas, by Carrie Hamilton, Alice Miles, and Richard Miles. Richard Miles, a vice-principal of a Houston-area high school, was married to Alice Miles, a registered nurse, and is the brother of both Hamilton, also a registered nurse, and Harold Miles, an APRO employee. When APRO obtained certification from the Texas Department of Health and a Medicare provider number, the company began to treat Medicare-covered patients and obtain reimbursement for in-home visits to such patients.

In this case, the government presented evidence that the defendants, through APRO, submitted cost reports that grossly inflated expenses for items ranging from mileage to employee salaries. For example, Hamilton was reimbursed for a whopping 282,000 travel miles from 1994 to 1996, a period when she also frequently visited Louisiana casinos. Alice Miles, another avid gambler, was reimbursed for 150,000 travel miles over three years, while her husband, whose primary job kept him occupied for most of the work day, was reimbursed for 180,000 miles over four years.

APRO also obtained reimbursement for costs that included personal expenses, such as renovations to the Hamiltons' home, renovations to the Miles's parents' residence, and various home appliances. Eventually, the amount of money coming in to APRO for fake charges became so large that in order to sustain the claimed level of expenses over the next year—so that APRO would not have to return overpayments to the federal government—the APRO principals began to use a variety of other methods to bilk Medicare out of taxpayer funds. These methods included their writing large-dollar checks to employees for "expenses" or "back pay" and then requiring the employees to cash the checks and hand the funds back to the APRO principals. Appellants billed expenses to Medicare for two or three times the actual cost incurred. At times, they engaged in more intricate schemes involving the splitting of large reimbursement checks into smaller cashier's checks that were then deposited into the APRO principals' bank accounts or used for personal expenses. On one occasion, Hamilton split an APRO check into cash and three cashier's checks at one bank. She deposited two of the cashier's checks into her own account at another bank and used a portion of the funds to obtain a fourth cashier's check to purchase a new Ford Mustang convertible. The third cashier's check from the original bank was cashed at the Star Casino.

In May 2012, more than 100 individuals were arrested in seven cities for a conspiracy involving more than $450 million in fraudulent medical billings.

MONEY LAUNDERING

Individuals involved in criminal fraud or drug or vice transactions confront the problem of accounting for their income. These individuals may want to live a high-profile lifestyle and buy a house or automobile that they could not afford based on the income reported on their tax forms. An obvious gap between lifestyle and income may attract the attention of the Internal Revenue Service or law enforcement. How can individuals explain their ability to purchase a million-dollar house when they report an income of only $30,000 a year? Where did the cash come from that they used to buy the house? Bank regulations require that deposits of more than $10,000 must be reported by the bank to the federal government. How can individuals explain to government authorities the source of the $50,000 that they deposit in a bank?

The solution is **money laundering**. This involves creating some false source of income that accounts for the money used to buy a house, purchase a car, or open a bank account. This typically involves schemes such as paying the owner of a business in cash to list a drug dealer as an employee of the individual's construction business. In other instances, individuals involved in criminal activity may claim that their income is derived from a lawful business such as a restaurant. Money laundering statutes are intended to combat the "washing" of money by declaring that it is criminal to use or transfer illegally obtained money or property. This is punishable by a fine of up to $500,000 and imprisonment for up to twenty years.

Money laundering includes the following elements:

- The defendant engaged or attempted to engage in a monetary transaction.
- The defendant knew the transaction involved funds or property derived from one or more of a long list of criminal activities listed in the statute.
- The transaction was intended to conceal or disguise the source of the money or property; or
- The transaction was intended to promote the carrying on of a specified unlawful activity.
- The money constituted the "proceeds" or "profit" from an unlawful activity.[18]

In *United States v. Johnson,* the defendant generated millions of dollars from a fraudulent scheme involving Mexican currency. The Tenth Circuit Court of Appeals ruled that the use of these funds to purchase an expensive home and a Mercedes violated the money laundering statute in that these purchases furthered the defendant's continued illegal activities by providing him with a legitimacy that he used to impress, attract, and ultimately victimize additional investors.[19]

In *United States v. Carucci,* a real estate agent was acquitted who assisted a well-known organized crime figure to purchase several homes. The federal appellate court agreed with the government prosecutors that the organized crime figure was unable to account for the source of the money that he used to buy real estate. The court, however, ruled that the government failed to clearly establish that these funds were derived from an illegal activity, such as illegal gambling, extortion, narcotics, or loan sharking.[20]

The next case in the text, *United States v. Jackson,* illustrates a money laundering scheme. Can you explain in a clear and straightforward fashion why Reverend Davis was convicted of money laundering?

Was Reverend Davis guilty of money laundering?

United States v. Jackson, 935 F.2d 832 (7th Cir. 1991). Opinion by: Flaum, J.

Issue

Mandell Jackson, Joseph Davis, and Romano Gines were each indicted and convicted of one count of conspiring to distribute over 50 grams of cocaine base in violation of 21 U.S.C. §§ 841(a)(1) and 846. In addition, Davis was indicted and convicted of one count of engaging in a continuing criminal enterprise in violation of 21 U.S.C. § 848 and three counts of money laundering in violation of 18 U.S.C. § 1956(a). Jackson and Gines were sentenced to 210 months each. Davis was sentenced to 30 years. Did the evidence support the conviction of Reverend Joseph Davis for money laundering?

Facts

The Reverend Joseph Davis describes himself as "a small-time, hellfire and brimstone country preacher." The evidence at trial, however, presented a more complete view of Mr. Davis' talents. It showed how he repaired a run-down East St. Louis church and revitalized its congregation, helping to restore the social fabric of a community in distress. Sadly, it also showed that Davis devoted his considerable skills to a variety of schemes that ranged from shady to downright illegal. One of these schemes was the ongoing distribution of late-twentieth century America's counterpart to brimstone, crack cocaine.

Davis became the preacher at the 15th Street Baptist Church in the mid-1980s. Shortly thereafter he began to sell drugs, and by mid-1987 was actively selling crack from two houses, the first at 735 Wabashaw and the second at 1479 Belmont. The Wabashaw house was managed by Dwayne Scruggs and the Belmont house by Mandell Jackson. Scruggs and Jackson directed teams of addicts who would sell crack on the streets around the houses as well as to passing motorists. They were paid for their efforts in crack. These addicts had occasional contacts with Davis, who would visit the houses to replenish the drug supply and collect cash. During the visits Davis would typically confer with the house manager privately. If the cocaine Davis supplied came in powder form, Jackson or Scruggs would cook it into cocaine base which they would then divide into smaller portions and distribute to the sellers for resale. At the Belmont house, Romano Gines helped with the cooking and otherwise assisted Jackson in running the house.

Davis deposited some of the cash he collected from the houses in bank accounts maintained in the name of the 15th Street Baptist Church Development Corporation ("Development Corporation account") and the 15th Street Baptist Church ("Church account") and at Illini Federal, a local savings and loan. Also deposited in the Development Corporation accounts were funds that Davis and the Corporation obtained from other activities. One of Davis' other activities was steering his parishioners and others to used-car outlets in the East St. Louis area in return for commissions from the dealers, a practice known as "bird-dogging." Davis would secure consumer credit for the cars and other purchases he helped to arrange through Sam Bennett, a loan officer at Jefferson Bank & Trust in St. Louis. Bennett, it is alleged, would turn a blind eye to the inability of many of the borrowers Davis sent his way to repay their obligations to Jefferson. In return, he and Davis would split the fees they received for arranging these loans. Davis also deposited in the Church and the Development Corporation accounts funds he received from more legitimate activities, including a contract for the Corporation to demolish a building in East St. Louis.

Davis could write checks on these accounts. Some of these checks were made out to cash, which Davis diverted to his personal use. Others were made out to local vendors who provided services such as beepers and mobile telephones. Still others were made out to the landlord who owned the Swansea, Illinois, residence where Davis lived. Davis also purchased numerous cars, spending over $79,000 on a variety of vehicles for personal and church use between October 1987 and November 1988.

Scruggs plead guilty to conspiracy and testified against his former confederates at their joint trial. At the close of this trial, Davis, Gines, and Jackson were each found guilty of conspiring to distribute cocaine base. The jury also found Davis guilty of engaging in a continuing criminal enterprise, and of three of the four counts of money laundering. . . .

Reasoning

The two provisions of 18 U.S.C. § 1956 under which Davis was charged provide, in relevant part, that:

(a) (1) Whoever, knowing that the property involved in a financial transaction represents the proceeds of some form of illegal activity, conducts or attempts to conduct such a financial transaction which in fact involves the proceeds of specified unlawful activity—

(A) (i) with the intent to promote the carrying on of specified unlawful activity; or . . .

(B) knowing that the transaction is designed in whole or in part—(i) to conceal or disguise the nature, the location, the source, the ownership, or the control of the proceeds of specified unlawful activity; . . . .

shall be sentenced to a fine of not more than $500,000 or twice the value of the property derived in the transaction, whichever is greater, or imprisonment for not more than twenty years, or both.

In a case brought under § 1956(a)(1)(B)(i), the government must prove that the transaction was designed to conceal one or another of the enumerated attributes of the proceeds involved. . . .

Agent Wehrheim testified as to Davis' sources of cash and established that Davis and the Development Corporation made bank deposits equal to approximately twice the amount that could be accounted for out of legitimate sources of income. Over half the total amount of these deposits was in cash. . . . The evidence at trial also established that Davis had access to large amounts of cash. For example, Marcie Rupert testified she had helped him count approximately $35,000 in currency. Pierre Manley, who worked as a runner transporting cash from the Wabashaw house to Davis, testified that he would often pick up $1,500 a day from the house. Reasonable jurors could certainly infer that the cash contained in the deposits Davis made or ordered to be made were derived to a large extent from Davis' drug operations and that Davis knew this.

Count three charged Davis with issuing seven Development Corporation checks to vendors providing beeper services and mobile telephone services. Pierre Manley testified that Davis gave him a beeper when he began to serve as one of Davis' runners, and that Davis would call Manley's beeper to tell him to contact Davis. When Manley called back, Davis instructed him to drive to the Wabashaw house to make pickups. This and other evidence of the use of beepers as an integral part of Davis' conduct of his continuing criminal enterprise suffice to establish that the use of the funds derived from Davis' drug activities to purchase beepers. . . . These transactions . . . fall within the second money laundering provision . . . The conversion of cash into goods and services as a way of concealing or disguising the wellspring of the cash is a central concern of the money laundering statute. . . . To convict under § 1956(a)(1)(B)(i) the government must prove not just that the defendant spent ill-gotten gains, but that the expenditures were designed to hide the provenance of the funds involved.

We believe that the government met this burden. Davis chose to place the proceeds of his drug sales not in a personal bank account but in bank accounts ostensibly maintained by the 15th Street Baptist Church and the 15th Street Baptist Church Development Corporation. He nevertheless treated the funds in these accounts as his own, using them to pay for cellular telephones he installed in his many cars and vans, his rent, and his credit card and home phone bills. The jury could reasonably infer that the use of the church accounts was an attempt to hide the ownership and source of Davis' drug money while preserving his ready access to the funds in the accounts, which were as close as the church's checkbook. Moreover, the evidence also established that Davis frequently removed himself still further from the funds in the church accounts by using the church secretary to present Development Corporation checks made out to cash at Illini Federal. She would then hand over cash she received to Davis. . . .

Holding

We affirm the defendant's criminal conviction.

Questions for Discussion

1. Describe Reverend Joseph Davis's criminal activity.

2. What are the elements of money laundering under 18 U.S.C. § 1956?

3. Why is Reverend Davis guilty of money laundering?

4. Explain the reason that money laundering is a crime.

ANTITRUST VIOLATIONS

The **Sherman (Antitrust) Act of 1890** is intended to insure a free and competitive business marketplace. The Sherman Act, according to former Supreme Court Justice Hugo Black, is designed to be a "comprehensive charter of economic liberty aimed at preserving free unfettered competition as the rule of trade." Imagine if every bar and restaurant in a college town agreed to sell beer at an inflated price rather than compete with one another for the business of students. The theory behind the Sherman Act, as explained by Justice Black, is that economic competition results in low prices and high quality and promotes self-reliance and democratic values. Can you explain why free competition leads to these benefits?[21]

The criminal provisions of the Sherman Act state that any person "who shall make any contract or engage in any combination or conspiracy" to interfere with interstate commerce is guilty of a felony. A corporation shall be punished with a fine of $10,000,000 and an individual by a fine of $1,000,000 or by imprisonment not exceeding three years, or both.[22]

A conviction requires proof that two or more persons or organizations

- knowingly entered into a contract or formed a combination or conspiracy; and
- the combination or conspiracy produced or potentially produced an unreasonable restraint of interstate trade.

In *United States v. Azzarelli Construction Co. and John F. Azzarelli,* the defendant was the owner of a construction company who agreed with the owners of other construction businesses to rig the process of bidding on state contracts to insure that each firm would receive state contracts. One firm would be designated to receive a contract and would submit an unreasonably high bid on the job. The competing firms would submit grossly inflated bids, insuring that the first firm would receive the contract. The firms that intentionally lost the job then would be compensated by receiving a kickback from the successful contractor. The court noted that this fraudulent practice interfered with interstate commerce by raising the cost of highway construction and resulted in less money being available to upgrade the highway system.[23]

You can find more cases on the study site: United States v. Allegheny Bottling Co., *www.sagepub.com/lippmanccl3e.*

CRIME IN THE NEWS

In December 2008, prominent New York investment broker Bernard L. Madoff called his sons into his office and announced that his business was a "big lie" and "basically a giant Ponzi scheme." Madoff sadly noted that there was "nothing left" and that he expected to "go to jail." He was arrested on December 11 and confessed to the FBI that he had looted investors of as much as $50 billion, making this the largest fraud in American history. Madoff's scheme was relatively unsophisticated. He would use the money provided by new investors to pay returns to old investors. This enabled Madoff to pay investors a consistent return of ten to fifteen percent a year. He was so successful that he could afford to turn investors away who lacked the "right background" and required most people who wanted to invest their money to provide at least $1 million. Other Wall Street brokers made millions of dollars by turning all their clients' funds over to Madoff for investment. This "house of cards" collapsed when the American economy took a downturn and a large number of Madoff's investors asked for the return of their money and found that their money had disappeared.

A portion of the money undoubtedly was used to support Madoff's quiet but luxurious lifestyle. This included memberships in most of the leading golf clubs in New York and Florida, partial ownership of two corporate jets and of two boats, and multiple homes, including one in France. Madoff reportedly was a frequent visitor to a luxury hotel in France where rooms rented at roughly $7,000 per night. Despite his affluent lifestyle, Madoff was respected for his public service and his charitable foundation and gave generously to worthy organizations in New York City including Carnegie Hall, the Public Theater, the Special Olympics, and the Gift of Life Bone Marrow Foundation. He also served on the boards of directors of various worthy causes, serving on the boards of Yeshiva University and of the Yeshiva business school. Yeshiva recognized his contribution with a special award of appreciation. Madoff's public profile and prominent role in the community attracted investors eager to claim that they were represented by Bernard Madoff.

Madoff defrauded his friends, his own sister, and the institutions that trusted him and demonstrated that even the most educated and sophisticated members of American society can be tricked by a skilled con man. Madoff's clients included the family that owned the New York Mets baseball team and the former owner of the Philadelphia Eagles football team. His victims included Yeshiva University, the institution that had embraced and honored him. New York Law School and Tufts University also suffered a loss of portions of their endowments. A number of charitable organizations lost most of their resources, including the foundation of Elie Weisel, the famed Holocaust survivor and commentator. The collapse of charitable foundations meant that many nonprofit foundations that had received donations now found that they had a shortage of money and confronted the prospect of closing their doors. The Picower Foundation lost roughly $268 million, threatening the foundation's ability to fund groups like the Children's Health Fund and the New York Public Library. The JEHT Foundation lost most of its assets and no longer would be able to fund social action organizations such as the Innocence Project,

Human Rights First, Center for Investigative Reporting, and Juvenile Law Center. The foundation of the famed film producer Steven Spielberg, which supported prominent hospitals, also lost a portion of its assets.

The collapse of Madoff's empire reverberated across the financial landscape and across the globe. A number of New York real estate developers were forced to cancel construction projects due to a lack of funds. In the past several years, Madoff had attracted significant investments from individuals, banks, and institutions in Austria, South America, Spain, Sweden, Switzerland, Asia, and the Middle East. Several days following the announcement that Madoff's firm was an empty shell, a French investment broker, R. Thierry Magnon de law Villehuchet, slashed his wrists and was found dead in his New York City office. The newspapers reported instances in which the elderly and infirm who counted on Madoff's investments to keep them economically secure found that they had been "wiped out" financially and faced the prospect of selling their homes or apartments in order to survive.

There is no obvious explanation for Madoff's corrupt conduct. He was from an extremely modest background, had lifted himself out of poverty through sheer intelligence, and had saved the money he earned as a young man from menial jobs and proceeded to build one of the most successful firms on Wall Street. Madoff pioneered the use of computers for investing and was a past president of a national organization of financial analysts and served on the group's board of governors. The salaries in his firm were modest by Wall Street standards, and yet he attracted a loyal group of employees who praised him for his personal kindness and concern for their welfare. He also made it a point to bring his brother, sons, and niece into the firm. Madoff prided himself on his integrity and accountability to his investors and proclaimed that his motto was that "the name is on the door." It was Madoff's prominence and his powerful clientele that may have intimidated the Securities and Exchange Commission and deterred the organization from investigating the performance of his investments, which most experts agreed was "too good to be true." Some observers have commented that investors were willing to tolerate Madoff's suspected illegalities so long as they were benefiting financially.

Madoff currently is serving a 150-year sentence and is in minimum security in Butner, North Carolina. Judge Denny Chin noted that Madoff's crimes were "extraordinarily evil" and that the sentence fit his "moral culpability." In an interview, Madoff accused the judge of making him into the "human piñata of Wall Street" and caustically observed that he was surprised that the judge did not "suggest stoning in the public square.'" He complained that "serial killers get a death sentence, but that's virtually what he gave me." Was Madoff's 150-year sentence disproportionate?

PUBLIC CORRUPTION

In 2004, the Corporate Crime Reporter released a report on public corruption in the United States. At the time of the report, three Connecticut mayors and the state treasurer had been sentenced to prison and the governor's former deputy chief of staff had pled guilty to accepting gold coins in return for government contracts. Governor John Rowland admitted that firms with state contracts had provided him with a hot tub and cathedral ceilings in his home. Connecticut, one of the cradles of American democracy, now was characterized by observers as the most corrupt state in the nation. Connecticut, however, is relatively "clean" compared to Mississippi, Louisiana, Illinois, Florida, and New York. In each of these states, a number of local and federal public officials have been accused of betraying the public trust by accepting or demanding money, campaign contributions, and other benefits in return for jobs, construction contracts, driver's licenses, and favorable judicial verdicts. These **crimes of official misconduct** are defined as the knowingly corrupt behavior by a public official in the exercise of his or her official responsibilities.[24]

The overwhelming number of prosecutions for public corruption are brought by federal authorities against state and federal officials. A study by the Corporate Crime Reporter of public corruption between 1993 and 2006 in states with populations of more than two million people concluded that the most corrupt states are Louisiana, Mississippi, Kentucky, Alabama, Ohio, Illinois, Pennsylvania, Florida, New Jersey, and New York. The least corrupt states are Washington, Utah, Kansas, Minnesota, Iowa, and Oregon. We all suffer when officials are corrupt because decisions are made based on monetary rewards rather than on the basis of what is best for the public.

The most frequently prosecuted state and federal official misconduct crime is the **bribery of a public official**. This offense, as we noted in discussing property offenses in Chapter 13, is committed by an individual who gives, offers, or promises a benefit to a public official as well as by a public official who demands, agrees to accept, or accepts a bribe. In other words, bribery punishes either giving or receiving a bribe and requires an intent to influence or to be influenced in carrying out a public duty. Bribery does not require a mutual agreement between the individuals. If you offer money to a police officer with the intent that he or she not charge you with a traffic offense, you are guilty of the bribery of a public official, regardless of whether the officer agrees to accept the bribe.

The offense of offering a bribe to a public official requires that

- the accused wrongfully promised, offered, or gave money or an item of value to a public official;
- the individual occupied an official position or possessed official duties; and
- the money or item of value was promised, offered, or given with the intent to influence an official decision or action of the individual.

The offense of soliciting a bribe requires that

- the accused wrongfully asked, accepted, or received money or an item of value from a person or organization;
- the accused occupied an official position or exercised official duties; and
- the accused asked, accepted, or received money or an item of value with the intent to have his or her decision or action influenced with respect to this matter.

Bribery is distinguished from **graft**. Graft does not require an intent to influence or to be influenced. Graft is defined as asking, accepting, receiving, or offering money or an item of value as compensation or a reward for an official decision. A builder that received a state contract to repair highways may express his or her appreciation and attempt to receive favorable consideration in the future by renovating a politician's summer home. Various state and federal statutes declare that it is a crime for a public official to ask for or to receive a reward for an official act. In such cases, the public is deprived of the right to receive "honest and faithful services."[25]

Why punish public bribery? Public officials are charged with acting in the interests of society rather than acting in the interest of individuals who provide a financial or other benefit. Corrupt behavior undermines confidence and trust in government, leads to dominance by the rich and powerful, and is contrary to the notion that every individual should be treated equally. In other words, government should be responsive to the majority of Americans rather than to the minority of millionaires.

For an international perspective on this topic, visit the study site.

The **Foreign Corrupt Practices Act** extends the concern with good government abroad and declares that it is illegal for an individual or U.S. company to bribe a foreign official in order to cause that official to assist in obtaining or retaining business. The statute makes an exception for "facilitating payments" to speed or insure the performance of a "routine governmental action," such as paying money to a foreign official to guarantee that an entry visa is quickly issued to a corporate employee. In 2005, the Titan Corporation in San Diego pled guilty and agreed to pay a fine of $28 million in settlement of various charges, including concealing a $2.1 million contribution to the election campaign of the president of the West African nation of Benin.[26]

The next case, *State v. Castillo*, discusses the basic elements of the law of bribery. Ask yourself whether this is clearly bribery.

Did the police officer demand and receive "unjust compensation" from the female motorist?

State v. Castillo, 877 So. 2d 690 (Fla. 2004). Opinion by: Cantero, J.

Facts

We present the facts in the light most favorable to the jury verdict. At about 4 a.m. on March 9, 2000, nineteen-year-old A.S., who had been drinking heavily, was traveling at about 55 m.p.h. in a 40 m.p.h. speed zone when a police cruiser drove up behind her with its overhead lights on. The respondent, Miami-Dade County Police Officer Fernando Castillo, on duty and in uniform, was driving. A.S. pulled over near a Burger King restaurant. Using the patrol car's loudspeaker, Officer Castillo ordered her out of her vehicle. A.S. feared she would be arrested because she was both drunk and speeding. As she walked toward the officer, she stumbled. Castillo remarked that "the party must have been good." After rummaging through her wallet, Castillo told A.S. to follow him into the empty Burger King parking lot. She complied. They both exited their cars and talked for awhile. Castillo was very friendly, smiling and touching A.S.'s shoulder as he stood close to her. Castillo noticed alcohol on her breath. At one point, Castillo asked her, "Do you want to follow me?" She said, "What?" and he replied, "You are going to follow me." Afraid not to obey, she complied. Castillo led her to a nearby deserted warehouse area. Again they exited their cars. He leaned her back on the hood of her car, pulled her pants and panties down, and mumbled "something like 'let me get that thing on.'" Commenting that she had the body of a stripper, he had vaginal intercourse with her. Because she was scared, A.S. did not look or say or do anything, and when he finished, she felt wetness on her lower stomach. As they dressed, Castillo smiled and told her that she was lucky he did not give her a ticket. He gave her his beeper number, and they each drove away.

Castillo did not report his over-forty-minute encounter with A.S. Instead, he reported that during that time he was engaged in various other patrol duties. Castillo's version of events, which the jury rejected, differed from A.S.'s. He testified that A.S. waved him over as he was passing her; that she suggested they talk at the Burger King; that she unexpectedly followed him from there and waved him down again to chat some more, and they did; that she met him a few hours later when his shift ended; and that at that time, they engaged in masturbatory sex. Castillo testified that he did not tell the officers investigating A.S.'s allegations that he engaged in a sexual act with her because they did not ask.

Castillo was charged with, and a jury found him guilty of, unlawful compensation and official misconduct. The trial court denied Castillo's motion for judgment of acquittal. On appeal, the district court focused on A.S.'s trial testimony that before she followed Castillo to the warehouse, he never specifically stated that he would arrest her if she did not have sex with him. The court concluded that because of "the absence of any spoken understanding," the State failed to establish an agreement to these terms. The court thus required direct evidence of a specific agreement to prove unlawful compensation.

During cross-examination, A.S. testified as follows:

Q: He never suggested he was going to arrest you for DUI?

A: No.

Q: He never said anything about along the lines of DUI, the entire encounter, did he?

A: No.

Q: It was never any quid quo pro [sic] that he wouldn't arrest you if you come with me, was there?

A: No.

The court affirmed Castillo's conviction for official misconduct. Castillo asks us to quash that part of the opinion. Because determination of this issue is not integral to the conflict issue presented for review, we do not address it.

The unlawful compensation statute provides in pertinent part as follows:

> It is unlawful for any person corruptly to give, offer, or promise to any public servant, or, if a public servant, corruptly to request, solicit, accept, or agree to accept, any pecuniary or other benefit not authorized by law, for the past, present, or future performance, nonperformance, or violation of any act or omission which the person believes to have been, or the public servant represents as having been, either within the official discretion of the public servant, in violation of a public duty, or in performance of a public duty.

The Florida Statutes define the terms *benefit* and *corruptly*. *Benefit* means gain or advantage, or anything regarded by the person to be benefited as a gain or advantage, including the doing of an act beneficial to any person in whose welfare he or she is interested. *Corruptly* means an act committed with a wrongful intent and for the purpose of obtaining or compensating or receiving compensation for any benefit resulting from some act or omission of a public servant which is inconsistent with the proper performance of his or her public duties.

Issue

We must decide two related issues concerning the statute: (A) whether a violation may be proved through circumstantial evidence; and (B) whether the State must prove a specific agreement. We discuss these issues below.

Reasoning

The district court in this case reversed Castillo's conviction because the State failed to establish a "spoken understanding" that if A.S. submitted to sexual intercourse with Castillo, he would not issue her a citation. Thus, the court required direct evidence of an agreement between the public official and the person unlawfully compensating him. The government, on the other hand, argues that "while the state must show a quid pro quo, it should be permitted to establish this element indirectly, through the use of circumstantial evidence."

The statute itself is silent on the type of proof required. It certainly does not require either a "spoken understanding" or any other direct evidence of a violation. In the absence of explicit statutory direction, it has long been established that circumstantial evidence is competent to establish the elements of a crime, including intent. If an express agreement were required to prove a violation of the statute, a public servant "could receive funds or other benefits from interested persons" and avoid prosecution "so long as he never explicitly promises to perform his public duties improperly." The element of intent, being a state of mind, often can only be proved by circumstantial evidence. Therefore, we hold that circumstantial evidence can establish a violation of the unlawful compensation statute. The district court's requirement of a "spoken understanding" imposes too high a burden on the State and would prohibit prosecution of all but the most blatant violations. Public corruption has become sophisticated enough at least to expect that public officials soliciting or accepting unlawful compensation ordinarily will not be so audacious as to explicitly verbalize their intent.

The second, related issue we must consider is whether the unlawful compensation statute requires evidence of an agreement or meeting of the minds. The district court held it did. It concluded that without direct evidence that Officer Castillo actually stated that he would arrest A.S. if she did not have sex with him, there was only evidence that A.S. believed this to be true. We respectfully disagree.

A "quid pro quo" refers to something exchanged for something else. It does not require an agreement. Harassment of this variety requires no "meeting of the minds."

On its face, the statute does not require an agreement. In fact, it criminalizes the mere solicitation of a "benefit not authorized by law," regardless of whether the solicited party accepts the offer. The statute expressly makes it unlawful for a public servant corruptly to request, solicit, or accept any pecuniary or other benefit not authorized by law. Such language implies that although evidence of an agreement is sufficient to prove a violation—the statute also prohibits agreeing to accept a benefit—it is not required. Section 838.016(1) further requires that the public servant must request, solicit, accept, or agree to accept the unlawful benefit "corruptly," which means "with a wrongful intent and for the purpose of obtaining or compensating or receiving compensation for any benefit resulting from some act or omission of a public servant which is inconsistent with the proper performance of his or her public duties." The statute thus focuses on the official's intent, not on an agreement. The statute does not require that the person from whom the public official requests or accepts a benefit agree to—or even understand—the exchange.

It is sufficient if the actor believes that he has agreed to confer or agreed to accept a benefit for the proscribed purpose, regardless of whether the other party actually accepts the bargain in any contract sense. . . . The evils of bribery are fully manifested by the actor who believes that he is conferring a benefit in exchange for official action, no matter how the recipient views the transaction. . . . Each defendant should be judged by what he thought he was doing and what he meant to do, not by how his actions were received by the other party. A specific agreement is not required. Only corrupt intent must be shown.

Holding

Applying our holdings that neither direct evidence nor evidence of a specific agreement is required to establish a violation of the statute, we conclude that competent, substantial evidence supports Castillo's conviction in this case. The evidence shows that Castillo, a uniformed officer in a marked patrol car, stopped A.S. while she was exceeding the speed limit. He recognized her intoxicated state when he remarked, after she stumbled, that "the party must have been good." He required A.S. to follow him to the nearby deserted restaurant parking lot where he was "very friendly" while they spoke. He smelled alcohol on her breath. He then required A.S. to follow him again, this time to a deserted warehouse area where he initiated and had intercourse with her. Afterwards, he told her she was lucky he did not ticket her, and he permitted her to leave. Castillo not only did not report his contact with A.S., but he misrepresented his activities during this almost hour-long period as official duties. Thus, the evidence of the officer's words and actions demonstrated his understanding that A.S. was violating the law when he stopped her, and his releasing A.S. without legal consequence after having sex with her demonstrates his corrupt intent in soliciting an unlawful quid pro quo.

The district court's conclusion that if Castillo thought that A.S. followed him to the warehouse voluntarily, then Castillo did not violate the statute, is groundless for two reasons. First, the evidence, taken in the light most favorable to the jury verdict, was that he required her to follow him. Second, as we explained above, the other participant's state of mind is irrelevant; it is the public servant's state of mind that matters. Although an agreement may be sufficient to prove a violation, it is not necessary. Accordingly, whether Castillo thought or believed A.S.'s actions were voluntary or whether her actions were in fact voluntary is irrelevant. Castillo demonstrated the causal relationship of his actions when he told A.S., after having intercourse with her, that she was lucky he did not give her a ticket. Thus, the competent, substantial evidence in this case demonstrates that Castillo acted with corrupt intent in accepting an unauthorized benefit—sex in exchange for his exercising his discretion not to issue a traffic citation.

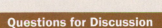

Questions for Discussion

1. What is the prosecution required to establish to convict a defendant of unjust compensation under Florida law?

2. Would Castillo's conviction have been affirmed by the Florida Supreme Court if an express agreement and direct (rather than circumstantial) evidence was required to establish unlawful compensation? Should an express agreement be required?

3. Would you have convicted Castillo, given that he did not mention that he had decided not to give A.S. a ticket until after the two engaged in sexual intercourse?

4. Would you charge and convict A.S. of offering to unjustly compensate Castillo?

5. Does the Florida statute punish both bribery and graft?

6. What is the societal interest in criminally convicting Castillo?

CHAPTER SUMMARY

Sociologist Edwin H. Sutherland pioneered the concept of white-collar crime, defining this as crime committed by an individual of respectability and high social status in the course of his or her occupation. The Justice Department, in contrast, focuses on the types of offenses that constitute white-collar crime as well as on the economic status of the offender. The Justice Department defines white-collar crime as offenses that employ deceit and concealment rather than the application of force to obtain money, property, or service; to avoid the payment or loss of money; or to secure a business or professional advantage. The definition notes that white-collar criminals occupy positions of responsibility and trust in government, industry, the professions, and civil organizations. In this chapter, we discussed some of the common white-collar criminal offenses.

Most white-collar crime prosecutions are undertaken by the federal government. These laws are based on the federal authority over interstate and foreign commerce and other constitutional powers. Environmental crimes threaten the health and natural resources of the United States and range from air and water pollution to the illegal transport and disposal of hazardous waste and pesticides. OSHA protects the health and safety of workers. Willful violations of the act that result in death are punished as a misdemeanor and may result in both a fine and imprisonment. A modest number of these violations have been pursued and cases are increasingly being referred to the EPA for prosecution.

The stock market is an important source of investment and retirement income for Americans. Securities law is designed to insure a free and fair market in order to maintain investor trust and confidence. The most active area of prosecution is insider trading, the use of information that is not publicly available to buy and sell stocks. Prosecutions for trading on the basis of inside information are brought against corporate insiders under a theory of "disclose or abate" and against corporate outsiders under a theory of "misappropriation."

A significant number of white-collar crime prosecutions are undertaken under the federal mail and wire fraud statutes. The mail fraud statute prohibits the knowing and intentional participation in a scheme or artifice to defraud money or property, the execution of which is undertaken through the use of the mails. The wire fraud statute requires the execution of an artifice or fraud through the use of wires that cross interstate or foreign boundaries. The mails or wires need only be used incident to an essential aspect of the scheme. A conspiracy to commit either of these offenses requires that the use of mails or wires will naturally occur in the course of the scheme or was reasonably foreseeable, although not actually intended.

The Travel Act is intended to assist state and local governments in combating organized crime and is directed at what ordinarily are considered the status offenses of illegal gambling, the shipment and sale of alcohol and controlled substances, extortion, bribery, arson, prostitution, and money laundering. This law punishes interstate or foreign travel or the use of the mails or any facility in interstate or foreign commerce with the intent to distribute the proceeds of one of the unlawful activities listed in the statute, to commit a crime of violence, or to commit a crime listed in the Travel Act. The law requires that a defendant thereafter attempt to commit or does commit a criminal offense.

Health fraud is an increasingly frequent and serious area of criminal activity. The federal health fraud statute addresses frauds against health insurance programs or the fraudulent obtaining of money or property from a health care benefit program. This typically involves claims by doctors to be reimbursed for unnecessary medical care or for medical care that was never provided.

A significant challenge confronting criminal offenders is to convert money or property obtained from illegal activity into what individuals can claim to be lawful income. The federal money laundering statute combats the "washing of money" by declaring that it is a crime to knowingly conduct a financial transaction involving the proceeds of a crime or to engage in a transaction involving property derived from criminal activity with the intent to conceal the source of the money or property or to promote an illegal enterprise.

The Sherman Antitrust Act is intended to insure free and competitive markets and declares that it is a crime to knowingly enter into a contract or to form a combination or conspiracy to interfere with interstate commerce. An example is price-fixing.

Public corruption costs the taxpayers a significant amount of money each year. These crimes of official misconduct are defined as the knowingly corrupt behavior by a public official in exercising his or her official responsibilities. An example is the bribery of a public official. Liability is imposed on both the individuals offering and accepting a bribe. Bribery is defined as wrongfully promising, offering, or giving money or a benefit with the intention of influencing an official decision or action. It is also bribery to ask for, accept, or receive money or a benefit with the intent to influence an official decision or action. Graft is defined as the unlawful compensation or reward for official action. The Foreign Corrupt Practices Act prohibits the bribery of overseas officials in order to assist in obtaining or retaining business.

In thinking about this chapter, consider whether sufficient attention is paid to the investigation and prosecution of white-collar crime. Would you punish white-collar criminals more or less severely than other offenders?

CHAPTER REVIEW QUESTIONS

1. What are the various approaches to defining white-collar crime?

2. Should criminal law textbooks devote a separate chapter to the discussion of white-collar crime?

3. Do you believe that the federal government should make it a priority to prosecute white-collar crime?

4. List some of the acts that are considered environmental crimes.

5. What is the purpose of criminal prosecutions under OSHA?

6. Define insider trading. Why is insider trading considered a crime?

7. What are the elements of mail and wire fraud? How do they differ from one another?

8. Discuss the purpose of the Travel Act.

9. What areas are covered under the federal Health Fraud Act?

10. Why is money laundering considered a criminal offense? Provide an example of money laundering.

11. Discuss the purpose of the Sherman Antitrust Act. Give an example of an antitrust violation.

12. What is public bribery? Why is bribery a crime? How is it distinguished from graft?

13. Should the federal government use its interstate commerce power to prosecute what are traditionally considered state criminal offenses? Why not let state government pursue criminal prosecutions?

14. Who poses a greater threat to society, white-collar or common criminals?

LEGAL TERMINOLOGY

bribery of a public official	disclose or abstain doctrine	Foreign Corrupt Practices Act
crimes of official misconduct	environmental crimes	fraud
	fiduciary relationship	graft

insider trading

mail fraud

misappropriation
doctrine

money laundering

Occupational Safety
and Health Act

Sarbanes-Oxley Act

Sherman
(Antitrust) Act of 1890

tippees

tippers

Travel Act

white-collar crime

wire fraud

CRIMINAL LAW ON THE WEB

Log on to the Web-based student study site at
www.sagepub.com/lippmanccl3e to assist you
in completing the Criminal Law on the Web
exercises, as well as for additional features such as
podcasts, Web quizzes, and video links.

1. Go to the FBI Web site on white-collar crime and
   summarize the criminal offenses that are listed and
   note some of the relevant criminal statutes.

2. Examine the corporate corruption across the globe.

3. Learn more about Bernie Madoff.

4. Read about recent trends in the sentencing of
   white-collar criminals.

BIBLIOGRAPHY

Ellen S. Podgar and Jerold H. Israel, *White Collar Crime in a Nutshell,*
3rd ed. (St. Paul, MN: Thomson/West, 1997). A brief overview
of the federal laws on white-collar crime.

Steven Rosoff, Henry Pontell, and Robert Tillman, *Profit Without
Honor: White-Collar Crime and the Looting of America,* 5th ed.
(Upper Saddle River, NJ: Prentice Hall, 2010). A good intro-
duction to white-collar crime.

Neal Shover and John Paul Wright, *Crimes of Privilege: Readings in White-Collar Crime* (New York: Oxford University

Press, 2000). An edited volume that provides a solid introduction to the academic literature on white-collar crime.
Essays cover the characteristics of white-collar crime and
of white-collar criminals, the motivation behind white-
collar crime, and approaches to controlling white-collar
crime.

J. Kelly Strader, *Understanding White Collar Crime,* 2nd ed. (New
York: Lexis, 2006). A clear and comprehensive discussion of
the laws and leading cases on white-collar crime.

15 CRIMES AGAINST PUBLIC ORDER AND MORALITY

Did the defendant's defense of the World Trade Center bombing constitute incitement to riot?

Viewed in context, defendant's words—IT'S GOOD THAT THE WORLD TRADE CENTER WAS BOMBED. MORE COPS AND FIREMEN SHOULD HAVE DIED. MORE BOMBS SHOULD HAVE DROPPED AND MORE PEOPLE SHOULD HAVE BEEN KILLED—were plainly intended to incite the crowd to violence, and not simply to express a point of view. But the allegations extend beyond mere words. It is further alleged that defendant accosted people in the crowd and shouted a threat— WE'VE GOT SOMETHING FOR YOUR ASSES—directly into the faces of some of the onlookers. It is also alleged that as the confrontation escalated, defendant and his accomplices refused police entreaties to disperse. This conduct went well beyond protected speech and firmly into the realm of criminal behavior. It was far more than "the mere abstract teaching . . . of the moral propriety or even moral necessity for a resort to force and violence"; under the circumstances, it constituted the very real threat of violence itself.

Learning Objectives

1. Know the definition of crimes against public order and morality.

2. Understand the elements of disorderly conduct and the types of acts punished by the crime of disorderly conduct.

3. Distinguish the crime of riot from the crime of disorderly conduct.

4. Understand crimes against the quality of life and the broken windows theory.

5. Discuss vagrancy and loitering and discuss homelessness and criminal law.

6. Understand the debate on the overreach of criminal law.

7. Know the definition of the crime of prostitution and various crimes related to prostitution.

8. Understand the definition of obscenity and the difference between obscenity and child pornography.

9. Discuss the reason for the crime of cruelty to animals.

INTRODUCTION

You undoubtedly have been walking down the street and have been approached by a "panhandler" asking for money in an annoying or aggressive manner. He or she might even have followed you down the street or blocked the sidewalk. You may have felt that your "personal space" was invaded or that you were being personally assaulted and might have vowed to avoid walking down this particular street again. Although you did not suffer physical harm, you may have developed a sense of insecurity or even fear. Did the panhandler in this example commit a crime or merely exercise his or her freedom of speech and assembly?

Crimes against public order and morality are intended to insure that individuals walking on sidewalks, traveling on the streets, or enjoying the public parks and facilities are free from harassment, fear, threat, and alarm. This category of crime includes a large number of seemingly unrelated offenses that threaten the public peace, quiet, and tranquility. The challenge presented by these offenses is to balance public order and morals with the right of individuals to exercise their civil liberties.

A prime example of a crime against public order is individual disorderly conduct. This broadly defined offense involves acts that create public inconvenience and annoyance by directly threatening *individuals'* sense of physical safety. Disorderly conduct entails offenses ranging from intentionally blocking the sidewalk and acting in an abusive and threatening manner to discharging a firearm in public. Group disorderly conduct (riot) entails tumultuous or violent conduct by three or more persons.

The second category of crimes against public order and morals covered in this chapter includes offenses against the public order that threaten the order and stability of a *neighborhood*. We focus on two so-called quality-of-life crimes. At common law, vagrancy was defined as moving through the community with no visible means of support. Loitering at common law was defined as idly standing on the corner or sidewalk in a manner that causes people to feel a sense of threat or alarm for their safety. These broad vagrancy and loitering statutes historically have been employed to detain and keep "undesirables" off the streets. The U.S. Supreme Court in recent years has consistently found these laws void for vagueness and unconstitutional. The same constitutional arguments are being used today to challenge ordinances directed against the homeless and gangs.

The last section of the chapter on crimes against public order and morals examines the overreach of the criminal law, or so-called victimless crimes. These are offenses against *morality*. The individuals who voluntarily engage in victimless crimes typically do not view their involvement as harmful to themselves or to others. We initially center our discussion of victimless crimes on prostitution and soliciting for prostitution. The next section on victimless crimes examines whether the prohibition on obscenity should be extended to violent video games that are thought to harm children or whether these games are protected under the First Amendment to the U.S. Constitution. We conclude our discussion of crimes against public order and morality by examining the use of the criminal law to protect domestic animals.

In this chapter, several issues arise. Ask yourself whether the statutes punishing crimes against public order and morality are, at times, employed to target certain undesirable individuals in order keep them off the streets rather than to protect society. Are some of these laws so broadly drafted that the police are provided with an unreasonable degree of discretion in determining whether to arrest individuals? Last, consider whether the criminal law reaches too far in punishing so-called victimless crimes. The overriding question is whether the enforcement of offenses against public order and morality is required in order to maintain social order and stability.

DISORDERLY CONDUCT

The common law punished a **breach of the peace**. This was defined as an act that disturbs or tends to disturb the tranquility of the citizenry. Blackstone notes that breaches of the peace included both acts that actually disrupted the social order, such as fighting in public, and what he terms constructive breaches of the peace or conduct that is reasonably likely to provoke or to excite others to disrupt social order. Blackstone cites as examples of constructive breaches of the peace both the circulation of material causing a person to be subjected to public ridicule or contempt and the issuing of a challenge to another person to fight.[1]

The common law crime of breach of the peace constituted the foundation for American state statutes punishing **disorderly conduct**. An example of a statutory definition of the misdemeanor of disorderly conduct is the Wisconsin law that punishes anyone who "in a public or private place, engages in violent, abusive, indecent, profane, boisterous, unreasonably loud or otherwise disorderly conduct under circumstances in which the conduct tends to cause or provoke a disturbance."[2] Other statutes specify the conduct constituting disorderly conduct. The Illinois statute defines disorderly conduct as any act knowingly committed in an unreasonable manner so as to "alarm or disturb another and to provoke a breach of the peace." The Illinois law then elaborates on this definition and lists specific acts that constitute a breach of the public peace, including a false fire alarm or false report of criminal activity to the police, a false report of an explosive device, a false report of child or elder abuse, and an annoying or intimidating telephone call made to collect a debt.[3] The Arizona statute requires an act committed with a specific intent to disturb the peace or quiet of a neighborhood, family, or person or committed with the knowledge that it is disturbing the peace. The Arizona law lists fighting, unreasonable noise, use of abusive or offensive language to any person likely to cause retaliation, commotion intended to prevent a meeting or procession, refusal to obey a lawful order to disperse, or recklessly handling, displaying, or discharging a deadly weapon or dangerous instrument.[4]

We can see that the *mens rea* of the Illinois statute is knowingly, while the *mens rea* of the Arizona statute is an intentional or knowing intent. Other state statutes extend disorderly conduct to include the reckless disturbance of the peace. The Illinois and Arizona statutes differ in one other respect. The Arizona law covers acts intended to disturb or that knowingly disturb the peace or quiet of a neighborhood, family, or person, while the Illinois statute is generally directed at threats or acts that alarm the community. The commentary to the Model Penal Code notes that defining disturbing the peace to include disturbing individuals authorizes the police to intervene and arrest individuals whose playing of their radio or television is considered unreasonably loud by their next-door neighbor. Disorderly conduct is a misdemeanor, although some states punish as felonies acts that create or threaten to create a significant disturbance.

As you can see, a broad range of conduct is punished under disorderly conduct statutes. For instance, a parent was convicted of disorderly conduct who loudly and aggressively disputed a referee's decision at his son's football game, refused the referee's request to leave the stadium, swore at spectators, placed his hands on the referee, and caused a halt in the game. A Minnesota court of appeals ruled that the parent's profanity combined with his aggressive acts caused anxiety and concern to the spectators and referees and constituted disorderly conduct. How does the court know that the defendant's behavior "disturbed the peace"? Do people assume the risk that they will be subjected to emotional outbursts by spectators attending a football game?[5] In an Illinois case, a defendant who declared to ticketing agents at the airport that he had a bomb in his shoe was sentenced to more than six months in prison. An appellate court affirmed the defendant's conviction for transmitting a false alarm relating to a bomb or other explosive device. The court noted that disorderly conduct under Illinois law requires only the uttering of a threat regardless of the response of other individuals. The court explained this strict standard by observing that the defendant must have known that there was a strong probability that the threat of a bomb in an airport would "cause alarm and mass disruption."[6]

Another challenge that is frequently presented by prosecutions for disorderly conduct is drawing the line between disorderly conduct and constitutionally protected speech. The Wisconsin Supreme Court affirmed that the statement by a thirteen-year-old that he was going to kill everyone at his school and make people suffer constituted disorderly conduct. The court explained that speech alone could constitute disorderly conduct where "a reasonable speaker . . . would foresee that reasonable listeners would interpret his statements as serious expressions of an intent to intimidate or inflict bodily harm."[7]

Disorderly conduct addresses relatively minor acts of criminality. Nevertheless, the commentary to the Model Penal Code stresses that this is an important area because disorderly conduct statutes affect a large number of defendants. Arrests for disorderly conduct in a given year are generally equal to the number arrested for all violent crimes combined. Thus, the enforcement of disorderly conduct statutes is critical in shaping public perceptions as to the fairness of the criminal justice system. A final point is that the concern of Americans with balancing crime control with civil liberties dictates that we take the time to consider whether disorderly conduct statutes intrude upon the rights of individuals and are in need of reform.

The next case in the book, *City of Cincinnati v. Summers,* illustrates the type of factual analysis that is engaged in by courts in determining whether a defendant is guilty of disorderly conduct. Ask yourself whether you agree with the court's decision.

Model Penal Code

Section 250.2. Disorderly Conduct

(1) A person is guilty of disorderly conduct if, with purpose to cause public inconvenience, annoyance or alarm, or recklessly creating a risk thereof, he

    (a) engages in fighting or threatening, or in violent or tumultuous behavior; or

(b) makes unreasonable noise or offensively coarse utterances, gesture or display, or addresses abusive language to any person present; or

    (c) creates a hazardous or physically offensive condition by any act which serves no legitimate purpose of the actor.

"Public" means affecting or likely to affect persons in a place to which the public or substantial group has access; among the places included are highways, transport facilities, schools, prisons, apartment houses, places of business or amusement, or any neighborhood.

(2) An offense under this section is a petty misdemeanor if the actor's purpose is to cause substantial harm or serious inconvenience, or if he persists in disorderly conduct after reasonable warning or request to desist. Otherwise disorderly conduct is a violation [subject to a fine].

Analysis

- The Model Penal Code limits disorderly conduct to specific acts likely to create what the code terms a public nuisance. The commentary notes that the proposed statute does not include conduct tending to corrupt or to annoy other individuals.
- The act must be committed with the purpose to cause public inconvenience, annoyance, or alarm or recklessly creating a risk thereof. Guilt cannot be based on the argument that an individual should have foreseen the risk of public annoyance or alarm; "nothing less than conscious disregard of a substantial and justifiable risk of public nuisance will suffice for liability. Conviction cannot be had merely on proof that the actor should have foreseen the risk of public annoyance or alarm."
- Disorderly conduct is directed at disturbing the peace and quiet of the community. The code excludes family disputes within the home.
- The section limits imprisonment to circumstances in which an individual's purpose is to cause significant harm or serious inconvenience or in which an individual continues the crime despite warnings or requests to halt.
- The Model Penal Code also includes specific sections on the abuse of a corpse; cruelty to animals; desecration of graves, monuments, and places of worship; disruption of meetings and processions; false public alarms; harassment; loitering or prowling; obstructing highways or other public passages and processions; public drunkenness; unlawful eavesdropping; surveillance; and breaching the privacy of messages.

Was Summers guilty of disorderly conduct?

City of Cincinnati v. Summers, 2003-Ohio-2773 (Ohio Ct. App. 2003). Per curiam.

Defendant-appellant Terry Summers appeals the judgment of the Hamilton County Municipal Court convicting him, following a bench trial, of disorderly conduct . . . . [W]e reverse the judgment of the trial court and discharge Summers.

Facts

The record reveals the following facts. Terry Summers is a member of a group called "Black Fist," which protests allegations of police misconduct. In the early evening of August 1, 2002, Black Fist was protesting at the intersection of Vine Street and Fifth Street in downtown Cincinnati. Summers was walking back and forth across the street at the crosswalk, dragging a sign

and shaking a small black baseball bat over his head. Upon observing Summers's actions while protesting, police officers arrested Summers for disorderly conduct because they perceived his actions as threatening to the passing motorists. Summers told the police that he was merely shaking the bat over his head and yelling "Black Power" to passing motorists. At trial, Police Officer David Johnston testified that Summers had been holding a bat over his head and that Officer Johnston believed that Summers's actions would provoke a violent response from passersby. Police Officer Pat Norton testified that Summers had been holding a small black baseball bat over his head and shaking it. But neither officer could hear what Summers was saying to the passing motorists.

At the conclusion of the testimony, the trial court found Summers guilty of disorderly conduct and ordered him to pay a $100 fine and court costs. In this appeal, Summers now brings forth . . . assignments of error.

Issue

In his first assignment of error, Summers asserts that the trial court's judgment was not supported by sufficient evidence. . . . R.C. 2917.11(A)(3) provides that "no person shall recklessly cause inconvenience, annoyance, or alarm to another, by doing any of the following: . . . (3) insulting, taunting, or challenging another, under circumstances in which such conduct is likely to provoke a violent response." Thus, we must determine if a reasonable trier of fact could have found that Summers had recklessly caused inconvenience, annoyance, or alarm to another by insulting, taunting, or challenging another under circumstances in which such conduct was likely to provoke a violent response.

Reasoning

There is evidence in the record upon which a reasonable trier of fact could have found beyond a reasonable doubt that Summers had in fact caused inconvenience and annoyance. The two police officers testified that the protest occurred during the afternoon rush hour, that there was heavy motorist and pedestrian traffic,

and that, despite the traffic, Summers was walking very slowly across the street. But the evidence was not sufficient to support the remaining elements of disorderly conduct beyond a reasonable doubt. There was no evidence that Summers had acted recklessly or had taunted or challenged any passing motorist. Summers stayed within the crosswalk when crossing the street and presumably crossed with the light in his favor, as there was no charge of jaywalking. Further, both officers testified that they had not heard what Summers was saying to the passing motorists. Although there was testimony that Summers had raised his bat in the air and shaken it, neither officer said that Summers had swung his bat at any passing car. Simply protesting within the limits of the law did not reasonably support the inference that Summers was insulting, taunting, or challenging passing motorists. Further, from our review of the record, we hold that peacefully protesting in a crosswalk while raising a small bat in the air and yelling "Black Power," without swinging the bat so as to hit a passing vehicle, was not something that was likely to provoke a violent response.

Holding

Accordingly, there was insufficient evidence to support the disorderly conduct conviction beyond a reasonable doubt. . . . We reverse the judgment of the trial court and discharge Summers. . . .

Questions for Discussion

1. Why did the appellate court acquit Summers?

2. Was the court's decision influenced by the fact that Summers was engaging in an act of political protest? Should the appellate

court reverse the decision of a lower court trial judge who was able to personally evaluate the credibility of the witnesses?

3. Do you agree with the appellate court's decision?

See more cases on the study site: In re Cesar, *www.sagepub.com/lippmanccl3e.*

RIOT

The common law punished group disorderly conduct as a misdemeanor. An **unlawful assembly** was defined as the assembling of three or more persons with the purpose to engage in an unlawful act. Taking steps toward the accomplishment of this common illegal purpose was punished as a **rout**. The law recognized a **riot** where three or more individuals engaged in an unlawful act of violence. The participants must have agreed to the illegal purpose prior to engaging in violence. However, the individuals were not required to enter into a common agreement to commit an illegal act prior to the assembly; the illegal purpose could develop during the course of the meeting. The English Riot Act of 1549 punished as a felony an assembly of twelve or more persons gathered together with an unlawful design that failed to obey an order to disperse within one hour of the issuance of the order to disband. The Riot Act was reintroduced in 1714. You may be interested to know that it is the reading of the act to an assembly that constitutes the basis of the popular phrase "reading the riot act."[8]

The American colonists were understandably reluctant to adopt a British statute that had been employed by the English Crown to punish people who gathered for purposes of political protest. However, all the states eventually adopted riot laws loosely based on the English statute. These laws continue to remain in force and, in effect, punish group disorderly conduct.

For a deeper look at this topic, visit the study site.

Why is there a separate offense of riot? After all, we could merely punish riot as aggravated disorderly conduct. A group has a mind of its own and both poses a greater threat to society and is less easily deterred than a single individual. Collective action also presents a problem for the police, who may have to resort to aggressive force to control the crowd. Courts have recognized that a clear distinction must be made between riots and the right of individuals to freely assemble to petition the government for the redress of grievances. In 1949, the U.S. Supreme Court upheld the constitutionality of an Arkansas riot statute, holding that it did not abridge free speech or assembly for the "state to fasten themselves upon one who has actively and consciously assisted" in the "promoting, encouraging and aiding of an assembly the purpose of which is to wreak violence."[9]

Under a New York statute, an individual is guilty of misdemeanor riot when, with four other persons, he engages in "tumultuous and violent conduct" and thereby "intentionally or recklessly causes or creates a grave risk of causing public alarm." The New York statutory scheme punishes riot as a felony when a group of ten or more persons engages in "tumultuous and violent conduct and thereby intentionally or recklessly causes or creates a grave risk of causing public alarm" and a person other than one of the participants suffers "physical injury or substantial property damage." What is the difference between misdemeanor and felony riot under the New York statute?[10]

New York also punishes the misdemeanor of unlawful assembly. An unlawful assembly is defined as the assembly of an individual with four or more others for the purpose of engaging or preparing to engage with them in "tumultuous and violent conduct likely to cause public alarm, or when, being present at an assembly which either has or develops such a purpose, he remains there with intent to advance that purpose." How does this differ from a riot? New York punishes incitement to riot when an individual "urges ten or more persons to engage in tumultuous and violent conduct of a kind likely to create public alarm."[11] Other states provide criminal penalties for the English statutory crime of a knowing failure to obey an order to disperse. An Ohio statute punishes five or more persons engaged in a course of disorderly conduct "who knowingly fail to obey an order to disperse where there are other persons in the vicinity whose presences creates the likelihood of physical harm to persons or property or of serious public inconvenience, annoyance, or alarm."[12]

Riot statutes are typically used when a conspiracy or accessoryship cannot be easily applied. The Utah riot statute provides that an individual is guilty of riot if "he assembles with two or more other persons with the purpose of engaging, soon thereafter, in tumultuous or violent conduct, knowing that two or more other persons in the assembly have the same purpose." In *J.B.A. v. State,* a Utah appellate court convicted a juvenile of riot and held him to be a delinquent. J.B.A. was determined to have been aware that his friends were collecting weapons and preparing to return to school to "settle some differences." The defendant voluntarily stood as part of a show of force in support of his friends as they fought with members of a rival group. The Utah court noted that J.B.A. was not an uninterested bystander and that he would have been expected to intervene in the event that his friends were in jeopardy of losing the fight. Does this situation fit your conception of participation in a riot? Why not merely punish this as conspiracy or as aiding and abetting disorderly conduct?[13]

In the next case, *People v. Upshaw,* the defendant was charged with inciting to riot and disorderly conduct after he and his friends confronted a crowd and praised the World Trade Center bombing and threatened crowd members. Consider whether the defendant was guilty of incitement to riot.

Model Penal Code

Section 250.1. Riot; Failure to Disperse

(1) A person is guilty of riot, a felony of the third degree, if he participates with two or more others in a course of disorderly conduct:

(a) with purpose to commit or facilitate the commission of a felony or misdemeanor;

(b) with purpose to prevent or coerce official action; or

(c) when the actor or any other participant to the knowledge of the actor uses or plans to use a firearm or other deadly weapon.

(2) Where three or more persons are participating in a course of disorderly conduct likely to cause substantial harm or serious inconvenience, annoyance or alarm, a peace officer or other public servant engaged in executing or enforcing the law may order the participants and others in the immediate vicinity to disperse. A person who refuses or knowingly fails to obey such an order commits a misdemeanor.

Analysis

- The Model Penal Code requires that an individual participate together with two or more other persons in a course of disorderly conduct with the required purpose or knowledge. It is not sufficient that an individual was present at the assembly or disturbance.
- The Model Penal Code also punishes a failure to disperse.

Did the defendant incite a riot?

People v. Upshaw, *741 N.Y.S.2d 664*
(Crim. Ct. N.Y.C. 2002). Opinion by: Harrington, J.

Facts

It is alleged that, within days of the September 11, 2001 terrorist assault on the World Trade Center, defendant and several alleged accomplices, on 42nd Street in the vicinity of Times Square, shouted at a gathering crowd of approximately fifty people in praise of the terrorist attack and the resulting deaths of police officers, firefighters, and civilians; vehemently expressed their shared disappointment that the carnage had not been greater; and accosted people in the crowd, yelling in the onlookers' faces, "We've got something for your asses." It is further alleged that arguments ensued between defendants and some of the crowd and that defendant and his alleged accomplices refused to disperse after police officers asked them to do so.

Defendant argues that the accusatory instrument [indictment], which charges him and two codefendants with inciting to riot and disorderly conduct is not facially sufficient and must be dismissed. Specifically, defendant argues that his actions, rather than criminal, were an exercise of his right to free speech under the First Amendment of the United States Constitution. . . . After reviewing the complaint, and after consideration of defendant's motion to dismiss and the People's opposition thereto, the court concludes that the accusatory instrument is facially sufficient. Therefore, and for the following reasons, defendant's motion is denied.

Penal Law section 240.08 provides that a person is guilty of inciting to riot "when he urges ten or more persons to engage in tumultuous and violent conduct of a kind likely to create public alarm." Although Penal Law section 240.08 does not expressly provide for the

element of intent, courts have recognized that in order to pass constitutional muster, the incitement statute necessarily includes the "elements of 'intent' and 'clear and present danger' before one's freedom of speech may be abridged under the First Amendment." "Thus, the People must prove not only that defendant's conduct . . . created a clear and present danger of riotous behavior, but also that by such conduct he in fact intended a riot to ensue." The complaint contains the following narrative of defendant's alleged criminal conduct:

> Deponent [Police Officer Charles Carlstrom] states that he observed each defendant at [234 W. 42nd Street in the County and State of New York] yelling and stating in substance: IT'S GOOD THAT THE WORLD TRADE CENTER WAS BOMBED. MORE COPS AND FIREMEN SHOULD HAVE DIED. MORE BOMBS SHOULD HAVE DROPPED AND MORE PEOPLE SHOULD HAVE BEEN KILLED. WE'VE GOT SOMETHING FOR YOUR ASSES.

> Deponent states that a total of 5 defendants (Eric White, Reggie Upshaw, Steven Murdock, Jesse Atkinson and Kyle Jones) where [sic] yelling the above statements to a crowd of approximately 50 people. Deponent states that said people gathered around defendants and some of said people yelled back at defendants.

> Deponent states that defendants did approach people in the crowd and yell in their faces.

> Deponent further states that defendants were asked to disperse and refused to do so.

Deponent states that defendants' conduct caused the crowd to gather and arguments to ensue.

Issue

Arguing that the complaint does not allege that he acted with the requisite intent to incite a riot, defendant contends that the complaint alleges merely that he "spoke in praise of the assault on the World Trade Center and stated that worse should have happened," but does not allege that "defendant urged or encouraged people to commit acts of terrorism or treason." . . . Defendant analogizes his conduct to "the mere abstract teaching . . . of the moral propriety or even moral necessity for a resort to force and violence" in contrast to "preparing a group for violent action and steeling it to such action." In defendant's view, "the language attributed to defendant was an expression of a political nature, intended to spur debate and thought, not to create the type of public harm contemplated by the statute."

Reasoning

In analyzing whether the allegations in the complaint evince defendant's intent that his alleged conduct led to riotous behavior, and whether his alleged conduct created a clear and present danger of riotous behavior, it is necessary to consider defendant's words and deeds in the context in which he and his alleged accomplices spoke and acted. The alleged crime took place only days after one of the greatest catastrophes this nation has suffered—the overwhelming brunt of which was felt most keenly here in New York—and within sight of the massive smoke plume emanating from the still-smoldering mass grave site that had been the twin towers of the World Trade Center. It took place while many New Yorkers were grieving for the loss of loved ones or praying in hope that the missing might yet be found, and as New Yorkers, indeed, all Americans, held their collective breath at what, at the time, appeared to be the likelihood, if not the inevitability, of additional terrorist attacks. It was under these circumstances that defendant and his cohorts allegedly chose a crowded 42nd Street near Times Square as their venue not merely to engage in what any reasonable person would consider to be a vile and morally reprehensible diatribe, but to intentionally confront the gathering crowd, at point blank range, for the purpose of inciting riotous behavior. It is estimated that approximately 3,000 people died in the World Trade Center attack. By comparison, 2,403 Americans were killed in the attack on Pearl Harbor.

There can be no doubt that the words and deeds alleged in the complaint make out the elements of the crime of inciting to riot. According to the complaint, defendant and his accomplices used extremely inflammatory language calculated to cause unrest in the crowd; praising the tragic deaths of thousands of innocents at the hands of terrorists and wishing for even more carnage while the threat of further attacks loomed over the city cannot be considered "an expression of a political nature, intended to spur debate and thought, not to create the type of public harm contemplated by the statute" to use defendant's words. The talismanic phrase "freedom of speech" does not cloak all utterances in legality. "It is one thing to say that the police cannot be used as an instrument for the suppression of unpopular views, and another to say that, when the speaker passes the bounds of argument or persuasion and undertakes incitement to riot, they are powerless to prevent a breach of the peace."

Holding

Viewed in context, defendant's words—IT'S GOOD THAT THE WORLD TRADE CENTER WAS BOMBED. MORE COPS AND FIREMEN SHOULD HAVE DIED. MORE BOMBS SHOULD HAVE DROPPED AND MORE PEOPLE SHOULD HAVE BEEN KILLED—were plainly intended to incite the crowd to violence, and not simply to express a point of view. But the allegations extend beyond mere words. It is further alleged that defendant accosted people in the crowd and shouted a threat—WE'VE GOT SOMETHING FOR YOUR ASSES—directly into the faces of some of the onlookers. It is also alleged that as the confrontation escalated, defendant and his accomplices refused police entreaties to disperse. This conduct went well beyond protected speech and firmly into the realm of criminal behavior. It was far more than "the mere abstract teaching . . . of the moral propriety or even moral necessity for a resort to force and violence"; under the circumstances, it constituted the very real threat of violence itself.

All that is required under Penal Law section 240.08 . . . is that defendant urge ten or more persons to engage in tumultuous and violent conduct of a kind likely to create public harm. Angrily confronting and threatening a crowd of onlookers with the intent to stir the crowd to violence is sufficient; the object of that tumultuous or violent conduct is irrelevant so long as the conduct defendant urges is of a kind likely to cause public harm. . . .

Penal Law section 240.20 provides, in pertinent part, that a person is guilty of disorderly conduct "when, with intent to cause public inconvenience, annoyance, or alarm, or recklessly creating a risk thereof . . . he engages in fighting or in violent, tumultuous or threatening behavior." . . . Defendant's words and deeds as alleged in the complaint demonstrate his intent to cause public inconvenience, annoyance, or alarm, or recklessly create a risk thereof by engaging in tumultuous or threatening behavior. Therefore, defendant's request to dismiss both counts in the accusatory instrument is denied. . . .

Questions for Discussion

1. What is the prosecution required to prove to convict a defendant of incitement to riot?

2. Did the defendant possess the necessary intent? Did his speech create the clear and present danger of a violent response? Was there any violence resulting from the defendant's statements?

3. Was the decision of the court influenced by the topic and timing of the defendant's statements?

4. Would you convict the defendant of incitement to riot or disorderly conduct?

5. Should the defendant's freedom of speech be limited by the "heckler's veto," the fact that others object to the defendant's expression? The leading Supreme Court cases addressing freedom of expression and disorderly conduct are *Hess v. Indiana*, 414 U.S. 105 (1973), and *Brandenburg v. Ohio*, 395 U.S. 444 (1969).

See more cases on the study site: People v. Upshaw, *www.sagepub.com/lippmanccl3e.*

PUBLIC INDECENCIES: QUALITY-OF-LIFE CRIMES

Criminal law texts traditionally devote very little attention to **public indecencies**. These offenses include public drunkenness, vagrancy, loitering, panhandling, graffiti, and urinating and sleeping in public. A significant number of arrests and prosecutions are devoted to these **crimes against the quality of life**, but for the most part, they receive limited attention because they are misdemeanors, are swiftly disposed of in summary trials before local judges, and disproportionately target young people, minorities, and individuals from lower socioeconomic backgrounds.

In the 1980s, scholars began to argue that seemingly unimportant offenses against the public order and morals were key to understanding why some neighborhoods bred crime and hopelessness while other areas prospered. This so-called **broken windows theory** is identified with criminologists James Q. Wilson and George Kelling. Why the name broken windows? Wilson and Kelling argue that if one window in a building is broken and left unrepaired, this sends a signal that no one cares about the house and that soon every window will be broken. The same process of decay is at work in a neighborhood. A home is abandoned, weeds sprout, the windows are smashed, and graffiti is sprayed on the building. Rowdy teenagers, drunks, and drug addicts are drawn to the abandoned structure and surrounding street. Residents find themselves confronting panhandlers, drunks, and addicts and develop apprehension about walking down the street and flee the area as property values drop and businesses desert the community. The neighborhood now has reached a tipping point and is at risk of spiraling into a downward cycle of crime, prostitution, drugs, and gangs. The solution, according to Wilson and Kelling, is to address small concerns before they develop into large-scale crimes.[14]

We can question, along with some researchers, whether small incidents of disorder inevitably lead to petty crime, then to serious offenses, and finally to neighborhood decay. Nevertheless, surveys indicate that most people are more concerned with the immediate threat to their quality of life posed by rowdy juveniles, drug dealers, prostitutes, and public drunkenness than they are with the more distant threats of rape, robbery, and murder.

A central focus of the broken windows theory in cities where it has been adopted is combating vagrancy and loitering.

Vagrancy and Loitering

Vagrancy is defined under the common law as wandering the streets with no apparent means of earning a living (without visible means of support). **Loitering** is a related offense defined as standing in public with no apparent purpose.

Vagrancy can be traced to laws passed in England as early as the thirteenth century. The early vagrancy statutes were passed in reaction to the end of the feudal system and required the vast army of individuals wandering the countryside to seek employment. These same laws were relied on during the labor shortage resulting from the Black Death in the fourteenth century to force individuals into the labor market. There was also the fear that these bands of men might loiter or gather together to engage in crime or rebellion.

The Statute of Laborers of 1349 authorized the imprisonment of males under sixty without means of financial support who refused to work. The Impotent Poor Act of 1530 stipulated that an impotent beggar who wandered from home and was engaged in begging was to be whipped or placed in stocks for three days and nights on bread and water. Able-bodied, but unemployed, wanderers were later subjected to harsh penalties, including branding and slicing off portions of the ear. Another provision stated that any person found begging or wandering shall be "stripped naked from the middle upwards, and be openly whipped until his or her body be bloody." An English law in force in the second half of the nineteenth century divided vagrants into three criminal classes: idle and disorderly persons (people who refuse to work), rogues and vagabonds (wanderers), and incorrigible rogues (repeat offenders). This descriptive language eventually found its way into the texts of American statutes.[15]

Statutes punishing vagrancy were adopted in virtually every American state. These laws typically punished a broad range of behavior, including wandering or loitering, living without employment and having no visible means of support, begging, failing to support a wife and child, and sleeping outdoors. Individuals were also punished for their status or lifestyle. Laws, for instance, condemned and categorized as criminal, prostitutes, drunkards, gamblers, gypsies engaged in telling fortunes, nightwalkers, corrupt persons, and individuals associating with thieves.[16]

The fear and distrust of the poor and unemployed led the U.S. Supreme Court to observe, in reference to an 1837 effort by New York City to prevent the inflow of the poor, that it is as necessary for a "state to provide precautionary measures against the moral pestilence of paupers, vagabonds, and possibly convicts; as it is to guard against the physical pestilence, which may arise from unsound and infectious articles imported or from a ship, the crew of which may be laboring under an infectious disease."[17]

By 1941, the U.S. Supreme Court had adopted a more sympathetic attitude toward the poor. California passed a statute in the early twentieth century preventing the influx of indigents. This was an unapologetic effort to limit budgetary expenditures for the poor and to prevent the introduction into the state of disease, rape, incest, and labor unrest. The Supreme Court criticized California and observed that it now was recognized in an industrial society that the "the relief of the needy has become the common responsibility and concern of the whole nation" and that "we do not think it will now be seriously contended that because a person is without employment and without funds he constitutes a 'moral pestilence.' Poverty and immorality are not synonymous." Justice Robert Jackson added that it is contrary to the history and tradition of the United States to make an individual's rights dependent on his or her economic status, race, creed, or color. As Justice Douglas noted, to hold a law constitutional that prevented those labeled as "indigents, paupers, or vagabonds" from seeking "new horizons" in California would be to reduce these individuals to an "inferior class of citizenship."[18]

In 1972, in the case of *Papachristou v. City of Jacksonville* (discussed in Chapter 2), the U.S. Supreme Court held a Jacksonville, Florida, ordinance unconstitutional that authorized the arrest of vagrants. This was a typical statute that classified a wide range of individuals as vagrants, including rogues, vagabonds, dissolute persons who go about begging, common gamblers, persons who use juggling or unlawful games, common drunkards, nightwalkers, pilferers or pickpockets, keepers of gambling places, common brawlers, habitual loafers, persons frequenting houses of ill fame, gambling houses, and places where alcoholic beverages are sold, and individuals living on the earnings of their wives or minor children.

The U.S. Supreme Court ruled that the statute was void for vagueness in that it failed to give a person of ordinary intelligence fair notice of the conduct that is prohibited by the statute and encourages the police to engage in arbitrary and erratic arrests and convictions. The Court explained that the true evil in the law was its employment by the police to target the young, the poor, and minorities. The "rule of law, evenly applied to minorities as well as majorities, to the poor as well as the rich, is the great mucilage that holds society together."[19]

Eleven years later, in *Kolender v. Lawson* (discussed in Chapter 2), the U.S. Supreme Court ruled a California loitering statute unconstitutional that authorized the arrest of persons who loiter or wander on the streets who fail to provide "credible and reliable" identification and to "account for their presence." The lack of a clear statement of what constitutes credible and reliable identification, according to the Court majority, left citizens uncertain how to satisfy the letter of the law and empowered the police to enforce the law in accordance with their individual biases and discretion.[20]

States now have amended their vagrancy and loitering statutes and have followed the Model Penal Code in punishing loitering or prowling under specific circumstances that "warrant alarm for the safety of persons or property."

Model Penal Code

Section 250.6. Loitering or Prowling

A person commits a violation if he loiters or prowls in a place, at a time, or in a manner not usual for law-abiding individuals under circumstances that warrant alarm for the safety of persons or property in the vicinity. Among the circumstances which may be considered in determining whether such alarm is warranted is the fact that the actor takes flight upon appearance of a peace officer, refuses to identify himself, or manifestly endeavors to conceal himself or any object. Unless flight by the actor or other circumstance makes it impracticable, a peace officer shall prior to any arrest for an offense under this section afford the actor an opportunity to dispel any alarm which would otherwise be warranted, by requesting him to identify himself and explain his presence and conduct. No person shall be convicted of an offense under this Section if the peace officer did not comply with the preceding sentence, if it appears at trial that the explanation given by the actor was true and, if believed by the peace officer at the time, would have dispelled the alarm.

Homelessness

City and local governments have increasingly relied on municipal ordinances to stem the tide of a growing homeless population. The National Law Center on Homelessness and Poverty and the National Coalition for the Homeless issued a report in 2009 titled *Homes Not Handcuffs: The Criminalization of Homelessness in U.S. Cities* that documents the increase and enforcement of laws prohibiting urban camping, sleeping in the parks and subways, aggressive panhandling, trespassing in areas under bridges and adjacent to parks, and blocking sidewalks. The report also finds laws against loitering, jaywalking, and open alcoholic containers. Several cities also prohibit charities, churches, and other organizations from serving food to the needy outside designated areas. The report concludes that these local ordinances have the effect of making it a crime to be homeless. The National Coalition for the Homeless singles out the ten "meanest cities" toward the homeless. These cities are Los Angeles, California; St. Petersburg, Florida; Orlando, Florida; Atlanta, Georgia; Gainesville, Florida; Kalamazoo, Michigan, San Francisco, California; Honolulu, Hawaii; Bradenton, Florida; and Berkeley, California.

In 1992, in *Pottinger v. City of Miami,* a federal district court ruled that the Miami police had employed the criminal law for the purpose of "eliminating or eradicating the presence of the homeless" or for "getting the homeless to move out of certain locations." One component of this strategy was to harass the homeless by preventing them from congregating in areas where pantries made free food available. The federal court found that the evidence supported the complainants' assertion that "there is no public place where they can perform basic, essential acts such as sleeping without the possibility of being arrested" and issued a judicial order directing the Miami police to halt this abuse of the criminal law. Another troubling trend is random violence by groups of young people against the homeless.[21]

The next case, *Joyce v. City and County of San Francisco,* involves a legal action seeking to prohibit San Francisco from enforcing the "Matrix Program" against the city's homeless population. Should a locality be prohibited from taking steps to control the homeless population?

Did San Francisco unconstitutionally target the homeless?

Joyce v. City and County of San Francisco,
846 F. Supp. 843 (N.D. Cal. 1994). Opinion by: Jensen, J.

Plaintiffs to this action seek preliminary injunctive relief on behalf of themselves and a class of homeless individuals alleged to be adversely affected by the City and County of San Francisco's (the "City's") "Matrix Program." While encompassing a wide range of services

to the City's homeless, the Program simultaneously contemplates a rigorous law enforcement component aimed at those violations of state and municipal law that arguably are committed predominantly by the homeless. Plaintiffs endorse much of the Program, challenging it not in its entirety but only insofar as it specifically penalizes certain "life sustaining activities" engaged in by the homeless. . . .

Facts

Institution of the Matrix Program followed the issuance of a report in April of 1992 by the San Francisco Mayor's Office of Economic Planning and Development, which attributed to homelessness a $173 million drain on sales in the City. In August of 1993, the City announced commencement of the Matrix Program, and the San Francisco Police Department began stringently enforcing a number of criminal laws. The City describes the Program as "initiated to address citizen complaints about a broad range of offenses occurring on the streets and in parks and neighborhoods." The program addresses offenses including public drinking and inebriation; obstruction of sidewalks; lodging, camping, or sleeping in public parks; littering; public urination and defecation; aggressive panhandling; dumping of refuse; graffiti; vandalism; street prostitution; and street sales of narcotics, among others.

An illustration of the enforcement efforts characteristic of the Program can be found in a four-page intradepartmental memorandum addressed to the Police Department's Southern Station Personnel. That memorandum, dated August 10, 1993, and signed by acting Police Captain Barry Johnson, defines "Quality of Life" violations and establishes a concomitant enforcement policy. Condemning a "type of behavior [which] tends to make San Francisco a less desirable place in which to live, work or visit," the memorandum directs the vigorous enforcement of eighteen specified code sections, including prohibitions against trespassing, public inebriation, urinating, or defecating in public; removal and possession of shopping carts; solicitation on or near a highway; erection of tents or structures in parks; obstruction and aggressive panhandling.

Pursuant to the memorandum, all station personnel shall, when not otherwise engaged, pay special attention and enforce observed "Quality of Life" violations. . . . One Officer . . . shall, daily, be assigned specifically to enforce all "Quality of Life" violations. . . .

In a Police Department Bulletin entitled "Update on Matrix Quality of Life Program," dated September 17, 1993, Deputy Chief Thomas Petrini paraphrased General Order D-6, the source of the intended nondiscriminatory policy of the Program's enforcement measures by stressing that all persons have the right to use the public streets and places so long as they are not engaged in specific criminal activity. Chief Petrini stressed that race, sex, sexual preference, age, dress, or appearance do not justify enforcement. The memorandum stated that the "rights of the homeless must be preserved," and included as an attachment

a Department Bulletin on "Rights of the Homeless," which stated that:

> [All members of the Department] are obligated to treat all persons equally, regardless of their economic or living conditions. The homeless enjoy the same legal and individual rights afforded to others. Members shall at all times respect these rights.

The Police Department has, during the pendency of the Matrix Program, conducted continuing education for officers regarding nondiscriminatory enforcement of the Program. In 1993, a police report recorded that since implementation of the Matrix Program, the City estimates that "according to unverified statistics kept by the Department," approximately sixty percent of enforcement actions have involved public inebriation and public drinking, and that other "significant categories" include felony arrests for narcotics and other offenses, and arrests for street sales without a permit. Together, enforcement actions concerning camping in the park, . . . sleeping in the park during prohibited hours, . . . and lodging . . . have constituted only approximately ten percent of the total.

Plaintiffs, pointing to the discretion inherent in policing the law enforcement measures of the Matrix Program, allege certain actions taken by police to be "calculated to punish the homeless." As a general practice, the Program is depicted by plaintiffs as "targeting hundreds of homeless persons who are guilty of nothing more than sitting on a park bench or on the ground with their possessions, or lying or sleeping on the ground covered by or on top of a blanket or cardboard carton." On one specific occasion, according to plaintiffs, police "cited and detained more than a dozen homeless people, and confiscated and destroyed their possessions, leaving them without medication, blankets or belongings to cope with the winter cold."

The City contests the depiction of Matrix as a singularly focused, punitive effort designed to move "an untidy problem out of sight and out of mind." Instead, the City characterizes the Matrix Program as "an inter-departmental effort . . . [utilizing] social workers and health workers . . . [and] offering shelter, medical care, information about services and general assistance. Many of those on the street refuse those services, as is their right; but Matrix makes the choice available." . . . The City claims it has attempted to conjoin its law enforcement efforts with referrals to social service agencies. One specific element of the Program seeks to familiarize the homeless with those services and programs available to them. This is accomplished by the dispersal of Department of Social Service social workers throughout the City in order to contact homeless persons.

Another element of the Matrix Program—the Night Shelter Referral Program—attempts to provide temporary housing to those not participating in the longer-term housing program. This effort was begun in December of 1993 and is designed to offer the option of shelter accommodations to those homeless individuals in violation of code sections pertaining to lodging, camping in public parks, and sleeping in parks during prohibited hours. . . . The City contends that of 3,820 referral slips offered to men, only 1,866 were taken and only 678 were actually utilized to obtain a shelter bed reserved for Police referrals. By the City's reckoning, "these statistics suggest that some homeless men may prefer to sleep outdoors rather than in a shelter."

The City emphasizes its history as one of the largest public providers of assistance to the homeless in the State, asserting that "individuals on general assistance in San Francisco are eligible for larger monthly grants than are available almost anywhere else in California." Homeless persons within the City are entitled to a maximum general assistance of $345 per month—an amount exceeding the grant provided by any of the surrounding counties. General assistance recipients are also eligible for up to $109 per month in food stamps. According to the City, some 15,000 City residents are on general assistance, of whom 3,000 claim to be homeless.

The City's Department of Social Services encourages participation in a Modified Payments Program offered by the Tenderloin Housing Clinic. Through this program, a recipient's general assistance check is paid to the Clinic, which in turn pays the recipient's rent and remits the balance to the recipient. The Clinic then negotiates with landlords of residential hotels to accept general assistance recipients at rents not exceeding $280 per month.

By its own estimate, the City will spend $46.4 million for services to the homeless for 1993–1994. Of that amount, over $8 million is specifically earmarked to provide housing and is spent primarily on emergency shelter beds for adults, families, battered women, and youths. An additional $12 million in general assistance grants is provided to those describing themselves as homeless, and free health care is provided by the City to the homeless at a cost of approximately $3 million.

The City contends that "few of the Matrix-related offenses involve arrest." Those persons found publicly inebriated, according to the City, are taken to the City's detoxification center or district stations until sober. "Most of the other violations result in an admonishment or a citation."

Since its implementation, the Matrix Program has resulted in the issuance of over 3,000 citations to homeless persons. Plaintiffs contend these citations have resulted in a cost to the City of over $500,000. Citations issued for encampment and sleeping infractions are in the amount of $76. . . . Those cited must pay or contest the citation within twenty-one days; failure to do so results in a $180 warrant for the individual's arrest, which is issued approximately two months after citation of the infraction. Upon the accrual of $1,000 in warrants, which equates roughly to the receipt of six citations, an individual becomes ineligible for citation release and may be placed in custody. The typical practice, however, is that those arrested for Matrix-related offenses are released on their own recognizance or with "credit for time served" on the day following arrest. Plaintiffs characterize the system as one in which "homeless people are cycled through the criminal justice system and released to continue their lives in the same manner, except now doing so as 'criminals.'" . . .

Plaintiffs have proffered estimates as to the number of homeless individuals unable to find nightly housing. Plaintiffs cite a survey conducted by Independent Housing Services, a nonprofit agency that among its aims seeks the improvement of access to affordable housing for the homeless. Begun in July of 1990 and conducted most recently in August of 1993, the survey tracks the number of homeless individuals turned away each night from shelters in the San Francisco area due to a lack of available bed space. Based on the data of that survey, plaintiffs contend that from January to July of 1993, an average of 500 homeless persons was turned away nightly from homeless shelters. That number, according to plaintiffs, increased to 600 upon the closing of the Transbay Terminal.

Issue

The named plaintiffs to this action have been exposed differently to the enforcement of the Matrix Program and to the rigors of a life of homelessness. Plaintiffs assert they are "homeless" individuals since they lack a "fixed, regular, and adequate nighttime residence." . . . On November 23, 1993, these plaintiffs filed a class action complaint seeking injunctive and declaratory relief against the City.

Plaintiffs have at this time moved the court to preliminarily enjoin the City's enforcement of certain state and municipal criminal measures that partially define the Matrix Program. Given this posture of the litigation, the court is called upon to decide whether to grant a preliminary injunction in the exercise of its equitable powers. Such relief constitutes an extraordinary use of the court's powers and is to be granted sparingly and with the ultimate aim of preserving the status quo pending trial on the merits. . . . Plaintiffs urge the court to implement an injunction under which the City shall be preliminarily enjoined from enforcing, or threatening to enforce, statutes and ordinances prohibiting sleeping, "camping" or "lodging" in public parks, or the obstruction of public sidewalks against the plaintiff class of homeless individuals for life-sustaining activities such as sleeping, sitting, or remaining in a public place. . . .

Reasoning

The injunction sought by plaintiffs at this juncture of the litigation must be denied . . . . [T]he proposed injunction lacks the necessary specificity to be enforceable and would give rise to enforcement problems. [Would the police be prevented from arrests for urinating and defecating in public or aggressive panhandling? How can the police determine who is a homeless person who is immune from arrest?] . . . The court cannot find at this time that upon conducting the required balance of harm and merit, plaintiffs have established a sufficient probability of success on the merits to warrant injunctive relief.

When asked by the court whether, under the proposed injunction, the City would be able to cite plaintiff Joyce for public camping, counsel for plaintiffs answered as follows: such citation might be permissible if Joyce had lodging available that night but would be otherwise impermissible. Counsel for plaintiffs suggested at the hearing that enforcement difficulties could be mitigated if police would merely ask questions to determine whether the person is "homeless" before citing him. This suggestion is no solution. . . . Responding to a question of the court, counsel for plaintiffs suggested at the hearing that anyone who placed and used three tents on San Francisco's Civic Plaza for a period of three days would have engaged in unpunishable activity under the proposed injunction. . . . The court cannot at this time sanction such a result. . . . The City's homeless have never been altogether immunized from enforcement of the various laws at issue here. . . .

Plaintiffs contend enforcement of the Matrix Program unconstitutionally punishes an asserted "status" of homelessness. The central thesis is that since plaintiffs are compelled to be on the street involuntarily, enforcement of laws that interfere with their ability to carry out life-sustaining activities on the street must be prohibited. . . . As a threshold matter, plaintiffs' citation to various court decisions striking down statutes criminalizing vagrancy is of indirect assistance to the present analysis. . . . The present efforts under the Matrix Program are dissimilar from those successfully challenged. . . . Matrix targets the commission of discrete acts of conduct, not a person's appearance as a vagrant. . . . Plaintiffs argue that the failure of the City to provide sufficient housing compels the conclusion that homelessness on the streets of San Francisco is cognizable as a status. This argument is unavailing at least for the fundamental reason that status cannot be defined as a function of the discretionary acts of others. . . . Although the plaintiffs' . . . challenge to the Matrix Program law enforcement activities will appropriately be subject to further scrutiny in this case, on this record, the plaintiffs have not demonstrated a probability of success on the merits of this claim. . . .

Predicate to an equal protection clause violation is a finding of governmental action undertaken with an intent to discriminate against a particular individual or class of individuals. In the present case, plaintiffs have not at this time demonstrated a likelihood of success on the merits of the equal protection claim, since the City's action has not been taken with an evinced intent to discriminate against an identifiable group. As discussed above, various directives issued within the Police Department mandate the nondiscriminatory enforcement of Matrix. . . . Further, the Police Department has, during the pendency of the Matrix Program, conducted continuing education for officers regarding nondiscriminatory enforcement of the Program.

It has not been proven at this time that Matrix was implemented with the aim of discriminating against the homeless. That enforcement of Matrix will . . . fall predominantly on the homeless does not in itself effect an equal protection clause violation. Notably, the absence of such a finding of intentional discrimination . . . .

Even were plaintiffs able at this time to prove an intent to discriminate against the homeless, the challenged sections of the Program might nonetheless survive constitutional scrutiny. Only in cases where the challenged action is aimed at a suspect classification, such as race or gender, or premised upon the exercise of a fundamental right, will the governmental action be subjected to a heightened scrutiny. . . . [T]he challenged Program would likely be tested only against a rational basis.

Plaintiffs contend the Matrix Program has been enforced in violation of the due process clauses of the United States . . . [in that] certain state codes are unconstitutionally vague.

Plaintiffs claim that San Francisco Park Code section 3.12 has been applied by police in an unconstitutional manner. That section provides that "no person shall construct or maintain any building, structure, tent or any other thing in any park that may be used for housing accommodations or camping, except by permission from the Recreation and Park Commission."

Plaintiffs contend the Police Department has impermissibly construed this provision to justify citing, arresting, threatening, and "moving along" those "persons guilty of nothing more than sitting on park benches with their personal possessions or lying on or under blankets on the ground." Plaintiffs have submitted declarations of various homeless persons supporting the asserted application of the San Francisco Park Code section. It appears, if plaintiffs have accurately depicted the manner in which the section is enforced, that the section may have been applied to conduct not covered by the section and may have been enforced unconstitutionally. . . .

Issuance of the remedy sought by plaintiffs would nevertheless be premature at this stage of the litigation. Police have been continually and increasingly educated in proper enforcement of the measures. Moreover, as it appears many of the alleged violations occurred prior to such instructions, it has not been sufficiently demonstrated by plaintiffs that the problematic enforcement continues, such that the continuing injury predicate to issuance of injunctive relief still exists. . . .

Plaintiffs also contend that San Francisco Park Code section 3.12 and California Penal Code section 647(i) are unconstitutionally . . . vague. Section 647(i) . . . punishes as a misdemeanor any person who "lodges in any building, structure, vehicle, or place, whether public or private, without the permission of the owner or person entitled to the possession or in control thereof."

The possible success of the vagueness challenge is . . . in doubt, as it seems readily apparent the measure is not "impermissibly vague in all of its applications. . . ." This likely failing follows from plaintiffs' inability to prove at this stage that police have been granted an excess of discretion pursuant to the statute. Plaintiffs assert vagueness in the San Francisco Park Code prohibition against maintaining "any other thing" that "may be used for . . . camping" and in enforcing the Penal Code prohibition against one "who lodges in . . . public." The plaintiffs claim that "the vagueness of these [Park Code] terms has apparently allowed San Francisco police officers to determine that blankets or possessions in carts are sufficiently connected to 'camping' to violate the ordinance." Plaintiffs have . . . submitted declarations of homeless person supporting these assertions and a concession by Assistant District Attorney Paul Cummins to the effect that the standards for enforcement are vague. . . .

The challenged Penal Code section cannot be concluded by the court at this time to be unconstitutionally vague. Read in conjunction with supplemental memoranda, the challenged measures appear, as a constitutional matter, sufficiently specific. Police officers were specifically cautioned in a September 17, 1993, memorandum that "the mere lying or sleeping on or in a bedroll of and in itself does not constitute a violation." While plaintiffs argue the additional memoranda were circulated too late to save the enforcement measures from vagueness and that they do not eliminate the confusion, it is far from clear that plaintiffs could meet the requisite showing that the measure was impermissibly vague in all its applications. Accordingly, even if the limits of permissible enforcement of these sections have not been perfectly elucidated, preliminary injunctive relief is inappropriate at this stage of the litigation.

Holding

In common with many communities across the country, the City is faced with a homeless population of tragic dimension. Today, plaintiffs have brought that societal problem before the court, seeking a legal judgment on the efforts adopted by the City in response to this problem. The role of the court is limited structurally by the fact that it may exercise only judicial power and technically by the fact that plaintiffs seek extraordinary pretrial relief. The court does not find that plaintiffs have made a showing at this time that constitutional barriers exist that preclude that effort. Accordingly, the court's judgment at this stage of the litigation is to permit the City to continue enforcing those aspects of the Matrix Program now challenged by plaintiffs.

The court therefore concludes that the injunction sought, both as it stands now and as plaintiffs have proposed to modify it, is not sufficiently specific to be enforceable. Further, upon conducting the required balance of harm and merit, the court finds that plaintiffs have failed to establish a sufficient probability of success on the merits to warrant injunctive relief. Accordingly, plaintiffs' motion for a preliminary injunction is denied.

Questions for Discussion

1. What is the Matrix Program? Why was it implemented by San Francisco? Did this program become necessary as a result of San Francisco's failure to provide adequate social programs for the homeless?

2. The lawyers for the homeless concede that portions of the Matrix Program are necessary. They are primarily concerned with freedom for the homeless to sleep in the parks and protecting the property and shopping carts of the homeless from seizure by the police. What do police arrest statistics indicate concerning whether this is a primary part of the Matrix Program? Are arrest statistics an accurate indicator of the conduct of the police?

3. Summarize the plaintiffs' contentions and the court's response concerning whether the Matrix Program unconstitutionally discriminates against the homeless and violates due process by targeting the homeless.

4. The lawyers for the homeless challenge as void for vagueness the law permitting arrest for "any other thing" that may be used for "camping" and the arrest of any person who "lodges in . . . public." They are particularly concerned that the police allegedly seize the blankets and possessions of the homeless. Does the police memorandum clarify these terms?

5. Explain why the plaintiffs' vagueness claim failed.

6. As a member of the San Francisco City Council, would you support the Matrix program? Why or why not?

See more cases on the study site: Joyce v. City and County of San Francisco, *www.sagepub.com/lippmanccl3e.*

Cases and Comments

1. **Panhandling.** A number of jurisdictions have adopted ordinances prohibiting street begging or panhandling. The Second Circuit Court of Appeals, in *Loper v. New York City Police Department,* 999 F.2d 699 (2d Cir. 1993), held that a statute that prohibited all panhandling on city sidewalks and streets violated the First Amendment. Indianapolis, in July 1999, responded by prohibiting "aggressive panhandling." This illegal activity was defined as touching the solicited person, approaching an individual standing in line and waiting to be admitted to a commercial establishment, blocking an individual's entrance to any building or vehicle, following a person who walks away from the panhandler, using profane or abusive language or statements or gestures that would "cause a reasonable person to be fearful or feel compelled," and panhandling in a group of two or more persons. The ordinance also prohibited soliciting at various locations, including bus stops, sidewalk cafes, vehicles parked or stopped on a public street, and within twenty feet of an automatic teller machine. Panhandling is also prohibited under the ordinance after sunset or before sunrise. Each act in violation of the ordinance is punishable by a fine of not more than $2,500. The court was authorized to issue an injunction or order prohibiting an individual convicted of violating the ordinance from repeating this violation. Violation of the injunction was punishable with imprisonment.

The ordinance was challenged by Jimmy Gresham, a homeless person who lived on a Social Security disability benefit of $417 per month. Gresham supplemented this income by panhandling. The Seventh Circuit Court of Appeals ruled that Indianapolis possessed a legitimate interest in promoting the safety and convenience of its citizens on the public streets, places, and parks and that the ordinance was a reasonable time, place, and manner restriction of speech. The court noted that the ordinance does not ban all panhandling, it merely restricts solicitations to situations that are considered "especially unwanted or bothersome" in which people would "feel a heightened sense of fear or alarm, or might wish especially to be left alone."

Panhandlers are also provided with reasonable alternative avenues to solicit funds. Panhandling is defined as a solicitation made in person upon any street, public place, or park in which a person requests an immediate donation of money or other gratuity. Individuals are free to directly ask for money during the day so long as they do not violate the ordinance. In addition, individuals have the right during the daytime or evening hours to engage in "passive panhandling" in which they display signs or engage in street performances during the evening.

The Seventh Circuit also dismissed the contention that the ordinance was void for vagueness. For instance, the contention that a polite request for a donation might be considered threatening by an unusually sensitive or fearful individual and that a panhandler would not be certain on how to conform to the ordinance according to the court was answered by the provision of a "reasonable person" standard in the ordinance. In another example, the court ruled that the prohibition on "following" an individual should be viewed as entailing a continuing request for a donation combined with following an individual. Walking in the same direction as the solicited person, as a result, would not be prohibited where the walking was "divorced from the request." Do you agree that it is possible to distinguish between "aggressive" and "passive" panhandling? What additional provisions might you propose to this ordinance? Would you vote for this ordinance as a member of the Indianapolis City Council? See *Gresham v. Peterson,* 225 F.3d 899 (7th Cir. 2000).

See more cases on the study site: Gresham v. Peterson, *www.sagepub.com/lippmanccl3e.*

2. **Homelessness and Status Offenses.** The 9th Circuit Court of Appeals held that homeless individuals arrested for "sitting, lying, or sleeping on a public street and sidewalks at all times and in all places within Los Angeles" had been subjected to cruel and unusual punishment in violation of their Eighth Amendment rights. The so-called Skid Row area of Los Angeles has the largest concentration of homeless in United States, and there generally is a shortage of over 1,000 beds each evening. The area is dominated by single residence hotels which charge $379 per month. The monthly welfare stipend for single adults in Los Angeles County is $221. Wait-lists for public housing and for housing assistance vouchers in Los Angeles are from three to ten years. The court of appeals held that "just as the Eighth Amendment prohibits the infliction of criminal punishment on an individual for being a drug addict, or for involuntary public drunkenness that is an unavoidable consequence of being a chronic alcoholic without a home," the Eighth Amendment prohibits the City from punishing "involuntary sitting, lying, or sleeping on public sidewalks that is an unavoidable consequence of being human and homeless without shelter in the City of Los Angeles." See *Jones v. City of Los Angeles,* 444 F.3d 1118 (9th Cir. 2006). (The decision was withdrawn following a settlement between the plaintiffs and Los Angeles.)

See more cases on the study site: **United States v. Jones,** *www.sagepub.com/lippmanccl3e.*

Gangs

It is estimated that that there are roughly 21,000 gangs active in the United States with an estimated 700,000 gang members. Gangs are no longer limited to large urban areas and today are active in nearly every city, suburb, and rural area. These gangs are involved in criminal activity ranging from drugs and prostitution, to extortion and theft, and some have members throughout the United States as well as in Mexico and Central America. The Illinois legislature made several legislative "findings" concerning the peril posed by gangs.

- Urban, suburban, and rural communities are being "terrorized and plundered by street gangs."
- Street gangs are often "controlled by criminally sophisticated adults" who manipulate or threaten young people into serving as drug couriers and into carrying out brutal crimes on behalf of the gang.
- Street gangs present a "clear and present danger to public order and safety."

An example is the Varrio Sureno Town gang that was described by the California Supreme Court, in the 1997 case of *Gallo v. Acuna,* as having converted the four-square-block neighborhood of Rocksprings in San Jose, California, into an "urban war zone." The gang was described as congregating on sidewalks and lawns and in front of apartment complexes at all hours. They openly drank, smoked dope, sniffed glue, snorted cocaine, and transformed the neighborhood into a drug bazaar. The court's opinion described drive-by shootings, vandalism, arson, and theft as commonplace. Garages were used as urinals; homes were "commandeered as escape routes" and served as storage sites for drugs and guns; and buildings, sidewalks, and automobiles were "turned into a . . . canvas of gang graffiti." The California Supreme Court concluded that community residents had become "prisoners in their own homes." Individuals wearing the color of clothing identified with rival gangs were at risk, and relatives and friends were reluctant to visit. Verbal and physical retaliation was directed against anyone who complained to the police or who served as an informant.[22]

States have adopted various legal approaches to controlling gangs. Special gang statutes make it a crime to solicit, to cause any person to join, or to deter any person from leaving a gang, and enhanced punishment is provided for crimes committed to further the interests of gangs. Gang members have also been prosecuted under organized crime statutes, and laws also provide for the vicarious civil liability of parents for the conduct of their children. Various school districts prohibit the display of gang paraphernalia and colors, and some correctional systems provide rewards for gang members who leave the gang and cooperate with authorities. In 2009, the City of Los Angeles successfully sued and collected a multimillion-dollar judgment for damages against individual gang members.

Gallo v. Acuna is an example of the type of innovative civil remedies that are being employed. In *Gallo,* the California Supreme Court affirmed an injunction (a court order halting certain acts) issued by a California trial court. The order declared the Varrio Sureno Town gang a "public nuisance," meaning that the gang's continued presence in the community prevented residents from the enjoyment of life and property, disrupted the quiet and security of the neighborhood, and interfered with the use of the streets and parks. Thirty-eight members of the gang were ordered to "abate" (end) the nuisance by halting conduct ranging from spray painting to the possession and sale of drugs and the playing of loud music, public consumption of alcohol, littering, urinating in a public place, communicating through the use of gang signals, and wearing gang insignia. An individual or group violating the injunction would be in contempt of court and subject to punishment by a fine or short-term incarceration.

Critics of these antigang efforts question whether we are sacrificing the civil liberties of both gang members and innocent young people in order to combat the violence perpetrated by a relatively small number of individuals. They point to the fact that young minority males who, in fact, may not be gang members are often targeted for harassment, detention, interrogation, and arrest by the police.

One of the most significant efforts to curb gang activity was the gang ordinance adopted by the Chicago City Council in 1992. This local law authorized the police to order suspected gang members who, along with at least two other individuals, were "loitering" in public to vacate the area. Between 1992 and 1995, the police issued 89,000 orders to disperse and arrested more than

42,000 people for disobeying an order to move on. In *City of Chicago v. Morales,* the U.S. Supreme Court considered the constitutionality of the ordinance. In reading the case, do you understand what type of behavior is prohibited under the ordinance? How do we balance the constitutional right of freedom of assembly against society's interest in combating gangs? Do you think that this type of ordinance is an effective approach to preventing gangs from terrorizing neighborhoods?

Should individuals reasonably believed by the police to be gang members have the right to assemble on street corners?

City of Chicago v. Morales, 527 U.S. 41 (1999). Opinion by: Stevens, J.

Issue

In 1992, the Chicago City Council enacted the Gang Congregation Ordinance, which prohibits "criminal street gang members" from "loitering" with one another or with other persons in any public place. The question presented is whether the Supreme Court of Illinois correctly held that the ordinance violates the Due Process Clause of the Fourteenth Amendment to the Federal Constitution.

Facts

Before the ordinance was adopted, the city council's Committee on Police and Fire conducted hearings to explore the problems created by the city's street gangs, and more particularly, the consequences of public loitering by gang members. Witnesses included residents of the neighborhoods where gang members are most active, as well as some of the aldermen who represent those areas. Based on that evidence, the council made a series of findings that are included in the text of the ordinance and explain the reasons for its enactment.

The council found that a continuing increase in criminal street gang activity was largely responsible for the city's rising murder rate, as well as an escalation of violent and drug-related crimes. It noted that in many neighborhoods throughout the city, "the burgeoning presence of street gang members in public places has intimidated many law abiding citizens." Furthermore, the council stated that gang members "establish control over identifiable areas . . . by loitering in those areas and intimidating others from entering those areas; and . . . members of criminal street gangs avoid arrest by committing no offense punishable under existing laws when they know the police are present. . . ." It further found that "loitering in public places by criminal street gang members creates a justifiable fear for the safety of persons and property in the area" and that "aggressive action is necessary to preserve the city's streets and other public places so that the public may use such places without fear."

Moreover, the council concluded that the city "has an interest in discouraging all persons from loitering in public places with criminal gang members."

The ordinance creates a criminal offense punishable by a fine of up to $500, imprisonment for not more than six months, and a requirement to perform up to 120 hours of community service. Commission of the offense involves four predicates. First, the police officer must reasonably believe that at least one of the two or more persons present in a "public place" is a "criminal street gang member." Second, the persons must be "loitering," which the ordinance defines as "remaining in any one place with no apparent purpose." Third, the officer must then order "all" of the persons to disperse and remove themselves "from the area." Fourth, a person must disobey the officer's order. If any person, whether a gang member or not, disobeys the officer's order, that person is guilty of violating the ordinance.

The ordinance states in pertinent part:

(a) Whenever a police officer observes a person whom he reasonably believes to be a criminal street gang member loitering in any public place with one or more other persons, he shall order all such persons to disperse and remove themselves from the area. Any person who does not promptly obey such an order is in violation of this section.

(b) It shall be an affirmative defense to an alleged violation of this section that no person who was observed loitering was in fact a member of a criminal street gang.

(c) As used in this section:

(1) "Loiter" means to remain in any one place with no apparent purpose.

(2) "Criminal street gang" means any ongoing organization, association in fact or group of three or more persons, whether formal or informal, having as one of its substantial activities the commission of

one or more of the criminal acts enumerated in paragraph

    (3) and whose members individually or collectively engage in or have engaged in a pattern of criminal gang activity.

. . .

    (5) "Public place" means the public way and any other location open to the public, whether publicly or privately owned.

. . .

  (e) Any person who violates this Section is subject to a fine of not less than $100 and not more than $500 for each offense, or imprisonment for not more than six months, or both.

In addition to or instead of the above penalties, any person who violates this section may be required to perform up to 120 hours of community service. . . .

Two months after the ordinance was adopted, the Chicago Police Department promulgated General Order 92-4 to provide guidelines to govern its enforcement. That order purported to establish limitations on the enforcement discretion of police officers "to ensure that the anti-gang loitering ordinance is not enforced in an arbitrary or discriminatory way." The limitations confine the authority to arrest gang members who violate the ordinance to sworn "members of the Gang Crime Section" and certain other designated officers, and establish detailed criteria for defining street gangs and membership in such gangs. In addition, the order directs district commanders to "designate areas in which the presence of gang members has a demonstrable effect on the activities of law abiding persons in the surrounding community," and provides that the ordinance "will be enforced only within the designated areas." The City, however, does not release the locations of these "designated areas" to the public.

During the three years of its enforcement, the police issued over 89,000 dispersal orders and arrested over 42,000 people for violating the ordinance. There were 5,251 arrests under the ordinance in 1993, 15,660 in 1994, and 22,056 in 1995. In the ensuing enforcement proceedings, two trial judges upheld the constitutionality of the ordinance, but eleven others ruled that it was invalid. In respondent Youkhana's case, the trial judge held that the "ordinance fails to notify individuals what conduct is prohibited, and it encourages arbitrary and capricious enforcement by police."

The City believes that the ordinance resulted in a significant decline in gang-related homicides. It notes that in 1995, the last year the ordinance was enforced, the gang-related homicide rate fell by twenty-six percent. In 1996, after the ordinance had been held invalid, the gang-related homicide rate rose eleven percent. However, gang-related homicides fell by nineteen percent in 1997, over a year after the suspension of the ordinance. Given the myriad factors that influence levels of violence, it is difficult to evaluate the probative value of this statistical evidence, or to reach any firm conclusion about the ordinance's efficacy. . . .

Reasoning

The basic factual predicate for the City's ordinance is not in dispute. As the City argues in its brief, "the very presence of a large collection of obviously brazen, insistent, and lawless gang members and hangers-on on the public ways intimidates residents, who become afraid even to leave their homes and go about their business. That, in turn, imperils community residents' sense of safety and security, detracts from property values, and can ultimately destabilize entire neighborhoods." The findings in the ordinance explain that it was motivated by these concerns. We have no doubt that a law that directly prohibited such intimidating conduct would be constitutional, but this ordinance broadly covers a significant amount of additional activity. Uncertainty about the scope of that additional coverage provides the basis for respondents' claim that the ordinance is too vague.

In fact, the City already has several laws that serve this purpose. . . . Deputy Superintendent Cooper, the only representative of the police department at the Committee on Police and Fire hearing on the ordinance before the Chicago City Council, testified that ninety percent of the conduct people complained that they were being arrested for was actually a criminal offense for which people could be arrested even absent the gang ordinance. These offenses included intimidation, criminal drug conspiracy, and mob action.

We are confronted at the outset with the City's claim that it was improper for the state courts to conclude that the ordinance is invalid on its face. . . . An enactment . . . may be impermissibly vague because it fails to establish standards for the police and public that are sufficient to guard against the arbitrary deprivation of liberty interests. . . . As the United States recognizes, the freedom to loiter for innocent purposes is part of the "liberty" protected by the Due Process Clause of the Fourteenth Amendment. We have expressly identified this "right to move from one place to another according to inclination" as "an attribute of personal liberty" protected by the Constitution. . . . Indeed, it is apparent that an individual's decision to remain in a public place of his choice is as much a part of his liberty as the freedom of movement inside frontiers that is "a part of our heritage" or the right to move "to whatsoever place one's own inclination may direct."

Vagueness may invalidate a criminal law for either of two independent reasons. First, it may fail to provide the kind of notice that will enable ordinary people to understand what conduct it prohibits; second, it may authorize and even encourage arbitrary and discriminatory enforcement. Accordingly, we first consider whether the ordinance provides fair notice to the citizen and then discuss its potential for arbitrary enforcement.

"It is established that a law fails to meet the requirements of the Due Process Clause if it is so vague and standardless that it leaves the public uncertain as to the conduct it prohibits. . . ." The Illinois Supreme Court recognized that the term *loiter* may have a common and accepted meaning, but the definition of that term in this ordinance—"to remain in any one place with no apparent purpose"—does not. It is difficult to imagine how any citizen of the city of Chicago standing in a public place with a group of people would know if he or she had an "apparent purpose." If she were talking to another person, would she have an apparent purpose? If she were frequently checking her watch and looking expectantly down the street, would she have an apparent purpose? . . . "The purpose simply to stand on a corner cannot be an 'apparent purpose' under the ordinance; if it were, the ordinance would prohibit nothing at all."

Since the city cannot conceivably have meant to criminalize each instance a citizen stands in public with a gang member, the vagueness that dooms this ordinance is not the product of uncertainty about the normal meaning of "loitering" but rather about what loitering is covered by the ordinance and what is not. The Illinois Supreme Court emphasized the law's failure to distinguish between innocent conduct and conduct threatening harm. Its decision followed the precedent set by a number of state courts that have upheld ordinances that criminalize loitering combined with some other overt act or evidence of criminal intent. However, state courts have uniformly invalidated laws that do not join the term *loitering* with a second specific element of the crime.

One of the trial courts that invalidated the ordinance gave the following illustration: "Suppose a group of gang members were playing basketball in the park, while waiting for a drug delivery. Their apparent purpose is that they are in the park to play ball. The actual purpose is that they are waiting for drugs. Under this definition of loitering, a group of people innocently sitting in a park discussing their futures would be arrested, while the 'basketball players' awaiting a drug delivery would be left alone."

The City's principal response to this concern about adequate notice is that loiterers are not subject to sanction until after they have failed to comply with an officer's order to disperse. "Whatever problem is created by a law that criminalizes conduct people normally believe to be innocent is solved when persons receive actual notice from a police order of what they are expected to do." We find this response unpersuasive for least two reasons.

First, the purpose of the fair notice requirement is to enable the ordinary citizen to conform his or her conduct to the law. "No one may be required at peril of life, liberty or property to speculate as to the meaning of penal statutes." Although it is true that a loiterer is not subject to criminal sanctions unless he or she disobeys a dispersal order, the loitering is the conduct that the ordinance is designed to prohibit. If the loitering is in fact harmless and innocent, the dispersal order itself is an unjustified impairment of liberty. . . . The police are able to decide arbitrarily which members of the public they will order to disperse. Because an officer may issue an order only after prohibited conduct has already occurred, it cannot provide the kind of advance notice that will protect the putative loiterer from being ordered to disperse. Such an order cannot retroactively give adequate warning of the boundary between the permissible and the impermissible applications of the law.

Second, the terms of the dispersal order compound the inadequacy of the notice afforded by the ordinance. It provides that the officer "shall order all such persons to disperse and remove themselves from the area." This vague phrasing raises a host of questions. After such an order issues, how long must the loiterers remain apart? How far must they move? If each loiterer walks around the block and they meet again at the same location, are they subject to arrest or merely to being ordered to disperse again? As we do here, we have found vagueness in a criminal statute exacerbated by the use of the standards of "neighborhood" and "locality." Both terms "are elastic and, dependent upon circumstances, may be equally satisfied by areas measured by rods or by miles."

Lack of clarity in the description of the loiterer's duty to obey a dispersal order might not render the ordinance unconstitutionally vague if the definition of the forbidden conduct were clear, but it does buttress our conclusion that the entire ordinance fails to give the ordinary citizen adequate notice of what is forbidden and what is permitted. The Constitution does not permit a legislature to "set a net large enough to catch all possible offenders, and leave it to the courts to step inside and say who could be rightfully detained, and who should be set at large." . . . This ordinance is therefore vague "not in the sense that it requires a person to conform his conduct to an imprecise but comprehensible normative standard, but rather in the sense that no standard of conduct is specified at all." . . .

The broad sweep of the ordinance also violates "the requirement that a legislature establish minimal guidelines to govern law enforcement." There are no such guidelines in the ordinance. In any public place in the city of Chicago, persons who stand or sit in the company of a gang member may be ordered to disperse unless their purpose is apparent. The mandatory language in the enactment directs the police to issue an order without first making any inquiry about their possible purposes. It matters not whether the reason that a gang member and his father, for example, might loiter near Wrigley Field is to rob an unsuspecting fan or just to get a glimpse of Sammy Sosa leaving the ballpark; in either event, if their purpose is not apparent to a nearby police officer, she may—indeed, she "shall"—order them to disperse.

Recognizing that the ordinance does reach a substantial amount of innocent conduct, we turn, then, to its language to determine if it "necessarily entrusts law-making to the moment-to-moment judgment of the policeman on his beat." As we discussed in the context of fair notice, the principal source of the vast discretion conferred on the police in this case is the definition of loitering as "to remain in any one place with no apparent purpose."

As the Illinois Supreme Court interprets that definition, it "provides absolute discretion to police officers to determine what activities constitute loitering." We have no authority to construe the language of a state statute more narrowly than the construction given by that State's highest court. "The power to determine the meaning of a statute carries with it the power to prescribe its extent and limitations as well as the method by which they shall be determined."

Nevertheless, the City disputes the Illinois Supreme Court's interpretation, arguing that the text of the ordinance limits the officer's discretion in three ways. First, it does not permit the officer to issue a dispersal order to anyone who is moving along or who has an apparent purpose. Second, it does not permit an arrest if individuals obey a dispersal order. Third, no order can issue unless the officer reasonably believes that one of the loiterers is a member of a criminal street gang.

Even putting to one side our duty to defer to a state court's construction of the scope of a local enactment, we find each of these limitations insufficient. That the ordinance does not apply to people who are moving—that is, to activity that would not constitute loitering under any possible definition of the term—does not even address the question of how much discretion the police enjoy in deciding which stationary persons to disperse under the ordinance. Similarly, that the ordinance does not permit an arrest until after a dispersal order has been disobeyed does not provide any guidance to the officer deciding whether such an order should issue. The "no apparent purpose" standard for making that decision is inherently subjective because its application depends on whether some purpose is "apparent" to the officer on the scene.

Presumably, an officer would have discretion to treat some purposes—perhaps a purpose to engage in idle conversation or simply to enjoy a cool breeze on a warm evening—as too frivolous to be apparent if he suspected a different ulterior motive. . . .

It is true, as the City argues, that the requirement that the officer reasonably believe that a group of loiterers contains a gang member does place a limit on the authority to order dispersal. That limitation would no doubt be sufficient if the ordinance only applied to loitering that had an apparently harmful purpose or effect, or possibly if it only applied to loitering by persons reasonably believed to be criminal gang members. But this ordinance, for reasons that are not explained in the findings of the city council, requires no harmful purpose and applies to nongang members as well as suspected gang members. It applies to everyone in the city who may remain in one place with one suspected gang member as long as their purpose is not apparent to an officer observing them. Friends, relatives, teachers, counselors, or even total strangers might unwittingly engage in forbidden loitering if they happen to engage in idle conversation with a gang member.

Ironically, the definition of loitering in the Chicago ordinance not only extends its scope to encompass harmless conduct but also has the perverse consequence of excluding from its coverage much of the intimidating conduct that motivated its enactment. As the city council's findings demonstrate, the most harmful gang loitering is motivated either by an apparent purpose to publicize the gang's dominance of certain territory, thereby intimidating nonmembers, or by an equally apparent purpose to conceal ongoing commerce in illegal drugs. As the Illinois Supreme Court has not placed any limiting construction on the language in the ordinance, we must assume that the ordinance means what it says and that it has no application to loiterers whose purpose is apparent. The relative importance of its application to harmless loitering is magnified by its inapplicability to loitering that has an obviously threatening or illicit purpose.

Finally, in its opinion striking down the ordinance, the Illinois Supreme Court refused to accept the general order issued by the police department as a sufficient limitation on the "vast amount of discretion" granted to the police in its enforcement. That the police have adopted internal rules limiting their enforcement to certain designated areas in the city would not provide a defense to a loiterer who might be arrested elsewhere. Nor could a person who knowingly loitered with a well-known gang member anywhere in the city safely assume that they would not be ordered to disperse no matter how innocent and harmless their loitering might be.

Holding

In our judgment, the Illinois Supreme Court correctly concluded that the ordinance does not provide sufficiently specific limits on the enforcement discretion of the police "to meet constitutional standards for definiteness and clarity." We recognize the serious and difficult problems testified to by the citizens of Chicago that led to the enactment of this ordinance. "We are mindful that the preservation of liberty depends in part on the maintenance of social order." However, in this instance, the City has enacted an ordinance that affords too much discretion to the police and too little notice to citizens who wish to use the public streets.

Accordingly, the judgment of the Supreme Court of Illinois is affirmed.

Dissenting, *Scalia, J.*

Until the ordinance that is before us today was adopted, the citizens of Chicago were free to stand about in public places with no apparent purpose—to engage, that is, in conduct that appeared to be loitering. In recent years, however, the city has been afflicted with criminal street gangs. As reflected in the record before us, these gangs congregated in public places to deal in drugs and to terrorize the neighborhoods by demonstrating control over their "turf." Many residents of the inner city felt that they were prisoners in their own homes. Once again, Chicagoans decided that to eliminate the problem, it was worth restricting some of the freedom that they once enjoyed. The means they took was similar to the second, and more mild, example given above rather than the first: Loitering was not made unlawful, but when a group of people occupied a public place without an apparent purpose and in the company of a known gang member, police officers were authorized to order them to disperse, and the failure to obey such an order was made unlawful. The minor limitation upon the free state of nature that this prophylactic arrangement imposed upon all Chicagoans seemed to them (and it seems to me) a small price to pay for liberation of their streets.

The majority today invalidates this perfectly reasonable measure . . . by elevating loitering to a constitutionally guaranteed right and by discerning vagueness where, according to our usual standards, none exists. . . . Respondent Jose Renteria—who admitted that he was a member of the Satan Disciples gang—was observed by the arresting officer loitering on a street corner with other gang members. The officer issued a dispersal order, but when she returned to the same corner fifteen to twenty minutes later, Renteria was still there with his friends, whereupon he was arrested. In another example, respondent Daniel Washington and several others—who admitted they were members of the Vice Lords gang—were observed by the arresting officer loitering in the street, yelling at passing vehicles, stopping traffic, and preventing pedestrians from using the sidewalks. The arresting officer issued a dispersal order, issued another dispersal order later when the group did not move, and finally arrested the group when they were found loitering in the same place still later. Finally, respondent Gregorio Gutierrez—who had previously admitted to the arresting officer his membership in the Latin Kings gang—was observed loitering with two other men. The officer issued a dispersal order, drove around the block, and arrested the men after finding them in the same place upon his return. . . .

In our democratic system, how much harmless conduct to proscribe is not a judgment to be made by the courts. So long as constitutionally guaranteed rights are not affected . . . all sorts of perfectly harmless activity by millions of perfectly innocent people can be forbidden—riding a motorcycle without a safety helmet, for example, starting a campfire in a national forest, or selling a safe and effective drug not yet approved. . . . All of these acts are entirely innocent and harmless in themselves, but because of the risk of harm that they entail, the freedom to engage in them has been abridged. The citizens of Chicago have decided that depriving themselves of the freedom to "hang out" with a gang member is necessary to eliminate pervasive gang crime and intimidation—and that the elimination of the one is worth the deprivation of the other. This Court has no business second-guessing either the degree of necessity or the fairness of the trade.

Dissenting, *Thomas, J.*

The duly elected members of the Chicago City Council enacted the ordinance at issue as part of a larger effort to prevent gangs from establishing dominion over the public streets. By invalidating Chicago's ordinance, I fear that the Court has unnecessarily sentenced law-abiding citizens to lives of terror and misery. The ordinance is not vague. . . . Nor does it violate the Due Process Clause. . . .

The human costs exacted by criminal street gangs are inestimable. In many of our Nation's cities, gangs have "virtually overtaken certain neighborhoods, contributing to the economic and social decline of these areas and causing fear and lifestyle changes among law-abiding residents." Gangs fill the daily lives of many of our poorest and most vulnerable citizens with a terror that the Court does not give sufficient consideration, often relegating them to the status of prisoners in their own homes. . . .

The city of Chicago has suffered the devastation wrought by this national tragedy. Last year, in an effort to curb plummeting attendance, the Chicago Public Schools hired dozens of adults to escort children to school. The youngsters had become too terrified of gang violence to leave their homes alone. The children's fears were not unfounded. In 1996, the Chicago Police Department estimated that there were 132 criminal street gangs in the city. Between 1987 and 1994, these gangs were involved in 63,141 criminal incidents, including 21,689 nonlethal violent crimes and 894 homicides. Many of these criminal incidents and homicides result from gang "turf battles," which take place on the public streets and place innocent residents in grave danger. . . . In 1996 alone, gangs were involved in 225 homicides, which was twenty-eight percent of the total homicides committed in the city. . . . Nationwide, law enforcement officials estimate that as many as 31,000 street gangs, with 846,000 members, exist. . . .

Following . . . [their] hearings, the Chicago City Council found that "criminal street gangs establish control over identifiable areas . . . by loitering in those areas and intimidating others from entering those areas." It further found that the mere presence of gang members "intimidates many law abiding citizens" and "creates a justifiable fear for the safety of persons and property in the area." It is the product of this democratic process—the council's attempt to address these social ills—that we are asked to pass judgment upon today.

As part of its ongoing effort to curb the deleterious effects of criminal street gangs, the citizens of Chicago sensibly decided to return to basics. The ordinance does nothing more than confirm the well-established principle that the police have the duty and the power to maintain the public peace and, when necessary, to disperse groups of individuals who threaten it. . . .

I do not suggest that a police officer enforcing the Gang Congregation Ordinance will never make a mistake. Nor do I overlook the possibility that a police officer . . . might enforce the ordinance in an arbitrary or discriminatory way. But our decisions should not turn on the proposition that such an event will be anything but rare. An individual is always free to challenge the constitutionality of an arrest in court.

The . . . conclusion that the ordinance fails to give the ordinary citizen adequate notice of what is forbidden and what is permitted is similarly untenable. There is nothing "vague" about an order to disperse. . . . It is safe to assume that the vast majority of people who are ordered by the police to "disperse and remove themselves from the area" will have little difficulty understanding how to comply.

Today, the Court focuses extensively on the "rights" of gang members and their companions. It can safely do so—the people who will have to live with the consequences of today's opinion do not live in our neighborhoods. Rather, the people who will suffer . . . are . . . people who have seen their neighborhoods literally destroyed by gangs and violence and drugs. They are good, decent people who must struggle to overcome their desperate situation, against all odds, in order to raise their families, earn a living, and remain good citizens. As one resident described, "There is only about maybe one or two percent of the people in the city causing these problems maybe, but it's keeping 98 percent of us in our houses and off the streets and afraid to shop." By focusing exclusively on the imagined "rights" of the two percent, the Court today has denied our most vulnerable citizens . . ."freedom of movement." And that is a shame. I respectfully dissent.

Questions for Discussion

1. Summarize Chicago's Gang Congregation Ordinance.

2. The majority concludes that the ordinance fails to provide both citizens and the police with clear guidelines and is unconstitutionally void for vagueness. By reading the law, would you know what conduct is prohibited? As a police officer, would it be clear under what circumstances you were authorized to order people to disperse?

3. Why is the ordinance void for vagueness?

4. Why does Justice Scalia take the time in his dissent to discuss the arrest of several individuals? How do Justices Scalia and Thomas balance civil liberties against public safety?

5. Do you believe that the Chicago Gang Congregation Ordinance is an effective approach to curbing gang activity?

6. As a member of the Chicago City Council, would you vote for the ordinance?

See more cases on the study site: **City of Chicago v. Morales,** *www.sagepub.com/lippmanccl3e.*

Cases and Comments

Chicago Gang Ordinance. On February 16, 2000, the Chicago City Council revised the 1992 Gang Congregation Ordinance. The amended ordinance defined "gang loitering" as remaining in any one place under circumstances "that would warrant a reasonable person" to believe that the purpose or effect of that behavior is to enable a criminal street gang "to establish control over identifiable areas, to intimidate others from entering these areas, or conceal illegal activities." The new ordinance authorizes the Superintendent of

Police to designate areas of Chicago in which enforcement is required because "gang loitering has enabled street gangs to establish control over identifiable areas, to intimidate others from entering the areas or to conceal illegal activities" (Municipal Code of Chicago sections 8-4-015, 8-4-017). The revised ordinance has not yet been tested in court. How does this differ from the 1992 Gang Congregation Ordinance? Do you think that the revised 2000 law is constitutional? Are there other amendments that you would make to the ordinance?

THE OVERREACH OF CRIMINAL LAW

Criminologists Norval Morris and Gordon Hawkins, writing in 1969, argue that the function of criminal law is to protect property and persons, particularly juveniles and those in need of special protection. They point out that roughly fifty percent of all arrests are for acts threatening public morality. These, for the most part, are acts that individuals engage in voluntarily and do not view as harmful to themselves or to others. In other words, people are arrested who do not believe that they should be treated as criminals or victims and who are not deterred by the threat of either arrest or punishment. This list of victimless crimes includes drunkenness, narcotics, gambling, prostitution, the possession of obscene materials, and various sexual offenses. Some may include under the heading of victimless offenses, seatbelt and motorcycle helmet laws, adultery and fornication, and the prohibition on assisted suicide. Morris and Hawkins criticize what they view as the moralist orientation of American criminal law and the long tradition of employing the law as an instrument for coercing men and women into acting in a virtuous fashion. In their view, people possess a complete right to choose a path that may lead to purgatory so long as they do not directly injure the person or property of another. Morris and Hawkins also point to the fact that criminalizing consensual, private behavior actually increases rather than decreases crime.[23]

- *Crime Tariff.* Making an activity illegal means that those engaged in the activity do not confront competition from legal businesses and will be free to charge a high price. This profit, in turn, is used to fund other organized crime activities. Also, addicts must resort to crime to support their expensive gambling, drug, and alcohol addictions.
- *Inconsistency.* The condemnation of activities such as gambling is undermined by the fact that there are legal lotteries and gambling casinos in Atlantic City, in Las Vegas, and on Native American reservations. This creates an inconsistency in legal rules and contributes to a lack of respect for the law.
- *Romanticism.* Declaring an activity illegal tends to make it appear romantic and appealing to younger people.
- *Law Enforcement.* Scarce law enforcement resources are devoted to enforcing these laws rather than more harmful offenses. Often, there are no complaining victims, and the police must resort to controversial undercover and sting operations. The amount of money involved in activities such as drug trafficking and the absence of complaining victims creates a situation with the potential for bribery, extortion, and corruption.
- *Criminal Subculture.* Making activities like gambling and prostitution illegal means that people involved in these activities are driven into a criminal environment and may be victimized or become involved in other crimes. Prostitutes often are exploited by pimps who offer protection and threaten customers.

This view is challenged by English Lord Patrick Devlin, who argues that society must be equally vigilant in protecting itself against threats from abroad and at home. Lord Devlin contends that the loosening of moral bonds is typically the first step toward the disintegration of the social order and that the maintenance of values is the proper concern of government. Lord Devlin argues that the notion that allegedly private behavior does not affect society is misguided. He concedes that while great social harm may not result from a single individual engaging in an alcoholic or gambling binge, society would crumble if the same activity is embraced by a quarter or more of the population. These so-called victimless crimes, according to Lord Devlin, impose hardships on the families of addicts, require society to spend money in treating addictions, corrupt the young, and ruin the lives of addicts. In short, a compassionate society does not permit individuals to "do their own thing." Lord Devlin concludes that we cannot, on the one hand, encourage people to live in a moral fashion and, on the other hand, tolerate immoral behavior. Where do you stand on the debate over so-called victimless crimes? Some studies have found an "epidemic" of illegal gambling by college students. Is this a "victimless crime"?[24]

The next section considers the **immorality crime** of prostitution, an activity that the famous British *Wolfenden Report* argued, in 1957, should not be subject to criminal punishment when carried on between consenting adults.

Prostitution and Solicitation

Prostitution is defined as engaging in sexual intercourse or other sexual acts in exchange for money or other items of value. You undoubtedly have heard someone refer to prostitution as "the world's oldest profession." Why is an activity that has been characteristic of both ancient and modern societies considered a crime? There are several reasons:

- *Disease*. Encourages transmission of infections such as AIDS.
- *Family*. Weakens marriage and the family.
- *Exploitation*. Exploits and degrades women.
- *Immorality*. Promotes social immorality and a culture tolerant of alcoholism, drug abuse, gambling, and acts of immorality.

Critics of laws punishing prostitution point out that the legitimacy of law enforcement is undermined by the fact that the police typically must resort to posing as "prostitutes" or "customers" in order to enforce prostitution laws and that this lowers respect for law enforcement. There is also an inconsistency in the fact that the police target street prostitutes while "call girls" who service the relatively wealthy are rarely arrested. Critics further note that despite the resources devoted to eliminating prostitution, the police have not been able to deter individuals from engaging in this activity. The argument is also made that categorizing prostitution as a crime insures that it will be controlled by organized crime and pimps (individuals who live off the proceeds of prostitution). This results in prostitutes being labeled as criminals, places them in danger, and deprives the government of tax revenues. Others argue that prostitution laws deprive women of the opportunity to utilize their bodies to advance their economic well-being. The most radical commentators point to the fact that prostitutes are no different than the large number of people who engage in sex with the intent of obtaining employment or material gain. Some favor decriminalization of prostitution and subjecting the practice to state regulation, the policy followed in the Netherlands. State regulation has the advantage of insuring that precautions are taken against the spread of HIV and other sexually transmitted diseases. A small number of commentators favor complete legalization.

Nevada is the only state in which prostitution is legal. Each of the smaller population counties in the state are free to determine whether to permit prostitution, and the practice is heavily regulated. Brothels must pay a licensing fee, and prostitutes are required to submit to monthly HIV tests. Condoms are required, and prostitutes must be at least twenty-one years old. Prostitution is not permitted anywhere other than in the brothels, and the brothels may not advertise in counties in which the practice is illegal. Nevada possesses roughly thirty legal brothels that employ roughly 300 prostitutes.

The Crime of Prostitution

Prostitution punishes both men and women who

- solicit or engage in
- any sexual activity
- in exchange for money or other consideration.

As you can see, prostitution is committed by exchanging sexual activity for money or other consideration or by **solicitation for prostitution**, asking or requesting another person to engage in prostitution. Note that it is the solicitation or actual exchange of money or value for sex that distinguishes prostitution from the legal act of approaching another person for consensual sexual activity. The crime of prostitution is not limited to sexual intercourse and encompasses all varieties of sexual interaction. Georgia's prostitution law provides that a person "commits the offense of prostitution when he or she performs or offers or consents to perform a sexual act, including but not limited to sexual intercourse or sodomy, for money or other items of value."[25]

Pennsylvania follows the Model Penal Code by providing that a person is guilty of prostitution who "is an inmate of a house of prostitution or otherwise engages in sexual activity as a business." An inmate is a person who engages in prostitution as a business in conjunction with a house of prostitution or as a "call girl," who makes use of an agency to obtain clients. An individual is guilty under this provision who engages in prostitution in affiliation with a house of prostitution. It is unnecessary to establish that the accused engaged in a specific act of prostitution.[26]

State statutes also commonly punish loitering for prostitution. California declares that it is "unlawful for any person to loiter in any public place with the intent to commit prostitution." An individual's *mens rea* is demonstrated by acts that indicate an intent to induce, entice, or solicit prostitution. The California statute notes that this intent may be established by the stopping and soliciting of pedestrians or of the occupants of passing automobiles. This type of provision recognizes that public loitering for solicitation is an essential step in engaging in the business of prostitution and that solicitation negatively impacts a neighborhood's sense of safety and stability.[27]

Prostitution statutes are gender neutral; prostitution may be committed by a male or female prostitute, and both prostitutes and customers may be guilty of soliciting or loitering for the purpose of prostitution. Several states explicitly punish a person who "hires a prostitute or any other person to engage in sexual activity . . . or if that person enters or remains in a house of prostitution for the purpose of engaging in sexual activity." Pennsylvania also provides that convictions and sentences for a second and all subsequent acts of prostitution shall be published in the newspaper.[28]

Another prostitution-related offense is **pimping**, which involves procuring a prostitute for another individual, arranging a meeting for the purpose of prostitution, transporting an individual to a location for the purpose of prostitution, receiving money or other thing of value from a prostitute knowing that it was earned from prostitution, or owning, managing, or leasing a house of prostitution or prostitution business. **Pandering** is the encouraging and inducing of another to become or remain a prostitute; this is punished more harshly when duress or coercion is employed.[29] **Living off prostitution** is committed by a person, other than a prostitute and the prostitute's minor child or other dependent, who is "knowingly supported in whole or substantial part by the proceeds of prostitution."[30] **Keeping a place of prostitution** involves "keeping a place of prostitution when [an individual] knowingly grants or permits the use of such place for the purpose of prostitution."[31] Pimping, pandering, and keeping a place of prostitution are also generally encompassed under the crime of **promoting prostitution**, which involves aiding or abetting prostitution by "any means whatsoever."[32] States also have extended their laws to criminally punish a "masseur or masseuse" who commits the offense of **masturbation for hire** when he or she "stimulates the genital organs of another, whether resulting in orgasm or not, by manual or other bodily contact exclusive of sexual intercourse or by instrumental manipulation for money or the substantial equivalent thereof."[33]

Prostitution is a misdemeanor. It is typically punished somewhat more severely for the third and subsequent offenses and is a felony in the event that an individual knew that he or she was infected with HIV. Georgia provides a sentence of between five and twenty years and a fine of up to $10,000 for keeping a place of prostitution, pimping, pandering, or solicitation involving an individual under eighteen years old.[34]

We should note in passing that there are several other misdemeanor sexual offenses that appear in various state statutes:

- *Adultery.* Consensual sexual intercourse between a male and a female, at least one of whom is married.
- *Bigamy.* Marrying another while already having a living spouse.
- *Fornication.* An unmarried person who engages in voluntary sexual intercourse with another individual.
- *Lewdness.* Public acts offending community standards, including the display of genitals, sexual intercourse, lewd sexual contact, and deviate sexual intercourse.

Legal Regulation of Prostitution

The difficulty of controlling prostitution through individual criminal prosecutions led the city of Milwaukee, Wisconsin, to obtain a court order declaring that prostitutes in designated areas of the city constituted a nuisance. A Wisconsin appellate court found that the police had received a high volume of complaints concerning prostitutes on the streets and private property in this neighborhood. The court further ruled that the enforcement of the laws against prostitution posed a danger to the police, who were forced to act undercover to apprehend prostitutes and that these officers were endangered by the fact that the prostitutes frequently carried sharpened objects, knives with long blades, and razors. The injunction issued by the court prohibited prostitutes from soliciting customers by stopping pedestrians and automobiles and from waiting at bus stops and pay phones and loitering in the doorways of businesses.[35]

The next case, *Harwell v. State,* illustrates the type of technical details that a court must examine to determine whether a defendant intentionally agreed to engage in sex for money.

Model Penal Code

Section 251.2. Prostitution and Related Offenses

(1) A person is guilty of prostitution, a petty misdemeanor, if he or she:

    (a) is an inmate of a house of prostitution or otherwise engages in sexual activity as a business; or

    (b) loiters in or within view of any public place for the purpose of being hired to engage in sexual activity.

    "Sexual activity" includes homosexual and other deviate sexual relations. A "house of prostitution" is any place where prostitution or promotion of prostitution is regularly carried on by one person under the control, management or supervision of another. An "inmate" is a person who engages in prostitution in or through the agency of a house of prostitution. "Public place" means any place to which the public or any substantial group thereof has access.

(2) Promoting Prostitution. . . . The following acts shall . . . constitute promoting prostitution:

    (a) owning, controlling, managing, supervising or otherwise keeping . . . a house of prostitution or prostitution business; or

    (b) procuring an inmate for a house of prostitution or a place in a house of prostitution for one who would be an inmate; or

    (c) encouraging, inducing, or otherwise purposely causing another to become or remain a prostitute; or

    (d) soliciting a person to patronize a prostitute; or

    (e) procuring a prostitute for a patron; or

    (f) transporting a person into or within this state with the purpose to promote that person engaging in prostitution, or procuring or paying for transportation with that purpose; or

    (g) leasing or otherwise permitting a place controlled by the actor . . . to be regularly used for prostitution or the promotion of prostitution, or failure to make reasonable efforts to abate such use by ejecting the tenant, notifying law enforcement authorities, or other legally available means; or

    (h) soliciting, receiving, or agreeing to receive any benefit for doing or agreeing to do anything forbidden by this Subsection.

(3) Grading of Offenses. An offense under Subsection (2) constitutes a felony of the third degree if:

    (a) the offense falls within paragraph (a), (b) or (c) of Subsection (2); or

    (b) the actor compels another to engage in or promote prostitution; or

    (c) the actor promotes prostitution of a child under 16 . . . ; or

    (d) the actor promotes prostitution of his wife, child, ward or any person for whose care, protection or support he is responsible.

Otherwise the offense is a misdemeanor.

(4) Presumption From Living off Prostitutes. A person, other than the prostitute or the prostitute's minor child or other legal dependent incapable of self-support, who is supported in whole or substantial part by the proceeds of prostitution is presumed to be knowingly promoting prostitution. . . .

(5) Patronizing Prostitutes. A person commits a violation if he hires a prostitute to engage in sexual activity with him, or if he enters or remains in a house of prostitution for the purpose of engaging in sexual activity.

(6) Evidence. On the issue whether a place is a house of prostitution the following shall be admissible evidence: its general repute; the repute of the persons who reside in or frequent the place; the frequency, timing and duration of visits by non-residents. Testimony of a person against his spouse shall be admissible to prove offenses under this Section.

Was there an agreement to engage in sex for money?

Harwell v. State, 821 N.E.2d 381 (Ind. Ct. App. 2004). Opinion by: Riley, J.

Appellant-defendant, Lisa Harwell (Harwell), appeals her conviction for Count I, prostitution, a Class D felony. . . . We affirm.

Issue

Harwell raises one issue on appeal, which we restate as follows: whether the State presented sufficient evidence to sustain her conviction for prostitution.

Facts

On September 12, 2003, Officer David Miller of the Indianapolis Police Department was investigating prostitution complaints in the College corridor, the area between Washington Street and 38th Street, in Indianapolis, Indiana. His undercover investigation consisted of driving around the area looking for prostitutes. At approximately 2:45 p.m., Officer Miller observed Harwell on the corner of 22nd Street and College Avenue. Upon stopping at the side of the road, he inquired if Harwell needed a ride. Without responding, Harwell entered the car and asked Officer Miller if he was a police officer. After denying he was a police officer, Officer Miller asked her if anything was going on, a question he uses to determine if women are looking to commit sexual acts. He further specified he was looking for fellatio. Although Harwell agreed to perform fellatio, she refused to discuss money when he asked about the price. Instead, she directed him towards an alley off the 2100 block of Yandes. When they arrived in the alley, Officer Miller again questioned Harwell about the cost, asking her if the act would be more than $20.00. Harwell simply responded "no." At that point, Officer Miller informed her that he was a police officer and that she was under arrest.

On January 26, 2004, a bench trial was held. At the close of the evidence, the trial court found Harwell guilty of prostitution, a Class D felony and sentenced her to 545 days of incarceration at the Indiana Department of Correction.

Reasoning

Harwell contends that the evidence presented at trial was insufficient to support her conviction. Specifically, Harwell argues that the State failed to prove that she offered or agreed to perform a sexual act in exchange for money or other property. . . . Prostitution as a Class D felony is defined as "[a] person who knowingly or intentionally performed, or offers or agrees to perform, sexual intercourse or deviate sexual conduct . . . for money or other property commits prostitution, a Class A misdemeanor. However, the offense is a Class D felony if the person has two (2) prior convictions under this section." Thus, in order to convict Harwell, the State was required to establish beyond a reasonable doubt that (1) she intentionally agreed to perform fellatio in exchange for money and that (2) she had two prior convictions for prostitution. Although Harwell now concedes that she intended to engage in a sexual act, she disputes that there was an agreement to perform fellatio for money. In particular, she asserts that despite Officer Miller's repeated inquiries about the cost of fellatio, she never indicated that she agreed to accept money. We are not persuaded.

As both parties correctly point out, there is no definition of "agreement" within the statute. Furthermore, neither party proffered nor did our research reveal any case law clarifying "agreement" as used in the charge of prostitution. . . . "Agreement" has a plain, and ordinary meaning: it is defined by Black's law dictionary as "a mutual understanding between two or more persons about their relative rights and duties regarding past or future performances; a manifestation of mutual assent by two or more persons." Analogizing to contract law, an agreement is considered to be a meeting of the minds between the parties, a mutual understanding of all terms of the contract. . . .

Here, the record shows that Officer Miller was conducting an undercover investigation in an area known for its high volume of prostitution. Testimonial evidence indicates that immediately after getting into Officer Miller's car, Harwell demanded to know whether he was a police officer. Further, Officer Miller testified that after Harwell agreed to perform fellatio, she was evasive about its cost. However, the record reflects that after directing him to an alley, Officer Miller again attempted to elicit a specific price for Harwell's services. At this point, Officer Miller stated that upon his question "if it was going to cost more than $20.00," Harwell responded, "no."

Holding

Being mindful that we "should not be ignorant as judges of what we know as men," we find that the State presented sufficient evidence to prove beyond a reasonable doubt that Harwell agreed to perform fellatio for money. Based on the evidence before us, we conclude that the agreement was implicit in the parties' words and actions when considered in the context in which they occurred. By indicating that the sexual service would not be more expensive than $20.00, Harwell emitted an inference that there was a cost involved and that she would accept money.

Moreover, Harwell's argument that a specific price has to be determined between the parties prior to there being a meeting of the minds is not supported by the statutory language, which only requires evidence of a performance, offer, or agreement to commit sexual services in exchange for money. The statute is silent as to the requirement of a preset price. Surely, it cannot be said that to constitute a violation of the statute, the agreement must be expressed and in precise statutory language. Therefore, we agree with the trial court that a meeting of the minds existed between Officer Miller and Harwell that she would perform fellatio for money, with a more specific price to be determined somewhere between 1 penny and $20.00, but definitely not more than $20.00. . . . Consequently, we hold that the State presented sufficient evidence to support Harwell's conviction for prostitution.

Questions for Discussion

1. Did Harwell possess the intent to exchange sex for money? Can there be a legal agreement given that Officer Miller did not possess an intent to exchange sex for money?

2. Absent an agreement as to price, are we able to conclude that Harwell intended to exchange sex for money? Was her consent conditional on the amount of money offered by Officer Miller?

3. Do you believe that there was a "meeting of the minds or an agreement to exchange sex for money"?

4. Was Harwell, in reality, convicted for being a prostitute rather than for engaging in an act of prostitution?

See more cases on the study site: Harwell v. State, *www.sagepub.com/lippmanccl3e.*

Cases and Comments

1. ***Decriminalization.*** Audrey James, twenty-one, and Laverne McCray, twenty-three, were first offenders, arrested and charged with prostitution. Both were enrolled in business school and worked as prostitutes in order to pay their tuition. They filed a petition under a provision of New York criminal procedure requesting the judge to dismiss the charges "in the interests of justice." This motion was opposed by the District Attorney, despite the fact that first offenders were traditionally released without punishment. Judge Stanley Gartenstein of the Criminal Court of New York City noted that prostitution is a victimless crime that the law had proven unable to control. He observed that this was a testimony to the fact that "morality cannot be legislated." In addition, Judge Gartenstein noted that the "the real victim of prostitution is the prostitute herself and the real criminal, her pimp, who keeps her virtually enslaved" and asked whether it "might . . . be more productive to focus society's efforts on the parasite rather than the host." Judge Gartenstein concluded that the time each of the defendants already had spent in jail was sufficient to communicate that prostitution was "self-destructive and degrading." The traditional punishment of the assessment of a fine with the alternative of jail time for nonpayment merely increased dependency on pimps and drove women back to the streets to earn even more money.

Releasing these two young women without penalty might save them from this self-perpetuating cycle of prostitution. Judge Gartenstein concluded by arguing that resources and money were being spent on prostitution and gambling offenses, which detracted from efforts to combat violent crime. The average cost for processing a prostitution case was $505.11, and for a gambling case, $625.78. This compared to $464.27 for a kidnapping and $470.49 for the average robbery. See *People v. James*, 415 N.Y.2d 342 (Crim. Ct. N.Y.C. 1979).

2. *Payment.* In *State v. Henderson,* the defendant approached a female undercover police officer who was working as a prostitution decoy and offered to exchange bubble gum for sex. The defendant was convicted and sentenced to six months' probation and fined $100. Henderson was convicted under a statute that prohibited solicitation of "another to engage . . . in sexual activity for hire." The appellate court concluded that "[i]n view of the statement Henderson made to [O]fficer St. Clair after he was arrested, the trial court could reasonably find that Henderson made a serious, if unusual, offer to [Officer] Briggs to engage in sex with him in exchange for some bubble gum." As the trial court noted, "bubble gum has economic value, even if that value is slight. . . . There is nothing . . . specifying the amount of compensation" required to constitute the offense of solicitation for prostitution. "Nor is there anything in the statute requiring that the proposed transaction be commercially reasonable." Do you agree? See *State v. Henderson,* 2007-Ohio-5367 (Ohio Ct. App. 2007).

3. *Sex Trafficking: Trafficking of Women.* International criminal gangs are estimated by the United Nations to generate $7 billion a year from the illegal trafficking of women. There are thought to be between 200,000 and 500,000 sex workers in Europe, two-thirds of whom are from Eastern Europe and the former Soviet Union and one-third from the developing world. The United States has estimated that between 14,500 and 17,500 individuals are trafficked into the country annually for sexual exploitation and forced labor. Globally, the estimate is that between 600,000 and 800,000 individuals are illegally transported across borders for various forms of exploitation. Studies indicate that roughly seventy percent of these individuals participate in the commercial sex trade. Europe, Asia, and the Pacific are the destination of roughly eighty percent of the victims of trafficking.

Women have experienced particularly high rates of unemployment and poverty in the collapsing economies and are susceptible to false promises of domestic and secretarial jobs in Canada, Western Europe, Japan, and the United States. Once the women reach their destination, their passports are confiscated, and they are told that they must work to pay off the cost of their travel and other expenses. The women are typically sold by the traffickers to pimps and bar owners who, in turn, may resell them to other individuals in the sex trade. According to the United Nations, Asian prostitutes in North America and Japan can sell for as much as $20,000. The United Nations reports that Russian prostitutes in Germany can earn about $7,500 a month, most of which is kept by the pimp or bar owner.

Women who resist are typically subjected to beatings and rape. Escape is difficult for the women. They typically lack language skills and money and may fear that in the event that they contact the police, they will be prosecuted for prostitution. Many of these women are from strict and conservative cultures and anticipate that in the event that the authorities send them home, they will suffer rejection and discrimination. The women find that there is little choice other than to cooperate with their captors and continue to risk contracting sexually transmitted diseases.

The American organization Human Rights Watch issued a report on the sex industry in Bosnia, formerly part of Yugoslavia. The report finds that women who managed to contact the police found themselves prosecuted for prostitution and that local authorities generally were connected with traffickers and shared in the profits and, as a consequence, took no action against the individuals who kidnapped and exploited the women.

In 2000, the United Nations adopted the Protocol to Prevent, Suppress and Punish Trafficking in Persons, Especially Women and Children. Countries that sign this agreement agree to prevent and combat trafficking, to protect and assist victims of trafficking, and to promote cooperation among nation-states to meet these goals. Signatories are to criminally punish traffickers and to consider providing support services for victims.

In 2000, the U.S. Congress passed the Victims of Trafficking and Violence Protection Act, which proclaims that the degrading institution of slavery continues in the world, including the sexual exploitation of women. The detailed law punishes by twenty years in prison individuals who provide or obtain the labor of another through threat of serious harm or restraint or who traffic in persons who are to be subjected to slavery or forced labor. Aggravated forced labor and sexual trafficking and the sexual trafficking of minors are subject to a punishment of up to life in prison. Various protections are afforded to the victims of sexual trafficking. The Justice Department reports that in the two years following passage of the law, the government has charged, convicted, or secured sentences against sixty-five traffickers in fourteen sex-trafficking and abuse cases. Several states have laws that enable the victims of sexual trafficking to petition a court to expunge their convictions for prostitution.

OBSCENITY

Until the eighteenth century, **obscenity** in England was punished before religious courts. In 1727, royal judges asserted jurisdiction over obscenity, asserting that possession of such material constituted an offense against the peace and weakened the "bonds of civil society, virtue, and morality."[36]

In *Roth v. United States,* the U.S. Supreme Court held that obscenity was not constitutionally protected speech or press within the First Amendment, reasoning that this form of expression is "utterly without redeeming social importance."[37] The lack of protection afforded to obscenity is based on several public policy considerations:

- *Protection of the Community.* A community is entitled to protect itself against threats to the moral fabric of society.
- *Antisocial Conduct.* Obscenity causes antisocial conduct.
- *Women.* Obscenity degrades women.
- *Communication.* Obscenity produces a sexual, rather than mental, response and is a form of sexual communication rather than the expression of ideas.

These assertions all have been challenged. For instance, the government *Report of the Commission on Obscenity and Pornography,* in 1970, concluded that exposure to explicit sexual materials does not play a significant role in causing delinquent or criminal behavior.

In *Miller v. California,* in 1973, the U.S. Supreme Court affirmed that obscene material is not protected by the First Amendment and established a three-part test for obscenity. Obscenity (as noted in Chapter 2) is defined as a description or representation of sexual conduct that taken, as a whole, by the average person applying contemporary community standards:[38]

- *Prurient Interest.* Appeals to the prurient interest in sex (an obsession with obscene, lewd, or immoral matters).
- *Offensive.* Portrays sex in a patently offensive way.
- *Value.* Applying a reasonable person standard, lacks serious literary, artistic, political, or scientific value when taken as a whole.

In *New York v. Ferber,* the U.S. Supreme Court held that **child pornography** could be prohibited despite the fact that the material did not satisfy the *Miller* standard. The Court upheld a New York law that prohibited the depiction of a child under sixteen years old in a "sexual performance," defined as engaging in actual or simulated sexual activity or the lewd display of the genitals. The sexual performance was considered criminal under the statute regardless of whether it possessed literary, artistic, political, or scientific value.[39]

It is illegal in every state to buy, sell, exhibit, produce, advertise, distribute, or possess with the intent to distribute obscene material or illegal child pornography. The offense of **lewdness** involves conduct that is obscene, such as willfully exposing the genitals of one person to another in a public place for the purpose of arousing or gratifying the sexual desire of either individual. **Indecent exposure** generally entails an act of public indecency, including sexual intercourse, exposure of the sexual organs, a lewd appearance in a state of partial or complete nudity, or a lewd caress or fondling of the body of another person.

Several municipalities have expanded the definition of obscenity to include other forms of communication that are considered harmful. An Indianapolis ordinance, for instance, was ruled unconstitutional that prohibited the portrayal of women as sexual objects who enjoy pain or humiliation or who experience sexual pleasure in being raped or who are presented as sexual objects for domination or violation. The ordinance was ruled unconstitutional by the Seventh Circuit Court of Appeals, which held that Indianapolis was improperly penalizing speech based on the content of the message: Communication depicting women in the approved way was lawful no matter how sexually explicit, and speech portraying women in the unapproved way was unlawful whatever the literary or artistic value.[40]

In the next case in the text, *Brown v. Entertainment Merchants Association,* the U.S. Supreme Court considered the constitutionality of a California law regulating violent and sexually explicit video games based on evidence that exposure to these games is related to harmful effects on minors. Does the First Amendment protect materials that California reasonably believes cause aggressive behavior by minors? Why must society wait until a young person commits a crime to intervene?

May California prohibit juveniles from purchasing or renting violent video games?

Brown v. Entertainment Merchants Association, __ U.S.___, 131 S. Ct. 2729 (2011). Opinion by: Scalia, J.

Issue

We consider whether a California law imposing restrictions on violent video games comports with the First Amendment.

Facts

California Assembly Bill 1179 (2005), Cal. Civ. Code Ann. §§1746–1746.5, prohibits the sale or rental of "violent video games" to minors unless accompanied by an adult, and requires their packaging to be labeled "18." The Act covers games "in which the range of options available to a player includes killing, maiming, dismembering, or sexually assaulting an image of a human being, if those acts are depicted" in a manner that "[a] reasonable person, considering the game as a whole, would find appeals to a deviant or morbid interest of minors," that is "patently offensive to prevailing standards in the community as to what is suitable for minors," and that "causes the game, as a whole, to lack serious literary, artistic, political, or scientific value for minors." Violation of the Act is punishable by a civil fine of up to $1,000.

Respondents, representing the video-game and software industries, brought a pre-enforcement challenge to the Act in the United States District Court for the Northern District of California. That court concluded that the Act violated the First Amendment and permanently enjoined its enforcement.

California correctly acknowledges that video games qualify for First Amendment protection. The Free Speech Clause exists principally to protect discourse on public matters, but we have long recognized that it is difficult to distinguish politics from entertainment, and dangerous to try. "Everyone is familiar with instances of propaganda through fiction. What is one man's amusement, teaches another's doctrine." Like the protected books, plays, and movies that preceded them, video games communicate ideas—and even social messages—through many familiar literary devices (such as characters, dialogue, plot, and music) and through features distinctive to the medium (such as the player's interaction with the virtual world). That suffices to confer First Amendment protection. Under our Constitution, "esthetic and moral judgments about art and literature . . . are for the individual to make, not for the Government to decree, even with the mandate or approval of a majority." And whatever the challenges of applying

the Constitution to ever-advancing technology, "the basic principles of freedom of speech and the press, like the First Amendment's command, do not vary" when a new and different medium for communication appears.

The most basic of those principles is this: "[A]s a general matter . . . government has no power to restrict expression because of its message, its ideas, its subject matter, or its content." There are of course exceptions. "'From 1791 to the present' . . . the First Amendment has 'permitted restrictions upon the content of speech in a few limited areas,' and has never 'include[d] a freedom to disregard these traditional limitations.'" These limited areas—such as obscenity, incitement, and fighting words—represent "well-defined and narrowly limited classes of speech, the prevention and punishment of which have never been thought to raise any Constitutional problem."

California does not argue that it is empowered to prohibit selling offensively violent works to adults. . . . Instead, it wishes to create a wholly new category of content-based regulation that is permissible only for speech directed at children. That is unprecedented and mistaken. "[M]inors are entitled to a significant measure of First Amendment protection" . . . . No doubt a State possesses legitimate power to protect children from harm, but that does not include a free-floating power to restrict the ideas to which children may be exposed. "Speech that is neither obscene . . . nor subject to some other legitimate proscription cannot be suppressed solely to protect the young from ideas or images that a legislative body thinks unsuitable for them."

California's argument would fare better if there were a longstanding tradition in this country of specially restricting children's access to depictions of violence, but there is none. Certainly the books we give children to read—or read to them when they are younger—contain no shortage of gore. Grimm's Fairy Tales, for example, are grim indeed. As her just deserts for trying to poison Snow White, the wicked queen is made to dance in red hot slippers "till she fell dead on the floor, a sad example of envy and jealousy." Cinderella's evil stepsisters have their eyes pecked out by doves. And Hansel and Gretel (children!) kill their captor by baking her in an oven.

High-school reading lists are full of similar fare. Homer's Odysseus blinds Polyphemus the Cyclops by grinding out his eye with a heated stake. In the *Inferno*, Dante and Virgil watch corrupt politicians struggle to stay submerged beneath a lake of boiling pitch, lest they be skewered by devils above the surface. And Golding's *Lord of the Flies* recounts how a schoolboy called Piggy is savagely murdered by other children while marooned on an island. . . .

California claims that video games present special problems because they are "interactive," in that the player participates in the violent action on screen and determines its outcome. The latter feature is nothing new: Since at least the publication of *The Adventures of You: Sugarcane Island* in 1969, young readers of choose-your-own-adventure stories have been able to make decisions that determine the plot by following instructions about which page to turn to. As for the argument that video games enable participation in the violent action, that seems to us more a matter of degree than of kind. As Judge Posner of the Seventh Circuit Court of Appeals has observed, all literature is interactive. "[T]he better it is, the more interactive. Literature when it is successful draws the reader into the story, makes him identify with the characters, invites him to judge them and quarrel with them, to experience their joys and sufferings as the reader's own."

Justice Alito has done considerable independent research to identify video games in which "the violence is astounding." "Victims are dismembered, decapitated, disemboweled, set on fire, and chopped into little pieces. . . . Blood gushes, splatters, and pools." Justice Alito recounts all these disgusting video games in order to disgust us—but disgust is not a valid basis for restricting expression. And the same is true of Justice Alito's description of those video games he has discovered that have a racial or ethnic motive for their violence—"'ethnic cleansing' [of] . . . African Americans, Latinos, or Jews." To what end does he relate this? Does it somehow increase the "aggressiveness" that California wishes to suppress? Who knows? But it does arouse the reader's ire, and the reader's desire to put an end to this horrible message. Thus, ironically, Justice Alito's argument highlights the precise danger posed by the California Act: that the ideas expressed by speech—whether it be violence, or gore, or racism—and not its objective effects, may be the real reason for governmental proscription.

Because the Act imposes a restriction on the content of protected speech, it is invalid unless California can demonstrate that it passes strict scrutiny—that is, unless it is justified by a compelling government interest and is narrowly drawn to serve that interest. The State must specifically identify an "actual problem" in need of solving, and the curtailment of free speech must be actually necessary to the solution. That is a demanding standard. "It is rare that a regulation restricting speech because of its content will ever be permissible."

California cannot meet that standard. . . . The State's evidence is not compelling. California relies primarily on the research of Dr. Craig Anderson and a few other research psychologists whose studies purport to show a connection between exposure to violent video games and harmful effects on children. These studies have been rejected by every court to consider them, and with good reason: They do not prove that violent video games cause minors to act aggressively. Instead, "[n]early all of the research is based on correlation, not evidence of causation, and most of the studies suffer from significant, admitted flaws in methodology." They show at best some correlation between exposure to violent entertainment and minuscule real-world effects, such as children's feeling more aggressive or making louder noises in the few minutes after playing a violent game than after playing a nonviolent game.

Even taking for granted Dr. Anderson's conclusions that violent video games produce some effect on children's feelings of aggression, those effects are both small and indistinguishable from effects produced by other media. In his testimony in a similar lawsuit, Dr. Anderson admitted that the "effect sizes" of children's exposure to violent video games are "about the same" as that produced by their exposure to violence on television. And he admits that the same effects have been found when children watch cartoons starring Bugs Bunny or the Road Runner, or when they play video games like *Sonic the Hedgehog* that are rated "E" or even when they "vie[w] a picture of a gun."

Of course, California has (wisely) declined to restrict Saturday morning cartoons, the sale of games rated for young children, or the distribution of pictures of guns. The consequence is that its regulation is wildly underinclusive when judged against its asserted justification, which in our view is alone enough to defeat it. Underinclusiveness raises serious doubts about whether the government is in fact pursuing the interest it invokes, rather than disfavoring a particular speaker or viewpoint. Here, California has singled out the purveyors of video games for disfavored treatment—at least when compared to booksellers, cartoonists, and movie producers—and has given no persuasive reason why.

The Act is also seriously underinclusive in another respect. . . . The California Legislature is perfectly willing to leave this dangerous, mind-altering material in the hands of children so long as one parent (or even an aunt or uncle) says it's OK. And there are not even any requirements as to how this parental or avuncular relationship is to be verified; apparently the child's or putative parent's, aunt's, or uncle's say-so suffices. That is not how one addresses a serious social problem.

California claims that the Act is justified in aid of parental authority: By requiring that the purchase of violent video games can be made only by adults, the Act ensures that parents can decide what games are appropriate. At the outset, we note our doubts that punishing third parties for conveying protected speech to children just in case their parents disapprove of that speech is a proper governmental means of aiding parental authority. Accepting that position would largely undermine the rule that "only in relatively narrow and well-defined circumstances may government bar public dissemination of protected materials to [minors]."

The video-game industry has in place a voluntary rating system designed to inform consumers about the content of games. The system, implemented by the Entertainment Software Rating Board (ESRB), assigns age-specific ratings to each video game submitted: EC (Early Childhood); E (Everyone); E10+ (Everyone 10 and older); T (Teens); M (17 and older); and AO (Adults Only—18 and older). The Video Software Dealers Association encourages retailers to prominently display information about the ESRB system in their stores; to refrain from renting or selling adults-only games to minors; and to rent or sell "M" rated games to minors only with parental consent. In 2009, the Federal Trade Commission (FTC) found that, as a result of this system, "the video game industry outpaces the movie and music industries" in "(1) restricting target-marketing of mature-rated products to children; (2) clearly and prominently disclosing rating information; and (3) restricting children's access to mature-rated products at retail." This system does much to ensure that minors cannot purchase seriously violent games on their own, and that parents who care about the matter can readily evaluate the games their children bring home. Filling the remaining modest gap in concerned-parents' control can hardly be a compelling state interest.

And finally, the Act's purported aid to parental authority is vastly overinclusive. Not all of the children who are forbidden to purchase violent video games on their own have parents who care whether they purchase violent video games. While some of the legislation's effect may indeed be in support of what some parents of the restricted children actually want, its entire effect is only in support of what the State thinks parents ought to want. . . .

Holding

We have no business passing judgment on the view of the California Legislature that violent video games (or, for that matter, any other forms of speech) corrupt the young or harm their moral development. Our task is only to say whether or not such works constitute a "well-defined and narrowly limited clas[s] of speech, the prevention and punishment of which have never been thought to raise any Constitutional problem," and if not, whether the regulation of such works is justified by that high degree of necessity we have described as a compelling state interest (it is not). Even where the protection of children is the object, the constitutional limits on governmental action apply.

As a means of protecting children from portrayals of violence, the legislation is seriously underinclusive, not only because it excludes portrayals other than

video games, but also because it permits a parental or avuncular veto. And as a means of assisting concerned parents it is seriously overinclusive because it abridges the First Amendment rights of young people whose parents (and aunts and uncles) think violent video games are a harmless pastime. . . .

Dissenting, *Breyer, J.*

California's law imposes no more than a modest restriction on expression. The statute prevents no one from playing a video game, it prevents no adult from buying a video game, and it prevents no child or adolescent from obtaining a game provided a parent is willing to help. All it prevents is a child or adolescent from buying, without a parent's assistance, a gruesomely violent video game of a kind that the industry itself tells us it wants to keep out of the hands of those under the age of 17. . . . .

The interest that California advances in support of the statute is compelling. As this Court has previously described that interest, it consists of both (1) the "basic" parental claim "to authority in their own household to direct the rearing of their children," which makes it proper to enact "laws designed to aid discharge of [parental] responsibility," and (2) the State's "independent interest in the well-being of its youth."

As to the need to help parents guide their children, the Court noted in 1968 that "parental control or guidance cannot always be provided." Today, 5.3 million grade-school-age children of working parents are routinely home alone. Thus, it has, if anything, become more important to supplement parents' authority to guide their children's development.

As to the State's independent interest, we have pointed out that juveniles are more likely to show a "lack of maturity'" and are "more vulnerable or susceptible to negative influences and outside pressures," and that their "character . . . is not as well formed as that of an adult." And we have therefore recognized "a compelling interest in protecting the physical and psychological well-being of minors. . . ."

There are many scientific studies that support California's views. Social scientists, for example, have found causal evidence that playing these games results in harm. Longitudinal studies, which measure changes over time, have found that increased exposure to violent video games causes an increase in aggression over the same period. Experimental studies in laboratories have found that subjects randomly assigned to play a violent video game subsequently displayed more characteristics of aggression than those who played nonviolent games. . . . .

I can find no "less restrictive" alternative to California's law that would be "at least as effective." The . . . voluntary system has serious enforcement gaps. . . . . [A]s of the FTC's most recent update to Congress, 20% of those under 17 are still able to buy M-rated video games, and, breaking down sales by store, one finds that this number rises to nearly 50% in the case of one large national chain. . . . The industry also argues for an alternative technological solution, namely "filtering at the console level." But it takes only a quick search of the Internet to find guides explaining how to circumvent any such technological controls. . . .

The upshot is that California's statute, as applied to its heartland of applications (*i.e.,* buyers under 17; extremely violent, realistic video games), imposes a restriction on speech that is modest at most.

I add that the majority's different conclusion creates a serious anomaly in First Amendment law. [A] State can prohibit the sale to minors of depictions of nudity; today the Court makes clear that a State cannot prohibit the sale to minors of the most violent interactive video games. But what sense does it make to forbid selling to a 13-year-old boy a magazine with an image of a nude woman, while protecting a sale to that 13-year-old of an interactive video game in which he actively, but virtually, binds and gags the woman, then tortures and kills her? What kind of First Amendment would permit the government to protect children by restricting sales of that extremely violent video game only when the woman—bound, gagged, tortured, and killed—is also topless? . . . [E]xtreme violence, where interactive, and without literary, artistic, or similar justification, can prove at least as, if not more, harmful to children as photographs of nudity. And the record here is more than adequate to support such a view. . . .

Questions for Discussion

1. Summarize Justice Scalia's reasons for concluding that the California statute is both "underinclusive." and "overinclusive."

2. Do you agree with Justice Breyer that the California law should be upheld as a constitutionally valid means of advancing the interest in protecting children from the uniquely destructive effects of violent video games?

3. Are the video games different than other media formats?

4. How do Justices Scalia and Breyer differ on their evaluation of the social science studies on the relationship between video games and violence?

5. Should the Supreme Court be involved in overturning a decision by the California legislature to regulate juvenile's access to video games?

See more cases on the study site: **Brown v. Entertainment Merchants Association,** *www.sagepub.com/lippmanccl3e.*

CRIME IN THE NEWS

In 2010, eighth grader Margarite held up her cell phone and snapped a photo that she then sent to her new boyfriend, Isaiah. They broke up a few weeks later, and Isaiah forwarded the photo to a classmate who pasted a text message on the photo. The message read: "Ho Alert!" . . . "If you think this girl is a whore then text this to all your friends." She pressed send, and in less than twenty-four hours, thousands of students in a suburb of Olympia, Washington, had received and, in turn, had forwarded the photo. In a brief period of time, several students found themselves in handcuffs and charged with the production and transmission of child pornography.

These were not the first arrests for sexting. In Florida, Phillip Albert, then eighteen, was convicted of transmitting child pornography after forwarding naked photos of his ex-girlfriend to his friends. Albert was required to register as a sex offender, lost friends, and was kicked out of school. The Florida law makes it a felony for any individual "who knew or reasonably should have known that he or she was transmitting child pornography" to text such material.

The Iowa Supreme Court upheld the conviction of Jorge Canal, age eighteen, for "knowingly disseminating obscene material to a minor" after he sent a fourteen-year-old a photo of his erect penis. Two teenagers in Florida were convicted of the production of child pornography after taking nude photos of themselves engaged in sexual behavior. A Florida court upheld a finding of delinquency against the female despite the fact that the photo had never been shared with anyone (*A.H. v. State,* 949 So. 2d 234 (Fla. Dist. Ct. App. 2007)). There is a fear that these photos will be widely disseminated and end up on pornographic websites.

The rhetoric surrounding sexting undoubtedly has become overheated; and the topic has become part and parcel of the debate over sex education, birth control, and teenage pregnancies. There are, nonetheless, at least two instances in which teenagers whose photos had been disseminated committed suicide.

Figures on the number of young adults engaged in sexting vary widely. A 2008 study by the National Campaign to Prevent Teen and Unplanned Pregnancy finds that twenty percent of thirteen-to nineteen-year-olds have sent or posted "nude or semi-nude photos or videos of themselves," and thirty-nine percent have sent or posted sexually suggestive messages via text, e-mail, or instant message.

The prosecutions of young adults for sexting is subject to criticism because child pornography laws are intended to protect children against being exploited by adults, and these prosecutions are labeling young people as "pornographers." Very few young people, of course, are jailed for sexting. They nonetheless may find themselves required to register as sex offenders, may be placed under judicial supervision for a number of years, and may have a felony conviction on their record. Commentators who support laws against the texting of child pornography argue that even if the law infrequently is enforced, it is important to keep these laws on the books to highlight that sexting is unacceptable behavior.

Opponents of prosecuting and punishing sexting among teenagers claim that the law is punishing young people for engaging in what most juveniles consider to be a culturally acceptable practice. Thousands of teenagers engage in this behavior, and it is impossible to prosecute every student who sends a sexual image over his or her cell phone or computer. Singling out a small number of students for prosecution is simply unfair.

More than twenty-one states have addressed sexting. Vermont and Illinois teenagers who sext will be subject to a delinquency hearing before a family court judge rather than being brought before a criminal court judge. Utah has reduced penalties on teenagers, and fourteen other states are considering laws that distinguish between young people who sext and adult pornographers. The Nebraska law removes criminal penalties from an individual under eighteen who sends his or her own photo to a willing recipient at least fifteen years old. An individual of whatever age who sends the photo of another person to other individuals is punished with felony child pornography charge and five years in prison. How should the criminal law address sexting by juveniles?

CRUELTY TO ANIMALS

For an international perspective on this topic, visit the study site.

The crime of cruelty to animals is recognized as an offense against public order and decency. These laws were originally based on the belief that respect for animals helped to teach people to act with sensitivity and regard for their fellow citizens, particularly the most vulnerable members of human society. Today, laws against cruelty to animals also reflect the emotional attachment that people have toward their pets and other animals and the increasingly common belief that animals experience pleasure and pain and possess rights. Violence toward animals is also thought to encourage aggression toward human beings. Prior to 1990, only six states punished cruelty to animals as a felony. As of 2011, forty-six states had felony provisions in their animal cruelty laws.

See more cases on the study site: Hall v. Indiana, *www.sagepub.com/lippmanccl3e.*

CHAPTER SUMMARY

Crimes against public order and morals have traditionally been viewed as of secondary importance. These misdemeanor offenses are disposed of in summary trials and carry modest punishments. Offenses such as disorderly conduct, however, constitute a significant percentage of arrests and prosecutions, and the treatment of these arrestees helps to shape perceptions of the criminal justice system.

Crimes against public order and morals were historically used to remove the unemployed and political agitators from cities and towns. Today, we are seeing a renewed emphasis on these offenses by municipalities. An increasing number of middle-class individuals are moving into urban areas and find themselves sharing their neighborhood with prostitutes, drug addicts, alcoholics, and gangs. The so-called broken windows theory reasons that the tolerance of small-scale, quality-of-life crimes leads to neighborhood deterioration and facilitates the growth of crime.

Individual disorderly conduct is directed at a broad range of conduct that (1) risks or causes public inconvenience, annoyance, or alarm and (2) risks causing or does cause a breach of the peace. The Model Penal Code punishes engaging in fighting or threatening violent behavior; creating unreasonable noise, offensive utterances, or gestures; or creating a hazardous or physically offensive condition. A riot is group disorderly conduct and entails participating with others in tumultuous and violent conduct with the intent of causing a grave risk of public alarm.

Two controversial quality-of-life crimes are vagrancy, defined in the common law as wandering the streets with no apparent means of earning a living, and loitering, a related offense that is defined at common law as standing in public with no apparent purpose. These broad statutes have historically been used against individuals based on their status as "undesirables." Vagrancy and loitering statutes have been found void for vagueness by the U.S. Supreme Court. Many states have responded by adopting the approach of the Model Penal Code and punish individuals whose conduct warrants "alarm for the safety of persons or property in the vicinity." Municipal ordinances directed against the homeless and gangs have been challenged as void for vagueness, and laws against the homeless also have been attacked as punishing individuals based on their economic status. These legal actions have generally proven unsuccessful.

Crimes against public order and decency have been criticized as punishing "victimless crimes," or consensual offenses that the individuals involved do not view as harmful. Other commentators argue that the law is properly concerned with private morality, and they challenge the notion that these offenses against public order and decency do not result in harm to individuals and to society. A particular object of debate is the criminalization of prostitution, the exchange of sexual acts for money or some other item of value. Obscenity is another offense that is claimed to be a victimless crime, which some claim creates social harm. There is particular controversy concerning efforts to extend obscenity to include depictions and descriptions of violence, particularly when directed to children. Another growing area of concern is the protection of animals.

CHAPTER REVIEW QUESTIONS

1. List specific acts constituting disorderly conduct.

2. What is the difference between disorderly conduct and riot?

3. Distinguish between vagrancy and loitering.

4. What constitutional objections have been raised to vagrancy and loitering statutes?

5. What was the constitutional basis for the Supreme Court's holding Chicago's Gang Congregation Ordinance unconstitutional? Explain the reasoning of the Supreme Court.

6. What are the elements of the crime of prostitution?

7. How does the U.S. Supreme Court define obscenity?

8. Considering the cases you read on homelessness and gangs, does the broken windows theory pose a threat to civil liberties? Support your answer with examples from the textbook.

9. Why do some commentators argue that the criminal law is overreaching?

10. Are prostitution and soliciting for prostitution victimless crimes?

11. Define the legal standard for obscenity and child pornography.

12. Should individuals be held criminally liable for cruelty to "wild" animals?

LEGAL TERMINOLOGY

adultery

bigamy

breach of the peace

broken windows theory

child pornography

crimes against public order and morality

crimes against the quality of life

disorderly conduct

fornication

immorality crimes

indecent exposure

keeping a place of prostitution

lewdness

living off prostitution

loitering

masturbation for hire

obscenity

pandering

pimping

promoting prostitution   riot   unlawful assembly

prostitution   rout   vagrancy

public indecencies   solicitation for prostitution

CRIMINAL LAW ON THE WEB

Log on to the Web-based student study site at www.sagepub.com/lippmanccl3e to assist you in completing the Criminal Law on the Web exercises, as well as for additional features such as podcasts, Web quizzes, and video links.

1. Read about the Chicago gang ordinance.
2. Consider the claim that homelessness is being "criminalized."
3. Compare cruelty-to-animal statutes in several states.

BIBLIOGRAPHY

American Law Institute, *Model Penal Code and Commentaries,* vol. 3, pt. 2 (Philadelphia: American Law Institute, 1980), pp. 309–523. A discussion of the common law background and various state statutes addressing offenses against public order and public decency.

Rollin M. Perkins and Ronald N. Boyce, *Criminal Law,* 3rd ed. (Mineola, NY: Foundation Press, 1982), pp. 453–498, 526–557. An introduction to the basic law of crimes against public order and morality.

16 CRIMES AGAINST THE STATE

Was American citizen John Walker Lindh an unlawful enemy combatant?

As part of his al Qaeda training, Lindh participated in "terrorist training courses in, among other things, weapons, orientating, navigation, explosives and battlefield combat." This training included the use of "shoulder weapons, pistols, and rocket-propelled grenades, and the construction of Molotov cocktails." During his stay at al Farooq, Lindh met personally with bin Laden, "who thanked him and other trainees for taking part in jihad." He also met with a senior al Qaeda official, Abu Mohammad Al-Masri, who inquired whether Lindh was interested in traveling outside Afghanistan to conduct operations against the United States and Israel. Lindh declined Al-Masri's offer in favor of going to the front lines to fight. It is specifically alleged that Lindh swore allegiance to jihad in June or July 2001.

Learning Objectives

1. Know the definition of treason and the constitutional requirements to convict an individual of treason.

2. State the definition of sedition.

3. Summarize the elements of the crime of sabotage.

4. Know the elements of the crime of espionage.

5. Distinguish between domestic and international terrorism.

6. Understand the reason for punishing material support for terrorists and material support for terrorist organizations.

7. Know the requirements for lawful combatants and the difference between lawful and unlawful combatants.

8. Understand the development of international criminal law and understand the concept of extraterritorial jurisdiction over international crimes.

INTRODUCTION

A significant percentage of the Europeans who settled in the American colonies were fleeing religious or political persecution and understandably developed a suspicion of government. This distrust was enhanced by the colonists' unhappy experiences with the often-repressive policies of the British authorities. There was almost uniform agreement to build the new American democracy on a foundation of strong limits on official authority along with a commitment to individual freedom. The colonists were also reluctant to adopt the type of harsh legislation that had been used by English monarchs to stifle dissent and criticism. Nevertheless, there was the reality that the United States confronted a threat from European countries eager to acquire additional territory in North America. A number of Americans and most Canadians also continued to harbor deep loyalties to England. This dictated that the United States put various laws in place to protect the government and people from attack. In this chapter, we examine these **crimes against the state**:

- *Treason.* Involvement in an attack on the United States.
- *Sedition.* A written or verbal communication intended to create disaffection, hatred, or contempt toward the U.S. government.

- *Sabotage.* Destruction of national defense materials.
- *Espionage.* Conveying information to a foreign government with the intent of injuring the United States.

Recent events have also led to the development of various counterterrorist laws, including the punishment of materially assisting terrorism. In reading about these counterterrorism offenses, you will see that these statutes are a modern and updated version of treason, sedition, sabotage, and espionage.

TREASON

English royalty prosecuted and convicted critics of the monarchy for **treason**. Monarchs intentionally avoided writing down the requirements of treason in a statute in order to permit the crime to be applied against all varieties of critics. Parliament was finally able to mobilize enough power in 1352 to limit the power of the king, and it forced Edward III to agree to a Declaration Which Offenses Shall Be Adjudged Treason.

British officials in the American colonies applied the law on treason against rebellious servants and government critics, who typically were punished by "drawing and hanging." The drafters of the U.S. Constitution harbored bitter memories of the abusive use of the law on treason against critics of the colonial regime. At the same time, the drafters of the U.S. Constitution were conscious of the need to protect the newly independent American states against the threat posed by individuals whose loyalty remained with England and against European states that desired to expand their territorial presence in North America.

How could these various concerns be balanced against one another? The decision was made for the Constitution to clearly set forth the definition of treason, the proof necessary to establish the offense, and the appropriate punishment. James Madison explained in the *Federalist Papers* that as "new-fangled and artificial treasons have been the great engines by which violent factions . . . have usually wreaked their . . . malignity on each other, the convention have, with great judgment opposed a barrier to this peculiar danger, by inserting a constitutional definition of the crime. . . ."[1] Article III, Section 3 of the U.S. Constitution provides that treason against the United States "shall consist only in levying War against them, or in adhering to their Enemies giving them Aid and Comfort. No Person shall be convicted of Treason unless on the Testimony of two Witnesses to the same overt Act or on Confession in open Court." Congress is also constitutionally prohibited from adopting the policy practiced in England of extending penalties beyond the individual offender to members of his or her family.

In summary, the United States adopted a law against treason while clearly limiting the definition of the offense in the Constitution.

Criminal Act and Criminal Intent

Treason is the only crime defined in the U.S. Constitution. Treason against the United States is a federal crime and may not be prosecuted by the states. Various states, such as California, prohibit treason against the state government.

The Constitution limits the *actus reus* of treason to individuals engaged in armed opposition to the government or in providing aid and comfort to the enemy.

- Levying war against the United States.
- Giving aid and comfort to the enemy.

Supreme Court Justice Robert Jackson, in *Cramer v. United States,* clarified that levying war consists of taking up arms and that giving aid and comfort involves concrete and tangible assistance. Justice Jackson stressed that a citizen may "intellectually or emotionally" favor the enemy or may "harbor sympathies or convictions disloyal" to the United States, but absent the required *actus reus,* there is no treason. He explained that an individual gives aid and comfort to the enemy by such acts as fomenting strikes in defense plants, charging exorbitant prices for essential armaments, providing arms to the enemy, or engaging in countless other acts that "impair our cohesion and diminish our strength."[2]

In a treason prosecution of sailors who seized, equipped, and armed a ship with the intent of attacking the federal government, Supreme Court Justice Joseph Fields observed that treason may be directed to the overthrow of the U.S. government throughout the country or only in certain states or in selected localities.[3] There is also no requirement that the enemy be shown to have benefited by the assistance provided by the accused.

The Constitution requires that treason be clearly established by the prosecution. This protects individuals against convictions based on passion, prejudice, or false testimony.

- Two witnesses must testify that the defendant committed the same overt act of treason, or
- the accused must make a confession in open court.

The *mens rea* of treason is an "intent to betray" the United States. Justice Jackson observed that "if there is no intent to betray there is no treason." Proof of the defendant's treasonous intent is not limited to the testimony of two witnesses. The required intent may be established by the testimony of a number of witnesses concerning the defendant's statements or behavior. Do not confuse motive with intent. An act that clearly assists and is intended to assist the enemy is treason, despite the fact that the defendant may be motivated by profit, anger, or personal opposition to war rather than by a belief in the justice of the enemy's cause.

In *Cramer,* Justice Jackson cautioned that treason is "one of the most intricate of crimes" and that the U.S. Constitution "gives a superficial appearance of clarity and simplicity which proves illusory when it is put to practical application. . . . The little clause is packed with controversy and difficulty."

Prosecuting Treason

The United States has brought only a handful of prosecutions for treason, and most offenders have had their death sentences modified or have received full pardons from the president. Courts have also been vigilant in insuring that the rights of defendants are protected. Justice Jackson observed that the United States has "managed to do without treason prosecutions to a degree that probably would be impossible except [where] a people was singularly confident of external security and internal stability."

Cramer v. United States, in 1945, is the most important treason case decided by the U.S. Supreme Court. A team of eight German saboteurs was transported across the Atlantic in two submarines and secretly put ashore in New York and Florida with the intent of engaging in acts of sabotage designed to impede the U.S. war effort and to undermine morale. Saboteurs Werner Thiel and Edward Kerling contacted a former friend of Thiel's in New York, Anthony Cramer. Cramer was subsequently charged and convicted of treason. His conviction was based on the testimony of two FBI agents who alleged that the three suspects drank together and engaged in long and intense conversation. The U.S. Supreme Court reversed Cramer's conviction based on the government's failure to establish an overt act that provided aid and comfort to the enemy. There was no indication that Cramer provided aid and comfort to the enemy by providing information; by securing food, shelter, or supplies; or by offering encouragement or advice. In summary, "without the use of some imagination it is difficult to perceive any advantage which this meeting afforded to Thiel and Kerling as enemies or how it strengthened Germany or weakened the United States in any way whatever."

See more cases on the study site: Iva Ikuko Toguri D' Aquino v. United States, *www.sagepub.com/lippmanccl3e.*

The Legal Equation

Treason **=** Overt act of levying war against the United States or giving aid and comfort to the enemies of the United States

+ intent to betray the United States

+ two witnesses to an overt act of levying war or giving aid or comfort or confession in open court.

SEDITION

Sedition at English common law was any communication intended or likely to bring about hatred, contempt, or disaffection with the king, the constitution, or the government. This agitation could be accomplished by **seditious speech** or **seditious libel** (writing). Sedition was punishable by imprisonment, fine, or pillory. In *The Case of the Seven Bishops* in 1688, English Justice Allybone pronounced that "[n]o man can take upon him to write against the actual exercise of the government . . . be what he writes true or false. . . . It is the business of the government to manage . . . the government; it is the business of subjects to mind their own properties and interests." Sedition was gradually expanded to include any and all criticism of the king or the government and the advocacy of reform of the government or church, as well as inciting discontent or promoting hostility between various economic and social classes.

During the debates over the U.S. Constitution, various speakers predicted that the effort to restrict the definition of treason would prove a "tempest in a teapot" because the government would merely resort to other laws to punish critics. This seemed borne out in 1798, when Congress passed the Alien and Sedition Acts. These laws punished any person writing or stating anything "false, scandalous and malicious" against the government, president, or Congress with the "intent to defame" or to bring them into "disrepute" or to "excite . . . the hatred of the . . . people of the United States, or to stir up sedition." The law differed from the common law in that the statute recognized truth as a defense. An individual convicted of sedition under the act was subject to a maximum punishment of two years in prison and by a fine of no more than $2,000. Although the law was defended as an effort to combat subversives who sought to sow the seeds of revolutionary violence, in fact, it was used to persecute political opponents of the government. For instance, a member of Congress from Vermont was sentenced to four months in prison for writing that President John Adams should be committed to a mental institution.

The modern version of the Alien and Sedition Acts is § 2383 in Title 18 of the U.S. Code. This statute punishes an individual who "incites, sets on foot, assists, or engages in any rebellion or insurrection against the authority of the United States or the laws thereof or gives aid or comfort thereto." This is punished by imprisonment for up to ten years, a fine, or both. An individual convicted of rebellion or insurrection under this law is prohibited from holding any federal office. Note that § 2383 prohibits incitement to sedition or criminal action against the United States or against a particular law.

The U.S. Code, in § 2384, punishes **seditious conspiracy**. This statute is directed at the use of force against the government, the use of force to prevent the execution of any law, or the use of force to interfere with governmental property and has been employed by prosecutors in recent years in terrorist prosecutions.

> If two or more persons in any state, territory, or in any place subject to the jurisdiction of the United States, conspire to overthrow, put down, or to destroy by force the government of the United States, or to levy war against them, or to oppose by force the authority thereof, or by force to prevent, hinder or delay the execution of any law of the United States, or by force to seize, take, or possess any property of the United States contrary to the authority thereof, they shall be fined . . . or imprisoned not more than twenty years or both.

In 1940, Congress adopted the Smith Act, which declared that it was a crime to conspire to teach or advocate the forcible overthrow of the U.S. government or to be a member of group that advocated the overthrow of the government. In *Dennis v. United States* in 1951, the Supreme Court upheld the constitutionality of this statute and affirmed the convictions of twelve leaders of the Communist Party. In 1957, the Supreme Court reconsidered the wisdom of this ruling in *Yates v. United States*. The Court, in that famous decision, held that individuals were free to advocate the overthrow of the U.S. government, although they may be prosecuted for urging individuals "to do something now or in the future." Congress was "aware of the distinction between the advocacy or teaching of abstract doctrine and the advocacy or teaching of action and . . . Congress did not intend to disregard this distinction."[4]

See more cases on the study site: Yates v. United States, United States v. Stone, *www.sagepub.com/ Lippmanccl3e.*

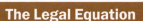

SABOTAGE

Sabotage is the willful injury, destruction, contamination, or infection of any war material, war premises, or war utilities with the intent of injuring or interfering or obstructing the United States or an allied country during a war or national emergency. Sabotage is punishable by imprisonment for not more than thirty years, a fine, or both.[5]

Whoever, when the United States is at war, or in times of national emergency as declared by the President or by Congress, with the intent to injure, interfere with, or obstruct the United States or any associate nation in preparing for or carrying on the war or defense activities, or with reason to believe that his act may injure, interfere with, or obstruct the United States or any associate nation in preparing for or carrying on the war or defense activities, willfully injures, destroys, contaminates or infects, or attempts to so injure, destroy, contaminate or infect any war material, war premises, or war utilities shall be fined under this title or imprisoned not more than thirty years, or both.

Sabotage may also be committed in peacetime against defense material, premises, or utilities.[6]

Whoever, with intent to injure, interfere with, or obstruct the national defense of the United States, willfully injures, destroys, contaminates or infects, or attempts to so injure, destroy, contaminate or infect any national-defense material, national-defense premises, or national-defense utilities, shall be fined under this title or imprisoned not more than 20 years, or both, and, if death results to any person, shall be imprisoned for any term of years or for life.

Other provisions punish the injury or destruction of harbors, premises, or utilities (e.g., transportation, water, power, electricity) and the production of defective national defense materials.[7]

Courts have held that sabotage requires a specific intent or purpose to damage the national defense of the United States. A defendant injuring property that he or she does not realize is part of the military defense may rely on the defense that although he or she intentionally damaged the property, there was a lack of a specific intent to injure the national defense. Several courts have taken the position that a knowledge standard satisfies the *mens rea* for sabotage. These judges reason that a defendant should be assumed to know that the destruction of defense material is practically certain to interfere with the national defense.

In *United States v. Kabat,* the defendants broke through a fence surrounding a missile silo in Missouri and used a jackhammer to slightly damage cables and chip a 100-ton lid covering the silo. The defendants were motivated by a desire to protest nuclear weapons and to educate the public concerning the mass destruction that would result from nuclear war. They hung banners and spray painted slogans that called attention to the fact that these weapons made the world less rather than more safe and were contrary to biblical teachings. Did the high-minded defendants who were motivated by a desire to save the planet from nuclear destruction possess a specific intent to injure, interfere with, or damage the national defense? The Eighth Circuit Court of Appeals ruled that the defendants' "intent to injure, interfere with or obstruct the national defense" was clear from their antinuclear statements and travel to Missouri for the specific purpose of damaging the missile silo. The damage to the silo clearly "interfered" with the defense of the United States, and to "allow citizens who thought they could further U.S. security to act on their theories at will could make it impossible for this country to maintain a coherent defense system." The issue remains whether the defendants intended to damage the national defense.[8]

The Legal Equation

Sabotage **=** Intentionally or knowingly

+ injures, interferes with, obstructs, destroys, contaminates, infects, or defectively produces

+ national defense materials, premises, utilities, or activities (in times of war, national emergency, or peacetime).

ESPIONAGE

The U.S. Code prohibits **espionage** or spying. The statute punishes espionage and espionage during war as separate offenses.[9]

For a deeper look at this topic, visit the study site.

Whoever, with intent or reason to believe that it is to be used to the injury of the United States or to the advantage of a foreign nation, communicates, delivers, or transmits, or attempts to communicate, deliver, or transmit, to any foreign government . . . faction or party or military or naval force within a foreign country . . . or to any representative or citizen thereof either directly or indirectly any document, writing, code book, signal book, sketch, photograph, photographic negative, blueprint, plan, map, model, note, instrument, appliance, or information relating to the national defense, shall be punished by death or by imprisonment for any term of years or for life, except that the sentence of death shall not be imposed unless [the jury or judge determines that the offense resulted in] the death of an agent of the United States . . . or directly concerned nuclear weaponry, military spacecraft or satellites, early warning systems, or other means of defense or retaliation against large-scale attack; war plans; communications intelligence or cryptographic information; or any other major weapons system or element of defense strategy.

Espionage in wartime is also defined in federal statutes.

Whoever, in time of war, with intent that the same shall be communicated to the enemy, collects, records, publishes, or communicates, or attempts to elicit any information with respect to the movement, numbers, description, condition, or disposition of any of the Armed Forces, ships, aircraft, or war materials of the United States, or with respect to the plans or conduct, or supposed plans or conduct of any naval or military operations, or with respect to any works or measures undertaken for or connected with, or intended for the fortification or defense of any place, or any other information relating to the public defense, which might be useful to the enemy, shall be punished by death or by imprisonment for any term of years or for life.

One of the most active areas of criminal activity in the new global economy is the theft by foreign governments of trade secrets from U.S. corporations. This may range from the formula for a new anticancer drug to the code for a computer program. Individuals involved in "industrial espionage" are subject to prosecution under the Economic Espionage Act of 1996.

In *Gorin v. United States,* the U.S. Supreme Court ruled that espionage during peacetime requires that the government establish that an individual acted in "bad faith" with the intent to injure the United States or to advantage a foreign nation. The foreign government that receives the information may be a friend or foe of the United States, because a country allied with the United States today may prove to be America's enemy tomorrow. The majority in *Gorin* held that "evil which the statute punishes is the obtaining or furnishing of this guarded information, either to our hurt or another's gain." Material that is stolen from American military files concerning British troop strength and armed preparedness may not directly harm the United States but may assist or advantage another country in protecting itself against the British and may constitute espionage.[10]

Note that espionage during wartime is easier for the prosecution to prove and requires the establishment of an intent to communicate information to the enemy along with a clear act toward the accomplishment of this goal.

National defense information that has not yet been officially released to the public but that has been reported by the press or generally referred to in government publications may be the subject of espionage because the government has not made the decision to release the specific details. Stealing plans for the design of a nuclear bomb in American defense files would be espionage, despite the fact that the broad outlines of the design are available on the Internet.

Information subject to espionage is not limited to the specific types of materials listed in the statutes. Under the espionage law, national defense is broadly interpreted to mean any information relating to the military and naval establishments and national preparation for war.

See more cases on the study site: United States v. Rosen, *www.sagepub.com/Lippmanccl3e.*

The Legal Equation

Espionage $=$ Communicates, delivers, transmits, or attempts to communicate, deliver, or transmit information

$+$ to a foreign government, faction, or military

$+$ purposely to injure or reason to believe that material will injure the United States or will be used to advantage a foreign nation.

Espionage during wartime $=$ Collects, records, publishes, communicates, or attempts to elicit information relating to national defense that might be useful to the enemy

$+$ with intent to communicate to the enemy

$+$ during time of war.

TERRORISM

The bombing of the Alfred P. Murrah federal building in Oklahoma City on April 19, 1995, and the attack on the United States on September 11, 2001, combined to push Congress to act to prohibit and to punish terrorism. Keep in mind that most acts of terrorism within the United States are prosecuted as ordinary murders, arson, kidnappings, and bombings rather than as acts of terrorism.

The central provisions of the U.S. law on terrorism are found in Title 18 of the U.S. Code, Chapter 113B, "Terrorism." These statutes have been amended and strengthened by the Anti-Terrorism and Effective Death Penalty Act (1996) and the Uniting and Strengthening America by Providing Appropriate Tools Required to Intercept and Obstruct Terrorism Act (2001), better known as the USA Patriot Act.

Definition of Terrorism

Various federal statutes use the term *terrorism* or *terrorist*. For instance, it is a crime to materially aid a terrorist or foreign terrorist organization. What is terrorism? Federal law divides terrorism into **international terrorism** and **domestic terrorism**.

International terrorism is distinguished by the fact that it occurs outside the United States. Both international and domestic terrorism are intended to intimidate or coerce the American population or are intended to influence or affect the public policy of the United States. We have several

For a deeper look at this topic, visit the study site.

tragic examples: the August 7, 1998, bombing of the U.S. embassies in Kenya (killing 213 and injuring more than 4,500) and in Tanzania (killing 11 and injuring 85); and the attack on a U.S. Navy warship, the *USS Cole,* in Yemen on October 12, 2000 (killing 17 U.S. sailors). Other clear examples are the acts of violence intended to force America to withdraw troops from Iraq. International terrorism is defined as

- violent acts or acts dangerous to human life that
- primarily occur outside the United States,
- would be criminal if committed in the United States, and
- appear to be intended to either
- intimidate or coerce a civilian population; or
- influence the policy of a government by intimidation or coercion; or
- affect the conduct of a government through mass destruction, assassination, or kidnapping.

Domestic terrorism is defined in the same fashion with the exception that it occurs "primarily within the territorial jurisdiction of the United States" rather than "outside the United States." Note that terrorism is defined in terms of the intent of the offender rather than by the target of the attack.

The U.S. Code defines the **federal crime of terrorism** as an offense "calculated to influence or affect the conduct of government by intimidation or coercion, or to retaliate against government conduct" that involves a violation of a long list of violent and dangerous federal offenses.[11] Chapter 113B of the U.S. Code provides for several specific terrorist crimes that are discussed in the following sections.

Terrorism Outside the United States

Section 2332 of the U.S. Code punishes various crimes against nationals of the United States that occur outside the United States. The killing of an American national outside the United States is punishable by imprisonment for a term of years or death as well as a fine. Voluntary manslaughter is subject to ten years' imprisonment along with a fine, and involuntary manslaughter is punished by a fine or imprisonment of not more than three years, or both. A conspiracy to kill a U.S. national is punished by up to life imprisonment as well as a fine, and a conspiracy leading to an attempt is subject to imprisonment for up to twenty years in prison as well as a fine. Physical violence with the intent to cause serious bodily injury to a U.S. national and physical violence that results in serious bodily injury are both subject to a fine as well as to imprisonment by up to ten years. The U.S. assertion of the right to prosecute and punish criminal acts that occur outside American territory is termed **extraterritorial jurisdiction**.[12]

Terrorism Transcending National Boundaries

U.S. law punishes as a felony acts of terrorism that occur within the United States but that are connected to foreign countries. Offenses involving conduct occurring both outside and within the United States are termed **terrorism transcending national boundaries**. In other words, the fact that conspirators meet in another country and plan to attack an American city would not remove the crime from the jurisdiction of the United States. Prime examples are the September 11, 2001, attacks on the World Trade Center and Pentagon, which were planned, directed, and funded from outside the United States. This statute permits the United States to prosecute individuals living in Afghanistan, England, Germany, Spain, and other countries who were involved in the 9/11 conspiracy. Federal law also provides that criminal attacks against U.S. agencies, embassies, or property abroad or in the air or at sea against property owned by the U.S. government or by an American citizen are considered to have taken place within American territory and are punishable by the United States.

Three crimes of violence are included within the statute on terrorism that transcends national boundaries:

- *Crimes Against the Person.* Killing, kidnapping, maiming, or committing an assault resulting in serious bodily injury or assault with a dangerous weapon against an individual within the United States.

- *Crimes Against Property Harming the Person.* Acts that create a substantial risk of serious bodily injury by destroying or damaging any structure, real estate, or object within the United States.
- *Inchoate Offense.* Threats, attempts, and conspiracies to commit either of these two offenses and accessories after the fact.

The offenses under this provision must meet one of several conditions. These include the following:

- *Federal Official.* The victim or intended victim is the U.S. government, a member of the uniformed services, or any official, officer, employee, or agent of the legislative, executive, or judicial branches or of any department or agency of the United States.
- *Property.* The building, vehicle, real estate, or object is owned or leased by the United States or a U.S. citizen.
- *Territory.* The offense is committed within the territorial sea or American airspace.
- *Interstate Commerce.* The offense makes use of the mail or any facility in interstate or foreign commerce.

Penalties include capital punishment for a crime resulting in death, life imprisonment for the crime of kidnapping, thirty-five years for maiming, thirty years for aggravated assault, and twenty-five years for damaging property.[13]

Weapons of Mass Destruction

The use, threat, attempt, or conspiracy to use a **weapon of mass destruction** is punishable by imprisonment for a term of years or life and, in the event of death, by life imprisonment.

The statute on weapons of mass destruction, 18 U.S.C. § 2332a, declares that it is a crime when such a weapon is used outside the United States against an American resident or citizen or within the United States against any person so as to affect commerce or against property within or outside the United States that is owned, leased, or used by the federal government. It is also an offense to threaten, attempt to use, or conspire to use a weapon of mass destruction. A weapon of mass destruction is defined as follows:

- Toxic or poisonous chemical weapons that are designed or intended to cause death or serious bodily injury (poison gas).
- Weapons involving biological agents (smallpox).
- Weapons releasing radiation or radioactivity at a level dangerous to human life (nuclear material).
- Explosive bombs, grenades, rockets, missiles, and mines.

Possession of a biological or toxic weapon or delivery system that cannot be justified by a peaceful purpose is subject to imprisonment for up to ten years or a fine, or both.[14] In April 2005, Zacarias Moussaoui, while denying involvement in the September 11, 2001, attacks against the Pentagon and World Trade Center, pled guilty to conspiring to use on an airplane a "weapon of mass destruction" to attack the White House.

Mass Transportation Systems

Subways, buses, and trains are some of the most vulnerable and potentially damaging terrorist targets. In 1995, a chemical gas attack on a subway train in Tokyo resulted in twelve deaths and more than 5,000 injuries. Federal law provides that it is a crime to willfully wreck, derail, set fire to, or disable a mass transportation vehicle or ferry or to damage or impair the operation of a signal or control system. It is also a crime to cause the death or serious bodily injury of an employee or passenger on a mass transportation vehicle. Other provisions prohibit using a weapon of mass destruction against a mass transportation system. These offenses are punishable by a fine and as many as twenty years in prison. An aggravated offense punishable by up to life in prison results when the mass transportation vehicle or ferry is carrying one or more passengers or the offense results in the death of any person.[15] The destruction of aircraft and air piracy are subject to punishment under separate provisions of the U.S. Code. Air piracy is defined in Title 49, § 46502, as "seizing or exercising control" over an aircraft by force, violence, threat of force or violence, or any form of intimidation with a wrongful intent.[16]

Harboring or Concealing Terrorists

It is a crime to harbor or conceal a person who an individual knows or has reasonable grounds to believe has committed or is about to commit various terrorist offenses or crimes posing a serious and widespread danger. Harboring a terrorist is subject to ten years in prison and to a fine.[17]

Material Support for Terrorism

The offense of providing **material support to a terrorist** is defined in the U.S. Code as providing material support or resources or concealing or disguising the nature, location, source, or ownership of material support or resources, knowing or intending that they are to be used in preparation for or in carrying out various terrorist acts or acts of violence or in the preparation of, concealment of, or escape from such crimes. This is punishable by imprisonment for no more than fifteen years, a fine, or both. In the event that death results, the accused is subject to imprisonment for a term of years or capital punishment.

There is a separate offense of knowingly providing **material support to a foreign terrorist organization** or an attempt or conspiracy to do so. This is also subject to imprisonment for up to fifteen years, a fine, or both. The death of a victim may result in imprisonment for up to life.

Material support or resources in support of terrorism includes money, financial services, housing, training, expert advice or assistance, personnel, safe houses, false documentation, communication equipment, facilities, weapons, lethal substances, explosives, transportation, and other "physical assets." Medical and religious materials are exempt from the prohibition on material support.[18] The U.S. Secretary of State is charged with determining whether a group is a foreign terrorist organization.[19]

These two statutes are the primary laws relied on by prosecutors in terrorist prosecutions in the United States, with close to sixty persons having been prosecuted between September 2001 and May 2004. Prosecutors explain that the material support statutes are central in combating terrorism because hardcore terrorists rely on the support provided by individuals willing to provide financial resources, passports, expertise in computer technology, weapons, and information. These laws also have been relied on by the government to prosecute individuals for "providing personnel to terrorist organizations" who have traveled abroad to undergo terrorist training. The material support statutes have the advantage of permitting the arrest of individuals before terrorist plots are carried out.

Critics caution that the material support provisions may be used to prosecute individuals who do not pose a threat. For instance, defendants in a New York case who had attended a terrorist training camp abroad and who had not been involved with terrorist activities after returning to the United States pled guilty to providing material support. It is also argued that these statutes are broadly written and that individuals may be criminally prosecuted who have merely donated money to a hospital or school in the Middle East run by a group labeled as terrorist. In *Holder v. Humanitarian Law Project,* the Supreme Court was asked to determine whether Congress may constitutionally criminally punish individuals who assist members of terrorist organizations to engage in lawful political activity.

The Legal Equation

Material support to terrorists = Provide material support to a terrorist or conceal or disguise the nature, location, source, or ownership of material

+ knowing or intending that support or resources are used in preparation for or in carrying out various terrorist acts.

Material support to foreign terrorist organizations = Provide material support to a foreign terrorist organization

+ knowingly provides material support.

Does the prohibition on material support to a foreign terrorist organization violate the first amendment protection on freedom of speech?

Holder v. Humanitarian Law Project,
___ U.S. ___, 130 S. Ct. 2705 (2010). Opinion by: Roberts, C.J.

Issue

Congress has prohibited the provision of "material support or resources" to certain foreign organizations that engage in terrorist activity. 18 U.S.C. section 2339B(a)(1). That prohibition is based on a finding that the specified organizations "are so tainted by their criminal conduct that any contribution to such an organization facilitates that conduct." The plaintiffs in this litigation seek to provide support to two such organizations. Plaintiffs claim that they seek to facilitate only the lawful, nonviolent purposes of those groups, and that applying the material-support law to prevent them from doing so violates the Constitution. In particular, they claim that the statute is too vague, in violation of the Fifth Amendment, and that it infringes their rights to freedom of speech and association, in violation of the First Amendment.

Facts

This litigation concerns 18 U.S.C. section 2339B, which makes it a federal crime to "knowingly provid[e] material support or resources to a foreign terrorist organization." Congress has amended the definition of "material support or resources" periodically, but at present it is defined as follows:

> [T]he term "material support or resources" means any property, tangible or intangible, or service, including currency or monetary instruments or financial securities, financial services, lodging, training, expert advice or assistance, safehouses, false documentation or identification, communications equipment, facilities, weapons, lethal substances, explosives, personnel (1 or more individuals who may be or include oneself), and transportation, except medicine or religious materials.

Section 2339A(b)(1); see also section 2339B(g)(4).

The authority to designate an entity a "foreign terrorist organization" rests with the Secretary of State. 8 U.S.C. sections 1189(a)(1), (d)(4). She may, in consultation with the Secretary of the Treasury and the Attorney General, so designate an organization upon finding that it is foreign, engages in "terrorist activity" or "terrorism," and thereby "threatens the security of United States nationals or the national security

of the United States." "'[N]ational security' means the national defense, foreign relations, or economic interests of the United States." An entity designated a foreign terrorist organization may seek review of that designation before the D. C. Circuit within 30 days of that designation.

In 1997, the Secretary of State designated 30 groups as foreign terrorist organizations. Two of those groups are the Kurdistan Workers' Party (also known as the Partiya Karkeran Kurdistan, or PKK) and the Liberation Tigers of Tamil Eelam (LTTE). The PKK is an organization founded in 1974 with the aim of establishing an independent Kurdish state in southeastern Turkey. The LTTE is an organization founded in 1976 for the purpose of creating an independent Tamil state in Sri Lanka. The District Court in this action found that the PKK and the LTTE engage in political and humanitarian activities. The Government has presented evidence that both groups have also committed numerous terrorist attacks, some of which have harmed American citizens. The LTTE sought judicial review of its designation as a foreign terrorist organization; the D. C. Circuit upheld that designation. The PKK did not challenge its designation.

Plaintiffs in this litigation are two U.S. citizens and six domestic organizations: the Humanitarian Law Project (HLP) (a human rights organization with consultative status to the United Nations); Ralph Fertig (the HLP's president, and a retired administrative law judge); Nagalingam Jeyalingam (a Tamil physician, born in Sri Lanka and a naturalized U.S. citizen); and five nonprofit groups dedicated to the interests of persons of Tamil descent. In 1998, plaintiffs filed suit in federal court challenging the constitutionality of the material-support statute, Section 2339B. Plaintiffs claimed that they wished to provide support for the humanitarian and political activities of the PKK and the LTTE in the form of monetary contributions, other tangible aid, legal training, and political advocacy, but that they could not do so for fear of prosecution under Section 2339B. As relevant here, plaintiffs claimed that the material-support statute was unconstitutional on two grounds: First, it violated their freedom of speech and freedom of association under the First Amendment, because it criminalized their provision of material support to the PKK and the LTTE, without requiring the Government to prove that plaintiffs had a specific intent to further the unlawful ends of those organizations. Second, plaintiffs argued that the statute was unconstitutionally vague.

[THIS SECTION OF THE DECISION RECOUNTS THE PROCEDURAL HISTORY OF THE CASE IN THE LOWER COURTS.]

In 2001, Congress amended the definition of "material support or resources" to add the term "expert advice or assistance." Uniting and Strengthening America by Providing Appropriate Tools Required to Intercept and Obstruct Terrorism Act of 2001 (Patriot Act), section 805(a)(2)(B).

On December 17, 2004 Congress again amended Section 2339B and the definition of "material support or resources." Intelligence Reform and Terrorism Prevention Act of 2004 (IRTPA), Section 6603. In IRTPA, Congress clarified the mental state necessary to violate Section 2339B, requiring knowledge of the foreign group's designation as a terrorist organization or the group's commission of terrorist acts. Section 2339B(a)(1). Congress also added the term "service" to the definition of "material support or resources," Section 2339A(b)(1), and defined "training" to mean "instruction or teaching designed to impart a specific skill, as opposed to general knowledge," Section 2339A(b)(2). It also defined "expert advice or assistance" to mean "advice or assistance derived from scientific, technical or other specialized knowledge." Section 2339A(b)(3). Finally, IRTPA clarified the scope of the term "personnel" by providing:

No person may be prosecuted under [Section 2339B] in connection with the term "personnel" unless that person has knowingly provided, attempted to provide, or conspired to provide a foreign terrorist organization with 1 or more individuals (who may be or include himself) to work under that terrorist organization's direction or control or to organize, manage, supervise, or otherwise direct the operation of that organization. Individuals who act entirely independently of the foreign terrorist organization to advance its goals or objectives shall not be considered to be working under the foreign terrorist organization's direction and control.

Section 2339B(h).

Reasoning

Given the complicated 12-year history of this litigation, we pause to clarify the questions before us. Plaintiffs challenge Section 2339B's prohibition on four types of material support—"training," "expert advice or assistance," "service," and "personnel." They raise three constitutional claims. First, plaintiffs claim that Section 2339B violates the Due Process Clause of the Fifth Amendment because these four statutory terms are impermissibly vague. Second, plaintiffs claim that Section 2339B violates their freedom of speech under

the First Amendment. Third, plaintiffs claim that Section 2339B violates their First Amendment freedom of association.

Plaintiffs . . . claim that Section 2339B is invalid to the extent it prohibits them from engaging in certain specified activities. With respect to the HLP and Judge Fertig, those activities are: (1) "train[ing] members of [the] PKK on how to use humanitarian and international law to peacefully resolve disputes"; (2) "engag[ing] in political advocacy on behalf of Kurds who live in Turkey"; and (3) "teach[ing] PKK members how to petition various representative bodies such as the United Nations for relief." With respect to the other plaintiffs, those activities are: (1) "train[ing] members of [the] LTTE to present claims for tsunami-related aid to mediators and international bodies"; (2) "offer[ing] their legal expertise in negotiating peace agreements between the LTTE and the Sri Lankan government"; and (3) "engag[ing] in political advocacy on behalf of Tamils who live in Sri Lanka."

Plaintiffs also state that "the LTTE was recently defeated militarily in Sri Lanka," so "[m]uch of the support the Tamil organizations and Dr. Jeyalingam sought to provide is now moot." Plaintiffs thus seek only to support the LTTE "as a political organization outside Sri Lanka advocating for the rights of Tamils." . . .

One last point. Plaintiffs seek preenforcement review of a criminal statute. We conclude that plaintiffs face "a credible threat of prosecution" and "should not be required to await and undergo a criminal prosecution as the sole means of seeking relief."

Plaintiffs claim that they provided support to the PKK and the LTTE before the enactment of Section 2339B and that they would provide similar support again if the statute's allegedly unconstitutional bar were lifted. The Government tells us that it has charged about 150 persons with violating Section 2339B, and that several of those prosecutions involved the enforcement of the statutory terms at issue here. The Government has not argued to this Court that plaintiffs will not be prosecuted if they do what they say they wish to do. Based on these considerations, we conclude that plaintiffs' claims are suitable for judicial review.

Plaintiffs . . . contend that we should interpret the material-support statute, when applied to speech, to require proof that a defendant intended to further a foreign terrorist organization's illegal activities. That interpretation, they say, would end the litigation because plaintiffs' proposed activities consist of speech, but plaintiffs do not intend to further unlawful conduct by the PKK or the LTTE.

We reject plaintiffs' interpretation of Section 2339B because it is inconsistent with the text of the statute. Section 2339B(a)(1) prohibits "knowingly" providing material support. It then specifically describes the type of knowledge that is required: "To violate this paragraph, a person must have knowledge that the organization is a designated terrorist organization . . . , that the organization has engaged or engages in terrorist activity . . . , or that the organization has engaged or engages in terrorism. . . ." Congress plainly spoke to the necessary mental state for a violation of Section 2339B, and it chose knowledge about the organization's connection to terrorism, not specific intent to further the organization's terrorist activities.

Plaintiffs' interpretation is also untenable in light of the sections immediately surrounding Section 2339B, both of which do refer to intent to further terrorist activity. See section 2339A(a) (establishing criminal penalties for one who "provides material support or resources . . . knowing or intending that they are to be used in preparation for, or in carrying out, a violation of" statutes prohibiting violent terrorist acts); Section 2339C(a)(1) (setting criminal penalties for one who "unlawfully and willfully provides or collects funds with the intention that such funds be used, or with the knowledge that such funds are to be used, in full or in part, in order to carry out" other unlawful acts). Congress enacted Section 2339A in 1994 and section 2339C in 2002. Yet Congress did not import the intent language of those provisions into section 2339B, either when it enacted Section 2339B in 1996, or when it clarified section 2339B's knowledge requirement in 2004.

Finally, plaintiffs give the game away when they argue that a specific intent requirement should apply only when the material-support statute applies to speech. There is no basis whatever in the text of section 2339B to read the same provisions in that statute as requiring intent in some circumstances but not others. It is therefore clear that plaintiffs are asking us not to interpret section 2339B, but to revise it. "Although this Court will often strain to construe legislation so as to save it against constitutional attack, it must not and will not carry this to the point of perverting the purpose of a statute." . . .

We cannot avoid the constitutional issues in this litigation through plaintiffs' proposed interpretation of Section 2339B. . . .

We turn to the question whether the material-support statute, as applied to plaintiffs, is impermissibly vague under the Due Process Clause of the Fifth Amendment. "A conviction fails to comport with due process if the statute under which it is obtained fails to provide a person of ordinary intelligence fair notice of what is prohibited, or is so standardless that it authorizes or encourages seriously discriminatory enforcement." We consider whether a statute is vague as applied to the particular facts at issue, for "[a] plaintiff who engages in some conduct that is clearly proscribed cannot complain of the vagueness of the law as applied to the conduct of others." We have said that when a statute "interferes with the right of free speech or of association, a more stringent vagueness test should apply." "But 'perfect clarity and precise guidance have never been required even of regulations that restrict expressive activity.'" . . .

Under a proper analysis, plaintiffs' claims of vagueness lack merit. Plaintiffs do not argue that the material-support statute grants too much enforcement discretion to the Government. We therefore address only whether the statute "provide[s] a person of ordinary intelligence fair notice of what is prohibited." As a general matter, the statutory terms at issue here are quite different from the sorts of terms that we have previously declared to be vague. We have in the past "struck down statutes that tied criminal culpability to whether the defendant's conduct was 'annoying' or 'indecent'—wholly subjective judgments without statutory definitions, narrowing context, or settled legal meanings." Applying the statutory terms in this action—"training," "expert advice or assistance," "service," and "personnel"—does not require similarly untethered, subjective judgments.

Of course, the scope of the material-support statute may not be clear in every application. But the dispositive point here is that the statutory terms are clear in their application to plaintiffs' proposed conduct, which means that plaintiffs' vagueness challenge must fail. Even assuming that a heightened standard applies because the material-support statute potentially implicates speech, the statutory terms are not vague as applied to plaintiffs.

Most of the activities in which plaintiffs seek to engage readily fall within the scope of the terms "training" and "expert advice or assistance." Plaintiffs want to "train members of [the] PKK on how to use humanitarian and international law to peacefully resolve disputes," and "teach PKK members how to petition various representative bodies such as the United Nations for relief." A person of ordinary intelligence would understand that instruction on resolving disputes through international law falls within the statute's definition of "training" because it imparts a "specific skill," not "general knowledge." Plaintiffs' activities also fall comfortably within the scope of "expert advice or assistance": A reasonable person would recognize that teaching the PKK how to petition for humanitarian relief before the United Nations involves advice derived from, as the statute puts it, "specialized knowledge." In fact, plaintiffs themselves have repeatedly used the terms "training" and "expert advice" throughout this litigation to describe their own proposed activities, demonstrating that these common terms readily and naturally cover plaintiffs' conduct. . . .

Plaintiffs also contend that they want to engage in "political advocacy" on behalf of Kurds living in Turkey and Tamils living in Sri Lanka. They are concerned that such advocacy might be regarded as "material support" in the form of providing "personnel" or "service[s]," and assert that the statute is unconstitutionally vague because they cannot tell.

As for "personnel," Congress enacted a limiting definition in IRTPA that answers plaintiffs' vagueness concerns. Providing material support that constitutes "personnel" is defined as knowingly providing a person "to work under that terrorist organization's direction or control or to organize, manage, supervise, or otherwise direct the operation of that organization." The statute makes clear that "personnel" does not cover independent advocacy: "Individuals who act entirely independently of the foreign terrorist organization to advance its goals or objectives shall not be considered to be working under the foreign terrorist organization's direction and control."

"[S]ervice" similarly refers to concerted activity, not independent advocacy. . . . Context confirms that ordinary meaning here. The statute prohibits providing a service "to a foreign terrorist organization." The use of the word "to" indicates a connection between the service and the foreign group. We think a person of ordinary intelligence would understand that independently advocating for a cause is different from providing a service to a group that is advocating for that cause.

Moreover, if independent activity in support of a terrorist group could be characterized as a "service," the statute's specific exclusion of independent activity in the definition of "personnel" would not make sense. Congress would not have prohibited under "service" what it specifically exempted from prohibition under "personnel." The other types of material support listed in the statute, including "lodging," "weapons," "explosives," and "transportation," section 2339A(b)(1), are not forms of support that could be provided independently of a foreign terrorist organization. We interpret "service" along the same lines. Thus, any independent advocacy in which plaintiffs wish to engage is not prohibited by section 2339B. On the other hand, a person of ordinary intelligence would understand the term "service" to cover advocacy performed in coordination with, or at the direction of, a foreign terrorist organization.

Plaintiffs argue that this construction of the statute poses difficult questions of exactly how much direction or coordination is necessary for an activity to constitute a "service." The problem with these questions is that they are entirely hypothetical. Plaintiffs have not provided any specific articulation of the degree to which they seek to coordinate their advocacy with the PKK and the LTTE. They have instead described the form of their intended advocacy only in the most general terms. . . . It is apparent with respect to these claims that "gradations of fact or charge would make a difference as to criminal liability," and so "adjudication of the reach and constitutionality of [the statute] must await a concrete fact situation."

We next consider whether the material-support statute, as applied to plaintiffs, violates the freedom of speech guaranteed by the First Amendment. Both plaintiffs and the Government take extreme positions on this question. Plaintiffs claim that Congress has banned their "pure political speech." It has not. Under the material-support statute, plaintiffs may say anything they wish on any topic. They may speak and write freely about the PKK and LTTE, the governments of Turkey and Sri Lanka, human rights, and international law. They may advocate before the United Nations. As the Government states: "The statute does not prohibit independent advocacy or expression of any kind." Congress has not, therefore, sought to suppress ideas or opinions in the form of "pure political speech." Rather, Congress has prohibited "material support," which most often does not take the form of speech at all. And when it does, the statute is carefully drawn to cover only a narrow category of speech to, under the direction of, or in coordination with foreign groups that the speaker knows to be terrorist organizations.

For its part, the Government takes the foregoing too far, claiming that the only thing truly at issue in this litigation is conduct, not speech. Section 2339B is directed at the fact of plaintiffs' interaction with the PKK and LTTE, the Government contends, and only incidentally burdens their expression. . . . Plaintiffs want to speak to the PKK and the LTTE, and whether they may do so under section 2339B depends on what they say. If plaintiffs' speech to those groups imparts a "specific skill" or communicates advice derived from "specialized knowledge"—for example, training on the use of international law or advice on petitioning the United Nations—then it is barred. On the other hand, plaintiffs' speech is not barred if it imparts only general or unspecialized knowledge.

The First Amendment issue before us is more refined than either plaintiffs or the Government would have it. It is not whether the Government may prohibit pure political speech, or may prohibit material support in the form of conduct. It is instead whether the Government may prohibit what plaintiffs want to do—provide material support to the PKK and LTTE in the form of speech.

Everyone agrees that the Government's interest in combating terrorism is an urgent objective of the highest order. Plaintiffs' complaint is that the ban on material support, applied to what they wish to do, is not "necessary to further that interest." The objective of combating terrorism does not justify prohibiting their speech, plaintiffs argue, because their support will advance only the legitimate activities of the designated terrorist organizations, not their terrorism.

Whether foreign terrorist organizations meaningfully segregate support of their legitimate activities from support of terrorism is an empirical question. When it enacted section 2339B in 1996, Congress made specific findings regarding the serious threat posed by international terrorism. One of those findings explicitly rejects plaintiffs' contention that their support would not further the terrorist activities of the PKK and LTTE: "[F]oreign organizations that engage in terrorist activity are so tainted by their criminal conduct that any contribution to such an organization facilitates that conduct."

Plaintiffs argue that the reference to "any contribution" in this finding meant only monetary support. There is no reason to read the finding to be so limited, particularly because Congress expressly prohibited so much more than monetary support in Section 2339B. Congress's use of the term "contribution" is best read to reflect a determination that any form of material support furnished "to" a foreign terrorist organization should be barred, which is precisely what the material-support statute does. Indeed, when Congress enacted section 2339B, Congress simultaneously removed an exception that had existed in section 2339A(a) for the provision of material support in the form of "humanitarian assistance to persons not directly involved in" terrorist activity. That repeal demonstrates that Congress considered and rejected the view that ostensibly peaceful aid would have no harmful effects.

We are convinced that Congress was justified in rejecting that view. The PKK and the LTTE are deadly groups. "The PKK's insurgency has claimed more than 22,000 lives." The LTTE has engaged in extensive suicide bombings and political assassinations, including killings of the Sri Lankan President, Security Minister, and Deputy Defense Minister. "On January 31, 1996, the LTTE exploded a truck bomb filled with an estimated 1,000 pounds of explosives at the Central Bank in Colombo, killing 100 people and injuring more than 1,400. This bombing was the most deadly terrorist incident in the world in 1996." It is not difficult to conclude as Congress did that the "tain[t]" of such violent activities is so great that working in coordination with or at the command of the PKK and LTTE serves to legitimize and further their terrorist means.

Material support meant to "promot[e] peaceable, lawful conduct," can further terrorism by foreign groups in multiple ways. "Material support" is a valuable resource by definition. Such support frees up other resources within the organization that may be put to violent ends. It also importantly helps lend legitimacy to foreign terrorist groups—legitimacy that makes it easier for those groups to persist, to recruit members, and to raise funds—all of which facilitate more terrorist attacks. "Terrorist organizations do not maintain organizational 'firewalls' that would prevent or deter . . . sharing and commingling of support and benefits." "[I]nvestigators have revealed how terrorist groups systematically conceal their activities behind charitable, social, and political fronts." "Indeed, some designated foreign terrorist organizations use social and political components to recruit personnel to carry out terrorist operations."

Money is fungible, and "[w]hen foreign terrorist organizations that have a dual structure raise funds, they highlight the civilian and humanitarian ends to which such moneys could be put." But "there is reason to believe that foreign terrorist organizations do not maintain legitimate financial firewalls between those funds raised for civil, nonviolent activities, and those ultimately used to support violent, terrorist operations." Thus, "[f]unds raised ostensibly for charitable purposes have in the past been redirected by some terrorist groups to fund the purchase of arms and explosives." There is evidence that the PKK and the LTTE, in particular, have not "respected the line between humanitarian and violent activities."

The dissent argues that there is "no natural stopping place" for the proposition that aiding a foreign terrorist organization's lawful activity promotes the terrorist organization as a whole. But Congress has settled on just such a natural stopping place: The statute reaches only material support coordinated with or under the direction of a designated foreign terrorist organization. Independent advocacy that might be viewed as promoting the group's legitimacy is not covered.

Providing foreign terrorist groups with material support in any form also furthers terrorism by straining the United States' relationships with its allies and undermining cooperative efforts between nations to prevent terrorist attacks. We see no reason to question Congress's finding that "international cooperation is required for an effective response to terrorism, as demonstrated by the numerous multilateral conventions in force providing universal prosecutorial jurisdiction over persons involved in a variety of terrorist acts, including hostage taking, murder of an internationally protected person, and aircraft piracy and sabotage." The material-support statute furthers this international effort by prohibiting aid for foreign terrorist groups that harm the United States' partners abroad: "A number of designated foreign terrorist organizations have attacked moderate governments with which the United States has vigorously endeavored to maintain close and friendly relations," and those attacks "threaten [the] social, economic and political stability" of such governments. "[O]ther foreign terrorist organizations attack our NATO allies, thereby implicating important and sensitive multilateral security arrangements."

For example, the Republic of Turkey—a fellow member of NATO—is defending itself against a violent insurgency waged by the PKK. That nation and our other allies would react sharply to Americans furnishing material support to foreign groups like the PKK, and would hardly be mollified by the explanation that the support was meant only to further those groups' "legitimate" activities. From Turkey's perspective, there likely are no such activities. (Turkey prohibits membership in the PKK and prosecutes those who provide support to that group, regardless of whether the support is directed to lawful activities).

In analyzing whether it is possible in practice to distinguish material support for a foreign terrorist group's violent activities and its nonviolent activities, we do not rely exclusively on our own inferences drawn from the record evidence. We have before us an affidavit stating the Executive Branch's conclusion on that question. The State Department informs us that "[t]he experience and analysis of the U.S. government agencies charged with combating terrorism strongly suppor[t]" Congress's finding that all contributions to foreign terrorist organizations further their terrorism. In the Executive's view: "Given the purposes, organizational structure, and clandestine nature of foreign terrorist organizations, it is highly likely that any material support to these organizations will ultimately inure to the benefit of their criminal, terrorist functions—regardless of whether such support was ostensibly intended to support non-violent, non-terrorist activities."

That evaluation of the facts by the Executive, like Congress's assessment, is entitled to deference. This litigation implicates sensitive and weighty interests of national security and foreign affairs. The PKK and the LTTE have committed terrorist acts against American citizens abroad, and the material-support statute addresses acute foreign policy concerns involving relationships with our Nation's allies. We have noted that "neither the Members of this Court nor most federal judges begin the day with briefings that may describe new and serious threats to our Nation and its people." It is vital in this context "not to substitute . . . our own evaluation of evidence for a reasonable evaluation by the Legislative Branch."

Our precedents, old and new, make clear that concerns of national security and foreign relations do not warrant abdication of the judicial role. We do not defer to the Government's reading of the First Amendment, even when such interests are at stake. We are one with the dissent that the Government's "authority and expertise in these matters do not automatically trump the Court's own obligation to secure the protection that the Constitution grants to individuals." But when it comes to collecting evidence and drawing factual inferences in this area, "the lack of competence on the part of the courts is marked," and respect for the Government's conclusions is appropriate.

One reason for that respect is that national security and foreign policy concerns arise in connection with efforts to confront evolving threats in an area where information can be difficult to obtain and the impact of certain conduct difficult to assess. The dissent slights these real constraints in demanding hard proof—with "detail," "specific facts," and "specific evidence"—that plaintiffs' proposed activities will support terrorist attacks. That would be a dangerous requirement. In this context, conclusions must often be based on informed judgment rather than concrete evidence, and that reality affects what we may reasonably insist on from the Government. The material-support statute is, on its face, a preventive measure—it criminalizes not terrorist attacks themselves, but aid that makes the attacks more likely to occur. The Government, when seeking to prevent imminent harms in the context of international affairs and national security, is not required to

conclusively link all the pieces in the puzzle before we grant weight to its empirical conclusions. . . . Congress and the Executive are uniquely positioned to make principled distinctions between activities that will further terrorist conduct and undermine United States foreign policy, and those that will not.

We also find it significant that Congress has been conscious of its own responsibility to consider how its actions may implicate constitutional concerns. First, section 2339B only applies to designated foreign terrorist organizations. There is, and always has been, a limited number of those organizations designated by the Executive Branch, and any groups so designated may seek judicial review of the designation. Second, in response to the lower courts' holdings in this litigation, Congress added clarity to the statute by providing narrowing definitions of the terms "training," "personnel," and "expert advice or assistance," as well as an explanation of the knowledge required to violate section 2339B. Third, in effectuating its stated intent not to abridge First Amendment rights, Congress has also displayed a careful balancing of interests in creating limited exceptions to the ban on material support. The definition of material support, for example, excludes medicine and religious materials. In this area perhaps more than any other, the Legislature's superior capacity for weighing competing interests means that "we must be particularly careful not to substitute our judgment of what is desirable for that of Congress." Finally, and most importantly, Congress has avoided any restriction on independent advocacy, or indeed any activities not directed to, coordinated with, or controlled by foreign terrorist groups.

At bottom, plaintiffs simply disagree with the considered judgment of Congress and the Executive that providing material support to a designated foreign terrorist organization—even seemingly benign support—bolsters the terrorist activities of that organization. That judgment, however, is entitled to significant weight, and we have persuasive evidence before us to sustain it. Given the sensitive interests in national security and foreign affairs at stake, the political branches have adequately substantiated their determination that, to serve the Government's interest in preventing terrorism, it was necessary to prohibit providing material support in the form of training, expert advice, personnel, and services to foreign terrorist groups, even if the supporters meant to promote only the groups' nonviolent ends.

We turn to the particular speech plaintiffs propose to undertake. First, plaintiffs propose to "train members of [the] PKK on how to use humanitarian and international law to peacefully resolve disputes." Congress can, consistent with the First Amendment, prohibit this direct training. It is wholly foreseeable that the PKK could use the "specific skill[s]" that plaintiffs propose to impart, as part of a broader strategy to promote terrorism. The PKK could, for example, pursue peaceful negotiation as a means of buying time to recover from short-term setbacks, lulling opponents into complacency, and ultimately preparing for renewed attacks. A foreign terrorist organization introduced to the structures of the international legal system might use the information to threaten, manipulate, and disrupt. This possibility is real, not remote.

Second, plaintiffs propose to "teach PKK members how to petition various representative bodies such as the United Nations for relief." The Government acts within First Amendment strictures in banning this proposed speech because it teaches the organization how to acquire "relief," which plaintiffs never define with any specificity, and which could readily include monetary aid. Indeed, earlier in this litigation, plaintiffs sought to teach the LTTE "to present claims for tsunami-related aid to mediators and international bodies," which naturally included monetary relief. Money is fungible, and Congress logically concluded that money a terrorist group such as the PKK obtains using the techniques plaintiffs propose to teach could be redirected to funding the group's violent activities.

Finally, plaintiffs propose to "engage in political advocacy on behalf of Kurds who live in Turkey," and "engage in political advocacy on behalf of Tamils who live in Sri Lanka." As explained above, plaintiffs do not specify their expected level of coordination with the PKK or LTTE or suggest what exactly their "advocacy" would consist of. Plaintiffs' proposals are phrased at such a high level of generality that they cannot prevail in this preenforcement challenge.

In responding to the foregoing, the dissent fails to address the real dangers at stake. It instead considers only the possible benefits of plaintiffs' proposed activities in the abstract. The dissent seems unwilling to entertain the prospect that training and advising a designated foreign terrorist organization on how to take advantage of international entities might benefit that organization in a way that facilitates its terrorist activities. In the dissent's world, such training is all to the good. Congress and the Executive, however, have concluded that we live in a different world: one in which the designated foreign terrorist organizations "are so tainted by their criminal conduct that any contribution to such an organization facilitates that conduct." One in which, for example, the United Nations High Commissioner for Refugees was forced to close a Kurdish refugee camp in northern Iraq because the camp had come under the control of the PKK, and the PKK had failed to respect its neutral and humanitarian nature. Training and advice on how to work with the United Nations could readily have helped the PKK in its efforts to use the United Nations camp as a base for terrorist activities.

If only good can come from training our adversaries in international dispute resolution, presumably it would have been unconstitutional to prevent American citizens from training the Japanese Government on using international organizations and mechanisms to resolve disputes during World War II. It would, under

the dissent's reasoning, have been contrary to our commitment to resolving disputes through "deliberative forces," for Congress to conclude that assisting Japan on that front might facilitate its war effort more generally. That view is not one the First Amendment requires us to embrace.

[W]e in no way suggest that a regulation of independent speech would pass constitutional muster, even if the Government were to show that such speech benefits foreign terrorist organizations. We also do not suggest that Congress could extend the same prohibition on material support at issue here to domestic organizations. We simply hold that, in prohibiting the particular forms of support that plaintiffs seek to provide to foreign terrorist groups, Section 2339B does not violate the freedom of speech.

Plaintiffs' final claim is that the material-support statute violates their freedom of association under the First Amendment. Plaintiffs argue that the statute criminalizes the mere fact of their associating with the PKK and the LTTE.

The Court of Appeals correctly rejected this claim because the statute does not penalize mere association with a foreign terrorist organization. As the Ninth Circuit put it: "The statute does not prohibit being a member of one of the designated groups or vigorously promoting and supporting the political goals of the group.... What Section 2339B prohibits is the act of giving material support...." Plaintiffs want to do the latter. Our decisions scrutinizing penalties on simple association or assembly are therefore inapposite.

Plaintiffs also argue that the material-support statute burdens their freedom of association because it prevents them from providing support to designated foreign terrorist organizations, but not to other groups. Any burden on plaintiffs' freedom of association in this regard is justified for the same reasons that we have denied plaintiffs' free speech challenge. It would be strange if the Constitution permitted Congress to prohibit certain forms of speech that constitute material support, but did not permit Congress to prohibit that support only to particularly dangerous and lawless foreign organizations. Congress is not required to ban material support to every group or none at all.

Holding

The Preamble to the Constitution proclaims that the people of the United States ordained and established that charter of government in part to "provide for the common defence." As Madison explained in Federalist No. 41, "[s]ecurity against foreign danger is... an avowed and essential object of the American Union." We hold that, in regulating the particular forms of support that plaintiffs seek to provide to foreign terrorist organizations, Congress has pursued that objective consistent with the limitations of the First and Fifth Amendments....

Questions for Discussion

1. Outline the process by which an organization is designated as a "foreign terrorist organization."

2. What types of activities do the plaintiffs wish to undertake on behalf of the PKK and LTTE?

3. Summarize the Court's holdings on the intent standard in the material-support statute and the Court's ruling on whether the law is unconstitutionally vague.

4. Why does Justice Roberts argue that it is constitutional to prohibit material support for a terrorist organization's criminal as well noncriminal activities. Explain how humanitarian or political or legal assistance to a terrorist organization furthers the organization's violent aims.

5. Will the Court's ruling subject individuals to criminal prosecution who write a letter to the editor of a newspaper in support of the PKK? Will the Court's ruling discourage this type of activity?

You can find more cases on the study site:
United States v. Shah, United States v. Khan, *www.sagepub.com/lippmanccl3e.*

CRIME IN THE NEWS

In November 2001, U.S. forces in Afghanistan captured a group of foreign fighters aligned with the Islamic fundamentalist Taliban regime. Among the captives was twenty-year-old John Walker Lindh, labeled by the media as the "American Taliban." The bearded Lindh, his face covered with dirt and grime, agreed to an interview with a cable news outlet, and Lindh's image filled the airwaves and the newspapers. Lindh related that he had entered Afghanistan to help the Taliban because they are the only government that is true to Islam and that he "definitely" believed that he was fighting for a just cause. The American public asked how a frail college-aged young person from northern California could end up fighting against the United States in Afghanistan. Were his parents correct that he had been brainwashed? His mother was quoted as observing that Lindh was a fearful and easily influenced young man who was "totally not streetwise." John's father insisted that "John loves America."

Lindh was the middle child and was named after the former member of The Beatles rock group John Lennon. His father was Catholic and worked as a government lawyer in Washington, D.C. His mother was a health care worker who became a Buddhist. The family moved north of San Francisco when John was ten, and he attended an alternative school for bright and independent students. His father, Frank Lindh, describes John as adept at music and languages and a serious student. John became interested in Islam at age twelve after watching the movie *Malcolm X,* which featured the religious pilgrimage (*Hajj*) to Mecca, Saudi Arabia, that is required of all Muslims. As John turned sixteen, he became a Muslim, regularly attended a mosque, adopted the names Suleyman al-Lindh and Suleyman al-Faris and occasionally wore a long white robe and turban. As his parents' marriage began to break apart in 1998, John asked for money to journey to Yemen to learn Arabic. He returned to California after a year and, still restless, returned to continue his studies in Yemen in February 2000. John e-mailed his parents following the terrorist attack on the *USS Cole* in Yemen in 2000 that the United States' docking of a naval ship in the foreign harbor was an act of war and that the *Cole*

had been properly targeted for attack. In October, John left for Pakistan and enrolled in an Islamic fundamentalist school (*madrasah*) where he joined the Harkadat-ul Mujahedeen-Al Almi, a terrorist group dedicated to liberating Kashmir from Indian control. John became disillusioned after twenty-four days of training and left to join the Taliban. He reportedly spent seven weeks at an al Qaeda camp where he was trained in weapons, explosives, battlefield technique, and map reading and reportedly met Osama bin Laden on several occasions and "swore allegiance to jihad" (armed struggle). In October 2001, Lindh was assigned with other foreign fighters to serve on the frontlines against the Northern Alliance, an opposition group aligned with the United States in seeking to forcefully remove the Taliban regime.

Lindh was discovered by American troops in a basement bunker with other captured Taliban soldiers. Lindh and the other fighters had retreated into the bunker after having launched a prison revolt that resulted in the death of American CIA agent Johnny Michael Spann.

Lindh entered into a plea agreement in which he agreed to plead guilty to supplying services to the Taliban in violation of an order issued by Presidents Clinton and Bush and of carrying a rifle and two hand grenades while engaged in a felony. In return, Lindh was sentenced to twenty years in prison. He also agreed to cooperate with investigators and to withdraw any claims of mistreatment. Some contend that the plea bargain reflected the fact that the government doubted whether it could obtain a conviction on treason. Others believed that the government did not want the public to learn of the physical abuse American interrogators allegedly directed toward Lindh.

In Lindh's fourteen-minute statement at sentencing, he condemned terrorism and denounced Osama bin Laden's attacks as contrary to Islamic teaching. Lindh explained that he had learned of the true nature of the Taliban and their repression of women only upon returning to the United States and that he had "made a mistake" by aligning himself with their cause. Are you persuaded by Lindh's apology? Did John Walker Lindh receive too lenient or too harsh a sentence?

Combat Immunity

American John Walker Lindh, the so-called American Taliban, was captured by American forces in Afghanistan and subsequently pled guilty to supplying services to the Taliban (the Islamic fundamentalist ruling party of Afghanistan) and of carrying an explosive during the commission of a felony. Lindh's lawyer made various arguments on his behalf, including **combat immunity**. This is the contention that as a member of the Afghan military, Lindh was immune from criminal prosecution for acts of lawful combat undertaken in the defense of Afghanistan against the United States. The U.S. government, however, contended that Lindh was not entitled to the status of a legal combatant and that his acts on behalf of the Taliban regime in Afghanistan were unlawful criminal offenses rather than acts of lawful warfare. The standard for determining whether an individual is a lawful or unlawful combatant is set forth in the **Geneva Convention of 1949**. The Geneva Convention is an international treaty regulating the law of war that the United States has signed and recognizes as part of American law.

The arguments discussed in *United States v. Lindh* raise the issue whether terrorists are entitled to be treated as prisoners of war rather than as unlawful fighters or as criminals. What is your view?

For an international perspective on this topic, visit the study site.

Is John Walker Lindh a lawful or unlawful combatant?

United States v. Lindh, *212 F. Supp. 2d 541 (E.D. Va. 2002). Opinion by: Ellis, J.*

John Phillip Walker Lindh is an American citizen who, according to the ten-count indictment filed against him in February 2002, joined certain foreign terrorist organizations in Afghanistan and served these organizations there in combat against Northern Alliance and American forces until his capture in November 2001. In seven threshold motions, Lindh sought dismissal of certain counts of the Indictment on a variety of grounds, including lawful combatant immunity.

Facts

The indictment's allegations may be succinctly summarized. In mid-2001, Lindh attended a military training camp in Pakistan run by Harakat ul-Mujahideen (HUM), a terrorist group dedicated to an extremist view of Islam. After receiving several weeks of training, Lindh informed HUM officials that "he wished to fight with the Taliban in Afghanistan." Thus, in May or June 2001, he traveled from Pakistan into Afghanistan "for the purpose of taking up arms with the Taliban," eventually arriving at a Taliban recruiting center in Kabul, Afghanistan—the Dar ul-Anan Headquarters of the Mujahideen. On his arrival, Lindh presented a letter of introduction from HUM and advised Taliban personnel "that he was an American and that he wanted to go to the front lines to fight." On October 8, 1997, HUM was designated by the Secretary of State as a foreign terrorist organization. . . .

According to the indictment, the Taliban is Afghanistan's dominant political force and its members, like the members of HUM, practice an extremist form of Islam. Specifically, members of the Taliban believe in conducting "jihad," or holy war, against those who they believe threaten their form of Islam, including the United States.

While at the Dar ul-Anan Headquarters, Lindh agreed to receive additional and extensive military training at an al Qaeda training camp. He made this decision "knowing that America and its citizens were the enemies of Bin Laden and al-Qaeda and that a principal purpose of al-Qaeda was to fight and kill Americans." In late May or June 2001, Lindh traveled to a bin Laden guest house in Kandahar, Afghanistan, where he stayed for several days, and then traveled to the al Farooq training camp, "an al Qaeda facility located several hours west of Kandahar." He reported to the camp with approximately twenty other trainees, mostly Saudis, and remained there throughout June and July. During this period, he participated fully in the camp's training activities, despite being told early in his stay that "Bin Laden had sent forth some fifty people to carry out twenty suicide terrorist operations against the United States and Israel." As part of his al Qaeda training, Lindh participated in "terrorist training courses in, among other things, weapons, orientating, navigation, explosives and battlefield combat." This training included the use of "shoulder weapons, pistols, and rocket-propelled grenades, and the construction of Molotov cocktails." During his stay at al Farooq, Lindh met personally with bin Laden, "who thanked him and other trainees for taking part in jihad." He also met with a senior al Qaeda official, Abu Mohammad Al-Masri, who inquired whether Lindh was interested in traveling outside Afghanistan to conduct operations against the United States and Israel. Lindh declined Al-Masri's offer in favor of going to the front lines to fight. It is specifically alleged that Lindh swore allegiance to jihad in June or July 2001.

The indictment alleges that al Qaeda is an organization, founded by Osama bin Laden and others, that is dedicated to opposing non-Islamic governments with force and violence. On October 8, 1999, al Qaeda was designated by the Secretary of State as a foreign terrorist organization, pursuant to section 219 of the Immigration and Nationality Act. . . .

When Lindh completed his training at al Farooq in July or August 2001, he traveled to Kabul, Afghanistan, where he was issued an AKM rifle "with a barrel suitable for long range shooting." Armed with this rifle, Lindh, together with approximately 150 non-Afghani fighters, traveled from Kabul to the front line at Takhar, located in northeastern Afghanistan, where the entire unit was placed under the command of an Iraqi named Abdul Hady. Lindh's group was eventually divided into smaller groups that fought in shifts against Northern Alliance troops in the Takhar trenches, rotating every one to two weeks. During this period, Lindh "carried various weapons with him, including the AKM rifle, an RPK rifle he was issued after the AKM rifle malfunctioned, and at least two grenades." He remained with his fighting group following the September 11, 2001 terrorist attacks, "despite having been told that Bin Laden had ordered the [September 11] attacks, that additional terrorist attacks were planned, and that additional al Qaeda personnel were being sent from the front lines to protect Bin Laden and defend against an anticipated military response from the United States." Indeed, it is specifically alleged that Lindh remained with his fighting group from October to December 2001, "after learning that United States military forces and United States nationals had become directly engaged in support of the Northern Alliance in its military conflict with Taliban and al Qaeda forces."

In November 2001, Lindh and his fighting group retreated from Takhar to the area of Kunduz, Afghanistan, where they ultimately surrendered to Northern Alliance troops. On November 24, 2001, he and the other captured Taliban fighters were transported to Mazar-e-Sharif and then to the nearby Qala-i-Janghi (QIJ) prison compound. The following day, November 25, Lindh was interviewed by two Americans—Agent Johnny Michael Spann from the Central Intelligence Agency (CIA) and another government employee. Later that day, it is alleged that Taliban detainees in the QIJ compound attacked Spann and the other employee, overpowered the guards, and armed themselves. Spann was shot and killed in the course of the uprising, and Lindh, after being wounded, retreated with other detainees to a basement area of the QIJ compound. The uprising at QIJ was eventually suppressed on December 1, 2001, at which time Lindh and other Taliban and al Qaeda fighters were taken into custody by Northern Alliance and American forces.

Following his capture, Lindh was interrogated, transported to the United States, and ultimately charged in this district with the following offenses in a ten-count indictment:

1. Conspiracy to murder nationals of the United States, including American military personnel and other governmental employees serving in Afghanistan following the September 11, 2001 terrorist attacks.

2. Conspiracy to provide material support and resources to HUM, a foreign terrorist organization.

3. Providing material support and resources to HUM.

4. Conspiracy to provide material support and resources to al Qaeda, a foreign terrorist organization.

5. Providing material support and resources to al Qaeda.

6. Conspiracy to contribute services to al Qaeda.

7. Contributing services to al Qaeda.

8. Conspiracy to supply services to the Taliban.

9. Supplying services to the Taliban.

10. Using and carrying firearms and destructive devices during crimes of violence.

Issue

Lindh claims that Count One of the indictment should be dismissed because, as a Taliban soldier, he was a lawful combatant entitled to the affirmative defense of lawful combatant immunity. Lindh makes no claim of lawful combatant immunity with respect to the indictment's allegations that he was a member or soldier of al Qaeda. Instead, Lindh focuses his lawful combatant immunity argument solely on the indictment's allegations that he was a Taliban member. This focus is understandable as there is no plausible claim of lawful combatant immunity in connection with al Qaeda membership. Thus, it appears that Lindh's goal is to win lawful combatant immunity with respect to the Taliban allegations and then to dispute factually the indictment's allegations that he was a member of al Qaeda.

Also worth noting is that the Government has not argued here that the Taliban's role in providing a home, a headquarters, and support to al Qaeda and its international terrorist activities serve to transform the Taliban from a legitimate state government into a terrorist institution whose soldiers are not entitled to lawful combatant immunity status. Put another way, the government has not argued that al Qaeda controlled the Taliban for its own purposes and that so-called Taliban soldiers were accordingly merely agents of al Qaeda, not lawful combatants.

Reasoning

Lawful combatant immunity, a doctrine rooted in the customary international law of war, forbids prosecution of soldiers for their lawful belligerent acts committed during the course of armed conflicts against legitimate military targets. Belligerent acts committed in armed conflict by enemy members of the armed forces may be punished as crimes under a belligerent's municipal law only to the extent that they violate international humanitarian law or are unrelated to the armed conflict. This doctrine has a long history, which is reflected in part in various early international conventions, statutes, and documents. But more pertinent, indeed controlling, here is that the doctrine also finds expression in the Geneva Convention Relative to the Treatment of Prisoners of War, August 12, 1949, to which the United States is a signatory. Significantly, Article 87 of the Geneva Principles of War [GPW] admonishes that combatants "may not be sentenced . . . to any penalties except those provided for in respect of members of the armed forces of the said Power who have committed the same acts." Similarly, Article 99 provides that "no prisoner of war may be tried or sentenced for an act which is not forbidden by the law of the Detaining Power or by international law, in force at the time the said act was committed." These articles, when read together, make clear that a belligerent in a war cannot prosecute the soldiers of its foes for the soldiers' lawful acts of war. The United States Constitution provides that "[t]his Constitution, and the laws of the United States which shall be made in pursuance thereof, and all treaties made, or which shall be made, under the authority of the United States, shall be the supreme law of the land. . . ."

Importantly, this lawful combatant immunity is not automatically available to anyone who takes up arms in a conflict. Rather, it is generally accepted that this immunity can be invoked only by members of regular or irregular armed forces who fight on behalf of a state and comply with the requirements for lawful combatants. Thus, it is well established that the law of war draws a distinction between the armed forces and the peaceful populations of belligerent nations and also between those who are lawful and unlawful combatants. Lawful combatants are subject to capture and detention as prisoners of war [POW] by opposing military forces. Unlawful combatants are likewise subject to capture and detention, but in addition, they are subject to trial and punishment by military tribunals for acts that render their belligerency unlawful.

The starting point in the analysis of Lindh's immunity claim is recognition that the President has unequivocally determined that Lindh, as a member of the Taliban, is an unlawful combatant and, as such, may not invoke lawful combatant immunity. On February 7, 2002, the White House announced the President's decision, as Commander-in-Chief, that the Taliban militia were unlawful combatants pursuant to GPW and general principles of international law, and therefore, they were not entitled to POW status under the Geneva Conventions. . . . The appropriate deference is to accord substantial or great weight to the President's decision regarding the interpretation and application of the GPW to Lindh, provided the interpretation and application of the treaty to Lindh may be said to be reasonable and not contradicted by the terms of the treaty or the facts. It is this proviso that is the focus of the judicial review here of the President's determination that Lindh is an unlawful combatant under the GPW.

The GPW sets forth four criteria an organization must meet for its members to qualify for lawful combatant status:

1. The organization must be commanded by a person responsible for his subordinates;

2. The organization's members must have a fixed distinctive emblem or uniform recognizable at a distance;

3. The organization's members must carry arms openly; and

4. The organization's members must conduct their operations in accordance with the laws and customs of war.

Lindh asserts that the Taliban is a "regular armed force," under the GPW, and because he is a member, he need not meet the four conditions of the Hague Regulations because only Article 4(A)(2), which addresses irregular armed forces, explicitly mentions the four criteria. This argument is unpersuasive; it ignores long-established practice under the GPW and, if accepted, leads to an absurd result. First, the four criteria have long been understood under customary international law to be the defining characteristics of any lawful armed force. Thus, all armed forces or militias, regular and irregular, must meet the four criteria if their members are to receive combatant immunity. Were this not so, the anomalous result that would follow is that members of an armed force that met none of the criteria could still claim lawful combatant immunity merely on the basis that the organization calls itself a "regular armed force." It would indeed be absurd for members of a so-called "regular armed force" to enjoy lawful combatant immunity even though the force had no established command structure and its members wore no recognizable symbol or insignia, concealed their weapons, and did not abide by the customary laws of war. Simply put, the label "regular armed force" cannot be used to mask unlawful combatant status.

The Taliban lacked the command structure necessary to fulfill the first criterion as it is manifest that the Taliban had no internal system of military command or discipline. Similarly, it appears the Taliban typically wore no distinctive sign that could be recognized by opposing combatants; they wore no uniforms or insignia and were effectively indistinguishable from the rest of the population. The requirement of such a sign is critical to ensure that combatants may be distinguished from the noncombatant, civilian population. Accordingly, Lindh cannot establish the second criterion.

Next, although it appears that Lindh and his cohorts carried arms openly in satisfaction of the third criterion for lawful combatant status, it is equally apparent that members of the Taliban failed to observe the laws and customs of war. Thus, because record evidence supports the conclusion that the Taliban regularly targeted civilian populations in clear contravention of the laws and customs of war, Lindh cannot meet his burden concerning the fourth criterion.

Holding

In sum, the President's determination that Lindh is an unlawful combatant and thus ineligible for immunity is controlling here because that determination is entitled to deference as a reasonable interpretation and application of the GPW to Lindh as a Taliban; because Lindh has failed to carry his burden of demonstrating the contrary; and because even absent deference, the Taliban falls far short when measured against the four GPW criteria for determining entitlement to lawful combatant immunity.

Questions for Discussion

1. What are the four standards in the Geneva Convention for determining lawful combatant status?

2. Why did the Taliban and Lindh not qualify as lawful combatants under the Geneva Convention?

3. What is the significance of whether an individual is a lawful or unlawful combatant?

4. Did Lindh commit treason against the United States? Was he punished too leniently or too harshly?

5. The four standards in the Geneva Convention seemingly preclude terrorists from being treated as prisoners of war. Should the standards be relaxed so that terrorists may be considered prisoners of war?

State Terrorism Statutes

Virtually every state has a terrorism statute to cover criminal acts that do not fall within federal jurisdiction. Consider the "beltway sniper" case from Virginia, *Muhammad v. Commonwealth*.

Were the shootings acts of terrorism?

Muhammad v. Commonwealth, 619 S.E.2d 16 (Va. 2005). Opinion by: Lemons, J.

In these appeals, we consider two capital murder convictions and two death sentences imposed upon John Allen Muhammad, along with his convictions for conspiracy to commit capital murder and the illegal use of a firearm in the commission of murder. This prosecution arose from the investigation of a series of sixteen shootings, including ten murders that occurred in Alabama, Louisiana, Maryland, Washington, D.C., and Virginia over a forty-seven-day period from September 5 to October 22, 2002. For the reasons discussed herein, the judgment of the trial court and the sentences of death will be affirmed.

Facts

On the morning of Wednesday, October 9, 2002, Dean H. Meyers was shot and killed while fueling his car at the Sunoco gas station on Sudley Road in Manassas, Virginia. Meyers was shot in the head by a single bullet. The bullet entered behind his left ear, where it

fragmented into multiple small pieces. The bullet fragments shattered the temporal bone, and the fragments of bullet and bone then traveled through his brain and caused multiple fractures of his skull. This gunshot wound was consistent with injuries from a bullet fired from a high-velocity rifle and was the cause of Meyers's death. Evidence at trial established that the bullet came from the .223 caliber Bushmaster rifle Muhammad possessed when he was arrested. An eyewitness testified that she saw Muhammad and Lee Boyd Malvo in the vicinity of the shooting approximately one hour beforehand. Police interviewed Muhammad immediately after the shooting in a parking lot across the street from where Meyers was shot. In both encounters, Muhammad was driving a Chevrolet Caprice in which he was later arrested. Muhammad's fingerprints were on a map police found in the parking lot where Muhammad had been interviewed. Meyers was killed during a forty-seven-day period, from September 5 to October 22, 2002, in which ten others were murdered and six more suffered gunshot wounds as a result of the acts of Muhammad and Malvo in concert. The murder of Meyers was the twelfth of these sixteen shootings.

The first shooting occurred in Clinton, Maryland, on September 5, 2002. Paul J. LaRuffa, the owner of Margellina's Restaurant, left the restaurant at closing and proceeded to his car with his briefcase and Sony portable computer. Inside the briefcase were bank deposit bags that contained $3,500 in cash and credit card receipts from that evening. LaRuffa placed the briefcase and laptop on the backseat of his car and then sat behind the steering wheel. He testified that almost immediately after he sat down, he saw a figure to his left and a flash of light. He heard gunshots and the driver's side window shattered. When he stepped out of his car, he realized he had been shot. The trauma surgeon who treated him testified that LaRuffa was shot six times: once in the back left side of his neck, three times in the left side of his chest, and twice in his left arm.

An employee who left the restaurant with LaRuffa, Paul B. Hammer, witnessed the shooting and called "911." Hammer testified that he saw a "kid" run up to LaRuffa's car, fire shots into it, and then open the rear door and take the briefcase and portable computer. He was unable to provide a detailed description because of lighting conditions but testified that the shooter was a male in his late teens or early twenties. The briefcase and empty bank deposit bags, along with a pair of pants and a shirt, were found six weeks later in a wooded area about a mile from the shooting. Hair on the clothing yielded DNA that was consistent with Malvo's DNA.

Four days later, on September 9, Muhammad purchased a 1990 Caprice automobile from Christopher M. O'Kupski in Trenton, New Jersey. O'Kupski testified that before the purchase, Muhammad got into the trunk and lay down. O'Kupski also testified that when Muhammad purchased it, the Caprice did not have a hole in the trunk or a passageway from the backseat to the trunk, the trunk was not spray-painted blue, and the windows were not tinted.

The second shooting occurred in Clinton, Maryland, on September 15, 2002. Muhammad Rashid was closing the Three Roads Liquor Store. Rashid testified that he noticed the Caprice outside the store shortly before closing. He testified that he was in the process of locking the front door from the outside when he heard gunshots from behind him. At the same time, a young man with a handgun rushed towards Rashid and shot Rashid in the stomach. At trial, Rashid identified Malvo as the person who shot him. Two bullets were removed from inside the store. The bullets had been shot through the front door and the trajectory of the bullets placed the shooter in a field across the street from the store.

The third and fourth shootings occurred in Montgomery, Alabama, on September 21, 2002. Claudine Parker and Kelly Adams closed the Zelda Road ABC Liquor Store and walked out. They were shot immediately. Parker died as a result of a single gunshot wound that entered her back, transected her spinal cord, and passed through her lung. Adams was shot once through her neck but lived. The bullet exited through her chin, breaking her jaw in half, shattering her face and teeth, paralyzing her left vocal cord, and severing major nerves to her left shoulder. Both gunshot wounds were consistent with injuries caused by a high-velocity rifle. Testing revealed that the bullet fragments recovered from the Parker shooting were fired from a Bushmaster rifle possessed by Muhammad when he was arrested.

As the rifle shots were fired, a young man, later identified as Malvo, ran up to Parker and Adams. A police car happened to pass the scene immediately after the shots were fired. A police officer observed Malvo with a handgun. He was going through the women's purses. The officer and another eyewitness chased Malvo. Although he escaped, Malvo dropped an "ArmorLite" gun catalogue during the chase. At trial, both the officer and the other eyewitness identified Malvo as the young man with the handgun who fled the scene. Additionally, Malvo's fingerprints were on the "ArmorLite" gun catalogue he dropped during the chase. The handgun Malvo carried that evening, a .22 caliber stainless steel revolver, was found in the stairwell of an apartment building that Malvo ran through during the chase. Forensic tests determined that this .22 caliber revolver was the same gun used to shoot both LaRuffa and Rashid.

The fifth shooting occurred in Baton Rouge, Louisiana, on September 23. Hong Im Ballenger, the manager of the Beauty Depot store, closed the store for the evening. As she was walking to her car, she was shot once in the head with a bullet fired from a high-velocity rifle. Ballenger died as the result of the single shot. The bullet entered the back of her head and exited through her jawbone. The wound caused massive bleeding and compromised her airway. Ballistic tests determined that the bullet fragments recovered from Ballenger were fired from the Bushmaster rifle possessed by Muhammad when he was arrested. An eyewitness saw a young man leave the scene with Ballenger's purse. At trial, this young man was identified as Malvo. Another eyewitness saw Malvo flee the scene with Ballenger's purse and get into the Caprice.

The sixth shooting occurred in Silver Spring, Maryland, on October 3, 2002. At approximately 8:15 a.m., Premkumar A. Walekar was fueling his taxicab. He was shot once with a bullet from a high-velocity rifle. The bullet passed through his left arm and then entered his chest, where it broke two ribs, shredded portions of his lungs, and damaged his heart. A physician, who was fueling her car next to Walekar, attempted CPR but was unsuccessful. Ballistic tests established that bullet fragments recovered from the Walekar shooting were fired from the Bushmaster rifle possessed by Muhammad when he was arrested.

The seventh shooting occurred in Silver Spring, Maryland, on October 3, 2002. At approximately 8:30 a.m., Sarah Ramos was sitting on a bench in front of the Crisp & Juicy Restaurant in the Leisure World Shopping Center. She was shot once with a bullet from a high-velocity rifle. The bullet entered the front of her head and exited through her spinal cord at the top of her neck. An eyewitness identified the Caprice at the scene prior to the shooting. Bullet fragments recovered from the Ramos shooting were fired from the Bushmaster rifle possessed by Muhammad when he was arrested.

The eighth shooting occurred in Kensington, Maryland, on October 3, 2002. At approximately 10:00 a.m., Lori Lewis-Rivera was vacuuming her car at the Shell gas station on the corner of Connecticut Avenue and Knowles Avenue. She was shot once in the back by a bullet from a high-velocity rifle as she vacuumed her car. An eyewitness testified that he saw the Caprice in the vicinity of the gas station approximately twenty minutes before the shooting. Bullet fragments recovered from the Lewis-Rivera shooting were fired from the Bushmaster rifle possessed by Muhammad when he was arrested.

The ninth shooting occurred in Washington, D.C. on October 3, 2002. At approximately 7:00 p.m., a police officer stopped Muhammad for "running" two stop signs. The police officer testified that the windows of the Caprice were heavily tinted and that he could not see anyone else in the car. The police officer gave Muhammad a verbal warning and let him go.

At approximately 9:15 p.m. on that day, Paschal Charlot was shot in the chest as he crossed the intersection of Georgia Avenue and Kalmia Road. This intersection was about thirty blocks from where the police officer stopped Muhammad. The bullet entered Charlot's chest and shattered his collarbone and three ribs before lacerating his lungs. Charlot died before emergency personnel arrived. Eyewitnesses testified that they saw the Caprice at the scene at the time of the shooting and that the driver drove away without its headlights on immediately after the shooting. It had been parked in a space on the street with its trunk positioned toward Georgia Avenue. One eyewitness testified that he saw a flash of light from the Caprice at the time the shot was fired. Ballistics tests determined that the bullet fragments recovered from the Charlot shooting were fired from the Bushmaster rifle possessed by Muhammad when he was arrested.

The tenth shooting occurred in Fredericksburg, Virginia, on October 4, 2002. Caroline Seawell had finished shopping at a Michael's Craft Store and was putting her bags in her minivan, when she was shot once in the back by a bullet from a high-velocity rifle. The bullet severely damaged her liver and exited through her right breast. Seawell survived the shooting. An eyewitness testified that he saw the Caprice in the parking lot at the time of the shooting. Ballistics tests determined that the bullet fragments recovered from the

Seawell shooting were fired from the Bushmaster rifle possessed by Muhammad when he was arrested.

The eleventh shooting occurred in Bowie, Maryland, on October 6, 2002. Tanya Brown ("Tanya") took Iran Brown ("Brown") to Tasker Middle School. As Brown was walking on the sidewalk to the school, he was shot once in the chest by a bullet from a high-velocity rifle. Tanya decided not to wait for emergency personnel and drove Brown to a health care center. Brown's lungs were damaged, there was a large hole in his diaphragm, the left lobe of his liver was damaged, and his stomach, pancreas, and spleen were lacerated by bullet fragments. Surgeons were able to save Brown's life, and he spent eight weeks recovering in the hospital.

Two eyewitnesses testified that they saw the Caprice in the vicinity of Tasker Middle School the day before the shooting and the morning of the shooting. One of these eyewitnesses positively identified both Muhammad and Malvo in the Caprice the morning of the shooting. They were seen in the Caprice, which was parked at an intersection with a line of sight to the school. Following the shooting, police searched the surrounding area and found a ballpoint pen and a shell casing in the woods next to the school. The pen and shell casing were located in an area that had been patted down like a hunting blind. This blind offered a clear line of sight to the scene of the shooting. Tissue samples from the pen matched Muhammad's DNA. The shell casing had been fired by the Bushmaster rifle possessed by Muhammad when he was arrested, and tests determined that the bullet fragments recovered from Brown were fired from that rifle. In the woods, police also found the first communication from Muhammad and Malvo. A tarot card, the one for death, was found with handwriting that stated, "Call me God." On the back of the card was handwriting that stated, "For you, Mr. Police. Code: Call me God. Do not release to the Press."

The twelfth shooting, discussed above, was the murder of Dean Meyers in Manassas, Virginia, on October 9, 2002.

The thirteenth shooting occurred in Massaponax, Virginia, on October 11, 2002. Kenneth Bridges was at an Exxon gas station on Jefferson Davis Highway. He was shot once in the chest by a bullet from a high-velocity rifle. The bullet damaged his lungs and heart, causing fatal internal injuries. Two eyewitnesses testified that they saw the Caprice at or near the Exxon station on the morning of the shooting. Ballistics tests determined that the bullet fragments recovered from the Bridges shooting were fired from the Bushmaster rifle possessed by Muhammad when he was arrested.

The fourteenth shooting occurred in Falls Church, Virginia, on October 14, 2002. Linda Franklin and her husband were shopping at a Home Depot store. As they loaded their purchases in their car, Franklin was shot and killed by a single bullet from a high-velocity rifle. The bullet entered the left side of her head, passed through her brain and skull, and exited from the right side of her head. An off-duty police officer testified that she saw Malvo driving the Caprice in the vicinity of the shooting immediately after it occurred. Tests determined that bullet fragments recovered from the Franklin shooting were fired from the Bushmaster rifle possessed by Muhammad when he was arrested.

On October 15, the day after Franklin was murdered, a Rockville, Maryland, police dispatcher received a telephone call in which the caller stated: "Don't say anything, just listen, we're the people who are causing the killings in your area. Look on the tarot card, it says, 'call me God, do not release to press.' We've called you three times before trying to set up negotiations. We've gotten no response. People have died." The dispatcher attempted to transfer the call to the Sniper Task Force, but the caller hung up.

Three days later, on October 18, Officer Derek Baliles, a Montgomery County, Maryland, Police Information Officer, received a telephone call. The caller told Officer Baliles to "shut up" and stated that he knew who was doing the shootings but wanted the police officer to verify some information before he talked further. The caller told Officer Baliles to verify information concerning a shooting at a liquor store near "Ann Street." The caller gave Officer Baliles the name and telephone number of a police officer in Alabama. Officer Baliles confirmed the shootings of Parker and Adams. The caller called Officer Baliles again. Officer Baliles told him that he had verified the information concerning the shootings of Parker and Adams. The caller then said that he had to find more coins for the call and had to find a telephone without surveillance and then hung up.

On the same day, William Sullivan, a priest in Ashland, Virginia, received a telephone call from two people. The first voice, a male, told him someone wanted to speak with him. Sullivan testified that a second male voice, told him that "the lady didn't have to die," and "it was at the Home Depot." The second voice also told him about a shooting at a liquor store in Alabama and then said, "Mr. Policeman, I am God. Do not tell the press." The second voice concluded by telling Sullivan to give this information to the police.

The fifteenth shooting occurred in Ashland, Virginia, on October 19, 2002. Jeffrey Hopper and his wife stopped in Ashland to fuel their car and eat dinner. They left the restaurant and were walking to their car when Hopper was shot in the abdomen. Hopper survived the shooting but underwent five surgeries to repair his pancreas, stomach, kidneys, liver, diaphragm, and intestines. In the woods near the shooting, police found a hunting-type blind similar to the one found at the Brown shooting. At the blind, police found a shell casing, a plastic sandwich bag attached to a tree with a thumbtack at eye level that was decorated with Halloween characters and self-adhesive stars, and a candy wrapper. Tests determined that the shell casing and bullet fragments recovered from the Hopper shooting came from the Bushmaster rifle possessed by Muhammad when he was arrested. Surveillance videotapes identified Muhammad in a Big Lots Store on October 19, 2002, near the shooting from which the plastic sandwich bag and decorations were likely obtained. The candy wrapper contained both Malvo's and Muhammad's DNA. Police also found a handwritten message in the plastic sandwich bag that read:

> For you Mr. Police. Call me God. Do not release to the Press. We have tried to contact you to start negotiation. . . . These people took our call for a Hoax or Joke, so your failure to respond has cost you five lives. If stopping the killing is more important than catching us now, then you will accept our demand which are non-negotiable. (i) You will place ten million dollar in Bank of america account. . . . We will have unlimited withdrawl at any atm worldwide. You will activate the bank account, credit card, and pin number. We will contact you at Ponderosa Buffet, Ashland, Virginia, tel. # . . . 6:00 am Sunday Morning. You have until 9:00 a.m. Monday morning to complete transaction. Try to catch us withdrawing at least you will have less body bags.
>
> (ii) If trying to catch us now more important then prepare you body bags.
>
> If we give you our word that is what takes place.
>
> Word is Bond
>
> P.S. Your children are not safe anywhere at anytime.

The note was not found until after the deadline had passed. The day after Hopper was shot at the Ponderosa, an FBI agent operating the "Sniper Tip Line" received a call from a young male who said, "Don't talk. Just listen. Call me God. I left a message for you at the Ponderosa. I am trying to reach you at the Ponderosa. Be there to take a call in ten minutes." On October 21, 2002, an FBI agent received a call to the FBI negotiations team, which had been rerouted from the Ponderosa telephone number referenced in the note left after the Hopper shooting. A recorded voice stated:

> Don't say anything. Just listen. Dearest police, Call me God. Do not release to the press. Five red stars. You have our terms. They are non-negotiable. If you choose Option 1, you will hold a press conference stating to the media that you believe you have caught the sniper like a duck in a noose. Repeat every word

exactly as you heard it. If you choose Option 2, be sure to remember we will not deviate. P.S.— Your children are not safe.

The sixteenth shooting occurred in Aspen Hill, Maryland, on October 22, 2002. At approximately 6:00 a.m., Conrad Johnson, a bus driver for the Montgomery County Transit Authority, was shot in the chest at the entrance to his bus. Johnson remained conscious until rescue workers arrived but died at the hospital. A single high-velocity rifle bullet killed Johnson. The bullet entered his right chest and caused massive damage to his diaphragm, liver, pancreas, kidneys, and intestines. Tests determined that the bullet fragments recovered from the Johnson shooting were fired from the Bushmaster rifle possessed by Muhammad when he was arrested. A hunting-type blind, similar to those found at the Brown and Hopper shootings, was found in the woods near where Johnson was shot. A black duffle bag and a left-handed glove were found. A hair from the duffle bag yielded DNA that matched Muhammad's DNA. The police also found another plastic sandwich bag, which contained a note and self-adhesive stars.

Muhammad and Malvo were captured and arrested on October 24, 2002, by agents of the FBI at a rest area in Frederick County, Maryland. They were asleep in the Caprice at the time of their capture. Inside the Caprice, police found a loaded .223 caliber Bushmaster rifle behind the rear seat. Tests determined that the DNA on the Bushmaster rifle matched the DNA of both Malvo and Muhammad. The only fingerprints found on the Bushmaster rifle were those of Malvo. The Caprice had been modified after Muhammad purchased it from O'Kupski. The windows were heavily tinted. The rear seat was hinged, providing easy access to the trunk from the passenger compartment. The trunk was spray-painted blue. A hole had been cut into the trunk lid, just above the license plate. The hole was blocked by a right-handed brown glove that matched the left-handed glove found in the woods near the Johnson shooting. The trunk also had a rubber seal that crossed over the hole.

Inside the Caprice, police found a global positioning system (GPS) receiver, a magazine about rifles, an AT&T telephone charge card, ear plugs, maps, plastic sandwich bags, a rifle scope, .223 caliber ammunition, "walkie-talkies," a digital voice recorder, a receipt from a Baton Rouge, Louisiana, grocery store dated September 27, 2002, an electronic organizer, a plastic bag from a Big Lots Store, a slip of paper containing the Sniper Task Force phone number, and a list of schools in the Baltimore area.

Police also found LaRuffa's portable computer in the Caprice. Muhammad had loaded software entitled "Microsoft Streets and Trips 2002" onto this computer on September 29, 2002. In this program, there were various maps showing particular routes and places

marked with icons, some with a skull and crossbones. Icons had been added to mark the places where Walekar, Lewis-Rivera, Seawell, Brown, Meyers, and Franklin were shot. There was also a Microsoft Word file titled "Allah 8.rtf" that contained portions of the text communicated to police in the extortion demands.

Subsequent to his arrest on October 24, 2002, Muhammad was indicted by a grand jury on October 28, 2002, for the capital murder of Meyers in the commission of an act of terrorism, capital murder of Meyers and at least one other person within a three-year period, conspiracy to commit capital murder, and illegal use of a firearm in the commission of capital murder.

From October 20 through November 17, 2003, Muhammad was tried before a jury in the Circuit Court of the City of Virginia Beach. The jury convicted Muhammad of all charges in the grand jury indictments. In a separate sentencing proceeding from November 17 through November 24, 2003, the jury sentenced Muhammad to two death sentences for the capital murder convictions, finding both the future dangerousness and vileness aggravating factors. The jury also sentenced Muhammad to thirteen years in prison upon the remaining convictions. At the conclusion of the sentencing proceeding, venue was transferred back to the Circuit Court of Prince William County. On March 9, 2004, the trial court imposed the two death sentences and the sentences of imprisonment as fixed by the jury. A final sentencing order was entered on March 29, 2004.

Issue

Muhammad maintains that the terrorism statutes, Code sections 18.2-31(13) and 18.2-46.4, are unconstitutionally overbroad and vague. We disagree.

Reasoning

A successful challenge to the facial validity of a criminal statute based upon vagueness requires proof that the statute fails to provide notice sufficient for ordinary people to understand what conduct it prohibits or proof that the statute "may authorize and even encourage arbitrary and discriminatory enforcement." But "one to whose conduct a statute clearly applies may not successfully challenge it for vagueness." Capital murder pursuant to Code is defined as the "willful, deliberate and premeditated killing of any person by another in the commission of or attempted commission of an act of terrorism as defined in Code § 18.2-46.4."

Act of terrorism means an act of violence as defined in clause (i) of subdivision A of section 19.2-297.1 committed with the intent to (i) intimidate the civilian population at large or (ii) influence the conduct or activities of the government of the United States, a state or locality through intimidation. Code section 18.2-46.4. The "act of violence" reference to Code section 19.2-297.1 includes a list of certain specific aggravated felonies including murder, voluntary manslaughter, mob-related felonies, malicious assault or bodily wounding, robbery, carjacking, sexual assault, and arson. The combination of these statutes defines criminal conduct that constitutes a willful, deliberate, and premeditated killing in the commission, or attempted commission, of one of the designated felonies with the intent to intimidate the civilian population or influence the conduct of government through intimidation. Additionally, under Code section 18.2-18, the General Assembly extended the reach of criminal conduct subject to the death penalty to include "a killing pursuant to the direction or order of one who is engaged in the commission of or attempted commission of an act of terrorism under the provisions of subdivision 13 of § 18.2-31."

Muhammad raises questions about the definition of *intimidation,* "civilian population at large," and "influence the conduct or activities of government." He suggests that failure to statutorily define these phrases renders the statutes unconstitutional. He further complains that "no distinction can be drawn between the newly defined crime and any 'base offense' which carries with it the same hallmarks of intimidation and influence" and that this allows "unguided and unbridled law enforcement discretion." Muhammad further maintains that extending the scope of the statute to reach those who order or direct a killing in the commission of or attempted commission of an act of terrorism somehow violates what he calls the "triggerman rule." In a particularly exaggerated statement, Muhammad claims that extending the scope of the statute "allows almost any violent criminal act to be classified as terrorism and thereby rendering any individual charged eligible for the death penalty." We disagree with each of Muhammad's contentions.

By referencing established criminal offenses as acts of violence subject to the statutory scheme, the legislature included clearly defined offenses. . . . Additionally, the term *intimidate* has been defined by case law (defining intimidation as unlawful coercion, extortion, duress, putting in fear). We have no difficulty understanding that *population at large* is a term that is intended to require a more pervasive intimidation of the community rather than a narrowly defined group of people. . . . We do not believe that a person of ordinary intelligence would fail to understand this phrase.

Similarly, we do not believe that a person of ordinary intelligence needs further definition of the phrase "influence the conduct or activities of government." . . . Muhammad claims that the statutes are designed "to address al-Qaeda type attacks—attacks motivated by a greater political purpose." . . . Nothing in the words of these statutes evinces an intent to limit its application to criminal actors with political motives.

Muhammad maintains that there is no distinction between the "base offense" and the capital offense based upon terrorism. What he appears to be arguing is that the terrorism statute is unnecessary on the one hand because a killing in the commission of one of the enumerated violent acts could result in the death penalty anyway, and on the other hand, its reach is extended too far by including those who order or direct such killings. Clearly, the General Assembly has the power to define criminal conduct even if statutes overlap in coverage. Whether a defendant can be simultaneously or successively charged with overlapping offenses implicates other questions not presented here.

Muhammad's quarrel with the expansion of the potential imposition of the death penalty to those who order or direct another in a killing in the commission of or attempted commission of an act of terrorism is a policy question well within the purview of legislative power so long as it is not otherwise unconstitutional. In that respect, Muhammad argues in Assignment of Error 18 that the provisions of Code section 18.2-18 allow the death penalty for a defendant with no demonstrated intent to kill the victim. Muhammad incorrectly characterizes the extension of the scope of the statute to reach traditional "aiders and abettors." The provisions of Code section 18.2-18 do not extend to "aiders and abettors"; rather, it extends only to those who "direct" or "order" the killing. The criminal actor who "orders" or "directs" the killing is not unlike the criminal actor who hires another to kill and is potentially subject to the death penalty under Code section 18.2-31(2). The criminal actor who "orders" or "directs" the killing shares the intent to kill with the one who carries out the murder. The provisions of Code section 18.2-18 do not have the effect imagined by Muhammad.

Holding

Muhammad's argument concerning vagueness does not focus on his conduct. Indeed, Muhammad does not claim in his brief that his actions and those of Malvo were not acts of terrorism under the statutory provisions. Rather, Muhammad hypothetically poses questions about the applicability of the statute in other circumstances. As discussed above, the statutes provide notice sufficient for ordinary people to understand what conduct they prohibit and do not authorize and/or encourage arbitrary and discriminatory enforcement. More importantly, Muhammad cannot and does not maintain that the statutes do not give him notice that his conduct and Malvo's conduct was prohibited. Nor does Muhammad allege that he has been subject to arbitrary or discriminatory enforcement of the statutes. One who engages in conduct that is clearly proscribed and not constitutionally protected may not successfully attack a statute as void for vagueness based upon hypothetical conduct of others.

Cases and Comments

1. ***The September 11 Attacks.*** On September 11, 2001, nineteen foreign nationals, functioning as separate terrorist teams, boarded and took control of four civilian aircraft. Two planes crashed into the twin towers of the World Trade Center in New York and a third careened into the Pentagon in Arlington, Virginia. The passengers on the fourth plane, realizing that the hijackers were intent on directing the plane into yet another government building, bravely resisted the hijackers, who responded by sending the plane spiraling into a Pennsylvania field. The terrorists' kamikaze attacks transformed the three aircraft and the 200,000 pounds of jet fuel into weapons of mass destruction and led to the death of roughly 3,000 people in the World Trade Center and hundreds of others at the Pentagon. Osama bin Laden, the leader of the al Qaeda terrorist organization, praised this as "good terror" and warned that the "battle has been moved inside America." He went on to proclaim that every American constituted the enemy and was to be killed.

The "suicide bombing" of September 11, 2001, ushered in what experts term "the new terrorism." This is characterized by religiously motivated groups who believe that the world must be destroyed in order to be saved from the "sex, drugs, and rock-and-roll" of American culture. The fear is that these terrorists may resort to the use of weapons of mass destruction.

The international community has been slow to respond to the threat of terrorism. The assassination of King Alexander I of Yugoslavia and Mr. Louis Barthou, Foreign Minister of France, on October 1, 1934, led the League of Nations (the early version of the United Nations) to formulate the 1937 Convention for the Prevention and Punishment of Terrorism. This required states to punish attacks committed on their territory against another country's leaders and officials. States that failed to prosecute the perpetrators were to send (extradite) the offenders to trial in the victim's state of nationality.

The international community failed to take action against terrorism until jolted into action by a series of attacks on airliners that threatened commercial air transportation. This led to two conventions, the Tokyo Convention on Offenses and Certain Other Acts Committed on Board Aircraft (1963) and the Montreal Convention for the Suppression of Unlawful Acts Against the Safety of Civil Aviation (1971), which required states to either prosecute or extradite individuals for trial who committed crimes of violence aboard aircraft or who intentionally destroyed key components of the aircraft. Most importantly, states agreed in the Hague Convention for the Suppression of Unlawful Seizure of Aircraft of 1970 to prosecute or send air hijackers to trial in nations claiming jurisdiction, including the state in which the plane was registered.

The next step was a series of treaties addressing other types of terrorism. Nations are bound to follow only the treaties that they agree to sign and make part of their own law. The United States has signed all of these counterterrorist agreements:

- *Convention on the Prevention and Punishment of Crimes Against Internationally Protected Persons, Including Diplomatic Agents (1973).* This protects Heads of State and other officials against attacks and requires states to prosecute assailants or to extradite them for trial in the victim's state of nationality.
- *International Convention Against the Taking of Hostages (1979).* States are to punish or extradite for trial any person who seizes or detains and threatens to kill, injure, or continue to detain any individual in order to compel a nation or international organization to perform or fail to perform any act as a condition of the release of a hostage.
- *International Convention for the Suppression of Terrorist Bombing (1989).* The treaty requires a

nation to punish an individual who intentionally delivers or detonates an explosive or other device into or against a government facility, public transportation system, or infrastructure facility with the intent to cause death or serious bodily injury or extensive destruction. Offenders are to be prosecuted or extradited for trial in the victim's state of nationality.

- *Convention for the Suppression of the Financing of Terrorism (1989).* This agreement obligates nations to prevent and to punish individuals who transmit money to terrorists across international borders.

The next conventions to be adopted will likely address the prevention and punishment of nuclear terrorism. A recent United Nations agreement urges nations to prevent and to punish incitement to terrorism.

Efforts over the past decade to formulate a comprehensive treaty on terrorism that obligates all states to prevent, to punish, and to cooperate in combating terrorist acts and organizations have proven unsuccessful. Such a treaty, for instance, might require states to refuse to permit terrorist groups to operate in their country or to establish bank accounts and might obligate states to adopt laws punishing terrorism and incitement to terrorism. The main obstacle to a comprehensive treaty is the disagreement between nations that view terrorism as a universal evil and nations that argue that terrorism is justified to resist the domination of powerful countries. The Hostage Convention, in fact, provides that the treaty shall not apply to an act of hostage taking committed in conflicts in which "peoples are fighting against colonial domination and alien occupation and against racist regimes in the exercise of their right of self-determination." Does the cause for which people are fighting justify the resort to terrorism? Another division is that some nations in the developing world believe that the terrorism carried out by big and powerful states in waging wars is far more serious than terrorism carried out by individual terrorists. Is terrorism a tactic that is employed by individuals, groups, and governments?

2. ***Crime on the Streets.*** We all are keenly aware that roughly 2,800 people died in the September 11, 2001, attack on the World Trade Center. This is the most costly terrorist incident in history as measured by dollars and the loss of human life. Terrorist statistics are notoriously unreliable because whether to categorize an act as terrorism can be fairly subjective. The statistics compiled by the National Counter Terrorism Center indicate that in 2004, we experienced 273 major terrorist attacks around the globe, the highest number of "significant terrorist incidents" since the United States started keeping statistics in 1968. This resulted in the death of 1,907 people, second only to the number killed in 2001 (due to September 11); the wounded numbered 6,704, and 710 people were taken hostage internationally. Eighty-three percent of attacks took place in the Middle East. Ten percent targeted U.S. citizens or property, and only one percent of all victims were Americans. Authors Clark R. Chapman and Alan W. Harris ask whether we might be overreacting to the September 11 attack and wasting time and money attempting to combat terrorism. They argue that the money could better be devoted to countering other types of accidents and improving social services. Consider the following observations by the authors:

- The number of deaths on September 11, 2001, are virtually identical to the monthly total of American traffic fatalities.
- September 11 deaths were ten times less than the number of annual deaths resulting from falls, from suicide, or from homicide.
- The Centers for Disease Control predicted in 2001 that 20,000 Americans would die from influenza during the winter.
- Twice as many people died in the floods stemming from the 1900 Galveston, Texas, hurricane as in the attack on the World Trade Center.
- The number of deaths on September 11, 2001, was equal to 1.5 percent of the worst epidemic in U.S. history; 1.5 percent of annual cancer deaths.

Are we devoting too many resources to countering terrorism? Are the authors overlooking the fact that a single successful terrorist attack could have a devastating impact? A nuclear bomb, for instance, would result in tens of thousands of casualties and contaminate and render uninhabitable a major American city. See Clark R. Chapman and Alan W. Harris, "How We Can Defeat Terrorism by Reacting to It More Rationally," *Skeptical Inquirer* (September/October 2002).

INTERNATIONAL CRIMINAL LAW

The origins of **international criminal law** can be traced to the prosecution of Nazi war criminals at Nuremberg in 1944. The international community subsequently agreed to a number of treaties that addressed crimes that are so serious that they are considered to be the concern of all nations and peoples. These treaties prohibit and punish acts such as genocide, torture, war crimes, and terrorism. A majority of countries in the world states have signed these treaties and have incorporated the provisions into their domestic criminal codes.[20]

International crimes typically are committed by individuals acting on behalf of the government. The exception of course is terrorism, which in most cases is committed by "non-state actors." Prosecutors in countries in which regimes carry out international crimes are reluctant or frightened to indict government officials for crimes, even after the officials have left office. As a result, the perpetrators of international crimes, in many instances, have not been brought to the bar of justice.

The international community periodically has convened international tribunals to prosecute and to punish government leaders who have carried out international crimes and who otherwise would have gone unpunished. In 1993, the United Nations established criminal tribunals to hear cases arising from genocide and war crimes in Rwanda and in Yugoslavia. The most significant step occurred in 2001 with the formation of the International Criminal Court (ICC). This court has jurisdiction over serious international crimes and is comprised of judges from countries that have joined the court. The United States, although a leading nation in the movement to prosecute and punish international crimes, is not a member of the ICC.

The United States, as part of its international obligation to punish international crimes, has claimed jurisdiction over international offenses committed outside America's territorial boundaries and has brought offenders to trial before U.S. domestic courts. The United States, for example, has prosecuted pirates for attacks on European and American ships off the coast of the African country of Somalia.

An example of **extraterritorial jurisdiction** is the United States' prosecution and conviction of Charles Emmanuel for acts of torture committed in Liberia between 1999 and 2003.

You can find more cases on the study site: United States v. Belfast, *www.sagepub.com/Lippmanccl3e.*

CHAPTER SUMMARY

The founding figures of the United States were fearful of a strong centralized government and provided protections for individual freedom and liberty. Nevertheless, they appreciated that there was a need to protect the government from foreign and domestic attack and accordingly incorporated a provision on treason into the Constitution. This was augmented by congressional enactments punishing sedition, sabotage, and espionage. In recent years, the United States has adopted laws intended to combat global terrorism.

Treason is defined in Article III, Section 3 of the U.S. Constitution. Treason requires an overt act of either levying war against the United States or providing aid and comfort to an enemy of the United States. The accused must be shown to have possessed the intent to betray the United States. The Constitution requires two witnesses or a confession in open court to an act of treason.

Sedition at English common law constituted a communication intended or likely to bring about hatred, contempt, or disaffection with the king, the constitution, or the government. This was broadly defined to include any criticism of the king and of English royalty. American courts have ruled that the punishment of seditious speech and libel may conflict with the First Amendment right to freedom of expression. As a result, judges have limited the punishment of seditious expression to the urging of the necessity or duty of taking action to forcibly overthrow the government. A seditious conspiracy requires an agreement to take immediate action.

Sabotage is the willful injury, destruction, contamination, or infection of war materials, premises, or utilities with the intent to injure, interfere with, or obstruct the United States or an allied nation in preparing for or carrying on war. Sabotage may also be committed in peacetime and requires that the damage to property is carried out with the intent to injure, interfere with, or obstruct the national defense of the United States.

An individual is guilty of espionage who communicates, delivers, or transmits information relating to the national defense to a foreign nation or force within a foreign nation with the purpose of injuring the United States or with reason to believe that it will injure the United States or advantage a foreign nation. Espionage in wartime involves collecting, recording, publishing, communicating, or attempting to elicit such information with the intent of communicating this information to the enemy.

The U.S. Code, Chapter 113B, punishes various terrorist crimes. This has been strengthened by the Anti-Terrorism and Effective Death Penalty Act of 1996 and by the USA Patriot Act (2001). Terrorism is divided into international and domestic terrorism. International terrorism is defined as violent acts or acts dangerous to human life that occur outside the United States; domestic terrorism is defined as violent acts or acts dangerous to human life that occur inside the United States. Terrorist acts transcending national boundaries are coordinated across national boundaries.

Terrorist acts are required to be intended to either intimidate or coerce a civilian population, to influence the policy of a government by intimidation or coercion, or to affect the conduct of a government by mass destruction, assassination, or kidnapping. The U.S. government has primarily relied on the prohibition against material assistance to terrorists and to terrorist organizations in order to prevent and punish terrorist designs before they are executed. This law has proven to be a powerful tool to deny terrorists the resources they require to carry out attacks. Federal law also punishes terrorist crimes involving attacks on mass transit systems and the use of weapons of mass destruction and prohibits the harboring or concealing of terrorists. The United States does not consider terrorists to be lawful combatants, and they are not viewed as prisoners of war under the Geneva Convention by American authorities. The detainees at Guantánamo are provided with the same protections as prisoners of war. Virtually every state has adopted a terrorism statute. Most terrorist offenses are based on the foundation offenses of treason, sedition, sabotage, and espionage. Terrorism is considered to be an international crime along with torture, genocide, war crimes, and other crimes that have been condemned by virtually all countries in the world.

CHAPTER REVIEW QUESTIONS

1. Treason is the only crime defined in the U.S. Constitution. What is treason? What type of evidence is required to establish treason?
2. Define sedition. Distinguish seditious speech from seditious libel. What constitutes a seditious conspiracy?
3. What is sabotage? Distinguish between sabotage and sabotage during wartime.
4. Explain the difference between espionage and espionage during wartime.
5. How do the definitions of international and domestic terrorism differ?
6. List some of the terrorist crimes set forth in the U.S. Code.
7. What are the elements of material support for terrorism? Provide examples of some provisions of the material support statute that courts have considered "void for vagueness." What are the arguments for and against prosecuting individuals for providing material support to terrorists or to foreign terrorist organizations?
8. Define and discuss combat immunity.
9. What factor is important in determining whether the federal government may assert jurisdiction over a terrorist act?
10. Discuss international crimes and extraterritorial jurisdiction.
11. Write a brief essay summarizing crimes against the state.

LEGAL TERMINOLOGY

combat immunity
crimes against the state
domestic terrorism
espionage
extraterritorial jurisdiction
federal crime of terrorism
Geneva Convention of 1949

international criminal law
international terrorism
material support to a foreign terrorist organization
material support to a terrorist
sabotage
sedition

seditious conspiracy
seditious libel
seditious speech
terrorism transcending national boundaries
treason
weapons of mass destruction

CRIMINAL LAW ON THE WEB

Log on to the Web-based student study site at www.sagepub.com/lippmanccl3e to assist you in completing the Criminal Law on the Web exercises, as well as for additional features such as podcasts, Web quizzes, and video links.

1. Read more about John Walker Lindh. Do you agree that John was a victim of media bias?

2. Examine the statistics on terrorist prosecutions in the U.S.

3. Read about "home grown" terrorism.

4. Consider the debate over the closing of Guantánamo Bay.

BIBLIOGRAPHY

Michael Byers, *War Law* (London: Atlantic Books, 2005). A good introduction to those aspects of the international humanitarian law of war that are relevant to the "war on terror."

Barton L. Ingraham, *Political Crime in Europe: A Comparative Study of France, Germany, and England* (Berkeley: University of California Press, 1979). A legal history of political crime in three European countries.

Nicholas N. Kittrie, *Rebels With a Cause* (Boulder, CO: Westview, 1999). A criminological analysis of political offenders and a discussion of the legal issues involved in prosecuting political offenders.

Nicholas N. Kittrie, *The War Against Authority* (Baltimore: Johns Hopkins University Press, 1995). A comprehensive legal history of political offenses in the United States.

Nicholas N. Kittrie and Eldon D. Wedlock, Jr., eds., *The Tree of Liberty* (Baltimore: Johns Hopkins University Press, 1986). A collection of historical documents on political crime in the United States.

Eric Lichtblau, *Bush's Law: The Remaking of American Justice* (New York: Pantheon Books, 2008). An account by a *New York Times* reporter of his experiences in covering several significant legal issues presented by the "war on terror."

Ronald C. Slye and Beth Van Schaack, *International Criminal Law: The Essentials* (New York: Aspen Publishers, 2009). An overview of international criminal law.

Steven T. Wax, *Kafka Comes to America: Fighting for Justice in the War on Terror: A Public Defender's Inside Account* (New York: Other Press, 2008). A federal public defender's personal account of representing defendants charged with violation of counterterrorism laws.

Benjamin Wittes, *Law and the Long War: The Future of Justice in the Age of Terror* (New York: Penguin, 2008). A balanced discussion of the major questions that arise in the investigation and prosecution of terrorist crimes.

NOTES

Chapter 1

1. Ala. Code § 13-12-5.
2. Fla. Stat. § 823.12.
3. R.I. Gen. Laws §§ 11-12-1–3.
4. Wyo. Stat. § 6-9-301(a).
5. Wyo. Stat. § 6-9-202.
6. *Idiot Laws.* (n.d.). Retrieved September 2011, from http://www.idiotlaws.com
7. La. Rev. Stat. § 14:67.13.
8. Henry M. Hart, Jr., "The Aims of the Criminal Law," *Law and Contemporary Problems* 23(3) (1958) 401–441.
9. *In re Winship,* 397 U.S. 358 (1972).
10. William Blackstone, *Commentaries on the Laws of England,* vol. 4 (Chicago: University of Chicago Press, 1979), p. 5.
11. *Kansas v. Hendricks,* 521 U.S. 346 (1997).
12. Tex. Penal Code Ann. § 1.02.
13. N.Y. Penal Law § 1.05.
14. Jerome Hall, *General Principles of Criminal Law,* 2nd ed. (Indianapolis, IN: Bobbs-Merrill, 1960), p. 18.
15. Wayne R. LaFave, *Criminal Law,* 3rd ed. (St. Paul, MN: West Publishing, 2000), p. 8.
16. Fla. Stat. §§ 796.04–796.045
17. LaFave, *Criminal Law,* pp. 70–71.
18. *Commonwealth v. Mochan,* 110 A.2d 788 (Pa. Super. Ct. 1955).
19. Lawrence M. Friedman, *Crime and Punishment in American History* (New York: Basic Books, 1993), pp. 64–65.
20. Fla. Stat. §§ 775.01–775.02.
21. Cal. Penal Code § 2-24-6.
22. Utah Code § 76-1-105.
23. LaFave, *Criminal Law,* pp. 72–73.
24. *Keeler v. Superior Court,* 470 P.2d 617 (Cal. 1970).
25. *Kovacs v. Cooper,* 336 U.S. 77 (1949).
26. *Village of Belle Terre v. Boraas,* 416 U.S. 1 (1974).
27. Joshua Dressler, *Understanding Criminal Law,* 3rd ed. (New York: Lexis, 2001), p. 31.
28. *United States v. Lopez,* 514 U.S. 549 (1995).
29. *United States v. Jones,* 529 U.S. 848 (2000).
30. *Oregon v. Gonzalez,* 546 U.S. 243 (2006).
31. *Texas v. Johnson,* 491 U.S. 397 (1989).
32. *Gonzalez v. Raich,* 541 U.S. 1 (2005).
33. *Koon v. United States,* 518 U.S. 81 (1996).

Chapter 2

1. Joshua Dressler, *Understanding Criminal Law,* 3rd ed. (New York: Lexis, 2001), p. 39.
2. American Law Institute, *Model Penal Code and Commentaries,* vol. 1, pt. 1 (Philadelphia: American Law Institute, 1980), § 1.05(1).
3. James Madison, "Federalist No. 44," in A. Hamilton, J. Madison, and J. Jay, *The Federalist Papers* (New York: New American Library, 1961), p. 282.
4. *United States v. Brown,* 381 U.S. 437, 442 (1965).
5. *United States v. Lovett,* 328 U.S. 303 (1946).
6. Alexander Hamilton, "The Federalist No. 84," in A. Hamilton, J. Madison, and J. Jay, *The Federalist Papers* (New York: New American Library, 1961), p. 512.
7. *Calder v. Bull,* 3 U.S. 386 (1798).
8. *Carmell v. Texas,* 529 U.S. 513 (2000).
9. *Stogner v. California,* 539 U.S. 607 (2003).
10. *Grayned v. City of Rockford,* 408 U.S. 104 (1972).
11. *Coates v. Cincinnati,* 402 U.S. 611 (1971).
12. *Kolender v. Lawson,* 461 U.S. 352 (1983).
13. *Gregory v. Chicago,* 394 U.S. 111 (1969).
14. *Papachristou v. Jacksonville,* 405 U.S. 156 (1971).
15. *Papachristou v. Jacksonville,* 405 U.S. at 156.
16. *Coates v. Cincinnati,* 402 U.S. at 611.
17. *Horn v. State,* 273 So. 2d 249 (Ala. App. 1973).
18. *Nebraska v. Metzger,* 319 N.W. 2d 459 (Neb. 1982).
19. Erwin Chemerinsky, *Constitutional Law Principles and Politics,* 2nd ed. (New York: Aspen, 2002), p. 642.
20. *Bolling v. Sharpe,* 347 U.S. 497 (1954).
21. *Buck v. Bell,* 274 U.S. 200, 208 (1927).
22. *Brown v. Board of Education,* 347 U.S. 483 (1954).
23. *McGowan v. Maryland,* 366 U.S. 420, 425–426 (1961).
24. *Westbrook v. Alaska,* 2003 WL 1732398 (Alaska Ct. App. 2003).
25. *Strauder v. West Virginia,* 100 U.S. 303 (1879).
26. *McLaughlin v. Florida,* 379 U.S. 184 (1964).
27. *Loving v. Virginia,* 388 U.S. 1 (1967).
28. *United States v. Virginia,* 518 U.S. 515 (1996).
29. *Michael M. v. Superior Court,* 450 U.S. 464 (1981).
30. *Michael M.,* 450 U.S. at 464.
31. *Gitlow v. New York,* 268 U.S. 652 (1925).
32. Thomas I. Emerson, *The System of Freedom of Expression* (New York: Vintage Books, 1970), pp. 6–7.
33. *Terminiello v. Chicago,* 337 U.S. 1 (1949).
34. *Chaplinsky v. New Hampshire,* 315 U.S. 568 (1942).
35. *Feiner v. New York,* 340 U.S. 315 (1951).
36. *Terminiello v. Chicago,* 337 U.S. 1 (1949).
37. *Watts v. United States,* 394 U.S. 705 (1969).

38. *Jacobellis v. Ohio*, 378 U.S. 184 (1974).
39. *Miller v. California*, 413 U.S. 15 (1973).
40. *New York v. Ferber*, 458 U.S. 747, 778 (1982).
41. *New York Times v. Sullivan*, 376 U.S. 254 (1964).
42. *Gertz v. Welch*, 418 U.S. 323 (1974).
43. *West Virginia v. Barnette*, 319 U.S. 624 (1943).
44. *New York v. Ferber*, 458 U.S. at 773.
45. *R.A.V. v. St. Paul*, 505 U.S. 377 (1992).
46. *Wisconsin v. Mitchell*, 508 U.S. 476 (1993).
47. *Virginia v. Black*, 538 U.S. 343 (2003).
48. Arnold L. Weinstein, *A Scream Goes Through the House: What Literature Teaches Us About Life* (New York: Random House, 2003), p. xxiii.
49. Samuel D. Warren and Louis D. Brandeis, "The Right to Privacy," *Harvard Law Review* 4(5) (1890) 1–23.
50. *Pavesich v. New England Life Ins. Co.*, 50 S.E. 68 (Ga. 1905).
51. *Griswold v. Connecticut*, 381 U.S. 479 (1965).
52. Arthur R. Miller, *The Assault on Privacy* (Ann Arbor: University of Michigan Press, 1971).
53. *Olmstead v. United States*, 277 U.S. 438 (1928).
54. *Eisenstadt v. Baird*, 405 U.S. 438 (1972).
55. *Carey v. Population Services International*, 431 U.S. 678 (1977).
56. *Roe v. Wade*, 410 U.S. 113 (1973).
57. *Planned Parenthood of Southeastern Pa. v. Casey*, 505 U.S. 833 (1992).
58. *Stanley v. Georgia*, 394 U.S. 557 (1969).
59. *Bowers v. Hardwick*, 478 U.S. 186 (1986).
60. *Lawrence v. Texas*, 529 U.S. 558 (2003).
61. *United States v. Miller*, 307 U.S. 174 (1939).
62. *District of Columbia v. Heller*, 552 U.S. 570 (2008).
63. *McDonald v. Chicago*, ___ U.S. ___, 130 S.Ct. 3020, 177 L.Ed.2d 894 (2010).

Chapter 3

1. Lawrence M. Friedman, *Crime and Punishment in American History* (New York: Basic Books, 1993), pp. 40–50.
2. Nicholas N. Kittrie, Elyce H. Zenoff, and Vincent A. Eng, *Sentencing, Sanctions, and Corrections: Federal and State Law, Policy, and Practice* (New York: Foundation Press, 2002), p. 736.
3. *Jackson v. Bishop*, 404 F.2d 571 (8th Cir. 1968).
4. Kittrie, Elyce, and Eng, *Sentencing, Sanctions, and Corrections*, p. 722.
5. George P. Fletcher, *Rethinking Criminal Law* (New York: Oxford University Press, 2000), pp. 408–410.
6. *Kennedy v. Mendoza-Martinez*, 372 U.S. 144 (1963).
7. 42 U.S.C. § 14071.
8. *Smith v. Doe*, 538 U.S. 84 (2003).
9. *United States v. Bergman*, 416 F. Supp. 496 (S.D.N.Y. 1976).
10. *United States v. Ursery*, 518 U.S. 267 (1996).
11. 18 U.S.C. §§ 3531–3626; 28 U.S.C. §§ 991–998.
12. *Apprendi v. New Jersey*, 530 U.S. 466 (2000).
13. *Blakely v. Washington*, 542 U.S. 296 (2004).
14. *Cunningham v. California*, 549 U.S. 270 (2007).
15. *United States v. Booker*, 543 U.S. 220 (2005).
16. *Rita v. United States*, 551 U.S. 338 (2007); *Gall v. United States*, 552 U.S. 38 (2007); *Kimbrough v. United States*, 552 U.S. 85 (2007).
17. *Simon & Schuster, Inc. v. New York Crime Victims Board*, 502 U.S. 105 (1992).
18. *Payne v. Tennessee*, 501 U.S. 808 (1991).
19. *Harmelin v. Michigan*, 501 U.S. 957 (1991).
20. Wayne R. LaFave, *Criminal Law*, 3rd ed. (St. Paul, MN: West Publishing, 2000), p. 189.
21. *Weems v. United States*, 217 U.S. 349, 369 (1910).
22. LaFave, *Criminal Law*, p. 187.
23. *In re Kemmler*, 136 U.S. 436 (1890).
24. *Delaware v. Cannon*, 190 A.2d 514 (Del. 1963).
25. *Trop v. Dulles*, 356 U.S. 86 (1958).
26. *Furman v. Georgia*, 408 U.S. 238 (1972).
27. *In re Kemmler*, 136 U.S. at 436.
28. *Wilkerson v. Utah*, 99 U.S. 130 (1878).
29. *In re Kemmler*, 136 U.S. at 436.
30. *State of Louisiana ex rel. Francis v. Resweber*, 329 U.S. 459 (1947).
31. *Hope v. Pelzer*, 536 U.S. 730 (2002).
32. *Brown v. Plata*, ___ U.S. ___, 131 S. Ct. 1910 (2011).
33. *In the Matter of R.B.*, 765 A.2d 396 (Pa. Super. Ct. 2000).
34. *Furman v. Georgia*, 408 U.S. 238 (1972).
35. *Furman v. Georgia*, 408 U.S. at 238.
36. *Woodson v. North Carolina*, 428 U.S. 280 (1976).
37. *Gregg v. Georgia*, 428 U.S. 153 (1976).
38. *Coker v. Georgia*, 433 U.S. 584 (1977).
39. *Kennedy v. Louisiana*, 554 U.S. 407 (2008).
40. *Baze v. Rees*, 533 U.S. 35 (2008).
41. *Kent v. United States*, 383 U.S. 541 (1966).
42. *Eddings v. Oklahoma*, 487 U.S. 815 (1982).
43. *Thompson v. Oklahoma*, 487 U.S. 815 (1988).
44. *Stanford v. Kentucky*, 492 U.S. 361 (1989).
45. *Lockyer v. Andrade*, 538 U.S. 63 (2003).
46. Cal. Penal Code, § 667.
47. *Ewing v. California*, 538 U.S. 11 (2003).
48. *Weems v. United States*, 217 U.S. 349 (1910).
49. *Rummel v. Estelle*, 445 U.S. 263 (1980).
50. *Hutto v. Davis*, 454 U.S. 370 (1982).
51. *United States v. Angelos*, 433 F.3d 738 (10th Cir. 2006).
52. Paul Butler, "Racially Based Jury Nullification: Black Power in the Criminal Justice System," *Yale Law Journal* 105 (1995) 677–725.
53. *Robinson v. California*, 370 U.S. 660 (1962).
54. *United States v. Edwardo-Franco*, 885 F.2d 1002 (2nd Cir. 1989).
55. *State v. Chambers*, 307 A.2d 78 (N.J. 1973).
56. *McCleskey v. Kemp*, 481 U.S. 279 (1987).

Chapter 4

1. Joshua Dressler, *Understanding Criminal Law*, 3rd ed. (New York: Lexis, 2001), p. 197.
2. Dressler, *Understanding Criminal Law*, p. 197.
3. Cal. Penal Code § 20 (1999).
4. Dressler, *Understanding Criminal Law*, p. 81.
5. Ind. Code § 35-41-2-1 (1993).
6. Wayne R. LaFave, *Criminal Law*, 3rd ed. (St. Paul, MN: West Publishing, 2000), p. 206.

7. *McClain v. State*, 678 N.E.2d 104 (Ind. 1997).
8. Markus D. Dubber, *Criminal Law: Model Penal Code* (New York: Foundation Press, 2002), p. 33.
9. Dressler, *Understanding Criminal Law*, p. 86.
10. LaFave, *Criminal Law*, p. 208.
11. *State v. Connell*, 493 S.E.2d 292 (1997).
12. *Fain v. Commonwealth*, 78 Ky. 183 (1879).
13. *People v. Decina*, 138 N.E. 799 (N.Y. 1956).
14. *People v. Newton*, 87 Cal. Rptr. 394 (Cal. Ct. App. 1970).
15. *Robinson v. California*, 370 U.S. 660, 661 (1962).
16. *Powell v. Texas*, 392 U.S. 514 (1968).
17. *People v. Beardsley*, 113 N.W. 1128 (Mich. 1907).
18. *People v. Beardsley*, 113 N.W. at 1132.
19. *Buck v. Amory Manufacturing*, 44 A. 809, 810 (N.H. 1897).
20. *Kuntz v. Montana Thirteenth Judicial District*, 995 P.2d 951 (Mont. 1999).
21. Dressler, *Understanding Criminal Law*, pp. 96–99.
22. *Hughes v. State*, 719 S.W.2d 560 (Tex. Crim. App. 1986).
23. 18 Pa. Cons. Stat. § 301(b).
24. *State v. Mally*, 366 P.2d 868 (Mont. 1961).
25. *Craig v. State*, 155 A.2d. 684 (Md. 1959).
26. *Commonwealth v. Pestinikas*, 617 A.2d 1339 (Pa. Super. 1992).
27. *People v. Oliver*, 258 Cal. Rptr. 138 (Cal. Dist. Ct. App. 1989).
28. *Jones v. State*, 43 N.E.2d 1017 (Ind. 1942).
29. Cal. Penal Code § 272.
30. *Williams v. Garcetti*, 853 P.2d 507 (Cal. 1993).
31. *Commonwealth v. Karetny*, 880 A.2d 505 (Pa. 2005).
32. *People v. Burton*, 788 N.E.2d 220 (Ill. App. 2003).
33. LaFave, *Criminal Law*, pp. 224–225.
34. *People v. Burton*, 788 N.E.2d at 220.
35. *Craig v. State*, 155 A.2d at 684.
36. *Craig v. State*, 155 A.2d at 684.
37. *State v. Walden*, 293 S.E.2d 780 (N.C. 1982).
38. *People v. Burton*, 788 N.E.2d at 220.
39. Arnold H. Lowey, *Criminal Law*, 4th ed. (St. Paul, MN: West Publishing, 2003), pp. 235–236.
40. Dressler, *Understanding Criminal Law*, pp. 92–93.
41. LaFave, *Criminal Law*, pp. 211–213.
42. *Hawkins v. State*, 89 S.W.3d 674 (Tex. App.–Houston 2002).
43. *People v. Mijares*, 491 P.2d 1115 (Cal. 1971).
44. *United States v. Byfield*, 928 F.2d 1163 (D.C. Cir. 1991).
45. *State v. Webb*, 648 N.W.2d 72 (Iowa 2002).
46. *State v. Rippley*, 319 N.W.2d 129 (N.D. 1982).
47. *Dawkins v. State*, 547 A.2d 1041 (Md. 1988).

Chapter 5

1. *Morissette v. United States*, 342 U.S. 246 (1952).
2. Oliver Wendell Holmes, Jr., *The Common Law* (Mark Howe ed.) (Boston: Little, Brown, 1963), p. 7.
3. *Dennis v. United States*, 341 U.S. 494 (1951).
4. *Morissette v. United States*, 342 U.S. at 246.
5. Jerome Hall, *General Principles of Criminal Law*, 2nd ed. (Indianapolis, IN: Bobbs-Merrill, 1960), pp. 106–107.
6. *People v. Conley*, 543 N.E.2d 138 (Ill. App. Ct. 1989).
7. *Alvarado v. State*, 704 S.W.2d 36 (Tex. Crim. App. 1986).
8. *United States v. Bailey*, 444 U.S. 394 (1980).
9. *United States v. Haupt*, 330 U.S. 631 (1947).
10. *United States v. Jewell*, 532 F.2d 697 (9th Cir. 1976).
11. *Tello v. State*, 138 S.W.3d 487 (Tex. 2004).
12. *Morissette v. United States*, 342 U.S. at 246.
13. *State v. York*, 2003-Ohio-7249, 2003 Ohio App. LEXIS 6532.
14. *Staples v. United States*, 511 U.S. 600 (1994).
15. Joshua Dressler, *Understanding Criminal Law*, 3rd ed. (New York: Lexis, 2001), pp. 180–181.
16. *United States v. Main*, 113 F.3d 1046 (9th Cir. 1997).
17. Dressler, *Understanding Criminal Law*, p. 191.
18. *Kibbe v. Henderson*, 534 F.2d 493 (2nd Cir. 1976).
19. *People v. Armitage*, 239 Cal. Rptr. 515 (Cal. Ct. App. 1987).
20. *People v. Schmies*, 51 Cal. Rptr. 2d 185 (Cal. Ct. App. 1996).
21. *People v. Saavedra-Rodriquez*, 971 P.2d 223 (Colo. 1998).
22. *United States v. Hamilton*, 182 F. Supp. 548 (D.D.C. 1960).
23. American Law Institute, *Model Penal Code and Commentaries*, vol. 1, pt. 1 (Philadelphia: American Law Institute, 1985), § 2.03.

Chapter 6

1. Rollin M. Perkins and Ronald N. Boyce, *Criminal Law*, 3rd ed. (Mineola, NY: Foundation Press, 1982), pp. 730–732.
2. Joshua Dressler, *Understanding Criminal Law*, 3rd ed. (New York: Lexis, 2001), p. 461.
3. *Pinkerton v. United States*, 328 U.S. 640 (1946).
4. *Wilcox v. Jeffery* (1951) 1 All E.R. 464.
5. Dressler, *Understanding Criminal Law*, p. 469.
6. *Brown v. State*, 864 So. 2d 1009 (Miss. Ct. App. 2004).
7. *State v. Mendez*, 2003 Tenn. Crim. App. LEXIS 790.
8. *Alejandres v. Texas*, 2004 Tex. Crim. App. LEXIS 257.
9. *State v. Jones*, 2004-Ohio-7280, Ohio Ct. App. LEXIS 1435.
10. *State v. Phillips*, 76 S.W.3d 1 (Tenn. Crim. App. 2001).
11. *State v. Doody*, 434 A.2d 523 (Me. 1981).
12. *State v. Noriega*, 928 P.2d 706 (Ariz. Ct. App. 1996).
13. *People v. Mullen*, 730 N.E.2d 545 (Ill. App. Ct. 2000).
14. *Bailey v. United States*, 416 F.2d 1110 (D.C. Cir. 1969).
15. *State v. Walden*, 293 S.E.2d 780 (N.C. 1982).
16. *People v. Perez*, 725 N.E.2d 1258 (Ill. 2000).
17. Dressler, *Understanding Criminal Law*, p. 472.
18. *United States v. Peoni*, 100 F.2d 401 (2d Cir. 1938).
19. *Backun v. United States*, 112 F.2d 635 (4th Cir. 1940).
20. *State v. Linscott*, 520 A.2d 1067 (Me. 1987).
21. *Wilson v. State*, 824 So. 2d 355 (Fla. Dist. Ct. App. 2002).
22. *Melahn v. State*, 843 So. 2d 929 (Fla. Dist. Ct. App. 2003).
23. *State v. Jordan*, 590 S.E.2d 424 (N.C. Ct. App. 2004).
24. Richard G. Singer and John Q. La Fond, *Criminal Law Examples and Explanations*, 2nd ed. (New York: Aspen, 2001), p. 108.
25. *People v. Travers*, 124 Cal. Rptr. 728 (Cal. Ct. App. 1975).
26. Wayne R. LaFave, *Criminal Law*, 3rd ed. (St. Paul, MN: West Publishing, 2000), p. 264.

27. Perkins and Boyce, *Criminal Law*, pp. 718–720.
28. *New York Cent. & H. R. R. Co. v. United States*, 212 U.S. 481 (1909).
29. *United States v. Dotterweich*, 320 U.S. 277 (1943).
30. *United States v. Park*, 421 U.S. 658 (1975).
31. *United States v. Dotterweich*, 320 U.S. at 281.
32. LaFave, *Criminal Law*, p. 275.
33. LaFave, *Criminal Law*, pp. 275–276.
34. *United States v. Allegheny Bottling Company*, 695 F. Supp. 856 (E.D. Va. 1988), *aff'd* 870 F.2d 655 (4th Cir. 1989).
35. *Commonwealth v. Penn Valley Resorts, Inc.*, 494 A.2d 1139 (Pa. Super. Ct. 1985).
36. National Governors Association, *Juvenile Justice Reform Initiatives in the States 1940–1996* (Washington, DC: Office of Juvenile Justice and Delinquency Prevention, 1997).
37. See Ala. Code § 12-15-13 (1975).
38. N.Y. Penal Law § 260.10.
39. *Williams v. Garcetti*, 853 P.2d 507 (Cal. 1993).

Chapter 7

1. Joshua Dressler, *Understanding Criminal Law*, 3rd ed. (New York: Lexis, 2001), pp. 373–374.
2. *People v. Miller*, 42 P.2d 308 (Cal. 1935).
3. Jerome Hall, *General Principles of Criminal Law*, 2nd ed. (Indianapolis, IN: Bobbs-Merrill, 1960), p. 559.
4. Hall, *General Principles of Criminal Law*, p. 560.
5. *Rex v. Scofield*, (1784) Caldecott 397.
6. *Rex v. Higgins*, (1801) 2 East 5.
7. Dressler, *Understanding Criminal Law*, pp. 384–385.
8. *People v. Gentry*, 510 N.E.2d 963 (Ill. App. Ct. 1987).
9. American Law Institute, *Model Penal Code and Commentaries*, vol. 11, pt. 1 (Philadelphia: American Law Institute, 1985), p. 305.
10. *State v. Reeves*, 916 S.W.2d 909 (Tenn. 1996).
11. *People v. Rizzo*, 158 N.E.2d 888 (N.Y. 1927).
12. *Commonwealth v. Gilliam*, 417 A.2d 1203 (Pa. Super. Ct. 1980).
13. Hall, *General Principles of Criminal Law*, p. 595.
14. *Attorney General v. Sillem*, (1863) 159 Eng. Rep. 178.
15. Colo. Rev. Stat. §18-1-201.
16. *People v. Dlugash*, 363 N.E.2d 1155 (N.Y. 1977).
17. *People v. Staples*, 85 Cal. Rptr. 589 (Cal. Ct. App. 1970).
18. Wayne R. LaFave, *Criminal Law*, 3rd ed. (St Paul, MN: West Publishing, 2000), pp. 563–564.
19. American Law Institute, *Model Penal Code and Commentaries*, pp. 358–362.
20. LaFave, *Criminal Law*, pp. 612–614.
21. *Krulewitch v. United States*, 336 U.S. 440 (1949).
22. *Commonwealth v. Azim*, 459 A.2d 1244 (Pa. Super. Ct. 1983).
23. *United States v. Brown*, 776 F.2d 397 (2nd Cir. 1985).
24. Jason Epstein, *The Great Conspiracy Trial* (New York: Vintage Books, 1971).
25. *Hyde v. United States*, 225 U.S. 347 (1912).
26. *Yates v. United States*, 354 U.S. 298 (1957).
27. *Cline v. State*, 319 S.W.2d 227 (Tenn. 1958).
28. Rollin M. Perkins and Ronald N. Boyce, *Criminal Law*, 3rd ed. (Mineola, NY: Foundation Press, 1982), pp. 685–687.
29. LaFave, *Criminal Law*, pp. 592–593.
30. *United States v. Falcone*, 311 U.S. 205 (1940).
31. *People v. Lauria*, 59 Cal. Rptr. 628 (Cal. Ct. App. 1967).
32. *Morrison v. California*, 291 U.S. 82 (1934).
33. Dressler, *Understanding Criminal Law*, pp. 439–440.
34. Dressler, *Understanding Criminal Law*, p. 440.
35. American Law Institute, *Model Penal Code and Commentaries*, p. 399.
36. *State v. Pacheco*, 882 P.2d 183 (Wash. 1994).
37. Richard G. Singer and John Q. La Fond, *Criminal Law Examples and Explanations*, 2nd ed. (New York: Aspen, 2001), p. 287.
38. *United States v. Bruno*, 105 F.2d 921 (2d Cir.), *rev'd on other grounds*, 308 U.S. 287 (1939).
39. *Kotteakos v. United States*, 328 U.S. 750 (1946).
40. *State v. McLaughlin*, 44 A.2d 116 (Conn. 1945).
41. *Rex v. Jones*, (1832) 110 Eng. Rep. 485.
42. *State v. Burnham*, 15 N.H. 396 (1844).
43. *People v. Fisher*, 14 Wend. 2 (N.Y. 1835).
44. *Shaw v. Director of Public Prosecutions*, (1962) A.C. 220.
45. *Musser v. Utah*, 333 U.S. 95 (1948).
46. 18 U.S.C. § 371.
47. *Pinkerton v. United States*, 328 U.S. 640 (1946).
48. Dressler, *Understanding Criminal Law*, pp. 455–456.
49. *Gebardi v. United States*, 287 U.S. 112 (1932).
50. *Harrison v. United States*, 7 F.2d 259 (2nd Cir. 1925).
51. LaFave, *Criminal Law*, pp. 569–573.
52. *Commonwealth v. Barsell*, 678 N.E.2d 143 (Mass. 1997).
53. *State v. Butler*, 35 P. 1093 (Wash. 1894).
54. *People v. Smith*, 806 N.E.2d 1262 (Ill. App. Ct. 2004).

Chapter 8

1. *In re Winship*, 397 U.S. 358 (1970).
2. *Victor v. Nebraska*, 511 U.S. 1 (1994).
3. *Jackson v. Virginia*, 443 U.S. 307 (1979).
4. *Commonwealth v. Webster*, 59 Mass. 295, 320 (1850).
5. Joshua Dressler, *Understanding Criminal Law*, 3rd ed. (New York: Lexis, 2001), pp. 63–65.
6. Wayne R. LaFave, *Criminal Law*, 3rd ed. (St. Paul, MN: West Publishing, 2000), p. 54.
7. Richard G. Singer and John Q. La Fond, *Criminal Law Examples and Explanations*, 2nd ed. (New York: Aspen, 2001), p. 374.
8. George P. Fletcher, *Rethinking Criminal Law* (New York: Oxford University Press, 2000), p. 759.
9. Dressler, *Understanding Criminal Law*, p. 205.
10. Dressler, *Understanding Criminal Law*, p. 216.
11. Dressler, *Understanding Criminal Law*, pp. 216–219.
12. Dressler, *Understanding Criminal Law*, pp. 206–209.
13. Singer and La Fond, *Criminal Law*, pp. 386–388.
14. *State v. Moore*, 689 N.E.2d 1 (Ohio 1998).
15. *United States v. Berrigan*, 417 F.2d 1002 (4th Cir. 1969).
16. *United States v. Dougherty*, 473 F.2d 1113 (D.C. Cir. 1972).
17. William Blackstone, *Commentaries on the Laws*

of England, vol. 4 (Chicago: University of Chicago Press, 1979), pp. 183–187.

18. *Brown v. United States,* 256 U.S. 335 (1921).

19. *Vigil v. People,* 353 P.2d 82 (Colo. 1960).

20. Dressler, *Understanding Criminal Law,* pp. 222–223.

21. *Harshaw v. State,* 39 S.W.3d 753 (Ark. 2001).

22. *State v. Schroeder,* 261 N.W.2d 759 (Neb. 1978).

23. *People v. Williams,* 205 N.E.2d 749 (Ill. 1965).

24. *State v. Pranckus,* 815 A.2d 678 (Conn. App. Ct. 2003).

25. Fletcher, *Rethinking Criminal Law,* pp. 870–875.

26. Dressler, *Understanding Criminal Law,* p. 227.

27. Dressler, *Understanding Criminal Law,* pp. 228–229.

28. *People v. Young,* 183 N.E.2d 319 (N.Y. 1962).

29. N.Y. Penal Law § 35.15.

30. Fletcher, *Rethinking Criminal Law,* p. 869.

31. *State v. Fair,* 211 A.2d 359 (N.J. 1965) (discussing the objective test).

32. *People v. Eatman,* 91 N.E.2d 387 (Ill. 1950).

33. Dressler, *Understanding Criminal Law,* pp. 263–265.

34. *State v. Anderson,* 972 P.2d 32 (Okla. Crim. App. 1998).

35. Dressler, *Understanding Criminal Law,* pp. 265–266.

36. Dressler, *Understanding Criminal Law,* pp. 276–277.

37. *Durham v. State,* 159 N.E. 145 (Ind. 1927).

38. Dressler, *Understanding Criminal Law,* pp. 276–277.

39. Paul Chevigny, *Edge of the Knife: Police Violence in the Americas* (New York: The New Press, 1995), p. 128.

40. American Law Institute, *Model Penal Code and Commentaries,* vol. 1, pt. 1 (Philadelphia: American Law Institute, 1985), p. 114.

41. Dressler, *Understanding Criminal Law,* p. 277.

42. *Commonwealth v. Chermansky,* 242 A.2d 237 (Pa. 1968).

43. *Queen v. Tooley,* (1909) 92 Eng. Rep. 349 (K.B.).

44. *John Bad Elk v. United States,* 177 U.S. 529 (1900).

45. *United States v. Di Re,* 332 U.S. 581 (1948).

46. Craig Hemmens, "Unlawful Arrest in Mississippi: Resisting the Modern Trend," *California Criminal Law Review* 2 (2000) 2 para. 1–33.

47. Arnold H. Lowey, *Criminal Law in a Nutshell,* 4th ed. (St. Paul, MN: West Publishing, 2003), pp. 82–83.

48. *Commonwealth v. French,* 611 A.2d 175 (Pa. 1992).

49. *State v. King,* 149 So. 2d 482 (Miss. 1963).

50. *State v. Valentine,* 935 P.2d 1294, 1312 (Wash. 1997).

51. *State v. Wiegmann,* 714 A.2d 841 (Md. 1998).

52. LaFave, *Criminal Law,* pp. 476–479.

53. Fletcher, *Rethinking Criminal Law,* p. 822.

54. *The Queen v. Dudley and Stephens,* (1884) 14 Q.B.D. 273.

55. Dressler, *Understanding Criminal Law,* p. 273.

56. *State v. Salin,* No. 0302016999, 2003 Del. Ct. C. P. LEXIS 39 (Kent County Ct. Common Pleas June 13, 2003).

57. Herbert L. Packer, *The Limits of the Criminal Sanction* (Palo Alto, CA: Stanford University Press, 1968), pp. 113–117.

58. Packer, *The Limits of the Criminal Sanction,* pp. 119–121.

59. Dressler, *Understanding Criminal Law,* p. 287.

60. *State v. Green,* 470 S.W.2d 565 (Mo. 1971).

61. *Humphrey v. Commonwealth,* 553 S.E.2d 546 (Va. Ct. App. 2001).

62. *Nelson v. State,* 597 P.2d 977 (Alaska 1979).

63. *United States v. Maxwell,* 254 F.3d 21 (1st Cir. 2001).

64. *Emry v. United States,* 829 A.2d 970 (D.C. 2003).

65. *People v. Brandyberry,* 812 P.2d 674 (Colo. Ct. App. 1991).

66. *United States v. Oakland Cannabis Buyers' Cooperative,* 532 U.S. 483 (2001).

67. *State v. Romano,* 809 A.2d 158 (N.J. Super. Ct. App. Div. 2002).

68. *Commonwealth v. Appleby,* 402 N.E.2d 1051 (Mass. 1980).

69. Fletcher, *Rethinking Criminal Law,* pp. 770–771.

70. Blackstone, *Commentaries on the Laws of England,* vol. iv, p. 5.

71. *State v. Brown,* 364 A.2d 27 (N.J. Super. Ct. Law Div. 1976).

72. *People v. Lenti,* 253 N.Y.S.2d 9 (N.Y. Co. Ct. 1964).

Chapter 9

1. Rollin M. Perkins and Ronald N. Boyce, *Criminal Law,* 3rd ed. (Mineola, NY: Foundation Press, 1982), pp. 950–951.

2. Jerome Hall, *General Principles of Criminal Law,* 2nd ed. (Indianapolis, IN: Bobbs-Merrill, 1960), p. 475.

3. Perkins and Boyce, *Criminal Law,* pp. 958–959.

4. Wayne R. LaFave, *Criminal Law,* 3rd ed. (St. Paul, MN: West Publishing, 2000), p. 331.

5. *Serritt v. State,* 582 S.E.2d 507 (Ga. Ct. App. 2003).

6. American Law Institute, *Model Penal Code and Commentaries,* vol. 2, pt. 1 (Philadelphia: American Law Institute, 1985), p. 166.

7. Arnold H. Loewy, *Criminal Law in a Nutshell,* 4th ed. (St. Paul, MN: West Publishing, 2003), p. 165.

8. *State v. Crenshaw,* 659 P.2d 488 (Wash. 1983).

9. *Clark v. Arizona,* 548 U.S. 735 (2006).

10. Joshua Dressler, *Understanding Criminal Law,* 3rd ed. (New York: Lexis, 2001), p. 349.

11. *Parsons v. State,* 2 So. 854 (Ala. 1887).

12. *United States v. Lyons,* 731 F.2d 243 (5th Cir. 1984).

13. *State v. Quinet,* 752 A.2d 490 (Conn. 2000).

14. 18 U.S.C. § 17.

15. *State v. Pike,* 49 N.H. 399 (1869).

16. *Durham v. United States,* 214 F.2d 862 (D.C. Cir. 1954).

17. *United States v. Brawner,* 471 F.2d 969 (D.C. Cir. 1972).

18. *Blocker v. United States,* 288 F.2d 853 (D.C. Cir. 1961).

19. LaFave, *Criminal Law,* pp. 374–377.

20. Dressler, *Understanding Criminal Law,* p. 356.

21. Idaho Code Ann. § 18-207(1)(2).

22. *State v. Bethel,* 66 P.3d 840 (Kan. 2003).

23. Dressler, *Understanding Criminal Law,* p. 360.

24. *United States v. Lyons,* 731 F.2d 243 (5th Cir. 1984).

25. *People v. Wells,* 202 P.2d 53 (Cal. 1949); *People v. Gorshen,* 336 P.2d 492 (Cal. 1959).

26. *People v. Gorshen,* 336 P.2d at 495–496, 504.

27. *State v. Skora,* 210 A.2d 193 (N.J. 1965).

28. Cal. Penal Code §§ 25–29.

29. American Law Institute, *Model Penal Code and Commentaries,* vol. 2, § 4.02.

30. *People v. Hood,* 462 P.2d 370 (Cal. 1969).

31. *Montana v. Egelhoff,* 518 U.S. 37, 44 (1996).
32. *Montana v. Egelhoff,* 518 U.S. at 44.
33. *People v. Hood,* 462 P.2d at 379.
34. Markus D. Dubber, *Criminal Law: Model Penal Code* (New York: Foundation Press, 2002), pp. 84–90.
35. Hall, *General Principles of Criminal Law,* p. 545.
36. *State v. Cameron,* 514 A.2d 1302 (N.J. 1986).
37. *Montana v. Egelhoff,* 518 U.S. at 44.
38. *City of Minneapolis v. Altimus,* 238 N.W.2d 851, 856 (Minn. 1976).
39. Perkins and Boyce, *Criminal Law,* pp. 936–938.
40. LaFave, *Criminal Law,* pp. 426–427.
41. LaFave, *Criminal Law,* pp. 427–429.
42. LaFave, *Criminal Law,* pp. 429–430.
43. *Kent v. United States,* 383 U.S. 541 (1966).
44. Hall, *General Principles of Criminal Law,* p. 438.
45. *Lynch v. D.P.P.,* (1975) A.C. 653.
46. *State v. Van Dyke,* 825 A.2d 1163 (N.J. Super. Ct. App. Div. 2003).
47. *People v. Anderson,* 50 P.3d 368 (Cal. 2000).
48. *Commonwealth v. Perl,* 737 N.E.2d 937 (Mass. App. Ct. 2000).
49. *State v. Keeran,* 674 N.W.2d 570 (Wis. Ct. App. 2003).
50. *United States v. Castro-Gomez,* 360 F.3d 216 (1st Cir. 2004).
51. *State v. Unger,* 338 N.E.2d 442 (Ill. App. Ct. 1975).
52. Wayne R. LaFave, *Principles of Criminal Law* (St. Paul, MN: West Publishing, 2003), p. 432.
53. Richard G. Singer and John Q. La Fond, *Criminal Law Examples and Explanations,* 2nd ed. (New York: Aspen, 2001), p. 79.
54. LaFave, *Criminal Law,* pp. 441–443.
55. LaFave, *Criminal Law,* pp. 442–443.
56. Singer and La Fond, *Criminal Law Examples and Explanations,* p. 83.
57. *Lambert v. California,* 355 U.S. 335 (1957).
58. *United States v. Cheek,* 498 U.S. 112 (1991).
59. *Cox v. Louisiana,* 379 U.S. 536 (1965).
60. LaFave, *Principles of Criminal Law,* pp. 433–434.

61. LaFave, *Principles of Criminal Law,* p. 434.
62. *Regina v. Morgan,* (1976) A.C. 182.
63. *Sorrells v. United States,* 287 U.S. 435 (1932).
64. *Sherman v. United States,* 356 U.S. 369, 372 (1958).
65. *Sherman v. United States,* 356 U.S. at 369.
66. LaFave, *Criminal Law,* pp. 463–464.
67. *United States v. Fusko,* 869 F.2d 1048 (7th Cir. 1989).
68. LaFave, *Criminal Law,* p. 458.
69. *Sherman v. United States,* 356 U.S. at 382–383.
70. *United States v. Russell,* 411 U.S. 423 (1973).
71. Alan M. Dershowitz, *The Abuse Excuse: And Other Cop-Outs, Sob Stories, and Evasions of Responsibility* (Boston: Little, Brown, 1994), p. 3.
72. George P. Fletcher, *Rethinking Criminal Law* (New York: Oxford University Press, 2000), pp. 801–802.
73. *Millard v. State,* 261 A.2d 227 (Md. Ct. Spec. App. 1970).
74. Dershowitz, *Abuse Excuse,* pp. 54–55.
75. *People v. Molina,* 249 Cal. Rptr. 273 (Cal. Ct. App. 1988).
76. *Commonwealth v. Garabedian,* 503 N.E.2d 1290 (Mass. 1987).
77. Singer and La Fond, *Criminal Law Examples and Explanations,* p. 486.
78. *State v. Phipps,* 883 S.W.2d 138 (Tenn. Ct. App. 1994).
79. Deborah Goldklang, "Post Traumatic Stress Disorder and Black Rage: Clinical Validity, Criminal Responsibility," *Virginia Journal of Social Policy & Law* 5 (1997) 213–243.
80. Wally Owens, "*State v. Osby,* The Urban Survival Defense," *American Journal of Criminal Law* 22 (1995) 821.
81. Patricia J. Falk, "Novel Theories of Criminal Defense Based Upon the Toxicity of the Social Environment: Urban Psychosis, Television Intoxication, and Black Rage," *North Carolina Law Review* 74 (1996) 731.
82. *United States v. Alexander,* 471 F.2d 923 (D.C. Cir. 1972).
83. *State v. Jerrett,* 307 S.E.2d 339 (N.C. 1983).
84. Doriane Lambelet Coleman, "Individualizing Justice Through Multiculturalism: The Liberals' Dilemma," *Columbia Law Review* 96 (1996) 1093, 1150.

Chapter 10

1. *Furman v. Georgia,* 408 U.S. 238 (1972).
2. *Coker v. Georgia,* 433 U.S. 584 (1977).
3. George P. Fletcher, *Rethinking Criminal Law* (New York: Oxford University Press, 2000), p. 236.
4. *State v. Jensen,* 417 P.2d 273 (Kan. 1966).
5. *Report of the Royal Commission on Capital Punishment,* 26–28 (London: HMSO, 1953).
6. Nev. Rev. Stat. § 200.010.
7. Nev. Rev. Stat. § 200.020.
8. Pa. Laws of 1794, ch. 257, §§ 1, 2 (1794).
9. American Law Institute, *Model Penal Code and Commentaries,* vol. 1, pt. 11 (Philadelphia: American Law Institute, 1980), pp. 14–15, 44–48.
10. William Blackstone, *Commentaries on the Laws of England,* vol. 4 (Chicago: University of Chicago Press, 1979), p. 196.
11. *Vesey v. Vesey,* 54 N.W.2d 385 (Minn. 1952).
12. *State v. Myers,* 81 A.2d 710 (N.J. 1951).
13. *Commonwealth v. Cass,* 467 N.E.2d 1324 (Mass. 1984).
14. *Roe v. Wade,* 410 U.S. 113 (1973).
15. Kans. Stat. Ann. § 77-202.
16. *State v. Fierro,* 603 P.2d 74 (Ariz. 1979).
17. Utah Code Ann. § 76-5-291.
18. *State v. Schrader,* 302 S.E.2d 70 (W. Va. 1982).
19. *State v. Bingham,* 719 P.2d 109 (Wash. 1986).
20. *State v. Moua,* 678 N.W.2d 29 (Minn. 2004).
21. Va. Code Ann. § 18.2-31.
22. Fla. Stat. § 921.141.
23. Wash. Rev. Code § 9A.32.050
24. Idaho Code Ann. § 18-4003.
25. La. Rev. Stat. Ann. § 14.30.1.
26. Ark. Code Ann. § 5-10-103.
27. *Alston v. State,* 643 A.2d 468 (Md. Ct. Spec. App. 1994).
28. *Commonwealth v. Malone,* 47 A.2d 445 (Pa. 1946).
29. Cal. Penal Code § 188.

30. *Commonwealth v. Malone,* 47 A.2d at 445.
31. *Banks v. State,* 211 S.W. 217 (Tex. 1919).
32. *Alston v. State,* 643 A.2d at 468.
33. *Hyam v. Director of Public Prosecutions,* (1975) A.C. 55.
34. Kan. Stat. Ann. § 21-3402.
35. *People v. Stamp,* 82 Cal. Rptr. 598 (Cal. Ct. App. 1969).
36. *People v. Fuller,* 150 Cal. Rptr. 515 (Cal. Ct. App. 1978).
37. *Regina v. Serne,* 16 Cox C.C. 311 (1887).
38. *Enmund v. Florida,* 458 U.S. 782 (1982).
39. Va. Code Ann. § 18.2-32
40. Mo. Rev. Stat. § 565.021.
41. 18 Pa. Cons. Stat. § 1102(d).
42. *People v. Burroughs,* 678 P.2d 894 (Cal. 1984).
43. *King v. Commonwealth,* 368 S.E.2d 704 (Va. Ct. App. 1988).
44. *Campbell v. State,* 444 A.2d 1034 (Md. 1982).
45. *Kinchion v. State,* 81 P.3d 681 (Okla. Crim. App. 2003).
46. Oliver Wendell Holmes, Jr., *The Common Law* (New York: Dover Press, 1991), pp. 57–58.
47. Francis T. Cullen and William J. Maakestad, *Corporate Crime Under Attack: The Ford Pinto Case and Beyond* (Cincinnati, OH: Anderson, 1987), pp. 145–308.
48. Wayne R. LaFave, *Criminal Law,* 3rd ed. (St. Paul, MN: West Publishing, 2000), pp. 272–278.
49. LaFave, *Criminal Law,* p. 717.
50. Joshua Dressler, *Understanding Criminal Law,* 3rd ed. (New York: Lexis, 2001), pp. 528–529.
51. Dressler, *Understanding Criminal Law,* pp. 529–530.
52. *Bedder v. Director of Public Prosecutions,* (1954) 1 W.L.R. 1119.
53. *Commonwealth v. Carr,* 580 A.2d 362 (Pa. Super. Ct. 1990).
54. Dressler, *Understanding Criminal Law,* p. 532.
55. *State v. Flory,* 276 P. 458 (Wyo. 1929).
56. *State v. Gounagias,* 153 P. 9 (Wash. 1901).
57. *People v. Bridgehouse,* 303 P.2d 1018 (Cal. 1956).
58. LaFave, *Criminal Law,* p. 705.
59. *State v. Flory,* 276 P. at 458.
60. *Commonwealth v. Mink,* 123 Mass. 422 (Mass. 1877).
61. Cal. Penal Code § 192(b).
62. *Todd v. State,* 594 So. 2d 802 (Fla. Dist. Ct. App. 1992).
63. *People v. Penny,* 285 P.2d 926 (Cal. 1955).
64. Cal. Penal Code §§ 191.5, 192.5.
65. Fla. Stat. § 782.07.

Chapter 11

1. Wayne R. LaFave, *Criminal Law,* 3rd ed. (St. Paul, MN: West Publishing, 2000), pp. 787–788.
2. *Coker v. Georgia,* 433 U.S. 584 (1977).
3. *People v. Liberta,* 474 N.E.2d 567 (N.Y. 1984).
4. *State v. Smith,* 426 A.2d 38 (N.J. 1981).
5. Joane Belknap, *The Invisible Woman: Gender, Crime, and Justice,* 3rd ed. (Belmont, CA: Thomson/Wadsworth, 2007), pp. 298-308.
6. Susan Estrich, *Real Rape* (Cambridge, MA: Harvard University Press, 1987).
7. National Center for Victims of Crime, *Male Rape* (1997). Retrieved November 30, 2008, from http://www.ncvc.org/ncvc/main.aspx?dbName=DocumentViewer&DocumentID=32361
8. LaFave, *Criminal Law,* p. 762.
9. *State of New Jersey in the Interest of M.T.S.,* 609 A.2d 1266 (N.J. 1992).
10. William Blackstone, *Commentaries on the Laws of England,* vol. 4 (Chicago: University of Chicago Press, 1979), p. 215.
11. LaFave, *Criminal Law,* p. 755.
12. Blackstone, *Commentaries on the Laws of England,* p. 213.
13. *State v. Rusk,* 424 A.2d 720 (Md. 1981).
14. *Brown v. State,* 106 N.W. 536 (Wis. 1906).
15. *State v. Schuster,* 282 S.W.2d 553 (Mo. 1955).
16. LaFave, *Criminal Law,* pp. 773–774.
17. LaFave, *Criminal Law,* pp. 775–777.
18. *People v. Minkowski,* 23 Cal. Rptr. 92 (Cal. Ct. App. 1962).
19. *Boro v. Superior Court,* 210 Cal. Rptr. 122 (Cal. Ct. App. 1985).
20. American Law Institute, *Model Penal Code and Commentaries,* vol. 1, pt. 2 (Philadelphia: American Law Institute, 1980), pp. 305–306.
21. LaFave, *Criminal Law,* pp. 755–756.
22. LaFave, *Criminal Law,* pp. 780–781.
23. Joshua Dressler, *Understanding Criminal Law,* 3rd ed. (New York: Lexis, 2001), pp. 574–576.
24. *Taylor v. State,* 12 N.E. 400 (Ind. 1887); *People v. Banks,* 552 N.E.2d 131 (N.Y. 1996).
25. Vt. Stat. Ann. tit. 13, § 3253.
26. American Law Institute, *Model Penal Code and Commentaries,* p. 346.
27. Cal. Penal Code § 263.
28. *State v. Kidwell,* 556 P.2d 20 (Ariz. Ct. App. 1976).
29. Cal. Penal Code § 261.
30. Mich. Comp. Laws § 750.520(b)(1)(e).
31. Tex. Penal Code Ann. § 22011(10).
32. 18 Pa. Cons. Stat. § 3101.
33. *Commonwealth v. Milnarich,* 542 A.2d 1336 (Pa. 1988).
34. Cal. Penal Code § 261.
35. *People v. Mayberry,* 542 P.2d 1337 (Cal. 1997).
36. *Director of Public Prosecutions v. Morgan,* (1976) A.C. 182, 2 All E.R. 347 (1975) 2 W.L.R. 913 (H.L.).
37. *State v. Plunkett,* 934 P.2d 113 (Kan. 1997).
38. *Tyson v. State,* 619 N.E.2d 276 (Ind. Ct. App. 1993).
39. LaFave, *Criminal Law,* p. 778.
40. LaFave, *Criminal Law,* pp. 778–779.
41. American Law Institute, *Model Penal Code and Commentaries,* p. 329.
42. *People v. Abbot,* 19 Wend. 192 (N.Y. 1838).
43. *State v. Colbath,* 540 A.2d 1212 (N.H. 1988).
44. *People v. Wilhelm,* 476 N.W.2d 753 (Mich. Ct. App. 1991).
45. Ga. Code Ann. § 16-5-22.
46. American Law Institute, *Model Penal Code and Commentaries,* § 211.1(1)(a)(b).
47. 720 Ill. Comp. Stat. 5/12-3.
48. Ga. Code Ann. § 16-5-23.1
49. Minn. Stat. § 609.226.
50. *State v. Humphries,* 586 P.2d 130 (Wash. Ct. App. 1978).
51. Ga. Code Ann. § 16-5-23.1.
52. Ga. Code Ann. § 16-5-24.
53. Cal. Penal Code §§ 244–245.
54. 720 Ill. Comp. Stat. 5/12-4–4.5.
55. S.D. Codified Laws § 22-18-1–3.
56. Minn. Stat. § 609.228.
57. 18 U.S.C. § 116.
58. Cal. Penal Code § 203.
59. Ga. Code Ann. § 16-5-20.
60. Cal. Penal Code § 240.

61. 720 Ill. Comp. Stat. 5/12-1.
62. Cal. Penal Code § 241.3.
63. Ohio Rev. Code Ann. § 2903.22.
64. LaFave, *Criminal Law,* pp. 376–377.
65. *Wilson v. State,* 53 Ga. 205 (1874).
66. Ga. Code Ann. § 16-5-21.
67. 720 Ill. Comp. Stat. 5/12-2.
68. 720 Ill. Comp. Stat. 5/12-2.5.
69. 720 Ill. Comp. Stat. 5/12-7.3.
70. 720 Ill. Comp. Stat. 5/12-7.4.
71. 720 Ill. Comp. Stat. 5/12-7.5.
72. S.D. Codified Laws § 22-18-1(4).
73. Rollin M. Perkins and Ronald N. Boyce, *Criminal Law,* 3rd ed. (Mineola, NY: Foundation Press, 1982), p. 282.
74. Mass. Gen. Laws ch. 265, § 26A.
75. Fla. Stat. § 787.01.
76. Tex. Penal Code Ann. § 20.04.
77. N.C. Gen. Stat. § 14-39.
78. *People v. Chessman,* 238 P.2d 1001 (Cal. 1951).
79. *People v. Daniels,* 459 P.2d 225 (Cal. 1969).
80. *People v. Lombardi,* 229 N.E.2d 206 (N.Y. 1967).
81. *State v. Goodhue,* 833 A.2d 861 (Vt. 2003).
82. Wis. Stat. § 940.31.
83. *Marbley v. State,* 100 S.W.3d 48 (Ark. Ct. App. 2003).
84. *People v. Majors,* 92 P.3d 360 (Cal. 2004).
85. Tex. Penal Code Ann. § 20.04.
86. Idaho Code Ann. § 18-2901.
87. Idaho Code Ann. § 18-2902.
88. Ark. Code Ann. § 5-11-104.
89. *People v. Cohoon,* 42 N.E.2d 969 (Ill. App. Ct. 1942).
90. *McKendree v. Christy,* 172 N.E.2d 380 (Ill. App. Ct. 1961).
91. Perkins and Boyce, *Criminal Law,* p. 224.
92. Wis. Stat. § 940.31.
93. Ala. Code § 13A-6-41.
94. *Shue v. State,* 553 S.E.2d 348 (Ga. Ct. App. 2001).
95. *State v. Overton,* N.C. Ct. App. LEXIS 915 (N.C. Ct. App. 2005).
96. *United States v. Bradley,* 390 F.3d 145 (1st Cir. 2004).

Chapter 12

1. American Law Institute, *Model Penal Code and Commentaries,* vol. 1, pt. 11 (Philadelphia: American Law Institute, 1980), p. 67.
2. William Blackstone, *Commentaries on the Laws of England,* vol. 4 (Chicago: University of Chicago Press, 1979), p. 223.

3. *Taylor v. United States,* 495 U.S. 575 (1990).
4. *State v. Miller,* 954 P.2d 925 (Wash. Ct. App. 1998).
5. *State v. Roberts,* 2004 Wash. App. LEXIS 255 (Wash. Ct. App. 2004).
6. *Sears v. State,* 713 P.2d 1218 (Alaska Ct. App. 1986).
7. *Dixon v. State,* 855 So. 2d 1245 (Fla. Dist. Ct. App. 2003).
8. *Stowell v. People,* 90 P.2d 520 (Colo. 1939).
9. 720 Ill. Comp. Stat. 5/19-1(a).
10. *State v. Wentz,* 68 P.3d 282 (Wash. 2003).
11. Cal. Penal Code § 459.
12. 18 Pa. Cons. Stat. § 3502(d).
13. Wayne R. LaFave, *Criminal Law,* 3rd ed. (St. Paul, MN: West Publishing, 2000), p. 890.
14. 18 Pa. Cons. Stat. § 3502(d).
15. Cal. Penal Code § 464.
16. Ariz. Rev. Stat. Ann. § 13-1506–1508.
17. Idaho Code Ann. § 18-1406.
18. 18 Pa. Cons. Stat. § 3502.
19. *State v. Stinton,* 89 P.3d 717 (Wash. Ct. App. 2004).
20. Tex. Penal Code Ann. § 30.05.
21. Tex. Penal Code Ann. §§ 30.05–30.06.
22. N.Y. Penal Law § 156.10.
23. Colo. Rev. Stat. § 18-5.5-102.
24. *Fugarino v. State,* 531 S.E.2d 187 (Ga. Ct. App. 2000).
25. Blackstone, *Commentaries on the Laws of England,* p. 220.
26. *State v. Braathen,* 43 N.W.2d 202 (N.D. 1950).
27. *State v. Spiegel,* 83 N.W. 722 (Iowa 1900).
28. *State v. Wyatt,* 269 S.E.2d 717 (N.C. 1980).
29. *Williams v. State,* 600 N.E.2d 962 (Ind. Ct. App. 1993).
30. N.J. Stat. Ann. § 2C:17-lb.
31. Fla. Stat. § 806.01(1).
32. 720 Ill. Comp. Stat. 5/201.1.
33. Fla. Stat. § 806.01(2).
34. Rev. Code Wash. 9A.48.030.
35. Rev. Code Wash. 9A.20.021.
36. Cal. Penal Code §§ 451–452.

Chapter 13

1. *People v. Hoban,* 88 N.E. 806, 807 (Ill. 1909).
2. *Anon. v. The Sheriff of London,* (1473) Year Book 13 Edw. IV, f. 9, p1.5.
3. Joshua Dressler, *Understanding Criminal Law,* 3rd ed. (New York: Lexis, 2001), pp. 550–551.
4. Rollin M. Perkins and Ronald N. Boyce, *Criminal Law,* 3rd ed.

(Mineola, NY: Foundation Press, 1982), p. 298.
5. Dressler, *Understanding Criminal Law,* p. 550.
6. Perkins and Boyce, *Criminal Law,* p. 300.
7. Dressler, *Understanding Criminal Law,* p. 553.
8. Dressler, *Understanding Criminal Law,* p. 554.
9. *Wilkinson v. State,* 60 So. 2d 786 (Miss. 1952).
10. *Smith v. State,* 74 S.E. 1093 (Ga. 1912).
11. *State v. Jones,* 65 N.C. 395 (N.C. 1871).
12. Tex. Penal Code Ann. § 31.03.
13. Cal. Penal Code §§ 484–502.9.
14. S.C. Code Ann. § 16-13-30.
15. Tex. Penal Code Ann. § 31.03(3).
16. 18 Pa. Cons. Stat. § 3903(c)(1).
17. 18 Pa. Cons. Stat. § 31.09(3).
18. Cal. Penal Code § 487(A).
19. Cal. Penal Code §§ 487(3), 489.
20. Tex. Penal Code Ann. § 31.03(E)(7).
21. Dressler, *Understanding Criminal Law,* p. 562.
22. American Law Institute, *Model Penal Code and Commentaries,* vol. 2, pt. 2 (Philadelphia: American Law Institute, 1980), p. 128.
23. Wayne R. LaFave, *Criminal Law,* 3rd ed. (St. Paul, MN: West Publishing, 2000), p. 850.
24. *Norton v. United States,* 92 F.2d 753 (9th Cir. 1937).
25. 18 U.S.C. § 1028(a)(7).
26. Utah Code Ann. §§ 76-6-1101–1104.
27. *State v. Green,* 172 P.3d 1213 (Kan. Ct. App. 2007).
28. *Lund v. Commonwealth,* 232 S.E.2d 745 (Va. 1977).
29. *Dawson's Case,* (1602) 80 Eng. Rep. 4.
30. *United States v. Monasterski,* 567 F.2d 677 (6th Cir. 1977).
31. Perkins and Boyce, *Criminal Law,* pp. 343–344.
32. Cal. Penal Code § 211.
33. *People v. Braverman,* 173 N.E. 55 (Ill. 1930).
34. LaFave, *Criminal Law,* pp. 869–870.
35. Fla. Stat. § 812.13.
36. American Law Institute, *Model Penal Code and Commentaries,* p. 104.
37. Fla. Stat. § 812.13.
38. Cal. Penal Code § 215.
39. N.J. Stat. Ann. § 2C:15-1.
40. Va. Code Ann. § 18.2-58.1.
41. Fla. Stat. § 812.133.

42. William Blackstone, *Commentaries on the Laws of England*, vol. 4. (Chicago: University of Chicago Press, 1979), p. 141.
43. Mich. Comp. Laws §§ 750.213–214.
44. N.Y. Penal Law § 155.05.
45. *State v. Crone*, 545 N.W.2d 267 (Iowa 1996).
46. *State v. Harrington*, 260 A.2d 692 (Vt. 1969).

Chapter 14

1. Edwin H. Sutherland, *White Collar Crime* (New Haven, CT: Yale University Press, 1983), p. 7.
2. Ellen S. Podgor and Jerold H. Israel, *White Collar Crime in a Nutshell*, 3rd ed. (St. Paul, MN: West Publishing, 1997), pp. 205–216.
3. 29 U.S.C. § 651.
4. 29 U.S.C. § 666.
5. David Barstow, "U.S. Rarely Seeks Charges for Deaths in Workplace," *New York Times*, Dec. 22, 2003.
6. 18 U.S.C. §§ 1348–1350.
7. *Diamond v. Oreamuno*, 248 N.E.2d 910 (N.Y. 1969).
8. *United States v. O'Hagan*, 521 U.S. 642 (1998).
9. 18 U.S.C. § 1341.
10. 18 U.S.C. § 1343.
11. 18 U.S.C. § 371.
12. *United States v. Jenkins*, 943 F.2d 167 (2d Cir. 1991).
13. *United States v. Goodman*, 945 F.2d 125 (6th Cir. 1991).
14. 18 U.S.C. § 1347.
15. *United States v. Baldwin*, 277 F. Supp. 2d 67 (D.D.C. 2003).
16. *United States v. Lucien*, 347 F.3d 45 (2d Cir. 2003).
17. *United States v. Miles*, 360 F.3d 472 (5th Cir. 2004).
18. *Cuellar v. United States*, 553 U.S. 550 (2008); *United States v. Santos*, 553 U.S. 507 (2008).
19. *United States v. Johnson*, 971 F.2d 562 (10th Cir. 1992).
20. *United States v. Carucci*, 364 F.3d 339 (1st Cir. 2004).
21. *Northern Pacific Railroad Co. v. United States*, 356 U.S. 1 (1958).
22. 15 U.S.C. § 1.
23. *United States v. Azzarelli Construction Co. and John F. Azzarelli*, 612 F.2d 292 (7th Cir. 1979).
24. Rollin M. Perkins and Ronald N. Boyce, *Criminal Law*, 3rd ed. (Mineola, NY: Foundation Press, 1982), pp. 528–530.

25. *United States v. Sun-Diamond Growers of California*, 526 U.S. 398 (1999).
26. 15 U.S.C. § 78dd-1.

Chapter 15

1. William Blackstone, *Commentaries on the Laws of England*, vol. 4 (Chicago: University of Chicago Press, 1979), pp. 142–152.
2. Wis. Stat. § 947.01.
3. 720 Ill. Comp. Stat. 5/26-1.
4. Ariz. Rev. Stat. Ann. § 13-2904.
5. *State v. McCarthy*, 659 N.W.2d 808 (Minn. Ct. App. 2003).
6. *People v. Barron*, 808 N.E.2d 1051 (Ill. App. Ct. 2004).
7. *State v. A.S.*, 626 N.W.2d 712 (Wis. 2001).
8. American Law Institute, *Model Penal Code and Commentaries*, vol. 3, pt. 2 (Philadelphia: American Law Institute, 1985), pp. 313–314.
9. *Cole v. Arkansas*, 338 U.S. 345 (1949).
10. N.Y. Penal Law §§ 240.05, 240.06.
11. N.Y. Penal Law §§ 240.10, 240.15.
12. Ohio Rev. Code Ann. § 2917.04.
13. *J.B.A. v. State*, 2004 UT App 450 (Utah Ct. App. 2004).
14. James Q. Wilson, *Thinking About Crime*, rev. ed. (New York: Vintage Books, 1983), pp. 75–89.
15. American Law Institute, *Model Penal Code and Commentaries*, p. 385.
16. American Law Institute, *Model Penal Code and Commentaries*, pp. 385–386.
17. *Mayor of the City of New York v. Miln*, 36 U.S. 102 (1837).
18. *Edwards v. California*, 314 U.S. 160 (1941).
19. *Papachristou v. City of Jacksonville*, 405 U.S. 156 (1972).
20. *Kolender v. Lawson*, 461 U.S. 352 (1983).
21. *Pottinger v. City of Miami*, 810 F. Supp. 1551 (S.D. Fla. 1992).
22. *Gallo v. Acuna*, 929 P.2d 596 (Cal. 1997).
23. Norval Morris and Gordon Hawkins, *The Honest Politician's Guide to Crime Control* (Chicago: University of Chicago Press, 1969).
24. Patrick Devlin, *The Enforcement of Morals* (New York: Oxford University Press, 1965).

25. Ga. Code Ann. § 16-6-9.
26. 18 Pa. Cons. Stat. § 5902(a).
27. Cal. Penal Code § 653.22.
28. 18 Pa. Cons. Stat. § 5902(e), (e.2).
29. Ga. Code Ann. §§ 16-6-11, 16-6-14.
30. 18 Pa. Cons. Stat. § 5902(d).
31. Ga. Code Ann. § 16-6-10.
32. N.C. Gen. Stat. § 14-204(7).
33. Ga. Code Ann. § 16-6-16.
34. Ga. Code Ann. § 16-6-13(b).
35. *City of Milwaukee v. Burnette*, 637 N.W.3d 447 (Wis. Ct. App. 2001).
36. *Rex v. Curl*, 93 Eng. Rep. 849 (K.B. 1727).
37. *Roth v. United States*, 354 U.S. 476 (1957).
38. *Miller v. California*, 413 U.S. 15 (1973).
39. *New York v. Ferber*, 458 U.S. 747 (1982).
40. *American Booksellers Ass'n v. Hudnut*, 771 F.2d 323 (7th Cir. 1985).

Chapter 16

1. James Madison, "The Federalist No. 43," in A. Hamilton, J. Madison, and J. Jay, *The Federalist Papers* (New York: New American Library, 1961).
2. *Cramer v. United States*, 325 U.S. 1 (1945).
3. *United States v. Greathouse*, 26 F. Cas. 18, 21 (C.C.N.D. Cal. 1863) (No. 15,254).
4. *Dennis v. United States*, 341 U.S. 494 (1951); *Yates v. United States*, 354 U.S. 298 (1957).
5. 18 U.S.C § 2153.
6. 18 U.S.C. § 2155.
7. 18 U.S.C. §§ 2152, 2156.
8. *United States v. Kabat*, 797 F.2d 580 (8th Cir. 1986).
9. 18 U.S.C. § 794.
10. *Gorin v. United States*, 312 U.S. 19 (1941).
11. 18 U.S.C. § 2332(b)(g)(3)5.
12. 18 U.S.C. § 2332.
13. 18 U.S.C. § 2332b.
14. 18 U.S.C. § 175.
15. 18 U.S.C. § 1992.
16. 49 U.S.C. § 46502.
17. 18 U.S.C. § 2339.
18. 18 U.S.C. § 2339A, 18 U.S.C. § 2339B.
19. 8 U.S.C. § 1189(a)(1).
20. Antonio Casssese, Guido Acquaviva, Mary Fan, and Alex Whiting, *International Criminal Law: Cases and Commentary* (New York: Oxford University Press, 2009).

GLOSSARY

abandonment: an individual who completely and voluntarily renounces his or her criminal purpose is not liable for an attempt. Abandonment as a result of outside or extraneous factors does not constitute a defense.

abuse excuse: criminal defense that claims a lack of criminal responsibility based on past abuse or experiences.

accessories: parties responsible for the separate and lesser offense of assisting a criminal offender to avoid apprehension, prosecution, or conviction.

accessories after the fact: individuals liable for assisting an offender to avoid arrest, prosecution, or punishment.

accessories before the fact: individuals under the common law who assist an individual prior to the commission of a crime and who are not present at the scene of the crime.

accomplices: parties liable as principals before and during the commission of a crime.

acquaintance rape: rape committed by a perpetrator who is known to the victim.

actual possession: an object within an individual's immediate physical control or on their person.

actus reus: a criminal act, the physical or external component of a crime.

adultery: consensual sexual intercourse between a male and a female, at least one of whom is married to someone else.

affirmative defenses: the burden of production and, in most cases, the burden of persuasion is on the defendant.

agency theory of felony murder: a felon is liable for a murder committed by a co-felon.

aggravated rape: a rape that is more harshly punished based on the use of force, injury to the victim, or the fact that the perpetrator is a stranger, or other factors.

aggravating factors: factors that permit enhancement of an offender's punishment, including an offender's prior record, nature of the offense, and identity of the victim.

aggressor: an individual initiating a physical confrontation is not entitled to self-defense unless he or she retreats.

alter ego rule: an individual intervening in defense of others possesses the rights of the person he or she is assisting.

American bystander rule: no legal duty to assist or to rescue an individual in danger.

American rule for resistance to an unlawful arrest: an individual may not resist an illegal arrest.

appellant: the individual appealing.

appellate courts: intermediate or supreme court of appeals.

appellee: the party against whom an appeal is filed.

arson: willful and malicious burning of the dwelling of another. Modified by statute to encompass any building or structure.

assault and battery: battery is the application of force to another person. An assault may be committed either by attempting to commit a battery or by intentionally placing another in fear of a battery.

assets forfeiture: seizure pursuant to a court order of the "fruits" of illegal narcotics transactions (along with certain other crimes) or of material that was used to engage in such activity.

attempt: an intent or purpose to commit a crime, an act or acts toward the commission of the crime, and a failure to commit the crime.

attendant circumstances: the conditions or context required for a crime.

bench trial: trial before a judge without a jury.

beyond a reasonable doubt: the standard of proof applied in a criminal case; requires that a judge or juror is convinced beyond a moral certainty.

bigamy: marrying another while already having a living spouse.

bilateral: there must be an agreement between at least two persons with the intent to achieve a common criminal objective.

bill of attainder: a legislative act directed against an individual or group of individuals imposing punishment without trial.

Bill of Rights: first ten amendments to the U.S. Constitution.

binding authority: a decision that establishes a precedent.

blackmail: taking property through the threat to disclose secret or embarrassing information.

brain death test: the irreversible function of all brain functions is the point at which an individual is legally dead.

breach of the peace: an act that disturbs or tends to disturb the tranquility of citizens.

bribery of a public official: offering an item of value to an individual occupying an official position to influence a decision or action.

brief: written legal argument submitted to an appellate court; also, to write a summary of a case.

broken windows theory: failing to prevent and punish misdemeanor offenses causes major crimes.

burden of persuasion: responsibility to convince the fact finder, usually beyond a reasonable doubt.

burden of production: responsibility to produce sufficient evidence for the fact finder to consider the merits of a claim.

burglary: breaking and entry of the dwelling house of another at night with the intention to commit a felony therein. Modified by statute to cover an illegal entry into any structure at any time, day or night, with a criminal intent.

capital felony: punishable by the death penalty or by life imprisonment in states without the death penalty.

capital murder: punishable by the death penalty or life imprisonment and, in non–capital punishment states, by life imprisonment. Also referred to in some states as aggravated murder.

carjacking: taking a motor vehicle in the possession of another, from his or her person or immediate presence, by force, and against his or her will.

case-in-chief: the prosecution's phase of the trial.

castle doctrine: individuals have no obligation to retreat inside their home.

causation: there must be a connection between an act and the resulting prohibited harm.

cause in fact: the defendant must be shown to be the "but for" cause of the harm or injury.

chain conspiracy: a conspiracy in which individuals are linked in a vertical chain to achieve a criminal objective.

child pornography: a juvenile engaged in actual or simulated sexual activity or in the lewd display of genitals.

choice of evils: the defense of necessity in which an individual commits a crime to avoid an imminent and greater social harm or evil.

circumstantial evidence: evidence that indirectly establishes that the defendant possessed a criminal intent or committed a criminal act.

civil commitment: a procedure for detaining psychologically troubled individuals who pose a danger to society.

civil law: protects the individual rather than societal interest.

clemency: an executive governmental official reduces a criminal sentence.

code jurisdiction: acts or omissions are only punishable that are contained in the state criminal code.

coincidental intervening act: a defendant's criminal act results in the victim being at a particular place at a particular time and being impacted by an independent intervening act. The defendant is responsible for foreseeable coincidental intervening acts.

collateral attack: a challenge to a conviction filed following the exhaustion of direct appeals.

combat immunity: individuals meeting standards set forth in the Geneva Convention are to be treated as prisoners of war when apprehended.

common law crimes: crimes developed by the common law judges in England and supplemented by acts of parliament and decrees issued by the king.

common law states: the common law may be applied where the legislature has not acted.

competence to stand trial: a defendant is competent to stand trial if he or she is able to intelligently assist his or her attorney and to follow and understand the trial.

complete attempt: an individual takes every act required to commit a crime and fails to succeed.

computer crime: crimes involving the computer, including unauthorized access to computers, computer programs, and networks; the modification or destruction of data and programs; and the sending of mass unsolicited messages and messages intended to trick and deceive.

concurrence: a criminal intent must trigger and coincide with a criminal act.

concurrent sentences: sentences for each criminal act are served at the same time.

concurring opinion: an opinion by a judge supporting a majority or dissenting opinion, typically based on other grounds.

consecutive sentences: sentences for each criminal act are served one after another.

conspiracy: an agreement to commit a crime. Various state statutes require an overt act in furtherance of this purpose.

constitutional democracy: a constitutional system that limits the powers of the government.

constructive intent: individuals who act in a gross and wantonly reckless fashion are considered to intend the natural consequences of their actions and are guilty of willful and intentional battery or homicide.

constructive possession: an individual who retains legal possession over property that is not within his or her actual control.

cooling of blood: the point at which an individual who has been provoked no longer is acting in response to an act of provocation.

corporate liability: the imposition of vicarious liability on a corporate officer or corporation.

corporate murder: a killing for which a business enterprise is held criminally liable.

corroboration: additional facts that support and lend credibility to the elements of a criminal charge or defense.

crime: conduct that, if shown to have taken place, will result in a formal and solemn pronouncement of the moral condemnation of the community.

crimes against public order and morality: offenses that threaten public peace, quiet, and tranquility.

crimes against the quality of life: misdemeanor offenses that diminish the sense of safety and security in a neighborhood.

crimes against the state: treason, sedition, sabotage, espionage, terrorism, and other offenses intended to harm the government.

crimes of cause and result: the intent to achieve a specific result.

crimes of official misconduct: knowingly corrupt behavior by a public official in exercise of his or her official responsibility.

criminal attempt: comprises an intent or purpose to commit a crime, an act or acts toward the commission of the crime, and a failure to commit the crime.

criminal homicide: all homicides that are neither justified nor excused.

criminal mischief: damage or destruction of tangible property.

criminal procedure: investigation and detection of crime by the police and the procedures used at trial.

criminal trespass: the unauthorized entry or remaining on the land or premises of another.

curtilage: the area immediately surrounding a dwelling that is considered part of the habitation.

custody: temporary and limited right to control property.

cyberstalking: stalking on the computer.

deadly force: use of physical force or a weapon likely to cause death or serious bodily harm.

defendant: individual charged with a crime and who is standing trial.

defiant trespass: entering or remaining on the property of another after receiving notice that an individual's presence is without the consent of the owner.

depraved heart murder: killing as a result of extreme recklessness, wanton unconcern, and indifference to human life with malice aforethought.

derivative liability: the guilt of a party to a crime based on the criminal acts of the primary party.

determinate sentencing: a sentence fixed by the state legislature.

diminished capacity: mental disease or defect admissible to demonstrate defendant's inability to form a criminal intent, typically limited to murder.

disclose or abstain doctrine: the obligation to either make corporate information public or refrain from trading in the corporation's stock.

disorderly conduct: intentionally or knowingly causing or risking public inconvenience, annoyance, or alarm.

disparity: sentences for a particular offense are not uniform and vary from one another.

dissenting opinion: an opinion by a judge disagreeing with the majority of a multiple judge court.

distinguishing precedents: showing why a case differs from existing precedents.

domestic terrorism: a violent or dangerous act occurring within the United States intended to intimidate or coerce the civilian population, to influence government policy by intimidation or coercion, or to affect the conduct of a government by mass destruction, assassination, or kidnapping.

double jeopardy: being prosecuted twice for the same crime.

dual sovereignty: sharing of power between federal and state governments. Each has different interests that permit both a state and a federal prosecution for the same crime.

duress: a crime is excused when committed to avoid what is reasonably believed to be the imminent infliction of serious physical harm or death.

Durham product test: a defendant's unlawful act is the product of a mental disease or defect.

duty to intervene: the legal obligation to act.

earnest resistance: a standard of resistance to rape under the common law.

Eighth Amendment: prohibits cruel and unusual punishment.

embezzlement: the fraudulent conversion of the property of another by an individual in lawful possession of the property.

English rule for resistance to an unlawful arrest: an individual may use reasonable force to resist an illegal arrest.

entrapment: defense based on governmental inducement of an otherwise innocent defendant to commit a defense (subjective test) or based on governmental conduct that falls below accepted standards and would cause an innocent individual to commit a criminal offense (objective test).

environmental crimes: crimes that threaten the natural environment, including harm to the air, water, land, and natural resources.

equal protection: the Fifth and Fourteenth Amendments to the U.S. Constitution guarantee individuals equal protection of the law.

espionage: deliver information to a foreign government with the intent or reason to believe that it is to be used to injure the United States or to advantage a foreign government.

European bystander rule: a rule in Europe imposing a legal duty on individuals to assist those in peril.

excusable homicide: individuals are relieved of criminal liability based on lack of criminal intent. This includes insanity, infancy, and intoxication.

excuses: defenses in which defendants admit wrongful conduct while claiming a lack of legal responsibility based on a lack of a criminal intent or the involuntary nature of their acts.

***ex post facto* law:** a law declaring an act criminal following the commission of the act.

extortion: taking property from another by threat of future violence or action, such as circulating secret or embarrassing information; by threat of a criminal charge; or by threat of inflicting economic harm.

extraneous factor: a circumstance that is not created by a defendant that prevents the completion of a criminal act.

extraterritorial jurisdiction: criminal jurisdiction outside the United States.

extrinsic force: an act of force beyond the effort required to accomplish penetration.

factual impossibility: a criminal act is prevented from being completed because of an extraneous factor.

false imprisonment: intentional and unlawful confinement or restraint of another person.

false pretenses: obtaining title and possession of property of another by a knowingly false representation of a present or past material fact with an intent to defraud that causes an individual to pass title to his or her property.

federal crime of terrorism: one or more violent federal offenses calculated to influence or affect the conduct of government by intimidation or coercion or to retaliate against government conduct.

federal criminal code: federal criminal statutes.

felony: crime punishable by death or by imprisonment for more than one year.

felony murder: a killing committed during the commission of a felony.

fiduciary relationship: a duty of care owed by a corporate official to the stockholders in a corporation.

fighting words: insulting words causing a breach of the peace.

First Amendment: protects freedom of expression, assembly, and free exercise of religion and prohibits the establishment of a religion.

first-degree murder: intentional and premeditated murder with malice afore-thought.

fleeing felon rule: the common law rule permitting deadly force against a felon fleeing the police.

fleeting possession: temporary dominion and control over an object; typically not considered possession for purposes of criminal liability.

Foreign Corrupt Practices Act: illegal for an individual or company to bribe a foreign official in order to gain assistance in obtaining or retaining business.

forgery: creating a false legal document or the material modification of an existing legal document with the intent to deceive or to defraud others.

fornication: name of the act when an unmarried person has voluntary sexual intercourse with another individual.

fraud: an intentional misrepresentation of a material existing fact with knowledge of its falsity, intended to induce another person to part with money, property, or a legal right.

fraud in inducement: misrepresentation in regard to the purpose or benefits of a sexual relationship does not constitute rape.

fraud in the *factum*: misrepresentation in regard to the act to which an individual consents constitutes rape.

Gebardi rule: an individual who is excluded from liability under a criminal statute may not be held legally liable as a conspirator to violate the law.

general deterrence: punishment intended to deter individuals other than the offender from committing a crime.

general intent: an intent to commit an *actus reus*.

Geneva Convention of 1949: international treaty on the law of war providing standards for lawful combatant status.

good motive defense: the fact that a defendant committed a crime for what he or she views as a good reason is not recognized as a defense.

Good Samaritan statute: legislation that exempts individuals from civil liability who assist individuals in peril.

grading: the categorization of homicide in accordance with the "moral blameworthiness" of the perpetrator.

graft: asking, accepting, receiving, or giving a thing of value as compensation or a reward for making an official decision.

grand larceny: a serious larceny, determined by the value of the property that is taken.

gross misdemeanor: punishable by between six and twelve months in prison.

guilty but mentally ill (GBMI): the defendant found to be guilty and mentally ill at the time of the criminal offense. The defendant is provided with psychiatric care while incarcerated. This is distinguished from a verdict of not guilty by reason of insanity (NGRI).

habeas corpus: an order issued by a court requiring the government to demonstrate that an individual is being legally detained.

hate speech: speech that denigrates, humiliates, and attacks individuals on account of race, religion, ethnicity, nationality, gender, sexual preference, or other personal characteristics and preferences.

head notes: short statements of the important points included in a legal decision.

heat of passion: acting in response to adequate provocation.

holding: the conclusion reached by a judge in a case.

identity theft: stealing of an individual's personal identifying information.

ignorantia lexis non excusat: ignorance of the law is no excuse.

immorality crimes: prostitution, obscenity, bigamy, and other offenses against the moral order.

imperfect self-defense: an honest, but unreasonable belief in the justifiability of self-defense that results in a conviction for manslaughter rather than murder.

incapacitation: a theory of punishment that protects the public by incarcerating offenders.

inchoate crimes: attempts, conspiracy, and solicitation. Each requires a specific purpose to accomplish a criminal objective and an act in furtherance of the intent. These offenses are punished to the same extent or to a lesser extent than the target crime.

incitement to violent action: words provoking individuals to breach the peace.

incomplete attempt: an individual abandons or is prevented from completing an attempt due to an extraneous or intervening factor.

incorporation theory: the Due Process Clause of the Fourteenth Amendment to the U.S. Constitution is interpreted to include most of the rights contained in the Bill of Rights and extends these protections to the states.

indecent exposure: an act of public indecency.

indeterminate sentencing: the state legislature provides judges with the ability within certain limits to set a minimum and maximum sentence. The offender is evaluated while imprisoned by a parole board.

infamous crimes: deserving of shame or disgrace.

infancy: at common law there was an irrebuttable presumption that children younger than seven lack criminal intent. In the case of children older than seven and younger than fourteen, there was a rebuttable presumption of a lack of capacity to form a criminal intent. Individuals older than fourteen were considered to possess the same capacity as an adult.

infractions: punishable by a fine.

inherent impossibility: an act that is incapable of achieving the desired result.

insanity defense: a legal excuse based on a mental disease or defect.

insider trading: use of confidential corporate information to buy or sell stocks.

intermediate level of scrutiny: classifications based on gender must be factually related to differences based on gender and must be substantially related to the achievement of a valid state objective.

international criminal law: criminal acts that violate international law.

international terrorism: a violent or dangerous act occurring outside the United States intended to intimidate or coerce the civilian population, to influence government policy by intimidation or coercion, or to affect the conduct of a government by mass destruction, assassination, or kidnapping.

Interstate Commerce Clause: constitutional power of U.S. Congress to regulate commerce among the states.

intervening cause: a cause that occurs between the defendant's criminal act and a social harm.

intervention in defense of others: the privilege to exercise self-defense on behalf of an individual in peril.

intrinsic force: the amount of force required to achieve penetration.

involuntary act: unconscious act or automatism.

involuntary intoxication: a defense to criminal offenses where the defendant meets the standard for mental illness in the state.

involuntary manslaughter: killing of another as a result of gross negligence, recklessness, or during the commission of an unlawful act.

irresistible impulse: mental disease that causes the defendant to lose the ability to choose between right and wrong and avoid engaging in criminal acts.

joint possession: several individuals exercise dominion and control over an object.

jury nullification: right of a jury to disregard the law and to acquit a defendant.

just deserts: offender receives sentence that he or she deserves.

justifiable homicide: murder is justified under the circumstances; this includes self-defense, police use of deadly force, and the death penalty.

justification: a defense based on the circumstances of a criminal act.

keeping a place of prostitution: the crime of using a building for prostitution.

kidnapping: the unlawful, nonconsensual, and forcible asportation of an individual.

knowingly: awareness that conduct is practically certain to cause a result.

knowing possession: individual awareness of criminal possession.

larceny: trespassory taking and carrying away of the personal property of another with the intent to permanently deprive the individual of possession of the property.

larceny by trick: obtaining possession by misrepresentation or deceit.

last step approach: common law approach to attempt that requires the last step to the completion of a crime.

legal impossibility: the defense that an individual's act does not constitute a crime as a matter of law.

legal reporters: books containing the published opinions of judges.

lewdness: willful exposing of the genitals of one person to another in a public place for purposes of sexual arousal or gratification.

libel: a civil action for words that harm an individual's reputation.

living off prostitution: being knowingly supported in whole or in substantial part by the proceeds of prostitution.

loitering: standing in public with no apparent purpose.

mail fraud: knowing and intentional participation in a scheme or artifice intended to obtain money or property through the use of the mails to execute the scheme.

majority opinion: the decision of a majority of the judges on a multiple judge panel.

make my day law: a statute that authorizes any degree of force against a trespasser who uses or threatens to use even slight force against the occupant of a home.

mala in se: crimes that are inherently evil.

mala prohibita: crimes that are not inherently evil.

malice aforethought: an intent to kill with ill will and hatred.

mandatory minimum sentences: the legislature requires judges to sentence an offender to a minimum sentence, regardless of mitigating factors. Prison sentences may be reduced by good-time credits while incarcerated.

manslaughter: killing of another without malice aforethought and without excuse or justification.

masturbation for hire: crime of stimulating the genitals of another.

material support to a foreign terrorist organization: providing material support or resources to a foreign terrorist organization or an attempt or conspiracy to do so.

material support to terrorists: providing support or resources or concealing the nature, location, source, or ownership of material support or resources, knowing or intending that the material is to be used in terrorist acts.

mayhem: depriving another individual of a member of his or her body or disfiguring or rendering it useless.

Megan's Law: sexually violent offender registration laws are named in memory and honor of Megan Kanka, a seven-year-old New Jersey child who was sexually assaulted and murdered by a neighbor in 1994.

mens rea: the mental element of a crime.

mere possession: unknowing possession.

mere presence rule: an individual's presence at the scene of a crime generally does not satisfy the *actus reus* requirement for accomplice liability.

minimum level of scrutiny test: law presumed constitutional so long as reasonably related to a valid state purpose.

misappropriation doctrine: an individual is prohibited from using inside information obtained from a firm or corporation to trade in another corporation's stock.

misdemeanant: individual charged with a misdemeanor.

misdemeanor: punishable by less than a year in prison.

misdemeanor manslaughter: the unintentional killing of another during the commission of a criminal act that does not amount to a felony.

mistake of fact: defense based on mistake of fact that negates a specific criminal intent, knowledge, or purpose.

mistake of law: an error of law, with isolated exceptions, is not a defense.

mitigating circumstances: factors that may reduce or moderate the sentence of a defendant convicted at trial.

M'Naghten test: a disease or defect of the mind that results in an individual's either not knowing what he or she was doing was right or wrong or not knowing what he or she was doing.

Model Penal Code: an influential criminal code drafted by prominent academics, practitioners, and judges affiliated with the American Law Institute to encourage state legislatures to adopt a uniform approach to the criminal law.

money laundering: financial transaction involving proceeds or property derived from unlawful activity.

multiple judge panel: a judicial tribunal with three or more judges.

murder: killing of another with malice aforethought and without excuse or justification.

natural and probable consequences doctrine: a person encouraging or facilitating the commission of a crime will be held liable as an accomplice for the crime he or she aided and abetted as well as for crimes that are the natural and probable outcome of the criminal conduct.

necessity defense: a criminal act is justified when undertaken to prevent an imminent, immediate, and greater harm.

negligently: a failure to be aware of a substantial and unjustifiable risk that constitutes a gross deviation from the standard of care that a reasonable person would observe in the actor's situation.

negligent manslaughter: arises when an individual commits an act that he or she is unaware creates a high degree of risk of human injury or death under circumstances in which a reasonable person would have been aware of the threat.

nolo contendere: a plea that has the legal effect of a plea of guilty, but does not constitute an admission of guilt in proceedings outside of the immediate trial.

nondeadly force: use of physical force or weapon that is not likely to cause death or serious injury.

nullum crimen sine lege, nulla poena sine lege: no crime without law, no punishment without law.

obiter dicta: observations from the bench.

objective approach to criminal attempt: requires an act that is very close to the completion of the crime.

objective test for intervention in defense of others: a person intervening in defense of others may intervene where a reasonable person would believe a person is in need of assistance.

obscenity: description or representation of sexual conduct that, taken as a whole by the average person applying contemporary community standards, appeals to the prurient interest in sex. Sex is portrayed in a patently offensive way and lacks serious literary, artistic, political, or scientific value when taken as a whole.

Occupational Safety and Health Act: a federal law protecting workplace safety.

omission: failure to act or to intervene to assist another.

oral argument: arguments before an appellate court.

overbreadth: a statute that is unconstitutionally broad and punishes both unprotected speech or conduct and protected speech or conduct.

overt act: an overt act in furtherance of an agreement is required under most modern conspiracy statutes.

pandering: encouraging and inducing another to remain a prostitute.

pardon: exempts an individual from additional punishment.

parental responsibility laws: statutory rule that parents are responsible for the criminal acts of their children.

parties to a crime: individuals liable for assisting another to commit a crime.

per curiam: an opinion not attributed to a particular judge, literally "for the court."

perfect self-defense: an honest and reasonable belief that constitutes a complete defense to a criminal charge.

persuasive authority: a decision that a court may consult to assist in a judgment that does not constitute binding authority.

petitioner: an individual filing a collateral attack on a verdict following the exhaustion of direct appeals.

petit larceny: a minor larceny, typically involving the taking of property valued at less than a designated monetary amount.

petty misdemeanor: punishable by less than six months in prison.

physical proximity test: an act constituting an attempt must be physically proximate to the completion of the crime.

pimping: procuring a prostitute for another.

Pinkerton rule: a conspirator is liable for all criminal acts taken in furtherance of the conspiracy.

plea bargain: negotiated agreements between the defense attorney and prosecutor and often approved by a judge.

plurality opinion: a judicial opinion that represents the views of the largest number of judges on a court, although short of a majority. The plurality opinion is typically combined with a concurring opinion to constitute the court majority.

plurality requirement: a conspiracy requires an agreement between two or more parties.

police power: duty to protect the well-being and tranquility of the community.

possession: physical control over property with the ability to freely use and enjoy the property.

precedent: a judicial opinion that controls the decision of a court presented with the same issue. A court may conclude that

a precedent does not fully fit the case it is adjudicating and distinguish the case before it from the existing precedent.

preemption doctrine: federal law is superior to state law in areas reserved to the national government.

premeditation and deliberation: the standard for first-degree murder involving planning and reflecting on a killing. Premeditation may occur instantaneously.

preparation: acts taken to prepare for committing a crime.

preparatory offense: a crime that is a step toward an even more serious offense.

preponderance of the evidence: the standard of proof in a civil case. The facts are probably more in favor of one side than the other.

presumption of innocence: an individual is presumed to be not guilty and the burden is on the government to establish guilt.

presumptive sentencing guidelines: a legislatively established commission establishes a sentencing formula based on various factors, including the nature of the crime and offender's criminal history. Judges may be strictly limited in terms of discretion or may be provided with some flexibility within established limits to depart from the presumptive sentence.

principals in the first degree: common law term for the actual perpetrator of a crime.

principals in the second degree: common law term for individuals who are present at the crime scene and assist in the crime.

privacy: the constitutional right to be free from unjustified governmental intrusion into the sphere of personal autonomy.

promoting prostitution: aiding or abetting prostitution.

prompt complaint: a rape victim at common law was required to lodge an immediate report of a rape.

proportionality: a sentence should "fit the crime."

prosecutrix: a victim or complainant in a rape prosecution.

prostitution: soliciting or engaging in sexual activity in exchange for money or other consideration.

proximate cause: the legally responsible cause of a criminal harm; may involve policy considerations.

proximate cause theory of felony murder: a felon is liable for all foreseeable results of the felony.

public indecencies: public drunkenness, vagrancy, loitering, panhandling, graffiti, and urinating and sleeping in public.

public welfare offense: regulatory offenses carrying fines that typically do not require a criminal intent.

purposely: a conscious intent to cause a particular result.

rape shield laws: the prosecution may not introduce evidence relating to the victim's sexual relations with individuals other than the accused and may not introduce evidence pertaining to the victim's reputation for chastity.

rape trauma syndrome: a psychological and medical condition common among victims of rape.

rational basis test: a law is presumed valid so long as it is reasonably related to a valid state purpose.

reasonable person: the ideal type of the balanced and fair individual.

reasonable resistance: resistance to rape that is objectively reasonable under the circumstances.

reasoning: an explanation of a judge's thinking in reaching a decision.

rebuttal: the defense case at trial.

receiving stolen property: accepting stolen property knowing it to be stolen with the intent to permanently deprive the owner of the property.

reception statutes: a state receives the common law as an unwritten part of a state's criminal law.

recklessly: conscious disregard of a substantial and unjustifiable risk that constitutes a gross deviation from the standard of conduct that a law-abiding person would observe in the defendant's situation.

recklessness: an individual is personally aware that his or her conduct creates a substantial risk of death or serious bodily harm.

rehabilitation: punishment intended to reform offenders and to transform them into law-abiding members of society.

relevant: evidence that assists in establishing a material fact of the crime.

res ipsa loquitur: "the thing speaks for itself." A test for attempt that asks whether an ordinary individual observing the acts of another would conclude that the individual intends to commit a crime.

resist to the utmost: at common law, a rape victim was required to demonstrate a determined resistance to the rape.

respondent: an individual against whom a collateral attack is directed.

responsive intervening act: a defendant's criminal act leads to an act undertaken by the victim in reaction to the threat. An unforeseeable and abnormal responsive act limits the defendant's criminal liability.

restoration: stresses the harm caused by crime to victims and requires offenders to engage in financial restitution and community service to compensate the victim and the community and to "make them whole once again."

result crime: requires that the act cause a very specific harm and requires a specific intent.

retreat: withdrawal from a conflict while indicating a desire to avoid a confrontation.

retreat to the wall: obligation to withdraw as fully as possible before resorting to self-defense.

retribution: offender receives the punishment that he or she deserves.

riot: group disorderly conduct by three or more persons.

robbery: taking personal property from an individual's person or presence by violence or intimidation.

rout: three or more persons taking steps toward the creation of a riot.

rule of legality: individual may not be punished for an act that was not criminally condemned in a statute prior to the commission of the act.

sabotage: during a time of war or national emergency, the willful injury to war material, premises, or utilities with the intent to injure, interfere with, or obstruct the United States or any associate nation in preparing for or carrying out the war or defense activities. During peacetime, sabotage requires an intent to injure, interfere with, or obstruct the national defense of the United States.

Sarbanes-Oxley Act: a securities fraud statute that requires corporate executive officers to certify that corporate financial statements are accurate.

scienter: guilty knowledge.

second-degree murder: intentional killing of another with malice aforethought.

sedition: any communication intended or likely to bring about hatred, contempt, or disaffection with the constitution or the government.

seditious conspiracy: an agreement to overthrow or destroy a government by force.

seditious libel: writing intended or likely to bring about hatred, contempt, or dissatisfaction with the constitution or the government.

seditious speech: verbal communications intended or likely to bring about hatred, contempt, or disaffection with the constitution or the government.

selective incapacitation: singles out repeat offenders and other dangerous individuals for lengthy incapacitation.

self-defense: a justification defense that recognizes the right of an individual to defend himself or herself against an armed attack.

Sherman (Antitrust) Act of 1890: criminal punishment of contracts, combinations, and conspiracies in restraint of interstate commerce.

simulation: punishes the creation of false objects with the purpose to defraud, such as antique furniture, paintings, and jewelry. Simulation requires proof of a purpose to defraud or proof that an individual knows that he or she is "facilitating a fraud."

social host liability laws: liability for serving or providing alcohol to minors in the event of an accident or injury.

Socratic method: use of question and answer technique in teaching.

solicitation: a written or spoken statement in which an individual intentionally advises, requests, counsels, commands, hires, encourages, or incites another person to commit a crime with the purpose that the other individual commit the crime.

solicitation for prostitution: requesting another person to engage in an act of prostitution.

Son of Sam laws: prohibits offenders from profiting from their crime.

specific deterrence: punishment intended to deter or discourage an offender from committing another crime.

specific intent: a mental determination to accomplish a specific result.

stalking: following another person or placing another person under surveillance.

stand your ground rule: no requirement to retreat.

stare decisis: precedent.

status offense: offense based on personal characteristics or condition rather than conduct that constitutes cruel and unusual punishment.

statutory rape: strict liability offense of intercourse with an underage individual.

strict liability: a crime that does not require a criminal intent.

strict scrutiny: the state has the burden of demonstrating that a law employing a racial or ethnic classification is strictly necessary to accomplish a valid objective.

subjective approach to criminal attempt: requires an act toward the commission of a crime that is sufficient to establish a criminal intent. The act is not required to be proximate to the completion of the crime.

substantial capacity test: a person is not responsible for criminal conduct, if, at the time of such conduct, as a result of mental disease or defect, the person lacks substantial capacity either to appreciate the criminality (wrongfulness) of his or her conduct or to conform his or her conduct to the requirements of law.

substantial step test: the Model Penal Code approach to determining attempt. There must be a clear step toward the commission of a crime that is not required to be immediately proximate to the crime itself. The act must be committed under circumstances strongly corroborative of an intent to commit a crime.

substantive criminal law: specific crimes, defenses, and general principles.

Supremacy Clause: the clause in the U.S. Constitution that provides that federal laws take precedence over state laws.

tactical retreat: an individual withdraws from a conflict while intending to continue the physical conflict.

tangible property: physical property, including personal property and real property. Distinguished from intangible property.

teen party ordinances: ordinances that make it an offense to hold a party at which minors are served alcohol.

terrorism transcending national boundaries: terrorism occurring partly within and partly outside the United States.

theft statute: consolidated state law punishing larceny, embezzlement, and false pretenses.

Three Strikes and You're Out law: provides mandatory sentences for individuals who commit a third felony after being previously convicted for two serious or violent felonies. Also, stringent penalties are typically provided for a second felony.

tippees: individuals who receive insider information.

tippers: individuals who provide insider information.

tort: civil action for injury to an individual or to his or her property.

transferred intent: what occurs when the intent to harm one individual is transferred to another.

Travel Act: interstate or foreign travel or use of the mails or of a facility in interstate or international commerce with the intent to distribute the proceeds of any specified unlawful activity or violence or to promote a specified unlawful activity and thereafter to commit or attempt to commit a crime.

treason: levying war or giving aid and comfort to the enemy.

trial de novo: a completely new trial conducted before an appellate court.

trial transcript: the written record of trial proceedings.

true man: an individual without fault who is able to rely on self-defense.

true threats: threats of bodily harm directed against an individual or a group of individuals.

truth in sentencing laws: laws that provide that offenders must serve a significant portion of their criminal sentences.

unequivocality test: a test for attempt that asks whether an ordinary individual observing a person's acts would

determine that the person intends to commit a crime.

unilateral: an individual with the intent to enter into a conspiratorial agreement is guilty regardless of the intent of the other party.

unlawful assembly: a gathering of at least three individuals for the purpose of engaging or preparing to engage in conduct likely to cause public alarm.

uttering: circulating or using a forged document.

vagrancy: wandering the street with no apparent means of earning a living.

vehicular manslaughter: killing resulting from the grossly negligent operation of a motor vehicle or resulting from driving while intoxicated.

vicarious liability: holding an individual or corporation liable for a crime committed by another based on the nature of the relationship between the parties.

victim impact statement: victim or victim's family may address the court at sentencing.

violation: minor crimes punishable by fines and not subject to imprisonment. Also called infractions.

void for vagueness: a law violates due process that fails to clearly inform individuals of what acts are prohibited and/or fails to establish clear standards for the police.

voluntary act: the individual is aware and fully conscious of acting. This is distinguished from an involuntary act, unconscious act, or automatism.

voluntary intoxication: defendant not held liable for an offense involving "knowledge or purpose." Increasingly not recognized as a defense.

voluntary manslaughter: instantaneous killing of another in the heat of passion in response to adequate provocation without a "cooling of blood."

weapons of mass destruction: toxic or poisonous chemical weapons, weapons involving biological agents, explosive bombs, or weapons releasing radiation or radioactivity at a level dangerous to human life.

Wharton's rule: an agreement by two persons to engage in a criminal act that requires the involvement of two persons cannot constitute a conspiracy.

wheel conspiracy: a conspiracy in which a single individual or individuals serve as a hub that is connected to various individuals or spokes.

white-collar crime: crimes committed by an individual of high status in the course of his or her occupation. The U.S. Justice Department defines white-collar crime as an illegal act that employs deceit and concealment rather than the application of force to obtain money, property, or service, to avoid the payment or loss of money, or to secure a business or professional advantage.

willful blindness: knowledge is imputed to individuals who consciously avoid awareness in order to avoid criminal responsibility.

wire fraud: knowing and intentional participation in a scheme or artifice intended to obtain money or property through the use of interstate wire communication.

withdrawal in good faith: individuals involved in a fight may gain the right of self-defense by clearly communicating that they are retreating from the struggle.

withdrawal of consent: an individual who initially consents to sexual penetration may change his or her mind.

writ of certiorari: a writ or order issued by the U.S. Supreme Court assuming jurisdiction over an appeal. Four judges must vote to review a case.

year-and-a-day rule: common law requirement, being abandoned by many states, that limits liability for homicide to a year and a day.

CASE INDEX

SUBJECT INDEX

Common law
accessory after the fact, 160
arson, 415
attempts, 177
burglary, 403–404
conspiracy, 192
criminal law from, 5–6
duty to intervene under, 98
history of, 6
insanity, 261–262
rape, 362–366
resisting unlawful arrests, 244–245
sedition, 532
self-defense, 214
states with, 6–7
transferred intent, 117
Comparative harm, 247, 249
Compassionate Use Act, 9
Competence to stand trial, 263
Complete attempt, 176
Complicity, 154
Computer crime, 426, 446–450, 465
Computer trespassing, 413
Concurrence, chronological, 134
Concurrent sentence, 57
Concurring opinion, 17
Congress
cocaine possession penalties, 77
description of, 8
Consecutive sentences, 57
Consensual crimes, 196
Consent
kidnapping and, 395
rape, 364, 373–374
withdrawal of, 378–380
Consent, as defense
description of, 253–254
exceptions to, 254–255
indications for, 254
legal capacity and, 254–255
rape, 365, 373–374
scope of, 254–255
in sports situations, 255, 257–258
statutory standard for, 255
Conspiracy
actus reus of, 192–193
agreement requirement for, 194, 195–196, 208
bilateral approach to, 194
chain, 195
common law description of, 192
criminal, 197–198
criminal objectives of, 195–196
definition of, 149, 192
federal laws against, 197
as felony, 192
Gebardi rule, 196
knowledge standard for, 193
mens rea of, 193–194
as misdemeanor, 192
overt act as proof of, 193
parties involved in, 194
Pinkerton rule, 149, 192
plurality requirement for, 194, 208
prosecution for, 196–197
punishment for, 192

purpose standard for, 193
reasons for punishing, 192
seditious, 532
specific intent, 194
structure of, 194–195
summary of, 175, 208
Supreme Court views on, 195–196
unilateral approach to, 194
Wharton's rule, 196
wheel, 195
Constitution
state, 6, 9
U.S. . See U.S. Constitution
Constitutional challenge, 16
Constitutional democracy, 21
Constructive intent, 117
Constructive possession, 105, 107–110, 427
Contract, obligation created by, 98–99
Controlled Substances Act, 9
Convention for the Suppression of the Financing of Terrorism, 557
Convention on the Prevention and Punishment of Crimes Against Internationally Protected Persons, Including Diplomatic Agents, 556
Cooling of blood, 347
Corporal punishment, 52
Corporate liability
cases involving, 341–345
criminal intent and, 167
criteria for, 167
history of, 166–167
Model Penal Code section, 167
murder, 341–344
respondeat superior, 167
summary of, 147
vicarious liability, 168–170
Correctional institutions, duress in, 288–289
Corroboration, 363
Corruption, public
bribery of a public official, 483–484
definition of, 468
prevalence of, 483
studies of, 483
summary of, 488
Counterterrorism agreements, 556–557
Court(s)
adversarial system, 15
appellate, 15, 16
trial, 15
Crack cocaine, 77
Credit card identity theft, 443–444
Crime(s). See also specific crime
broken windows theory of, 499, 526
categories of, 4–5
of cause and result, 117
common law, 6
computer, 426, 446–450, 465
condemnation of, 2
consensual, 196
corporate, 166–167
of criminal conduct causing a criminal harm, 136
definition of, 2

environmental. See Environmental crimes
felony. See Felony
hate, 36
infamous, 5
infractions, 4
mala in se, 4–5, 130, 357
mala prohibita, 5, 130, 357
misdemeanors. See Misdemeanor
of moral turpitude, 4
against nature, 25
of official misconduct, 483
parties to. See Parties to a crime
against public order and morality. See Public order and morality, crimes against
punishment for, 4–5
victimless. See Victimless crimes
violations, 4
white-collar. See White-collar crimes
Crime Victims Rights Act, 59
Crimes against the state
espionage, 530, 534–535, 558
sabotage, 530, 533–534, 558
sedition, 529, 532–533, 558
summary of, 529–530, 558–559
terrorism. See Terrorism
treason. See Treason
types of, 529–530
Criminal act(s). See also Actus reus
attendant circumstances, 87–88
bystander rule, 97–98
concurrence with criminal intent, 134–135
criminal intent and, 3–4, 134–135
definition of, 3
duress and, 287
failure to act as. See Omission
harm from, 247
omission as. See Omission
status offenses as, 92–93
summary of, 87–88, 112
voluntary requirement. See Voluntary act
Criminal attempt. See Attempt
Criminal conspiracy, 197–198
Criminal intent. See also Mens rea
accessory after the fact and, 161, 163–164
cases involving, 181
concurrence with criminal act, 134–135
constructive, 117
corporate liability and, 167
criminal act and, 3–4, 134–135
definition of, 3
description of, 130
duress and, 287
general, 117
kidnapping, 394
knowingly standard, 121–122, 144
larceny, 428
negligently standard, 127–128, 145
purposely standard, 118–119, 144
recklessly standard, 124, 144
specific, 117

About the Author

Matthew Lippman has taught criminal law and criminal procedure in the Department of Criminology, Law, and Justice at the University of Illinois at Chicago (UIC) for more than twenty years and has served in every major administrative position in that department, including Department Head and Director of Undergraduate and Graduate Studies. He has also taught courses on civil liberties, law and society, terrorism, and genocide and teaches international criminal law at John Marshall Law School in Chicago. He earned a doctorate in political science from Northwestern University and a master of law from Harvard Law School and is a member of the Pennsylvania Bar. He has been voted by the graduating seniors at UIC to receive the Silver Circle Award for outstanding teaching on six separate occasions and has also received the UIC Flame Award from the University of Illinois Alumni Association, as well as the Excellence in Teaching Award, Teaching Recognition (Portfolio) Award, and Honors College Fellow of the Year Award. The university chapter of Alpha Phi Sigma, the criminal justice honor society, recognized Professor Lippman as "criminal justice professor of the year" on three occasions. In 2008, Professor Lippman was named College of Liberal Arts and Sciences Master Teacher. He was honored by the College of Liberal Arts and Sciences, which named him Commencement Marshal at the May 2012 graduation. Professor Lippman is also recognized in *Who's Who Among America's Teachers*.

Professor Lippman is author of more than a hundred articles and four books. These publications focus on criminal law and criminal procedure, international human rights, and comparative law. He is the author of the Sage text, *Criminal Procedure* (2011). His work is cited in hundreds of academic publications and by international courts and organizations. He has served on legal teams appearing before the International Court of Justice in The Hague, and has participated in a number of international human rights cases before federal district and appellate courts and before the U.S. Supreme Court. Professor Lippman also has testified as an expert witness on international law before numerous state and federal courts, and has consulted with both private organizations and branches of the U.S. government. He regularly appears as a radio and television commentator and is frequently quoted in leading newspapers.

$SAGE researchmethods

The essential online tool for researchers from the world's leading methods publisher

Find exactly what you are looking for, from basic explanations to advanced discussion

More content and new features added this year!

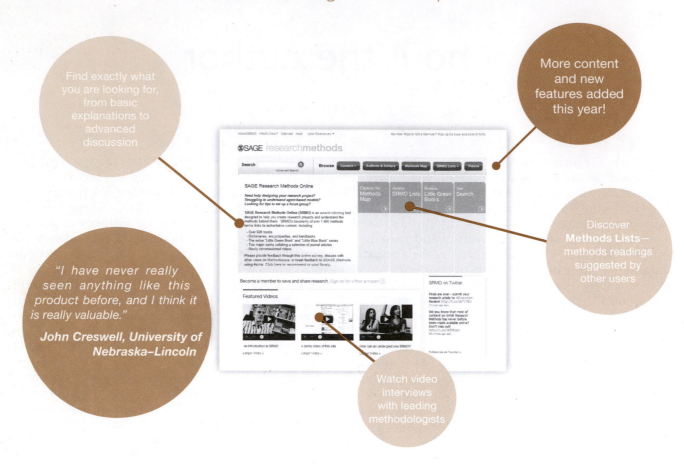

Discover **Methods Lists**—methods readings suggested by other users

"I have never really seen anything like this product before, and I think it is really valuable."

John Creswell, University of Nebraska–Lincoln

Watch video interviews with leading methodologists

Explore the **Methods Map** to discover links between methods

Search a custom-designed taxonomy with more than 1,400 qualitative, quantitative, and mixed methods terms

Uncover more than 120,000 pages of book, journal, and reference content to support your learning

Find out more at
www.sageresearchmethods.com